Developments in Clay Science

Volume 8

# Infrared and Raman Spectroscopies of Clay Minerals

Developments in Clay Science

Volume 8

# Infrared and Raman Spectroscopies of Clay Minerals


Edited by

**W.P. Gates**

Institute for Frontier Materials, Deakin University, Victoria, Australia

**J.T. Kloprogge**

Department of Chemistry, College of Arts and Sciences University of the Philippines, Visayas, Miagao, Iloilo, Philippines

**J. Madejová**

Slovak Academy of Sciences, Institute of Inorganic Chemistry, Bratislava, Slovakia

**F. Bergaya**

CNRS, Interfaces, Confinement, Matériaux et Nanostructures (ICMN) Orléans, France



ELSEVIER

Elsevier
Radarweg 29, PO Box 211, 1000 AE Amsterdam, Netherlands
The Boulevard, Langford Lane, Kidlington, Oxford OX5 1GB, United Kingdom
50 Hampshire Street, 5th Floor, Cambridge, MA 02139, United States



ISBN: 978-0-08-100355-8
ISSN: 1572-4352



*Publisher:* Candice Janco
*Acquisition Editor:* Amy Shapiro
*Editorial Project Manager:* Tasha Frank
*Production Project Manager:* Maria Bernard
*Designer:* Victoria Pearson

Typeset by SPi Global, India

# Dedication

**In loving memory of my son Andrew Olav Kloprogge**

**Feb. 2000–May 2016**
**J.T. Kloprogge**

# Contents

# Contributors

**D. Bain**, Middleton Steading, North Deeside Road, Cults, Aberdeen, AB15 9PL, United Kingdom; derekcbain@gmail.com

**E. Balan**, Sorbonne Universités, Paris, France; etienne.balan@impmc.jussieu.fr

**F. Bergaya**, CNRS, Interfaces, Confinement, Matériaux et Nanostructures (ICMN) Orléans, France; f.bergaya@cnrs-orleans.fr

**J.L. Bishop**, SETI Institute, Mountain View, CA, United States; jbishop@seti.org

**J. Carter**, Institut d'Astrophysique Spatiale, CNRS, University de Paris - Sud, Orsay, France; john.carter@ias.u-psud.fr

**G.D. Chryssikos**, National Hellenic Research Foundation, Athens, Greece; gdchryss@eie.gr

**W.P. Gates**, Institute for Frontier Materials, Deakin University, Melbourn, Victoria, Australia; will.gates@deakin.edu.au

**H.P. He**, Guangzhou Institute of Geochemistry, Chinese Academy of Sciences, Guangzhou, China; hehp@gig.ac.cn

**C.T. Johnston**, Purdue University, West Lafayette, IN, United States; clays@purdue.edu

**J.T. Kloprogge**, Department of Chemistry, College of Arts and Sciences University of the Philippines, Visayas, Miagao, Iloilo, Philippines; t.kloprogge@qut.edu.au

**J. Madejová**, Institute of Inorganic Chemistry, Slovak Academy of Sciences, Bratislava, Slovakia; jana.madejova@savba.sk

**J.R. Michalski**, Department of Earth Sciences and Laboratory for Space Research, University of Hong Kong, Hong Kong, China; jmichal@hku.hk

**S. Petit**, Institut de Chimie des Milieux et Matériaux de Poitiers, Université de Poitiers, Poitiers, France; sabine.petit@univ-poitiers.fr

**H. Pálková**, Institute of Inorganic Chemistry, Slovak Academy of Sciences, Bratislava, Slovakia; uachpalk@savba.sk

**J. Zhu**, Guangzhou Institute of Geochemistry, Chinese Academy of Sciences, Guangzhou, China; zhujx@gig.ac.cn

# Acknowledgements


**(1)** *Faïza Bergaya* acknowledges her co-editors for accepting this non-trivial task and for the heavy work they provided during these past 2 years. The author is particularly thankful to Jana Madejová for her enthusiastic participation as co-author of several chapters, and as co-editor of the book. The author also extends her gratitude to Helena Pálková, for help with the conception and development of the cover. She also expresses her thanks to Will Gates who acted as co-author of several chapters and also, later during its final stages, as first main editor of this book. His help in sharing this last responsibility of managing this book was particularly invaluable. A special thought should be addressed to Theo Kloprogge who, despite a personal setback (a sad event that happened while editing this book), continued to be active writing important chapters for this book. Both Faïza and Will are very grateful for his effort and valuable contributions.

**(2)** *Will P. Gates* thanks the co-editors for asking him to join the team and his co-authors for three of the chapters. As co-author of Chapter 7, he acknowledges that some of the data were collected at the Australian Synchrotron during commissioning of the far-IR beamline. The authors of this chapter acknowledge *D. Appadoo* and *J.D. Cashion* for their able assistance in collecting the data displayed in Fig. 7.9.

**(3)** *Jana Madejová* acknowledges her co-authors for three chapters of this book, and *P. Komadel*, for many years of close collaboration in the IR study of clays and clay minerals.

**(4)** *Theo Klopproge* extends his thanks, first of all, to *Faïza Bergaya* for making a long-held dream come true after his forced retirement from the Queensland University of Technology. He specially acknowledges the co-editors for their tremendous help and support in what was for him personally an extremely difficult year. Finally, Theo wants to thank, in particular, *Will Gates* as he took over so much of the work of the main editor in getting this book to the finish line.

**(5)** All four co-editors express their thanks to the external authors and co-authors of the different chapters of this volume for their useful and timely contributions which made this book possible. Last but not least, several anonymous reviewers are thanked for their precious help in improving the scientific quality of this book. The hard work of the authors and reviewers has resulted in a hopefully excellent addendum to Farmer's book.

**(6)** Co-author of Chapter 4, *Georgios Chryssikos*, thanks *V. Gionis* (NHRF) for many years of close collaboration in the study of clays, and *G. Kacandes* (Geohellas S.A.) and *M. Stefanakis* (S&B Industrial Minerals, now Imerys) for setting the challenge and providing the means.

**(7)** The authors of Chapter 14, *Janice Bishop*, *Joseph Michalski* and *John Carter* are grateful to *R.E. Arvidson* and *J.J. Wray* for their helpful comments that greatly improved this chapter.

Chapter 1

# General Introduction

F. Bergaya, W.P. Gates, J. Madejová, J.T. Kloprogge and D. Bain

## 1.1 ORIGIN AND CONTENT OF THE BOOK

The initiative to prepare a book on *Infrared and Raman Spectroscopies of Clay Minerals* was originally conceived during the XVth International Clay Conference held in Rio de Janeiro (Brazil) in 2013, by one of the co-editors (*F. Bergaya*, who acts also as series editor for the Developments in Clay Science). In fact, she realised that the famous book *Infrared Spectra of Minerals*, by V.C. Farmer (Farmer, 1974a), used as a reference for infrared (IR) studies of clay minerals throughout the world, had not been updated for 40 years. Not surprisingly, this proves that Farmer's book was, and continues to be, 'a fundamental piece of work'. The idea for this volume was to update the available information on IR spectroscopy of the clay minerals, and quickly evolved to include data on Raman spectroscopy. Both techniques are non-destructive, are often used as complementary tools, and critically in the case of Raman, require an up-to-date assessment of the state of the science as applied to studies for the clay minerals.

As series editor, *F. Bergaya* invited *J. Madejová* who accepted without hesitation to contribute as co-editor of this volume. During initial discussions on the framework of this book, she proposed inviting also as co-editor, *W.P. Gates*, who immediately accepted the invitation with great pleasure. On invitation to lead the Raman spectroscopy sections, *J.T. Kloprogge* also accepted to join the editorial team. All co-editors discussed the final table of contents with the hope that it would cover the broad field of IR and Raman research involving clay minerals.

This book contains 14 chapters, including this chapter, which introduces its origin and a short biography of Farmer, as well as brief concluding comments of the editors. Chapter 2 introduces the theoretical basis of IR and Raman spectroscopies. Advancement in instrumentation with the most appropriate configuration and the sampling methods for IR and Raman spectroscopic study of the clay minerals, since Farmer's 1974 book, are revisited in

Developments in Clay Science, Vol. 8. https://doi.org/10.1016/B978-0-08-100355-8.00001-1

Chapter 3. In further recognition of modern IR and Raman studies, the spectral manipulation of large datasets and application of multivariate analysis is critically presented in Chapter 4. Chapters 5 and 6 present, respectively, IR (mid-infrared and near-infrared) and Raman spectral libraries of representative clay minerals, with band assignments allowing their differentiation and identification.

The next eight chapters present critical reviews on focused topics, on the application of IR and/or Raman spectroscopy. Determination of the site occupancy in clay minerals is covered in Chapter 7. Characterisation of products in the synthesis of clay minerals is presented in Chapter 8. Studies of clay mineral-water interaction are presented in Chapter 9. Applications towards determining interlayer environments of organoclay intercalates (Chapter 10), intercalated kaolins (Chapter 11), pillared clays (Chapter 12) and modifications of smectites by different treatments (Chapter 13) are covered in detail. Finally, Chapter 14 describes the remote detection of clay minerals in the visible/NIR regions on Earth and how it is often more easily performed on Mars and other extraterrestrial entities.

As this book was conceived to be in part the continuity of Farmer's contribution to IR spectroscopy of clay minerals, it seems opportune to say at least a few words about this eminent scientist and his role in developing the science of IR spectroscopy of minerals. Colin Farmer, whom F. Bergaya met personally at each International Clay Conference from the 1970s till the 1990s, is remembered as an elegant, pleasant man, always very approachable despite his high scientific notoriety. *D. Bain*, Colin's long-time coworker at Macaulay Institute, kindly provided the following brief biography on Farmer.

## 1.2 VICTOR COLIN FARMER (1920–2006)

Although born in Ireland, Colin Farmer was brought up in Scotland and graduated with first class honours in chemistry from the University of Glasgow in 1943. This was followed by his PhD in 1947 from the University of Aberdeen for a thesis entitled "Spectroscopic investigations on the minor element content of plants and soils," the research having been carried out mainly at the Macaulay Institute for Soil Research in Aberdeen. He was then appointed to a research post at the Macaulay Institute to work on applying chemical analytical techniques to the study of plants and soils and spent his whole career at the Macaulay Institute, retiring in 1983.

In 1955 a decision was taken to purchase an infrared (IR) spectrometer for the Macaulay Institute, and Colin was given the responsibility of investigating the possible applications of IR spectroscopy to soil science. This was an inspired decision that was to prove the start of establishing IR spectroscopy as a powerful tool for the study of minerals and establish Colin as a research scientist of outstanding international repute. He was ably assisted by Jim

Russell and Tony Fraser, and this team produced an amazing flow of publications of high quality. Their reputation and willingness to collaborate resulted in a steady flow of visiting researchers to the Macaulay Institute from all over the world. Working with Colin and his team was highly sought after as they established themselves as the world's leading experts on IR spectroscopy.

Colin in effect established IR spectroscopy as a powerful tool for investigating the chemistry of minerals, particularly clay minerals. He showed that it could be used for studying mineral surfaces and their reactions in addition to simply identifying the mineral phases. In relation to soil science, it had the advantage over the more commonly used techniques such as X-ray diffraction of being able to identify and characterise X-ray amorphous phases. Much pioneering work on amorphous soils constituents was carried out in Colin's lab, and much attention was paid to the tubular paracrystalline mineral imogolite. This had first been recorded in Japanese soils derived from volcanic material but Colin found that it occurred, albeit in smaller amounts, in podzolic soils in Scotland derived from other parent materials. Subsequently it was discovered in other similar soils in other countries, and Colin proposed a new theory of podzolisation based on the transport of iron and aluminium down through soil profiles as inorganic complexes as opposed to organic complexation as in the conventional theory. This new theory was never generally accepted by the soil science community but kindled much discussion and debate in the scientific literature. Colin believed in the serendipity approach to research and was fortunate to have been active when this was possible, but science itself benefited from the research he conducted as a result.

As recognition of his outstanding ability in scientific research, Colin received a special merit promotion, a rare award in agricultural research in the United Kingdom, to enable him to pursue his research at a higher grade without having to be promoted to an administrative position such as head of a department.

One of Colin's characteristics was a real passion and profound interest in scientific matters, and he was not afraid to point out, through publications or verbally at meetings, anything that he considered to be incorrect or unsubstantiated. He never did this as a means of embarrassing anyone, but only in pursuit of the scientific truth.

After retiring from his post at the Macaulay Institute in 1983, Colin was in demand as a scientific researcher and spent time as a visiting professor or research fellow in several research establishments across the world such as the University of Adelaide and the CSIRO Division of Soils in Australia; Instituto di Chimica Agraria, Portici in Italy; University of Saskatchewan in Canada and INRA, Versailles, in France. He also continued research at the Macaulay as an honorary research fellow as long as he was able and was involved in approximately 40 publications after his retirement, mostly concerned with amorphous and poorly ordered materials in soils.

Of all Colin's publications, the one which ranks the highest is the monograph published by the Mineralogical Society of Great Britain and Ireland in 1974 on "*Infrared Spectra of Minerals*." Colin wrote many of the chapters in this book as well as being the overall editor. For many researchers, this is the 'bible' of IR spectroscopy, and it is nothing less than astonishing that it is still widely used over 40 years later.

Colin's work was recognised with a number of honours and awards. In the United Kingdom he was made a Distinguished Member of the Clay Minerals Group at a special dinner in Aberdeen organised in his honour for his lifetime research into clay mineralogy and IR spectroscopy. In the United States, the Clay Minerals Society granted him the Pioneer in Clay Science Award for his research contributions that led to important new directions in clay mineral science.

## 1.3 PERSPECTIVES AND CONCLUDING REMARKS

The chapters presented in this book clearly indicate the actually wide range of applications that IR and Raman spectroscopies bring to the study of clay minerals and clay polymer nanocomposites. While perhaps the development of vibrational theory has not advanced much since the 70's, our ability to use it to interpret and predict IR and Raman spectra of clay minerals has; largely because of the advances in computing over the intervening decades, the potential of this technique is not exhausted yet. Certainly improvements in instrumentation, computing and methods for presenting samples to IR or Raman instrumentation have changed markedly since Farmer's book. These advances have resulted in the availability of a rather complete set of instruments for probing nearly any sample in any condition, anywhere.

Clay scientists have not been slow to employ these new instruments and methods and have adapted them into their research as they have become available. Detailed interpretation of the IR and Raman data on clay minerals obtained from the spectra taken in the traditional mid-IR (MIR) region, but also in the less common near- and far-IR (NIR and FIR) regions, supported by advanced theoretical studies, applications of different accessories for specific purposes, or real-time IR measurements are allowing us to continue to probe unanswered questions. Significant progress in instrumentation in connection with the fiberoptic probes for portable Fourier transform infrared (FTIR) spectrometers operating in both NIR and MIR regions, allowing complex characterisation of soils and clay deposits, has occurred. Utilisation of synchrotron radiation for far-infrared (FIR) region or for IR microscopy has significantly improved the detection limit of the method. Combination of FTIR and Raman spectrometers with a variety of other analytical instruments will provide more complex information on the clay minerals alone and in materials with other substances. Enormous progress in orbiter-carried visible and near-infrared (VIS-NIR) spectrometers performing global mapping of the surface of Mars and elsewhere has brought, nearly daily, marvellous details on the occurrence of clay minerals on the red planet. It follows clearly that vibrational spectroscopy will remain one of the most important experimental methods for clay minerals investigation over the next 40 years.

The editors hope that this volume fulfils its promise to update and expand on Farmer's landmark book. The editors recognise that this is a hard task to achieve, given the almost universal uptake of Farmer's book by clay scientists worldwide, but are hopeful that it will be as useful for clay researchers and teachers.

Chapter 2

# Theoretical Aspects of Infrared and Raman Spectroscopies

E. Balan and J.T. Kloprogge

## 2.1 INTRODUCTION

Infrared and nonresonant Raman spectroscopies are based on the excitation of specific vibrational modes of the system under study. The frequency, intensity and width of the measured signals are highly sensitive to the local structure, composition and microstructure of the investigated samples, leading to a wide range of applications in mineralogy (e.g. Farmer, 1974c,d) and in clay science (e.g. Farmer, 1974c,d, 1979; Russell and Fraser, 1994; Madejová and Komadel, 2001; Madejová, 2003; Kloprogge, 2005). In this chapter, the physical relations between the sample structure and the measured spectrum are exposed with a special attention to the specificity of clay minerals. When possible, the presentation will be based on classical models, which provide a reasonable understanding of the involved phenomena and parameters. The relation with their quantum mechanical counterparts will be discussed when needed.

The chapter begins with an introduction to lattice dynamics, aiming at drawing direct relationships between the crystal structure of the clay minerals and their vibrational modes. Infrared (IR) and Raman spectroscopies will then be presented successively. The interaction of a vibrational mode with an oscillating electric field in the IR frequency range will be presented using the phenomenological 1D Drude-Lorentz model, illustrating how the low-frequency dielectric tensor of a crystal can be built from atomic-scale parameters. The specificities of IR spectroscopy applied to powder materials will then be discussed in the light of popular effective medium models. The principles of the nonresonant Raman spectroscopy will be exposed using a simplified picture. The final paragraph deals with some aspects of the atomic-scale modeling approaches applied to the prediction of vibrational spectra, a field that has greatly expanded during the last 30 years.

Developments in Clay Science, Vol. 8. https://doi.org/10.1016/B978-0-08-100355-8.00002-3

## 2.2 LATTICE DYNAMICS IN THE HARMONIC APPROXIMATION

### 2.2.1 Classical Model of Crystal Vibrations

The main features of the vibrational properties of crystals within the framework of classical mechanics are summarised in this section. The objective is to quantitatively describe the collective oscillatory motion of atoms in the vicinity of their equilibrium position. Various approaches can be used to expose this subject. For example, 1D models consisting in infinite chains of atoms can be used to capture the main physical ingredients determining the vibrational properties of an extended system and to display the specific features related to its translational invariance (Hadni, 1974; Bruesch, 1982). These models have the merit they can be solved in an analytic fashion, leading to explicit relations between vibrational frequencies and microscopic parameters such as atomic mass and force constant of bonds. In the present chapter, the three-dimensional crystal structure will be explicitly considered. Although mathematically less straightforward relations are obtained, this facilitates the application of theory to real systems. Extensive presentations of the lattice dynamic theory can be found in, for example, Born and Huang (1954), Bruesch (1982) and Böttger (1983).

In practical terms, the position of any atom in a crystal is defined by the vector $\vec{R}_{ak}$ (Fig. 2.1):

$$\vec{R}_{ak} = \vec{R}_a + \vec{r}_k + \vec{u}_{ak} \tag{2.1}$$

where $\vec{R}_a$ is a lattice vector corresponding to the cell of integer index $a$, $\vec{r}_k$ is defined by the reduced atomic coordinates of atom $k$ at equilibrium position, and $\vec{u}_{ak}$ is its displacement from equilibrium position.

Within the Born-Oppenheimer approximation, electron and nuclear dynamics are decoupled. The force experienced by an atom $k$ in cell $a$ along the Cartesian direction $\alpha$ can be obtained from a potential energy term depending on atomic positions:

$$F_{ak\alpha} = -\frac{\partial E_{pot}}{\partial R_{ak\alpha}} \tag{2.2}$$

where $E_{pot} = E_{pot}\left(\left\{\vec{R}_{ak}\right\}\right)$ is the potential energy surface of the system. By definition, the potential energy is minimal and the forces on atoms are nil for the equilibrium geometry.

The potential energy surface contains contributions from the electronic energy and nuclear–nuclear repulsion energy. Alternatively, assuming that the system under study can be described using an ionic bond formalism, the potential energy surface can be described as a balance between short-range dominantly repulsive interactions (incorporating smaller van der Waals

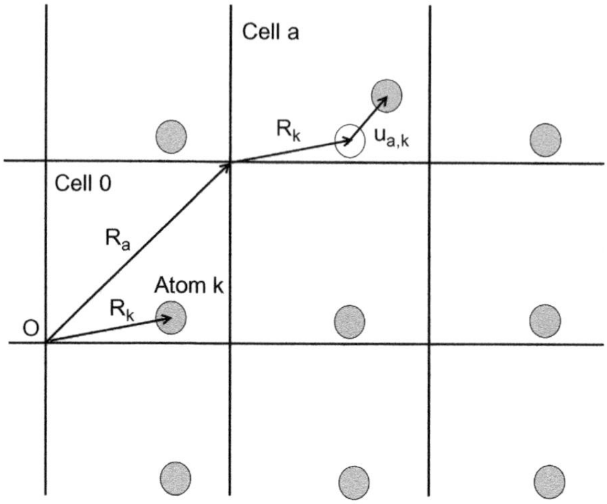


**FIG. 2.1** Equilibrium position vector and displacement vector describing the displacement of atom k in cell a of a periodic crystal structure.

attractive contributions) and long-range Coulombic interactions between atoms. An explicit expression of the potential energy surface as a function of atomic positions in the vicinity of the equilibrium geometry can be obtained using a Taylor expansion:

$$E_{pot} = \sum_{ak\alpha} \left( \frac{\partial E_{pot}}{\partial R_{ak\alpha}} \right)_0 u_{ak\alpha} + \frac{1}{2} \sum_{ak\alpha} \sum_{bk'\beta} \left( \frac{\partial^2 E_{pot}}{\partial R_{ak\alpha} \partial R_{bk'\beta}} \right)_0 u_{ak\alpha} u_{bk'\beta} + \cdots \quad (2.3)$$

where the partial derivatives are taken at the equilibrium positions and the minimum of the potential energy is taken equal to zero. The indexes $a$ and $b$ span over the crystal cells, $k$ and $k'$ over the atoms in a single cell, and $\alpha$ and $\beta$ over the three Cartesian coordinates. The potential energy of the system being minimal for the equilibrium structure, the first-order terms of the development are zero and the second-order terms are positive:

$$\left( \frac{\partial E_{pot}}{\partial u_{ak\alpha}} \right)_0 = 0 \ , \qquad \left( \frac{\partial^2 E_{pot}}{\partial u_{ak\alpha} \partial u_{bk'\beta}} \right)_0 \geq 0 \quad (2.4)$$

The second-order derivatives thus define the coefficients of a symmetric matrix, referred to as the force-constant matrix:

$$C_{k\alpha,k'\beta}(a,b) = \left( \frac{\partial^2 E_{pot}}{\partial u_{ak\alpha} \partial u_{bk'\beta}} \right)_0 \quad (2.5)$$

In the harmonic approximation, the Taylor expansion is restricted to second-order terms, leading to the following expression of the potential energy of the system:

$$E_{pot}^{harm} = \frac{1}{2} \sum_{ak\alpha} \sum_{bk'\beta} C_{k\alpha,k'\beta}(a,b)\ u_{ak\alpha} u_{bk'\beta} \tag{2.6}$$

This approximation holds for atomic displacements small enough such that third and higher orders can be neglected. The harmonic restoring force experienced by atom $k$ in cell $a$ along direction $\alpha$ when it is displaced from its equilibrium position is thus:

$$F_{ak\alpha} = -\sum_{bk'\beta} C_{k\alpha,k'\beta}(a,b)\, u_{bk'\beta} \tag{2.7}$$

From this expression, the application of Newton's second law leads to a system of coupled differential equations that relate the acceleration of atom $k$ in cell $a$ along direction $\alpha$ to the displacements of atoms $k'$ in cell $b$ along direction $\beta$:

$$M_k \frac{\partial^2 u_{ak\alpha}(t)}{\partial t^2} = -\sum_{bk'\beta} C_{k\alpha,k'\beta}(a,b)\, u_{bk'\beta}(t) \tag{2.8}$$

Adopting a complex notation for oscillatory quantities (i.e. the corresponding oscillating real quantities are obtained by adding to the complex quantity its conjugate value and dividing the sum by two), the discrete oscillatory solutions of the system given in Eq. (2.8) that are consistent with the translational invariance of the crystal are obtained under the form:

$$u_{ak\alpha}(t) = \eta_{k\alpha} e^{i\vec{q}\vec{R}_a} e^{-i\omega t} \tag{2.9}$$

where $\eta_{k\alpha}$ is the coordinate of the vector (referred to as the polarisation vector) describing the displacement of atom k in the cell origin along the direction $\alpha$, $\vec{q}$ is a vector belonging to the reciprocal space and $\vec{R}_a$ the position of the crystal lattice node $a$, and $\omega$ is the angular frequency describing the time periodicity. These solutions have the form of plane waves defined over the nodes of the crystal lattice. It can be shown that the solutions of Eq. (2.8) for different wave vectors are decoupled. For a given wave vector, $\vec{q}$, the oscillatory motion of atoms around their equilibrium positions is thus fully described by a system of $3N$ equations, where $N$ is the number of atoms in a single crystal unit cell. In addition, it is not useful to consider values of $\left|\vec{q}.\vec{R}_a\right|$ larger than $\pi$ because of the periodicity of the function $e^{ix}$, which restricts the choice of $\vec{q}$ vectors to the first Brillouin zone of the crystal.

To describe the collective motion of atoms, Eq. (2.10) is then obtained by substituting Eq. (2.9) into Eq. (2.8) and by multiplying both sides by $e^{i\omega t}$ and $e^{-i\vec{q}\vec{R}_a}$:

$$M_k \omega^2 \eta_{k\alpha} = \sum_{bk'\beta} C_{k\alpha,k'\beta}(a,b)\, \eta_{k'\beta} e^{i\vec{q}\left(\vec{R}_b - \vec{R}_a\right)} \tag{2.10}$$

Because of the translational invariance, the force-constant coefficients only depend on the relative positions of cells $a$ and $b$:

$$C_{k\alpha,k'\beta}(a,b)=C_{k\alpha,k'\beta}(0,b-a) \tag{2.11}$$

As the sum is performed over all crystal cells, the index a can be arbitrarily chosen as corresponding to the origin cell; that is, $a=0$, and Eq. (2.10) can be rewritten as:

$$M_k\omega^2\eta_{k\alpha}=\sum_{k'\beta}\eta_{k'\beta}\tilde{C}_{k\alpha,k'\beta}\left(\vec{q}\right) \tag{2.12}$$

where $\tilde{C}_{k\alpha,k'\beta}\left(\vec{q}\right)$ is the discrete Fourier transform of the force constants:

$$\tilde{C}_{k\alpha,k'\beta}\left(\vec{q}\right)=\sum_{b}C_{k\alpha,k'\beta}(0,b)e^{i\vec{q}\vec{R}_b} \tag{2.13}$$

The system of $3N$ equations expressed in Eq. (2.12) can be rewritten in a more tractable form using parameters obtained by multiplying the displacement of each atom $k$ with the square root of its mass:

$$\gamma_{k\alpha}=\sqrt{M_k}\eta_{k\alpha} \tag{2.14}$$

Eq. (2.12) is then rewritten as:

$$\omega^2\gamma_{k\alpha}=\sum_{k'\beta}\gamma_{k'\beta}\tilde{D}_{k\alpha,k'\beta}\left(\vec{q}\right) \tag{2.15}$$

where the coefficients $\tilde{D}_{k\alpha,k'\beta}\left(\vec{q}\right)$ define the dynamical matrix $\tilde{D}\left(\vec{q}\right)$:

$$\tilde{D}_{k\alpha,k'\beta}\left(\vec{q}\right)=\frac{\tilde{C}_{k\alpha,k'\beta}\left(\vec{q}\right)}{\sqrt{M_k}\sqrt{M_{k'}}} \tag{2.16}$$

It can be shown that the $3N\times3N$ dynamical matrix $\tilde{D}\left(\vec{q}\right)$ is Hermitian. Thus the nontrivial solutions of the system of Eq. (2.15), referred to as vibrational modes, can be obtained by diagonalising $\tilde{D}\left(\vec{q}\right)$. The eigenvalues $\lambda_m\left(\vec{q}\right)$, where the index m spans from 1 to $3N$, are real, and the mechanical stability of the structure requires them to be positive. This leads to $3N$ real angular frequencies $\omega_m\left(\vec{q}\right)=\sqrt{\lambda_m\left(\vec{q}\right)}$. The eigenvectors are orthogonal and can be normalised to form an orthonormal and complete basis set. Because of the orthogonality of eigenvectors, each vibrational mode can be

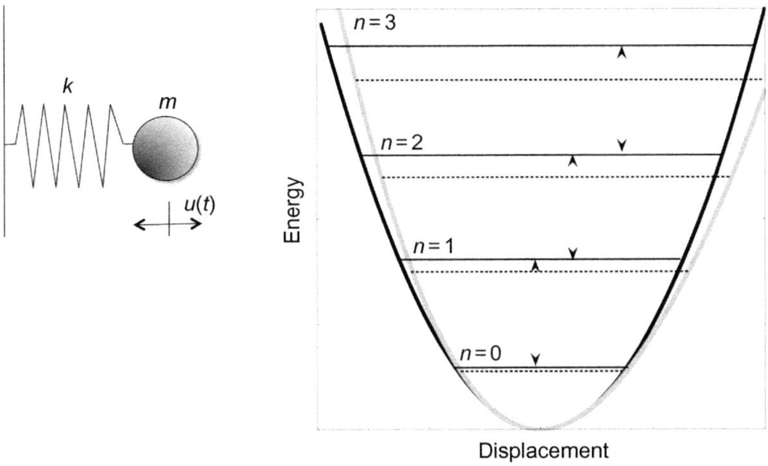


**FIG. 2.2** Left: Mechanical analog of a vibrational mode made of a mobile mass linked by a spring to a fixed point. Right: Potential energy curve for a displacement of the mass. The anharmonic potential corresponds to the grey curve, whereas the harmonic one is in black and displays a parabolic shape. The corresponding vibrational quantum levels are indicated by the horizontal lines and the corresponding quantum number $n$ indicated. The harmonic levels are equidistant whereas the anharmonic ones are progressively downshifted. Note that the position of the fundamental state does not coincide with the minimum of the potential well, contributing to the zero-point energy.

considered as an independent harmonic oscillator (Fig. 2.2). The corresponding atomic displacement patterns can be obtained from the corresponding $3N$ eigenvectors $\vec{\gamma}_m\left(\vec{q}\right)$ using Eq. (2.14). When the atoms are displaced from their equilibrium position along the pattern corresponding to a given mode $(u_{m,k\alpha}\left(\vec{q}\right) = \lambda\,\eta_{m,k\alpha}\left(\vec{q}\right)$, where $\lambda$ is a real number and the mode polarisation is normalised such as $\sum_{k\alpha}\eta^2_{m,k\alpha}\left(\vec{q}\right) = 1)$, the harmonic potential energy curve follows a parabolic law as a function of $\lambda$:

$$E_p(\lambda) = \frac{M_m\left(\vec{q}\right)\omega_m\left(\vec{q}\right)^2}{2}\lambda^2 \tag{2.17}$$

where $M_m\left(\vec{q}\right)$ is the mode effective mass defined by $M_m\left(\vec{q}\right) = \sum_{k\alpha} M_k\left(\eta_{m,k\alpha}\left(\vec{q}\right)\right)^2$.

Accordingly, each vibration mode behaves as an isolated harmonic oscillator and can be described using its simple mechanical system counterpart, constituted by a mobile effective mass and a spring (Fig. 2.2).

The vibrational frequencies in crystals typically range between $10^{12}$ and $10^{14}$ Hz. It is often convenient to relate them to wavenumbers, denoted as $\bar{\nu}$, and defined as the inverse of the corresponding electromagnetic wavelength $\lambda$ ($\lambda = c/\nu$, with $c$ the speed of light). The wavelength is usually expressed in cm and the corresponding "$cm^{-1}$" unit leads to numbers typically between 10 and 4000 $cm^{-1}$.

### 2.2.2 Categorisation and Symmetry of Vibrational Modes

The vibrational frequencies depend on the $\vec{q}$ vectors. The curves reporting the frequency as a function of the $q$ value along selected directions are referred to as dispersion curves. For $q=0$, corresponding to the centre of the first Brillouin zone of the crystal, all crystal cells vibrate in phase, displaying identical atomic displacements. These zone-centre modes play a key role in IR and Raman spectroscopies because momentum conservation rules restrict the interaction of vibrational modes with exciting IR or visible radiation (whose wavelength is significantly larger than typical crystal cell dimensions, i.e. few Å) to the long wavelength limit ($\vec{q} \rightarrow 0$). For three vibrational modes, referred to as acoustic modes, the vibrational frequency tends to zero for vanishingly small values of $q$. The corresponding displacement patterns reduce to rigid crystal translation along the three Cartesian directions.

The other vibrational modes display finite frequencies at $q=0$ and are categorised as optic modes. The oscillating motion of ions in some of these modes can be associated with an oscillating dipole moment in the unit cell, which can induce a uniform (macroscopic) electric field in the crystal when $\vec{q} \rightarrow 0$. This oscillating field affects the ionic displacements (and in turn is affected by them) and modifies the corresponding vibrational frequency (Born and Huang, 1954; Baroni et al., 2001). Its occurrence depends on the direction of the $q$ vector that has been chosen for the limit, $\vec{q} \rightarrow 0$. The macroscopic electric field is zero when the electrostatic polarisation vector and phonon wave vectors are perpendicular, which corresponds to transverse optic (TO) modes. In a cubic crystal, the macroscopic electric field is maximal when the polarisation vector and phonon wave vector are parallel, leading to the higher vibrational frequency of the longitudinal optic (LO) modes. The distinction between transverse and longitudinal modes still applies in crystals with lower symmetry.

Beside the lattice translational invariance, crystal structures are invariant by the application of the finite number of operations belonging to their space group. The potential energy surface describing the motion of atoms in the vicinity of their equilibrium positions must also remain unchanged under these symmetry operations. This property imposes mathematical relations between the coefficients of the dynamical matrix that in turn impose specific symmetry properties to the related eigenvectors. Accordingly, the vibrational modes can be categorised by their transformation properties under the crystal symmetry operations. This procedure is efficiently achieved using the principles of group theory (e.g. Farmer and Lazarev, 1974). In highly symmetric structures, the relations imposed by symmetry constraints limit the number of independent force constants and facilitate their determination from experimental data.

The vibrational patterns of solids displaying molecular ions with strong bonds (e.g. silicates, hydroxyls) can also be conveniently described by relating them to the vibrational modes of the isolated molecule (e.g. Lazarev, 1974). For such minerals, some vibrational patterns may involve the quasirigid motion of the whole molecular group with respect to the crystal lattice (external translational or rotational modes) whereas other patterns correspond to the deformation of the molecular groups (internal modes). For example, stretching or bending modes involve changes in bond lengths or angles, respectively. This type of description is particularly useful when dealing with low-symmetry systems or disordered systems. In general, internal modes are observed at relatively high frequency and qualified as hard modes, such as Si-O stretching (typically $\overline{\nu} \approx 1000\ \mathrm{cm}^{-1}$) or O-H stretching (typically $\overline{\nu} \approx 3600\ \mathrm{cm}^{-1}$) vibrations. They are useful probes of structural modifications of crystals, including phase transitions and varying composition of solid solutions (Salje and Bismayer, 1997; Salje et al., 2000).

In hydrous minerals, OH stretching modes are usually observed at a significantly higher frequency than the other modes because of the small mass of the hydrogen atom (Eq. (2.17)). In this case, the OH stretching frequency is sensitive to the molecular environment of OH groups. It depends on the H-bond strength (e.g. Ryskin, 1974; Libowitzky, 1999; Johnston et al., 2008; Welch et al., 2012) but is also sensitive to nearby cationic substitutions (e.g. Farmer, 1974b; Besson and Drits, 1997a, 1997b; Sainz-Diaz et al., 2000; Martínez-Alonso et al., 2002; Botella et al., 2004; Petit et al., 1999b, 2004a). Note that the vibrational modes of the crystal can involve coupling between the vibrational motion of nonequivalent molecular groups (Fig. 2.3). For example, the OH stretching modes of the kaolinite polymorphs (kaolinite, dickite, nacrite) display significant variations that can be ascribed to a different coupling scheme between the nonequivalent OH groups of their structure (Farmer, 1974b; Balan et al., 2001, 2005).

### 2.2.3 Relation to the Quantum Mechanical Description of Vibrational Properties

Although they have been obtained using a classical mechanics framework, the vibrational properties described can be related to their quantum mechanical counterparts. The energy carried by each vibrational mode is quantised (Fig. 2.2). The quantified energy of a harmonic oscillator, characterised by an angular frequency $\nu_m$ is

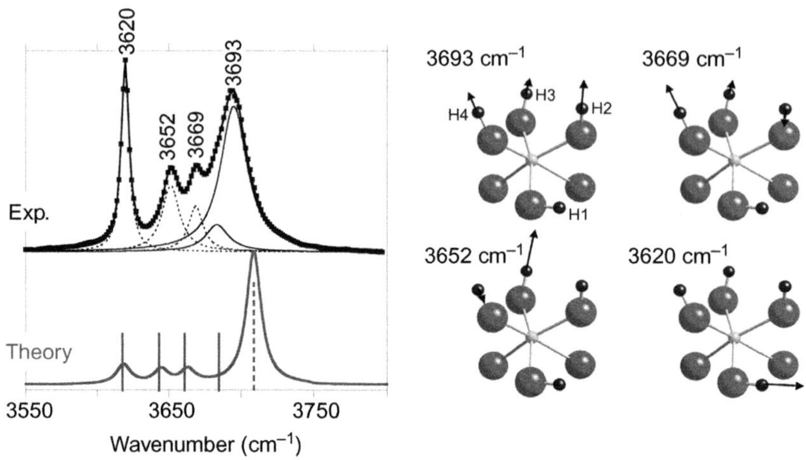


**FIG. 2.3** Infrared absorption spectrum of OH stretching modes in kaolinite (KGa-1 sample, upper curve). The corresponding wavenumber is indicated above each band. The theoretical spectrum computed for small platy particles is reported below. Note the shift of the absorption bands with respect to the vibrational mode frequencies of bulk kaolinite (vertical bars, dotted bar: LO frequency), due to long-range electrostatic interactions in the particle. The vectors describing the atomic displacements of each mode are reported on the right (large spheres: oxygen atoms, small black spheres: H atoms, central sphere: aluminum, only a small cluster belonging to the kaolinite structure is represented). Note the coupled nature of the inner-surface OH stretching mode (involving H2, H3, H4 atoms) in kaolinite (From Balan et al. (2014)).

$$E_n = \left(n + \frac{1}{2}\right) h\nu_m(q) \tag{2.18}$$

where $n$ is a positive integer and $h$ the Planck constant. The number $n$ is the quantum number indexing the discrete excited states of the harmonic oscillator. It can be equivalently understood as an occupation number of a vibration state by quasiparticles referred to as phonons (Bruesch, 1982). At thermal equilibrium, the mean occupation number is described by the Bose-Einstein distribution:

$$\overline{n} = \frac{1}{e^{\frac{h\nu_m(q)}{kT}} - 1} \tag{2.19}$$

The progressive increase in the mean occupation number with temperature corresponds to heat storage by the material. At very low temperature, the thermal energy is small with respect to that involved in vibrational excitations and only the vibrational ground state, defined by n=0, is occupied. Its energy $\left(\frac{h\,\nu_m(q)}{2}\right)$ does not coincide with the minimum of the potential energy surface and depends on atomic masses and force constants. This zero-point energy can significantly contribute to the low-temperature thermodynamic properties of the system. The corresponding ground-state nuclear wave function describes the quantum delocalisation of atoms at very low temperature.

### 2.2.4 Anharmonic Vibrational Properties

These developments have been obtained using the harmonic approximation, that is, by restricting the Taylor expansion of the potential energy surface to the second-order terms (Eq. 2.3). However, important properties also depend on higher-order terms of the expansion, among which are the thermal expansion of crystals and the finite lifetime of vibrational excitations (e.g. Bruesch, 1982; Menéndez and Cardona, 1984). These properties are referred to as "anharmonic" and are, in most insulating materials, efficiently described using the third- (cubic) and fourth- (quartic) order terms in the Taylor expansion of the potential energy.

The effect of anharmonic terms can be envisioned by considering the basis formed by the eigenvectors of the harmonic dynamical matrix. Because of the anharmonic terms, the system dynamics is no longer described by a perfectly diagonal matrix in this basis. The finite out-of-diagonal terms describe the anharmonic interaction between different phonons which contribute to the temperature dependence of spectroscopic parameters, including the position and width of vibrational signals (e.g. Lazzeri et al., 2003; Zhang et al., 2007). The temperature dependence of IR and Raman spectra can bring rich information on their vibrational properties. For example, the low-temperature shift and narrowing of OH stretching bands have been used to unravel OH signatures ascribed to specific stacking schemes in disordered kaolinite group minerals (Prost et al., 1989; Johnston et al., 2008; Balan et al., 2010, 2014).

The diagonal values also differ from the harmonic ones contributing to modifications of the vibrational frequencies. In many cases, however, the anharmonic features only correspond to minor variations of the harmonic scheme. An important exception concerns the OH stretching modes. Because of the small H mass, even the vibrational ground state is far above the minimum of the potential energy (Eq. 2.17), and the harmonic approximation appears a very crude one: The anharmonic stretching frequencies of OH groups are downshifted by about 200 $cm^{-1}$ with respect to their harmonic counterparts. A quantitative assessment of the frequency shift can be obtained by using models based on a Morse potential description of the OH bond (Pascale et al., 2004; Tosoni et al., 2005) or by applying perturbation theory to the harmonic oscillator (Balan et al., 2007).

The significant anharmonicity of the OH stretching potential also leads to the detection of specific multiphonon bands in the near-infrared (NIR) range. The overtone bands correspond to absorptions from the ground vibrational state to the second or the third excited state of the OH stretching oscillator (Fig. 2.2). Because of the anharmonicity of the potential, these vibrational states are no longer equally spaced, and the position of the corresponding bands in the NIR range provides quantitative information on the anharmonicity parameters. In addition, bands corresponding to the combination of different vibrational modes can be observed (e.g. Petit et al., 2004a,b; Petit, 2005). The combination between OH stretching ($\nu$(OH) ~ 3600 cm$^{-1}$) and bending ($\delta$(OH) ~ 1600 cm$^{-1}$) vibrations in the water molecule leads to a characteristic band at ca. 5200 cm$^{-1}$, whereas combination bands involving the stretching of structural OH groups and libration modes are observed at ca. 4500 cm$^{-1}$. This makes it possible to make a better distinction between structural OH groups and molecular water, whose stretching spectra usually overlap in the mid-infrared (MIR) range. Applications of NIR spectroscopy range to the remote sensing of planetary surfaces (e.g. Bishop et al., 2008c) and ore prospection using portable spectrometers (e.g. Marsh and McKeon, 1983).

## 2.3 PROBING THE VIBRATIONAL MODES WITH IR LIGHT

### 2.3.1 Drude-Lorentz Model Applied to IR Spectroscopy

The interaction between an electric field oscillating at IR frequencies and a crystal can be described by the induced macroscopic polarisation of the material due to the displacement of the charged particles from their equilibrium position. In turn, the polarisation can be decomposed in an electronic contribution at fixed ionic positions and an ionic contribution due to ion displacements. In the range of IR frequencies, far from the frequencies typical of electronic excitations, the electronic contribution can be considered to be frequency independent. In contrast, the ionic contribution in a polar material exhibits a strong dependence on the IR light frequency. This dependence can be understood within a classic mechanical framework, by considering the interaction of a damped harmonic oscillator of fundamental angular frequency, $\omega_0$, with an oscillating electric field, $E$. The oscillator is characterised by a mobile material point of effective mass, $M_{eff}$, its electrostatic charge, $Z_{eff}$, and a restoring force constant, $k = M_{eff}\omega_0^2$. In this model, the natural angular frequency corresponds to the transverse frequency of the vibrational mode. In case of a system that can be described in terms of individual localised oscillators (e.g. OH groups), the model frequency is not that of the free oscillator but takes into account the long-range electrostatic interactions at $q = 0$. A phenomenological damping parameter, conveniently expressed as $M_{eff}\,\gamma$, is also introduced to account for the anharmonic interaction of the oscillator with the other vibrational states of the crystal. The damping parameter $\gamma$ is usually much smaller than transverse frequencies, typically few cm$^{-1}$.

The second-order linear differential equation describing the dynamical properties of this system results from Newton's second law and Hooke's law:

$$M_{eff}\frac{d^2u}{dt^2} = -ku - M_{eff}\gamma\frac{du}{dt} + Z_{eff}E \tag{2.20}$$

where $u$ is the displacement of the mass with respect to its equilibrium position. Assuming a sinusoidal variation of the electric field $E = E_0 e^{-i\omega t}$, a general solution of Eq. (2.17) will be of the form:

$$u = u_0 e^{-i\omega t} \tag{2.21}$$

where $u_0$ is a complex displacement amplitude, which can account for a potential phase lag between the excitation and the response of the system.

Eq. (2.20) leads to the following expression of the displacement amplitude:

$$u_0 = \frac{Z_{eff}E_0}{M_{eff}}\left(\frac{1}{\omega_0^2 - \omega^2 - i\gamma\omega}\right) \tag{2.22}$$

This expression can be split into real and imaginary parts:

$$u_0 = \frac{Z_{eff}E_0}{M_{eff}}\left(\frac{\omega_0^2 - \omega^2}{\left(\omega_0^2 - \omega^2\right)^2 + \gamma^2\omega^2} + i\frac{\gamma\omega}{\left(\omega_0^2 - \omega^2\right)^2 + \gamma^2\omega^2}\right) \tag{2.23}$$

As expected for a driven oscillator, the system displays a resonance when the exciting frequency coincides with the eigen frequency of the oscillator. Such resonances typically occur in the far-IR and MIR range. Because of the damping term, the displacement amplitude does not diverge, but a $\pi/2$ phase lag occurs between the excitation and the displacement (Eq. 2.23).

To relate the mechanical resonance to an absorption process, it is relevant to compute the time-averaged electromagnetic power dissipated in the system. This quantity is obtained by considering the time average of the work done by the electric field on the moving charge:

$$\langle P\rangle_t = \left\langle Z_{eff}\mathrm{E}\frac{du}{dt}\right\rangle_t \tag{2.24}$$

Using the real expression of the electric field and mass displacement (Eq. 2.24) leads to:

$$\langle P\rangle_t = \frac{1}{2}\omega Z_{eff}E_0\,\mathrm{Im}(u_o) \tag{2.25}$$

For a weak damping, combining Eqs. (2.23) and (2.25) shows that for frequencies close to the resonance one, the electromagnetic absorption will have a nearly Lorentzian shape centered on the fundamental frequency of the oscillator. This explains why spectroscopic line shapes are most often fitted using symmetric Lorentzian functions. Voigt functions can be used to model inhomogeneously broadened bands, that is, related to a distribution of vibrational frequencies (e.g. Brauns and Meier, 2009). A different approach based on spectral autocorrelation has also been developed to extract width parameters from absorption bands displaying a more complex line profile (Salje et al., 2000).

From this model, it is also apparent that the absorption is closely connected to the polarisation induced by the ionic displacements. A straightforward consequence of this observation is that when the polarisation is nil for symmetry reasons, for example, for atomic displacements symmetric with respect to an inversion center, no absorption occurs and the mode is not IR active.

### 2.3.2 Low-Frequency Dielectric Permittivity Tensor of a Crystal and Born Effective Charge Tensors

The response of a crystal to an applied oscillating electric field can be modeled by considering that the crystal consists in a collection of damped oscillators characterised by their fundamental frequency, effective charge $Z_{eff,m}$, and effective mass $M_m$. Each oscillator corresponds to an IR active mode of the crystal. The total polarisation is obtained by adding the ionic and electronic contributions, here considered as isotropic:

$$P = \frac{1}{\Omega}\sum_m Z_{eff,m} u_m + \varepsilon_0 \chi_\infty E = \varepsilon_0 \chi E \tag{2.26}$$

where $\Omega$ is the volume of the unit cell, $\chi_\infty$ is the electronic electric susceptibility, considered as frequency independent in the IR range, and $\chi$ the complex electric susceptibility of the crystal. Eq. (2.26) defines the frequency-dependent susceptibility that is related to the complex and dimensionless dielectric permittivity $\varepsilon$, via:

$$\varepsilon = 1 + \chi \tag{2.27}$$

where the frequency dependence is implicitly assumed.

A more realistic expression of the dielectric permittivity should, however, take into account the tridimensional nature of the crystal structure, leading to the dielectric permittivity tensor (Gonze and Lee, 1997):

$$\varepsilon_{\alpha\beta}(\omega) = \varepsilon_{\alpha\beta}(\infty) + \frac{4\pi}{\Omega}\sum_m \frac{S_{m,\,\alpha\beta}}{\omega_m^2 - \omega^2 - i\omega\gamma_m} \tag{2.28}$$

where $\alpha$ and $\beta$ vary over the Cartesian coordinates, and $S_{m,\alpha\beta}$, $\gamma_m$ and $\omega_m$ are the oscillator strength, damping coefficient and transverse angular frequency of mode $m$, respectively. The mode oscillator strength is defined by:

$$S_{m,\,\alpha\beta} = \left[\sum_{k,\alpha'} Z^*_{k,\alpha\alpha'} \eta_{m,k\alpha'}\right]\left[\sum_{k,\beta'} Z^*_{k,\beta\beta'} \eta_{m,k\beta'}\right] \tag{2.29}$$

where $Z_{k,\beta\beta'}$ is the Born effective charge tensors of atom $k$ and $\eta_{m,k\alpha}$ is the displacement of atom $k$ along the direction $\alpha$ for the mode $m$. The atomic displacement patterns $\eta_{m,k\alpha}$ verify the following condition due to the orthonormality of the eigenvectors of the dynamical matrix:

$$\sum_{k,\alpha} M_k \eta^*_{m,k\alpha} \eta_{n,k\alpha} = \delta_{nm} \tag{2.30}$$

where $M_k$ is the mass of atom $k$, and $\delta_{nm}=1$ if $n=m$, $\delta_{nm}=0$ if $n \neq m$.

The macroscopic dielectric properties of a crystal, here described by its dielectric permittivity tensor, can thus be obtained from microscopic parameters, such as electrostatic charges, atomic masses and force constants. At this stage, the interaction of IR light with the system, including propagation, absorption and scattering, can be determined knowing the complex dielectric permittivity tensor of the material and using Maxwell's laws of electromagnetism. In Eq. (2.29), the expression of the charges differs from the simplified static and isotropic effective charge used to characterise the coupling of the mechanical oscillator with the oscillating electric field in Eq. (2.20). The static ionic charges already differ from the formal ones because of the partially covalent character of chemical bonds in most clay minerals. In addition, the anisotropic and dynamical nature of the charges also have to be considered for a more realistic description of the low-frequency dielectric permittivity of crystals. The Born effective charge tensors (Born and Huang, 1954; Pasquarello and Car, 1997; Gonze and Lee, 1997) are defined at the centre of the Brillouin zone ($q=0$) and describe the linear relation between the polarisation per unit cell along the direction $\beta$, and the displacement of the atom $k$ along the direction $\alpha$:

$$Z^*_{k,\alpha\beta} = \frac{\Omega}{e} \frac{\partial P_\beta}{\partial \eta_{k\alpha}} \tag{2.31}$$

where $e$ is the electron charge. Equivalently, they describe the linear relation between the force acting on the atom $k$ in the direction $\beta$ and a homogeneous (i.e. constant over the crystal cell) electric field along the direction $\alpha$. The Born effective charges are dynamical quantities because they incorporate the changes of the electronic structure induced by the atomic displacements and related to the local polarisability of atoms or interatomic charge transfers (Ghosez et al., 1998). When a positive atom is displaced toward a negative atom, the electronic shell around a polarisable negative atom can move toward the positive atom or a change in the bond hybridisation can cause the transfer of electrons from the negative to the positive atom. As a consequence, the polarisation due to the ion displacement is dynamically augmented with respect to static charges. This property has strong consequences on IR absorption intensity (e.g. Pasquarello and Car, 1997) because the absorption scales as the squared electric charge.

Note that a different formulation of the dielectric tensor can be adopted to account for a different damping of TO and LO modes (e.g. Gervais and

Piriou, 1974). Such type of formulation is required in materials exhibiting a large difference between the LO and TO frequencies, which implies that the anharmonic mode coupling to the vibrational density of states of the TO and LO modes can be significantly different.

### 2.3.3 IR Spectroscopy of Powder Materials

#### *2.3.3.1 Light Reflection and Transmission by an Isotropic Dielectric Slab*

Various experimental geometries can be used to record an IR spectrum, depending on the nature of the sample. The measured signal can correspond, for example, to the fraction of incident light intensity reflected or transmitted by a macroscopic sample (Fig. 2.4). In the case of powder material, such as clay minerals, the spectroscopic information is usually not recorded on a single crystal. Transmission measurements are performed on composite samples consisting in a small quantity of powder diluted in a nonabsorbing matrix (Fig. 2.5), whereas the IR spectrum of pure powders can be recorded using other geometries (e.g. attenuated total reflection (ATR) spectroscopy;

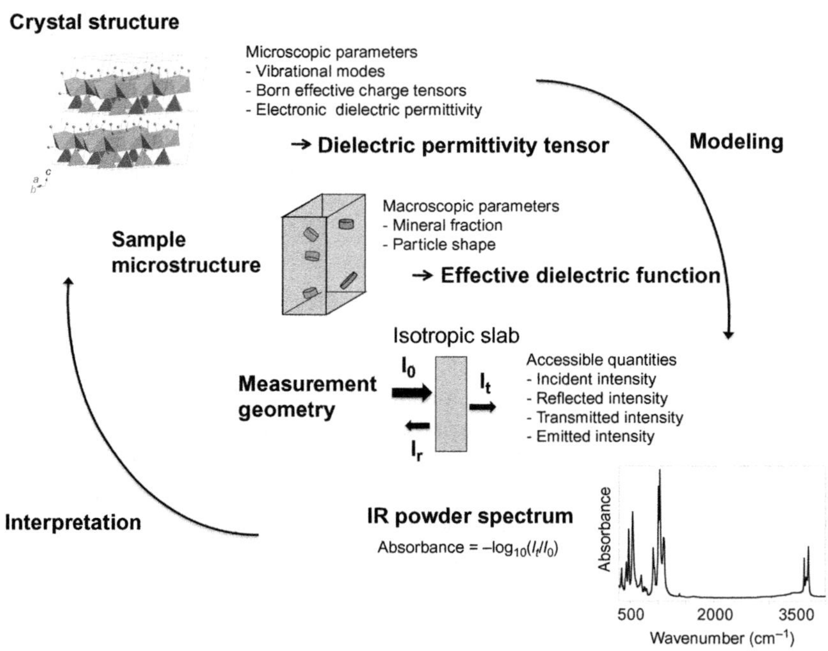


**FIG. 2.4** Flowchart relating the atomic-scale structure of a mineral to its measured powder infrared spectrum. Note that all the atomic-scale parameters are in the dielectric permittivity tensor of the crystal whereas the effective dielectric function of the measured sample also incorporates macroscopic parameters. The interpretation of experimental spectra has to consider these various aspects and can be sustained by direct modeling strategies.

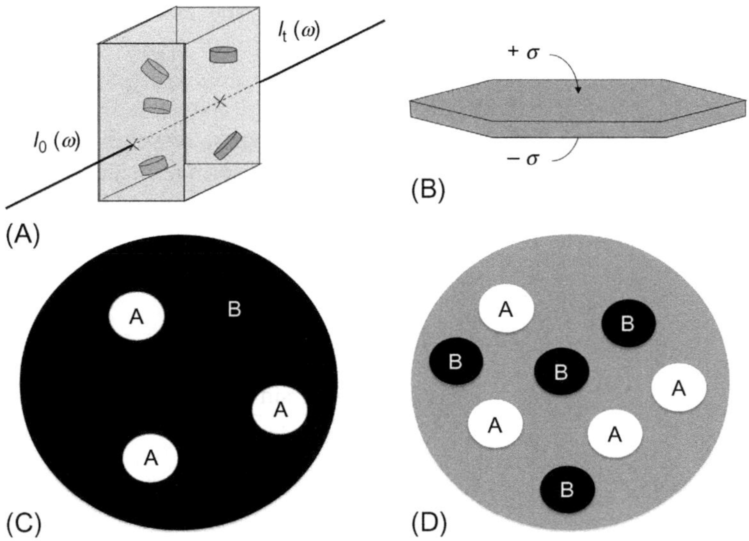


**FIG. 2.5** (A) Usual geometry of powder transmission measurements. The sample consists in dilute particles in a nonabsorbing matrix. (B) For platy particles, the influence of the particle shape on powder IR spectra is related to the oscillating charges appearing at the surface of the polarised particles when the polarisation field is perpendicular to the plate. In this case, the absorption spectrum can be related to the orientation and time average of the electromagnetic power dissipated in a particle. Models explicitly accounting for the fraction of substance in a composite sample are illustrated below. In the Maxwell Garnett model (C), the particles of one substance A are considered as embedded in a matrix B whereas in the Bruggeman model (D) the two substances, A and B, are considered as embedded in the effective medium (in grey).

photoacoustic spectroscopy). In this latter case, the powder sample displays some porosity and actually consists in a composite material made of mineral and air. Therefore, although individual particles are characterised by a significant anisotropy, the actual macroscopic sample usually displays isotropic properties, provided that the particle size can be considered as small with respect to the IR wavelength and sufficient statistical orientation homogeneity is ensured.

Considering an electromagnetic wave perpendicularly incident on a slab of nonopaque isotropic body with parallel faces, a part of the radiation will be reflected off the surface, whereas another one is transmitted through the slab (Fig. 2.4). Both parts can be actually detected and measured. Neglecting any light scattering, a balance of energy can relate these two parts to a last one, which corresponds to the radiation absorbed on its way through the slab:

$$r_\nu + \alpha_\nu + t_\nu = 1 \tag{2.32}$$

where $r_\nu$ = reflectance, $t_\nu$ = transmittance, $\alpha_\nu$ = absorbance. All these quantities are frequency-dependent "apparent quantities" which depend on the thickness of the slab. They are quantitatively determined by the imaginary refractive index $\hat{\eta}(\omega)$, which describes the propagation of the electromagnetic wave through the material:

$$\hat{\eta}(\omega) = n_1 + in_2 \tag{2.33}$$

where $n_1$ and $n_2$ are the real and imaginary parts, respectively, and the frequency-dependence implicitly considered. For a normal incidence on a slab of thickness $d$ with isotropic properties, the transmittance $t$ and reflectance $r$ are given by (Bruesch, 1986):

$$t = \frac{I_t}{I_0} = \frac{\left[(1-R)^2 + 4R\sin(\phi)^2\right]T}{(1-TR)^2 + 4TR\sin(\phi+\alpha)^2} \tag{2.34}$$

$$r = \frac{I_r}{I_0} = R\frac{(1-T)^2 + 4T\sin(\alpha)^2}{(1-TR)^2 + 4TR\sin(\phi+\alpha)^2} \tag{2.35}$$

In these expressions, $R$, $\phi$, $\alpha$ and $T$ are frequency-dependent quantities defined by:

$$R = \frac{(1-n_1)^2 + n_2^2}{(1+n_1)^2 + n_2^2} \tag{2.36}$$

$$\tan(\phi) = \frac{-2n_2}{1 - n_1^2 - n_2^2} \tag{2.37}$$

$$\alpha = \frac{2\pi n_1 d}{\lambda_0} \tag{2.38}$$

$$T = e^{-Kd} \tag{2.39}$$

In this last relation, the absorption coefficient $K$ describes the absorption of the electromagnetic wave propagating through the sample and is related to the imaginary part of the refractive index by:

$$K = 2n_2\omega/c \tag{2.40}$$

The IR spectrum of a material measured in transmission or reflection reports the fraction of transmitted or reflected intensity as a function of the light frequency, usually expressed in wavenumber units. It is thus fully determined at a macroscopic level by the experimental geometry and the refractive index of the sample. In turn, Maxwell's equations imply that the refractive index is related to the dielectric function of the material $\varepsilon(\omega)$ (which links the macroscopic polarisation of the material to the macroscopic electric field). For an isotropic material, this relation is

$$\hat{\eta}^2(\omega) = \varepsilon(\omega) \tag{2.41}$$

In the case of a crystalline material, its dielectric function is directly determined by the microscopic (atomic-scale) properties of the crystal (Eq. (2.28)).

In the case of a powder crystalline sample, the long-range nature of electrostatic interactions and the finite size of particles can lead to a significant dependence of the measured spectra on the microstructure of the actual sample. Thus an additional step has to be considered to relate the dielectric function, and corresponding IR spectrum, of the composite sample to the bulk dielectric properties of the crystal (Fig. 2.4). Different strategies can be used to achieve this goal, depending on the characteristics of the powder and related composite sample.

#### 2.3.3.1.1 IR Absorption by Isolated Small Particles in a Nonabsorbing Matrix

Pellets used for transmission measurements contain a small fraction of powder diluted in a nonabsorbing matrix, most often KBr (Fig. 2.5). In the limit of high dilution, the electrostatic interaction between particles can be neglected, and the sample absorption is obtained as an orientation average of the electromagnetic power dissipated in a single particle. Although not taking into account the spectral dependence on the substance fraction in the KBr pellet, this approach accounts for the relations between the absorption spectrum and the shape of particles.

For a particle size smaller than the IR wavelength and an ellipsoidal particle shape, the internal macroscopic electric and polarisation fields are homogeneous. The internal electric field differs from the external one because of the polarisation of the particle, which behaves as a capacitor (Fig. 2.5). The transitory accumulation of charges at the surface of particles, related to the oscillating motion of ions, induces a macroscopic electric field opposing their displacement. This contribution adds to the microscopic restoring force related to the interatomic potential, thus increasing the vibrational frequency. Such electrostatic effect is closely related to the electrostatic interactions giving birth to the LO-TO splitting in the bulk infinite crystals. The observed frequencies will range between the transverse and the longitudinal ones (Fig. 2.3), depending on the boundary conditions that control the charge accumulation at the particle surface. Larger effects will be observed on the strong absorption bands, corresponding to a large ionic polarisation. Interestingly, these electrostatic effects not only involve the vibrational properties of minerals and the shape of particles but also the various electronic contributions to the sample polarisability, that is, the dielectric permittivity of the nonabsorbing matrix and the high-frequency electronic permittivity of the mineral.

The influence of these macroscopic electrostatic effects on powder IR absorption spectra has been investigated in simple oxides and relatively symmetric minerals such as periclase, hematite, zircon or apatite (e.g. Rendon and Serna, 1981; Serna et al., 1987; Iglesias et al., 1990; Pecharroman et al., 1994; Pecharroman and Iglesias, 1996, 2000; Blanchard et al., 2008; Balan et al., 2011b). They also have to be considered for the interpretation of the spectrum of a number of clay minerals and related minerals, such as kaolinite group minerals (Farmer, 1998, 2000; Balan et al., 2001, 2005), serpentine group minerals (Balan et al., 2002a,b; Trittschack et al., 2012), aluminium and iron (oxy) hydroxides (Balan et al., 2006, 2008; Delattre et al., 2012; Blanchard et al., 2013). A specific dependence of the IR spectrum on the high-frequency electronic permittivity of the mineral has been demonstrated on irradiated kaolinite samples (Fourdrin et *al.*, 2008). In this case, irradiation induces the creation of electronic defects and increases the electronic polarisability of the material, indirectly affecting the vibrational frequencies and the position of specific IR absorption bands.

#### 2.3.3.1.2 Effective Dielectric Functions of Composite Samples

Various models provide the approximate dielectric function of a composite sample from its individual constituents in a quasistatic approach (e.g. Sihvola and Kong, 1988; Spanier and Herman, 2000; Kendrick and Burnett, 2016). They neglect scattering losses and can be extended to frequency-dependent dielectric properties when the characteristic size of individual particles is much smaller than the wavelength of the electromagnetic radiation, a condition usually approximately verified in MIR spectroscopy ($\lambda_{IR} > 10$ µm). The effective dielectric function of the isotropic composite sample can then be related to the measured spectrum, either through the related absorption or reflection coefficients depending on the geometry of the experiment.

One of the simplest models was proposed by Maxwell-Garnett (Garnett, 1904), in which the particles of one of the constituents are considered as embedded in a homogeneous medium, corresponding to the other constituent for binary samples. In this model, the two constituents are therefore not treated in a symmetrically equivalent way. The Maxwell Garnett model is closely related to the Clausius-Mossotti derivation of the dielectric constant of molecular substances from the polarisability of molecules (e.g. Landauer, 1978) and is well suited to describe the dielectric properties of a sample in which a low amount of one of the constituents is diluted in a nonabsorbing matrix. In the limit of high dilution, it leads to results equivalent to those obtained by considering isolated particles. Note, however, that it also leads to the average properties of the pure material for a sample fraction tending to one. The correct asymptotic behavior at both high and low dilution limits explains its wide use for the interpretation of the electromagnetic properties of composite samples.

However, for arbitrary microstructures and intermediate volume fractions, the particles of one of the constituents cannot be considered as isolated and embedded in a homogeneous matrix of the other constituent. Correspondingly, an alternative model could be used in which each individual particle of a mixture is considered as embedded in the effective medium (Bruggeman, 1935; Landauer, 1978). At variance with the MG model, all constituents are treated in a symmetrically equivalent way and the effective dielectric properties of the sample are obtained self-consistently.

The two effective medium approaches display very distinct properties as a function of the volume fraction of the constituents (Aufort et al., 2016). This is particularly evident when considering a model isotropic substance displaying a single resonance in its frequency-dependent complex dielectric function $\varepsilon_s = \varepsilon_s' + i\,\varepsilon_s''$ (Fig. 2.6). In the limit of a vanishingly small damping coefficient, the maximum of $\varepsilon_s''$ defines the TO frequency, whereas the LO frequency is defined by $|\varepsilon_s| = 0$. The second medium of the binary composite is characterised by its dielectric constant $\varepsilon_h$, taken to be equal to unity in Fig. 2.6. The corresponding Maxwell Garnett effective dielectric function is defined for spherical particles by:

$$\frac{\varepsilon_{MG} - \varepsilon_h}{\varepsilon_{MG} + 2\varepsilon_h} = f\frac{\varepsilon_S - \varepsilon_h}{\varepsilon_S + 2\varepsilon_h} \tag{2.42}$$

It presents a single resonance, which progressively shifts toward the resonance frequency of the pure material when the volume fraction increases. In the limit of low volume fraction and vanishingly small damping parameter, this resonance coincides with that of the isolated sphere in the host matrix. The intensity of the resonance is not linearly related to the volume fraction of the composite.

Under the same assumptions, the Bruggeman model (Bruggeman, 1935) leads to the following relation:

$$(1-f)\frac{\varepsilon_h - \varepsilon_{Br}}{\varepsilon_h + 2\,\varepsilon_{Br}} + f\frac{\varepsilon_S - \varepsilon_{Br}}{\varepsilon_S + 2\,\varepsilon_{Br}} = 0 \tag{2.43}$$

In the limit cases $f \to 0$ or $f \to 1$, the Bruggeman dielectric function is identical to that obtained using the Maxwell Garnett model, with a single resonance at the characteristic frequency of the isolated sphere or of the pure bulk material, respectively. In contrast, the line shape of the imaginary part of the dielectric function at intermediate volume fractions significantly broadens and becomes asymmetric (Fig. 2.6). For volume fractions above 1/3, the lines extend between their TO and LO frequencies with a relatively higher intensity on the TO frequency side.

More sophisticated models can also be used to better account for the specificities of actual samples (Spanier and Herman, 2000). However, the better agreement with experiment is obtained at the cost of an increasing number of parameters. In many cases, the simple models exposed here already provide a good approach to disentangle the various microscopic and macroscopic effects contributing to the measured IR spectra.

### 2.3.3.2 *Infrared Emission Spectroscopy (IES)*

Beside usual reflection or transmission measurements mentioned in the preceding paragraphs, it is also possible to measure the IR light emitted by a

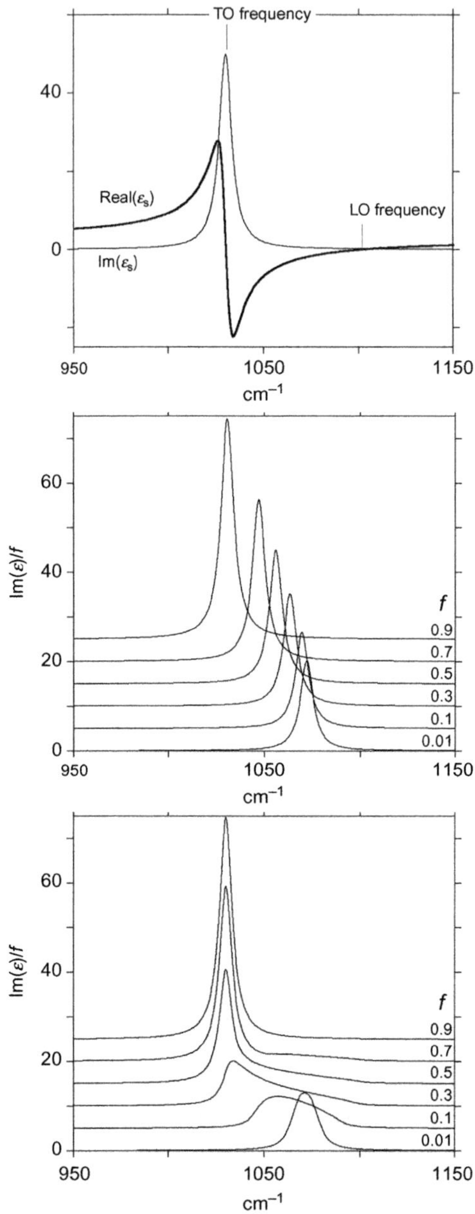


**FIG. 2.6** Real and imaginary parts of a model single-resonance dielectric function (top) and corresponding imaginary part of the effective medium dielectric function in the Maxwell Garnett (middle) and Bruggeman (bottom) models for a composite sample made of the substance (volume fraction f ) and a second substance of dielectric constant equal to unity. The effective dielectric functions have been divided by corresponding volume fraction, reported on the right. Note the significant shift of the resonance of the effective dielectric function with respect to that of the pure substance, as well as the specific lineshape in the Bruggeman model for intermediate fractions (From Aufort et al., 2016).

sample in thermal equilibrium with a heat source. This is particularly useful to investigate the high-temperature properties of materials. The emitted light in the IR range displays characteristic peaks arising from the radiative decay of the thermally populated vibrational modes. The IR emission spectra thus bring information of the same nature as those recorded with a more conventional setup.

Dehydration, dehydroxylation and phase changes of clay minerals have mostly been studied using thermal analytical techniques such as thermal gravimetric analysis (TGA), differential thermal analysis (DTA) and differential scanning calorimetry (DSC) (Baraldi, 1990; Mackenzie, 1957; Stucki and Bish, 1990). More recently other analytical methods have become available to study these changes in inorganic materials in situ, such as heating stage X-ray diffraction, heating stage scanning electron microscopy and heating stage transmission electron microscopy. The use of in situ spectroscopic methods such as heating stage Raman spectroscopy and IR emission spectroscopy (IES) has been generally limited. In order to study dehydration and dehydroxylation of clay minerals, the clay mineral powders are heated to a certain temperature, then rapidly cooled down (quenched), followed by conventional spectroscopic analysis at room temperature. Though in many cases this provides useful data, there is no guarantee that the clay mineral structure does not change upon cooling before the spectral measurement. Smectites, for example, are known to quickly rehydrate after dehydration when exposed to moist atmospheric conditions. Therefore, IES forms a very useful spectroscopic technique that can be applied in situ during the heat treatment of clay minerals (e.g. Kloprogge and Frost, 2005b).

IES has been shown for nearly 50 years to be a useful technique for gaining spectral information at increased temperatures of a variety of compounds, such as organic and inorganic materials including coal, clay minerals, polymers and even liquids or melts (Baraldi, 1990; DeBlase and Compton, 1991). Probably one of the very first papers on the theory of IES can be ascribed to Max Planck (Planck, 1901) in which he developed the distribution law for emitted radiation as a function of temperature and wavelength. For a rather long period this technique received essentially no attention as it was believed that no useful information could be obtained. Coblentz (1908, 1912) was one of the very few who reported on the emission spectra at increased temperatures of semitransparent minerals. Kapff (1946; 1948) defined some of the parameters affecting the emitted intensity of a sample. McMahon (1950) proposed a theory for IR emission through a semitransparent material that was later refined by Gordon to comprise hemispherical emitted radiation from the surface (Gordon, 1956).

Low (1965) used Fourier transform infrared (FTIR) spectroscopy to study the IR emission spectra of clay mineral surfaces. The advancement from dispersive IR to FTIR with its high sensitivity to low energy sampling has been very significant for the increasing interest in IES (Baraldi, 1990; Fabbri and Baraldi, 1972; Griffiths, 1972; Jonson et al., 1986; Mink and Keresztury, 1993). Vassallo et al. (1992) described a number of phase changes observed by IES such as the $\alpha$-quartz-$\beta$-quartz transition and the formation of cristobalite, the transformation of calcite to aragonite and of kaolinite via metakaolinite to mullite.

The single-beam reference spectrum obtained during the recording of a normal IR transmission spectrum is in fact the emission spectrum of the instrument source, modified by the instrument response function. Transmission spectra are obtained by taking the ratio of the signals in the absence and presence of the sample. Likewise, emission spectra can be obtained by taking the ratio of the emission signal of the sample and that of a reference, typically a blackbody source that emits radiation according to Planck's radiation formula (i.e. a continuum). The energy radiated by a blackbody is dependent on the frequency and temperature according to the equation:

$$\rho(v,T) = \frac{8\pi h v^3}{c^3}\left[\frac{1}{\exp\left(\frac{hv}{kT}\right) - 1}\right] \tag{2.44}$$

where $\rho(v, T)$ is the density of radiation of frequency v at temperature T. The frequency corresponding to the maximum of $\rho(v, T)$ increases and shifts to higher frequencies with temperature, so that at 300°C a blackbody is emitting energy over the whole of the MIR range of 400–4000 $cm^{-1}$. Consistent with Kirchoff's law for a body in thermal equilibrium with its surroundings, the absorbed and emitted energies are equal, and since a perfect blackbody will absorb all incident radiation, it will also be the perfect emitter. Hence real samples will always emit less energy than a blackbody. At a given temperature $T$, the ratio of the energy emitted by the sample ($E_s$) to that of a blackbody ($E_b$) at frequency ($v$) defines its emissivity $\varepsilon_v(v)$:

$$\varepsilon_v(v) = E_s(v,T)/E_b(v,T) \tag{2.45}$$

The emissivity of a sample will consequently always be less than or equal to 1. It is not constantly required to ratio the emitted energy to a blackbody radiator in order to obtain a useful emission spectrum. A body for which $\varepsilon_v(v)$ is constant and less than 1 for all frequencies is known as a grey body. The energy distribution as a function of the frequency is the same as for a blackbody at the same temperature. Nevertheless, in many materials, the emittance fluctuates strongly with the frequency, as materials absorb and consequently emit strongly at their characteristic vibrational frequencies.

McMahon (1950) established a model for the unidirectional emission normal to the front surface of a semitransparent slab. For such sample, the total emittance can be related to the other measurable quantities characterising the slab, that is, reflectance and transmittance (Eqs. 2.34 and 2.35):

$$\varepsilon_v(v) = \frac{(1-r)(1-t)}{1-rt} \tag{2.46}$$

Later, Hvistendhal et al. (1983) derived a similar equation for emittance of a sample in contact with a highly reflective backplate:

$$\varepsilon_v(v) = \frac{(1-r)(1-t^2)}{1-rt^2} \tag{2.47}$$

In the two models the emittance depends on both the transmittance and reflectance of the sample. The emittance spectrum in the limiting case of an opaque sample (transmittance = 0) is totally determined by the difference in surface reflectance with frequency. In semitransparent samples, strong emission bands can show some distortions or false splitting because the reflectance as well as the transmittance is fluctuating strongly where the absorption coefficient is large. These reflectance effects may be removed by taking the ratio of the emission with that of the emission of a thick sample of the same material rather than to that of a blackbody.

When very thin sample layers are measured (about 1 μm), reflectance effects are seldom a problem. Furthermore, disproportionate temperature gradients are also avoided by using very thin samples, thus minimising the readsorption of intense frequencies by cooler, outer layers. As the sample thickness increases past a certain point, the spectral contrast between an actual band and the background noise decreases. Both reflectance and reabsorption affects results in attenuation of the stronger bands. Therefore the best spectra are obtained with the thinnest samples.

When comparing emission spectra with absorption spectra, the difference in scales must be kept in mind. Absorbance is measured as $-\log_{10}$(transmittance) and varies from 0 to ∞ as absorbance and emittance vary between 0 and 1. Accordingly, weaker bands appear to be enhanced in emission spectra relative to absorbance spectra. Emittance values should be converted as $-\log(1-\alpha)$ in order to obtain units that are linearly proportional to concentration. Care should be taken, however, in quantitative applications because of the reflectance and reabsorption effects.

Generally, it is more practical to take the ratio of "grey body samples" to those of "grey body references." This method is used to improve the signal to noise ratio of the emission spectrum. When a grey body radiator is used instead of a blackbody radiator, the measured sample emissivity $\varepsilon'(v)$ is equal to:

$$\varepsilon_v'(v) = E_s(v,T)/E_{gb}(v,T) \tag{2.48}$$

and is related to the true emissivity by a function $K(v,T)$:

$$K(v,T) = E_{gb}(v,T)/E_b(v,T) \tag{2.49}$$

In the normal course of events, three sets of spectra are obtained: (1) the blackbody radiation (in reality the grey body radiation) over the temperature range selected at the various temperatures, (2) the platinum plate radiation is obtained at the same temperatures, and (3) the spectra from the platinum plate covered with the sample. Normally only one set of blackbody and platinum radiation is required. The emittance spectrum at a particular temperature can then be calculated by subtraction of the single-beam spectrum of the platinum sample holder from that of the platinum sample holder +sample, and finally the ratio is taken of this resulting spectrum to that of the single-beam spectrum of an approximate blackbody (graphite). This spectral manipulation can be carried out after all the spectral data have been collected.

## 2.4 RAMAN SPECTROSCOPY

Conventional Raman spectroscopy probes the vibrational modes of a material through a nonresonant process (Raman and Krishnan, 1928; Long, 1977; McMillan and Hofmeister, 1988; Nasdala et al., 2004). It has been widely used in the investigation of clays and clay minerals (e.g. Johnston et al., 1985, 1998; Frost, 1995; Frost et al., 1996; Frost and Van Der Gaast, 1997; Frost and Kloprogge, 2000a,b; Ruan et al., 2001). Raman contributions are usually detected in the light scattered by a sample illuminated by an intense monochromatic laser source with a frequency larger than fundamental vibrational frequencies but still smaller than that corresponding to the energy gap between valence and conduction bands.

In a classical picture, the scattered light is related to the time-dependent polarisation of the system induced by the incident electric field $E = E_i cos(\omega_i t)$. Far from the vibrational excitations, the macroscopic electrostatic polarisation of the solid is dominated by the electronic contribution:

$$P(t) = \varepsilon_0 \chi_\infty E_i \cos(\omega_i t) \tag{2.50}$$

Although the electronic susceptibility was assumed to be constant in the preceding expressions, it actually depends on atomic positions. Limiting the development to first-order derivatives, the oscillatory motion of atoms in a zone-center vibrational mode characterised by the displacement pattern $Q_m$ and angular frequency $\omega_m$ induces a modulation of the electronic susceptibility at the corresponding frequency:

$$\chi_\infty(t) = \chi_\infty(0) + \left(\frac{\partial \chi_\infty}{\partial Q_m}\right)_0 Q_m \cos(\omega_m t) \tag{2.51}$$

where the first derivative of the electronic susceptibility is computed for atomic displacements along the pattern corresponding to the mode $Q$m. This leads to the following expression for the induced polarisation:

$$P(t) = \varepsilon_0 \mathrm{E}_i \left( \chi_\infty(0)\cos(\omega_i t) + \frac{1}{2}\left(\frac{\partial \chi_\infty}{\partial Q_m}\right)_0 Q_m[\cos((\omega_i+\omega_m)t) + \cos((\omega_i-\omega_m)t)] \right) \tag{2.52}$$

Accordingly, the scattered light will contain a dominant contribution at a frequency equal to that of the laser light. As this contribution does not involve any energy exchange between the electromagnetic wave and the system, it is referred to as the elastic (or Rayleigh) contribution. Two weaker contributions in the scattered light occur at higher (anti-Stokes) or lower (Stokes) frequency with respect to that of the incident light. Their intensity is about three orders of magnitude lower than that of the elastically scattered light and six orders of magnitude lower than that of the incident light. The frequency difference between the incident light and these weak contributions in the scattered light corresponds to that of the vibrational mode $Q_m$. As for the IR absorption, the Raman activity is strongly dependent on the system symmetry. Only modes inducing a modification of the electronic polarisability of the crystal are Raman active. For example, the Raman-active modes in structures presenting the inversion symmetry are symmetric with respect to inversion, whereas the IR-active modes are antisymmetric with respect to inversion. In this specific case, Raman and IR activities are mutually exclusive. For other symmetries, a given mode can display both Raman and IR activities. Note also that some zone-center modes referred to as silent modes can be inactive both in Raman and IR spectroscopies.

A more detailed theoretical treatment leads to the expression of the Raman intensity in insulating crystals. The intensity of the Stokes contribution is proportional to (Lazzeri and Mauri, 2003; Veithen et al., 2005):

$$I_m \propto \frac{(\omega_0-\omega_m)^4}{c}\left|\vec{E}_i.\overleftrightarrow{A}_m.\vec{E}_s\right|^2 \frac{h}{4\pi\omega_m}(n_m+1) \tag{2.53}$$

where $\vec{E}_i$ and $\vec{E}_s$ are the polarisation of the incident and scattered light, $\omega_0$ is the pulsation of the incident laser light, $\omega_\mathrm{m}$ the angular frequency of the vibrational mode $m$, $n_m = (\exp(\hbar\omega_m/k_B T) - 1)^{-1}$, and T is the temperature. The Raman tensor $\overleftrightarrow{A}_m$ is defined by:

$$A_{m,kj} = \sum_{i\alpha} \frac{\partial \chi_{\infty,kj}}{\partial r_{i\alpha}} \frac{\gamma_{m,i\alpha}}{\sqrt{M_i}} \tag{2.54}$$

where $E_k$ is the $k^{\text{th}}$ Cartesian component of a uniform electric field, $r_{i\alpha}$ the displacement of the $i^{\text{th}}$ atom in the $\alpha^{\text{th}}$ direction, and $\gamma_{m,i\alpha}$ the displacement of this atom in the orthonormal vibrational eigenmode m.

Raman intensity in crystals depends on the polarisation of the incident and scattered light (Eq. 2.53). The experimental geometry is usually reported using the Porto notation $A(BC)D$, where $A$ and $D$ are the propagation directions and $B$ and $C$ are the polarisation directions of the incident and scattered light, respectively (Damen et al., 1966). In general, only relative Raman intensities are considered because absolute experimental intensities depend on other processes, including light absorption, which are difficult to monitor. The ratio of Stokes to anti-Stokes intensities depends on temperature (Long, 1977):

$$\frac{I_{mS}}{I_{mA}} = \frac{(\omega_0 - \omega_m)^4}{(\omega_0 + \omega_m)^4} e^{\frac{h\omega_m}{2\pi k_B T}} \tag{2.55}$$

The Stokes spectrum is thus preferentially recorded because it is generally more intense than the anti-Stokes spectrum.

As in IR spectroscopy, the intrinsic linewidth of Raman spectra is related to the anharmonic damping of phonons. Electrostatic effects can also shift and broaden the bands related to polar modes (e.g. Farmer, 1998, 2000). In nanometer-sized particles, phonons with wave vectors larger than zero can contribute to the spectrum. This confinement effect leads to an asymmetric broadening and shift of the Raman lines, which can be related to the phonon dispersion curves (e.g. Zhang et al., 2000).

## 2.5 MODELING OF VIBRATIONAL SPECTRA FROM FIRST PRINCIPLES

All the ingredients required to compute the harmonic vibrational spectra of solids correspond to derivatives of the total energy of the system (i.e. the sum of the electronic energy and nuclear repulsion energy) with respect to atomic displacements and/or the application of an external electric field. These quantities include the force constants, Born effective charge tensors, dielectric permittivity tensor, and Raman tensors (Gonze and Lee, 1997; Baroni et al., 2001; Lazzeri and Mauri, 2003; Veithen et al., 2005). In the past 10 years, the electronic structure calculations based on density functional theory have reached a level of accuracy high enough to reproduce with an excellent agreement the vibrational properties of systems with a chemical complexity and unit cell size relevant for mineralogical studies, that is, of the order of one hundred atoms per unit cell. Theoretical vibrational frequencies are usually observed within a few percents from experimental values, whereas Born effective charge and Raman tensors are typically reproduced within 10%.

Alternative approaches based on molecular dynamics have also been used to obtain the vibrational properties of condensed systems at finite temperature, but they often rely on parameterised potentials to minimise the computation time (e.g. Kirkpatrick et al., 2005).

First-principles methods thus provide the raw ingredients required to compute IR or Raman spectra of crystals (e.g. Balan et al., 2001, 2002a,b, 2005, 2006, 2008, 2011b; Blanchard et al., 2008, 2013; Lazzeri and Mauri, 2003; Delattre et al., 2012; Fourdrin et al., 2008; Aufort et al., 2016). The comparison with experimental data recorded on powder samples, such as clay minerals or hydroxides, can be significantly improved by taking into account parameters related to the experimental geometry (reflection or transmission) or to specific properties of the investigated sample (particle shape, diluting matrix). Note, however, that the agreement between theoretical and experimental spectra is never perfect as the approach significantly differs from empirical fitting schemes. The discrepancies arise from the various approximations done during the calculation. They are related to the implementation of density functional theory, to the harmonic approximation, to the neglect of temperature effects and to the macroscopic modeling of the sample.

Several codes are now routinely used to compute the vibrational properties of minerals, such as the Quantum-Espresso package, which includes the PWSCF and PHONON codes (Giannozzi et al., 2009). In addition, ABINIT (Gonze et al., 2005), CASTEP (Clark et al., 2005), CPMD (Marx and Hutter, 2009), CRYSTAL (Dovesi et al., 2014), SIESTA (Soler et al., 2002), VASP (Kresse and Furthmüller, 1996) have all been successfully applied to modeling the vibration properties and spectra of minerals.

Chapter 3

# Modern Infrared and Raman Instrumentation and Sampling Methods

G.D. Chryssikos

## 3.1 INTRODUCTION

The second chapter in Farmer's famous infrared (IR) book is about IR 'Instrumentation and Techniques,' written by Russell (1974). The chapter was composed at the closing of the era of dispersive IR spectrometers and the dawn of computationally efficient Fourier transform IR (FTIR) instruments or attenuated total reflectance (ATR) cells. Russell addressed good practise in IR data acquisition by transmittance and reflectance methods, sample preparation, variable-temperature and orientation cells, etc., essentially capturing the technical know-how behind the enormous contribution of the 'Farmer group.' More than 40 years later, Russell's chapter is still recommended to those wishing to study clay minerals, clays, clay-based materials and soils by vibrational spectroscopy. The present chapter is an attempt to both update the state-of-the-art and to pay tribute to these pioneers.

Since the early 1990s, the number of vibrational studies of clay minerals has been growing exponentially. IR spectroscopy and, to a lesser extent, Raman spectroscopy are now affordable and routinely available in the modern (geo)chemical laboratory. In times of fast expansion, the meticulous care for collecting high-quality data may be jeopardised by the apparent simplicity of their acquisition. This is an additional reason for revisiting the ways available for collecting vibrational spectra, the advantages of the various sampling techniques and their limitations.

The early motivations for recording the vibrational spectra of clays were within the fields of soil science and planetary research. These are now enriched and cover additionally a range of fields from advanced materials or environmental applications, to deposit management and process mineralogy.

Developments in Clay Science, Vol. 8. https://doi.org/10.1016/B978-0-08-100355-8.00003-5

In all these cases, it is desirable to study the sample 'as is,' in the least invasive manner. The sample may come in various forms: bulk rocks, drill cores, powders, dispersions or oriented films. The clay mineral can be pure, a constituent of a rock or soil, or a component of a hybrid material. The size of the sample can range from isolated particles to tons. In some cases, data acquisition may be needed as a function of temperature, pressure, relative humidity or reaction time. Can these requirements be met by vibrational spectroscopy? What are the techniques of choice?

Aiming at providing answers to these questions, this chapter outlines the key features of the main instruments and measurement configurations. The necessary background in physics is kept to a minimum, and specific examples are drawn from the literature to illustrate the theory, the technical capabilities and the good practises that are specifically relevant to the study of clay minerals. The focus is on the measurement itself and not on the interpretation of the data, which will be covered in detail in other chapters. Some important vibrational techniques such as step-scan IR, near-field vibrational microscopy (excluding ATR), and surface enhancement Raman spectroscopy (SERS) are not included because their use in clay-related problems has been scarce so far. Others, like remote sensing for aerial survey or extraterrestrial application, are covered in dedicated chapters of this book. The author apologises for unwittingly, but inevitably, omitting important contributions.

## 3.2 INSTRUMENTATION

### 3.2.1 IR Spectroscopy

IR spectroscopy involves the resonance of IR photons (typically in the 12,500–10 $cm^{-1}$, 0.8–1000 μm, range) with the vibrational energy differences of the fundamental electronic configuration of a material. For reasons stemming mostly from instrumentation and sampling, the IR range is traditionally split into three subranges: near- (12,500–4000 $cm^{-1}$, 0.8–2.5 μm), mid- (4000–400 $cm^{-1}$, 2.5–25 μm), and far-IR (400–10 $cm^{-1}$, 25–1000 μm). A typical polyatomic system exhibits many vibrational modes, but only those involving a change in dipole moment $\mu$ are active in the IR, $\delta\mu/\delta Q \neq 0$. Transitions between the lowest energy and the first excited vibrational state (*fundamentals*) typically result in strong absorption that is proportional to $\delta\mu/\delta Q$, but higher-order transitions (*overtones* and *combinations*) are also permitted due to anharmonicity.

Nearly all laboratory IR spectrometers today are of the Fourier transform (FT) type (Griffiths and de Haseth, 2007). Instead of a monochromator and its slits, an interferometer is located between the source and the detector. Broadband IR light emitted from the source is modulated by the interferometer, absorbed in part by the sample, detected in the length (time) domain,

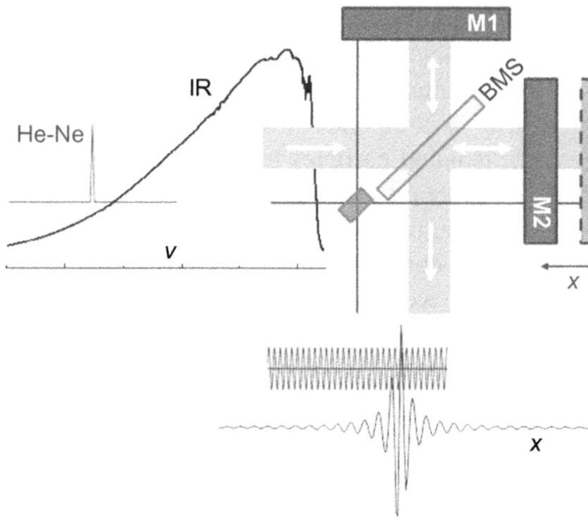


**FIG. 3.1** Schematics of a Michelson interferometer with coupled IR (broadband) and laser (monochromatic) sources. Arrows indicate the light path from the sources *(left)*, through the mirrors *(top and right)*, to the detectors *(bottom)*, where an interferogram is produced.

digitised and Fourier-transformed back to the wavenumber (frequency) domain. The same procedure without the sample provides the reference spectrum.

The operational principle of a simple Michelson interferometer is well known (Fig. 3.1). Light emitted from the source is directed to a beamsplitter (BMS) where it separates evenly into two branches or paths, which in the case of the Michelson type are perpendicular to each other. One path is of fixed length determined by the position of a fixed mirror (M1); the other is variable length because mirror M2 moves (*scans*) very precisely. The two beams recombine at the beamsplitter and are directed to the detector. The as-produced recombined beam (*interferogram*) is then modulated by the optical path difference of the two branches or, equivalently, by the displacement of M2 ($x$). As such, it is recorded by the detector versus $x$, that is, in the length domain.

In the simplest case of a monochromatic source with wavelength $\lambda$, the interferogram is a simple $\cos(2\pi x/\lambda)$ or $\cos(2\pi \nu x)$ wave, where $x$ is the displacement, $\lambda$ is the wavelength and $\nu = 1/\lambda$ is the wavenumber (Fig. 3.1). In the real case of a broadband source, the interferogram is the superposition of all the cosine functions of the wavenumber content of the source which are in phase only at zero path difference, hence the observed centerburst. In the complex shape of its 'wings', the interferogram expresses in the time domain its entire polychromatic input (*multiplexing*). The full spectral information, collected typically within seconds during one scan of the moving mirror, can be converted back to the wavenumber domain by a fast Fourier transform (FFT) computation.

In the case of dispersive spectrometers, achieving higher spectral resolution would require narrowing the slits and would have a detrimental effect on the signal to noise (S/N) ratio. FTIR spectrometers have no slits (*throughput advantage*). Instead, they have a relatively large circular aperture after the source to eliminate oblique rays from reaching the interferometer, as these would reduce resolution in the high wavenumber side of the spectrum. Spectral resolution is varied by changing the maximum path difference (i.e. allowing the moving mirror to travel over longer distances, $x$) and tuning the digitisation (*sampling*) of the interferogram.

For this purpose, FTIR spectrometers combine a broadband IR source with a monochromatic visible laser source of accurately known wavelength (typically HeNe at $\lambda \approx 633$ nm or $\nu \approx 15{,}878\ \text{cm}^{-1}$), both coupled to the same interferometer (although with dedicated beamsplitters and detectors) as in Fig. 3.1. The simultaneous collection of the cosine interferogram of the laser provides a set of regularly spaced zero crossing points (*Dirac delta comb*) that serves as a very accurate ruler for sampling the interferogram at intervals of $\lambda/2$. As a result, FTIR spectrometers offer much higher wavenumber accuracy and reproducibility than dispersive instruments (*Connes' advantage*) and enable much finer spectral comparisons. Increasing the optical path difference spreads the interferogram and results in improved spectral resolution due to denser sampling. For a given speed of the scanner, the spectral resolution is inversely proportional to the scanning time. For a given experimental setup, the signal to noise (S/N) ratio of the spectra varies as the square root of the number of single interferograms (*scans*) that are collected and averaged prior to the FT; hence S/N varies with the square root of the time needed to collect the final spectrum and increases by about 41% for each doubling of time required. This dependence sets a practical time limit for the improvement of S/N by simply increasing the number of scans and justifies investment in efficient instruments and sampling accessories.

Compared to the simple Michelson case, modern instruments incorporate significant improvements in interferometer design. As the need for higher resolution requires moving the mirror M2 over longer distances, even minute off-axis tilts occurring during this movement are undesirable. There are many ways to overcome this problem such as dynamic alignment or the use of cube corner mirrors at M1 and M2. In some configurations, both M1 and M2 mirrors are attached to a swinging double pendulum mechanism. The as-resulting interferometer is small and very robust, suitable for both fundamental research and industrial applications provided that the optical resolution requirements are not very high (e.g. such as those needed for the study of gasses). The reader interested in interferometer design can consult Jackson (2002).

In broad terms, the same types of interferometer are suitable for covering the whole IR region, from the near- to the far-IR. The operating wavenumber range of the spectrometer is then defined by the choice of source, beamsplitter material and detector. Sophisticated research instruments allow for selecting various combinations of prealigned sources, beamsplitters and detectors and enable optimised data acquisition over the desired wavenumber range, as well as the merging of different ranges into a continuous spectrum.

The most common type of source in the mid-IR region is the silicon carbide-based ceramic Globar, which is thermally emitting in the ~8000–100 $cm^{-1}$ range (and therefore can be used to cover also parts of the far- and near-IR). The full near-IR range (~12,000–4000 $cm^{-1}$) is usually covered by an incandescent tungsten-halogen bulb. These thermal emission sources are blackbody radiators. As an exception, thermally emitting samples occupy the place of the blackbody source in the case of IR emission spectroscopy (Section 3.3.6). In the far-IR, especially below ~100 $cm^{-1}$, a high-pressure mercury arc lamp is used. The latter operates via the secondary emission of low-energy photons from its quartz enclosure, excited by the discrete UV and visible lines of the mercury arc. Demanding experiments requiring narrow, collimated and very bright IR beams can benefit from the available synchrotron beamlines (e.g. Pawley and Jones, 2011).

In terms of beamsplitters, the mid-IR (MIR) range is usually covered by a Ge-coated KBr element down to ~400 $cm^{-1}$. $SiO_2$ or $CaF_2$ beamsplitters are typically used in the near-IR (NIR). The far-IR (FIR) region has been traditionally covered by a series of polyethylene terephthalate (Mylar) films with thickness varying from 3 to 100 μm. Increasing the thickness of the film moves the operational window of the beamsplitter to lower wavenumbers, but decreases its width. As a result, the full optimal coverage of the FIR (e.g. 400–30 $cm^{-1}$) would require changing 4 or 5 different beamsplitters, which can be cumbersome. Recent developments of multilayer (e.g. Ge on Mylar) or silicon-based elements are beginning to enable the continuous coverage of the MIR and FIR ranges by a single beamsplitter and are expected to break the barrier between MIR and FIR spectroscopies.

With the exception of the special case of photoacoustic spectroscopy (Section 3.3.7), the most common detectors in both the MIR and FIR ranges (down to ~10 $cm^{-1}$) are pyroelectric devices (e.g. DLATGS, standing for deuterated L-alanine-doped triglycine sulphate) sealed with an appropriate nonabsorbing front window (e.g. KBr, CsI or polyethylene). Pyroelectric detectors operate at ambient temperature and are reasonably fast, but their efficiency drops with increasing wavenumber. For this reason, room temperature diode, PbS or InGaAs detectors are used in the NIR. An appealing alternative to pyroelectric detection in the MIR is offered by mercury cadmium telluride (MCT, $Hg_xCd_{1-x}Te$) photo-conducting elements. These are inherently more sensitive than pyroelectrics and faster, therefore suitable for detecting weak signals or achieving fast scanning rates (in many cases fast scanning rates are required to avoid detector saturation). MCT detectors come in three basic types—narrow-, medium- and wide-band—with low-wavenumber cutoff (750–450 $cm^{-1}$), but also with sensitivity decreasing along the series. Further, they can be fabricated in two-dimensional array forms (e.g. $64 \times 64$ to $256 \times 256$ individually interrogated pixels) for IR imaging applications. Their main drawback is the need for liquid $N_2$ cooling to suppress thermal noise.

Measurements in the low-wavenumber end of the FIR (100–3 $cm^{-1}$) are now possible by the so-called time-domain terahertz (TD-THz) spectroscopy, which offers improved S/N compared to normal FT acquisition (Yan et al., 2001). The description of this technique, which involves the propagation of femtosecond laser pulses through the sample, is beyond the scope of this chapter. The interested reader can follow the first applications of TD-THz to clay samples in Janek et al. (2009, 2010) and in Wilke et al. (2014).

### 3.2.2 Raman Spectroscopy

Raman is an inelastic light scattering technique, fundamentally different from the resonance phenomena measured in the IR (Long, 2002). Monochromatic laser light (with a frequency $\nu_o$) is directed to the sample where most of it is scattered elastically (Rayleigh, also at $\nu_o$), but a tiny fraction ($\sim 10^{-6}$) interacts with the vibrational states and is scattered inelastically at lower (Stokes, $\nu_o - \nu_{vib}$) or higher (anti-Stokes, $\nu_o + \nu_{vib}$) energies, where $\nu_{vib}$ is the vibrational energy of a given vibrational mode. In the absence of electronic resonance effects, the intensity of the Raman scattering is proportional to the fourth power of its absolute wavenumber, $(\nu_o - \nu_{vib})^4$. The Raman intensity is not, therefore, determined solely by the molecular characteristics of the sample which are expressed as the square of the electronic polarisability changes associated with the vibrational mode, $(\vartheta\alpha/\vartheta Q)^2$, but is also very strongly dependent on the selection of excitation laser. Lasers with the highest $\nu_o$, for example, emitting in the blue part of the visible spectrum, even in the ultraviolet, would be the obvious choice (Fig. 3.2, upper). The problem with such choice is that high-energy excitations may activate higher electronic states and induce photoluminescence at $\nu_L < \nu_o$. Photoluminescence can be more efficient than Raman scattering by several orders of magnitude and, even if due to admixtures at impurity levels, it can fully mask the Raman spectrum. A solution to this problem can come by choosing among more than one laser source if installed on the same spectrometer, each with optimised filters and focusing elements. By changing the excitation line $\nu_o$, one can move the Raman spectrum (observed at fixed $\nu_o - \nu_{vib}$) away from the photoluminescence (observed at fixed $\nu_L$), as shown schematically in the lower part of Fig. 3.2. Alternatively, the unwanted excitation of photoluminescence can be suppressed by choosing a low-$\nu_o$ near-IR laser. In practise, this leads to two different designs of Raman spectrometers, single- or multilaser dispersive with visible excitation and FT with NIR excitation.

Top-of-the-line dispersive instruments combine up to three relatively long monochromator modules in various configurations to disperse the scattered light so efficiently that, with the use of slits, the Rayleigh line at $\nu_o$ can

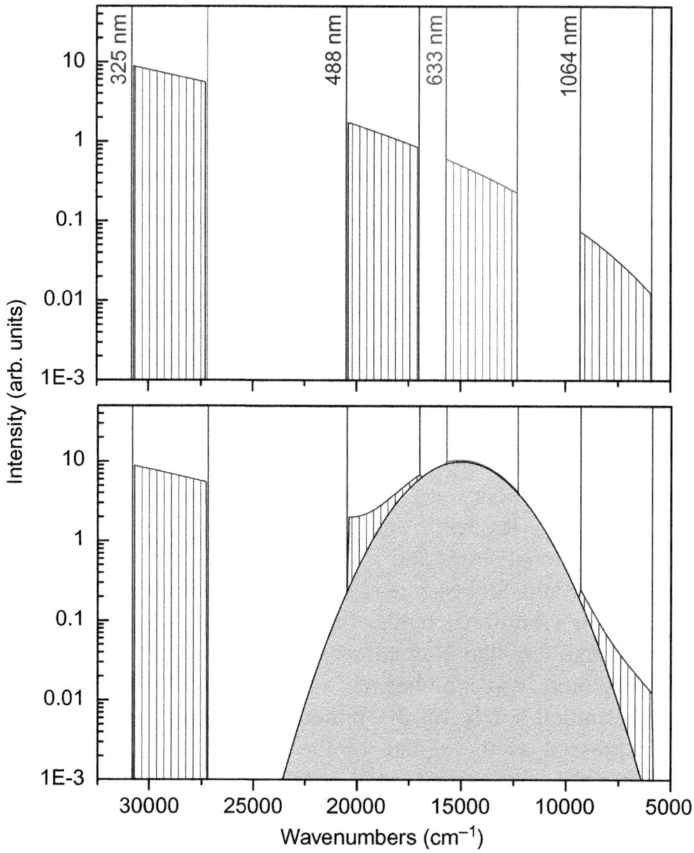


**FIG. 3.2** (*Upper*) Illustration of the dependence of Raman intensity on the absolute wavenumber, $(\nu_o - \nu_{vib})^4$, for four different excitation lines $\nu_o$, and $0 < \nu_{vib} < 3500\ cm^{-1}$. (*Lower*) The same Raman intensities superimposed on a photoluminescence band located at $\nu_L = 15{,}000\ cm^{-1}$. The response of the detector is assumed to be independent of wavenumber. Note the log scale of the *y*-axis.

be approached within $\pm 10\ cm^{-1}$ or less. More compact dispersive instruments employ a single grating configuration and use notch filters with a cut-off of $\sim\pm 50\ cm^{-1}$ from the laser, or one-sided (long pass) edge filters ($-75\ cm^{-1}$) to suppress the excitation line. Various levels of automation enable the change of excitation line (which involves changing filters, optics and, perhaps, gratings), the collection geometry (typically 90 or 180 degrees), the switching between a macrosampling chamber, a fiberoptical sampling head, or a microscope. Ultaviolet (UV) Raman spectrometers need to employ sophisticated and expensive components (coatings, lenses, gratings, etc.) to handle the harsh UV photochemistry.

Typical laser lines for the excitation of Raman spectra in dispersive spectrometers are the commonly available $Ar^+$ (strongest lines at 488.0 and

514.5 nm; 20,492 and 19,436 $cm^{-1}$), $Kr^+$ (530.9 and 674.1 nm; 18,836 and 14,826 $cm^{-1}$), He-Ne (632.8 nm; 15,803 $cm^{-1}$), Nd:YAG (doubled, 532 nm; 18,797 $cm^{-1}$) and several diode lasers emitting in the near-IR (e.g. 785 and 830 nm; 12,739 and 12,048 $cm^{-1}$). UV excitation is possible by using commercial UV lasers such as the HeCd (325 nm, 30,770 $cm^{-1}$) or frequency-doubled $Ar^+$ and $Kr^+$ lines.

Since the mid-1980s, silicon-based charge-coupled devices (CCD) took over from photomultiplier tubes as the most common detectors in dispersive Raman spectrometers. These are array detectors, designed to provide spectral and/or spatial resolution. Their quantum efficiency is high, which is important in recording weak Raman scattering signals. They have a response curve ranging typically between ~450 and 1000 nm (~22,000–10,000 $cm^{-1}$), which can be extended on the UV side but not towards near-IR. Given that fundamental vibrations can be observed up to $\nu_o - \nu_{vib} \approx 3600\ cm^{-1}$, CCDs are not suitable for recording the full Stokes Raman spectrum excited by near-IR lasers and fail progressively beyond $\nu_o - \nu_{vib} \approx 2600\ cm^{-1}$ for 785-nm excitation, or 1900 $cm^{-1}$ for 830-nm excitation energies.

In many cases, near-IR excitation at wavelengths beyond 830 $cm^{-1}$ is required to suppress photoluminescence, despite the concomitant dramatic decrease of the $(\nu_o - \nu_{vib})^4$ factor. This requirement has set the ground for the advent of FT Raman spectroscopy (Hirschfeld and Chase, 1986).

FT Raman instruments do not differ much from their IR analogues and can be designed as an accessory of the latter. They have the same advantages over dispersive instruments (with precautions concerning multiplexing, see below) that compensate in part for the $(\nu_o - \nu_{vib})^4$ drawback. Excitation is provided typically by a continuous-wave, diode-laser-pumped Nd:YAG laser emitting at 1064 nm (~9400 $cm^{-1}$), which is particularly sturdy and requires no cooling. Alternative lines at 976 nm (~10,250 $cm^{-1}$) or 1339 nm (~7470 $cm^{-1}$) have been employed, but are less common. The light scattered both elastically (Rayleigh) and inelastically (Raman) by the sample is collimated and directed to the interferometer through an entrance aperture. A key element of an FT Raman spectrometer is the requirement for very efficient filtering of the Rayleigh and HeNe beams. This is because, in the absence of such filtering, their modulated signal (and especially the noise associated with it) spreads over the whole interferogram and covers that of the weak Raman signal in a manner that turns the multiplex advantage to a disadvantage. With the proper choice of filters, the acquisition of both the Stokes and anti-Stokes branches of the spectrum is possible. Detection is typically by InGaAs or Ge detectors at ambient or liquid-$N_2$ temperatures.

### 3.2.3 IR and Raman Microscopies

Both IR and Raman spectrometers can be coupled to appropriate microscopes with reflective (Schwarzschild, Cassegrain) and refractive optics, respectively.

Both instruments aim at increasing the spatial resolution of the measurement by condensing the beam on the sample. In the case of weak Raman scattering materials one can additionally benefit from the increased energy flux of the excitation line at the focal point (provided that the sample is not absorbing). In both microscopes, the spatial resolution is diffraction limited and not better than $\lambda/2$. In practise, unless a synchrotron source is employed, resolution is $\sim 2\lambda$ for a beam transmitted in air. As a result of its smaller absolute $\lambda$, Raman microscopy offers considerably better spatial resolution than IR ($\sim$1 μm for 514.5-nm excitation vs $\sim$5–40 μm in the 4000–500 $cm^{-1}$ IR range). Further, a suitable Raman microscope used in the confocal mode allows for depth profiling of thick ($\sim$100–150 μm) samples with a resolution of a few μm, provided that the sample does not absorb light at $\nu_o$. Both types of IR and Raman microscope employ collinear white-light systems for sample viewing.

Raman microscopes are relatively similar to their common optical counterparts. The sample is measured in backscattering geometry (180 degrees) and the scattered beam is Rayleigh filtered, analysed and detected by a normal or miniaturised Raman spectrometer. The design of IR microscopes is quite different (Fig. 3.3). The modulated IR beam is transmitted and focused on the sample by spherical mirror optics. Their use is mandatory not only because of the light absorption properties of common lens materials, but also because of the need to minimise aberrations due to the broad wavenumber range of operation. After the sample, the beam is directed to dedicated detectors in one or more compartments of the microscope. Liquid $N_2$-cooled MCT elements are typically employed in the mid-IR range. Focal plane array (FPA) detectors provide additional hyperspectral imaging capabilities. Perhaps the most distinguished feature of IR microscopes is their ability to take multiple objective heads with different optical configurations (Fig. 3.3). Whereas the various optical configurations for IR spectroscopy and their application in the study of clay-related samples are detailed in the following section, these objectives are listed here for completeness. The standard Schwarzschild mirror lenses allow for transmission, near-normal incidence specular reflection or reflection-absorption. Additional devices can be used for grazing or diffuse reflectance. Of great interest is an objective incorporating miniature attenuated total reflectance (ATR) element. The ATR element is retractable to enable sample visualisation, and its contact with the sample can be controlled by a pressure sensor for efficient micro-ATR acquisition. In this latter case the beam is directed to the sample via a high refractive index medium and the spatial resolution is improved by a factor up to 4 ($\sim$5 μm at 1000 $cm^{-1}$ for a Ge ATR element). Further improvement of spatial resolution, exceeding the diffraction limit and down to $\sim$50 nm, is only possible by near-field and/or plasmon resonance enhancement techniques, which are not covered by this chapter.

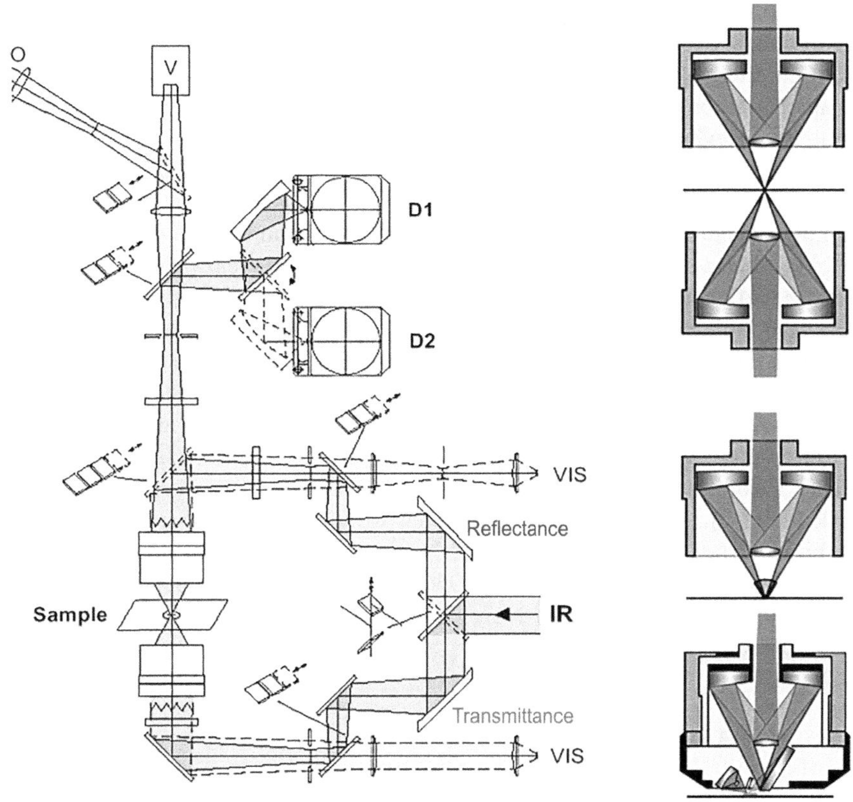


FIG. 3.3 *(Left)* Optical layout of a modern infrared microscope. The incoming modulated beam (IR) can be switched between prealigned pathways for transmittance and reflectance measurements. Visible light beams (VIS) run along the transmittance and reflectance lightpaths for sample viewing. Detector compartments (D), oculars (O) and a CCD camera (V), are seen in the top part of the instrument. *(Right)* Schematics of the mirror objectives for transmittance and near-normal specular reflectance *(top)*, micro-ATR *(middle)* and grazing-angle reflectance measurements *(bottom)*. *(Adapted and reproduced with permission from Bruker Optics, Ettlingen, Germany.)*

## 3.2.4 Portable and Miniature Instruments

The design of portable and miniature instruments is driven by the need for small, robust systems having a minimum number of moving parts, which can be handheld or remotely operated in order to be used for field or process mineralogical monitoring on Earth or other planets. These instruments are inevitably inferior in overall capabilities compared to their research grade analogues, but they can be tuned to the spectral range, resolution and sensitivity requirements of specific applications and can come integrated with built-in chemometric or library-search algorithms. Interferometry is frequently avoided in an effort to minimise moving parts, and near-IR sources are common because of their small size, robustness and lower cost. Portable instruments exploit various optoelectronic and mechanical components and configurations, the full coverage of which is beyond the scope of this chapter. Among these components, diode lasers, lamps, arrays of LED sources or even sunlight are used as the source. Wavelength selection in these types of instruments relies on passive interference filters including variable ones, tunable acousto-optical devices, as well as stationary or rapidly scanning gratings. Detection is by multichannel arrays that can be cooled thermoelectrically, and sampling is by optical fibres or other diffuse reflectance heads. Mobile or handheld Raman spectrometers are also available for in situ measurements. The interested reader can consult the relevant general (e.g. Kawata, 2002; Vandenabeele et al., 2014) or instrument-specific (e.g. Hargreaves et al., 2008; Vítek et al., 2012) literature. A very recent example of miniaturisation and portability is an NIR micro-spectrometer with embedded illumination and a thin-film linear variable Fabry-Perot filter monochromator bonded directly to an array detector (Alcalà et al., 2013). The spectrometer weighs ~60 g, is smaller than a Rubik's Cube, and performs diffuse reflectance or transmittance measurements over the 1150–2150 nm or 950–1650 nm ranges (~8700–4650, and ~10,500–6050 $cm^{-1}$, respectively), albeit with a modest optical resolution (~1.25% of the centre wavelength).

## 3.3 IR SAMPLING TECHNIQUES

### 3.3.1 Transmission Through Dispersions in Transparent Media

A fine dispersion of ~1 wt% clay in spectroscopic-grade alkali halide (typically KBr) powder, sintered to a pellet form under pressure was, perhaps up to the mid 1990s, the most common way of preparing samples for mid-IR measurements of clay minerals and is still widely used. Preparing a good KBr pellet requires the proper grinding of the sample in a manner that does not damage the crystallites, its thorough mixing with the matrix, as well as consistent drying of the sample, matrix or the final pellet at temperatures in the range of 100–200°C for extended periods of time. Although simple in principle, the preparation of good-quality dispersions can be time-consuming and limit high-throughput applications. Detailed and standardised protocols for KBr pellet preparation are an essential part of collecting IR transmission data aiming at mineral identification (e.g. Farmer and Russell, 1964, 1967; Russell and Farmer, 1964; Madejová and Komadel, 2005), quantification of mineralogical composition (e.g. Joussein et al., 2001; Kaufhold et al., 2012), or analysis of the octahedral composition of clay minerals (e.g. Wilkins and Ito, 1967; Slonimskaya et al., 1986; Madejová et al., 1994; Besson and Drits, 1997a,b; Petit et al., 2002; Bishop et al., 2002a,c; Zviagina et al., 2004, 2015), and should not be overlooked.

Regardless of efforts in standardising the preparation of pellets, the use of the alkali halide matrices induces inevitable perturbations to the clay composition and structure. A main reason for these perturbations is the need to dry the pellet, usually by a combination of prolonged heating at 100–200°C and/or vacuum. Concerns stem mostly from the uncontrolled redistribution of water and mobile cations between the matrix and the clay (Russell, 1974; Pelletier et al., 1999), but additional effects due to the changes of the covalent framework of the layer are not uncommon. Water, cations, termination defects and guest organics in close association with the clay minerals layers can be permanently altered by the presence of KBr. The surface, the responsive interface that links the mineral to its environment and determines its uses, becomes unavailable for direct investigation. A striking example are the silent terminal silanol groups of dry sepiolite and palygorskite dispersed in KBr, which are abundantly observed in the corresponding spectra of self-supported clay minerals films (Ahlrichs et al., 1975).

As alkali halides have limited transparency in the low wavenumber range (KBr absorbs strongly below ~400 $cm^{-1}$), 10%–25% dispersions in polyethylene, melted between two flat plates and quenched to a film, are typically used for FIR measurements in transmission (e.g. Laperche and Prost, 1991; Madejová et al., 2006; Diaz et al., 2010).

### 3.3.2 Transmission Through Thin Films

This is the sampling technique of the early pioneers (Serratosa, 1960; Russell and Farmer, 1964; Farmer, 1968): The spectra are measured in transmission through oriented, self-supported (>0.7 $mg/cm^2$) clay mineral films. These films are made by depositing clay mineral/water dispersions on aluminium foil, polyethylene or polyester sheets, evaporation and subsequent peeling of the dry film. Before the advent of ATR sampling, this sampling technique was the only option available for studying adsorption and surface reactivity phenomena, for example, in the general context of soil science and agricultural applications. Further, this type of sample preparation was relevant to companion X-ray diffraction (XRD) measurements on the same oriented films and enabled comparisons between spectroscopic and diffraction data.

Several examples of homemade IR transmission cells from the earlier literature can be found in Stevens and Anderson (1996), Farmer (1968) and Russell (1974). Their design would enable the control of clay hydration by heating, vacuum, or saturated salt solutions, as well as the tilting of the sample against the beam. The latter would enable the measurement of dichroism for determining the orientation of the structural OH groups with respect to the oriented layers (Serratosa, 1960; Fripiat et al., 1960; Calvet and Prost, 1971; Farmer and Russell, 1971; Suquet et al., 1977). Heating was typically provided by a suitable mantle or sleeve, but Fripiat et al. (1960) proposed casting the clay mineral film on a platinum screen that could be used for resistive heating. Other transmission cells were developed specifically for studying the adsorption of organic vapours following a degassing treatment (e.g. Eberly, 1967; Annabi-Bergaya et al., 1980). Adsorption/desorption studies called for the design of cells enabling the simultaneous acquisition of IR and gravimetric data. An early version of such a cell was employed by Swoboda and Kunze (1964) in their study of pyridine adsorption on clay mineral films. In the early 1970s, Prost (1975b) developed a transmission cell with two interconnected environmental chambers for collecting sample and reference spectra in a double beam spectrometer, as well as accessories for varying the orientation of the film, for heating (up to 400°C), cooling (by liquid $N_2$), or under controlled humidity atmosphere, and also for gravimetric measurements in situ by a spring balance. This system served a number of pioneering studies regarding the hydration, H/D exchange and dehydration of clay minerals and zeolites (Prost, 1975a,b; Suquet et al., 1977; Sposito and Prost, 1982; Sposito et al., 1983). Later, Johnston et al. (1991, 1992a) advanced a controlled-environment MIR transmission cell fitted with a Cahn microbalance to study the adsorption of organics and $H_2O$ in clay minerals. The latest version of this cell is described by Xu et al. (2000). A figure depicting the schematics of this cell is provided in Fig. 9.12 of Chapter 9.

One drawback of transmission through self-supported films is that the minimum practical thickness of the sample is still too large for recording the strong Si—O stretching envelope at ~1000 $cm^{-1}$ (Hofmeister and Bowey, 2006). This drawback is typically addressed by the deposition of thinner films (~0.25 $mg/cm^2$) on IR transmitting windows, for example, ZnS (Lerot and Low, 1976), or ZnSe (Fialips et al., 2002). On the other hand, sample thickness is not an issue when measuring the weak FIR bands in transmission (Laperche and Prost, 1991; Schroeder, 1992).

IR microscopy has enabled transmission studies with spatial resolution of the order of ~20 μm and is valuable when studying single particles or multimineralic samples (e.g. Johnston et al., 1990; Shewring et al., 1995; Diaz et al., 2000; Robin et al., 2013). Microscope stages with controlled atmosphere and programmable heating or freezing capabilities are now commercially available. As a result, most of the features of transmission IR cells can now be miniaturised. Further, diamond anvil cells (Butler and Frost, 2006) offer the possibility of extending the measurements to high pressures, typically up to 20–25 GPa (e.g. Alabarse et al., 2011; Pawly and Jones, 2011).

### 3.3.3 External Specular Reflection

Specular reflection applies ideally to absorbing bulk samples with infinite thickness and one optically flat surface (*semiinfinite*). The reflected ($I_R$) beam lies in the plane of incidence and at an angle that is identical to the incidence angle. The specular reflectance spectrum, $R$, is recorded as the ratio $I_R/I_o$, where $I_R$ is the reflected signal from the sample and $I_o$ is the reference

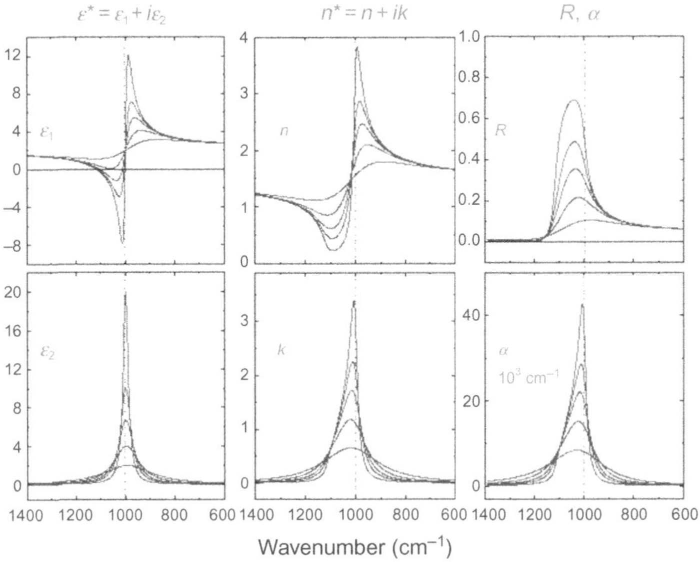


**FIG. 3.4** Simulated normal incidence angle reflectance ($R$), extinction coefficient ($\alpha = 4\pi\nu k$), real and imaginary parts of the complex dielectric permittivity ($\varepsilon^* = \varepsilon_1 + i\varepsilon_2$) and refractive index ($n^* = (\varepsilon^*)^{1/2} = n + ik$) as a function of wavenumber $\nu$, of a Lorentz oscillator with $\nu_o = 1000\ \text{cm}^{-1}$ (marked with a vertical dashed line), variable damping coefficient (width) $\Gamma = 25$, 50, 75, 125 and 250 cm$^{-1}$ increasing from top to bottom, $\varepsilon_\infty = 2.5$, oscillator strength ($\varepsilon_s - \varepsilon_\infty = 0.5$). Based on the value of $\alpha(\nu_o)$, a transmission measurement of a sample with $\Gamma = 25\ \text{cm}^{-1}$ and thickness of $d \approx 0.7\ \mu\text{m}$, or with $\Gamma = 250\ \text{cm}^{-1}$ and $d \approx 2.3\ \mu\text{m}$ would yield a value of maximum measured absorbance $A = -\log I/I_o = 1$. Note that, for a given oscillator strength, sharp bands ($\Gamma/\nu_o \leq 0.05$ for $\varepsilon_s - \varepsilon_\infty = 0.5$) exhibit negative values of $\varepsilon_1$ on the high-wavenumber side of the peak maximum and pronounced asymmetries of $n$. Also note that among the loss curves shown, only $\varepsilon_2$ is symmetric around $\nu_o$, whereas both $k$ and $\alpha = 4\pi\nu k$ are skewed on the high-wavenumber side. At fixed $\Gamma$, high-frequency distortions decrease upon decreasing oscillator strength (not shown). The detailed account of the optical dispersion formalism can be found in many reference books and articles (e.g. Hofmeister et al., 2003 or Kamitsos, 2015).

spectrum collected from a front-coated Al or Au mirror, placed instead of the sample. Regardless of the angle of incidence, spectra produced by specular reflectance are very different than those collected from transmission measurements (Fig. 3.4). The reason for this difference is that $R$ depends strongly on both the real ($n$) and imaginary ($k$) parts of the complex reflective index ($n^* = n + ik$) of the sample, as described by the Fresnel equations, whereas transmittance, $T$, is solely a function of the imaginary part. The Fresnel equations define reflectance as a function of the complex refractive index of the sample ($n^*$), which is a function of the relative (dielectric) permittivity ($\varepsilon^* = \varepsilon_1 + i\varepsilon_2 = n^{*2}$), the trigonometric functions of the angle $\theta$ and the polarisation of the light ($p$- or $s$-, in and perpendicular to the plane of incidence, respectively). In the simplest case of near-normal incidence ($\theta_o < 15$ degrees, commonly 11 degrees), the functions of $\theta$ cancel out and the polarisation of the electric field becomes irrelevant if the sample is isotropic perpendicularly to the plane of incidence. Then $R$ can be described as $R=[(n-1)^2+k^2]/[(n+1)^2+k^2]$ (Fig. 3.4).

The main advantage of specular reflectance studies is that the same sample and sampling accessory can be employed to different spectral ranges. The resulting spectra can be merged to provide the full response of the material over very broad wavenumber ranges that can, in some cases, extend from the FIR to the visible or UV (e.g. Hofmeister et al., 2003; Kitamura et al., 2007). The Kramers-Kronig transformation or the analysis of dispersion in terms of a sum of damped harmonic oscillators can then be employed to calculate both the complex refractive index ($n^*$) and the dielectric permittivity ($\varepsilon^*$) of the sample (e.g. Hofmeister and Bowey, 2006; Kamitsos, 2015).

Clay minerals are not expected to fulfil the requirement for a strongly absorbing, bulk semiinfinite system that is needed to perform quantitative specular reflectance analysis. Nevertheless, some near-normal reflectance studies have been reported on pressed pellets to study the Si—O stretching envelope of montmorillonite subjected to ion-exchange and heating treatments (Karakassides et al., 1997, 1999). In addition, Che et al. (2011) have reported the MIR and FIR reflectance spectra (at 30 degrees incidence angle) of selected clay minerals and zeolites at various stages of dehydration and/or dehydroxylation, as a reference database for planetary research.

### 3.3.4 Reflection-Absorption of Thin Films on Mirror Substrates

The same fixed- or variable-angle specular reflectance accessories can be employed to study thin, oriented films of clay minerals deposited on mirror surfaces instead of bulk samples. This sampling technique is sometimes called IR reflection-absorption spectroscopy (IRRAS) because the resulting spectra are almost identical to those obtained in transmission from a sample of double thickness. Two extreme cases can be distinguished (Fig. 3.5): Near-normal reflectance ($\theta<15$ degrees) provides almost exclusively the spectrum of the in-plane vibrations of the layers, and grazing-angle reflectance ($\theta\approx70$–$85$ degrees) probes ideally the vibrations having a dipole moment vector perpendicular to the layer. At near-normal incidence, a node is formed at the front mirror surface (i.e. at the mirror/sample interface), and therefore the maximum amplitude of the electric field is observed at a distance of $\lambda/4$ from the mirror. Thus near-normal reflection is most sensitive to samples that have a thickness of $\sim\lambda/4$, that is, of the order of microns, $\sim2.5\ \mu m$ at $\nu=1000\ cm^{-1}$. On the contrary, a beam at grazing incidence results in an electric field that exhibits the maximum amplitude of its $p$-component exactly on the surface of the mirror, whereas the corresponding $s$-components of the incident and reflected beams cancel out (Fig. 3.5). As a result, grazing-angle reflectance is restricted to vibrational modes probed by the $p$-component and is inherently more sensitive than near-normal reflectance when the study

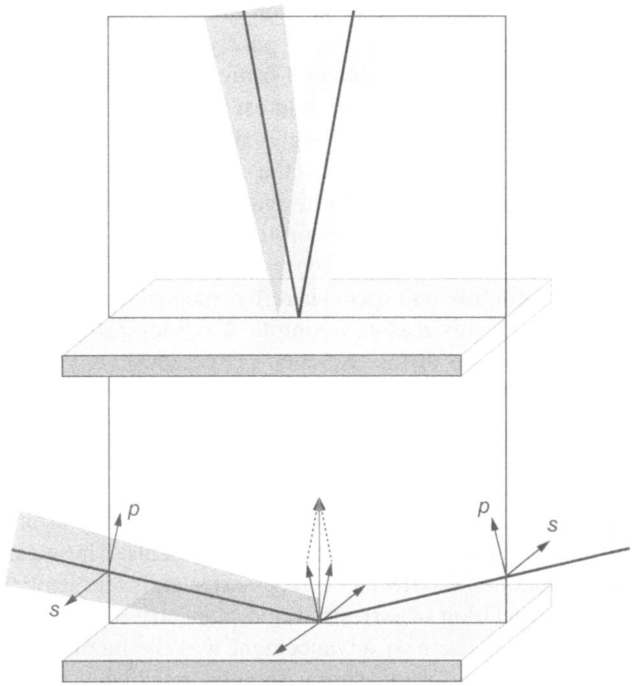


**FIG. 3.5** Near-normal *(upper)* and grazing-angle *(lower)* reflectance geometries for reflection-absorption (IRRAS) measurements. The sample is a thin oriented film deposited on the mirror.

of nm-thick layers is of interest. The ability of grazing-angle reflectance to measure the IR spectra of extremely thin clay mineral films, including single-platelet layers and layer-by-layer hybrid assemblies, has been demonstrated by Ras et al. (2003, 2004, 2006).

### 3.3.5 Diffuse Reflectance MIR and NIR Spectroscopies

This is the technique that measures the light that penetrated, scattered, absorbed and reflected in a diffuse (i.e. nonspecular) manner from loose, powdered samples. As such, it is a natural choice for studying clay minerals, clays, soils, etc., with minimum or no sample preparation. Radiation transmits, reflects, refracts through many particles leading to the enhancement of weak bands (including those due to higher-order transitions), but strong bands in the MIR can be saturated unless the sample is diluted in a transparent medium like KBr, for example. Of major importance in diffuse reflectance spectroscopy is the reduction of the inevitable specular contribution, which, as discussed in Section 3.3.3, distorts the spectra in a manner that cannot be compensated by a suitable background, nor handled explicitly. Specular distortions (*reststrahlen*) are considerably more pronounced in the vicinity of strong bands due to the dispersion of the real part of the refractive index. They are also more pronounced in poorly ground, undiluted samples. The background spectrum ($I_o$) is obtained from an ideal 'matte' surface with no bands in the wavenumber range of interest (*Lambertian reflector*), which can be approximated by NaCl, KBr, MgO, $BaSO_4$, fluoropolymers such as Spectralon. Several functions (such as the Kubelka-Munk transformation, see Yang and Miklavicic, 2005, and references therein) have been proposed to account for light scattering and absorption effects in diffuse reflectance spectra, and the way these effects influence spectral intensity. However, scattering is difficult to control, and specular reflectance is not taken into account by these functions. For this reason, a simple $A = -\log(I_R/I_o)$ relation is frequently used for presenting diffuse reflectance spectra. In any case, quantitative analysis of the data based on band intensities is only possible within sample sets of identical scattering coefficients, that is, with the same particle size distribution and concentration.

Historically, diffuse reflectance IR spectroscopy was first applied on samples of mineralogical interest in the NIR and not in the MIR, as an extension to the visible spectra obtained by dispersive instruments. These early studies were based on bidirectional reflectance accessories and were explicitly meant to enable the 'interpretation of terrestrial and planetary remote sensing data' (Hunt and Ross, 1967). The next advancement was the modification of integrating sphere accessories for measuring samples positioned in a horizontal, instead of vertical, ~30-mm diameter pan (Lindberg and Snyder, 1972; Hunt, 1980). Similar integrating spheres were used in the pioneering work of Cariati et al. (1981, 1983a,b) and the remarkable compilation of high-resolution spectra of minerals by Clark et al. (1990).

The transition to FT spectrometers brought MIR and NIR into a common platform, triggered the design of new accessories for diffuse reflectance (Yang, et al., 1986) and made the technique known as DRIFT, diffuse reflectance IR FT spectroscopy (e.g. Leyden and Murthy, 1988). Different designs of diffuse reflectance cells fitting the sample compartment of an FT spectrometer are commercially available. These cells employ a biconical design (e.g. Fig. 3.6) with wide-angle incidence and collection mirrors and can differ considerably in sensitivity to specular distortions (Hembree and Smyrl, 1989). As a result of these advances and also because of the minimal sample preparation requirements, diffuse reflectance became a mainstream technique in the MIR study of clay minerals (e.g. Parker and Frost, 1996; Madejová et al., 2000a; Madejová and Komadel, 2001; Gates, 2005, 2008), usually applied on 10 wt% dilutions of sample in KBr or NaCl. These studies were immediately extended to the NIR, by employing the same types of accessory and with appropriate changes in source, beamsplitter and/or detector. The relatively broad and weak NIR-active combination and overtone bands could be recorded from undiluted, neat samples in a highly reproducible manner with minimum or no distortions from specular reflectance artefacts

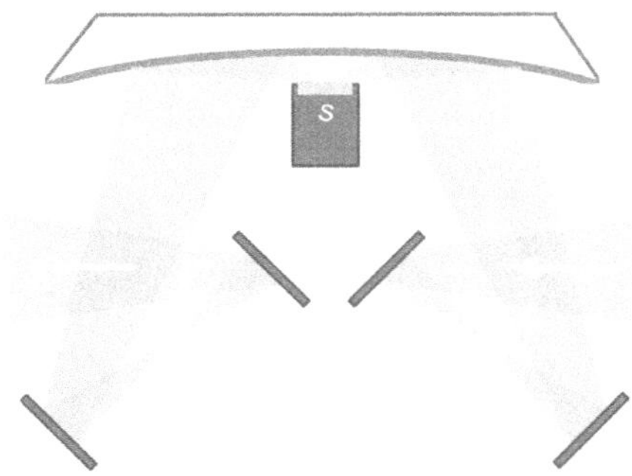


**FIG. 3.6** Schematic optical diagram of an infrared diffuse reflectance accessory with biconical incidence and collection optics. The position of the sample is marked (S). Accessories of this type fit inside the sample compartment of Fourier-transform mid- and near-infrared instruments and have been widely used to study clay minerals and related materials.

(e.g. Post and Noble, 1993; Petit et al., 1999b, 2002, 2004b; Madejová, 2005; Madejová et al., 2000b, 2006, 2015; Gates, 2005). Remarkably, applications of simultaneous mid-IR DRIFTS and Extended X-ray Absorption Fine Structure (EXAFS) or other X-ray techniques as a function of temperature have been demonstrated recently, albeit not in the study of clay minerals (Newton, 2009; Newton and van Beek, 2010; Beyer et al., 2013).

Besides standard biconical DRIFT accessories, integrating spheres have also been developed for dedicated FT NIR spectrometers (Fig. 3.7, centre). Contrary to the early designs that required fitting the sphere on top of the sample (Hunt, 1980), modern accessories measure from below via an NIR transparent window (e.g. quartz or sapphire) which absorbs light in the MIR. Modern designs optimised for NIR spectroscopy are typically gold-coated and employ a rough gold reflector for measuring the background spectrum. Incident light enters the sphere from a small opening located opposite to the sample opening, performs multiple reflections interacting with the sample or reference material and is detected by a dedicated detector that is placed on the side of the sphere. The sample (~50 mg or more) can be placed directly on the window of the sphere and covered with an environmental cup, or inside a flat-bottom container (vial, reaction vessel) made of clear glass. Due to this containment, the sample can be conveniently isolated from the environment following a thermal or chemical treatment, or measured in situ during dynamic changes such as purging, hydration or hydrogen for deuterium (H/D) exchange (e.g. Gionis et al., 2006; Chryssikos et al., 2009; Sánchez del Río, et al., 2009; Stathopoulou et al., 2011; Pentrák et al., 2012a; Skoubris et al., 2013; Petit et al., 2015; Tsampodimou et al., 2015; Pálková et al., 2015). Powdered samples can be allowed to settle in a suitable solvent (e.g. $CCl_4$ which is transparent in the NIR) enabling the real-time study of the adsorption of organics dissolved in the liquid phase by measuring the precipitated clay through the bottom of the vial (Tsiantos et al., 2012).

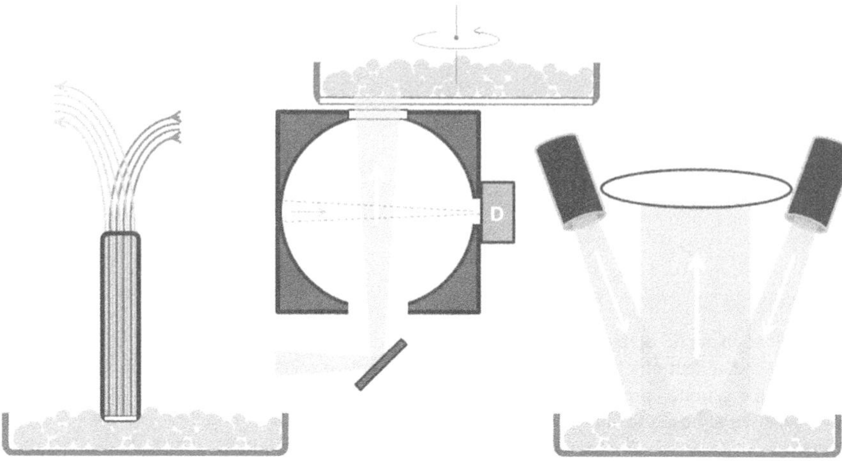


**FIG. 3.7** Principle of operation for three types of 'external' diffuse reflectance accessories for near-infrared measurements: *(left)* bifurcated optical fibre bundle, *(centre)* integrating sphere for stationary or rotating samples, also enclosed in colourless flat-bottomed glass vessels, reactors, etc., and *(right)* external sample illumination head with remote collection optics.

The typical diameter of the integrating sphere sampling window (~20 mm) is not suitable for the statistically representative measurement of samples that are both coarse (>10 mm) and heterogeneous. Instead, relatively large (~50–100 g) amounts of such samples can be placed in a larger diameter, quartz-bottomed vessel (Fig. 3.7) and rotated off-centre above the integrating sphere during the measurement to enable statistical averaging (e.g. Wight et al., 2016). A hemispherical diffuse reflectance design has been incorporated in field NIR spectrometers and used for drillcore logging and geochemical exploration (Kruse, 1996; Yang et al., 2000, 2001; Sun et al., 2001; Thomas and Walter, 2002).

A third option for diffuse reflectance measurements in the NIR is the use of optical fibre probes. Their design is based on low-OH silica fibres that have negligible absorption in the NIR. A diffuse reflectance fibre probe consists of a number of such fibres which are bundled together at the sample end and bifurcate as two separate bundles on the spectrometer side, one for illumination and the other for collection (Fig. 3.7, left). The typical sampling diameter of the bundle is ~5 mm, and the physical length of the fibre can range between ~1 and 2 m for handheld operation and >50 m for industrial installation (the round-trip pathlength for light is twice as large). The low wavenumber cutoff of these fibres depends on their length and the concentration of OH in the silica. A 1.5-m-long fibre transmits above ~4000 $cm^{-1}$, but a 65-m-long probe does not transmit below 4500 $cm^{-1}$. Measurements require the contact (or near contact) of the fibre tip to the sample. Background is typically provided by a fluoropolymer (Spectralon) disk. Although diffuse reflectance fibre probes have a lower throughput than integrating spheres, they do enable the collection of high-quality spectra with handheld operation and acquisition times of <1 min (e.g. Gionis et al., 2007; Stathopoulou et al., 2008).

Optical fibre probes eliminate the need for introducing the sample to the spectrometer. As such, they enable the study of samples in bulk rock form, drill cores, or powders in bulk bags with a spatial resolution of ~5 mm, as well as field applications. Further, optical fibres can be conveniently incorporated into cells for other types of measurement, thereby enabling the design of instrumentation for synchronous, hyphenated, data acquisition (see Section 3.3.9).

A last type of diffuse reflectance probes for NIR measurements involves the external illumination of the sample and the remote collection and analysis of the reflected light (Fig. 3.7, right). These probe heads operate at a distance of ~15–30 cm from the surface of the sample and have a relatively large sampling area ~100 cm$^2$. They are therefore convenient for measuring large samples, for example, as they are transported in conveyor belts for process mineralogy applications (Fig. 3.8). A miniature version of this type has been described in Section 3.2.4 (Alcalà et al., 2013).

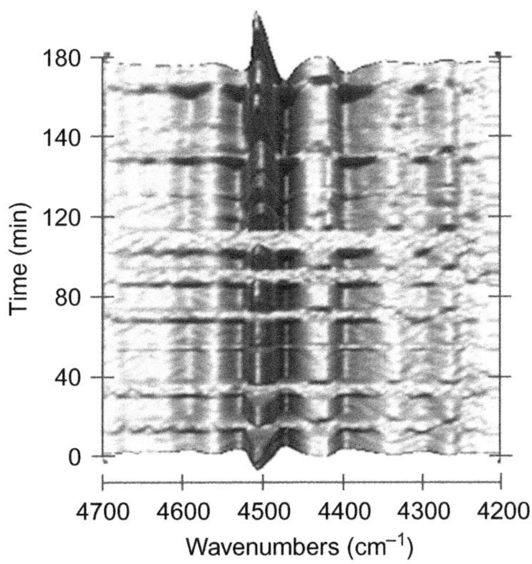


**FIG. 3.8** Real-time spectroscopic monitoring of a bulk dioctahedral smectite ore during transportation on an open-air conveyor belt. Data acquisition is by Fourier-transform NIR at 8-cm$^{-1}$ resolution, employing an external illumination head and diffuse reflectance collection optics (as in Fig. 3.7, right) at a distance of 30±5 cm from the irregular surface of the moving sample. The fingerprinting OH combination region is shown in the second derivative mode to demonstrate spectral quality. Each spectrum corresponds to 2-min acquisition time and represents an average of ~5 tons of ore moving under the beam. The head performed a periodic lateral movement, transverse to the transportation direction, to avoid granular convection (*Brazil nut*) effects in sampling. Horizontal 'stripes' correspond to empty belt.

### 3.3.6 IR Emission

The IR emission technique involves the measurement of the thermal emittance of the sample, as opposed to transmittance or reflectance. The sample, normally in the form of a thin film on a platinum plate, occupies the place of the source in the spectrometer and is thermally excited by a suitable heater. As a result of heating, the system populates its excited vibrational states and emits in the IR upon return to the ground state. Sample heaters can be external to the FTIR instrument, in which case condensing mirrors are necessary to focus and amplify signal to spectrometer, or internal thermal elements located within the sample compartment of the spectrometer.

The IR emittance spectrum is obtained from the intensity ratio between the emission from the sample and the emission from a blackbody reference, and is identical to transmittance in terms of band positions, resolution, etc. In principle, IR emission requires the population of the excited vibrational state and occurs at any temperature above the absolute zero, but is too weak to be explored unless the temperature is sufficiently high. This needs to be taken into account for the study of clays and clay minerals because of reactions of these materials at elevated temperatures (water loss, dehydroxylation, $Li^+$ fixation and charge reduction, etc.).

The technique is mostly adapted to strongly absorbing and weakly reflecting samples at high temperatures which are normally difficult to study by other techniques. It is, therefore, ideally suited for the study of high-temperature chemisorption phenomena on solid catalysts. The IR emission spectra of clay minerals have been extensively studied by Frost's group at temperatures in the ~250–800°C range, which is relevant to dehydroxylation events and higher-temperature transformations. A lower temperature (~80°C) study of clays has been reported by Friedlander et al. (2015). The field was reviewed by Kloprogge and Frost (2005a).

### 3.3.7 Photoacoustic Spectroscopy

The sample (~200 mg) is enclosed in a sealed cell filled with an inert, IR transparent gas, irradiated by a modulated IR beam via a transparent window and excited to its higher vibrational levels. Nonradiative decay to the ground state results in a temperature fluctuation. The latter is translated to a pressure fluctuation of the gas (He is preferred due to its superior thermal transport properties) and detected as a sound wave by a sensitive microphone. Particulate samples with large surface area (as in diffuse reflectance spectroscopy) favour the efficient transfer of the signal from the temperature to the pressure domain. The sampling depth is of the order of ~10–100 μm and is inversely related to the modulation frequency of the interferometer (McClelland et al., 1993). Photoacoustic spectroscopy requires little or no sample preparation and is, therefore, applied in the study of strongly absorbing systems. Its application to soils and clay minerals has been demonstrated by Schmidt

(1980). With appropriate cells, the technique is applicable over broad energy ranges, including the NIR (e.g. Donini and Michaelian, 1986). To the author's knowledge, applications of photoacoustic IR spectroscopy to the crystallochemical investigation of clay minerals have not been published, but they have reached a level of accessibility and reproducibility that should allow the sampling of large datasets of, for example, soil samples, suitable for chemometric analysis (Du et al., 2008).

### 3.3.8 Internal Reflection IR and Attenuated Total Reflectance (ATR) Spectroscopy

Light propagating in an optically dense transparent medium (with refractive index $n_1$) and reaching an interface with a less dense medium ($n_2 < n_1$) at an angle $\theta$ to the normal, undergoes total (*internal*) reflection if $\theta$ is greater than a critical angle $\theta_c = \arcsin(n_2/n_1)$, (Fig. 3.9). Under such supercritical angle conditions ($\theta > \theta_c$), a standing wave is created within the optically thin medium. The amplitude of this wave decreases exponentially with the distance from the interface and reaches $1/e$ of its original value (~37%) at some distance, $d_p$, that is defined as the penetration depth. The dependence of $d_p$ on wavelength, $\lambda$, the angle of incidence, $\theta$, and the refractive indexes of the two media, $n_1$ and $n_2$, is given by:

$$d_p = \frac{\lambda}{2\pi\sqrt{n_1^2 \sin^2\theta - n_2^2}}$$

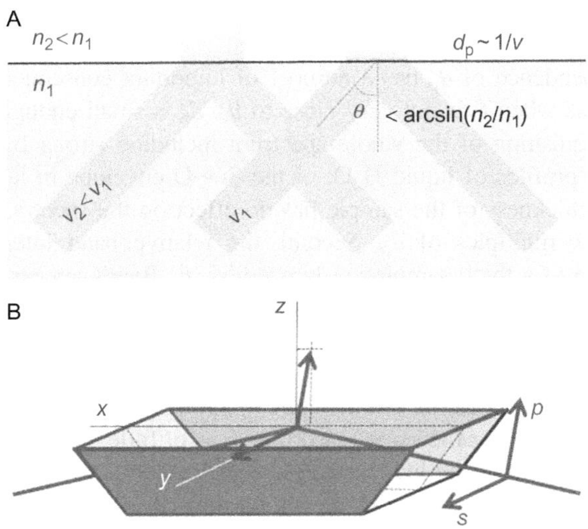


**FIG. 3.9** (A) Principle of internal reflection, evanescent wave and penetration depth in the optically thin medium. (B) Radiation polarisation vs crystal (and oriented sample) coordinates in ATR spectroscopy.

This relationship is of paramount importance for internal reflection IR spectroscopy: for any given set of $n_1$, $n_2$ and $\theta$, $d_p$ becomes a linear function of the wavelength, $\lambda$. Also, for any $\lambda$, $d_p$ becomes imaginary for $\theta < \theta_c$, it is infinite at $\theta = \theta_c$ and then decreases asymptotically to a small, but nonzero, value as $\theta > \theta_c$ approaches 90 degrees (grazing incidence).

If neither of the two media absorbs light ($n^*_1 = n_1$ and $n^*_2 = n_2$ are real numbers over the wavelength range of the experiment), the supercritical internal reflection is elastic and can be called total. However, if the optically thin medium is absorbing, that is, has a nonzero imaginary refractive index ($n^*_2 = n_2 + i\kappa_2$, $\kappa_2 \neq 0$), part of the incident beam crossing the interface will be absorbed by the sample. As a result, reflection will no longer be total, and the reflected beam will be attenuated selectively by the wavenumber-dependent loss in the sample. If recorded as a function of wavenumber and appropriately ratioed to the reference state, this beam will provide the so-called attenuated total reflectance (ATR) spectrum of the sample.

Although $d_p$ is properly defined only in the case of nonabsorbing media with real $n_1$, $n_2$, its dependence on $\theta$ and $\lambda$ provides a useful approximation for choosing the appropriate internal reflection accessory for ATR measurements. The requirement for transparent optically dense media over broad wavenumber ranges, and with sufficiently high $n_1$ (so that the condition $n_1 > n_2$ can be valid for a range of samples), limits the choice of internal reflecting media to a few: Ge ($n_1 = 4$), Si (3.4), ZnSe (2.4) and diamond (2.4) are the most common crystals for ATR spectroscopy (note the index matching of ZnSe and diamond). Assuming $n_2 = 1.5$, these media satisfy the supercritical angle conditions for $\theta_c(\text{Ge}) > 22^\circ$, $\theta_c(\text{Si}) > 26^\circ$, $\theta_c(\text{ZnSe}) = \theta_c(\text{diamond}) > 39$ degrees. The corresponding values of $d_p$ at $\theta = 45^\circ$ and at $\lambda = 10\ \mu\text{m}$ ($1000\ \text{cm}^{-1}$) are ~0.66, 0.85 and 2.0 μm. These decrease linearly with wavelength and at $\lambda = 2.5\ \mu\text{m}$ ($4000\ \text{cm}^{-1}$) they drop to 1/4 of these values.

The $\lambda$-dependence of $d_p$ has a number of important consequences. First, in the typical case when $\theta$ is not very close to $\theta_c$, $d_p$ is small enough (<5 μm) to ensure the acquisition of the whole spectrum including strong bands (such as the stretching profiles of liquid $H_2O$, or the Si—O envelope in layer silicates). The physical thickness of the sample has no effect on the spectra, as long as it exceeds a few multiples of $d_p$. Second, the relative band intensities in the ATR spectrum of a thick sample are biassed by $d_p$: Bands appear weaker upon increasing wavenumber because $d_p$ becomes shallower. For this reason, the spectra of thick samples are typically intensity-normalised by $\lambda$ in order to yield relative intensities similar to those in a transmission spectrum. No such correction is needed for samples that are thinner than $d_p$ throughout the wavenumber range of the measurement. For the same reason, ATR is of little use in the NIR, where the combined effects of the low extinction coefficient of the higher-order vibrational modes and the negligible $d_p$ are prohibiting. As a last remark, achieving very good contact between the sample and the crystal is critical, as minute air gaps would cause loss of the ATR signal, especially at high wavenumbers.

These general considerations are based on the interactions of IR radiation with nonabsorbing materials and become qualitative even in the case of weakly absorbing samples. In this case, an effective penetration depth of the radiation in the sample, $d_{eff}$, can be defined as the sample thickness that would produce the same absorbance as in a transmission experiment. Parameter $d_{eff}$ is linearly related to $d_p$ and of similar magnitude but depends on the polarisation of the incident beam. It is larger when radiation is polarised within (*p-polarised*) rather than perpendicular (*s-polarised*) to the plane of incidence. For $\theta = 45^o$, $d^p{}_{eff} = 2d^s{}_{eff}$. The relevant equations are given in reference books on ATR spectroscopy (e.g. Harrick, 1967; Milosevic, 2012).

Depending on application, ATR elements can be prisms with trigonal, parallelogram or trapezoidal cross section, or cylindrical with conical ends, or semicylindrical, or semispherical. These are fitted to single or multiple reflection configurations, the latter used mainly for very thin or low absorption samples. They can be placed slightly below the level of their optical bench to serve as a trough for liquids, or above in order to handle large specimens. The operational wavenumber window for ATR spectroscopy depends on the nature of the internal reflection element and on its size. Typical multiple reflection Si (or Ge) crystals are not useful below $\sim 1300\ cm^{-1}$ ($\sim 700\ cm^{-1}$), but miniature (<250 μm diameter) single reflection elements, such as those in microsampling sample-compartment accessories or ATR microscope objectives (Sections 3.2.3, Fig. 3.3), can be used down to $\sim 500\ cm^{-1}$. Diamond, most frequently used as a single-reflection element, has the broadest window extending to the far-IR ($\sim 35\ cm^{-1}$), albeit with some strong absorption bands that limit its use in the 2300–1900 $cm^{-1}$ range, especially in multiple reflection cells. Some of the most popular ATR elements combine diamond on the sample side with ZnSe on the spectrometer side (both with $n \approx 2.4$) for producing larger elements with higher throughput. Due to the ZnSe absorption, these elements are limited to $\nu > 550\ cm^{-1}$. Additional considerations in choosing an ATR unit are the chemical and mechanical durability of the crystal (both very high in the case of diamond), the ability to withstand temperature changes, etc.

In the ATR configuration, the samples can be studied as dispersions (which offer ideal contact with the element), thin oriented films adhering to the crystal made by drying dispersions, or loose powders. In the latter case, a suitable press is needed to establish good contact and single-reflection elements are employed. Commercially available anvil tips exert pressures in the range of ~10–100 MPa, which may, in some cases, affect the exact position and shape of the vibrational bands (Friedrich and Weidler, 2010).

The ATR spectra are never identical to those obtained in transmission, and this goes beyond the relatively simple dependence of $d_p$ or $d_{eff}$ (hence, band intensity) on $\lambda$. The reason is that a peak in the imaginary part of the refractive index, $\kappa$, is accompanied by the abrupt variation (dispersion) of the real part, $n$. As a result, $d_{eff}$ is different on the two sides of the peak, and the latter appears shifted to lower wavenumbers in the ATR spectra compared to transmission.

Comparative spectroscopic investigations of clay minerals by ATR versus transmission from self-supported films or KBr pellets as well as diffuse reflectance have demonstrated the ease of operation and reproducibility of ATR, which is particularly valuable when studying the interaction of $H_2O$ (e.g. Madejová et al., 2002a; Madejová, 2003) or organics (e.g. Gates et al., 2016) with mineral surfaces. As a result, ATR spectroscopy can be conveniently employed for the routine mineralogical analysis of powdered samples (e.g. Klinkenberg et al., 2006) and for collecting spectral libraries suitable for chemometric analysis (e.g. Müller et al., 2014).

Multiple reflection ATR cells have enabled the study of the difficult-to-measure Si—O envelope in aqueous systems with variable smectite/$H_2O$ content, exchangeable cations, and electrolytes present or as a function of Fe-reduction (Shewring et al., 1995; Yan et al., 1996a,b; Yan and Stucki, 1999), more accurately and conveniently than in earlier transmission work (e.g. Gan and Low, 1993). As a follow-up to these studies, the introduction of a wire polariser in the incident beam path enabled the polarisation-dependent study of dilute aqueous dispersions of smectite demonstrating the orientation of the smectite layers (Johnston and Premachandra, 2001), the orientation of organics adsorbed on smectite (e.g. Nguyen, 1986), and the formation of trioctahedral domains upon $Fe^{3+}$ reduction (Bzdek and McGuire, 2009). The optical configuration of these ATR cells is such that the *s*-polarised component of the incident beam (i.e. the one perpendicular to the plane of incidence) interacts solely with the in-plane modes of the oriented sample, whereas its *p*-polarised counterpart probes both in-plane and out-of-plane modes (Fig. 3.9, lower). This is to be compared with near-normal ($<15$ degrees) incidence in transmittance or external reflectance configurations that would probe exclusively the in-plane modes, regardless of beam polarisation, or with grazing-angle external reflectance ($>75$ degrees) that would be sensitive to the out-of-plane components (Fig. 3.5). The extreme sensitivity of 25-reflection, 45 degrees ZnSe or Ge ATR elements in recording the spectra of Langmuir-Blodgett smectite-based monolayers in both polarisations has been demonstrated (Ras et al., 2003).

A modified commercial high-temperature diamond ATR cell operating up to 300°C was employed to study in situ the intercalation of molten polymer into layer silicates by recording the changes of the Si—O band envelope (Cole, 2009). The interaction of $CO_2$ gas with clay minerals was studied by a twin cylindrical ATR cell sealed in a high-pressure chamber (Krukowski et al., 2015). Twin cells have two identical crystals (one with sample, the other as reference) exposed to the same conditions. These crystals are alternated mechanically in the light path, to provide properly referenced spectra that are suitable for sensing very subtle changes (Goodman et al., 2005).

As the studied sample is in intimate contact with the ATR crystal, simultaneous IR and gravimetric measurements during water uptake, for example, are no longer possible. To circumvent this problem, Schuttlefield et al. (2007a,b) coupled a quartz microbalance (QCM) module in series with the ATR module. The two modules were exposed to the same variable environmental conditions, hosted different specimens of the same sample and were used to provide two parallel sets of synchronised spectroscopic and gravimetric data.

### 3.3.9 Combined Acquisition in the MIR and NIR

There are many cases where the vibrational spectra of clays need to be collected in both the MIR and NIR ranges and analysed in a common frame (e.g. Cariati et al., 1981; Madejová et al., 2000a,b; Post and Borer, 2000; Madejová and Komadel, 2001; Frost et al., 2001b; Bishop et al., 2002a,b; Petit et al., 2004b; Gates, 2005; Gionis et al., 2006; Skoubris et al., 2013). This type of study is typically motivated by the need to assign vibrational bands in a self-consistent manner, to distinguish between the overlapping vibrational signatures of $H_2O$ and structural OH, to measure anharmonicity, to transfer laboratory techniques to field (or space) application, etc.

Instrumentation is not an obstacle for this type of combined study, because dedicated FT MIR and NIR instruments are common. Further, a single pyroelectric or MCT detectors can now cover both the mid-IR and a good part of the NIR (e.g. 6000–400 $cm^{-1}$). This extended range is sufficient for probing the higher-order modes of structural OH and $H_2O$ in addition to the fundamental stretching spectrum (e.g. Gates, 2005; Zhang et al., 2006). Last, the full MIR and NIR spectral coverage ($\sim$12,000–400 $cm^{-1}$) can be obtained on instruments which combine more than one source and detector, and support exchangeable beamsplitters (e.g. Bishop et al., 1994; Madejová and Komadel, 2001; Zhang et al., 2007).

The problems with the continuous MIR and NIR acquisition stem, instead, from the properties of the sample itself, namely the great difference of the extinction coefficient between fundamental and higher order modes. A single sampling technique, such as transmission, diffuse reflectance or ATR, cannot handle these differences. The common way to overcome the problem is by performing two separate experiments with the same sampling accessory (e.g. diffuse reflectance as in Fig. 3.6), and two different sample concentrations (e.g. neat powder in the NIR and 10 wt% in KBr for the MIR). If the samples of interest are equilibrated and stable at ambient or other well-defined conditions, the two independent sets of data are comparable. On the other hand, 'synchronising' experiments that require spectral acquisition upon varying conditions or as a function of time may not be straightforward.

To address this problem, a simple environmental cell was designed to couple the single-reflection ATR cell of an MIR spectrometer, with the diffuse reflectance fibre optical probe of an NIR spectrometer (Bukas et al., 2013). The cell (Fig. 3.10) is fitted with a gas input and output, which enable purge-drying, hydration or $D_2O$ enrichment, the introduction of gaseous reactants, etc. The sample is deposited first on the ATR element from a

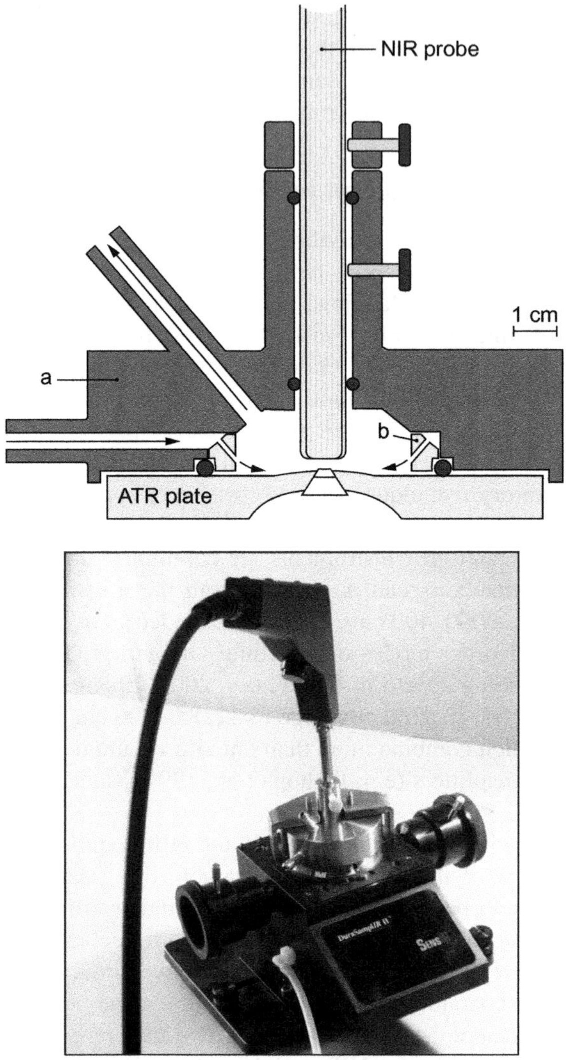


**FIG. 3.10** Schematics of an environmental chamber (a) for synchronous mid-and near-infrared measurements by a single-reflection diamond ATR element and a diffuse-reflectance fibre optic bundle probe, respectively. The sample (~100 μm thick, not shown) is deposited on the ATR crystal. The input gas is distributed evenly around the sample via a perforated metal flange (b) (Bukas et al., 2013 *reproduced with permission*). The assembled cell, outside the sample compartment of the mid-infrared spectrometer, is also shown.

concentrated suspension in $H_2O$ to form a well-adhering film of ~8-mm diameter and ~100 μm thickness, which is assumed to be uniform. The cell is then assembled and the NIR probe is positioned at a distance of <1 mm from the top of the sample. The key idea is that the thickness of the sample is sufficient for the acquisition of high-quality NIR spectra, whereas ATR is probing a much thinner layer ($d_{eff} \approx \lambda$) from the lower side of the film. The two spectrometers are set to perform sequences of synchronised spectral measurements which can be maintained for several hours. To ensure that the near-IR beam does not 'contaminate' the MIR interferogram by leaking through the sample and the diamond, the two spectrometers need to be operated with different scanner velocities. Kinetic processes can be studied at a time resolution of ~60 s with spectral quality that suffices for 2nd derivative analysis. Background spectra for the two accessories are collected in the usual manner prior to the assembly of the cell, and also at the end of each experiment to compensate for changes of the atmospheric gases by interpolation, see Chapter 4, Section 4.2.3. The same cell can be used without the NIR probe, that is, exclusively for ATR measurements under controlled atmosphere as in Kuligiewicz et al. (2015a,b), or fitted to a temperature-controlled stage instead of the ATR cell for variable-temperature NIR measurements.

## 3.4 RAMAN SAMPLING TECHNIQUES

Raman spectroscopy requires no sample preparation and provides the full vibrational spectrum (conforming to the $\vartheta\alpha/\vartheta Q)\neq 0$ selection rule) in the Stokes range. The scattering geometry is simple (180 or 90 degrees), and with a dedicated microscope small samples can be studied with a spatial resolution of a few μm. Sample heating could be a problem especially when NIR excitation is employed, but can be avoided by placing the sample in a miniature heat sink made of aluminium.

Despite these seemingly simple requirements and despite advances in instrumentation, there are few Raman studies of clay minerals and a very small subset of these is on smectites. Photoluminescence is an important limitation. Iron oxides and hydr(oxides) have electronic absorption bands in the visible and NIR parts of the spectrum that could be pumped by the excitation beam (Sherman and Waite, 1985). Further complications may arise from the instability of iron oxides and hydr(oxides) even under moderate laser powers (Hanesch, 2009). Broad luminescence peaks due to these species and also due to trace organic carbon in natural samples have been observed even with NIR excitation (1064 nm). Luminescence due to organic carbon could be eliminated by oxidation (30% $H_2O_2$), but a standard iron-removing, citrate-dithionite-bicarbonate treatment had little or no effect on the spectra (Schroeder et al., 2003). Rare earths are also known to produce strong and sharp fluorescence lines, should they be present (e.g. as ppm traces in phosphate admixtures).

But the main reason for the low Raman scattering efficiency of clay minerals stems from their chemical composition per se. Clay mineral layers consist of light atoms and low electron density bonds, arranged in defect-rich structures. As such, most clay minerals lack the well-defined symmetric 'breathing' modes that give strong peaks in Raman spectroscopy because they are associated with high changes in polarisability. Molecular $H_2O$ in the interlayer space of hydrous phyllosilicates and structural OH groups are also very weak Raman scatterers. Further, the fine particles of clay minerals scatter the excitation line in a manner that prevents the deep penetration of the laser into the sample and reduces the scattering volume (Farmer, 1998). For this reason, it is often preferable to measure the Raman spectra of bulk samples and not fine powders (e.g. Frost and Rintoul, 1996a). The obvious risk is that coarser particles in the sample (which may be due to crystalline accessory phases with a simpler structure) are likely to dominate the spectrum.

Kaolinite and its polytypes are by far the most systematically studied clay minerals by Raman spectroscopy. This is related to their low layer substitutions and to their relatively high crystal perfection compared to the smectites. The Raman studies of kaolinite in the literature involved excitation from the NIR to the UV of samples in bulk, or in aqueous dispersion forms (e.g. Johnston et al., 1985; Farmer, 1998; Frost et al., 2001a). These studies placed emphasis on the analysis of the structural OH stretching range. Diamond anvil cells have been used to study the pressure dependence of the spectra and detect pressure-induced phase transitions (e.g. Holtz et al., 1993; Johnston et al., 2002b; Butler and Frost, 2006). Single particles of talc were also studied in one of the first applications of Raman microscopy (Rosasco and Blaha, 1980).

Smectites have been much less systematically studied. Frost and Rintoul (1996a) reported the FT-Raman spectra of several montmorillonites in the low frequency range (50–1150 $cm^{-1}$). A recent study by Ritz et al. (2016) compares Raman scattering and fluorescence in the spectra of four reference montmorillonites, measured with 532-, 780- and 1064-nm excitation. Some data on synthetic and natural saponites are also available (e.g. Kloprogge and Frost, 2000; Pelletier et al., 2003; Bisio et al., 2008). Raman studies of clay mineral hydration are more demanding because they require the sealing of the sample in a suitable cell, but have been demonstrated in the case of synthetic clay minerals with 514.5-nm excitation (Rinnert et al., 2005).

The weak Raman scattering cross section of clay minerals is turned to an advantage when the focus is at the accessory phases present in clays. As an example, anatase admixtures in kaolin have been quantified by FT-Raman spectroscopy down to $\sim$0.3 wt% on the basis of the intensity of the very strong anatase peak at 144 $cm^{-1}$ (Schroeder et al., 2003). In another example, micro-Raman spectroscopy permitted the identification of the corrosion products of iron in contact with bentonite (Wersin et al., 2015).

For the same reason, Raman spectroscopy is a very valuable tool when studying the structure of species adsorbed on clay minerals, with little or no spectral interference from the latter (e.g. Rana et al., 2009; Sánchez del Río et al., 2009; Ruiz-Garcia et al., 2013; Siqueira et al., 2014;

Kubota et al., 2015). This type of study is very relevant to the emerging field of clay-based, organic-inorganic hybrid materials, and can be combined with synchronous XRD measurements (Sánchez del Río et al., 2009) or other X-ray techniques (Newton and van Beek, 2010). In the particular case of organic dyes interacting with clay minerals, one must take into account (and, ideally, explore) the preresonance or resonance effects that originate from the interaction of the excitation lines with the electronic spectrum of the chromophore. These effects influence considerably the number and relative intensity of the bands via the resonance Raman effect. The interpretation of these spectra is often perplexed by the fact that the electronic spectrum of the dye may be changing upon adsorption on the clay mineral, therefore leading to different resonance enhancement profiles (e.g. Teixeira-Neto et al., 2012; Tsiantos et al., 2012).

Whereas the study of clay-based hybrid systems by Raman spectroscopy is expected to flourish, the systematic study of clay minerals per se, especially natural smectites, may remain scarce compared to study by IR. The field seems to be turning towards NIR excitation and FT acquisition to reduce photoluminescence (see Chapter 6). The opposite strategy may also be worth pursuing. UV excitation may prove to be an appealing alternative once UV-Raman spectrometers become more affordable because of the enormous $(\nu_o - \nu_{vib})^4$ Raman enhancement. A recent Raman investigation of the UV-Raman spectrum of montmorillonite in the O—H stretching range (325-nm excitation, 3000–3750 $cm^{-1}$) with good quality and at several levels of hydration may be showing the way (de Ligny et al., 2013).

## 3.5 EPILOGUE

Farmer got his hands on an IR spectrometer in 1954. Immediately afterwards ‘in his typical open-door fashion he was prepared to look at any sample that was presented to him. And steadily a very large collection of many thousands of spectra was amassed in a well organised filing system which proved invaluable in future investigations’ (Russell, 2008). Farmer’s paradigm shows the way for advancing science via the meticulous collection and rational organisation of observations (cf. Hazen, 2014). Systematics and concepts are emerging from such technique-specific observations based on a sufficiently large and diversified set of clays and clay minerals. Vibrational spectroscopy, if properly mastered, is a powerful toolbox for collecting rich data on the chemical structure and (some) properties of clays and clay minerals. X-ray diffraction, chemical analysis and several other experimental techniques are as powerful as vibrational spectroscopy, but yield different, complementary information, because they are based on different principles and are subject to different limitations. Joining efforts to tune these different method-specific glossaries into a common language remains an ultimate goal. The name of the game has not changed since Farmer’s time.

Chapter 4

# Spectral Manipulation and Introduction to Multivariate Analysis

G.D. Chryssikos and W.P. Gates

## 4.1 INTRODUCTION

The vibrational spectrum of any clay mineral is complicated for a number of reasons. First, it represents a complex and polyatomic unit cell with both covalent (layer) and ionic (interlayer) features. In addition, many clay minerals of interest are formally defined over a range of compositions with variable type and degree of chemical substitutions: The species montmorillonite, for example, cannot be associated with a unique infrared (IR) spectrum. Further, a smectite-enriched sample, such as bentonite, may contain many clay mineral phases (e.g. montmorillonite, beidellite, illite and kaolinite) and variable amounts of accessory minerals (e.g. quartz, carbonates and sulphates). All these factors of chemical and mineralogical complexity are convoluted in the spectrum of a single sample and need to be distinguished prior to or during analysis (e.g. Balan et al., 2001; Madejová et al., 2002).

Another form of complexity related to modern IR and Raman spectral analysis comes from the spatial distribution of minerals or other species within samples which may have been collected to represent the larger object of study. Modern IR and Raman instrumentation offers the possibility of collecting hundreds or thousands of spatially resolved spectra efficiently and often noninvasively (see Chapter 3). Spatial complexity, for example, in thin sections studied in reflectance by mid-IR (MIR) microscopy, or in a set of drill cores from the exploration of a deposit studied by near-IR (NIR) spectroscopy, is recorded in large databases. These are called hyperspectral because they contain a (two- or, often, multiple–dimensional) grid of sampling points expressed in spatial coordinates and an extra dimension of a full spectrum


Developments in Clay Science, Vol. 8. https://doi.org/10.1016/B978-0-08-100355-8.00004-7

for each point. As expected, hyperspectral data contain large amounts of information that can be difficult to handle, much less interpret. But they can be analysed to yield the key spectral components that describe the variance of the system of interest, as well as the spatial distribution of these components within the macrosample. This leads to applications like hierarchical classification schemes and identification algorithms for the improved understanding of complex mineralogy. The same spectral databases can also be correlated with independent chemical, mineralogical and physical parameters of the sample, and trained to provide quantitative predictions on unknown samples. The development and application of such qualitative and quantitative prediction using multivariate analysis and chemometric tools can reduce drastically the time, cost and environmental footprint associated with the analysis of a large number of individual samples by conventional methods.

The terms *multivariate analysis* and *chemometrics* are often used interchangeably in the literature, but they are not equivalent. Multivariate analysis is a method for revealing underlying latent descriptors in a dataset and can be called chemometric only when calibrated to predict properties quantitatively. Multivariate analysis can be done independently of chemometric methods, but the opposite is not true. For example, principal component analysis (PCA) (see Section 4.4.1) is often used purely as a data exploration tool. Multivariate analysis and chemometric methods are not limited to vibrational techniques and clay minerals, and can be applied to any large datasets obtained by any technique that can provide an information-rich output regarding the materials of interest and their properties. Among vibrational techniques, multivariate analysis and chemometric methods have been historically based first on NIR spectra, primarily because these were the first that could be collected noninvasively and with sufficiently high signal-to-noise ratio, but also because NIR features were too complicated to assign and analyse using a simple Beer–Lambert approach. Chemometric analysis of MIR data came only after the advent of diffuse and attenuated total reflectance (ATR) techniques. Vibrational chemometric tools are widely applied in diverse fields including food science (e.g. González-Martín et al., 2002; Jie et al., 2013), biology and medicine (e.g. Grootveld, 2015), pharmaceutics (e.g. Reich, 2005), soil science (e.g. Tauler et al., 1996; Viscarra Rossel et al., 2006; Samuels et al., 2006; Reeves and Smith, 2009; Waruru et al., 2014) and mineralogy and ore processing (e.g. Andrés and Bona, 2005). The application of multivariate or chemometric methods to the processing of vibrational (mostly NIR and MIR) datasets has proven to be highly effective in extracting useful information, and this chapter is an attempt to introduce the reader to the salient aspects of these methods. As a prerequisite to multivariate analysis, but also as an independent part of spectral analysis, the following section covers the most common methods used in the mathematical treatment of vibrational spectra to enable their semiquantitative (identification) and quantitative analysis.

## 4.2 OVERVIEW OF POSTCOLLECTION SPECTRAL PROCESSING

The use of any spectroscopic technique based on resonance absorption is underpinned by an understanding of the Beer–Lambert Law, namely:

$$A = -\log(T) = -\log(I/I_o) = \epsilon \cdot b \cdot c \tag{4.1}$$

where $A$ is the absorbance expressed as the negative logarithm of the transmittance ($T = I/I_o$) of light through a sample ($I$) to the initial light intensity ($I_o$), $\epsilon$ is the wavelength-dependent molar absorptivity (also known as the absorption or extinction coefficient), $b$ is the path length of light through the sample and $c$ is the molar concentration. The Beer–Lambert law describes the experimentally observed exponential attenuation of electromagnetic radiation through a homogeneous absorbing (and nonscattering) medium as a function of both distance and concentration (Griffiths, 2002). As such, the Beer–Lambert law is applicable to IR absorption but not to Raman scattering. In IR spectroscopy, the absorbing medium can be a molecule (e.g. interlayer $H_2O$), a polyatomic ion (e.g. $CO_3{}^{2-}$ or $NH_4{}^{+}$) or a functional group (e.g. structural OH bonded to different octahedral cations in montmorillonite) or, collectively, a crystalline or amorphous network of covalent bonds. The exact energy and intensity profile of IR absorption, expressed by the wavenumber-dependent coefficient $\epsilon$, is a useful fingerprint property specific to the vibrating species involved and depends on the dipole moment changes associated with its symmetry-dependent vibrational transitions (selection rules), as well as its local and crystal symmetry (see Chapter 2).

The Beer–Lambert law applies to multicomponent systems provided that there are no chemical interactions or matrix effects. In the ideal case where the sample is of known thickness and the observed vibrational transitions are well separated from each other by a zero baseline, the law allows for both the identification and the quantification of the absorbing species via their unique wavenumber dependence of $\epsilon$. If $\epsilon$ is not known, it can be obtained from a series of standards of variable concentration $c$, or estimated by theoretical analysis (e.g. Balan et al., 2009).

The Beer–Lambert relation typically behaves linearly over the range $0.2 < A < 1$, and for many oscillators of interest in clay science (such as the O—H or Si—O bonds) this range may impose unrealistic constraints on concentration and/or sample thickness. Depending on the sample, these constraints may be critical in determining the proper IR measuring technique (see Chapter 3). Additional reasons causing nonlinearity, such as light scattering, electronic absorption, reflection phenomena and associated optical dispersion effects, may be difficult to be explicitly accounted and call for determining the linearity regime of the law, prior to its application.

The ideal spectrum that can be fully assigned and quantified as described in the previous paragraph is rarely encountered, even less so in clay minerals. In addition to the chemical and mineralogical complexity that yields spectra with many strongly overlapping features, noise and broadly varying non zero baselines are common features of these spectra. In fact, this situation applies to both IR and Raman spectroscopies. As a result, the analysis of the vibrational spectra requires mathematical pretreatments to reduce or eliminate non vibrational features, as well as to identify and assign discrete vibrational bands, prior to extracting quantitative (or semiquantitative) information. Unless otherwise indicated, these pretreatments apply to the analysis of both, single spectra and large spectral datasets.

### 4.2.1 Smoothing

Smoothing aims at reducing the noise level in a spectrum without reducing the number of spectroscopically significant variables. Usually smoothing is applied in cases where noise reduction by increasing acquisition time is impractical (as is the case in high-throughput measurements) or not possible. Application of smoothing is based on the assumption that noise peaks are much sharper than the vibrational bands of interest, and smoothing thus serves as a low-pass filter. Smoothing routines use different equation forms where the values of the points on either side of a datum may be 'averaged' or applied algorithmically to estimate the central value in some way, and this average then replaces the original central datum. The most common smoothing routines used in spectroscopy are based on the Savitzky–Golay filter (Savitzky and Golay, 1964), which is also used in conjunction with differentiation (see Section 4.3.2). The filter employs polynomial functions, fit to either side of each successive individual datum point within a spectrum. The functions are calculated from a regression fit of each successive polynomial to a moving window of user-chosen filter-width (e.g. $n=1$, 2, 3 or more) points on either side of each successive datum in the spectrum, to project calculated intensities for each datum in turn. The user should ensure that the filter width (i.e. the total number of points used in the smoothing, $2n+1$, multiplied by the spacing of the data, $\Delta\nu$, in $cm^{-1}$) and the order of the polynomial are chosen carefully. The filter should be of the same or smaller width than the nominal width of non noise features present in the spectrum, and the polynomial order should be adjusted to remove only noise. The larger the filter width and the lower the polynomial order chosen, the greater will be the resulting smoothing effect.

Smoothing results in decreasing spectral resolution and can affect the position, width and intensity of vibrational bands, as well as the ability to resolve components of vibrational envelopes. Therefore smoothing should be applied only if needed, and with the appropriate adaptation of its parameters to the specific spectrum and its acquisition conditions. Any reference to the number of smoothing points employed should be accompanied by the frequency spacing, $\Delta\nu$, of the data, which is not necessarily identical to the spectral resolution setting of the experiment.

### 4.2.2 Baseline Corrections

The baseline of IR and Raman spectra are often excessively non zero in some regions that may make the detection of weak vibrational bands difficult. Baseline correction is used to minimise the presence of non informative background in the measurements, thereby enhancing the spectral features of interest. A strongly sloping background is very common in diffuse reflectance spectra due to variations in light scattering from irregularities in the sample: non uniform particle size, shape, sample packing or irregular surface texture (Martens et al., 2003), but also in the case of NIR, from the high-frequency tails of strong MIR bands. Similar effects are common in Raman spectra due to luminescence (fluorescence). Consistent data acquisition can minimise some of these effects, but it is usually not possible to remove all baseline issues. Such unwanted baseline effects can propagate in the spectra during subsequent treatment (e.g. normalisation, see the following) unless they are properly compensated for.

The simplest baseline correction is an offset correction applied to bring relatively flat (non sloping) baselines to zero. Similarly, a sloping baseline can be corrected by subtracting a suitable straight-line segment from the experimental spectrum over the wavenumber range of interest. More typical is the situation where the features of interest are located on a curved baseline, usually the tail of a broad and strong band of vibrational or electronic origin. This is, for example, the case of the OH deformation modes of smectites that appear on the low-frequency side of the strong Si—O stretching envelope (e.g. Vantelon et al., 2001; Gates, 2005), or the corresponding, NIR-active, OH combination modes (Post and Noble, 1993), which are observed on the high-frequency side of the strong, MIR-active, O—H stretching envelope (e.g. Madejová et al., 1994; Bishop et al., 1999; Gates et al., 2002; Zviagina et al., 2004; Petit et al., 2015). This unwanted contribution from the broad overlapping features is typically removed by subtracting background Gaussian functions from one or both sides of the wavenumber range of interest (e.g. Fig. 4.1). Obviously, this type of correction is critical in estimating the occupancy of octahedral sites in clay minerals from their IR spectra (e.g. Petit et al., 2015). Another common approach to baseline correction has been the simple subtraction of an artificial baseline 'spectrum' created to simulate the wavenumber dependence of the baseline over the spectral range or ranges of interest (e.g. Petit et al., 2015). Whereas these baseline correction

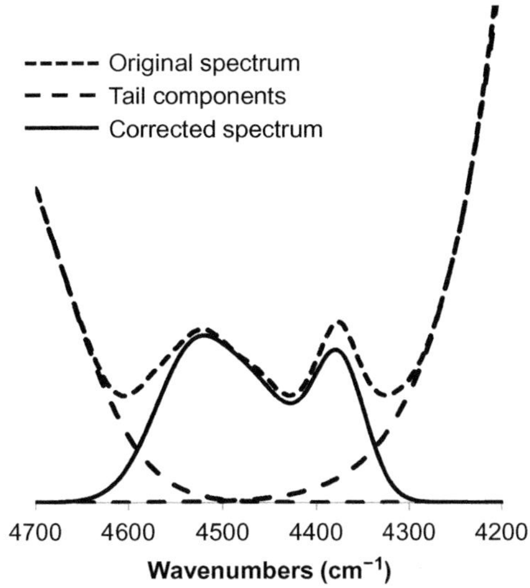


**FIG. 4.1** Baseline correction by removing Gaussian-shaped tails from the raw NIR spectrum of a synthetic smectite. Such correction enables least squares quantitative decomposition analysis of bands of interest. Spectral treatment applied by the authors to the IR spectrum of smectite synthesised and characterised by Andrieux and Petit (2010).

approaches tend to be somewhat empirical and subjective, they have served sufficiently for most applications, as long as they have been employed consistently across a spectral dataset and documented in detail.

### 4.2.3 Atmospheric Compensation

Compensation for the residual spectra of atmospheric gases is a special form of baseline correction in IR spectroscopy. Unless specifically removed, air within the optical path of an experiment often contains enough $H_2O$ vapour to produce a very rich vibrational-rotational spectrum consisting of a series of sharp bands at ~1900–1300 cm$^{-1}$ ($\delta$), 4000–3400 cm$^{-1}$ ($\nu$), 5600–5100 cm$^{-1}$ ($\nu+\delta$) and 7400–7000 cm$^{-1}$ ($2\nu$). Similarly, residual atmospheric $CO_2$ yields a characteristic doublet at 2360, 2340 cm$^{-1}$ and a sharp bending mode at 668 cm$^{-1}$. The latter features are usually of a lesser concern, except perhaps in the case of deuteration studies or in the detailed analysis of the OH bending and lattice deformation modes of smectites. The bands of atmospheric gasses are very prominent in the single-beam IR spectra and should ideally cancel out when intensity ratios of the sample and background spectra are calculated. In many cases, however, this cancellation is incomplete, either because the optical path is open or, more seriously, because process monitoring measurements need to be performed over long times and cannot be interrupted for updating the background single-beam measurement

(Li et al., 2000). Despite overlap with the vibrational features of structural OH and $H_2O$ in clay minerals, the positive or negative residual rotational-vibrational bands of $H_2O$ vapour are easily discerned in high-resolution (e.g. 1 $cm^{-1}$ or higher) spectra because they are much sharper than the vibrational features of the sample and often can be corrected by application of an appropriate smoothing filter. Under more routine resolution settings (e.g. 4 $cm^{-1}$ or lower), however, the bands of $H_2O$ vapour appear broader and may therefore be confused with the sharpest bands of the sample (e.g. the structural $Mg_3OH$ stretch of talc) and as such, should not be remedied with smoothing. Unfortunately, the incorrect assignment of $H_2O$ vapour bands to all sorts of fictitious chemical species in the sample is not at all rare in the literature.

Modern commercial spectroscopic software offers general-purpose atmospheric compensation options that can be satisfactory for the routine presentation of absorbance and reflectance spectra, but are unsuitable for subsequent derivative analysis. The best practise to avoid interference from $H_2O$ vapour is to isolate the optical path as much as possible from the atmosphere (which is more easily achieved in ATR spectroscopy) and to equilibrate by desiccants and/or purging with dry $CO_2$-free air or $N_2$. Any remaining positive or negative changes of $H_2O$ vapour within the instrument itself are best compensated by measuring a second background spectrum after the completion of the experiment and using it to remove by subtraction any residual contribution for the vapour (e.g. Bukas et al., 2013). This procedure essentially interpolates two reference background spectra collected before and after sample measurement to compensate in full the individual spectra of a series of successive measurements. The only requirement for the successful vapour band removal by subtraction is that temperature remains constant during measurement to avoid shifting the $H_2O$ vapour spectrum.

All of this described here refers to IR spectroscopy. Raman spectroscopy with excitation in the visible spectrum is generally immune to such atmospheric effects because the absolute frequency of the measurement is well away from the vibrational spectrum of $H_2O$ (see Chapter 3). FT-Raman spectra, however, collected with 1064-nm excitation (~9400 $cm^{-1}$) can exhibit atmospheric interference, for example, in the $\Delta\nu = 2400$–$2000$ $cm^{-1}$ range from the $2\nu$ vibrations of $H_2O$. Such interference is in the form of negative peaks on a nonzero, sample-dependent background (due to luminescence or blackbody radiation, for example), and its compensation is difficult.

### 4.2.4 Normalisation

Normalisation is used to adjust the intensity of a spectrum to the same scale as other spectra in a dataset in order to assist visual comparisons or quantitative analysis. Normalisation helps to remove non systematic effects, for example when sample concentrations may differ in dilutions, or when the path length of a transmission cell may differ between experiments, and serves to give all sample spectra equal impact for comparison. Another important use of normalising spectra is when the signal is a function of source power rather than sample concentration, as may occur in some instances in synchrotron-based IR or Raman spectroscopy or when a comparison between data obtained by different techniques over the same energy range is required (e.g. IR and Raman). Several procedures exist for the normalisation of IR spectra, the most common of which are area normalisation, vector normalisation and multiplicative scatter correction (e.g. Randolph, 2006). Other normalisation procedures exist, but are less commonly used in the vibrational spectroscopy of clay minerals and will not be discussed here. Normalisation can be performed on a specific band, or envelope of bands, even the entire spectral range, depending on needs. It can be applied to spectra in all forms, including absorbance, transmittance, derivatives, and others.

Most normalisation procedures transform the spectral features within a set of spectra to represent the same number of oscillators distributed over two or more component bands within the frequency range of interest. This is approximated by converting the spectra to the same integrated intensity (area) over the range of interest and explains why this should be done on baseline-corrected spectra. The critical assumption underlying this approximation when normalised data are used for quantitative analysis is that the absorptivity, $\epsilon_i$, of the component bands (scattering cross-sections in Raman experiments) are identical or, at least, similar to each other.

*Area and intensity normalisation.* Area normalisation transforms the observed data so that the area under the baseline-corrected spectrum in the frequency range of interest is the same for all samples. Intensity normalisation on the same band (usually the maximum of the strongest) within the region of interest can be applied instead of area normalisation, provided that the band position and width remain fixed across a series of spectra. This is the typical normalisation applied in cases where the sample is spiked with a suitable non-interacting, IR- or Raman-active admixture that serves as an internal standard. Maximum-weighted normalisation divides each spectrum by the maximum intensity observed, and therefore all spectra have vector norms that are scaled by their maximum values.

Vector normalisation refers to a group of normalisation methods that are based on linear algebra rather than empirical as is often the case for intensity or area normalisations. These normalisations attempt to give each spectrum equal impact on any model subsequently developed, and also assist with visualisation. Vector normalisation is based on the vectorial definition and properties of a spectrum $\boldsymbol{x}$, where $\boldsymbol{x}$ is a $1 \times k$ matrix row vector described by:

$$\boldsymbol{x} = [x_1, x_2, \ldots x_n] \tag{4.2}$$

where $x_i$ is the measured intensity and $n$ is the number of points over the wavenumber range of interest. The method transforms spectrum $\boldsymbol{x}$ so that its vector norm becomes equal to 1. This is done by subtracting first the average spectral intensity from the spectrum, leading to a new spectrum, $\boldsymbol{x}' = [x'_1, x'_2, \ldots x'_n]$, with both positive and negative intensities so that $\Sigma x' = 0$. The sum of the squares of the intensities $\Sigma(x'^2)$ is then calculated and the spectrum $\boldsymbol{x}'$ is normalised by the square root of this sum. Similarly, standard normal variate (SNV) normalisation scales $\boldsymbol{x}'$ to the standard deviation of the data points in the spectrum (Barnes et al., 1989; Rinnan et al., 2009). Absolute value vector normalisation divides each spectrum by the sum of the absolute value of all intensities, $|x_{ij}|$, in the spectrum. This returns a vector norm with a unit area, $w_i$, under the spectrum equal to 1. An example is given in Fig. 4.2, where spectra (Fig. 4.2A) have been vector normalised (Fig. 4.2B).

*Multiplicative scattering correction.* Within the general frame of normalisation transformations, multiplicative scatter correction (MSC) is a special type of compensation for variable individual baseline contributions within a set of spectral data. As discussed earlier, baseline variability within the dataset can be caused by particle size, sample packing and density variations, inhomogeneous particle distribution as a function of depth of sample or sample surface roughness, and is particularly common in diffuse reflectance spectroscopy (Afseth and Kohler, 2012). All these poorly controlled factors influence the optical path length, $b$ (Eq. 4.1), and create additive or multiplicative effects on the spectral baseline (Rinnan et al., 2009; Huang et al., 2010). Some of these effects can be taken into account by application of the Kubelka–Munk theory (Yang and Kruse, 2004), but further normalisation within the spectral dataset is frequently needed. MSC cannot be applied to individual spectra and is mostly relevant to the analysis of large spectral datasets by, for example, PCA (see Section 4.4.4). MSC operates on the mean of the entire dataset, and it scales a group of spectra to an equalised reflectance for quantitative comparison. In its simplest form MSC relies on the assumption that any measured IR or Raman spectrum can be successfully approximated by the sum of a baseline offset and a variably amplified 'chemical' Beer–Lambert absorbance. Mathematically, this is expressed (Huang et al., 2010) as:

$$x_{ij}(\mathrm{MSC}) = a_{ij} + b_{ij}x_{ij} \tag{4.3a}$$

where $x_{ij}$ is the absorbance value of the spectrum $i$ at wavelength $j$; $a_{ij}$ is the corresponding baseline offset; $b_{ij}$ is the reference absorbance value for $i$ at each wavelength $j$ (Rinnan et al., 2009) and accounts for signal amplification effects due to variations in optical pathlength; and $x_{ij}$(MSC) is then the absorbance associated with the sample chemistry. Extended-MSC (EMSC) allows for further compensation of effects like sample mass differences

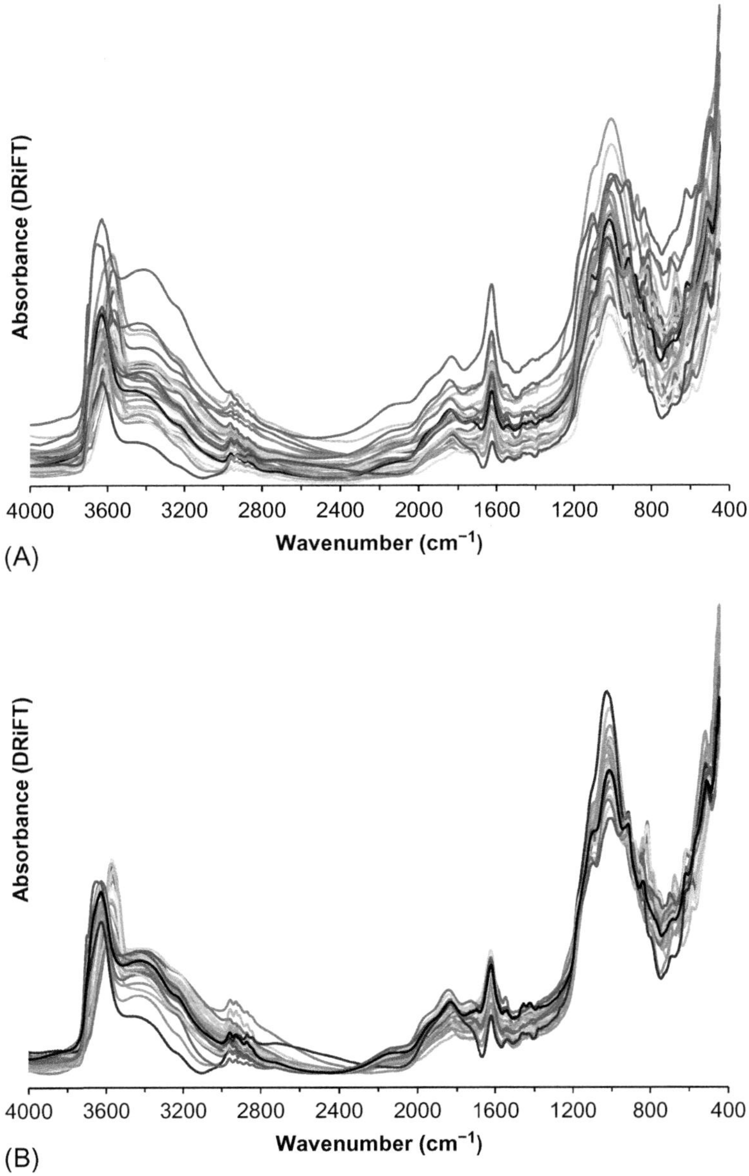


**FIG. 4.2** (A) MIR spectra of 32 dioctahedral smectites (neat air-dried, $Ca^{2+}$ form). (B) The same dataset after unit vector normalisation and offset correction. Absorbance values are the Kubelka–Munk transformations of diffuse reflectance (DRiFT) spectra.

and wavelength-dependent spectral effects (e.g. interference and path length variations) in order to remove their variance from the spectral information of interest (Martens et al., 2003). In EMSC a spectrum is expressed linearly as:

$$x_{ij}(\text{EMSC}) = a_{ij} + b_{ij}x_{ij} + d_{ij}\lambda + e_{ij}\lambda^2 \tag{4.3b}$$

where additional $d_{ij}$ and $e_{ij}$ terms, calculated from the dataset, are introduced to describe wavelength-dependent, $\lambda$, variations.

Both MSC and EMSC methods regress each individual spectrum, $i$, of a set of spectra against a reference, $b_i$, typically the average spectrum of the set, and it is important to realise that the (E)MSC corrected output depends on the set of spectra where this reference spectrum belongs (Maleki et al., 2007). In mathematical terms, correction has the form:

$$x_{ij\,\text{MSCcor}} = \left(x_{ij} - a_{ij}\right) \Big/ b_i \tag{4.4a}$$

$$x_{ij\,\text{EMSCcor}} = \left(x_{ij} - a_{ij} - d_i\lambda - e_i\lambda^2\right) \Big/ b_i \tag{4.4b}$$

In both expressions, $x_{i(\text{E})\text{MSCcor}}$ returns the frequency-dependent absorbance of the spectrum containing the chemical information of interest with minimal interference from nonchemical sources of variability (Li-Chan et al., 2011). The corrections are most commonly performed using a first-order polynomial (Rinnan et al., 2009). Since each spectrum is treated as such, direct intensity (area) comparisons become fully quantitative. Examples of both MCS and EMSC corrected spectra are shown in Fig. 4.3.

## 4.3 IDENTIFICATION AND SEPARATION OF OVERLAPPING VIBRATIONAL TRANSITIONS

In most cases, the mere identification, much less quantification, of the mineral components in a sample by vibrational spectroscopic techniques cannot generally be separated from the consideration of the chemical (compositional) variability of these components and its effect on the spectra. In fact, the complex mineralogy and crystal chemistry of the clay minerals call for the separation of overlapping vibrational bands, as a prerequisite to their assignment (identification) and subsequent quantification. Solution of the problem involves ideally the precise and reproducible determination of the exact number of component bands underlying a complex vibrational envelope, as well as the position, width and intensity of each component band. This determination is usually based on band decomposition and/or derivative procedures. The two approaches, each with its own advantages and deficiencies, will be outlined in the following sections.

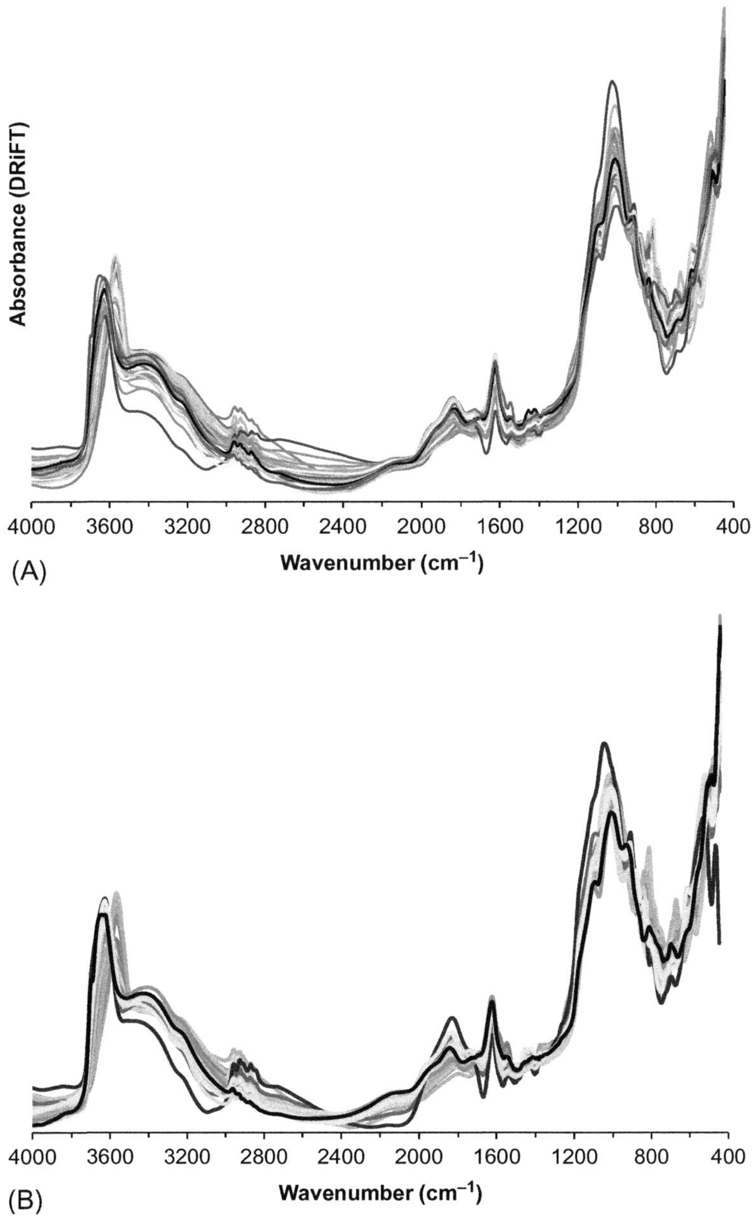


**FIG. 4.3** The (A) MSC-corrected and (B) EMSC-corrected Kubelka–Munk transformed MIR DRiFT spectra shown in Fig. 4.2A. Note the differences from vector normalisation (Fig. 4.2B).

### 4.3.1 Decomposition of Overlapping Bands

The compositional and structural variations inherent in smectites and other clay minerals result in band broadening and overlap, which negates the applicability of a simple intensity measurement for quantification. Instead, the broadened absorption envelopes can be analysed by separating into a number of overlapping components, where each component is made up of a distribution of intensities over a discrete frequency range. This process, called spectral band decomposition fitting, or simply decomposition analysis, provides a means to quantify spectra if the resulting component integrated intensities associated with each particular transition are summed. The goal in decomposition analysis is to simulate the experimental spectrum (typically in absorbance and after correction for baseline effects) over the frequency range of interest with a number of component bands (Fig. 4.4).

Decomposition analysis is an iterative curve-fitting problem, solved by minimising the least squares difference between the simulated (sum of components) and experimental spectra. Most available software allows for choosing the minimisation algorithm (e.g. damped or local least squares) and the least squares threshold for terminating the iteration. Many researchers have performed decompositions using spreadsheet formulations (e.g. Gates et al., 2002; Petit et al., 2015, 2016). Most fitting routines require initial 'guesses' about the number, shape (typically Lorentz, Gauss or Voigt), position, width and intensity of the component bands to produce an initial 'calculated'

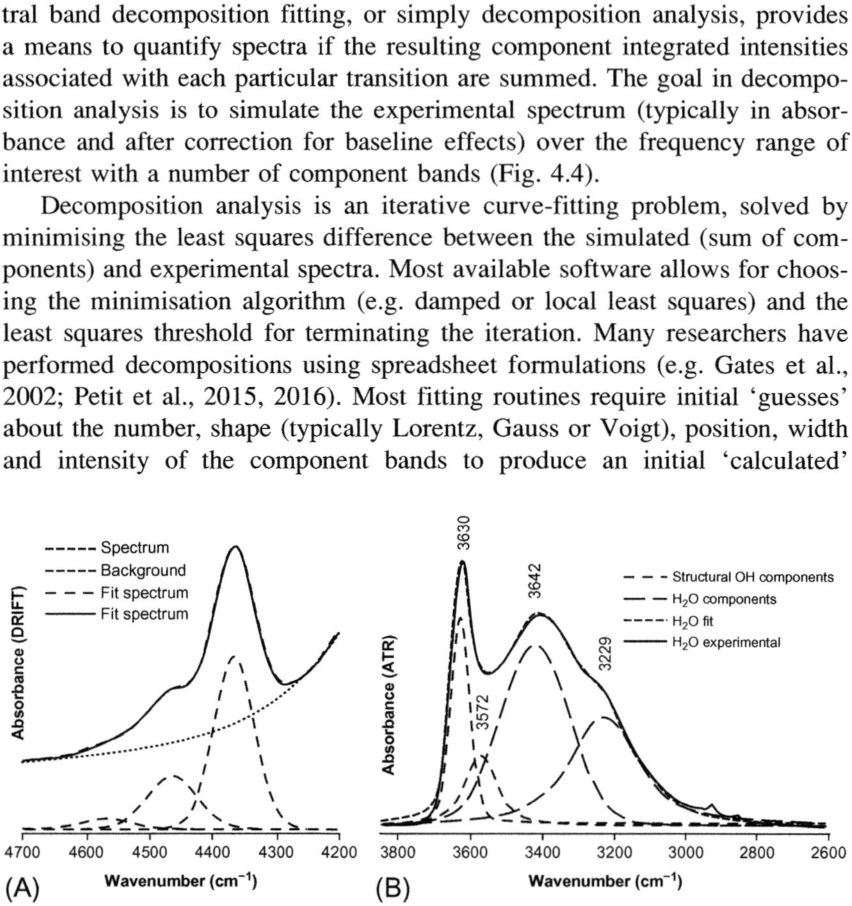


**FIG. 4.4** Representative decompositions of the IR spectra of smectites. (A) Decomposition of the OH combination ($\nu+\delta$) bands of NG-1 nontronite as shown in Gates et al. (2002) following baseline correction as presented in Fig. 4.1. (B) Decomposition of the complex OH stretch ($\nu$) bands of a montmorillonite following the procedure outlined by Madejová et al. (2002a). In (A) baseline correction by removing the tail on the low frequency side of the region of interest had little impact on the sensitivity of the fit and quantification of the two strongest bands. In (B), however, the fitting uncertainty of the 3572 cm$^{-1}$ component (which is not evident in the absorbance spectrum) can have a significant effect on the parameters of the 3422 cm$^{-1}$ component.

spectrum. The calculated spectrum is then least-squares fit to approximate the experimental spectrum by iterative changes of these initial parameters.

Each component is defined by three parameters (position, width and intensity) to be fitted (four in the case a Voigt function having variable Lorentz character), leading to a multidimensional global minimisation problem. In such problems, defining the single global minimum, rather than one among several possible local minima, is not straightforward and may require several optimisation runs starting from different initial conditions. Often it is also necessary for the operator to fix individual parameters in a controlled iterative fashion to ensure that the calculated minimum produces a realistic model of the experimental spectrum. Some programmes enable setting limits, or boundary conditions, to the parameters to minimise the opportunity for the fits to fall into 'false minima.' Obviously, for large spectral datasets, spectral decomposition fitting can become quite cumbersome even with modern software packages.

One pitfall to the method is that the introduction of more component bands into the model spectrum produces better fits without improving necessarily the physical meaning of the outcome (Gates, 2005). To achieve meaningful results, intuition and experience are important for deciding what parameters to constrain, fix, or allow to evolve freely during the fitting routine, as well as when to do so (e.g. Zviagina et al., 2004; Petit et al., 2015). It probably cannot be stressed strongly enough that a consistent approach is key to meaningful application of decomposition fitting.

Despite the difficulties in controlling peak fitting algorithms, several authors have utilised the method effectively to the study of clay minerals. Among the first successful applications of this method to smectites was an adaptation by Muller et al. (1991) of the breakthrough approach of Slonimskayá et al. (1986) on micas. Since then, peak fitting for spectral decomposition has been applied to the study of a variety of smectite samples for various purposes (e.g. Madejová et al., 1994, 2002a; Yan et al., 1996a; Besson and Drits, 1997a,b; Petit et al., 1999b, 2015; Frost et al., 2001b; Vantelon et al., 2001; Gates et al., 2002; Zviagina et al., 2004, 2015; Bishop et al., 2011). The popularity of spectral decomposition analysis stems from the fact that the relative integrated intensities (areas) of resolved components with common origin (e.g. structural OH) can be fed to a Beer–Lorentz-type quantification. Converting relative intensities to relative concentrations is based on the assumption that the absorptivity coefficients, $\epsilon_i$, of the resolved component IR bands (or the scattering cross sections of Raman components) are identical. The validity of this assumption should always be tested against independent analytical data (e.g. Gates, 2005; Petit et al., 2016). An overview on the applications of band decomposition analysis to the structural characterisation of dioctahedral clay minerals is discussed in greater detail in Chapter 7.

Spectral decomposition by least squares peak fitting approaches reaches its limits in cases where the experimental spectral envelope to be fitted contains both sharp and broad overlapping bands, especially in cases where the relative integrated intensity of the former is weak. Examples of such cases are encountered in the O—H stretching and overtone regions of the IR spectra of smectites, due to the complex overlap of the sharper bands of structural OH with the broader (and possibly variable) bands of $H_2O$, or to the coexistence of dioctahedral and trioctahedral clay minerals. In these cases a small acceptable error in the optimised solution of the strong broad components can induce large and unacceptable errors in resolving the weak sharp components. This is because fitting is usually done on spectra with fixed frequency intervals, $\Delta\nu$. Thus sharp features are represented by fewer experimental points than broad ones and have less influence on the minimisation criterion. For this reason, some software routines offer the option to use different frequency intervals within segments of the range of interest, but this option remains scarcely explored in the field.

### 4.3.2 Derivative Analysis

Some of the drawbacks of band decomposition analysis are conveniently remedied by the use of spectral derivatives, $d^nA/d\nu^n$. Derivatization is a multifunctional mathematical tool that can be used simultaneously for enhancing spectral resolution and filtering out broad spectral features, including sloping and curved baselines. This explains why derivative analysis has become very popular among NIR spectroscopists, although its use on other vibrational techniques is equally important. A deeper insight about the properties of derivatives and their use in spectroscopy can be found in Mark and Workman (2003).

For Lorentzian band shapes the amplitude of the $n^{\text{th}}$ derivative decreases with $n$, but also varies inversely as the $n^{\text{th}}$ power of the bandwidth. The dependence of Gaussian band shape amplitudes on the order of the derivative is weaker, but still very significant (Maddams and Mead, 1982a; Maddams and Southon, 1982b). This behaviour is the basis for the high discriminating power of derivative spectra for weak sharp peaks superimposed on intense broad peaks or background (Fig. 4.5), which increases sharply with derivative order. The main drawback of using higher-order derivatives is that the derivative amplitude of any random high-frequency noise decreases with increasing order of the derivative at a slower rate than the corresponding amplitude of the broader vibrational bands of interest. As a result, the existing noise in the spectrum becomes more pronounced, and the signal-to-noise ratio deteriorates significantly with increasing derivative order. The compromise between desired resolution and signal-to-noise ratio sets a practical limit on the maximum order of the derivative, which is usually the second derivative. In comparison to second derivatives, first derivatives have a better

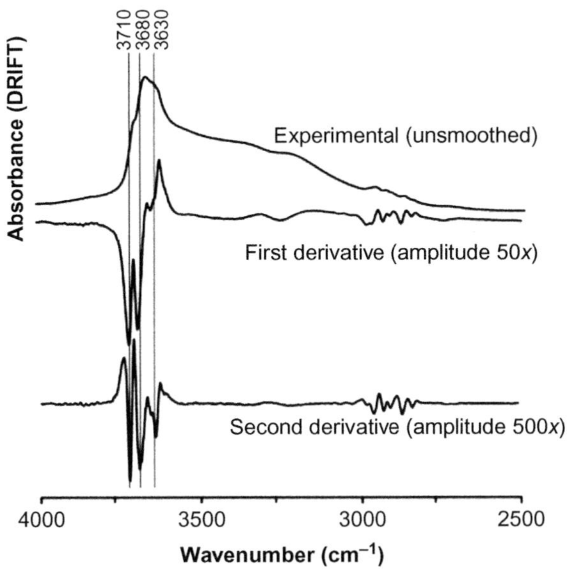


**FIG. 4.5** DRIFT MIR spectrum (2 $cm^{-1}$ resolution with $\Delta v = 0.96\ cm^{-1}$) of the OH stretching region of a smectite sample detailing how absorbance features within complex bands of experimental spectra can be better viewed after derivatization. Application of first derivative with Savitzky–Golay function over $2n+1=11$ points (50×); Application of second derivative $2n+1=15$ points (500×).

signal-to-noise ratio, but a smaller resolving power and a band shape that can be difficult to interpret (zero-crossings at peak maxima).

Derivatives can be computed in various ways, for example, as simple differences of adjacent raw data points, or over fixed-gap distances. Spectral derivatives are usually computed over a finite differentiation interval using an algorithm such as that parametrised by Savitzky and Golay (1964) and corrected by Steinier et al. (1972). This algorithm is applicable on spectra with equidistant data points, $\Delta v$, and computes derivatives at any point of the spectrum (except a few points at either end) on the basis of a range of $n$ points on each side of the central point, that is, over a spectral range of $(2n+1)\cdot \Delta v\ cm^{-1}$, similarly to smoothing routines described previously. Both $n$ and $\Delta v$ need to be selected judiciously because, although the increase of their product results in an increase of the signal-to-noise ratio, it also results in loss of spectral detail.

In practical terms, the aforementioned considerations regarding the properties of the derivative and their computation imply that all the spurious sharp features in a spectrum (noise, spikes, spectrum of atmospheric gases) must be eliminated prior to differentiation. Smoothing should not be used prior to differentiation because, as has been described, it is embedded in the Savitzky–Golay derivative algorithm. Similarly, offset, sloping or broadband backgrounds will be eliminated or damped significantly by differentiation, so these corrections prior to performing derivatives are unnecessary. For a given experimental frequency spacing $\Delta v$ (which is typically controlled by the spectral resolution setting and any zero-filling factor employed), the smallest number of Savitzky–Golay points, $(2n+1)$, that can resolve the spectral bands of interest with satisfactory signal-to-noise ratio should be employed (Fig. 4.6). If the spectrum of interest contains peaks with different bandwidths, more than one value of $(2n+1)$ can be chosen for analysing different parts of the spectrum. At the expense of effort to acquire high-quality spectra, this procedure can yield a very clean picture of the sharper vibrational features of interest, with high discriminating power and free of interference from broader features, either spurious or intrinsic (e.g. bands of variable intensity due to physisorbed $H_2O$) (Fig. 4.7). Properly applied derivatization has been shown to be highly suitable for the qualitative assessment of the mineralogical or chemical composition of clay mineral samples and can be applied quickly on large sets of relevant spectra (Gionis et al., 2006, 2007). In contrast to band decomposition, derivatives can be applied to preprocess large sets of spectra in multivariate analysis where they assist in removing nonvibrational effects to create robust training sets (see Section 4.4.2).

On the other hand, the direct exploitation of derivatives for quantitative determinations on clay minerals is rare, because of the resulting lower signal-to-noise ratio of the derivative spectrum. Nevertheless, such quantitative analysis applications have been demonstrated on the basis of second derivative amplitudes as an extension to Beer–Lambert's law (Gionis et al., 2007; Chryssikos et al., 2009), or on the basis of the position of sharp $H_2O$ stretching bands (Kuligiewicz et al., 2015a,b).

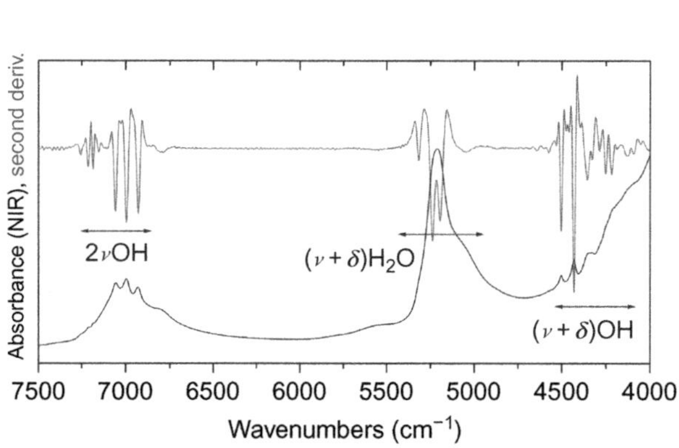


**FIG. 4.6** Vector-normalised NIR absorbance and second derivative spectra of a palygorskite sample demonstrating the excellent resolving power and baseline compensation of a properly tuned second derivative. The spectrum was collected by averaging 100 scans at 4 $cm^{-1}$ resolution ($\Delta v = 2$ $cm^{-1}$), and the derivative was computed by application of a Savitzky–Golay, $2n+1=13$ function.

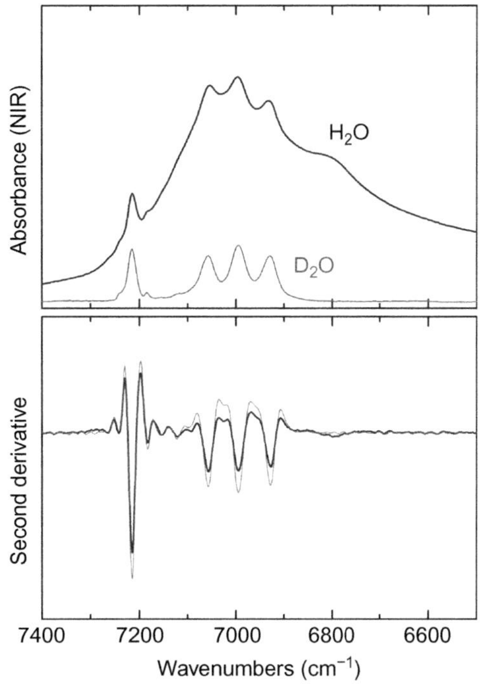


**FIG. 4.7** Experimental NIR absorbance data *(upper spectrum)* of a palygorskite specimen with mixed dioctahedral–trioctahedral character in its natural $H_2O$ form, and after exchange with $D_2O$ to remove the broad and complex spectrum of $H_2O$. The corresponding second derivative spectra *(lower panel)* are nearly identical to each other indicating that the accurate identification of structural OH modes can be obtained directly from the $H_2O$ form, without the need for deuteration. Sample drying was in this case out of the question because it affects the structure of the layers. This is a case where peak fitting the envelope of the $H_2O$ form (as exemplified in Fig. 4.4B) would introduce serious uncertainty, especially in cases of variable relative humidity. For more examples and technical details, see Bukas et al. (2013).

## 4.4 MULTIVARIATE ANALYSIS AND CHEMOMETRIC QUANTIFICATION

Multivariate analysis and chemometrics are a group of methods that provide a statistically meaningful basis for describing complex multiparametric phenomena (Kramer, 1998). When applied to large sets of vibrational spectra (e.g. Fig. 4.2), multivariate analysis enables the extraction of qualitative information that would otherwise be too complex to obtain by other means. Similarly, chemometric techniques allow for quantitative predictions in cases where the straightforward application of Beer–Lambert law would be impossible. These methodologies for qualitative and quantitative analysis rely on good-quality spectral data (the independent variables in the multivariate analysis) as well as good quality chemical or physical data (the dependent variables in chemometric analysis). Application of such methods to NIR and MIR spectra of materials has proven to be an inexpensive method for rapidly identifying the mineral and chemical components in soils and rocks (Karstang et al., 1991; Sudduth and Hummel, 1993; Chang et al., 2001; Viscarra Rossel et al., 2006; Reeves, 2010; Reeves et al., 2010) as well as establishing predictive correlations against mineralogical, physical and chemical datasets associated with the same materials (Reich, 2005; Viscarra Rossel et al., 2006; Bona and Andrés, 2007; Gomez et al., 2008; Reeves and Smith, 2009; Mouazen et al., 2010; Kerr et al., 2011; Filgueiras et al., 2014). Recent advances in specialised commercial software packages enable rapid incorporation into the spectroscopy laboratory.

A brief introduction to matrix or vector approaches which form the basis for multivariate analysis is presented. This introduction is intended to provide a starting point for researchers interested in utilising the chemometric approach to the study of clays and clay minerals. For greater detail the reader is referred to many specialised literature sources including Mardia et al. (1979), Naes and Martens (1988), Naes et al. (2002) and Schlens (2014).

### 4.4.1 Introduction to PCA and PLS

Given the specificity and sensitivity of vibrational spectroscopy to chemical composition and structure, any large sample set can be measured and represented by a hyperspectral database (e.g. absorbance spectra) that is very rich in information content. The database can be represented as an $N \times K$ matrix, $\boldsymbol{X}$, where $K$ is the number of discrete frequency points (independent parameters) in each spectrum, and each of the $N$ rows represents a spectrum as a point in a $K$-dimensional space (see Section 4.2.4 on vector normalisation). This matrix is usually called *training* or *calibration set* in the context of multivariate analysis and chemometrics, respectively.

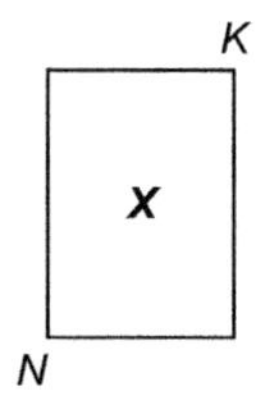


In the typical case of interest, the number of spectra, $N$, is much larger than the number of chemical or mineral components that can possibly be present in the sample set, $n$. There also is sufficient variability between the spectra (rows of the matrix) at most frequencies (columns). Each individual spectrum is considered to represent a different (for simplicity, linear) combination of this limited number of chemical components (i.e. the spectra are highly correlated). Of course, each spectrum also contains errors from random noise and, possibly, instrumental artefacts.

For the simplest analysis of a *multivariate* set of data, one should apply a classical least squares approach in order to express the matrix $\boldsymbol{X}$ as a linear combination of a finite number $n \ll N$ of reference spectra (row vectors $\boldsymbol{r}_n$), one for each of the chemical components of interest, such that the residuals, $\boldsymbol{E}$, are small and contain only error. In this way, every spectrum of the training set can be mapped on a predefined $n$-dimensional space formed by the reference vectors $\boldsymbol{r}_n$. This would be a *supervised* solution (the reference spectra $\boldsymbol{r}_n$ are chosen by the user) expressed in the form:

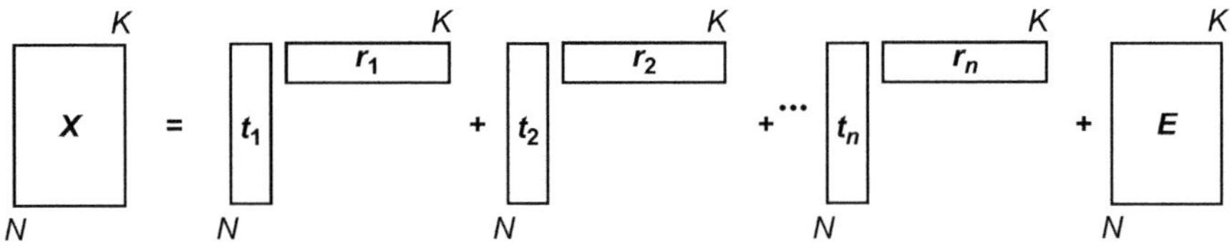


(4.5)

Eq. (4.5) contains both the qualitative (spectra $\boldsymbol{r}_n$) and quantitative (coefficients $\boldsymbol{t}_n$) information that fully describes database $\boldsymbol{X}$. Nonlinear variations to Eq. (4.5) also occur. It was already argued (see Section 4.3) that this approach can be problematic because of the lack of properly defined reference mineral spectra and the uncertainty surrounding how many spectra would be necessary to describe a database. The problem, therefore, calls for a *multivariate analysis* solution to identify and extract the main features underlying the dataset $\boldsymbol{X}$ in an *unsupervised* manner, without any input of the user. There are several methods available to perform this extraction based on multivariate statistics and linear algebra, and one of the most popular for decoding spectroscopic data is Principal Component Analysis (PCA) (e.g. Cowe and McNicol, 1985; Geladi and Kowalski, 1986; Bro and Smilde, 2014; Schlens, 2014).

PCA expresses the training matrix $\boldsymbol{X}$ (or rather, its mean-centred version) as a linear combination of calculated independent one-dimensional (*rank* 1) variables, called *loadings, components,* or more specifically *principal components*, $\boldsymbol{p}_n$, weighted by appropriate coefficients called *scores*, $\boldsymbol{c}_n$. The principal components represent calculated, *latent,* reference spectra and the scores represent their intensity changes within $\boldsymbol{X}$. The original set of spectra containing, for example, absorbance versus wavenumber data is now expressed into an equivalent set of scores on principal components.

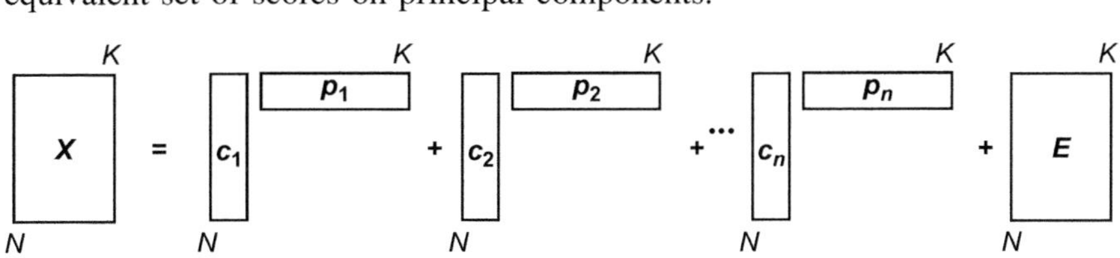


(4.6)

PCA differs from the supervised modelling that was previously based on spectra of reference compounds, $\boldsymbol{r}_n$ (Eq. 4.5), in that the linear terms in the series are not preselected by the user. The linear terms are computed from the data in decreasing order of importance for explaining the variability in $\boldsymbol{X}$. The elements of $\boldsymbol{p}_n$, are linear combinations of the independent variables in $\boldsymbol{X}$, and the vectors $\boldsymbol{p}_n$ are the new, *latent*, independent variables. They are computed to be fully uncorrelated from each other, and they constitute the eigenvectors of the covariance in $\boldsymbol{X}$. As such they form an orthonormal vectorial basis suitable for the projection of any spectrum within $\boldsymbol{X}$. The scores, $\boldsymbol{c}_n$, are the corresponding eigenvalues, and they are scaled so that the sum of their weight coefficients squared is unity.

The PCA calculation starts by mean-centring the spectra in $\boldsymbol{X}$, through subtraction of the means across all dimensions of the data:

$$x'_{ij} = \left(x_{ij} - \bar{x}_{ij}\right) \tag{4.7}$$

where $x'_{ij}$ is the mean-centred result and $\bar{x}_{ij}$ is the mean value. This operation centres the variability along each of the original $K$ dimensions (frequency points) at zero. The first principal component, $\boldsymbol{p}_1$, is determined as the (eigen)vector along which $\boldsymbol{X}$ shows the highest covariance. The second principal component, $\boldsymbol{p}_2$, accounts for most of the remaining variability that is orthogonal (independent) to the first. The iteration continues, always maintaining orthogonality, until the variability that remains unexplained after the calculation of $\boldsymbol{p}_n$ approaches the size of the measurement error. The eigenvalues of the principal component eigenvectors constitute the covariance matrix.

The successful application and usability of PCA relies on the extent to which the dimensionality of the original matrix $\boldsymbol{X}$ can be reduced. In other words it depends on the *rank* of the solution, which is the minimum number of principal components in Eq. (4.6) that explain most of the variance in the original data (without explaining noise). As the principal components are computed in order of decreasing explained variance, the cumulative explained variance approaches 100% asymptotically at $n = N$ (Fig. 4.8). Fortunately, this convergence usually occurs at a number of ranks that is significantly smaller than the original number of spectra in $X$, $n \ll N$. In contrast to Eq. (4.5), where the rank is predefined (and limited) by the number of reference minerals used or available, in unsupervised PCA, a decision must be made about truncating the terms of Eq. (4.6). This decision depends in part on the contents of the training set, the amount and type of information that needs to be explained, and the identification of the principal component(s) along which this information can be retrieved most clearly. General guidelines regarding these aspects and specific examples on the development and application of PCA will be provided in subsequent sections.

Any spectrum within $\boldsymbol{X}$ can be uniquely located in the $n$-dimensional space of the principal components by its scoring on each component. The location

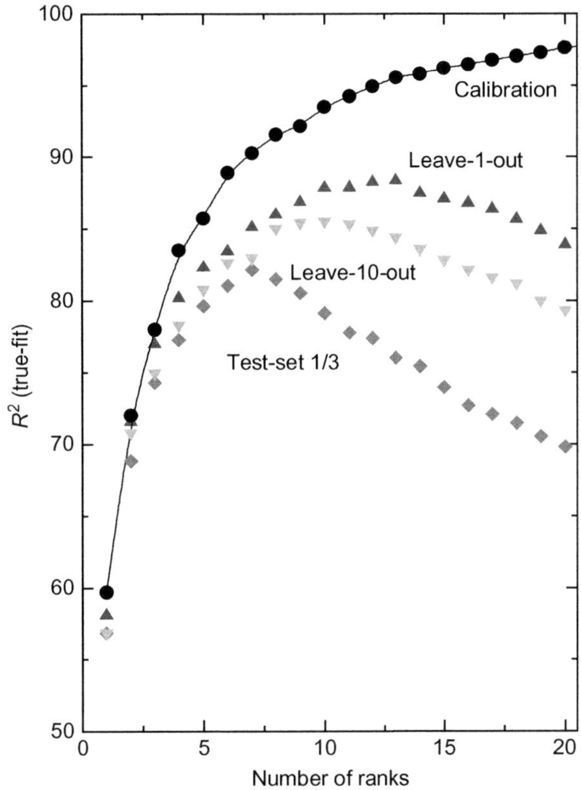


**FIG. 4.8** True-fit correlation coefficient $R^2$ of a case-study PLS calibration and three different internal validations for predicting a property of bentonite based on the NIR spectra of a set of 150 samples. All methods have been optimised in terms of the same spectral pretreatment and frequency ranges. Opposite to the trivial behaviour of calibration, all validations exhibit maxima with position (rank) and performance ($R^2$) that depend on the validation method. Leave-one-out cross-validation is clearly optimistic (overfitting), and the recommended number of ranks (~12) is high in comparison to the relatively small size of the dataset. On the other hand, a test set validation based on 50 spectra (1/3 of the dataset) affords a significantly smaller calibration set and may be underfitting at 7 ranks.

of several spectra can be projected on any plane defined by two principal components and be used for the unsupervised discrimination among groups of spectra, should the latter internal structure exist in the set. By the same means, unknown spectra that are relevant to the spectra in $\boldsymbol{X}$ can be identified as members of these groups. Although the common names of the groups may be coming from other techniques (X-ray diffraction or chemical analysis), the multivariate analysis of vibrational spectroscopic data from clay minerals and related materials is a powerful, independent and self-consistent data exploration tool for discrimination and identification applications.

Besides identification, there is often additional or independent need for quantitative predictions, as multidimensional extensions of the Beer–Lambert law. Such quantitative multivariate tools are commonly called chemometric due to their relevance in analytical chemistry, but they are important in many fields, even outside the physical sciences. The basic operating principles of two of the most popular chemometric methods will be presented here: the Principal Component Regression (PCR) which builds on the results of PCA (e.g. Naes and Martens, 1988), and PLS (standing for Projection to Latent Structures but commonly referred to as Partial Least Squares) which is wholly independent of PCA (e.g. Geladi and Kowalski, 1986; Wold et al., 2001). A detailed presentation of the two methods with common examples and notation can be found in Hasegawa (2006).

Training (more specifically, *calibrating*) a method to provide quantitative predictions for the sample set represented by the spectra within $\boldsymbol{X}$, requires that a set of independent, accurate property measurements $\boldsymbol{Y}$ is available for exactly the same sample set by other techniques (see Section 4.4.2).

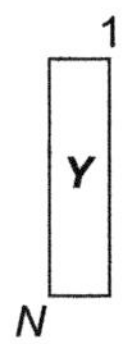


PCR is based on calculating the correlations between the scores of each principal component derived by PCA and the property $\boldsymbol{Y}$. The user can then construct a predictive algorithm including only those principal components that correlate with the property and excluding all others as irrelevant. This approach is conceptually identical to testing how the absorbance values in a set of spectra at each individual wavenumber correlate with the property of interest and selecting the region of interest for a predictive model, but it is performed on an orthogonal vectorial basis with reduced dimensionality. Although the predictive principal components of a PCR method are by definition identical to those derived by PCA, they are not necessarily employed in the same order. This is because the principal components of PCA were determined by order of explaining the overall latent structure of the spectra, and not the subset of this structure that may specifically be related to any property $\boldsymbol{Y}$. One would ideally prefer that the $\boldsymbol{Y}$-predictive principal components are also among those that explain a large part of the overall spectral variability (meaning that they have a rather strong spectral signature), but this may not always be the case.

In the PLS method, however, the stage is set from the beginning for predicting $\boldsymbol{Y}$:

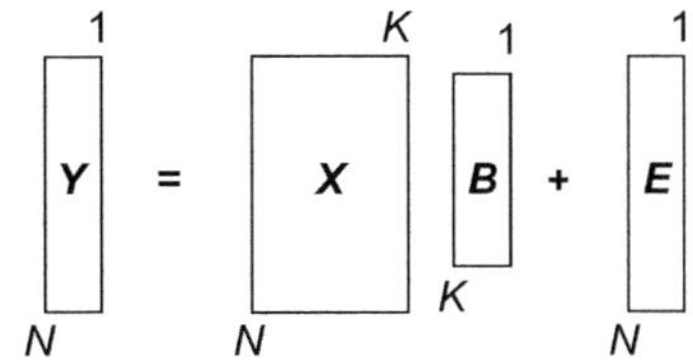


(4.8)

where a vector $\boldsymbol{B}$ is sought such that, when applied to $\boldsymbol{X}$, yields a prediction of $\boldsymbol{Y}$ with minimum error $\boldsymbol{E}$ (Geladi and Kowalski, 1986). In this sense, PLS is a method of dimensional reduction which is *a priori* supervised by the property data $\boldsymbol{Y}$. The computation of $\boldsymbol{B}$ involves the usual expression of $\boldsymbol{X}$ as a linear series of scores on loadings (Eqs. 4.5 and 4.6), but the loadings are now different than the principal components of PCA, because they are specifically calibrated to describe the variability in $\boldsymbol{Y}$ and obtained in order of decreasing ability to describe this variability. Further, the PLS loadings are not forced to be orthogonal, but their scores are. Similar to PCA, an important decision must be made about the minimum number of terms (dimensions) that are needed to fit the property without fitting the error (*overfitting*).

### 4.4.2 Training (Calibration) and Property datasets

As elegantly stated by DiFoggio (2000), the fundamental assumption behind any multivariate regression or chemometric method is that the (training or calibration) set of spectra $\boldsymbol{X}$ must contain all the information that is sought by the method. This assumption sets specific requirements about both the contents of the spectral matrix $\boldsymbol{X}$, and the kind of properties, $\boldsymbol{Y}$, that can be reasonably extracted. These requirements will be discussed in this section.

The training set $\boldsymbol{X}$ can be selected from a larger set of spectra (the usual case) or be developed wholly separately from reference materials and then applied to the analysis of another dataset. The quality of the training set controls to a large degree the performance of any multivariate or chemometric application. The number of spectra to be included in $\boldsymbol{X}$ is rarely an issue because matrices of very large size can now be manipulated with commonly available computing power. The number of spectra chosen for the training set, however, must be much larger than the number of anticipated latent variables $N \gg n$ usually by a factor of 10–100. Ideally, every spectrum in $\boldsymbol{X}$ should be a linear combination of many, if not all, anticipated latent variables: The spectra in $\boldsymbol{X}$ must be correlated. In fact, the whole normal distribution of each anticipated variable should be ideally represented in $\boldsymbol{X}$, and this is the main reason behind the need for large sample datasets. A *design of experiments* approach is frequently useful in selecting spectra for $\boldsymbol{X}$ that meet these requirements among even larger groups of available data (Kjeldhal and Bro, 2010).

As a general rule, it is inadvisable to include the spectrum of a pure reference compound in $\boldsymbol{X}$, if the latter is meant to represent a population of real-world mixtures that are too different from this reference. In such a case, the reference compound would be singled out as the first principal component in PCA (because it would be explaining a very large part of the variance in the whole set), but would be too different (distant, in the geometrical sense) from the rest of the samples and hence of little value in describing the finer details of their latent structure. In fact, including the spectra of all anticipated reference compounds in $\boldsymbol{X}$ would reduce the problem to the supervised approach expressed in Eq. (4.5). The same situation is encountered if the variance of nonvibrational features in the spectra (e.g. strong and variable sloping baselines or atmospheric gas signatures) dominates the overall variance in $\boldsymbol{X}$. These features would be promoted to the early principal component ranks, despite the fact that they are completely irrelevant to the chemical structure or mineralogical composition of the samples. This explains why the removal of such non vibrational sources of variance by appropriate mathematical pretreatments (see Section 4.2) is of paramount importance in multiple regression applications.

In practise, the statistics of the scores of each loading do reveal whether the representation of this loading in $\boldsymbol{X}$ is close to the desirable normal distribution or, instead, is mostly determined by a small isolated minority of spectra. If a particular loading is lacking such statistically distributed scores, one should decide whether this loading represents an outlier (that should be rejected prior to redoing the analysis), a spurious effect in the spectra (that should be corrected by an appropriate pretreatment), or a real chemical component which is rarely (but truly) present in the spectra of $\boldsymbol{X}$. In this latter case, one may decide to exclude the minority of spectra that represent the suspect component in $\boldsymbol{X}$, or enrich the spectral set with additional spectra bridging the compositional (or property) gap between the majority and minority populations. The decision depends on the anticipated significance of the chemical component with respect to the problem of interest. Tuning the contents of the spectral set $\boldsymbol{X}$ and deciding on the necessary mathematical pretreatment of the spectra that removes spurious effects and provides a satisfactory compromise between spectral resolution and signal quality typically requires a few trial analyses.

As an additional general rule, the spectra to be included in $\boldsymbol{X}$ should be of the same quality as those that are anticipated for future analysis. Methods based on high-quality laboratory data often lose their robustness when applied to field measurements. This is because the latter may introduce additional sources of spectral variability that are not encountered in the former and have not been considered by the model. In cases where this situation cannot be avoided, a *desensitisation* approach can be adopted, involving the *a posteriori* addition of the missing variability in the calibration spectra (DiFoggio, 2000).

These guidelines are sufficient for exploring the latent structure of spectral sets by multivariate analysis, but additional considerations are needed for chemometric (quantitative) applications. For example, what are the properties, ***Y***, that can be considered for chemometric analysis based on the spectra of a set of mineral samples, ***X***? In principle, ***Y*** can be any property that can be related fundamentally or empirically to the composition and local chemical structure (bonding and symmetry) of the samples; that is, to the latent features normally found in vibrational spectra. The elements of the matrix ***Y*** must come from independent measurements, and these measurements must be as accurate and precise as possible, because their quality characteristics will be transferred at large to the chemometric prediction tool (DiFoggio, 2000).

In practise, chemometrics serve to substitute for the routine measurement of properties that would otherwise be determined by time-consuming, expensive or environmentally problematic standardised measurement protocols (Viscarra Rossel et al., 2006; Waruru et al., 2014). In many cases, chemometrics are called to be based on (and eventually substitute for) existing quality control schemes. These may include mineralogical (e.g. % quartz in bentonite), chemical (e.g. moisture content, cation exchange capacity [CEC], or methylene blue [MB] adsorption, Al occupancy in the tetrahedral sheet, total organic carbon), physical (e.g. refractive index, viscosity of a dispersion) or engineering (e.g. swelling capacity, Atterberg limit) properties. One must be fully aware of the limitations and assumptions inherent to the measurement protocols that are used to produce ***Y***. One should also make sure that these protocols are applied exactly to the same samples that produced the spectra in ***X***, are conducted in the exact fashion for all samples and preferably to the same type of samples, and in the same fashion, anticipated for future analysis. In cases where field, rather than laboratory, applications are sought, the requirement for matching data in ***X*** and ***Y*** can set major difficulties in assembling these matrices. For example, in designing a chemometric application for the field measurement of moisture in soil, one must make sure that the natural moisture of the calibration samples is fully preserved until both spectroscopic and thermogravimetric measurements are completed.

One must also be aware that a chemometric prediction is based on the correlation between the spectra of a particular set of samples and their properties. For this reason, chemometric predictions may perform very well, but only within the domain set by their latent structures. Chemometrics are not direct measurement techniques and therefore cannot be used to predict the properties of any set of samples that lie outside that range of properties of the set they have been calibrated against. Two subtle issues regarding the information content of the spectra in ***X*** may be added at this point:

Can the quality of the chemometric predictions surpass that of the independent measurement? The intuitive answer is 'no', but the regression mathematics used for establishing the chemometric tool average out random error in the reference values and can, in principle, improve the precision (DiFoggio, 2000). In addition, the study of the same sample set by two independent experimental methods feeding ***X*** and ***Y*** often reveals the presence of *outliers* (wrong samples, wrong measurements, typographical errors, etc.) in either or both sets of data. Outliers are samples that are (1) described badly by ***X*** or ***Y*** or (2) outside the range of predictive capability of the method. Elimination of these outliers improves the accuracy of the measurement.

Can one employ chemometrics to measure concentrations of a species that are unobserved in spectra? Again, the intuitive answer is 'no', and is correct unless the concentration of the invisible species is (positively or negatively) correlated with that of other constituents that have a pronounced spectral signature. A typical example is the prediction of % quartz in clay mineral samples by NIR spectroscopy, despite the fact that quartz is transparent in the NIR. Such predictions are based on the assumption that quartz is the only NIR-transparent mineral in the samples and that its presence and concentration can be inferred from the absence of the remaining, nontransparent components. Such indirect determinations should be used with great caution.

### 4.4.3 Validation and Optimum Dimensionality

The single purpose of any multivariate analysis or chemometric methodology is the satisfactory reduction of the dimensionality (rank) of the problem from the initial large number of $N$ correlated spectra in ***X***, to a much smaller number of noncorrelated latent loadings, $n \ll N$. Recall that the linear decomposition of a matrix ***X*** consisting of $N$ spectra can involve asymptotically up to $N$ linear terms (e.g. Eq. 4.6) and that the information content of these terms is expected to progressively drift from significance to triviality. The fundamental question is how one can decide where to truncate the series, thereby defining the *optimum rank* of the method, in order to capture most of the significant latent structure in ***X***, or the best possible correlation with the external variable ***Y***, without fitting irrelevant variables and errors.

Some insight about the anticipated dimensionality of the solution can be obtained from a general knowledge of the sample set under investigation. For example, the octahedral sheet of palygorskite contains trioctahedral magnesian and dioctahedral aluminous-ferric components which are manifested in the O—H stretching spectra by discrete $Mg_3OH$, $Al_2OH$, AlFeOH and $Fe_2OH$ bands (Chryssikos et al., 2009; Stathopoulou et al., 2011). The palygorskite content and composition is then defined by three independent variables, that is, the trioctahedral fraction, the $Fe^{III}$ occupancy in the dioctahedral domains and the extent to which mixed Al-$Fe^{III}$ pairs can form. Indeed, both a Beer–Lambert approach based on the intensities of the O—H stretching overtones and a 3-rank chemometric model yield the same $R^2=0.94$ performance in predicting the palygorskite content in samples containing other clay and nonclay minerals (Gionis et al., 2007). In the more general case of modelling all species in a mixture, the number of significant principal components would ideally be the same as the number of those mineral components that can be observed by the spectroscopic technique employed. Similarly, in a predictive model for CEC, the rank of the solution should be similar to the number of mineral species with exchangeable cations present, multiplied by the number of spectroscopically active chemical substitutions that are encountered in each of these species and may be determining their layer charge. Such considerations do not take into account the possible presence of other variables or noise that can be strong enough to obscure some of the anticipated loadings.

For these reasons, the optimum rank as well as the figures of merit of any multivariate or chemometric methodology at this rank (e.g. the percent explained variance in PCR, or the root mean square error of prediction in PLS) should never be derived from data that have been used in the calibration of this methodology. As with any scientific methodology, multivariate analysis and chemometrics should be evaluated on the basis of independent, but plausibly similar, data by a process called *validation*. For details about the significance of validation, the various validation strategies and their potential pitfalls, the reader is referred to the specialised literature (e.g. Shao, 1993; DiFoggio, 2000; Wold et al., 2001; Steyerberg et al., 2003; König et al., 2007; Esbensen and Geladi, 2010; Westad and Marini, 2015).

There are many types of validation depending on the way the validation samples are chosen. The strictest and most conservative approach is *external validation* which uses an independent dataset—collected as similarly as possible to the parent dataset, but different from the training set—to test the statistical significance and predictive capability of a chemometric model. The external validation set must be representative and large enough to enable a statistically relevant assessment of the robustness of the method. Depending on the sources of irrelevant variability that must be accounted for, this independent set may arise from different operators, with data obtained over different time periods, or even from a (geological) setting that is analogous, but not identical, to the one used for training or calibration.

Very frequently, the conditions for constructing a true external validation dataset cannot be met, or the externally validated methodology turns out to be too conservative for the specific application of interest. In such cases, validation is based on a subset of data that are removed from the parent dataset and is called *internal validation*. As with its external counterpart, the internal validation dataset must be sufficiently large and well spread over the latent space, but still constitute a small fraction of the parent dataset so that the structure of the remaining calibration set is not disrupted. A common scheme of internal validation, *test set validation*, involves the selection of 10%–35%

samples for validation purposes. The method of selecting validation samples for the test set (random, in blocks, or evenly spaced across $\boldsymbol{X}$) can be critical if there are hidden systematics in the numbering of the spectra in $\boldsymbol{X}$ (time series, blocks provided by different operators or equipment, blocks added to fill-in gaps in original data, etc). There are many possible permutations of test set over parent data, and therefore several test set validations of the same method can be performed to optimise the rank, identify gaps in the calibration set, remove outliers, etc.

Another popular method of internal validation is *cross-validation*, also known as the *leave-$N_v$-out* method. The method removes a set of spectra $N_v$ from the original set of $N$ spectra in $\boldsymbol{X}$, calibrates on the basis of the remaining $N$–$N_v$ spectra and applies this calibration to the excluded $N_v$ spectra for method evaluation. The process is iterated over all spectra in $\boldsymbol{X}$. In some cases, $N_v=1$ and the method is called a *leave-one-out validation*. Cross-validations tend to be optimistic, especially in their leave-one-out version, and should be used with caution.

An example about the use of validation in determining the optimum rank of a chemometric method is shown in Fig. 4.8. The example is taken from the PLS modelling of a property based on the NIR diffuse reflectance spectra of ~150 bentonite samples in powder form. The true-fit correlation coefficient $R^2$ is shown as a function of dimensionality (rank) for calibration as well as three different validation runs. Upon increasing number of ranks, the $R^2$ of the calibration run tends asymptotically (and trivially) to 100%. There is no way to define the optimum rank from such data. On the contrary, the $R^2$ of any one of the three validations first increases with increasing dimensionality (albeit remaining always inferior to calibration), reaches a maximum and then decreases, which is a clear sign that the model is now fitting the property on the basis of irrelevant variables (*overfitting*). The optimum dimensionality (and associated performance) of the method can be evaluated from the rank at maximum $R^2$. The case study in Fig. 4.8, however, indicates that the suggested optimum dimensionality can vary between the optimistic leave-one-out validation and the conservative test-set validation. If encountered, the PLS correlation should be repeated after increasing the size of the representative calibration spectra to support a proper test-set validation or reducing possible data clustering in the dataset (which can bias leave-one-out validations).

### 4.4.4 PCA and PCR Chemometrics in the Study of Clay Minerals

This section illustrates the use of PCA and PCR to extract and explore the variability present in a dataset consisting of the MIR diffuse reflectance (DRIFT) IR spectra of 32 dioctahedral smectites. This is a well-studied dataset (Gates, 2005) with accompanying chemical and crystallographic data and has been already shown in Figs. 4.2 and 4.3 to demonstrate the effect of various mathematical pretreatments. The dataset is very limited in size compared to the anticipated dimensionality of the problem (see Section 4.4.3), but it does cover most of the chemistry range of the dioctahedral smectites. As such it will serve to guide the reader through the PCA process from training to validation and the development of PCR predictions.

*PCA calibration.* The set of 32 vector-normalised and baseline-offset MIR data presented in Fig. 4.2A was subjected first to PCA analysis. Employing the Pearson's correlation coefficient,

$$R = \frac{\sum (x - x_i)}{\left(\sum (x - x_i)^2\right)^{\frac{1}{2}}} \tag{4.9}$$

as a measure of variance, the effect of each calculated component in explaining the total variance of the dataset can be deduced in Fig. 4.9A. As explained in Section 4.4.1, components are calculated in order of decreasing explained variance and their incorporation to the fit results in increasing cumulative explained variance. In the vector-normalised dataset, the first component, PC1, accounts for ~55% of the explained variance in the dataset, PC2 accounts for ~45% of the residual variability, and each successive component describes approximately 33% of the successive residual. The effect of derivatization treatments as well as multiplicative scattering corrections on the resulting explained variance of the training set is also shown in Fig. 4.9. Derivatization (Fig. 4.9A) results in an increased contribution by PC1, but successive components have less influence on the explained variance. Conducting MSC or EMSC corrections on the data prior to derivatization has little influence on the resulting explained variance (Fig. 4.9B). Two remarks can be made on the basis of the data in Fig. 4.9: First, as this is a calibration run,

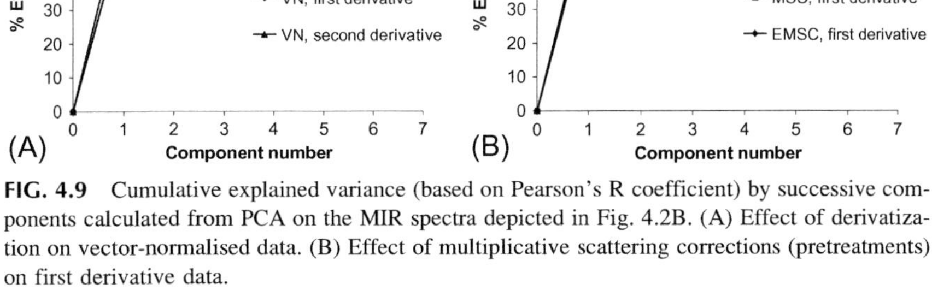


**FIG. 4.9** Cumulative explained variance (based on Pearson's R coefficient) by successive components calculated from PCA on the MIR spectra depicted in Fig. 4.2B. (A) Effect of derivatization on vector-normalised data. (B) Effect of multiplicative scattering corrections (pretreatments) on first derivative data.

the explained variance will be based on increasingly trivial information beyond a certain number of ranks until reaching 100% at 32 principal components (cf. Fig. 4.8). Second, this slowly converging solution is typical for an information-rich system. The description of such systems requires a relatively large number of independent latent variables and should be based on datasets that are significantly larger than the one used in the present example.

Nevertheless, the minimum mathematical preprocessing of the vector-normalised spectral set allows for rationalising the shape of the first principal components (Fig. 4.10) in terms of smectite chemistry. For example, PC1 (Fig. 4.10A) is dominated by positive correlations with the $\nu$(OH) and $\delta$(OH) spectrum of nontronite (e.g. strong positive bands at ~3580, 830 cm$^{-1}$). A sharp positive feature at 1620 cm$^{-1}$ also occurs in PC1, associated with interlayer $H_2O$. Capturing the largest amount of the remaining variability (20% of the total), PC2 (Fig. 4.10B) exhibits an opposite response in the $\nu$(OH) range (positive near 3665, negative near 3570 cm$^{-1}$), as well as a strong negative response associated with ν(SiO) stretch near 1180 and 1060 cm$^{-1}$. Negative features associated with bound water near 3350 cm$^{-1}$ are also present in PC2.

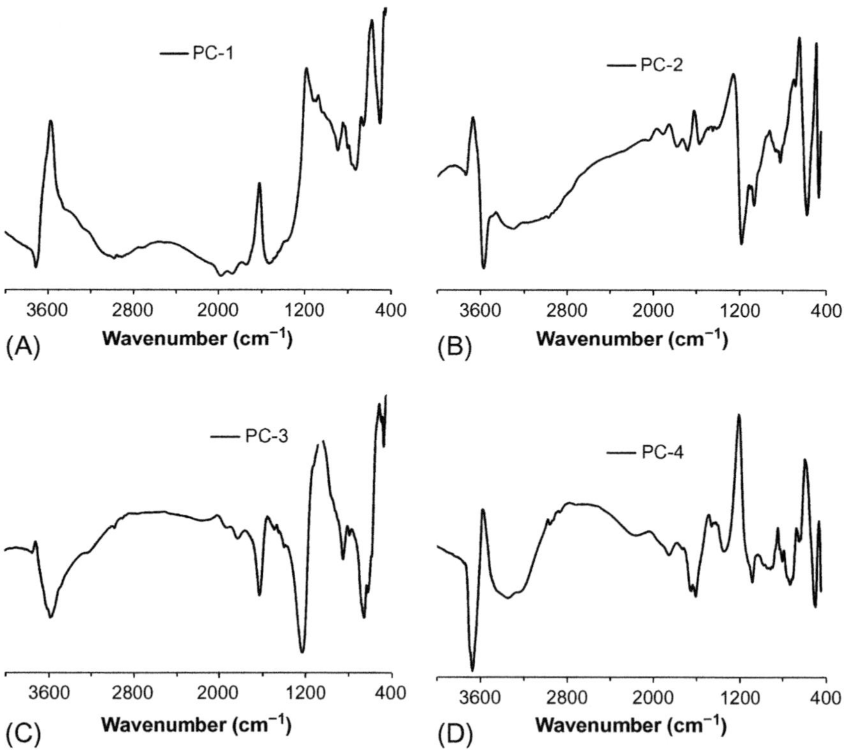


**FIG. 4.10** Spectral loading plots of the first four components (A: PC1; B: PC2; C:PC3 and D: PC4) computed from vector-normalised MIR spectra of 32 dioctahedral smectites. The plots are all differently and arbitrarily scaled.

PC3 (Fig. 4.10C), which captures the next largest proportion (12%) of the remaining variability, indicates a broader negative feature at 3580 $cm^{-1}$ than PC1, as well as strongly negative features at 1620, 1222, and 650 $cm^{-1}$, and a strongly positive feature at 1055 $cm^{-1}$. PC4 (Fig. 4.10D) amounts to only 5% of the captured variability, and appears to convey issues associated with scattering (broad features centred near 2800 $cm^{-1}$).

Score plots (Fig. 4.11A) showcase what physical parameters may be associated with the calculated components. For the 32-sample dataset, nontronites are arrayed negatively with PC1 and both positively and negatively with PC2, indicating that PC1 captures well the variability in Fe content of the samples. This relationship extends to the ferrian smectites (dashed circle), which are positive in PC1 and negative in PC2. Beidellites and montmorillonites are arrayed positively with PC1, but interspersed with each other and mostly positively, in PC2. PC2, which only captures 13% of the variability of the dataset, appears to be associated with the magnesium content of the samples.

The locus of the 32 samples in the score plot of Fig. 4.11A is maintained with small deviations upon performing a leave-two-out internal validation. About one-third of the samples deviate slightly along PC2 and of these, most deviate positively. One sample (circled) is poorly validated along PC2. Such a sample is a candidate outlier and should be checked thoroughly using, for example, a Hotelling statistic (Hotelling, 1931).

The percent explained variance of all the components calculated is shown in Fig. 4.11B, where the validation set returns a lower percentage than

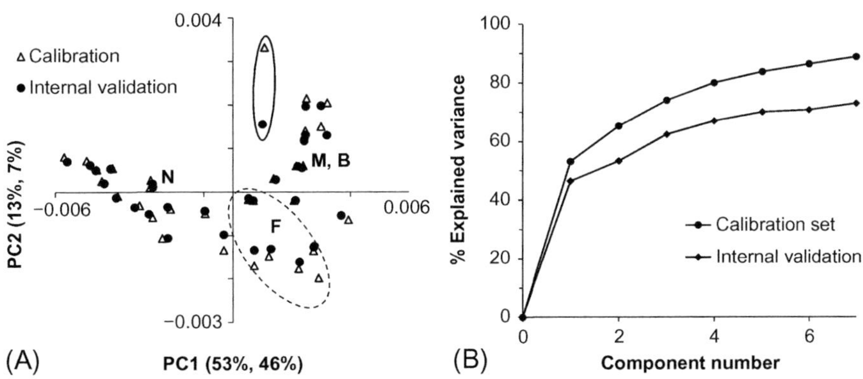


**FIG. 4.11** (A) Scatter plot of the first two principal component scores (which account for 65% of the explained variance) resulting from PCA determined on the 32-sample training set. The internal validation was performed by applying the PCA model iteratively to the data with two randomly selected samples omitted (leave-two-out). The pair of values circled indicates the sample with the strongest deviation of the validation from the calibration. (B) Percentage of explained variance of components (based on Pearson's R) for the 32-sample calibration set and the randomly assigned leave-two-out internal validation. Symbols: *N*, nontronite; *F*, ferruginous smectite; *M*, montmorillonite; *B*, beidellite. The ferruginous smectite grouping is encircled by a dashed line.

the calibration set, without showing a maximum at a certain optimum rank (cf. Fig. 4.8). This indicates that considerable variability remains uncaptured by PCA.

*PCR model validation.* Armed with an example description of the system in terms of its principal components, it is now possible to examine whether the principal components calculated by PCA regress, by least squares, against known chemical or physical properties of the samples. Validations will be presented on the basis of a representative and randomly chosen 8-sample internal test set, calibrated against the remaining 24 spectra. Both calibration and validation sets have identically measured property sets and were subject to the same pretreatments. For the following PCR model development, the PCA components derived from vector-normalised datasets (Fig. 4.10) were used without any other corrections, and applied to the octahedral Fe content given in Gates (2005).

For PCR applied to the octahedral Fe content, PC1 and PC2 component loadings describe the same features of the dataset as PCA (Fig. 4.12), and thus indicate that these components remain the most influential in describing octahedral Fe for this particular dataset. However, the first two components are inverted in importance with respect to PCA (not shown). Among the 24 calibration samples, PC1 accounts for about 11% of the explained variance in the regression on octahedral Fe whereas PC2 accounts for 83% (Fig. 4.13), resulting in a cumulative capture of 94%. Subsequent principal components have no influence on the resulting regressions. Note also that the 8-sample validation set tracks very well the 24-sample calibration set, amounting to 94% cumulative explained variance (16% from PC1 and 78% from PC2).

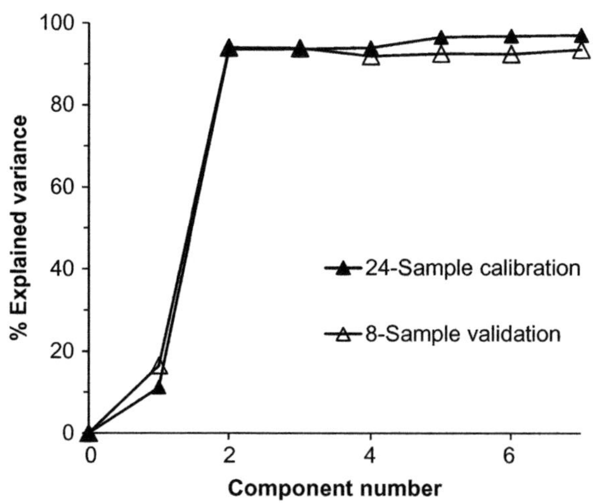


**FIG. 4.12** Percentage of explained variance in the PCR treatment of 32 dioctahedral smectites. The internal validation consisted of a randomly selected set of 8 samples from the original 32-sample dataset used.

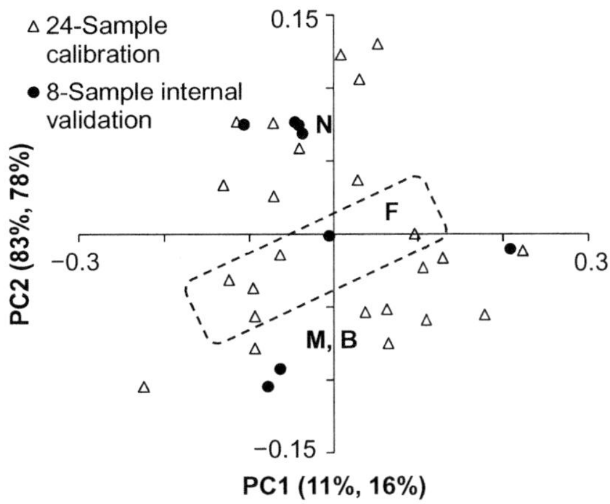


**FIG. 4.13** Scatter plots of the first two principal component scores resulting from PCR on octahedral Fe determined on the vector-normalised, 24-sample training set *(open symbols)*. The internal validation was performed by applying the PCR model to the octahedral Fe contents of eight randomly selected samples comprising the validation set *(filled symbols)*. Percentage values for each PC correspond to, respectively, the PCA calibration and the PCR internal validation. Symbols: *N*, nontronite; *F*, ferruginous smectite; *M*, montmorillonite; *B*, beidellite. The ferruginous smectite grouping is encircled by a dashed line.

A scatter plot for the scores resulting from regression of the first two principal components onto octahedral Fe content (Fig. 4.13) reveals that the samples distribute uniformly along both PC1 and PC2, and do not show the strong distribution along PC1 that was observed previously in Fig. 4.11A. The different clay minerals (nontronite, ferrian (ferruginous) smectite, montmorillonite, and beidellite) are still largely preserved, but with a different orientation in comparison to Fig. 4.11A. In fact, the locus of the groups indicates that PC2 captures most of the passage between montmorillonite-beidellite, ferrian smectite and nontronite, that is, most of the variability in octahedral Fe content. The 8-sample validation set, composed of randomly selected nontronites (4), montmorillonites (2), beidellite (1) and ferrian smectite (1) are well dispersed within the calibration set, indicating that any predictions resulting for this PCR model will be robust.

The 24-sample calibration predictions and the 8-sample validation predictions are depicted in Fig. 4.14 for octahedral Fe content. The prediction of octahedral Fe by PC1 is poor (Fig. 4.14A), which is not surprising given that it accounts for 11% of the variability. Prediction based solely on PC2 highly underestimates octahedral Fe (Fig. 4.14B), but correlates well with the data. The PCR model that combines linearly the orthonormal PC1 and PC2 loadings (Fig. 4.14C) significantly improves the prediction of octahedral Fe, again decreasing significantly the root mean square error (RMSE) and also increasing the resulting coefficient of correlation ($R^2$). As expected, the 8-sample validation set tracks the calibration in the same fashion.

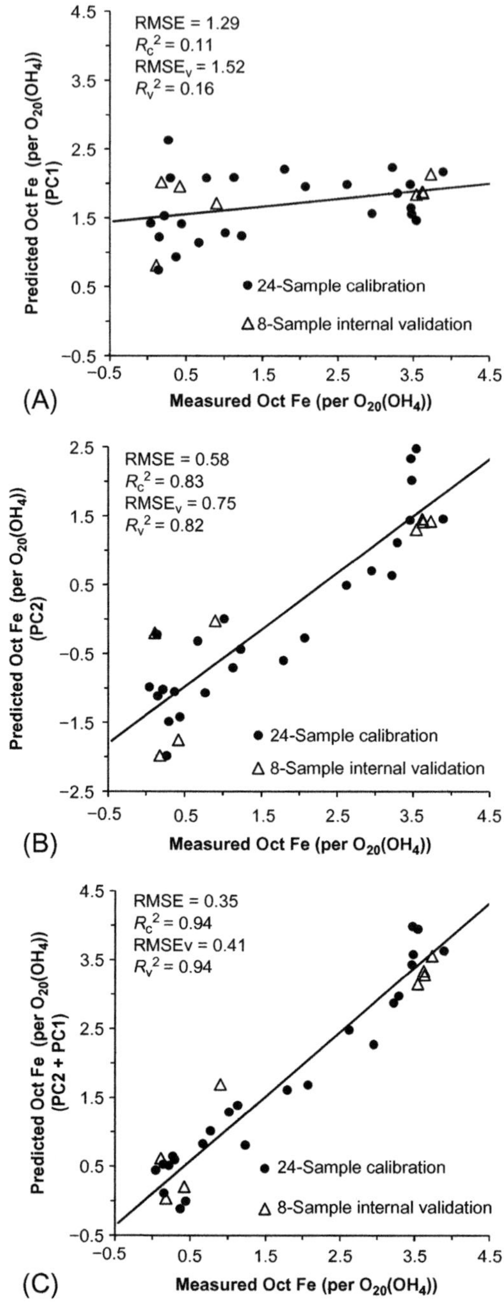


**FIG. 4.14** PCR regression results for (A) PC1, (B) PC2 and (C) a two-component (PC2+PC1) PCR model to predict octahedral Fe content. Root mean square error (RMSE) correlation coefficient ($R^2$) reported is reported for the regression on the calibration and internal validation set for each model.

In summary, the PCA and PCR treatment on the MIR spectra of a series of well-studied dioctahedral smectites indicates the capacity of this multivariate analytical approach to produce valid and useful predictions of their physical or chemical properties. Increasing the number of samples in the calibration set could capture better the latent structure of their spectra over the same, or broader, ranges. It could also enable a more thorough validation, adding to the robustness of the predictive models and offering an appealing alternative for the high-throughput analysis of similar samples.

### 4.4.5 PLS Chemometrics for Clay Mineral Processing Applications

This section highlights the application of PLS chemometrics as a platform technology for quantitative analysis at the clay mineral processing plant. The perspective is both appealing and challenging: Assuming that good-quality spectra can be collected at suitable points of the processing line, these can be used to produce instantaneous predictions on several prevalidated properties of the sample. The latter can then be used for optimising subsequent processing steps or for the quality assurance of the final products. In addition, this approach would create over time a spectral record of all processed materials at the selected sampling points. If properly managed, such detailed records of production can be revisited at later stages for additional data mining regarding the identification of unknown components, for troubleshooting, or for setting up new methodologies.

There are several issues that are critical for the implementation of such projects and require close collaboration and mutual understanding between the spectroscopist and the end user. These issues concern the selection of sampling points, the choice of the spectroscopic technique, the properties to be targeted for fitting among the characterisation data that can be independently available, and the operational maintenance of the application. In addition, they concern the way the spectroscopic chemometric tool will be developed, tested and installed in a 'transparent' way that does not obstruct or delay normal operation. Few of these aspects are detailed in the peer-reviewed literature (but see Goetz et al., 2009; Konrad et al., 2015), because each application is custom made for a specific deposit and plant. Thus this section provides general guidelines and examples rather than detailed accounts of specific applications.

A deposit is mined selectively on the basis of exploratory data, for example on drill-core samples. A typical clay mineral processing plant accepts the mined ore in coarsely crushed rock form, with near-natural moisture content and with variable mineralogical composition. What follows is a series of physical (mixing, drying, milling, beneficiation), and sometimes chemical (ion exchange, calcination) modification steps that are necessary for producing the final product(s). All these steps from ore mining to the final product(s) eliminate progressively the original variability of the material, resulting in a product with a relatively narrow range of compositions and consistent specified properties. Each of these steps down the process line provides a potential sampling point, with its own requirements and limitations.

The overall geological structure of the deposit may be diverse, but the samples that are isolated from drill cores can be relatively uniform at the medium (100 g–5 kg) scale. In addition they can be available dry after equilibration to ambient conditions and in homogenised powder or bulk rock forms. Such samples are relatively easy to measure by many vibrational techniques, although their potentially large number calls for high-throughput data acquisition (Herrmann et al., 2001; Yang et al., 2001). The opposite holds for the agglomerate rock that is accepted as raw material for further processing, usually after coarse crushing. Critical screening decisions may need to be made at this early stage of production. The material, however, can be very diverse on the medium scale of single lumps. Their surface is created by fracture and cannot be assumed to be identical to the bulk, and their moisture content can vary locally. Such material needs to be sampled and measured in a manner that represents large volumes, of the order of many tons (e.g. Goetz et al., 2009). Subsequent processing steps typically produce finer and more uniform material but may require measurements under dynamically changing conditions, for example, during the evolution of equilibration to ambient reactions and other diffusion-dependent processes. In the final products the properties may require precise measurement over narrow ranges of variability.

Among the various vibrational sampling techniques discussed previously in Chapter 3, diffuse reflectance in the NIR is perhaps the most suitable in fulfilling the majority of the diverse sampling requirements of the processing plant and in providing high-quality and throughput spectra that are suitable for chemometric modelling (see Section 3.3.5). Optical fibres, integrating spheres and external illumination diffuse reflectance probes (see Fig. 3.7) have progressively increasing sampling spot sizes which are useful for averaging the signal from the surface of heterogeneous samples. Fibres and spheres are more suitable for laboratory, off-line applications, whereas external illumination probes allow additionally for the contactless, quasicontinuous, real-time measurement of large or moving samples, for example, during belt transportation (see Fig. 3.8).

Besides noninvasive sampling, NIR spectroscopy has additional advantages that are specific to the study of clay-containing ores. Due to its sensitivity to vibrational modes involving the hydrogen atom, NIR can provide spectra that may be diagnostic for natural samples that contain significant amounts of nonhydrous associated minerals. Further, due to the activity of O—H stretching-bending combination modes, and also due to anharmonicity effects, NIR typically offers better resolution between structural OH and $H_2O$ in the sample than MIR. On the other hand, ATR-MIR is superior to NIR in modelling the nonhydrous components (Müller et al., 2014), the content of which can be among the critical specifications of the final product. Robust, industrial-grade FT instruments are preferable due to their higher resolution and accuracy, but simpler instruments can be employed if portability and/or the cost associated with the number of sampling points are the dominant considerations (Herrmann et al., 2001; Yang et al., 2001).

The next decision concerns the choice of properties that need to be validated for the chemometric modelling of the vibrational spectra (see Section 4.4.2). A good (and often the only possible) starting point is the modelling of the properties that are already implemented in the plant and used for decision making. With some precautions, these established quality control operations can be used for providing matching calibration spectra and independent property data. The advantage is that sampling and on-site analytics are already available and based on standardised procedures. The usual drawback is that, due to time and cost considerations, the on-site materials characterization can be coarse in terms of fundamental understanding. It may, therefore, be lacking the specificity in terms of structure and composition that would be desirable for correlation with vibrational spectroscopic data. As an example, industrial determinations of smectite content may be based on titrations with a cationic dye (Lagaly, 1981). Such methods can provide the average CEC or layer charge but are unsuitable for distinguishing the type and content of the various smectite species in the sample. Besides that, they can be sensitive to the aggregation of the dye molecule (Budják et al., 2002). As another example, several problems can be anticipated when the exchange of the interlayer cations of the parent material is required in order to develop performance. First, the effect of the nature of interlayer cations in the NIR (or MIR) spectra of clay minerals is indirect and weak. Second, the ion-exchange routines applied on site may be lacking the accuracy of those performed in an academic laboratory. More seriously, the chemometric model may be required to predict the anticipated performance of the raw material, prior to the processing that is needed for the development of this performance.

Despite these concerns, properly validated PLS chemometrics can perform very well at the clay mineral processing plant. The predictions shown in Fig. 4.15 are selected among the results of two independent industrial feasibility projects on dioctahedral bentonite and palygorskite-smectite systems. The spectra have been collected by FT NIR instruments in the diffuse reflectance mode with an external illumination unit and a powder probe, respectively. In both cases, the samples were in a $<250\ \mu m$ homogenised powder form, dried at ambient conditions without further treatments. They were selected from company depositories of reference production and deposit exploration samples, respectively, and their properties were determined by in-house standardised procedures applied after a common preparation step. PLS chemometrics were developed with the QUANT2 package of the OPUS software by Bruker Optics. In most cases, spectral preprocessing involved Savitzky–Golay

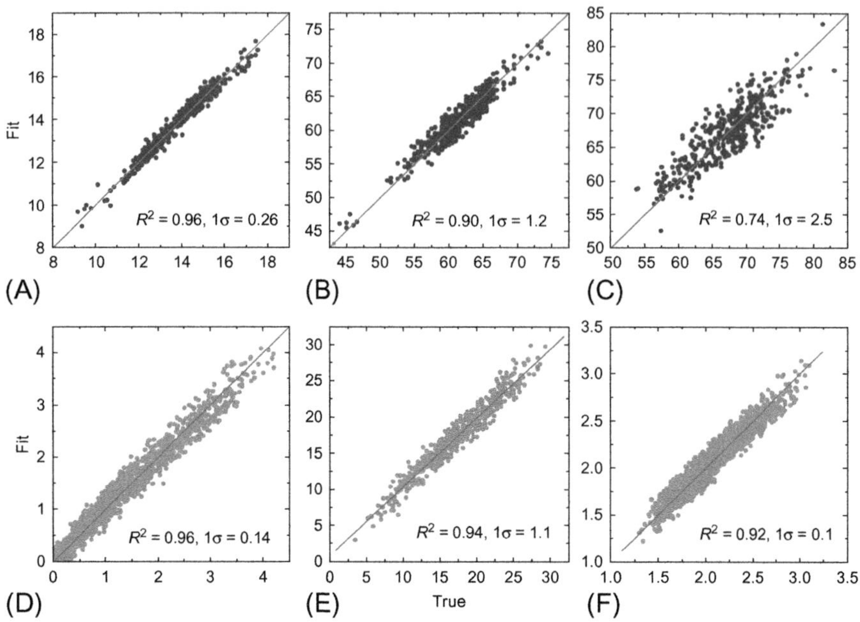


**FIG. 4.15** True-fit performance of NIR-based PLS chemometrics on industrial clay mineral sample sets obtained by cross-validation leaving out 10% of the original samples. The upper series (A–C) is from a dioctahedral bentonite system represented here by ~400 samples, and the lower series (D–F) is from a palygorskite-smectite system (~1200 samples). Property data cannot be disclosed, but some details are provided in the text.

second derivatives, for filtering out variable baselines or broad vibrational bands, and enhancing the resolution of sharp features (see Section 4.3.2). The methods were typically cross-validated by applying a leaving-10%-out method after randomising the order of the spectra in $\boldsymbol{X}$ to remove possible time-series systematics. The properties shown in Fig. 4.15 (arbitrary units) include moisture content (A) estimated gravimetically, 'smectite content' (B, E) estimated by titration methods, palygorskite content (D) estimated from the intensity of the $d_{110}$ XRD reflection at ~10.4 Å as in Gionis et al. (2007), and two proprietary sets of data on anticipated performance (C, F). At this level of development, PLS chemometrics can complement or substitute seamlessly the existing laboratory quality control procedures on site. In addition, a number of knowledge-based parameters, typically based on peak positions or relative intensities (e.g. Post and Noble, 1993; Yang et al., 2001; Chryssikos et al., 2009; Stathopoulou et al., 2011) can be extracted from the spectra and included in the characterisation report of each sample as additional proxies of crystal structure and composition.

As discussed in Section 4.4.2, the successful development of any chemometric correlation method relies on the similarity between the calibration samples and the anticipated unknowns. This is because predictions should be

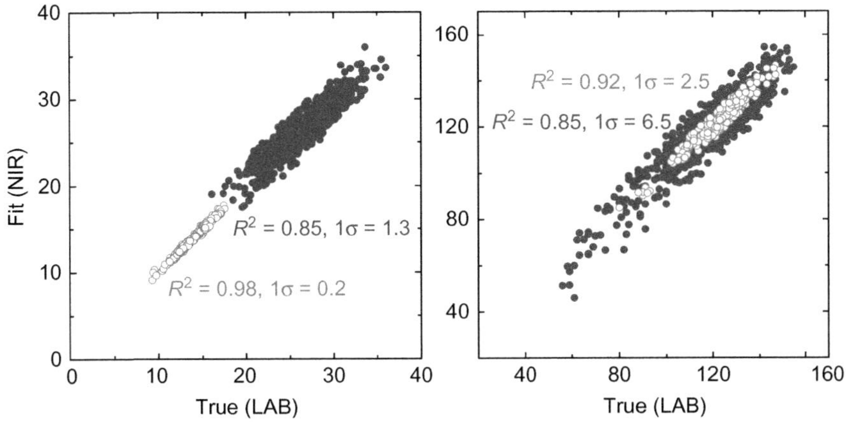


**FIG. 4.16** Comparison of cross-validated PLS predictions at two different sampling points of the same plant during the same time period. Solid symbols (blue in online version) correspond to real-time NIR measurements of the wet, coarse material over the conveyor (~800 samples). Open symbols (red in online version) represent laboratory spectral measurements on <250 μm powders dried at ambient conditions (~300 samples) performed with the same diffuse reflectance head. The left panel is on moisture content validated against the results of an industrial gravimetric method; the right panel represents smectite content proxied by a titration (arbitrary units).

based on interpolation rather than extrapolation, but also because diffuse reflectance techniques depend strongly on the particle size of the sample. For these reasons, the chemometric monitoring of the same material at a different sampling point along the process pipeline requires in most cases independent calibration and validation. In Fig. 4.16 are compared two such calibrations for moisture and smectite content (arbitrary units) based on the incoming wet, coarse material from the conveyor belt against laboratory measurements on powdered samples equilibrated at ambient conditions (as in Fig. 4.15). The two sets of data are obtained from the same production line over the same time period of several months. The data should, therefore, represent the same distributions of the essential properties of the material (except moisture content and particle size), which is in agreement with the results shown in Fig. 4.16. On the other hand, the RMSE of the predictions based on wet, coarse calibrants is clearly inferior to that based on dry powders. The reason for this must be sought in the way calibration and property data (***X*** and ***Y***, respectively, see Section 4.4.2) are obtained. For practical and safety reasons, the characterisation of the coarse material may require taking a sample from the conveyor, sealing it to avoid loss of natural moisture, and measuring it at a later time in the laboratory by a sampling system that simulates spectral acquisition over the belt (e.g. a rotating platform as in Goetz et al., 2009). Measuring the independent properties of the same sample by the usual standard methods requires sample drying and homogenisation. Obtaining such ***X*** and ***Y*** data on samples that exhibit local variations in

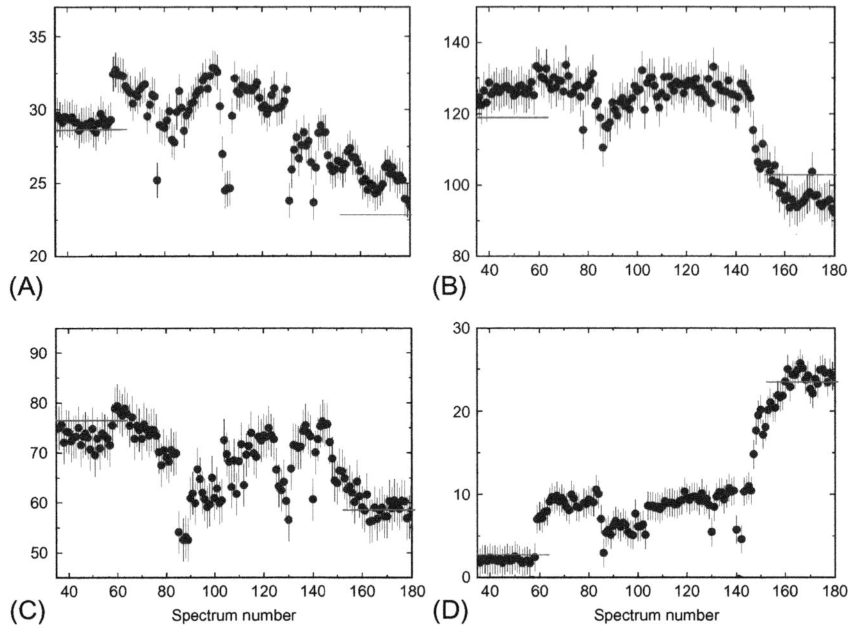


**FIG. 4.17** Real-time PLS chemometric predictions of four different properties during the monitoring of wet, coarse clay-mineral ore during transportation on an open-air conveyor. The spectral time series, similar to that shown in Chapter 3 (Figure 3.8) was collected by an external illumination diffuse reflectance probe. The predictions (arbitrary units) concerned the moisture (A), smectite (B), interlayer (C) and one common impurity (D) contents of a 2-min average sample, over a period of ~5 h of continuous operation. Lines on the left and right side of each panel (red in online version) correspond to the results of independent laboratory analysis from the routine periodic sampling of the transported material.

composition is by no means trivial. Calibration is prone to higher sampling errors in both the spectral and property measurements, as well as in matching exactly the samples of the two sets.

Overcoming these obstacles leads to chemometric tools that can be applied on-line to characterise the incoming raw material over, for example, a conveyor belt in real time. This approach is illustrated in Fig. 4.17. It is based on a representative time series of diffused reflectance NIR data (similar to that in Chapter 3, Fig. 3.8) and a set of suitable chemometric validations as in Fig. 4.16. In this manner, the time dependence of several sample properties can be obtained simultaneously. These properties can feed-back decisions on the potential use of the incoming materials and, if applicable, on the parameters of subsequent treatments. The data are equispaced in time but can be converted to tonnage by considering the time-dependent loading capacity of the conveyor. The time resolution of the graphs in Fig. 4.17 is sufficiently sensitive to sudden changes, but average predictions can be obtained over longer periods, if needed. The performance of each prediction algorithm can be accessed in real time by the quality figures of the PLS fit, and can also be externally evaluated (and validated) by existing protocols for the regular sampling and characterisation of the raw material (horizontal ticks in each panel of Fig. 4.17).

It needs to be emphasised that any quality control process based on chemometric correlations requires regular maintenance. Modern spectrometers have internal validation standards that can be used regularly for detecting and correcting drifts with time. Systematic changes regarding the state of the material at the sampling points need to be accounted for. More importantly, it cannot be assumed that the unknown incoming samples will remain within the calibrated range of latent variables as mining advances with time. The problem is typically addressed by the periodic update of the sample and property datasets and their use for the revalidation of the methods. It has been proposed that the necessary adjustments and chemometric updates can be performed in an unsupervised manner (Li et al., 2000; Benndorf, 2015). Including the exploration samples as a separate stage of chemometric assessment can provide an early external validation warning about anticipated material changes in production. The need for method maintenance implies that the in-house conventional quality control procedures should not be abandoned. Chemometrics should be employed for increasing the number of characterised samples by orders of magnitude, also for providing an independent assessment of the material, but not as a full substitute of conventional characterisation.

## 4.5 CONCLUDING REMARKS

When applying mathematical treatments to interrogate the data, it is important to keep in mind that statistics do not generate information. Instead, they just help extract relevant information from signals that contain noninformative content. The less noninformative content, the greater is the accuracy of information extraction. Emphasis should therefore be placed on the quality of the raw data *per se*, and this implies decisions regarding the most suitable sampling technique, the spectral range, the optical and digital resolution, the acquisition time, the elimination of (or compensation for) spectral artifacts. With appropriate adaptations, the same holds for the various types of independent data that are considered for chemometric correlation with the spectral data.

Deciding what part(s) of the spectral information is(are) informative is question dependent: It is determined by what information is sought from the spectra. Vibrational spectroscopy is extremely rich in information concerning the composition, structure and bonding of clay minerals and related samples. Yet one should be prepared to encounter situations where the information content of the spectra that is relevant to the specific question of interest is below the detection limit of a particular (or any) vibrational spectroscopic technique. Further, the threshold between significance and triviality can be vague and should only be approached statistically by validation procedures.

Vibrational spectra are fingerprints of samples (Farmer and Russell, 1964, 1967). From the perspective of multivariate analysis, spectra (or samples) are pixels of a more general picture and acquire most of their significance as specific parts of that general picture. Thus a single spectrum of a specific montmorillonite has little to offer if observed outside the context of the IR spectra of other montmorillonites or smectites in general. The situation calls for a large number of observations that are needed to create the general picture.

Multivariate analysis is a toolbox designed for identifying and analyzing systematics within large sets of observations, or among large sets of observations obtained by different 'sensors'. It is in their latter capacity that multivariate analysis and especially chemometrics can bridge between disciplines and open the way towards scientific discovery.

Yet it should always be remembered that the human brain has amazing multivariate analytic capabilities in handling and rationalising observations. The visual inspection of large numbers of spectra on screen, subjected to various pretreatments (derivatives, normalisations, etc.) has always been rewarding in identifying important trends and subtle systematics.

A final remark is deserved regarding multivariate analysis and chemometrics: Their performance can be properly judged during application, but its reporting in the literature is often received with scepticism. This is because the models are built from datasets that may not be available for inspection, even less for review, but also because the physical meaning of the latent variables is rarely straightforward. Further, it is easy to use commercial software and modest computing resources for producing solutions of no value. Poorly validated 'results' appear more impressive the more mathematically trivial they are. For these reasons, multivariate analysis and chemometric modelling should not be developed, nor reported as 'black box' applications. They should, instead, be discussed in their real context, as tools for classifying, identifying and predicting the behaviour of individual objects that are part of a complex, poorly understood cosmos.

Chapter 5

# IR Spectra of Clay Minerals

J. Madejová, W.P. Gates and S. Petit

## 5.1 INTRODUCTION

Over the past 60 years spectroscopic analysis in the mid-infrared (MIR) region has provided a great deal of information on chemical composition and crystal-chemical aspects of clays and clay minerals. An IR spectrum can serve as a fingerprint for mineral identification and can also provide unique information concerning the mineral structure, including the family of minerals to which the specimen belongs, the nature of isomorphic substituents and the presence of both crystalline and noncrystalline impurities. The early IR studies of clays were closely connected with a tremendous body of work by V.C. Farmer and his coworkers, performed at the Macaulay Institute for Soil Research in Aberdeen, Scotland. Farmer's scientific contribution to the clay/soil community included (co)authoring a total of 164 refereed journal papers, many of them related to the IR characterisation of clay minerals (e.g. Farmer, 1958, 1968, 1971, 1979; Farmer and Russell, 1964, 1967) or to the specific problems associated with clay minerals structure (e.g. Farmer, 1964, 1998, 2000; Russell and Farmer, 1964; Farmer and Russell, 1966, 1971; Farmer et al., 1967, 1971, 1988, 1994; Russell et al., 1970; Farmer and Velde, 1973; Diaz et al., 2000). Farmer's most important publication was the monumental monograph *Infrared Spectra of Minerals* (Farmer, 1974a), which he not only edited, but to which he also contributed several chapters, including the excellent one devoted to layer silicates (Farmer, 1974b).

In this chapter the IR spectra of the representatives of the individual families of the layer silicates are discussed in relation to their structural characteristics (i.e. 1:1 vs 2:1 clay minerals and their trioctahedral vs dioctahedral character), including symmetry groups and chemical composition (i.e. isomorphous substitutions in tetrahedral and octahedral sheets). The main emphasis is devoted to the vibrations of the basic constituent units and/or functional groups, that is, the hydroxyl groups and the silicon–oxygen network. Most of Farmer's assignments still remain valid today, being confirmed or

Developments in Clay Science, Vol. 8. https://doi.org/10.1016/B978-0-08-100355-8.00005-9

improved by other studies on natural (e.g. Besson and Drits, 1997a,b; Zviagina et al., 2004; Gates, 2005; Kloprogge, 2006) as well as on synthetic clay minerals (e.g. Decarreau et al., 1992, 2008; Petit et al., 1999b, 2004a,b, 2015, 2016; Andrieux and Petit, 2010; Baron et al., 2016a). Importantly, these assignments have also been largely confirmed by computational approaches (e.g. Chatterjee et al., 1996; Sainz-Diaz et al., 2000; Balan et al., 2001, 2005, 2007; Martínez-Alonso et al., 2002; Botella et al., 2004; Scholzová et al., 2003; Larentzos et al., 2007; Smrčok et al., 2010; Escamilla-Roa et al., 2014; Szczerba et al., 2016).

Since the publication of Farmer's famous book 40 years ago, several important review chapters related to the characterisation of clay minerals by IR spectroscopy have been published (e.g. Johnston and Aochi, 1996; Johnston and Wang, 2002; Schroeder, 2002; Petit, 2006; Madejová et al., 2010, 2011; Petit and Madejová, 2013). Russell and Fraser, Farmer's coworkers at the Macaulay Institute, presented in 1994 spectra of many types of clay and associated minerals. They also included some representative soil clays, indicating how IR spectra may be used to assess the constituent minerals in the samples. In 2005, Kloprogge edited a volume of the CMS Workshop Lectures entitled "The Application of Vibrational Spectroscopy to Clay Minerals and Layered Double Hydroxides," which included a series of contributions devoted mainly to IR spectroscopy of clay minerals (Gates, 2005; Kloprogge and Frost, 2005; Madejová, 2005; Madejová and Komadel, 2005; Petit, 2005). Due to improved instrumentation (see Chapter 3), application of IR spectroscopy to clay science has greatly increased. A number of research articles have reported results in which IR spectroscopy was applied to investigate synthesis, surface reactivity, dissolution, reduction–oxidation and organo-modification, to name a few. Moreover, recent advances by researchers using the near-infrared (NIR) region also offer the possibility of successful study of the structural chemistry of clay minerals (e.g. Madejová and Komadel, 2001; Bishop et al., 2002a, 2008a; Frost et al., 2002b; Petit et al., 2002, 2004a,b; Gates, 2005; Gionis et al., 2006; Cheng et al., 2010; Madejová et al., 2011; Bukas et al., 2013; Cuadros et al., 2015a; Tsampodimou et al., 2015; Baron and Petit, 2016). Among the advantages offered by NIR spectroscopy, the speed, simplicity of sample preparation and nondestructive nature of the technique should be mentioned.

It is not possible to cite all research employing the IR technique for clay minerals; for example, for the last 20 years (1995–2015) using the keywords 'IR spectroscopy' and 'clay' displays more than 4300 papers that have been published in scientific journals registered in the Web of Science (WOS) database. Some of them used IR spectroscopy only as a subsidiary or complementary technique, but for many it was the principal method for the identification and characterisation of clay minerals.

The purpose of this chapter is to present the MIR and NIR spectra of selected minerals from the important clay minerals families and to give the assignment of their characteristic functional groups. The assignments of the bands appearing in the MIR spectra are based on those suggested by Farmer but take into account new information obtained by more recent research in the field of clay mineralogy. The intention is to underpin the importance of IR spectroscopy as a stand-alone method for identifying and characterising clay minerals.

## 5.2 EXPERIMENTAL

Raw materials were obtained from different deposits and collections worldwide or from the Source Clay Repository of the Clay Minerals Society (Table 5.1). For IR analysis, fine fractions (<2 μm) were separated from clay ores by sedimentation.

The IR spectra were obtained on a Nicolet 6700 Fourier transform IR (FTIR) spectrometer from Thermo Scientific by coaddition of 128 scans at a resolution of 4 $cm^{-1}$. To obtain MIR spectra, the KBr pressed disk technique (1 mg sample diluted in 200 mg of KBr) was used. After pressing, a disc was heated in an oven overnight at 120°C to minimise the amount of water adsorbed onto KBr and the sample. The NIR spectra were collected by the diffuse reflectance technique using the Smart Diffuse Reflectance Accessory from Thermo Scientific. Spectra manipulations were performed using the Thermo Scientific OMNIC software package. The second derivative procedure using the Savitsky–Golay approach (Savitzky and Golay, 1964) was used to find the exact position of the vibrational modes appearing as shoulder in the IR spectra.

## 5.3 CHARACTERISTIC VIBRATIONS OF CLAY MINERALS

The IR spectra of clay minerals are rich with information that reveals much about their structure and chemical composition. Clay minerals have a layered structure formed by two basic units: tetrahedral and octahedral sheets (Brigatti et al., 2013). Each tetrahedron consists of a central cation, mostly $Si^{4+}$, coordinated to four oxygen atoms, and linked to neighbouring tetrahedra by sharing three basal oxygen atoms. The fourth oxygen atom (the apical oxygen) is a constituent of a plane connecting the tetrahedral with the adjacent octahedral sheet. The octahedral sheet is composed of octahedrons containing a central cation, mostly by $Al^{3+}$, $Fe^{3+}$ or $Mg^{2+}$, coordinated with apical oxygen (from the tetrahedral sheet) and also with hydroxyl. Two types of octahedral sheets occur in clay minerals: the dioctahedral type, where two octahedral sites are occupied mainly by trivalent cations (e.g. $Al^{3+}$ or $Fe^{3+}$) and the third position is vacant, and the trioctahedral type, with all octahedral sites occupied by divalent cations (e.g. $Mg^{2+}$ or $Fe^{2+}$).

**TABLE 5.1** List of Clay Minerals (Including Their Location and Reference, If Available)

| Mineral Name/Label | Sample Location | Reference |
|---|---|---|
| *1:1 minerals* | | |
| Kaolinite/KGa-1 | Washington County, Georgia, United States | CMS Source Clay[a] |
| Kaolinite/KGa-2 | Warren County, Georgia, United States | CMS Source Clay |
| Halloysite | Biela hora, Michalovce, Slovakia | Madejová et al., 1997 |
| Dickite | St. Claire, Pennsylvania, United States | Balan et al., 2010 |
| Nacrite | Lodève, France | Bautier et al., 1996 |
| Chrysotile | Dobšiná, Slovakia | Putiš et al., 2012 |
| Lizardite | Dobšiná, Slovakia | Putiš et al., 2012 |
| *2:1 minerals* | | |
| Pyrophyllite | Vígľašská Huta, Slovakia | Uhlík et al., 2000 |
| Talc/T111 | Trimouns, France | Petit et al., 2004b |
| Talc/ANT | Sterling mine, Antwerp, New York, United States | Petit et al., 2004b |
| Talc/GP | Gemerská Poruba, Slovakia | Čavajda et al., 2015 |
| Beidellite | Glen Silver Pit, De Lamar, Idaho, United States | Gates, 2005 |
| Montmorillonite/SAz-1 | Apache County, Arizona, United States | CMS Source Clay |
| Montmorillonite/JP | Jelšový Potok, Slovakia | Madejová et al., 1994 |
| Montmorillonite/SWy-2 | Crook County, Wyoming, United States | CMS Source Clay |

| | | |
|---|---|---|
| Fe-smectite/RH | Redhill, England | Gates, 2005 |
| Fe-smectite/BLS | Blšany, Doupovske hory, Czech Rep. | Čičel et al., 1992 |
| Fe-montmorillonite | Ölberg, Germany | Petit et al., 2002 |
| Ferruginous smectite/SWa-1 | Grant County, Washington, United States | CMS Special Clay[a] |
| Nontronite/NAu-1 | South Australia | Keeling et al., 2000 |
| Nontronite/NG-1 | Germany | Gates, 2005 |
| Nontronite/NAu-2 | South Australia | Keeling et al., 2000 |
| Nontronite | Spokane, Washington, United States | Gates, 2005 |
| Saponite | Ballarat, California, United States | Post, 1984 |
| Hectorite/SHCa-1 | San Bernardino County, California, United States | CMS Source Clay |
| Griffithite | Griffith Park, California, United States | Komadel et al., 2000 |
| Illite | Ward's, Morris # 36, Illinois, United States | Środoń, 1984 |
| Muscovite | Křiženec, Czech Republic | www.mindat.org |
| Celadonite | Zavalye, Russia | Slonimskaya et al., 1986 |
| Vermiculite | Santa Olalla, Spain | Maqueda et al., 2008 |
| Biotite | Mineralogical collection, University of Poitiers, France | |
| Phlogophite | Sludjanka Lake, Bajkal, Russia | www.mindat.org |
| Clinochlore | Mineralogical collection, University of Poitiers, France | |
| Ripidolite/CCa-2 | Flagstaff Hill, El Dorado County, California United States | CMS Special Clay |

*Continued*

**TABLE 5.1** List of Clay Minerals (Including Their Location and Reference, If Available)—cont'd

| Mineral Name/Label | Sample Location | Reference |
|---|---|---|
| Sudoite | Athabasca Basin, Saskatchewan, Canada | Billault et al., 2002 |
| Palygorskite/PFl-1 | Gadsden County, Florida, United States | CMS Source Clay |
| Palygorskite/Gr-1 | Western Macedonia, Greece | Gionis et al., 2006 |
| Sepiolite/SepSp-1 | Valdemore, Spain | CMS Special Clay |

[a]*CMS Source/Special Clay: Physicochemical data and corresponding references on The Clay Minerals Society Source Clays and Special Clays are available at: http://www.clays.org.*

Tetrahedral and octahedral sheets form a layer, consisting of one tetrahedral and one octahedral sheet (1:1 type) or of one octahedral sheet sandwiched between two tetrahedral sheets (2:1 type).

The absorption bands present in the MIR spectra (4000–400 $cm^{-1}$) of clay minerals can be related to the fundamental stretching ($\nu$) and bending ($\delta$) vibrations of their basic functional groups, that is, OH and Si—O groups. The stretching and bending vibrations of OH groups absorb in the 3700–3500 and 950–600 $cm^{-1}$ regions, respectively, and are affected by the octahedral atoms to which the OH groups are coordinated, by interlayer cations in the micas, and by the configuration of, and charge distribution in, the surrounding tetrahedral lattice (Farmer, 1974b; Gates, 2005; Petit, 2005; Gates et al., 2017 and the references therein). The OH stretching vibrations are only weakly coupled with other vibrations of the structure, similar to the Si—O stretching modes occurring in the 1200–700 $cm^{-1}$ region. In contrast, the Si—O bending bands lying between 600–400 $cm^{-1}$ are strongly coupled with vibrations of the octahedral atoms and/or with translator vibrations of hydroxyl groups (Farmer, 1974b).

The bands present in the NIR spectra (8000–4000 $cm^{-1}$) of clay minerals are related to overtones of fundamental stretching vibrations or combinations involving stretching and bending modes of the structural OH groups. The first overtone of the OH stretching vibration mode ($2\nu$(OH)) occurs near 7060 $cm^{-1}$. The appearance of the first overtones, at wavenumbers slightly lower than twice the fundamentals, is due to the anharmonic character of vibrations (Workman and Weyer, 2008). The combination band(s) of OH stretching and bending fundamental modes ($\nu$(OH) + $\delta$(OH)) are present in the 4600–4300 $cm^{-1}$ region (Bishop et al., 1994; Madejová and Komadel, 2001).

The positions of the individual vibrational bands of OH and Si—O groups significantly depend on the structural characteristics and chemical composition of the clay mineral under study. For example, the frequency of vibrations of hydroxyl groups allows us to probe their local (neighbouring) atomic environments and therefore determine the types and amounts of octahedral atoms coordinated by the hydroxyls (Slonimskaya et al., 1986; Madejová et al., 1994; Besson and Drits, 1997a,b; Gates et al., 2002; Zviagina et al., 2004; Gates, 2005; Chryssikos et al., 2009). Detailed analysis of the relation of the OH vibration bands positions and chemistry of (mostly) dioctahedral smectites is presented in Chapter 7.

## 5.4 THE 1:1 CLAY MINERALS

### 5.4.1 Dioctahedral 1:1 Clay Minerals: The Kaolin Group

The kaolin group consists of dioctahedral 1:1 layer structures with the general composition of $Al_2Si_2O_5(OH)_4$. Kaolin minerals show different polytypes: kaolinite, dickite, nacrite and halloysite. These minerals are stacked with different positions of octahedral vacancies in adjacent layers (Smrčok et al., 1990; Bish, 1993; Brigatti et al., 2013). Halloysite, containing water molecules within the interlayers, also belongs to this group. Kaolin minerals can form a wide variety of ordered and disordered polytypes, which differ in stacking faults or defects occurring within the structure. All these structural variations significantly affect the shape of their IR pattern, mainly in the OH stretching region. The different OH groups present in kaolin minerals have attracted considerable attention as sensitive indicators of their structural and chemical characteristics (Farmer and Russell, 1964; Farmer, 1974b, 1998, 2000; Prost et al., 1989; Johnston et al., 1990; Bish and Johnston, 1993; Frost, 1998; Joussein et al., 2005; Welch et al., 2012; Cuadros et al., 2015).

The MIR spectra of selected representatives of this group are given in Fig. 5.1a–e. The OH stretching region of well-ordered kaolinite KGa-1 shows four resolved bands at 3694, 3669, 3652 and 3620 $cm^{-1}$ (Fig. 5.1a). The assignment of the bands follows well-established band assignments proposed by Farmer (1974b) and confirmed later by quantum-mechanical calculations (Balan et al., 2001, 2005, 2010; Benco et al., 2001a). The 3620 $cm^{-1}$ band arises from the stretching vibrations of single 'inner' hydroxyl groups ($\nu$(OH)) bonded to octahedral cations, but residing largely within the plane of apical oxygen atoms that link the tetrahedral and octahedral sheets. The other three bands are due to coupled stretching vibrations of nonequivalent 'inner-surface' OH groups located at surface of the dioctahedral sheet of the layers. These hydroxyls form strong H-bonds with the surface oxygen atoms of the tetrahedra of an adjacent layer, and are therefore located on the inner surface of the unit cell. The 3694 $cm^{-1}$ band corresponds to the in-phase coupled stretching vibration of perpendicular-surface-OH, whereas 3669 and 3652 $cm^{-1}$ bands are due to out-of-phase coupled vibrations of these groups (Balan et al., 2001, 2005).

The spectra of kaolin minerals of different geological origin (hydrothermal or sedimentary), as well as different degree of stacking disorder, show considerable variation in the relative intensities of the $\nu$(OH) bands (Parker, 1969; Prost et al., 1989; Madejová et al., 1997). The spectrum of KGa-2 (Fig. 5.1b) is typical for a disordered kaolinite showing only one broad complex $\nu$(OH) band at 3652 $cm^{-1}$ with a shoulder near 3670 $cm^{-1}$ instead of two bands. Low-temperature (15 K) transmission measurements, however, may help to resolve the spectrum of disordered kaolinites (Prost et al., 1989; Johnston et al., 2008).

The IR spectra of the two kaolinite polymorphs, dickite (St. Clair, Pennsylvania, United States) and nacrite (Lodève, France) show only three resolved bands in the OH stretching region as a consequence of different geometry of OH bonds. In dickite, the three nonequivalent, inner-surface OH groups have a different orientation from that in kaolinite (Bish and Johnston, 1993). Instead of being almost perpendicular to the layers, one OH group (OH3) is inclined and shares a longer H-bond with the next layer. The two other inner-surface OH groups (OH2 and OH4) remain perpendicular (Bish

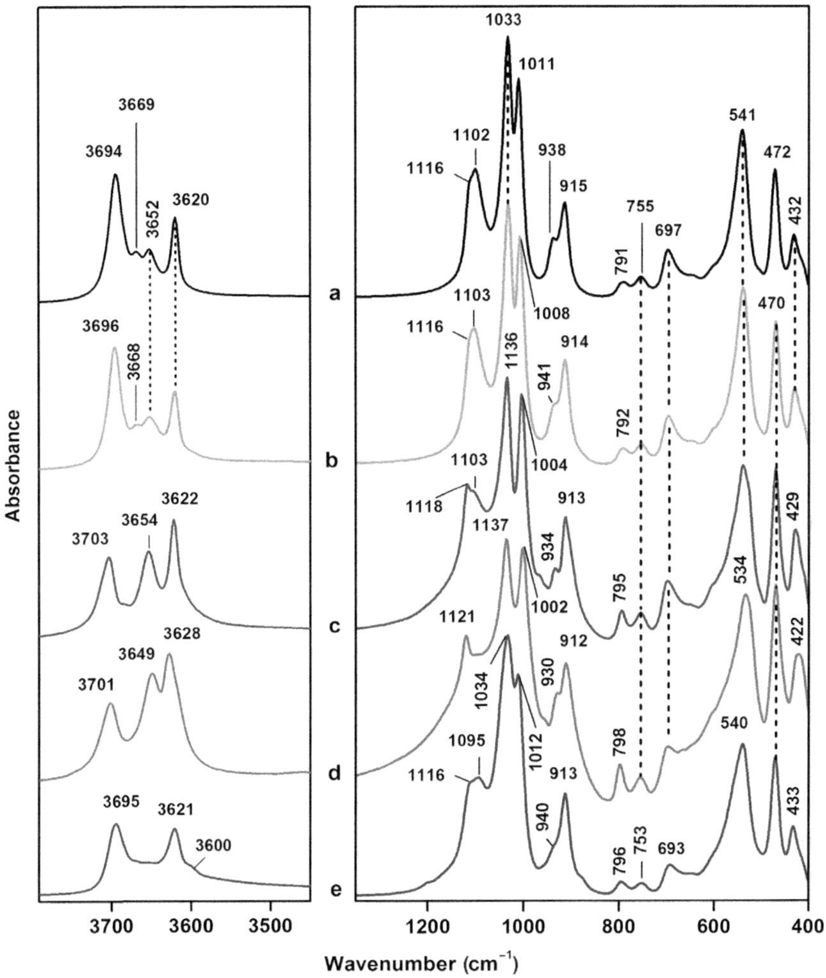


**FIG. 5.1** MIR spectra of (a) well-ordered kaolinite KGa-1, (b) disordered kaolinite KGa-2, (c) dickite, (d) nacrite, (e) halloysite.

and Johnston, 1993). The spectrum of dickite displays three absorption bands at 3703, 3654 and 3622 cm$^{-1}$ (Fig. 5.1c). These bands are classically observed in ordered dickite samples (Prost et al., 1989; Beaufort et al., 1998; Balan et al., 2010). Following the interpretation proposed by Balan et al. (2005), the band observed at 3622 cm$^{-1}$ is unambiguously ascribed to inner OH1 groups. The band at 3654 cm$^{-1}$ contains two components corresponding to the in-phase vibration of OH2 and OH4 groups, split by interlayer coupling. The high-frequency band at 3703 cm$^{-1}$ corresponds to the vibration of the inclined OH3 group.

The IR spectrum of nacrite shows similar OH pattern to those of dickite (Fig. 5.1d). Based on computational analysis by Balan et al. (2005), the 3628 $cm^{-1}$ band is assigned to in-plane stretching vibration of inner OH1, whereas that of 3701 $cm^{-1}$ band is due to vibrations of inclined inner-surface OH2 group. The band observed at 3649 $cm^{-1}$ is related to the in-phase stretching motion of the OH2 and OH3 groups.

Though the structure and chemical composition of halloysite is similar to that of kaolinite, the IR spectrum of halloysite from Michalovce (Slovakia) exhibits only two bands at 3695 $cm^{-1}$ and 3621 $cm^{-1}$ in the OH stretching region (Fig. 5.1e). As reported by Farmer (1974b), a different type of disorder, involving more than simply a random stacking sequence, is necessary to account for the type of IR spectrum given by most halloysites, which often have no distinct features between these two bands. The weaker intensity of the 3695 $cm^{-1}$ band compared to kaolinite reflects hydrogen bonding between surface OH groups and interlayer water molecules (Joussein et al., 2005). A weak shoulder near 3600 $cm^{-1}$ indicates a presence of some $Fe^{3+}$ in the octahedral positions (Petit et al., 1999b).

In contrast to Joussein et al. (2005) reporting only two OH stretching bands at 3695 and 3620 $cm^{-1}$ for halloysites, Hillier et al. (2016) documented very recently that the IR spectra of some tubular halloysites can also show one or two weak intermediate OH stretching bands near 3670 and 3653 $cm^{-1}$. Based on detailed X-ray diffraction (XRD), scanning electron microscopy (SEM) and transmission electron microscopy (TEM) analysis, they concluded that two main OH stretching bands are characteristic for cylindrical tubular halloysites while the IR pattern with intermediate bands, resembling well-ordered kaolinite, are evident in prismatic tubular specimens. The important point was that studied prismatic halloysites showed no obvious evidence for kaolinite as a separate phase. Most probably the intermediate bands observed for prismatic halloysite are an indication of a generally more ordered pattern of hydrogen bonding of the layers, presumably a reflection of the fact that many layers are planar in the prismatic examples whereas in the cylindrical halloysites there is added disorder due to curvature of the layers which probably results in uncoupling of the surface OH bonds (Farmer, 1974b).

The IR spectra of all kaolin minerals have a characteristic band near 3700 $cm^{-1}$, which is frequently used for identification of kaolinites in ores or raw materials. This band is well separated from absorption bands of most other minerals, and therefore allows identification of very low concentrations of kaolinite in mixtures (Madejová et al., 2002). Further, the difference in position and relative intensities of the OH stretching bands between kaolinite and dickite allows, using an IR microscope, differential mapping of these kaolin minerals in petrographic thin sections of rocks (Robin et al., 2013).

The spectra of all kaolin minerals are similar over the 1400–400 $cm^{-1}$ region (Fig. 5.1). The Si—O stretching vibrations give several well-resolved strong and sharp bands in the 1120–1000 $cm^{-1}$ region. The antisymmetric in-plane Si—O—Si stretching vibrations give two strong bands at 1037–1033 $cm^{-1}$ and 1012–1002 $cm^{-1}$, while the in-phase stretching band of apical Si—O bonds appears near 1100 $cm^{-1}$. The high-frequency band at 1121–1116 $cm^{-1}$ was assigned to the symmetric stretching Si—O vibrations involving the basal oxygens (Balan et al., 2001). The $Al_2OH$ bending bands of kaolin minerals near 915 and 935 $cm^{-1}$ arise from vibrations of inner and inner-surface OH groups, respectively. Well-ordered KGa-1, dickite and nacrite display a resolved inner-surface OH deformation band, whereas only a shoulder or an inflexion is present in the KGa-2 and halloysite spectra. The weak bands at the 755 and 697 $cm^{-1}$ in the spectra of kaolin minerals are related to the perpendicular Si—O vibrations involving probably the surface hydroxyl layer (Farmer, 1974b). Two strong bands near 540 and 470 $cm^{-1}$ are due to the bending vibrations of Si—O—$Al^{VI}$ ($Al^{VI}$ indicates aluminium in the octahedral positions) and Si—O—Si groups, respectively (Farmer, 1974b).

The NIR spectra of kaolin minerals are presented in Fig. 5.2. In the OH overtone region four bands are observed for well-ordered KGa-1 kaolinite (Fig. 5.2a). Petit et al. (2004b) showed that a strong band at 7065 is assigned to $2\nu(OH)$ overtones of OH stretching modes of inner hydroxyl groups ($\nu(OH) = 3620\ cm^{-1}$), while a complex band with several more or less resolved peaks in the 7200–7100 $cm^{-1}$ region corresponds to overtones of the inner-surface OH groups. The 7172 and 7142 $cm^{-1}$ bands are due to the first overtones of the 3669 and 3652 $cm^{-1}$ bands, respectively. A very broad shoulder located near 7220 $cm^{-1}$ is supposed to correspond to the overtone of the 3694 $cm^{-1}$ band. The 7113 $cm^{-1}$ band, however, cannot be considered as an overtone of any observable $\nu(OH)$ band. On the other hand, the combination of the 3620 and 3669-$cm^{-1}$ bands accounts well for the 7113 $cm^{-1}$ value (Petit et al., 2004b).

The shape of the OH overtone bands, similar to that of the OH stretching region, reflects the different degree of disorder of kaolin minerals. The spectrum of disordered KGa-2 shows a sharp band at 7065 $cm^{-1}$ and a broad band at 7171 $cm^{-1}$ with hardly resolved shoulder near 7117 $cm^{-1}$ (Fig. 5.2b). The spectrum of halloysite exhibits a strong band at 7069 $cm^{-1}$ related to the first overtone of the inner hydroxyl groups and a broad band at 7195 $cm^{-1}$ involving contributions of OH overtones of the inner-surface OH groups (Fig. 5.2e). NIR spectra of dickite and nacrite show two sharp bands in the OH overtone region; however, in comparison with kaolinites their positions are shifted to higher wavenumbers (Fig. 5.2c and d) corresponding more closely to their theoretical positions. The 7070 and 7085 $cm^{-1}$ bands found for dickite and nacrite, respectively, are due to the first overtone of the 3622 and 3628 $cm^{-1}$ bands of the inner OH groups. The bands at 7239 and 7227 $cm^{-1}$ are related to the first OH overtone of the inner surface OH groups. The first overtones of the 3654 $cm^{-1}$ band for dickite and the 3649 $cm^{-1}$ band for nacrite do not occur or are too weak to be observed (Petit et al., 2004b).

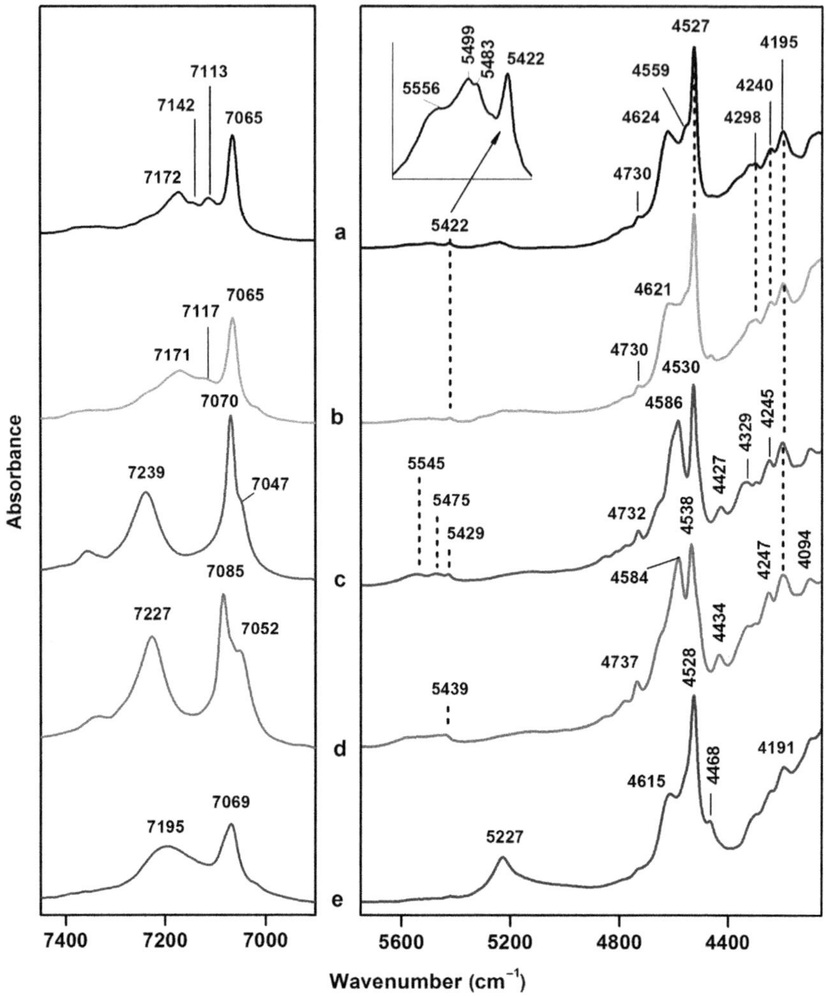


**FIG. 5.2** NIR spectra of (a) well-ordered kaolinite KGa-1, (b) disordered kaolinite KGa-2, (c) dickite, (d) nacrite, (e) halloysite.

The NIR spectra of kaolin minerals show a series of weak bands near 5500 cm$^{-1}$ which, while probably observed previously, have not been reported or discussed in the literature so far. For example, 5556, 5499, 5483 and 5422 cm$^{-1}$ bands can be clearly distinguished in the spectrum of KGa-1 kaolinite (inset in Fig. 5.2a). Taking into account the positions observed for OH stretching and bending in the MIR region, these bands can be related to the ($\nu+2\delta$)OH combination modes (i.e. 3694+2*938=5570, 3669+2*938=5545, 3652+2*938=5528 and 3620+2*915=5448). The calculated values are in reasonable agreement with the observed experimental data.

Spectrum of halloysite shows also a band at 5227 $cm^{-1}$ related to the combination OH stretching and bending modes of water ($(\nu+\delta)H_2O$) which confirms the presence of adsorbed water in this clay mineral. The NIR spectra of all kaolin minerals reveal bands in the 4700–4400 $cm^{-1}$ region corresponding to combinations of OH stretching and bending OH vibrations. The major band at 4527 $cm^{-1}$ observed in the NIR spectrum of KGa-1 is associated with the $(\nu+\delta)$OH of inner OH groups, the bands at 4624 and 4559 $cm^{-1}$ correspond to similar combination modes of inner surface hydroxyls. The weak band at 4730 $cm^{-1}$ is likely due to a combination of the inner OH stretching vibrations (3620 $cm^{-1}$) with lattice vibration (Si—O stretching near 1100 $cm^{-1}$). The bands observed between 4300 and 4000 $cm^{-1}$ are probably combinations of OH stretching bands with lattice deformation vibrations involving octahedral atoms (Hunt and Salisbury, 1970; Hunt et al., 1973; Petit et al., 1999b; Baron and Petit, 2016).

## 5.4.2 Trioctahedral 1:1 Clay Minerals: The Serpentine Group

The trioctahedral 1:1 clay minerals chrysotile and lizardite have the same chemical composition $Mg_3Si_2O_5(OH)_4$, but they are structurally different. Lizardite shows ideal layer topology, whereas chrysotile is bent (Brigatti et al., 2013). The structural differences affect also the shape of their IR spectra (Balan et al., 2002a,b; Šontevska et al., 2007; Anbalagan et al., 2010). The MIR spectra of chrysotile and lizardite both obtained from the Dobšiná (Slovakia) deposit are reported in Fig. 5.3. The OH stretching region of chrysotile displays two well-resolved bands at 3693 and 3647 $cm^{-1}$, while a band

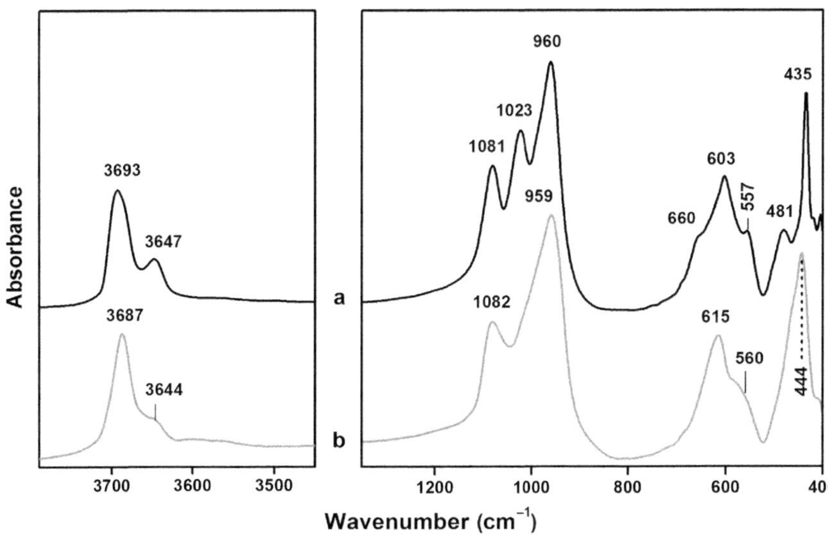


FIG. 5.3 MIR spectra of (a) chrysotile, (b) lizardite.

at 3687 $cm^{-1}$ with a shoulder near 3644 $cm^{-1}$ is observed in the spectrum of lizardite. The assignment of the OH stretching bands of trioctahedral 1:1 clay minerals is not as well understood as for the kaolin minerals. Farmer (1974b) reported that it was not possible to distinguish the IR absorptions of inner and inner surface hydroxyls. In contrast, Anbalagan et al. (2010) ascribed the 3689 and 3648 $cm^{-1}$ bands of chrysotile to in-phase outer and inner Mg—OH stretch, respectively. The assignment of the OH bands of lizardite presented by Balan et al. (2002b) is based on first-principle calculations, as was performed for kaolin minerals. They assigned the 3684 $cm^{-1}$ band, observed in the experimental spectrum, to the in-phase stretching of the three inner-surface OH groups, and the shoulder near 3645 $cm^{-1}$ to the two degenerate out-of-phase stretching modes of the inner surface OH groups. The positions of the OH stretching vibrations of lizardite shown in Fig. 5.3b are in agreement with those reported by Balan et al. (2002b). The same assignments can be made also for chrysotile; that is, the band at 3693 $cm^{-1}$ is due to the in-phase stretching of the three inner-surface OH groups and the band at 3647 $cm^{-1}$ to the two degenerate out-of-phase stretching modes of the inner-surface OH groups (Fig. 5.3a). The absorption due to the vibrations of the inner OH groups, reported by Balan et al. (2002a) near 3700 $cm^{-1}$, is most probably overlapped with the 3690 $cm^{-1}$ band.

The Si—O stretching region of both minerals shows a band near 1080 $cm^{-1}$ due to the out-of-plane (perpendicular to the layers) vibration and the band near 960 $cm^{-1}$ assigned to the two degenerate in-plane (parallel to the layer) Si—O vibrations (Balan et al., 2002b). The IR spectrum of chrysotile, compared to that of lizardite, displays also a strong absorption band at 1023 $cm^{-1}$ due to an out-of-plane Si—O vibration. A strong, broad band at 603 and 615 $cm^{-1}$ observed in the spectrum of chrysotile and lizardite, respectively, involves contributions of several vibrational modes including $Mg_3OH$ bending vibrations, whereas the absorption observed near 440 $cm^{-1}$ arises from the overlapping in-plane Mg—OH vibrations and bending Si—O—Si vibrations.

Structural differences between chrysotile and lizardite influence also the shape of the bands observed in the NIR spectra (Fig. 5.4a and b). In contrast to the OH stretching region, where only two OH bands are seen (Fig. 5.3), three bands can be distinguished in the first OH overtone region. Recently, Baron and Petit (2016) reported a detail analysis of the vibrational modes for synthetic lizardite in both MIR and NIR regions. Taking into account the anharmonicity constant ($X = -85.6$ $cm^{-1}$) calculated for lizardite, the band at 7208 $cm^{-1}$ is attributed to the overtone of the 3687 $cm^{-1}$ band, that is, the in-phase stretching vibrations of the three inner-surface OH groups. Similarly, the 7207 $cm^{-1}$ band of chrysotile should correspond to the 3693 $cm^{-1}$ band. The calculation is less accurate for the 7167 $cm^{-1}$ band of lizardite and 7181 $cm^{-1}$ of chrysotile (Fig. 5.4). If the same anharmonicity constant (Petit et al., 2004b) is used, then the calculated values are 3669 $cm^{-1}$ for lizardite (experimental value is 3644 $cm^{-1}$) and 3676 $cm^{-1}$

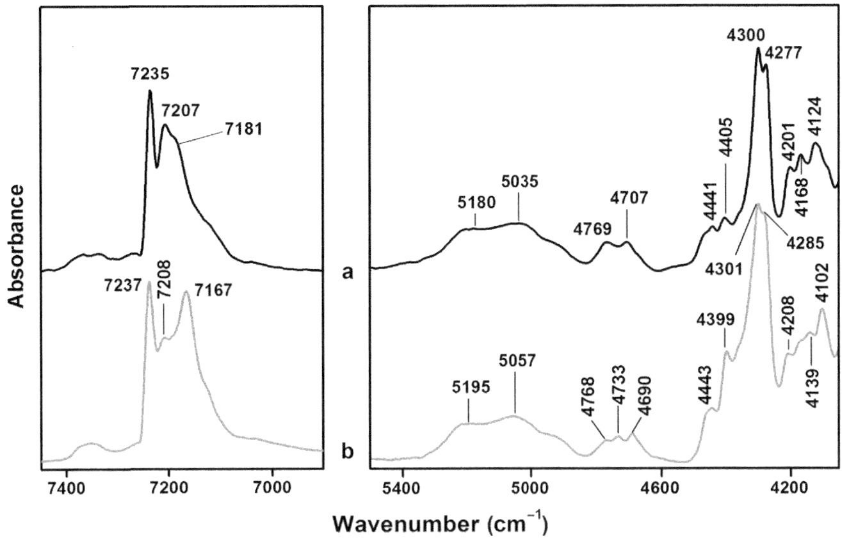


**FIG. 5.4** NIR spectra of (a) chrysotile, (b) lizardite.

for chrysotile (experimental value is 3647 cm$^{-1}$), even though both bands (i.e. at 7167 and 7181 cm$^{-1}$) are related to the out-of-phase stretching modes of the inner surface OH groups. The NIR spectra of both trioctahedral 1:1 clay minerals show also a sharp band near 7235 cm$^{-1}$ (Fig. 5.4). The calculation using the $X = -85.6$ cm$^{-1}$ gives value of 3704 cm$^{-1}$ for lizardite and 3703 cm$^{-1}$ for chrysotile, which is in very good agreement with the value reported by Balan et al. (2002b) for the vibrations of the inner OH groups of lizardite. It is evident that the NIR region allows identification of not only the vibrational modes related to the inner surface OH groups, but also that of inner OH groups.

The NIR spectra of both clay minerals show very broad and complex bands in the 5200–5000 cm$^{-1}$ region corresponding to the combination modes of water molecules bonded to their surface with H-bonds of different strength. The region below 5000 cm$^{-1}$ shows two strong bands near 4300 and 4280 cm$^{-1}$ corresponding to the combination modes $(\nu + \delta)$OH of the structural OH groups. The other less intensive bands arise from different combinations of two or more fundamental vibrations ($\nu$(OH), $\delta$(OH) and lattice vibrations) which are observed in the MIR region. Detailed analysis of all bands present in the NIR spectrum of synthetic lizardite is given in Baron and Petit (2016).

## 5.5 THE 2:1 CLAY MINERALS

### 5.5.1 Pyrophyllite, Talc

Pyrophyllite is a dioctahedral clay mineral containing $Al^{3+}$ in octahedral positions while talc is a trioctahedral clay mineral with mainly $Mg^{2+}$ in octahedral sheets. Only $Si^{4+}$ occupies the tetrahedral sites in both minerals. These nonexpanding layer silicates are often considered as parent members of the 2:1 group from which the other minerals like smectites or micas can be derived by heterovalent isomorphous substitution in octahedral and/or tetrahedral sheets. Spectra of pyrophyllite and talc show differences in peak positions due to di- and trioctahedral occupancy of the octahedral sheets. The absorption bands of these minerals are sharp as anticipated for well-crystallised, ordered clay minerals (Farmer, 1974b; Russell and Fraser, 1994; Kloprogge et al., 2000b; Reddy et al., 2016).

The MIR spectra of pyrophyllite and talc from different deposits are shown in Fig. 5.5. The OH stretching region of pyrophyllite from Vígľašská Huta (Slovakia) reveals only one relatively sharp band at 3674 $cm^{-1}$ due to the stretching vibrations of hydroxyl groups lying nearly in the plane of the layers (Fig. 5.5a). The Si—O stretching region of pyrophyllite shows well-resolved bands at 1122 and 1071 $cm^{-1}$ due to in-plane, and at 1052 $cm^{-1}$ due to out-of-plane, Si—O vibrations (Farmer, 1974b). The $Al_2OH$ bending band, which is remarkably well resolved for pyrophyllite, appears at

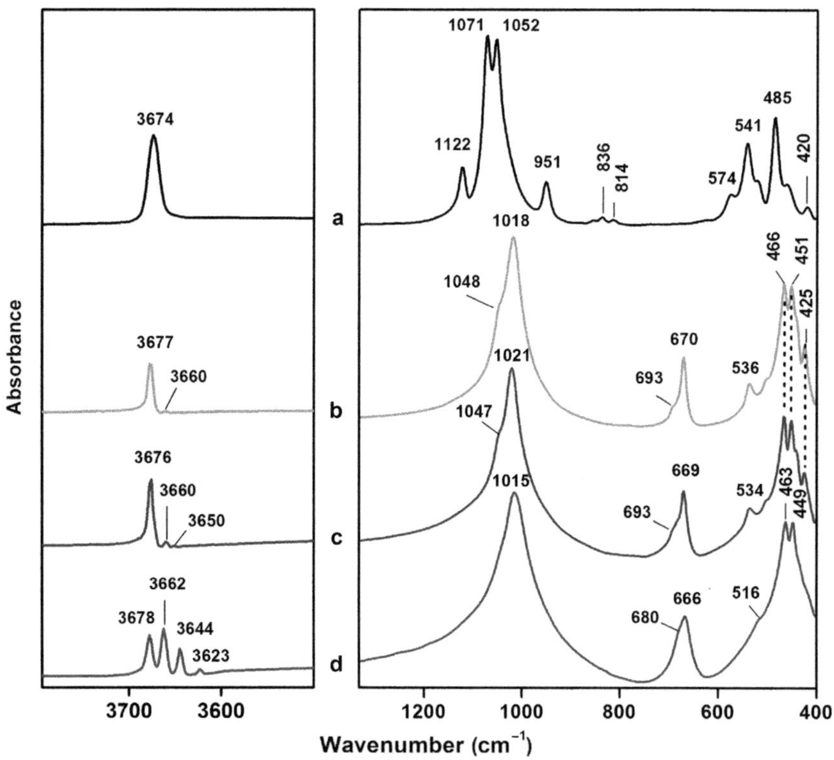


**FIG. 5.5** MIR spectra of (a) pyrophyllite, (b) talc T111, (c) talc GP, (d) talc ANT.

951 $cm^{-1}$. The Si—O bending vibrations contribute to the strong absorption in the 550–400 $cm^{-1}$ region with Si—O—Al bending band at 541 $cm^{-1}$.

In talc, the full octahedral occupancy, mainly by Mg, results in symmetrical environments for OH groups and this fact is reflected in a narrower OH stretching band compared to pyrophyllite. The IR spectra have been often used to identify a talc structure and chemical composition (Farmer, 1974b; Martin et al., 1999a; Ferrage et al., 2003b; Petit et al., 2004a; Parry et al., 2007).

The MIR spectra of talcs from different localities are presented in Fig. 5.5b–d. Talc T111 comes from Trimouns (France), talc GP from Gemerská Poruba (Slovakia) and talc ANT from Antwerp (United States). In trioctahedral 2:1 clay minerals each OH group is bonded to the three central octahedral cations and the position of the OH stretching band is affected by the nature of these cations. Up to four $\nu$(OH) bands can be distinguished when Mg is partially substituted by other divalent cations. The spectra of all studied talcs (Fig. 5.5b–d) show a sharp band near 3677 $cm^{-1}$ assigned to the stretching mode of the $Mg_3OH$ unit (Wilkins and Ito, 1967; Farmer, 1974b; Petit et al., 2004a). The spectrum of talc T111 displays also a very weak band at 3660 $cm^{-1}$ due to the $Mg_2Fe^{2+}OH$ stretching vibration indicating a small amount of iron in this mineral, confirmed also by Martin et al. (1999a) based on different spectroscopic and chemical analyses. A weak band at 3660 $cm^{-1}$ due to $Mg_2Fe^{2+}OH$ appears also in the spectrum of talc GP; moreover, a zoomed spectrum revealed also a weak band at 3650 $cm^{-1}$ related to $MgFe^{2+}{}_2OH$ stretching vibrations. The presence of iron in talc GP was confirmed also by chemical analysis reported in Čavajda et al. (2015). The intensities of the 3660 and 3650 $cm^{-1}$ bands indicate the very low content of iron in the octahedral positions of T111 and GP talcs. However, the IR spectrum of talc ANT indicates a greater content of $Fe^{2+}$. The intensities of the $Mg_3OH$ (3678 $cm^{-1}$) and $Mg_2Fe^{2+}OH$ (3662 $cm^{-1}$) bands are similar, the $MgFe^{2+}{}_2OH$ (3644 $cm^{-1}$) is obvious and in addition a new, though weak, band at 3623 $cm^{-1}$, assigned to $Fe_3OH$ (Farmer, 1974b; Petit et al., 2004a), appears in the spectrum of talc from Antwerp. The IR pattern of the talc ANT in the OH stretching region is in good agreement with chemical analysis of this sample, indicating relatively high content of $Fe^{2+}$ in the octahedral positions (Petit et al., 2004a).

In the 1200–400 $cm^{-1}$ region the spectra of studied talc samples show well-defined absorption bands related to the vibrations of Si—O and OH groups. Detailed analyses of all vibrational modes of talc are given in Farmer (1974b). The spectra presented in Fig. 5.5b–d show the most intense band at 1015–1021 $cm^{-1}$, assigned to the in-plane Si—O stretching vibration, while a shoulder near 1048 $cm^{-1}$, appearing for talcs with low iron content, that is, T111 and GP talcs, belongs to the perpendicular Si—O stretching vibration involving apical oxygens of the tetrahedral sheet. Another perpendicular Si—O vibration appears as a shoulder near 690 $cm^{-1}$ overlaps with the complex band centred near 670 $cm^{-1}$, which is assigned to the bending vibration of OH groups coordinated mainly to the octahedral Mg (i.e. $Mg_3OH$). A slightly lower position (666 $cm^{-1}$) observed for talc ANT reflects the greater octahedral $Fe^{2+}$ content in this sample, and it is broadened due to the absorptions of $Mg_2Fe^{2+}OH$, $MgFe^{2+}{}_2OH$ and $Fe^{2+}{}_3OH$ groups. The absorption band near 535 $cm^{-1}$ is related to the perpendicular Mg—O vibration, the band at 466 $cm^{-1}$ (463 $cm^{-1}$ for ANT) to the translation vibrations of the OH groups. The band near 450 $cm^{-1}$ belongs to the Si—O—Si bending vibrations. Comparison of the positions of the bands found in the 1200–400 $cm^{-1}$ region for all talc samples shows that due to the higher occupancy of the octahedral positions with Fe in talc ANT, all observed bands are shifted to lower wavenumbers.

The NIR spectrum of pyrophyllite shows a single band at 7175 $cm^{-1}$ corresponding to the first OH stretching overtone and at 4615 $cm^{-1}$ to the OH combination mode of the OH stretching and bending vibrations of the structural hydroxyls (Fig. 5.6a). Small bands below 5000 $cm^{-1}$ are supposed

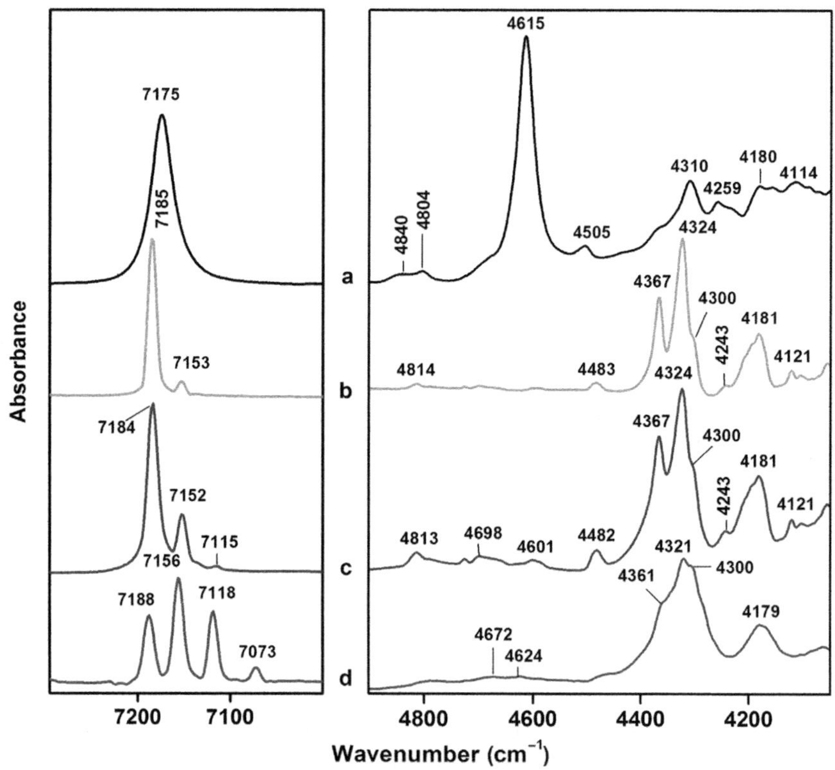


**FIG. 5.6** NIR spectra of (a) pyrophyllite, (b) talc T111, (c) talc GP, (d) talc ANT.

to arise from the combination of fundamental OH and Si—O modes (Baron and Petit, 2016).

The NIR spectra of talc samples in the first OH overtone region (Fig. 5.6b–d) are similar to those obtained in the MIR region (Fig. 5.5b–d). Based on Petit et al. (2004a) the band at 7184–7188 $cm^{-1}$ was assigned to $2\nu(Mg_3OH)$, at 7152–7156 $cm^{-1}$ to $2\nu(Mg_2Fe^{2+}OH)$, at 7115–7118 $cm^{-1}$ to $2\nu(MgFe^{2+}{}_2OH)$, and that at 7073 to $2\nu(Fe^{3+}{}_2OH)$. Better resolution of the individual OH bands, as well as their higher intensities, in the OH overtone compared to the OH stretching region makes the NIR spectroscopy particularly useful for studying the chemistry of talcs.

In the 5000–4000 $cm^{-1}$ region, the spectra of T111 and GP talcs exhibit two main bands at 4367 and 4324 $cm^{-1}$ while the positions are slightly shifted to lower wavenumbers (4361 and 4321 $cm^{-1}$) for ANT talc containing high Fe content. Taking into account the positions of the OH stretching and bending bands (Fig. 5.5), the 4367 $cm^{-1}$ band can be assigned to the combination of the 3677 $cm^{-1}$ band ($\nu$(OH)) and the 693 $cm^{-1}$ band related to the Si—O vibration involving the apical oxygens. The 4324 $cm^{-1}$ is related to $\nu$(OH) + $\delta$(OH), that is, the combination of 3677 and 670 $cm^{-1}$ bands, respectively. The third intense and complex band in this region near 4180 $cm^{-1}$ is supposed to involve the combinations of 3677 with 536 $cm^{-1}$ band (i.e. perpendicular Mg—O) and 3677 with 466 $cm^{-1}$ (i.e. translation OH). The other weak bands most probably result from different combinations of OH with lattice vibrations (Baron and Petit, 2016.)

## 5.5.2 Smectites

Smectites are the most widely studied of the 2:1 layer-type clay minerals. Isomorphic substitution within the tetrahedral and/or octahedral sheets generates a negative charge on the layers, which is balanced by hydrated exchangeable cations in the interlayer spaces or on the outer mineral surfaces. Smectites are usually classified according to the occupancy of the octahedral sheet to dioctahedral and trioctahedral. The general structural formula (SF) for dioctahedral smectites per unit cell is

$$\left(M^{+}{}_{x+y}\cdot nH_2O\right)\ \left(Si_{8-x}Al_x\right)\ \left(R^{3+}{}_{4-y}R^{2+}{}_{y}\right)\ O_{20}(OH)_4$$

and that for trioctahedral species is

$$\left(M^{+}{}_{x+y}\cdot nH_2O\right)\ \left(Si_{8-x}Al_x\right)\ \left(R^{2+}{}_{6-y}R^{+}{}_{y}\right)\ O_{20}(OH)_4$$

where $x$, $y$ are the layer charge resulting from the substitutions in the tetrahedral and octahedral sites, respectively; $R^{3+}$, $R^{2+}$ are generic octahedral cations; and $M^+$ is the generic interlayer cation. For these idealised formulae, substitution of other cations can occur (i.e. $Fe^{3+}$) and, particularly for the trioctahedral type, some trivalent substitution may occur in the octahedral sites, resulting in an excess of positive charge in that sheet.

### 5.5.2.1 Dioctahedral Smectites

Different cations can occupy the central positions in the tetrahedral and octahedral sheets of dioctahedral smectites. Beidellites are Al-rich smectites containing only $Al^{3+}$ in the octahedral positions and $Si^{4+}$ and $Al^{3+}$ in the tetrahedral positions. The layer charge is located on the tetrahedral sheets. Montmorillonites also belong to the Al-rich smectites, but their layer charge originates mainly due to the substitutions in octahedra where $Al^{3+}$ is more or less substituted with $Fe^{3+}$ and $Mg^{2+}$, and the range of $Al^{3+}$ for Si substitution in the tetrahedral positions is lower. If the $Al^{3+}$ is the main octahedral cation, but the content of octahedral $Fe^{3+}$ is greater than 1 atom per unit cell, the clay minerals are frequently named Fe-smectites. Nontronites and ferruginous smectite are also Fe-rich smectites, but their octahedral positions are dominated by $Fe^{3+}$ ($\geq$3 atoms per unit cell), and their substitution with $Al^{3+}$ is low or is missing. Layer charge in nontronites is created due to the $Al^{3+}$ (or $Fe^{3+}$) for $Si^{4+}$ replacement in the tetrahedral positions. Several Fe-rich smectites, however, can have considerable (and some dominant) octahedral layer charge resulting from $Mg^{2+}$ for $Fe^{3+}$ substitution (Petit et al., 2002; Gates, 2005; Cashion et al, 2008).

As a result of the different kinds and amounts of substitutions within smectite layers, the crystalline order is reduced and structural imperfections arise, leading to diverse IR absorption bands due to lowering of symmetry, but also considerable broadening of those bands compared to, for example, unsubstituted pyrophyllite (Russell and Fraser, 1994). The following section will present MIR spectra of some characteristic smectites with basic assignment of their absorption bands. More detailed discussion on the effect of the crystal chemistry of smectites on the position of the bands is given in Chapter 7 (and the references therein).

The MIR spectra of dioctahedral smectites having aluminium as the main octahedral cation but differing in the occupancy of octahedral and tetrahedral sheets are reported in Fig. 5.7. Structural formulae of the samples, calculated conventionally from the chemical analysis, are given in Table 5.2. The OH stretching region shows a typically broad OH stretching band in the 3620–3650 $cm^{-1}$ range, involving the contributions of the OH groups coordinated to different octahedral cations, mainly $Al_2OH$, AlMgOH and $AlFe^{3+}OH$. The spectrum of beidellite displays the OH band with two maxima at 3652 and 3623 $cm^{-1}$. While the position at higher wavenumber is characteristic for beidellite, that is, smectite dominated by tetrahedral substitution, the band at 3623 $cm^{-1}$ indicates that the sample contains montmorillonite domains, that is, smectite with prevailing octahedral substitution, possibly interstratified with beidellite (Fig. 5.7a). Spectra of all other samples contain the maximum intensity of the OH stretching band at 3621–3622 $cm^{-1}$ (Fig. 5.7b–f). A weak shoulder at 3698 $cm^{-1}$ indicates a kaolinite admixture in the Blšany sample (Fig. 5.7f). A broad band near 3420 $cm^{-1}$ is due to OH stretching vibrations

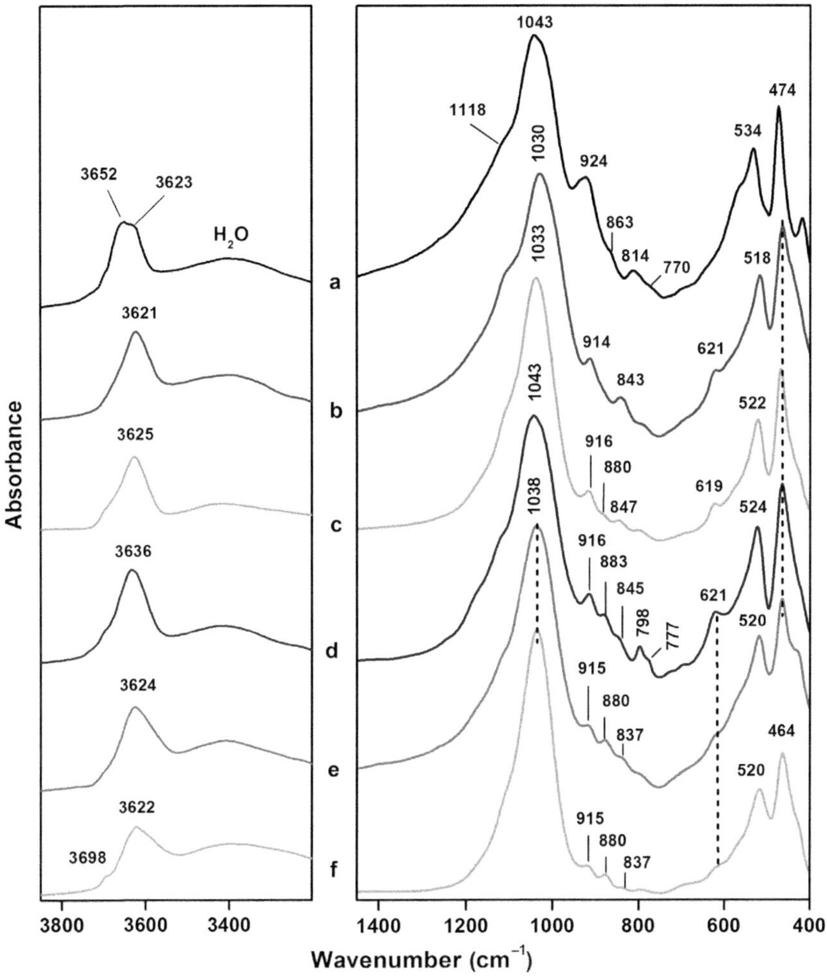


**FIG. 5.7** MIR spectra of Al-/Fe-smectites: (a) beidellite, (b) SAz-1, (c) JP, (d) SWy-2, (e) RH, (f) BLS.

of water molecules that were not released upon heating of KBr pellet at 120°C (Farmer, 1974b).

The OH bending region (950–800 $cm^{-1}$) is more sensitive to the occupancy of the octahedral sheets showing well-resolved OH bands for individual pairs of cations. A single band at 924 $cm^{-1}$ ($\delta(Al_2OH)$) in the spectrum of beidellite confirms the dominance of Al within octahedral sites (Fig. 5.7a). A weak inflexion near 863 $cm^{-1}$ was assigned to AlMgOH bending vibration by Gates (2005), which supports that some montmorillonite is present in this sample. A strong band at 814 $cm^{-1}$ with an inflexion near 770 $cm^{-1}$ was assigned to Al—O out-of-plane and Al—O—Si in-plane vibrations

**TABLE 5.2 Structural Formulae Per $O_{20}(OH)_4$ for the Smectites Given in Fig. 5.7 Calculated From the Chemical Analyses**

| | Tetrahedral | | Octahedral | | | |
|---|---|---|---|---|---|---|
| **Sample** | *Si* | *Al* | *Al* | *Fe* | *Mg* | **Reference** |
| Beidellite | 7.29 | 0.71 | 3.73 | 0.05 | 0.23 | Gates, 2005 |
| SAz-1 | 7.98 | 0.02 | 2.75 | 0.15 | 1.15 | Gates, 2005 |
| JP | 7.80 | 0.20 | 3.01 | 0.40 | 0.60 | Calculated from CA |
| SWy-2 | 7.80 | 0.20 | 3.05 | 0.42 | 0.51 | Gates, 2005 |
| RH | 7.87 | 0.13 | 2.50 | 0.77 | 0.74 | Gates, 2005 |
| BLS | 7.26 | 0.74 | 2.31 | 0.93 | 0.76 | Číčel et al., 1992 |

(Farmer, 1974b; Russell and Fraser, 1994). Two peaks at 914 ($\delta(Al_2OH)$) and 843 $cm^{-1}$ ($\delta$(AlMgOH)) for SAz-1 and at 916 and 847 $cm^{-1}$ for JP reflect substantial substitution of octahedral Al by Mg in these montmorillonites (Table 5.2). The spectrum of JP montmorillonite shows also a weak inflexion near 880 $cm^{-1}$ related to the $AlFe^{3+}OH$ (Fig. 5.7c). The slightly higher intensity of the $\delta(AlFe^{3+}OH)$ at 883 $cm^{-1}$, and the presence of $\delta$(AlMgOH) as only a shoulder near 845 $cm^{-1}$ (Fig. 5.7d), reflects the higher Fe and lower Mg content in SWy-2 compared to JP montmorillonite. A doublet at 798 $cm^{-1}$ and 777 $cm^{-1}$ indicates presence of quartz admixture in SWy-2. The spectra of smectites containing higher amounts of octahedral $Fe^{3+}$ (RH and BLS) have correspondingly lower intensity of the $\delta$(AlAlOH) band and higher intensity of the $\delta(AlFe^{3+}OH)$ band (Fig. 5.7e and f), which is in agreement with their structural formulae derived from chemical analysis (Table 5.2).

The Si—O stretching region of dioctahedral smectites, in contrast to pyrophyllite, exhibits only one broad complex band in the 1040–1030 $cm^{-1}$ range involving the contributions of in-plane Si—O—Si stretching vibrations (1070–970) and perpendicular Si—O vibration near 1100 $cm^{-1}$ (Farmer, 1974b). The broadness of this complex band reflects the increased bonding disorder in the smectite layers compared to the more crystalline pyrophyllite. The $Al^{VI}$—O—Si bending band ($Al^{VI}$–Al in octahedral positions) appears near 520 $cm^{-1}$ and Si—O—Si near 470 $cm^{-1}$ (Fig. 5.7). The band near 620 $cm^{-1}$ due to coupled Al—O out-of-plane and Si—O vibrations is characteristic for montmorillonites, while for smectites with higher $Fe^{3+}$ content, such as RH and BLS, only a weak shoulder can be recognised.

The MIR spectra of the dioctahedral ferruginous smectites and nontronites with dominant $Fe^{3+}$ in the octahedral positions (Table 5.3) show a quite

**TABLE 5.3 Structural Formulae Per $O_{20}(OH)_4$ for the Nontronites Given in Fig. 5.8 Calculated From the Chemical Analyses**

| | Tetrahedral | | | Octahedral | | | |
|---|---|---|---|---|---|---|---|
| **Sample** | *Si* | *Al* | *Fe* | *Al* | *Fe* | *Mg* | **Reference** |
| Ölberg | 7.84 | 0.16 | | 0.74 | 1.96 | 1.42 | Köster et al., 1999 |
| SWa-1 | 7.40 | 0.60 | | 1.10 | 2.62 | 0.25 | Gates, 2005 |
| NAu-1 | 6.98 | 0.96 | 0.06 | 0.35 | 3.62 | 0.04 | Gates, 2005 |
| NG-1 | 7.12 | 0.26 | 0.62 | 0.72 | 3.22 | 0.05 | Gates, 2005 |
| NAu-2 | 7.55 | 0.16 | 0.29 | 0.34 | 3.54 | 0.05 | Gates, 2005 |
| Spokane | 7.26 | 0.04 | 0.70 | 0.03 | 3.89 | 0.04 | Gates, 2005 |

different pattern of absorption bands (Fig. 5.8) than the smectites discussed previously (Fig. 5.7). The OH stretching band of nontronites appears in the 3573–3563 $cm^{-1}$ range, reflecting that $Fe_2^{3+}OH$ grouping is dominant in these Fe-rich samples. Broadening to higher frequency for Ölberg and SWa-1 (Fig. 5.8a and b) indicates that $Al_2OH$, AlMgOH and $AlFe^{3+}OH$ stretching vibrations contribute to this complex band (Madejová et al., 1994; Petit et al., 2002; Gates, 2005).

A high content of Fe affects also the shape of the OH bending region (Petit et al., 2002, 2015; Gates, 2005, 2008). Spectra of all nontronites show a sharp band in the 821–816 $cm^{-1}$ range assigned to $Fe^{3+}{}_2OH$ bending vibration. This band, together with OH stretching near 3570 $cm^{-1}$, is an important criterion for identification of nontronites in clay materials (Goodman et al., 1976; Stucki and Roth, 1976; Russell and Fraser, 1994; Gates et al., 2002; Gates, 2005). The OH bending region of most natural nontronites studied displays also a band near 870 $cm^{-1}$ due to $AlFe^{3+}OH$ vibration. Its intensity differs depending on the content of Al in the octahedral positions, and its position is related to the octahedral Fe content (Gates, 2008). The spectrum of Ölberg sample shows also a band at 765 $cm^{-1}$ assigned to $Fe^{3+}MgOH$ bending vibration, indicating a presence Mg in the octahedral positions. Moreover, Petit et al. (2002) found that the layer charge in this smectite arises mainly in the octahedral sheets. Therefore this smectite, in spite of having Fe as dominant octahedral cation, should be referred to as a Fe-montmorillonite (Petit et al., 2002). The spectra of nontronites with high $Fe^{3+}$ content, that is, NAu-1, NG-1, NAu-2 and Spokane, also show a band near 839–846 $cm^{-1}$. This band has been previously assigned to $Fe^{3+}{}_2OH$ bending vibration (Goodman et al., 1976; Gates, 2005), and its presence and intensity appears to be related to tetrahedral $Fe^{3+}$ in natural nontronites (Gates, 2008). However, Baron et al. (2016a)

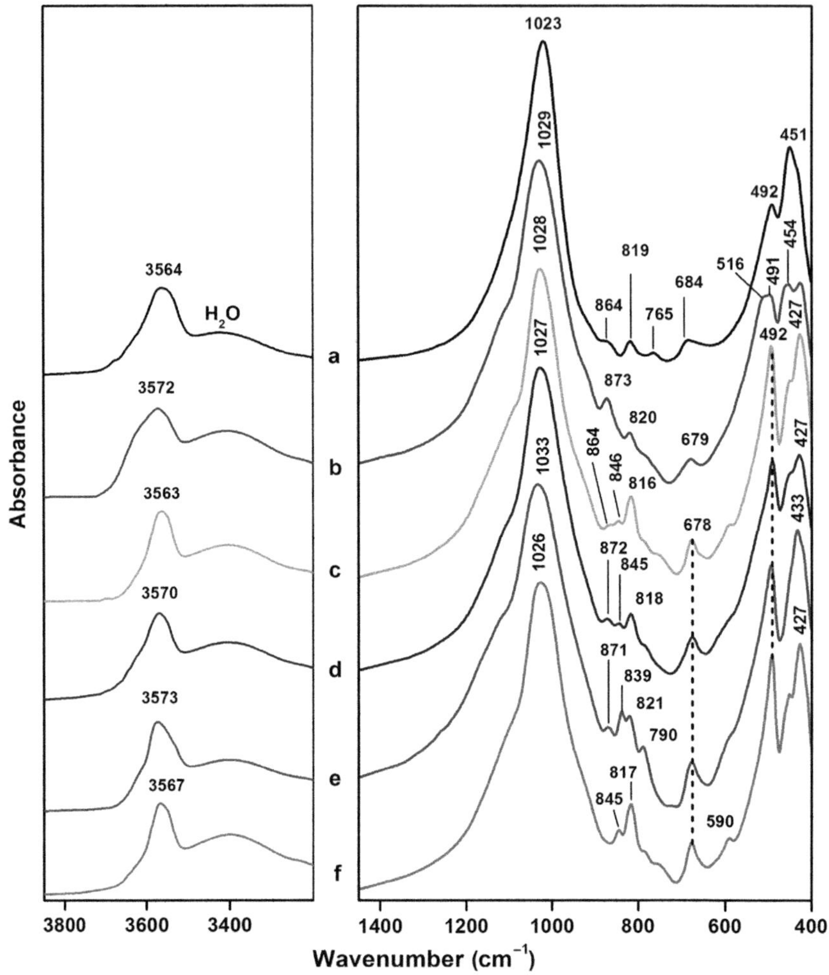


**FIG. 5.8** MIR spectra of nontronites: (a) Ölberg, (b) SWa-1, (c) NAu-1, (d) NG-1, (e) NAu-2, (f) Spokane.

have shown that this band is absent in the spectra of synthetic nontronites with more than 1 $Fe^{3+}$ per 8 Si. As such, its exact assignment is still questioned.

The Si—O stretching band of nontronites appears at slightly lower positions than in montmorillonites, for most samples below 1030 $cm^{-1}$, due to higher Si by $Al^{3+}$ substitution on the tetrahedral sheets (Fig. 5.8). The band at 684 $cm^{-1}$ for Ölberg Fe-montmorillonite and near 678 $cm^{-1}$ for all other nontronites is associated with Fe—O out-of-plane vibration (Farmer, 1974b). A band at 590 $cm^{-1}$, clearly resolved only in the spectrum of SPO nontronite, that is, a natural nontronite with the highest total Fe content per unit cell thus far known, is assigned to coupled $Fe^{3+}{}_{oct} - O_{ap} - Si_{tet}$

deformation (Fig. 5.8f). In the Si—O bending region, the $Fe^{3+}$—O—Si band near 490 $cm^{-1}$ occurs together with Si—O—Si bending bands near 450 and 430 $cm^{-1}$. A shoulder at 516 $cm^{-1}$ observed in the spectrum of SWa-1 is due to Al—O—Si bending vibration and confirms rather high content of Al in the octahedral positions (Fig. 5.8b).

The NIR spectra of Al-rich dioctahedral smectites show a complex band in the 7100–7050 $cm^{-1}$ range assigned to the first OH overtone ($2\nu$(OH)) of the structural OH groups and the first overtone of water molecules involved in weak hydrogen bonding (Fig. 5.9). If the $2\nu(H_2O)$ band is assumed to have

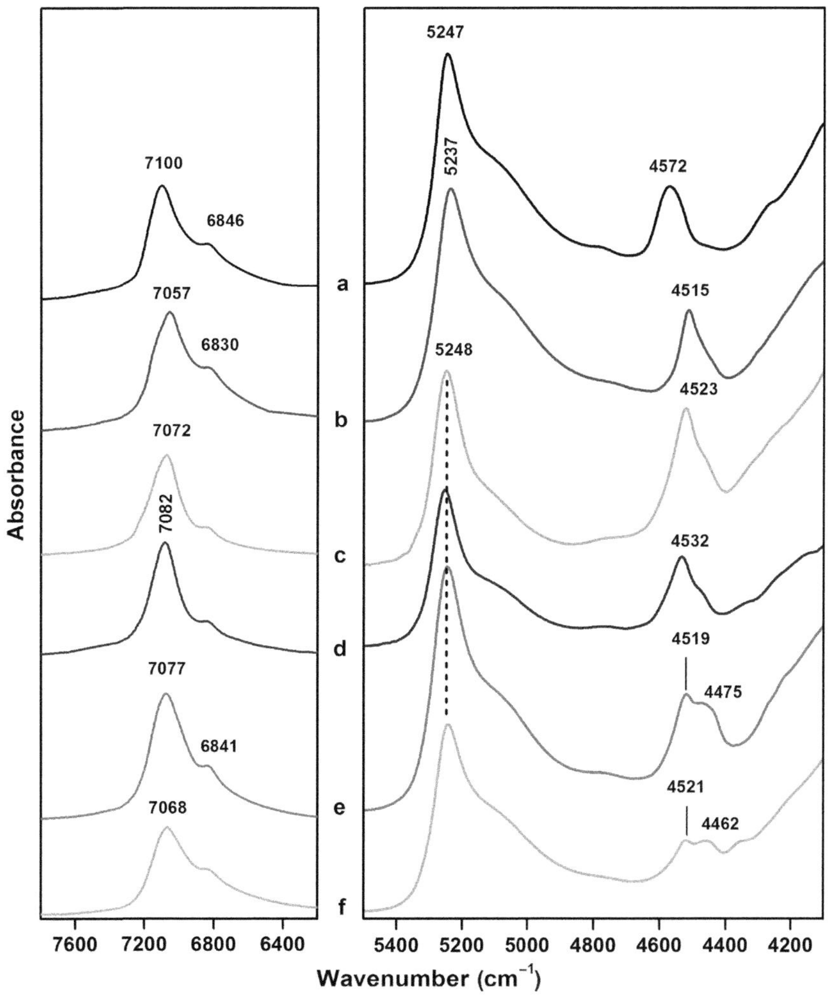


**FIG. 5.9** NIR spectra of Al-/Fe-smectites: (a) beidellite, (b) SAz-1, (c) JP, (d) SWy-2, (e) RH, (f) BLS.

the same (or very close) position, then the distinctions in the overtone band position reflect differences observed also in the positions of their $\nu$(OH) bands (Fig. 5.7). Therefore, for example, the $2\nu$(OH) position of beidellite is seen at 7100 $cm^{-1}$, while 7057 $cm^{-1}$ is observed for SAz-1 montmorillonite. A shoulder near 6840 $cm^{-1}$ is related to the first overtone of interlayer water molecules involved in strong hydrogen bonding. The broad and complex band near 5250 $cm^{-1}$ corresponds to the combination mode of the stretching and bending vibrations of water molecules engaged in both weak and strong hydrogen bonding.

The combination modes of the stretching and bending vibrations of the structural OH groups of dioctahedral smectites are observed in the region below 5000 $cm^{-1}$ (Fig. 5.9). Similar to that observed for OH bending, the combination region is very sensitive to the type and distribution of the cations in the octahedral sites. The relatively high position of the $(\nu+\delta)Al_2OH$ band of beidellite (4572 $cm^{-1}$) results, as expected, from relatively high values found for $\nu$(OH) (3652 $cm^{-1}$) and $\delta(Al_2OH)$ (924 $cm^{-1}$). The combination bands of SAz-1, JP and SWy-2 montmorillonites at 4515, 4523 and 4532 $cm^{-1}$, respectively, are complex envelopes involving overlapping contributions of $(\nu+\delta)Al_2OH$ near 4520 $cm^{-1}$, $(\nu+\delta)$AlMgOH near 4440 $cm^{-1}$, and $(\nu+\delta)AlFe^{3+}OH$ near 4470 $cm^{-1}$. In the NIR spectra of smectites containing more octahedral $Fe^{3+}$, that is, RH and BLS (Fig. 5.10e and f), the $AlFe^{3+}OH$ band at 4475–4462 $cm^{-1}$ is clearly resolved from 4520 $cm^{-1}$, confirming the increased content of Fe in octahedral sheets.

The position of the $2\nu$(OH) band appearing in the NIR spectra of nontronites (Fig. 5.10) reflects, similarly as observed for $\nu$(OH) band in the MIR region (Fig. 5.8), the high content of Fe in the octahedral sites. Samples with more Al, Ölberg and SWa-1 show the first OH overtone at 7050 and 7064 $cm^{-1}$, respectively, that is, positions closer to those found for Al-smectites. The broadening of the band towards lower wavenumbers, however, reveals the large contribution of $Fe^{3+}{}_2OH$ grouping. The band near 6960 $cm^{-1}$ is assigned to $2\nu(Fe^{3+}{}_2OH)$ (Madejová et al., 2000b), and it dominates in the spectra of nontronites for which the content of Fe is higher than 3.0 per structural unit (Table 5.3). The combination region of Fe-rich smectites shows a strong band near 4375 $cm^{-1}$ due to $(\nu+\delta)Fe^{3+}{}_2OH$. A shift of the main combination band to 4351 $cm^{-1}$ for Ölberg (Fig. 5.10a) results from the contribution of $(\nu+\delta)Fe^{3+}MgOH$ vibration absorbing near 4340 $cm^{-1}$. Gradually decreasing intensity of the $(\nu+\delta)AlFe^{3+}OH$ band near 4470 $cm^{-1}$ correlates with decreasing content of octahedral Al (Table 5.3).

#### *5.5.2.2 Trioctahedral Smectites*

Hectorite and saponite represent the trioctahedral subgroup of the smectites. Though Mg dominates the octahedral sites in these minerals, a partial substitution of $Li^+$ for $Mg^{2+}$ occurs in hectorite. Saponite has $Al^{3+}$ for $Si^{4+}$

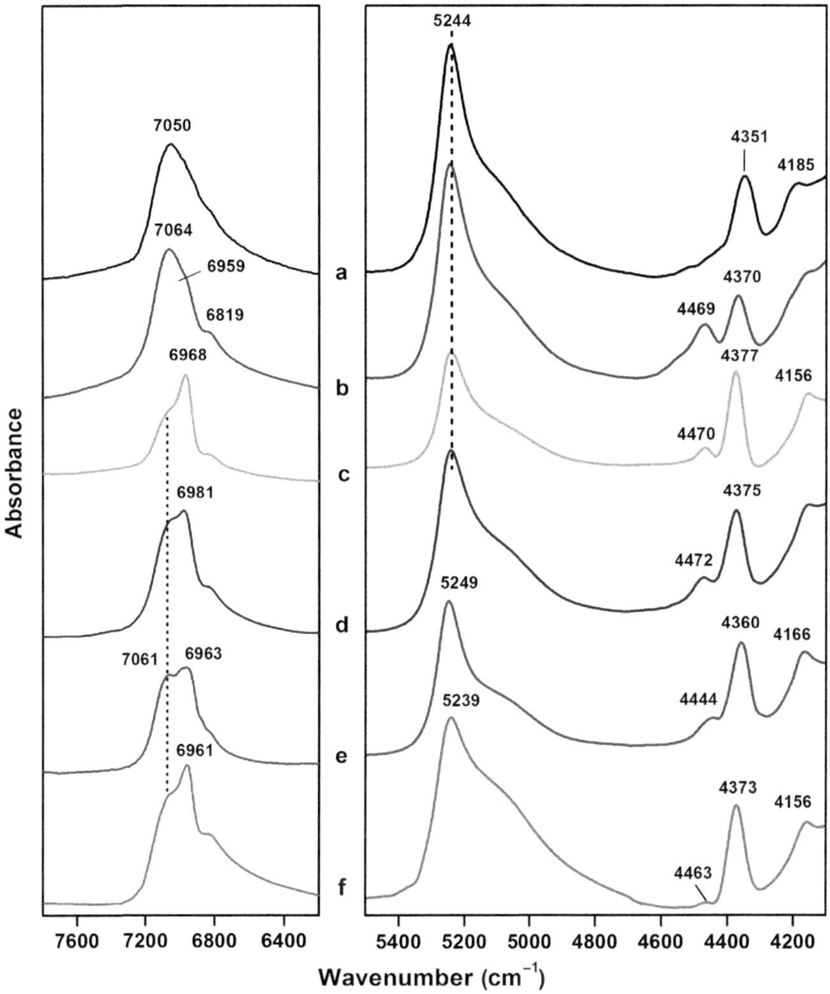


**FIG. 5.10** NIR spectra of nontronites: (a) Ölberg, (b) SWa-1, (c) NAu-1, (d) NG-1, (e) NAu-2, (f) Spokane.

substitutions in the tetrahedral sheets, as well as some $Al^{3+}$ for $Mg^{2+}$ substitution in the octahedral sheet. In part as a result of substitutions, the MIR spectra of both minerals show marked perturbation from that of talc; that is, the bands are broader and slightly shifted (Fig. 5.11a and b). The OH stretching region of hectorite (SHCa-1) and saponite (Ballarat) shows characteristically weak $\nu(Mg_3OH)$ band near 3680 cm$^{-1}$. The spectrum of saponite, in addition, reveals also a weak absorption near 3720 cm$^{-1}$. When a KBr pellet with saponite is heated (to remove adsorbed water), the highly localised negative charge on the Si—O surface attracts $K^+$ ions from KBr, which disturbs the

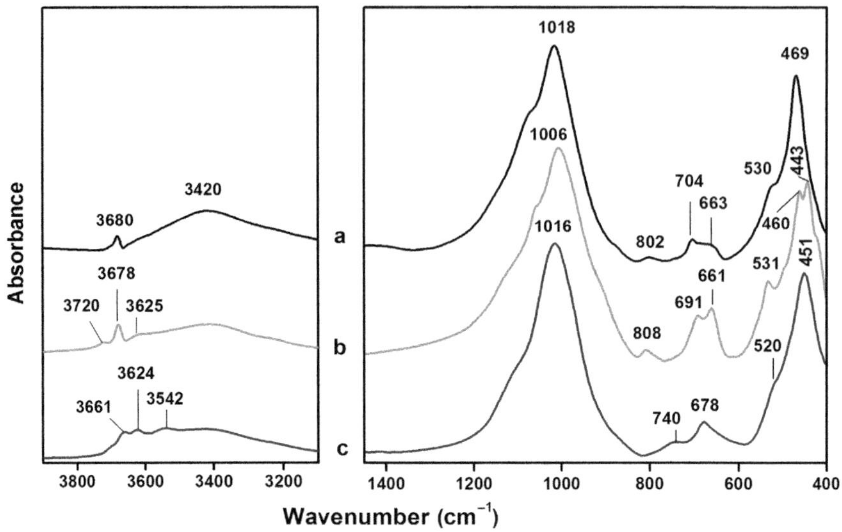


**FIG. 5.11** MIR spectra of (a) hectorite, (b) saponite, (c) griffithite.

orientation of the OH groups, and a high-frequency $\nu(Mg_3OH)$ satellite appears (Russell and Fraser, 1994; Pelletier et al., 1999). The main Si—O stretching band of hectorite is observed at 1018 cm$^{-1}$, but its position is shifted to a lower frequency for saponite (1006 cm$^{-1}$) due to Al for Si substitution (Fig. 5.11a and b). Similarly, as for talc (Fig. 5.6) the complex band with maxima at 704 and 663 cm$^{-1}$ for hectorite (Fig. 5.11a), and at 691 and 661 cm$^{-1}$ for saponite (Fig. 5.11b), involves the contributions of the perpendicular Si—O and bending $Mg_3OH$ vibrations. These bands are broadened as for the dioctahedral smectites due to local disorder. The absorption near 530 cm$^{-1}$ corresponds to the perpendicular Mg—O vibration. While the spectrum of saponites shows resolved bands at 460 and 443 cm$^{-1}$ due to the translation OH and Si—O—Si bending vibrations, respectively, the spectrum of hectorite reveals only one complex band.

The MIR spectrum of $<6$ μm fraction of the Fe-saponite 'griffithite' (from Griffith Park, California, United States, described in detail in Komadel et al., 2000) is also shown in Fig. 5.11. Griffithite is primarily an iron-rich saponite with dioctahedral domains with SF:

$$Li_{0.86}[Si_{6.67}Al_{1.33}]\,[Al_{0.09}Fe^{3+}{}_{1.48}Fe^{2+}{}_{0.15}Mg_{3.64}Mn_{0.07}]\,O_{20}(OH)_4.$$

The OH stretching region shows several well-resolved components (Fig. 5.11c). The absorption at 3661 cm$^{-1}$ due to $\nu(Mg_2Fe^{2+}OH)$ reflects a partial substitution of Mg by $Fe^{2+}$ in the octahedra. The absorption band at 3624 cm$^{-1}$ is supposed to arise from OH groups coordinated with $Mg_2Fe^{3+}$ atoms. However, Farmer and Russell (1971) reported for saponite and hectorite that OH stretching vibrations of water molecules weakly hydrogen bonded to the oxygens of the silicate lattice give a 3630 $cm^{-1}$ band. Thus both $\nu(Mg_2Fe^{2+}OH)$ and $\nu(H_2O)$ may contribute to the 3624 $cm^{-1}$ band in the spectrum of griffithite. The absorption band at 3542 $cm^{-1}$ is assigned to dioctahedral-like $\nu(MgFe^{3+}OH)$; however, vibrations of $Fe^{3+}{}_2OH$ groups, if present, would also contribute to this band. The spectrum of griffithite shows two strong bands at 1016 and 451 $cm^{-1}$ due to Si—O stretching and bending vibrations, respectively. The absorption band near 740 $cm^{-1}$ is probably associated with vibrations of Si—O—$Al^{IV}$ bonds. The asymmetric absorption band at 678 $cm^{-1}$ is the sum of two components corresponding to perpendicular Si—O and OH bending vibrations (Farmer, 1974b).

The NIR spectra of saponite and hectorite, and partly also griffithite, show that the trioctahedral character of these smectites can be clearly distinguished (Fig. 5.12). While in spectra of dioctahedral Al-rich smectites (Fig. 5.9) the structural OH stretching overtone cannot be resolved from the water band because they both appear near 7060 $cm^{-1}$, in the spectra of trioctahedral smectites the $2\nu(OH)$ occurs near 7190 $cm^{-1}$ (7178 $cm^{-1}$ for griffithite) and the water band in the 7085–7080 $cm^{-1}$ region, so they are easily distinguished from one another. The broad complex band in the 5252–5237 $cm^{-1}$ range corresponds to water molecules involved into H-bonds of different strength. The sharp band at 4328 $cm^{-1}$ in the spectrum of hectorite (Fig. 5.12a) corresponds to the combination of OH stretching (3678 $cm^{-1}$) and deformation (663 $cm^{-1}$) vibrations of $Mg_3OH$ groups. Two combination bands at 4361 and 4320 $cm^{-1}$, occurring in the spectrum of saponite (Fig. 5.12b), arise from combination of 3678+691 and 3678+660 $cm^{-1}$ vibrational modes,

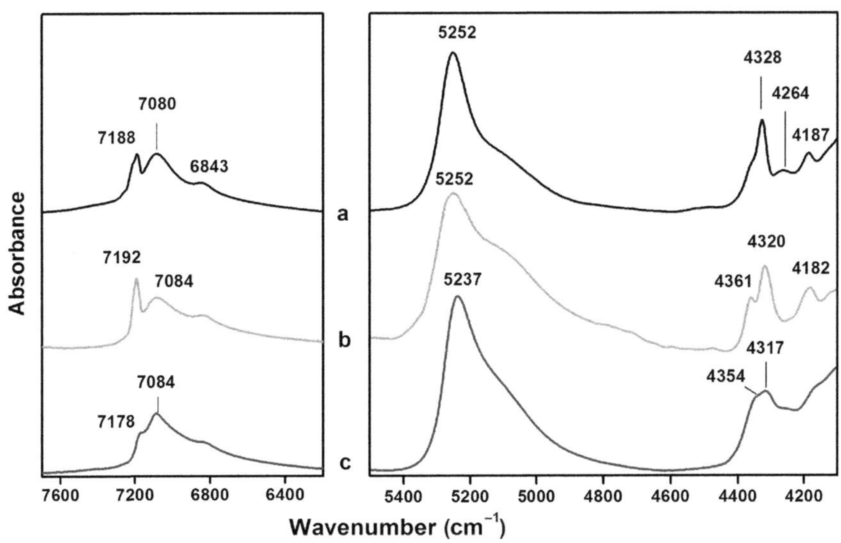


**FIG. 5.12** NIR spectra of (a) hectorite, (b) saponite, (c) griffithite.

respectively. The broad complex band near 4317 $cm^{-1}$ involves the variety of structural OH groups coordinated to different octahedral cations of griffithite (Fig. 5.12c).

### 5.5.3 Vermiculite, Illite and Micas

The 2:1 family of clay minerals, in addition to smectites, includes clay minerals with nonexpandable and nonhydrated interlayer spaces, including illites and dioctahedral or trioctahedral micas or vermiculites with expandability significantly reduced compared to smectites. Different chemistry, as well as the di- or trioctahedral occupancy, affects significantly the shape of their spectra (Farmer and Russell, 1964; Farmer et al., 1971; Farmer and Velde, 1973; Farmer, 1974b; Slonimskaya et al., 1986; Besson and Drits, 1997a,b; Kloprogge et al., 1999a; Zviagina et al., 2015; Heller-Kallai and Lapides, 2015; Singha and Singh, 2016).

Illites are dioctahedral minerals with variable chemical composition in the octahedral and tetrahedral sheets. The interlayer space of illite is mainly occupied by unhydrated potassium cations (~80%) responsible for the absence of expandability (Gualtieri et al., 2008). Some illites contain also $NH_4^+$ ions replacing $K^+$, the presence of which can be clearly identified by the $\delta$(N–H) vibration of $NH_4^+$ ions in the 1400–1430-$cm^{-1}$ region (Šucha et al., 1994, 1998; Petit et al., 1999a; Pironon et al., 2003). The MIR spectrum of illite presented in Fig. 5.13a resembles that of dioctahedral smectite (Fig. 5.7). Only a single band at 3623 $cm^{-1}$ without any shoulder is present in the OH stretching region. The 1400–400 $cm^{-1}$ region shows an intense complex band at 1030 $cm^{-1}$ related to Si—O stretching vibrations. The position of $\delta(Al_2OH)$ band at 916 $cm^{-1}$ is similar to those found for montmorillonites. A characteristic, though relatively weak band of illites at 831 $cm^{-1}$, was assigned to $Al^{IV}$—O out of plane vibration and/or to octahedral AlMgOH bending vibration and the 756 $cm^{-1}$ one to Al—O—Si in plane vibration (Farmer, 1974b). The doublet at 800 and 780 $cm^{-1}$ indicates quartz admixture in the sample.

Muscovite, a spectrum of which is given in Fig. 5.13b, belongs to dioctahedral Al-rich mica group of clay minerals. The octahedral positions are occupied by Al, while a pronounced Al for Si substitution also occurs in the tetrahedral sheets. Negative layer charge is compensated by potassium cations present mainly in the interlayer space. Muscovite, like all micas, is a nonexpandable clay mineral with no water in the interlayer space, thus having a fixed interlayer spacing. The OH stretching vibrations give a slightly asymmetric band at 3623 $cm^{-1}$ reflecting particular variability in the OH group environment. The high-frequency shoulder near 3659 $cm^{-1}$ is supposed to be related to OH groups close to $AlO_4$ tetrahedron while the peak at 3627 $cm^{-1}$ corresponds to OH groups with neighbouring $SiO_4$ tetrahedron. High extent of the tetrahedral substitution in muscovite broadens the Si—O stretching (1028 $cm^{-1}$) and OH bending (933 $cm^{-1}$) bands and generates the $Al^{IV}$—O (831 $cm^{-1}$) and $Al^{IV}$—O—Si (753 $cm^{-1}$) vibrations. The intensity of the latter

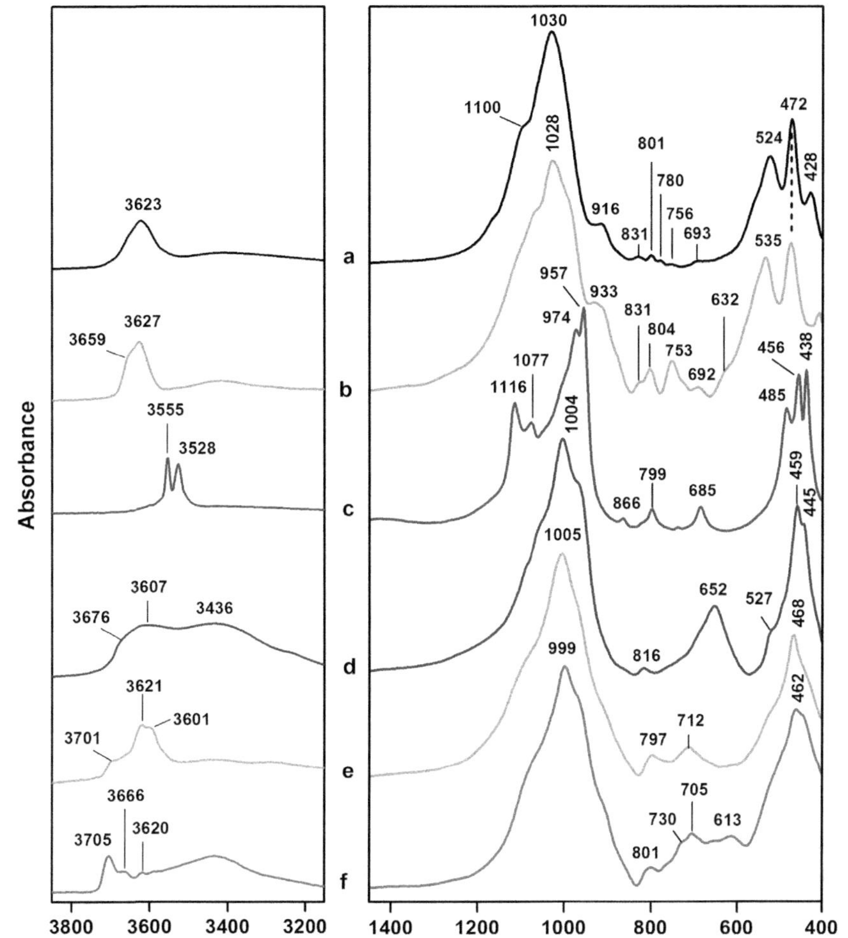


**FIG. 5.13** MIR spectra of (a) illite, (b) muscovite, (c) celadonite, (d) vermiculite, (e) biotite, (f) phlogopite.

band is frequently diagnostic for muscovite-like clay minerals (e.g. illite). The 524 $cm^{-1}$ band is due to $Al^{VI}$—O—Si bending vibration, and both 472 and 428 $cm^{-1}$ bands originate from the Si—O—Si bending vibrations.

Celadonite is a dioctahedral iron-rich mica with a layer charge derived from replacement of trivalent Al and Fe with divalent Mg and Fe cations in the octahedral sheet. Slonimskaya et al. (1986) showed that celadonite Zavalye (Fig. 5.13c) has the following structural formulae per half unit:

$$(K_{0.89}Ca_{0.10})\ (Si_{3.96}Al_{0.04})\ \left(Al_{0.05}Fe^{3+}{}_{0.96}Fe^{2+}{}_{0.26}Mg_{0.73}\right)\ O_{10}(OH)_2$$

The sharp bands observed in the MIR spectrum of celadonite point clearly to a highly ordered cation distribution in the octahedral sheets.

The OH stretching region shows two sharp bands at 3555 and 3528 $cm^{-1}$ assigned to $\nu(Fe^{3+}MgOH)$ and $\nu(Fe^{3+}Fe^{2+}OH)$ vibrations. Slonimskaya et al. (1986) used these bands together with OH stretching bands of other dioctahedral micas providing for the first time revised crystalochemical formulae of celadonites and glauconites based on quantitative IR analysis. The 1400–400 $cm^{-1}$ region (Fig. 5.13c) displays four well-resolved Si—O stretching bands at 1116, 1077, 974 and 957 $cm^{-1}$, the OH bending bands at 866 $cm^{-1}$ ($\delta(AlFe^{3+}OH)$) and 799 $cm^{-1}$ ($\delta(Fe^{3+}MgOH)$), the Fe—O vibration at 685 $cm^{-1}$ and the bands below 500 $cm^{-1}$ related to the mixed vibrations involving silicon–oxygen network and octahedral cations (Farmer, 1974b; Russell and Fraser, 1994).

Vermiculites are 2:1 clay minerals containing hydrated exchangeable cations compensating the negative layer charge. In comparison with smectites, the expandability of vermiculites is rather limited due to their high tetrahedral layer charge. The MIR spectrum of trioctahedral vermiculite shows a broad OH stretching band near 3607 $cm^{-1}$ (Fig. 5.13d). The width of the band as well as its position indicates substantial diversity of the OH group environment. The presence of $Mg_2Fe^{2+}OH$ and $MgFe^{2+}{}_2OH$ groups is highly probable. A shoulder near 3676 $cm^{-1}$ reflects the $Mg_3OH$ grouping. The broad band at 3436 $cm^{-1}$ corresponds to stretching vibrations of water molecules. The spectrum in the 1400–400 $cm^{-1}$ region resembles trioctahedral smectite–saponite. The Si—O stretching band is downward shifted in comparison with dioctahedral clay minerals, and the OH bending vibrations absorb at 652 $cm^{-1}$.

Biotite and phlogopite are trioctahedral micas with substantial variability in the composition of the octahedral and tetrahedral sheets. In phlogopites all octahedral sites are generally occupied, but in biotite the octahedra occupied with $Fe^{2+}$, Mg and Al are present together with some vacant octahedra. The MIR spectrum of biotite shows a complex pattern of the OH vibrations (Fig. 5.13e). The broad shoulder near 3700 $cm^{-1}$ involves contributions of $\nu(Mg_3OH)$, the inflexion at 3665 $cm^{-1}$ (revealed only by 2nd derivative approach) corresponds to $\nu(Mg_2AlOH)$ or $\nu(Mg_2Fe^{2+}OH)$. A peak at 3621 $cm^{-1}$ may be related to $\nu(MgAlOH)$, while the band at 3601 $cm^{-1}$ corresponds to stretching vibrations of the $MgFe^{3+}{}_2OH$ and/or $AlFe^{2+}OH$ groups. The OH stretching region of phlogopite (Fig. 5.13f) displays bands at 3705 $cm^{-1}$ ($\nu(Mg_3OH)$), 3666 $cm^{-1}$ ($\nu(Mg_2AlOH)$) and 3620 $cm^{-1}$ ($\nu(MgAlOH)$) which confirm presence of Al in the octahedral positions. The spectra of biotite and phlogopite are similar in the 1400–400 $cm^{-1}$ region with typical features of trioctahedral minerals. The substitutions of Al for Si in phlogophite and $Fe^{2+}$ for Mg in biotite have little effect on the position of the main Si—O stretching band near 1000 $cm^{-1}$, but there are significant differences in the 900–500 $cm^{-1}$ region where bands related to OH bending overlap with bands resulting from vibrations of the tetrahedral sheets (Fig. 5.13e and f).

The NIR spectrum of illite shows the $2\nu(OH)$ overtone band at 7071 $cm^{-1}$ and the structural OH combination band at 4500 $cm^{-1}$ (Fig. 5.14a). The assignment of the 4257 $cm^{-1}$ band has not been reported so far; however, taking into account results reported by Baron and Petit (2016), this band can be related to the combination of $\nu(OH)$ and $\nu(SiOAl)$ vibrations. The simple calculation 4257–3623 $cm^{-1}$ gives 634 $cm^{-1}$. No band at this position can be

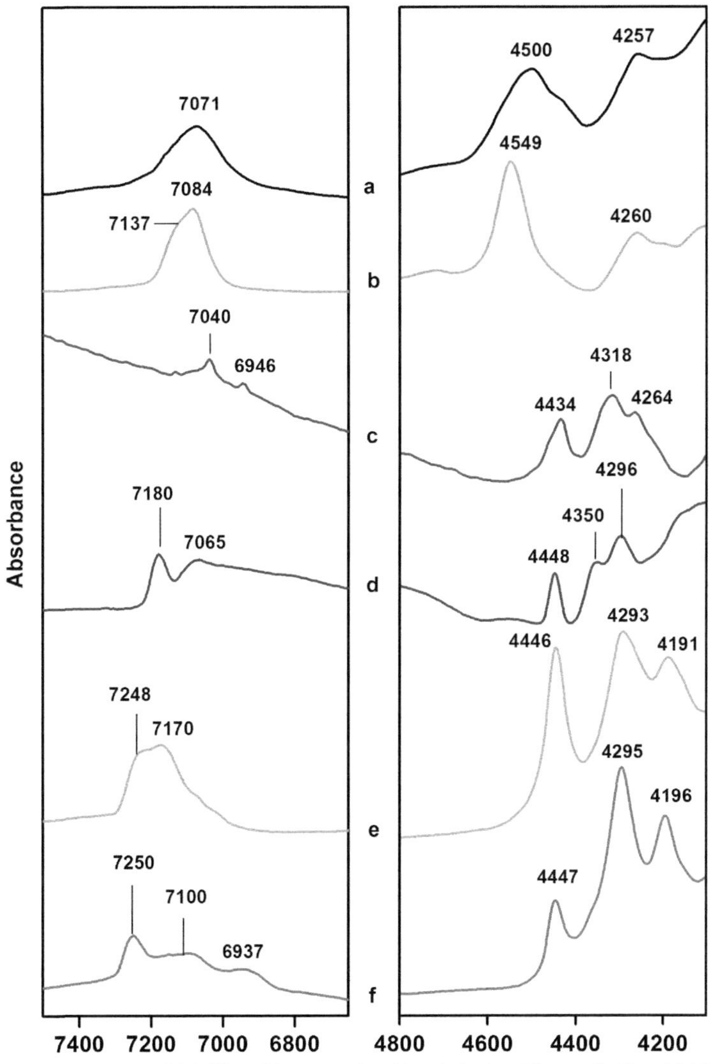


**FIG. 5.14** NIR spectra of (a) illite, (b) muscovite, (c) celadonite, (d) vermiculite, (e) biotite, (f) phlogopite.

recognised in the MIR spectrum of illite; however, the coupled Si—$O_{ap}$ and Al—O vibration near 620 $cm^{-1}$ was reported for dioctahedral Al-rich smectites (Farmer, 1974b). The first OH overtone (2$\nu$(OH)) of muscovite (Fig. 5.14b) appears at slightly higher position compared to illite, which correlates with higher $\nu$(OH) value observed in MIR region (Fig. 5.13b). The shoulder at 7137 $cm^{-1}$ represents the contribution of OH groups close to $AlO_4$ tetrahedra. The structural $(\nu+\delta)$OH combination mode gives rise to 4549 $cm^{-1}$ band. Similarly as for illite the 4260 $cm^{-1}$ band (Fig. 5.14b) is supposed to be due to combination of 3627 and 632 $cm^{-1}$-bands (Fig. 5.13b). High content of Fe and Mg in dioctahedral celadonite modified significantly the shape of its NIR spectrum (Fig. 5.14c). The sloping baseline due to the low-energy tail of the charge transfer bands occurring in the visible range in $Fe^{2+}$-minerals hinders to some extent a clear identification of the structural OH overtones. Nevertheless the weak bands at 7040 and 6946 $cm^{-1}$ are supposed to be dioctahedral 2$\nu$($Fe^{3+}$MgOH) and 2$\nu$($Fe^{3+}Fe^{2+}$OH) overtones, respectively, and the combination bands of structural OH groups appears below 4500 $cm^{-1}$.

The NIR spectra of trioctahedral vermiculite, biotite and phlogopite are reported in Fig. 5.14d–f. Presence of water molecules in vermiculite (Fig. 5.14d) is confirmed by the wide band at 7065 $cm^{-1}$ due to 2$\nu$($H_2O$) and the strong broad band at 5193 $cm^{-1}$ (not shown) related to the $(\nu+\delta)$ $H_2O$. The first overtone band of the structural OH groups appears at 7180 $cm^{-1}$. Using the calculations suggested by Petit et al. (2004b), the corresponding $\nu$($Mg_3$OH) should be at about 3675 $cm^{-1}$. In the MIR region of vermiculite, however, only one very broad band centred near 3607 $cm^{-1}$ can be observed; the position of weak inflexion at 3676 $cm^{-1}$ can be obtained only by 2nd derivative. In contrary, the corresponding overtone to the 3607 $cm^{-1}$ band should occur near 7043 $cm^{-1}$. This band, however, cannot be readily identified in the NIR due to it being strongly overlapping with a broad water band. Thus analysis of both the MIR and NIR regions are required to recognise the occupancy of the octahedral sheets of vermiculite. The OH combination region of vermiculite shows several overlapped bands of which the most intensive are bands at 4448 and 4296 $cm^{-1}$.

The OH stretching overtone region of biotite gives a broad band with two almost unresolved maxima at 7248 and 7170 $cm^{-1}$ (Fig. 5.14e). The corresponding calculated $\nu$(OH) values are 3700 $cm^{-1}$ ($\nu$($Mg_3$OH)) and 3667 $cm^{-1}$ ($\nu$($Mg_2$AlOH) and/or $\nu$($Mg_2Fe^{2+}$OH)), respectively, which fit quite well with the pattern in the MIR spectrum (Fig. 5.13e). On the other hand, the vibrations of dioctahedral $\nu$(MgAlOH) (3621 $cm^{-1}$) and $\nu$(Mg$Fe^{3+}$OH) (3601 $cm^{-1}$) resulting in well-resolved peaks in MIR (Fig. 5.13e) are not visible in the NIR region due to a broad low-frequency shoulder superimposed on the 2$\nu$(OH) overtone. These results show how complementary data are available when using both the MIR and NIR spectroscopy.

The OH stretching overtone region of phlogophite reveals a band at 7250 $cm^{-1}$ corresponding to $2\nu(Mg_3OH)$ overtone and a broad band centred near 7100 $cm^{-1}$ involving the contributions of $2\nu(Mg_3OH)$ and $2\nu(Mg_2AlOH)$ and $2\nu(MgAlOH)$ overtones (Fig. 5.14f). The positions of the bands present in the combination region of both micas (and partly also in vermiculite) are similar, but differences are obvious in their relative intensities. The exact assignment of the individual bands, however, has not yet been established, but they result from various combinations of $\nu(OH)$ with $\delta(OH)$ and/or $\delta(SiO)$.

## 5.5.4 Chlorites

The structure of chlorite is composed of regularly stacked, negatively charged 2:1 layers with a single layer of positively charged interlayer octahedra connected by H-bonds. Chlorites are usually trioctahedral, with $Mg^{2+}$, $Al^{3+}$, $Fe^{3+}$, and $Fe^{2+}$ on octahedral sites, tetrahedral cations are mainly $Si^{4+}$ and $Al^{3+}$. The interlayer octahedral sheet is composed of either trioctahedral (Mg or Mn) or dioctahedral arrangement of oxyhydroxides. Chlorites are divided into four subgroups: (1) trioctahedral chlorites, the most common, where both the interlayer and the 2:1 octahedral sheets are trioctahedral (e.g. clinochlore); (2) dioctahedral chlorites, where both the interlayer and the 2:1 octahedral sheets are dioctahedral; (3) ditrioctahedral chlorites, where the 2:1 octahedral sheet is dioctahedral and the interlayer sheet is trioctahedral (e.g. sudoite) and (4) tridioctahedral chlorites, where the interlayer sheet is dioctahedral and the 2:1 octahedral sheet is trioctahedral (Bailey, 1980a).

MIR spectra of trioctahedral Mg-chlorite—clinochlore, with general structural formula $(Si_3Al)(Mg_5Al)O_{10}(OH)_8$, ripidolite—Fe variety of clinochlore with SF

$$(Si_3Al)(MgFeAl)_6O_{10}(OH)_8$$

and ditrioctahedral Al-rich chlorite–sudoite with SF

$$(Si_3Al)(Mg_2Al_3)O_{10}(OH)_8$$

are presented in Fig. 5.15. The absorption bands in the OH stretching region arise from vibrations of OH groups belonging to both the 2:1 layer and interlayer hydroxide sheet (Farmer, 1974b; Russell and Fraser, 1994). The spectrum of clinochlore shows a broad OH stretching band at 3628 $cm^{-1}$ with weak inflexion near 3680 $cm^{-1}$ and a wide shoulder centred near 3450 $cm^{-1}$ (Fig. 5.15a). The absorption near 3680 $cm^{-1}$ is supposed to be related to the $Mg_3OH$ stretching vibration in the trioctahedral 2:1 layer while two lower-frequency bands arise from the OH stretching vibrations of the interlayer hydroxide sheet (Russell and Fraser, 1994). The $(Mg_3OH)$ absorption is well resolved at 3671 $cm^{-1}$ in the MIR spectrum of ripidolite (Fig. 5.15b). As a

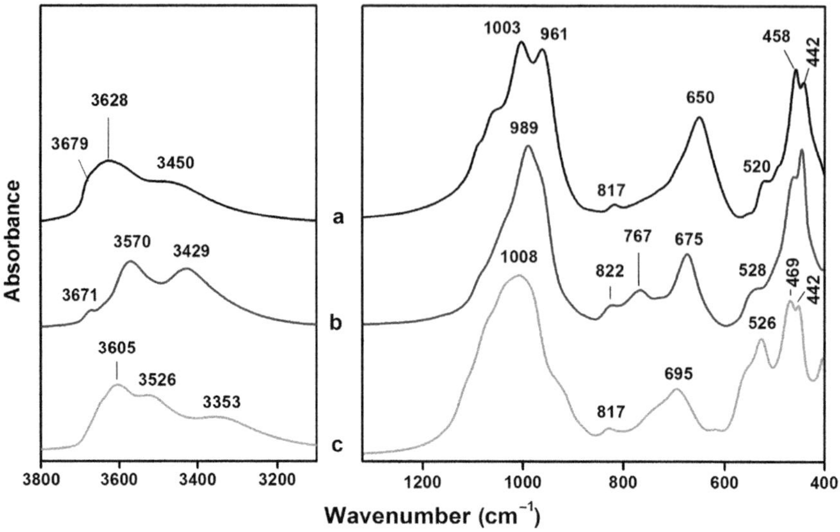


**FIG. 5.15** MIR spectra of chlorites: (a) clinochlore, (b) ripidolite, (c) sudoite.

result of higher octahedral iron content, the bands associated with the interlayer oxy-hydroxyl sheets are shifted to lower positions, that is, to 3570 and 3429 cm$^{-1}$. The MIR spectrum of sudoite gives a distinctive pattern with three bands at 3605, 3526 and 3353 cm$^{-1}$ (Fig. 5.15c). The high-frequency band at 3605 cm$^{-1}$ resembles that of vermiculite (Fig. 5.13d), while the other two bands correspond to OH vibrations of the interlayer hydroxide sheet. The presence of two bands associated with the interlayer hydroxyl sheets is clear evidence for two different OH—O distances (Farmer, 1974b).

The Si—O stretching region of clinochlore, in contrast to ripidolite and sudoite, shows the marked splitting of the Si—O band with components at 1003 and 961 cm$^{-1}$, reflecting the relatively low level of Al for Si substitution (Russell and Fraser, 1994). The absorption band near 820 cm$^{-1}$, observed in the spectra of all three chlorites, is related to the tetrahedral Al—O vibrations. A strong band at 700–600 cm$^{-1}$ arises from OH bending vibrations in both 2:1 layer and in the hydroxide sheets, and its position appears to be sensitive to octahedral composition (Farmer, 1974b).

The NIR spectrum of clinochlore shows only one sharp band at 7198 cm$^{-1}$ assigned to the first overtone of the structural OH stretching mode of $Mg_3OH$ (Fig. 5.16a). A similar position for $Mg_3OH$, at 7185 cm$^{-1}$, is observed for trioctahedral talc (Fig. 5.6b). In contrast, the spectrum of ripidolite, containing also Fe, reveals three bands at 7178, 7110 and 6488 cm$^{-1}$ (Fig. 5.16b). Comparable positions (7185, 7115 and 6500 cm$^{-1}$) are reported by Ferrage et al. (2003a) for chlorite samples containing also iron in the octahedral positions. Based on their results, the 7178 cm$^{-1}$ band is assigned to $2\nu(Mg_3OH)$, the

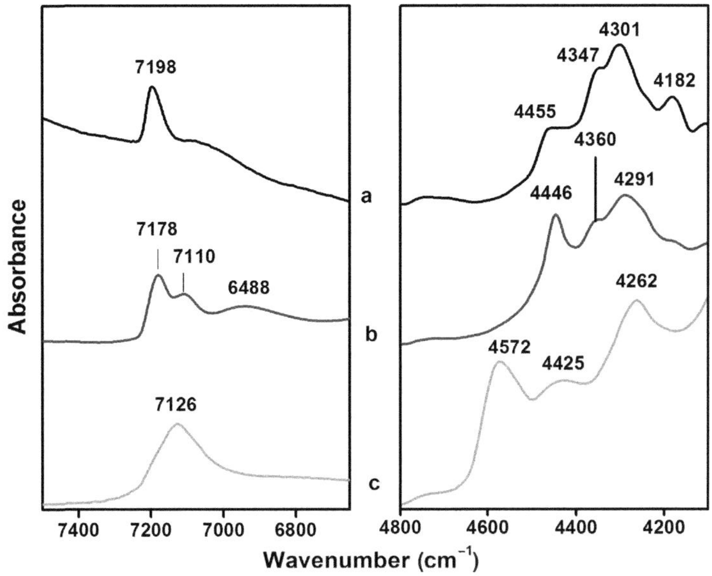


**FIG. 5.16** NIR spectra of chlorites: (a) clinochlore, (b) ripidolite, (c) sudoite.

7110 cm$^{-1}$ to the 2$\nu$($Mg_2AlOH$) of the interlayer hydroxide sheet and the 6488 cm$^{-1}$ band is attributed to overtones of OH stretching fundamental modes of (SiAl)O—OH. The position of the 2$\nu$(OH) band at 7126 cm$^{-1}$ in the spectrum of sudoite may reflect the presence of Al in the dioctahedral positions (Fig. 5.16c). The shape of the bands observed in the spectra of clinochlore and ripidolite below 5000 cm$^{-1}$ resemble to some extent those of trioctahedral micas, while dioctahedral sudolite is closer to dioctahedral illite or muscovite. The exact assignments of the individual bands, however, are still disputed.

## 5.6 PALYGORSKITE, SEPIOLITE

Clay minerals from the palygorskite–sepiolite group contain a modulated structure. The modulation is due to the periodic inversion of apical oxygen atoms of the tetrahedra in every two (palygorskite) or three (sepiolite) silicate chains, causing a periodic discontinuity of the octahedral sheet and the formation of alternating ribbons (modules) and tunnels/channels running along the fibre axis. These clay minerals exhibit high internal and external specific surface areas decorated by various types of $H_2O$ such as bound water molecules directly coordinated to Al or Mg ions at the edges of the 2:1 ribbon layers, unbound zeolitic water molecules within the channels, $H_2O$ adsorbed on the external surface and physisorbed water. While these clay minerals are non-swelling, because of their modular structure, they can fold on desiccation.

The structures of palygorskite and sepiolite have been described in detail in many papers (e.g. Brindley and Brown, 1980; Jones and Galan, 1988; Guggenheim and Eggleton, 1988). Comprehensive review of different structural modes has been provided by Brigatti et al. (2013). Galán and Carretero (1999) reviewed the literature on their chemical analyses, including bulk chemical analyses, and energy dispersive X-ray (EDX) analyses of selected sepiolite and palygorskite samples. Their assessment indicates that sepiolite is a true trioctahedral mineral with eight octahedral positions filled by $Mg^{2+}$ and negligible structural substitutions. The structure of palygorskite, on the other hand, has both di- and trioctahedral character. The octahedral sheet of palygorskite contains mainly $Mg^{2+}$, $Al^{3+}$ and $Fe^{3+}$, with an $R^{2+}/R^{3+}$ ratio close to 1, in which four of the five structural positions are occupied. Chahi et al. (2002) found the evidence of both dioctahedral and trioctahedral features for Mg-rich palygorskite from Ganntour (Morocco). In contrast, Serna et al. (1977a), basing their work on IR spectra, described Georgia palygorskite as completely dioctahedral with octahedral edges of the ribbons occupied only by Mg cations.

IR spectroscopy has an important role in structural understanding (site occupancy, distribution of cations, di- or trioctahedral character) of different palygorskite and sepiolite samples (Serna et al., 1977a; Blanco et al., 1989; Russell and Fraser, 1994; Madejová and Komadel, 2001; Chahi et al., 2002). Second derivative NIR spectroscopic analysis of the bands position has been successfully used to study in detail the octahedral site occupancies in palygorskites and sepiolites of different origin, as well as to follow the structural changes upon minerals drying (Gionis et al., 2006, 2007; Chryssikos et al., 2009; Stathopoulou et al., 2011; Bukas et al., 2013). Therefore in the following part only a basic description of the MIR and NIR bands of palygorskite and sepiolite is given.

The MIR spectra of aluminous palygorskite from Florida (PFl-1), iron-rich palygorskite from Macedonia, Greece (GR-1) and sepiolite from Valdmore, Spain (SepSp-1) are presented in Fig. 5.17. The spectra were measured in KBr pellets at room temperature (Fig. 5.17a–c) and after heating KBr pellets for 24 h at 150°C (Fig. 5.17a*–c*). The OH stretching region of unheated PFl-1 reveals absorption bands at 3614, 3546 and 3423 $cm^{-1}$ and a shoulder at 3581 $cm^{-1}$ (Fig. 5.17a). Similar positions were reported previously for palygorskites (e.g. Russell and Fraser, 1994; Chahi et al., 2002; Gionis et al., 2006). After heating the sample at 150°C, the complex OH stretching band significantly decreases in intensity due to removal of physisorbed and partly also zeolitic water (Fig. 5.17a*). Moreover, the shape of the $\nu$(OH) band is also changed. A sharp doublet at 3623 and 3610 $cm^{-1}$ is observed in the spectrum instead of the 3614 $cm^{-1}$ band. Weak shoulder at 3581 $cm^{-1}$ is transformed into a well-resolved band at 3580 $cm^{-1}$, the intensity of the band near 3540 $cm^{-1}$ decreases and a new band appears at 3517 $cm^{-1}$ (Fig. 5.17a*). In agreement with chemical composition of PFl-1

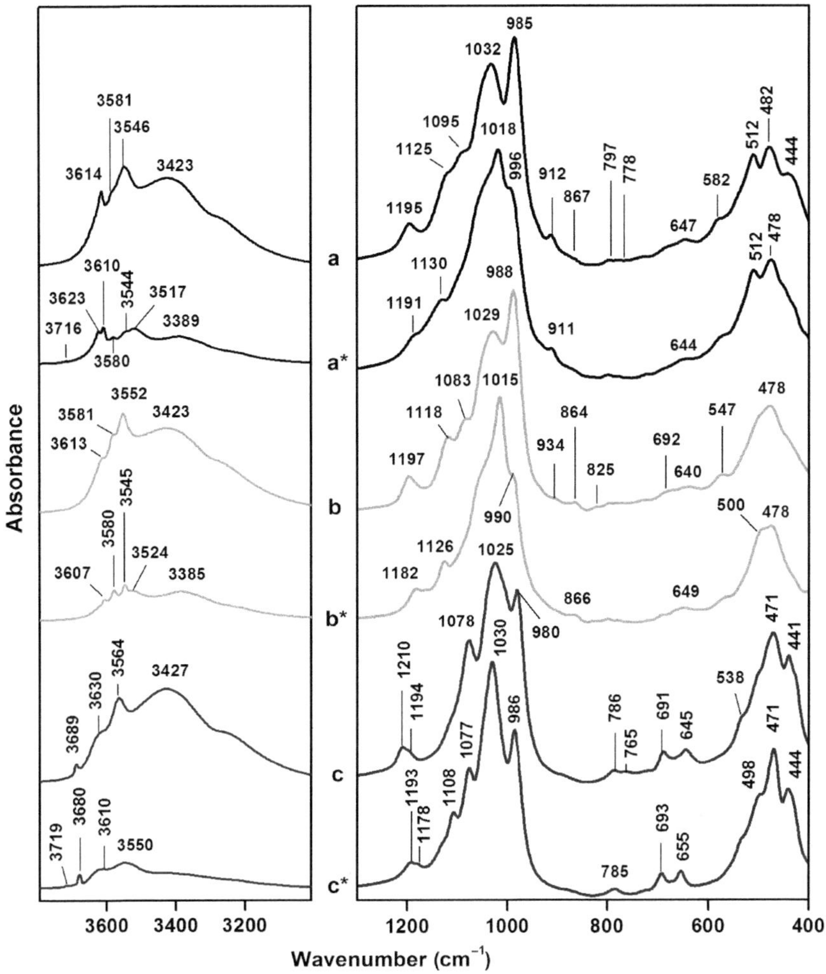


**FIG. 5.17** MIR spectra of unheated: (a) palygorskite PFl-1, (b) palygorskite GR-1, (c) sepiolite SepSP-1 and MIR spectra of (a*) palygorskite PFl-1, (b*) palygorskite GR-1, (c*) sepiolite heated for 24 h at 150°C.

(Mermut and Cano, 2001a), and the assignments suggested by Russell and Fraser (1994), Chahi et al. (2002), and Gionis et al. (2006), the 3614 $cm^{-1}$ band is related to the $\nu(Al_2OH)$. The splitting of the $\nu(Al_2OH)$ band into 3623 and 3610 $cm^{-1}$ components is due to the structural modifications of the palygorskite fibres upon heating ('folding structure') due to the loss of zeolitic and coordinated water (Presinger, 1963). The 3580 $cm^{-1}$ band can be assigned to $\nu(AlFe^{3+}OH)$; however, because its shape is modified upon heating, the $\nu(OH)$ of bound water is likely to occur close to (or at) this wavenumber

(Chahi et al., 2002). The bands observed in the 3550–3500 $cm^{-1}$ region for unheated and heated PFl-1 are mainly because of bound water molecules; however, contributions from $\nu(Fe^{3+}{}_2OH)$ cannot be excluded (Gionis et al., 2006). The unbound zeolitic and physisorbed water molecules absorb below 3450 $cm^{-1}$ (Russell and Fraser, 1994). In addition to the previously discussed bands, the zoomed spectrum of heated PFl-1 supported by the 2nd derivative analysis (not shown) reveals also a new weak band at 3716 $cm^{-1}$, which is related to free (i.e. not bonded to water molecules) surface silanol groups (Serna et al., 1977a; Gionis et al., 2006).

The OH stretching region of unheated iron-rich palygorskite GR-1 (Fig. 5.17b) displays the complex OH absorption band with overlapping absorptions at similar positions as for PFl-1. Their intensities, however, are different due to higher Fe content in GR-1 than PFl-1 (Gionis et al., 2006). As a result of different chemistry, the intensity of the $\nu(Al_2OH)$ band at 3613 $cm^{-1}$ of GR-1 is lower and those of at 3581 and 3552 $cm^{-1}$ higher than PFl-1, supporting their assignment to $\nu(AlFe^{3+}OH)$ and $\nu(Fe^{3+}{}_2OH)$, respectively. Upon heating, the overall intensity of the complex $OH/H_2O$ stretching band of GR-1 decreases due to the removal of bound and unbound (zeolitic and physisorbed) $H_2O$ (Fig. 5.17b*). Compared to PFl-1, no splitting of the $\nu(Al_2OH)$ band is observed; instead it shifts to 3607 $cm^{-1}$ indicating partial modification of the GR-1 structure. Decrease of the complex OH band intensity together with the appearance of the weak bands at 3545 and 3524 $cm^{-1}$ (similarly as found for PFl-1) reflects the reduction of the bound water molecules content.

The MIR spectrum of unheated trioctahedral sepiolite (Fig. 5.17c) shows a sharp $\nu(Mg_3OH)$ band at 3689 $cm^{-1}$, slightly higher position than that of talc (Fig. 5.5). A shoulder near 3630 $cm^{-1}$ and a strong band at 3564 $cm^{-1}$ are due to water molecules bound to Mg at the edges of the discontinuous octahedral sheet inside the tunnels (Russell and Fraser, 1994; Bukas et al., 2013). In the spectrum of heated sepiolite, similar changes as for palygorskites are observed, that is, pronounced reduction of the complex OH stretching band intensity followed by partial modifications of the individual bands positions. The appearance of the weak band at 3719 $cm^{-1}$, unambiguously confirmed by the 2nd derivative, confirms the creation of free SiOH groups upon SepSP-1 drying.

The 2:1 ribbon layer structure of palygorskite and sepiolite affects also the shape of their MIR spectra in the 1300–400-$cm^{-1}$ region. The Si—O stretching region of both unheated palygorskites reveals a complex band with five more or less resolved absorptions at 1195, 1125, 1095, 1032 and 985 $cm^{-1}$ for PFl-1 and at 1197, 1118, 1083, 1029 and 988 $cm^{-1}$ for GR-1 (Fig. 5.17a and b). The band near 1200 $cm^{-1}$ is characteristic for ribbon structure, and is one of the highest recorded for clay minerals. The ~1095 and ~640-$cm^{-1}$ bands are attributed to the asymmetric and symmetric stretching vibration of Si—O—Si bonds linking the 2:1 ribbons along their edges (Russell and Fraser, 1994; Gionis et al., 2006). The bands in the 1100–980 $cm^{-1}$ region are related to the antisymmetric stretching modes of the tetrahedral silicate sheet. The $\nu(Al_2OH)$, and $\nu(AlFe^{3+}OH)$, appear at 912 and 867 cm$^{-1}$ for PFl-1 and at 934 and 864 cm$^{-1}$ for GR-1. The spectrum of unheated iron-rich palygorskite GR-1 shows also shoulders near 825 and 692 cm$^{-1}$. The absorption bands of similar, though slightly lower, wavenumbers are observed for iron-rich smectites (Fig. 5.8) and are associated with $Fe^{3+}{}_2OH$ bending and $Fe^{3+}$—O out-of-plane vibrations. Some quartz (797 and 778 cm$^{-1}$) and possibly feldspar (582 cm$^{-1}$) admixtures are revealed in the MIR spectrum of PFl-1. The vibration bands in the 550–400 cm$^{-1}$ region correspond to the mixed Al—O—Si and Si—O—Si bending vibrations of Al-rich palygorskite (Fig. 5.17a), while the complex band centred at 478 cm$^{-1}$, due to lattice vibrations, reflects higher Fe content in GR-1.

The removal of water molecules upon heating the PFl-1 and GR-1 KBr pellets at 150°C has a pronounced effect on the shape of the complex Si—O stretching band (Fig. 5.17a*,b*). The individual components are slightly shifted to higher or lower wavenumbers, and the resolved peaks near 1030 and 985 cm$^{-1}$ are merged into one broad band near 1015 cm$^{-1}$. In the region below 950 cm$^{-1}$, only negligible changes in the MIR spectra of both palygorskite samples are observed.

The MIR spectrum of the unheated SepSP-1 (Fig. 5.17c) has several similar features in the 1300–400 region as found for palygorskites (Fig. 5.17a and b), but due to trioctahedral character some differences can be recognised as well. The strongest IR bands of sepiolite observed around 1000 cm$^{-1}$ are attributed to the antisymmetric stretching modes of the silicate tetrahedra. The in-plane Si—O modes are supposed to be located at 1025 and 980 cm$^{-1}$, out-of-plane mode at 1078 cm$^{-1}$. In addition, a doublet at 1210 and 1194 cm$^{-1}$ is associated with the Si—O—Si linkages between the ribbons of the modulated 2:1 layer (Bukas et al., 2013). The origin of the 786 and 765 cm$^{-1}$ bands is unclear yet; most probably they are related to Si—O—Si modes. The IR bands in the 700–400 cm$^{-1}$ region are typical for trioctahedral saponite (Fig. 5.11b). The 691 cm$^{-1}$ band is attributed to an out-of-plane Si—O mode, the 645 cm$^{-1}$ to bending vibration of the $Mg_3OH$ groups. By analogy with saponite, the absorption near 538 cm$^{-1}$ corresponds to the perpendicular Mg—O vibration, while the bands at 471 and 441 cm$^{-1}$ to the translation OH and Si—O—Si bending vibrations, respectively.

The release of water upon SepSP-1 heating imparts small shifts of the $\nu$(Si—O) bands positions. In contrast to PFl-1 and GR-1, two most intense Si—O bands of sepiolite are not merged into one complex band, but importantly, their intensities invert. No significant changes in the bands positions or intensities are visible for heated SepSP-1 (Fig. 5.17c*) compared to the unheated one in the region below 700 cm$^{-1}$ (Fig. 5.17c).

The NIR spectra of palygorskites and sepiolite are presented in Fig. 5.18. The assignment of the bands is based mainly on Gionis et al. (2006) and Bukas et al. (2013). As mentioned in the experimental part, to obtain the positions of the weak shoulders or bands, the second derivative was used. Several

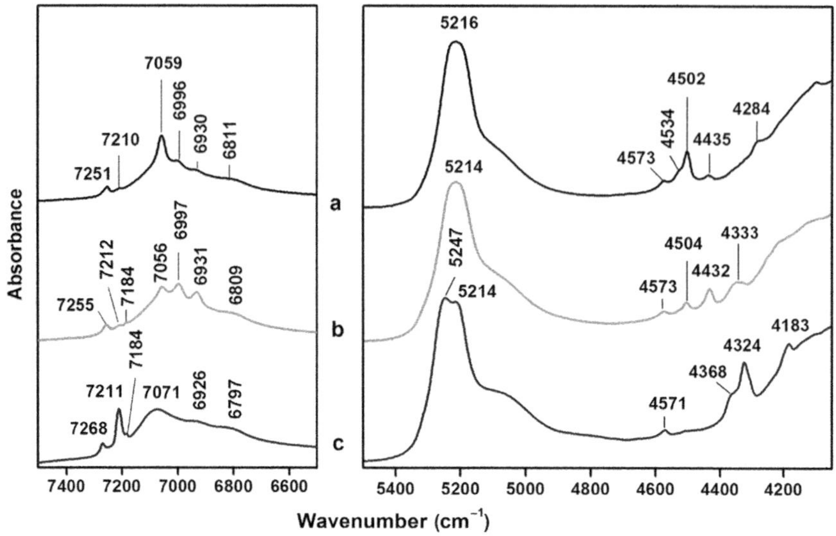


**FIG. 5.18** NIR spectra of unheated: (a) palygorskite PFl-1, (b) palygorskite GR-1, (c) sepiolite.

sharp features due to structural OH stretching overtones are superimposed with a broad and complex envelope of $H_2O$ stretching overtones in the 7500–6500 cm$^{-1}$ region. The spectrum of PFl-1 (Fig. 5.18a) shows a dominant band at 7059 cm$^{-1}$ and two weak bands at 6996, and 6930 cm$^{-1}$. In the case of $Fe^{3+}$-rich Gr-1 these bands are of nearly the same intensity (Fig. 5.18b). These bands reflect the occupancy of the dioctahedral sheets of palygorskites by Al and $Fe^{3+}$. Thus the band near 7059 cm$^{-1}$ was assigned to $2\nu(Al_2OH)$, the 6996 cm$^{-1}$ to $2\nu(AlFe^{3+}OH)$, and 6930 cm$^{-1}$ to $2\nu\left(Fe^{3+}{}_2OH\right)$ and zeolitic water. The weak band near 6810 cm$^{-1}$ is due to the first $H_2O$ overtone of water ($2\nu(H_2O)$). Both samples exhibit also a weak but sharp band near 7250 cm$^{-1}$, weak shoulder near 7210 cm$^{-1}$, whereas GR-1 shows also a small shoulder at 7184 cm$^{-1}$. Gionis et al. (2006) assigned both weak absorptions to the $2\nu(Mg_3OH)$. While the position at 7188 cm$^{-1}$ was found to be diagnostic for $Mg_3OH$ group in trioctahedral hectorite (Fig. 5.12a) or talc (Fig. 5.6b), a sharp band at 7211 cm$^{-1}$ is characteristic for trioctahedral sepiolite (Fig. 5.18c) and should be also related to $2\nu(Mg_3OH)$. Based on the NIR spectra solely, however, it is not possible to determine whether the $Mg_3OH$ features detected in both PFl-1 and GR-1 are related to clay admixtures or to trioctahedral domains present in palygorskites. The high-frequency region of all three samples shows a weak but well-resolved band at 7251–7268 cm$^{-1}$ corresponding to the first stretching overtone of the SiOH groups occurring at outer surfaces. The only mineral showing the $\nu$(SiOH) also in the MIR region is sepiolite (Fig. 5.17c). Broad overlapping bands below 7100 cm$^{-1}$ present in the spectrum of sepiolite correspond to the first overtones of various forms of water molecules.

The most intense band observed in the 5500–4000 $cm^{-1}$ region of palygorskites and sepiolite belongs to the combination modes of water molecules (Fig. 5.18). While palygorkites exhibit only one broad band near 5216 (5214) $cm^{-1}$, the sepiolite spectrum reveals two bands at 5247 and 5214 $cm^{-1}$ related to coordinated and zeolitic water (Gionis et al., 2006; Bukas et al., 2013). The combination modes of the structural hydroxyl groups appear below 4600 $cm^{-1}$. All three samples showed a weak band at almost the same position, that is, at 4573 $cm^{-1}$ for palygorskites and 4571 $cm^{-1}$ for sepiolite, corresponding to the $(\nu+\delta)$SiOH combination (Fig. 5.18). The spectrum of dioctahedral PFl-1 is dominated by a rather strong band at 4502 $cm^{-1}$ with shoulder at 4534 $cm^{-1}$. Weaker bands are observed at 4435 and 4284 $cm^{-1}$ (Fig. 5.18a). Similar features can also be observed in the spectrum of GR-1 with the strongest band at 4432 $cm^{-1}$, and comparing to PFl-1 less intense band at 4504 $cm^{-1}$. Moreover, a broad band near 4333 $cm^{-1}$ is clearly visible for GR-1 (Fig. 5.18b). The 4502 (4504) $cm^{-1}$ band can be related to the $(\nu+\delta)Al_2OH$, and at 4435 (4432) $cm^{-1}$ to the $(\nu+\delta)AlFe^{3+}OH$. A broad band at 4333 $cm^{-1}$ represents the envelope involving most probably the $(\nu+\delta)Fe^{3+}{}_2OH$ and $(\nu+\delta)$ $MgFe^{3+}OH$ modes found near 4370 and 4330 $cm^{-1}$ for dioctahedral smectites (Gates, 2005). Due to its trioctahedral character, sepiolite exhibits a rather different pattern of the bands in the OH combination region. The positions, as well as intensities, of the bands are very similar to those observed for trioctahedral saponite (Fig. 5.12b). Two combination bands at 4368 and 4324 $cm^{-1}$, occurring in the spectrum of sepiolite, arise from combination of (3680+693 $cm^{-1}$) and (3680+655 $cm^{-1}$) vibrational modes, respectively. Finally, more or less resolved bands appearing in the NIR spectra of palygorskites and sepiolite below 4300 $cm^{-1}$ result from different combinations of the structural OH groups and Si—O—Si vibrations.

## 5.7 CONCLUSIONS

This chapter demonstrated the potential usefulness of IR spectroscopy for the identification and chemical and structural characterisation of the clay minerals. Detailed interpretation of the IR spectra of clay minerals measured in traditional MIR, but also in the less common NIR region, is a relatively simple way to gain insight into octahedral, tetrahedral and hydroxyl structural units of clay minerals.

Thanks to pronounced and continuous development in instrumentation and advanced computational methods, the understanding of spectral response to changes in clay mineral crystal chemistry is currently more fully developed. The potential of IR spectroscopy in assisting in solving problems related to clay mineral chemistry, as well as the numerous industrial applications of clay minerals, is certainly not exhausted. Continued progress in experimental as well as theoretical/computational methods will further extend its utilisation in clay minerals studies and can resolve yet unanswered questions.

Chapter 6

# Raman Spectroscopy of Clay Minerals

J.T. Kloprogge

## 6.1 INTRODUCTION

Raman spectroscopy is a relatively old technique based on theoretical predictions for the existence of inelastic light scattering reported in 1923 by Smekal (1923) and experimentally proven by Sir C.V. Raman in 1928 (Raman and Krishnan, 1928). However, it took much longer before Raman spectroscopy became a tool used routinely by clay mineralogists, as compared to infrared (IR) spectroscopy. The first studies of clay minerals by Raman spectroscopy were published during the latter quarter of the last century (e.g. Blaha and Rosasco, 1978; Wiewiora et al., 1979; Haley et al., 1982; Johnston et al., 1984, 1985, 1998; Michaelian, 1986). Useful Raman spectra of materials such as clay minerals were unavailable until powerful laser spectrometers became accessible. In recent decades both Fourier transform (FT)-Raman spectroscopy and Raman microscopy have become mainstream spectroscopic methods, either as stand-alone techniques or in combination with IR spectroscopy. This also applies to the study of minerals in general, but clay minerals have received less attention due to a number of problems associated with obtaining useful spectra. The main reasons why Raman spectroscopy has received relatively less attention by clay scientists are the weakness of the Raman scattered signal, the photo-degradation of the sample, and the occurrence of fluorescence overlapping the Raman signal. Either the sample or impurities contained within it may absorb the laser radiation and re-emit it as fluorescence. If this occurs, Raman spectra can be obscured by a broad, strong fluorescence band, the intensity of which could be as much as $10^4$ larger than the Raman signal (Frost and van der Gaast, 1997; Ferraro et al., 2003; Ritz et al., 2016). More recently, the advent of FT Raman spectroscopy (Chase, 1986; Cutler, 1990), which uses near IR (NIR) radiation as the excitation source and a FT spectrometer, has enabled the study of clay mineral

Developments in Clay Science, Vol. 8. https://doi.org/10.1016/B978-0-08-100355-8.00006-0

structure. The use of FT Raman spectroscopy offers the advantages of reduced fluorescence, improved signal-to-noise by the co-adding of scans, and the longer wavelength of light reduces sample degradation.

The first compendious Raman study of the five major groups of phyllosilicates, such as kaolinite-serpentine, pyrophyllite-talc, mica, smectite, and chlorite groups was reported Wang et al. (2015). These authors demonstrated that phyllosilicates have a unique Raman spectral pattern that can be used to distinguish them from other groups of silicates (orthosilicate, chain silicate, ring silicate, and framework silicate with different degrees of silicate polymerization). Moreover, they distinguished characteristics of Raman spectral features associated with phyllosilicates of different stacking sequences and showed how minor changes in chemical compositions and structural details in some phyllosilicates can affect their Raman spectral patterns, peak positions, and peak widths. Several empirical rules were extracted from this study that can help to identify and characterize phyllosilicates based on in situ Raman spectroscopic measurements. The results are significant for planetary surface exploration, especially Mars, where the existence of phyllosilicates has been indicated by Vis-Near-IR spectroscopy used in orbital remote sensing (see Chapter 14 in this volume) and by X-ray diffraction in surface exploration. Small portable remote Raman systems, which are suitable for planetary rovers and landers, are currently being developed for use in future missions not only to Mars (Bishop and Murad, 2004; Wang et al., 2015) but also to Venus (Sharma et al., 2010; Villar and Edwards, 2005). There has been no attempt made to give detailed analysis of the Raman spectroscopic results of the major clay mineral groups until now. This chapter aims to provide such an overview based on the different structural units in the clay mineral structures, such as the hydroxyl groups and the layer structure itself.

## 6.2 HYDROXYL STRETCHING REGION

### 6.2.1 The 1:1 Clay Minerals

#### *6.2.1.1 The Kaolin Group Minerals*

Kaolin group polymorphs (kaolinite, dickite, nacrite and halloysite) are dioctahedral 1:1 clay minerals (Brigatti et al., 2013). The structure can be described as a repetition of units consisting of an $AlO_2OH$ octahedral sheet, where 2/3 of the octahedra are occupied by $Al^{3+}$ and coupled to a single $SiO_4$ tetrahedral sheet. Within this structure a differentiation can be made between the four hydroxyl (OH) groups. Three of these groups reside at one basal surface of the layers and form hydrogen bonds (H-bonds) to the next layer. These OH groups are referred to as the inner surface (or sometimes outer) OH groups. Another OH group, known as the inner OH group, is located within the ditrigonal cavity between the tetrahedral and octahedral sheets. The FT-Raman spectra in the OH stretching region (3750–3550 $cm^{-1}$) shown for kaolin group

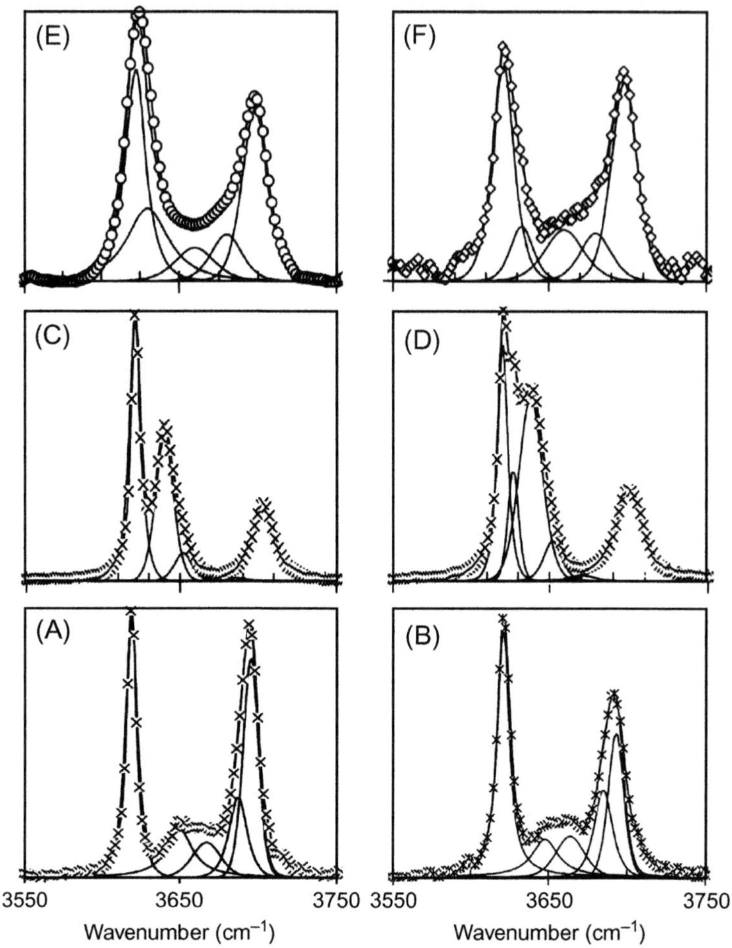


**FIG. 6.1** The band component analysis of the FT-Raman spectrum of (A) kaolinite (KGa-1); (B) kaolinite (KGa-2); (C) dickite, San Juanito; (D) dickite, Sainte Claire; (E) halloysite, New Zealand; and (F) halloysite, Eureka, in the 3550–3750 cm$^{-1}$ spectral region. *Reproduced with permission from Frost (1997).*

minerals (Fig. 6.1, Frost, 1997), resemble the mid-infrared (MIR) spectra of corresponding minerals (see Chapter 5). The positions of the Raman OH stretching bands observed for kaolins by different authors are given in Table 6.1.

The Raman band observed at 3619 cm$^{-1}$ for well-crystalline KGa-1 kaolinite (Kaol), and at 3621 cm$^{-1}$ for poorly crystalline KGa-2 Kaol has been assigned to the stretching vibrations of the inner OH ($\nu(Al_2OH)_I$). A similar band near 3620 cm$^{-1}$ exists for dickites and ordered halloysites, but less crystalline halloysites may exhibit this band even at 3627 cm$^{-1}$ (Frost, 1995). Dickite (Saint Claire) shows a complex spectrum in the 3620 cm$^{-1}$ region.

**TABLE 6.1** Raman Positions (in $cm^{-1}$) of the Inner ($\nu(Al_2OH)_I$ and $\nu(Mg_3OH)_I$) and Inner Surface ($\nu(Al_2OH)_{IS}$ and $\nu(Mg_3OH)_{IS}$) Hydroxyl Stretching Modes for Dioctahedral and Trioctahedral Clay Minerals 1:1 Clay Minerals

| Dioctahedral[1] | | | | | | | | |
|---|---|---|---|---|---|---|---|---|
| | *Kaolinite* | | *Dickite* | | *Halloysite* | | | |
| | *KGa-1* | *KGa-2* | *San Juanito* | *Sainte Claire* | *NZ* | *Eureka* | | |
| $\nu(Al_2OH)_I$ | 3619 | 3621 | 3621 | 3620 | 3622 | 3621 | | |
| $\nu(Al_2OH)_{IS}$ | | | | 3627 | | | | |
| | 3650 | 3648 | 3639 | 3638 | 3630 | 3631 | | |
| | 3667 | 3664 | 3652 | 3652 | 3660 | 3655 | | |
| | 3686 | 3685 | | | 3680 | 3679 | | |
| | 3695 | 3693 | 3703 | 3700 | 3698 | 3712 | | |
| Trioctahedral | | | | | | | | |
| | Lizardite | | Chrysotile | | | | Antigorite | |
| | (2) | (3) | (4) | (2) | (3) | (5) | (2) | (3) |
| $\nu(Mg_3OH)_I$ | 3697 | 3703 | 3697 | 3698 | 3698 | 3696 | 3698 | 3695 |
| | 3706 | | | 3701 | | | | |
| $\nu(Mg_3OH)_{IS}$ | | | | | | | 3619 | |
| | 3654 | 3660 | 3649 | 3649 | 3651 | 3643 | 3641 | 3665 |
| | 3670 | | 3681 | 3684 | | 3678 | 3652 | |
| | 3683 | 3683 | 3688 | 3689 | 3691 | 3686 | 3661 | |
| | 3690 | | | 3694 | | | 3686 | |

(1) Frost (1995).
(2) Auzende et al. (2004).
(3) Petriglieri et al. (2015).
(4) Trittschack and Grobety (2013).
(5) Kloprogge et al. (1999c).

Similarly, halloysite (Eureka) shows a similar although less well-resolved pattern for this OH group (Frost, 1997).

The Raman bands of KGa-1 Kaol related to the stretching vibrations of the inner surface OH ($\nu(Al_2OH)_{IS}$) at 3650, 3667, 3686 and 3695 $cm^{-1}$ (Table 6.1) are in good agreement with the data published for KGa-1 by Johnston et al. (1985). The positions of the $\nu(Al_2OH)_{IS}$ bands for disordered KGa-2 are due to the disordering slightly shifted to lower wavenumbers as a result of increased or more available H-bonding (Frost, 1997). The Raman spectra of halloysite also shows four $\nu(Al_2OH)_{IS}$ bands; however, their positions differ somewhat from Kaol. The Raman spectra of dickite reveal only three bands near 3639, 3652 and 3700 $cm^{-1}$ related to the inner surface OH groups (Table 6.1; Frost, 1997).

The kaolins have been extensively studied using Raman spectroscopy. Frequently the problem with the assignment of the four Raman bands near 3652, 3668, 3686 and 3695 $cm^{-1}$ to the three inner surface OH has been discussed. Most often the bands were attributed to the coupled in-phase and out-of-phase $\nu(Al_2OH)_{IS}$ oscillators. Shoval et al. (2002) obtained oriented single crystal polarized Raman spectra of Kaol with five bands at 3694, 3683, 3668, 3650 and 3620 $cm^{-1}$. Their intensities depended on the orientation of the crystal and the scattering geometry. Frost and Kloprogge (2001), in a similar fashion, measured the Raman spectra of an oriented Kaol single crystal cooled to 77 K. In agreement with the results reported by Johnston et al. (1990) an absence of complete extinction of the $\nu(Al_2OH)_{IS}$ bands was observed. Frost and Kloprogge (2001) gave an explanation that considered the possibility of having leakage of the collection optics over a wide angle, resulting in the observation of some intensity in both the transverse optical (TO) and longitudinal optical (LO) modes (see below for further discussion). They also concluded that upon cooling the in-phase modes became partially or wholly uncoupled, but the out-of-phase modes became more narrow.

Pajcini and Dhamelincourt (1994) published the first low temperature (between 293 and 80 K) Raman spectra of Kaol employing a Raman microscope. Compared to the room temperature (RT) spectrum the $\nu(Al_2OH)_{IS}$ at 3621 $cm^{-1}$ was shifted slightly to 3618 $cm^{-1}$ and a new shoulder appeared at 3628 $cm^{-1}$, while two $\nu(Al_2OH)_{IS}$ at 3670 $cm^{-1}$ and 3697 $cm^{-1}$ were shifted significantly to the higher wavenumbers. The remaining $\nu(Al_2OH)_{IS}$ band at 3652 $cm^{-1}$ split into two components at 3650 $cm^{-1}$ and 3657 $cm^{-1}$ as reported also by Frost (1995). In contrast, Johansson et al. (1998) did not observe splitting of this OH stretching mode, but instead a shift from 3620 $cm^{-1}$ to 3615 $cm^{-1}$. The $\nu(Al_2OH)_{IS}$ modes all moved to higher wavenumbers by about 7 $cm^{-1}$ and the resolution between the bands improved significantly upon cooling. They attributed these shifts to the decreased strength of H-bonds formed between the inner surface OH groups and the oxygen atoms of the adjacent siloxane sheet of the Kaol layer. The reasoning behind this interpretation was that upon cooling to 77 K the Kaol layers would be closer to each other and the inner surface OH groups, which are oriented at angles of 60–73 degrees to the *ab* plane (Giese, 1988), would be pushed flatter against the surface. This phenomenon then results in weaker H-bonds between adjacent layers resulting in increased frequency of the $\nu(Al_2OH)_{IS}$ bands.

Raman spectra of kaolinites, in addition to the four $\nu(Al_2OH)_{IS}$ bands observed also in the IR spectra, show a band at 3686 $cm^{-1}$ (e.g. Wiewiora et al., 1979; Johnston et al., 1985, 1990; Michaelian, 1986; Frost, 1995; Frost and van der Gaast, 1997; Frost et al., 1996). Johnston et al. (1985) described that the intensity of this fifth band depended on the nature and origin of the kaolinite. Farmer (1998, 2000) offered a theoretical explanation for the presence of this band based on the presence of TO and LO modes. If a unit cell vibration develops an oscillating dipole, two corresponding long-wavelength modes exist: one with the dipole moment parallel to the wave vector (LO) and one perpendicular (TO). The $\nu(Al_2OH)_{IS}$ band at 3686 $cm^{-1}$, observed in the Raman spectra of Kaol samples containing crystals thicker than the exciting wavelength (usually ~500 nm), is that of a TO mode involving the in-phase vibration of the three other inner surface OH groups. Crystals much thinner than the wavelength of the exciting radiation exhibit a higher frequency for this $\nu(Al_2OH)_{IS}$ vibration (3697 $cm^{-1}$) which corresponds to that of the LO mode. IR studies are usually made on clay-size (<2 μm) particles, which have mean thicknesses of 30–150 nm and IR radiation of wavelength near 2.7 μm can thus only excite vibrations perpendicular to the plates, such as the TO mode. However, the position of this TO mode (~3697 $cm^{-1}$) in IR spectra is at the same frequency as that of the LO mode in Raman spectra of Kaol.

The precise assignment of the OH stretching vibrations and the orientation of OH groups in the kaolin group minerals is still debated. In addition to experimental studies the quantum mechanical calculations were conducted to determine the structure and vibrational properties of these minerals (Balan et al., 2001, 2005, 2011a; Benco et al., 2001a,b; Tosoni et al., 2006). Balan et al. (2005) computed the OH-stretching modes of kaolin group minerals (i.e. kaolinite, dickite and nacrite) using the density functional theory and the results were compared with experimentally obtained IR and Raman spectra. Good agreement between experimental IR and Raman data and theoretical derived structures was obtained. The experimental Raman spectrum of well-crystallized Kaol displayed four well-developed bands at 3618, 3649, 3664, 3679 $cm^{-1}$ and a shoulder at 3688 $cm^{-1}$. The calculated Raman values were very similar, that is, 3618, 3643, 3661 and 3684 $cm^{-1}$. Consistent with previous observations the 3618 $cm^{-1}$ band was assigned to $\nu(Al_2OH)_I$, the 3649, 3664 $cm^{-1}$ band to the coupled out-of-phase $\nu(Al_2OH)_{IS}$ modes and the band at 3679 $cm^{-1}$ to perpendicular in-phase $\nu(Al_2OH)_{IS}$ modes. The shoulder at 3688 $cm^{-1}$ was interpreted as being due to the presence of thin domains within the particles as many large kaolinite crystals are actually stacks or booklets of thin crystals with very similar orientations (Neder et al., 1999).

The experimental Raman spectrum of dickite reported by Balan et al. (2005) showed bands at 3622, 3642 and 3708 $cm^{-1}$, the band component analysis of the OH stretching region revealed also weak bands at 3624 and 3655 $cm^{-1}$. The first-principle calculation provided Raman bands at 3607, 3625, 3635 and 3750 $cm^{-1}$ (Balan et al., 2005). Based on previous experimental and calculated spectra the 3622 $cm^{-1}$ band was assigned to the inner OH1 stretching mode (OH groups labelled based on the data by Bish, 1993). The high frequency band 3708 $cm^{-1}$ (3750 $cm^{-1}$ in the calculated spectrum) corresponded to the in-phase motion of the inner surface OH3 groups and the band at 3642 $cm^{-1}$ to the LO and TO frequencies of the in-phase mode of the two OH2 and OH4 groups. Weak band at 3624 $cm^{-1}$ superimposed to the OH1 band was attributed to the out-of-phase interlayer motion of the OH2 and OH4 groups.

Balan et al. (2005) also analyzed the OH vibrational modes of nacrite. The experimental Raman spectrum showed three strong bands at 3629, 3636, and 3701 $cm^{-1}$, and shoulders at 3620 and 3644 $cm^{-1}$. The calculated positions were at 3571, 3591, 3655 and 3690 $cm^{-1}$. The 3629 $cm^{-1}$ was related to the in-plane $\nu(Al_2OH)_I$, the band at 3701 $cm^{-1}$ (3690 $cm^{-1}$ in the calculated spectrum) to $\nu(Al_2OH)_{IS}$ of OH2 groups. The band observed at 3636 $cm^{-1}$ was attributed to the in-phase inter- and intra-layer $\nu(Al_2OH)_{IS}$ of the OH3 and OH4 groups. The weak band at 3644 $cm^{-1}$ was supposed to be due to the depolarization field occurring in thin domains of the nacrite particles.

#### *6.2.1.2 The Serpentine Group Minerals*

The serpentine group minerals (mainly lizardite, antigorite and chrysotile) form the Mg-rich trioctahedral equivalent of the kaolinite group with the same 1:1 layer structure but all the positions in the octahedral sheet are filled with $Mg^{2+}$. Lizardite shows an ideal layer topology owing to shifts of the octahedral and tetrahedral cations away from their ideal positions and to the limited $Al^{3+}$ for $Si^{4+}$ substitution in the tetrahedral sites (Rinaudo et al., 2003), whereas antigorite is modulated and chrysotile is bent (Brigatti et al., 2013). The positions of the Raman OH stretching modes reported for different serpentine minerals are summarized in Table 6.1.

The first Raman spectra of lizardite and chrysotile were reported by Bard et al. (1997). The OH stretching region of lizardite showed an intense band near 3700 $cm^{-1}$ with shoulders near 3685 and 3650 $cm^{-1}$. For chrysotile, a less intense OH band at 3700 and a shoulder near 3685 $cm^{-1}$ was reported. No assignment of the bands was provided. Kloprogge et al. (1999c) used polarized Raman microscopy to study tubular chrysotile. Polarized spectra of a single chrysotile crystal for different crystal faces are shown in Fig. 6.2. The letters (A, B and C) refer to the Porto notation indicating the propagation directions of the incident and scattered light and their polarizations described in terms of the crystallographic axes *a*, *b* and *c*. The band

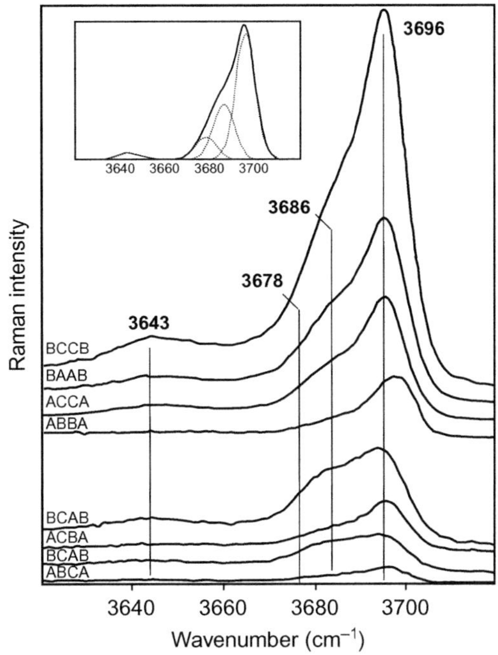


**FIG. 6.2** Raman spectra of the hydroxyl-stretching region of tubular chrysotile from the different crystal faces following the Porto notation. Inset: band component analysis of the hydroxy-stretching region (BCCB), modified after Kloprogge et al. (1999c).

component analysis revealed four bands around 3696, 3686, 3678 and 3643 $cm^{-1}$ (insert in Fig 6.2, Table 6.1). Their intensities were strongly dependent on the crystal orientation and the directions of the polarization of the laser and the scattered radiation. The band around 3678 $cm^{-1}$ was not detected before. The weak band at 3643 $cm^{-1}$ was also not described in the Raman spectrum of chrysotile by Bard et al. (1997), but only in the IR spectra as a band around 3650 $cm^{-1}$. Based on the analogy with kaolinite it was reasonable at that time to assign the three highest wavenumber bands to $\nu(Mg_3OH)_{IS}$ and the band around 3643 $cm^{-1}$ to $\nu(Mg_3OH)_I$ located in the lower shared plane of the octahedral sheet (Kloprogge et al. (1999c)).

More recently the precise attribution of the Raman bands in the high-wavenumber part of the spectrum of the 1:1 clay minerals, in terms of vibrations of the different OH groups occurring in the serpentine polymorphs, was revised by Auzende et al. (2004). They investigated the behavior of serpentine minerals lizardite, antigorite and chrysotile at increasing pressures. The positions of the OH stretching modes observed at low-pressure are summarized in Table 6.1. Upon compression, all Raman bands shift toward higher wavenumbers. The positive shifts can be expected for the inner OH groups because no strong H-bonding occurs in these hydroxyl groups. Thus the positive shifts indicate that the O–H bond is strengthening under compression. Positive shifts were unexpected in the case of the inner surface OH groups, which in serpentine minerals are considered to be involved in weak H-bonding in order to link adjacent 1:1 layers. The positive shifts of the O–H symmetric stretching vibrations in serpentine minerals indicate the absence of pressure-induced enhancement of H-bonding. Auzende et al. (2004) considered that the inner surface OH groups could be expected to behave as ionic bonds, whose strength increases with pressure, and thus leads to a new assignment for the Raman OH-stretching modes. The stretching modes with the greatest pressure-dependence arise from the inner surface OH groups, whereas the less pressure-sensitive stretching modes, located close to 3700 $cm^{-1}$, can be assigned to $\nu(Mg_3OH)_I$ modes. This assignment is in close agreement with the ab initio calculations of the IR spectrum of lizardite by Balan et al. (2002b), who calculated that the positions of the IR $\nu(Al_2OH)_{IS}$ bands occurred at lower wavenumbers than bands due to inner hydroxyl groups. First-principle calculations of Balan et al. (2002b) and Prencipe et al. (2009) give three Raman active modes plus a contribution of one longitudinal-optical (LO) mode for lizardite-1*T* ($C_{3v}$ factor group).

A higher number of modes than in lizardite can be expected due to the curved layers and a higher degree of disorder in chrysotile (Auzende et al., 2004). Trittschack and Grobety (2013) ascribed narrow bands at 3681 and 3688 $cm^{-1}$ to LO-TO splitting of the in-phase $\nu(Mg_3OH)_{IS}$ mode as proposed by Prencipe et al. (2009), while the bands at 3649 and 3697 $cm^{-1}$ were ascribed to the out-of-phase $\nu(Mg_3OH)_{IS}$ and $\nu(Mg_3OH)_I$ modes, respectively (Table 6.1).

Petriglieri et al. (2015) showed a Raman spectrum of antigorite with the main band maxima located at 3665 and 3695 $cm^{-1}$, plus a weak inflection at 3619 $cm^{-1}$, far out of range compared to the positions of the structural OH groups. No explanation was presented for the origin of this latter band. Schwartz et al. (2012) indicated that antigorite presented a spectrum broader than that observed for lizardite with a primary band near 3665 $cm^{-1}$ and a secondary band at 3697 $cm^{-1}$ with weak inflection around 3682 $cm^{-1}$. No attempt was made to deconvolute the spectrum in either of these two papers. The Raman spectrum for synthetic antigorite was simpler than for a natural sample, with two main bands located at 3672 and 3698 $cm^{-1}$ (Reynard and Wunder, 2006). The position of the first band ($\nu(Mg_3OH)_I$) and its relative intensity varies slightly, while the $\nu(Mg_3OH)_{IS}$ band varies significantly for different orientations. This is possibly due to the existence of many modes with slightly split wavenumbers in the modulated and low symmetry antigorite structure and to variations in the modulation of antigorite in the experimental charge (Wunder et al., 2001).

These structural complexities can also explain the larger bandwidths observed for OH modes in the complex antigorite structure with respect to the higher symmetry lizardite. The band positions of the synthetic antigorite were in good agreement with those obtained from ab initio calculations (Balan et al., 2002b) for the in-phase $\nu(Mg_3OH)_{IS}$ and the $\nu(Mg_3OH)_I$ modes of lizardite, at 3693 and 3726 cm$^{-1}$, respectively. The contribution for the degenerate out-of-phase outer OH stretching mode, expected at 3655 cm$^{-1}$ from ab initio calculation, was not clearly observed in the raw experimental spectra, but deconvolution was not performed to determine the exact number of bands and their positions. It could correspond to a weak broad shoulder near 3650 cm$^{-1}$ on the low wavenumber side of the intense OH bands. A shoulder in the highest wavenumber OH band is observed at 3693 cm$^{-1}$. A similar shoulder was observed in the natural antigorite spectrum and was interpreted as probably being due to positional disorder of the inner OH in the distorted silica rings of the tetrahedral layer. A small shoulder at 3701 cm$^{-1}$ was sometimes observed in the synthetic product and ascribed to a trace amount of chrysotile (Wunder et al., 1997).

## 6.2.2 The 2:1 Clay Minerals

### 6.2.2.1 *Pyrophyllite and Talc*

Pyrophyllite ($Al_4Si_8O_{20}(OH)_4$) is the end member of dioctahedral 2:1 clay minerals containing no isomorphic substitution and therefore no layer charge. All tetrahedral sites are filled with $Si^{4+}$, two-thirds of the octahedral sites are filled with $Al^{3+}$, and the remaining one-third of the sites is vacant. The electrically neutral sheets in pyrophyllite are bonded together by relatively weak dipolar and van der Waals forces (Evans and Guggenheim, 1988). For both the Raman and IR spectra of pyrophyllite, the extremely sharp $\nu(Al_2OH)$ mode observed by Zhai et al. (2008) at 3670 cm$^{-1}$ (Fig. 6.3) and at 3674–3675 cm$^{-1}$ by Farmer (1974c), Kloprogge and Frost (1999e), Wang et al. (2002), Lantenois et al. (2007) is very characteristic. The second sharp band found at 3647 cm$^{-1}$ (Fig. 6.3; Zhai et al., 2008) in the Raman and near 3645 cm$^{-1}$ in the IR spectra were associated with $\nu(AlFe^{3+}OH)$ mode (Farmer, 1974c). Lantenois et al. (2007) also observed an IR band at 3645 cm$^{-1}$ in natural and synthetic samples (some with Fe substitution).

The 2:1 layer, trioctahedral equivalent of the serpentine group minerals is talc, with structural formula $Mg_3Si_4O_{10}(OH)_2$. As for pyrophyllite there are two different OH groups related through an approximate or exact center of symmetry. Therefore, two $\nu(OH)$ modes are possible, one active principally in the Raman spectrum, in which both OH groups vibrate in phase, and the other active principally in the IR, in which the two OH groups vibrate in anti-phase (Farmer, 1974c). Because coupling between the OH groups is weak, the wavenumbers are nearly the same in the IR and the Raman spectra (Petit, 2005). The $\nu(OH)$ modes of talc are affected by the octahedral cations. As a result up to four OH stretching modes can be expected upon substitution

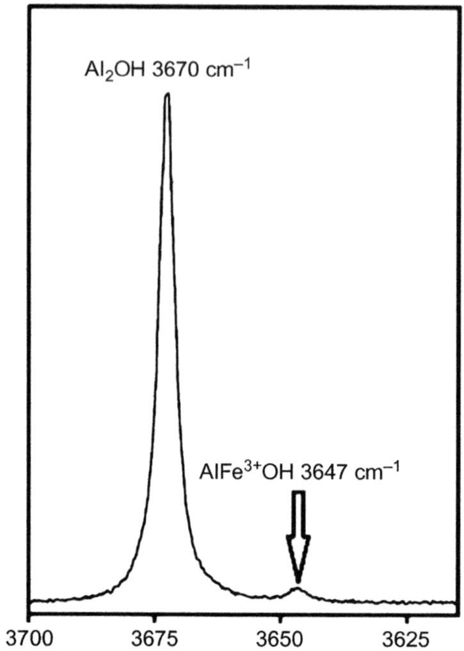


**FIG. 6.3** Raman spectrum of natural pyrophyllite in the hydroxyl stretching region showing the strong mode associated with $\nu(Al_2OH)$ at 3670 $cm^{-1}$ and the $\nu(AlFe^{3+}OH)$ at 3647 $cm^{-1}$ due to substitution of $Al^{3+}$ by $Fe^{3+}$ in the octahedral sheet.

of $Mg^{2+}$ by another divalent cation (3Mg, 2Mg, $M^{2+}$, Mg, $2M^{2+}$ and $3M^{2+}$ in the case of only one other divalent cation). This has been shown in a number of studies in the IR spectra for cations such as $Ni^{2+}$, $Co^{2+}$, $Zn^{2+}$, $Fe^{2+}$, $Mn^{2+}$ and $Cu^{2+}$ (Petit (2005) and references therein). Only a limited number of papers have reported the Raman spectrum of talc and only a few of these have included the $\nu$(OH) region. In general, a single band is reported between 3672 and 3677 $cm^{-1}$ due to $\nu(Mg_3OH)$ vibration (Blaha and Rosasco, 1978; Rosasco and Blaha, 1980; Holtz et al., 1993, 1994; Sharma et al., 2010). Wang et al. (2015) showed also a Raman spectrum of talc (AW, 08) with main peak at 3678 $cm^{-1}$ and a second small peak at 3663 $cm^{-1}$ related to the presence of $Fe^{2+}$ in the talc structure.

### *6.2.2.2 Smectites*

Smectites are abundantly occurring clay minerals of the 2:1 layer type. Isomorphic substitution within the tetrahedral and/or octahedral sheets generates a negative charge on the layers, which is balanced by hydrated exchangeable cations located in the interlayer spaces (mostly $Ca^{2+}$, $Mg^{2+}$ and $Na^{+}$). The dioctahedral smectites can be divided into two principal groups: (1) the

Al-rich smectites, such as montmorillonite (Mt) and beidellites (Bd), and (2) the Fe-rich varieties including ferrian (also ferruginous) smectites and nontronite. However, all these species are related by systematic changes in chemical composition. Trioctahedral smectites include for example saponites (Sap) and hectorite (Ht) (Brigatti et al., 2013). For smectites the $\nu$(OH) region in the IR spectra is dominated by not only the bands related to OH groups of the octahedral sheets but also by physi-sorbed and interlayer water. In the corresponding Raman spectra the influence of this water is much more difficult to observe. The observation of the Raman spectra of these clay minerals is complicated by the generally strong fluorescence, and thus careful experimentation is needed to obtain useful data.

Ritz et al. (2016) recently presented the comprehensive study of the Mt standards (SAz-2, SWy-2, STx-1b) from the Clay Mineral Society (CMS) and synthetic Mt (Kunipia-F) by dispersive Raman spectroscopy with laser excitation wavelengths ($\lambda$) of 532 nm and 780 nm, dispersive Raman microscopy with laser excitation $\lambda = 532$ nm and 100 × magnifying lens, and Fourier Transform-Raman spectroscopy with laser excitation $\lambda = 1064$ nm. The spectra of Mt differed mainly in the type of interlayer cation and substitution of octahedral aluminium by magnesium or iron. Almost all measured spectra of Mt were significantly affected by strong fluorescence; only SAz-2 was only slightly (or not at all) influenced by fluorescence. The band at 3620 $cm^{-1}$ present in Raman spectra of all Mt, corresponded to the $\nu(Al_2OH)$ vibrations of the structural OH groups. The study of Ritz et al. (2016) was focused on the effects of laser $\lambda$ on fluorescence behaviour in the spectra, and as such, the authors did not comment on the structure of the OH stretching band in the Raman spectrum of SAz-2. This spectrum clearly shows a weak shoulder at the high side of the 3620 $cm^{-1}$ band and a very broad band centred around 3450 $cm^{-1}$, probably due to the OH-stretching modes of interlayer water ($\nu$(HOH)).

The presence of the Fe in the octahedra of the NG-1 nontronite (CMS standard) shifted the OH peak in the Raman spectrum down to 3570 $cm^{-1}$ compared to Mt (Wang et al., 2015). Frost and Kloprogge (2000c,d) studied the IR and Raman spectra of the ferrian smectite and nontronites. The results of the band component analyses are summarized in Table 6.2. The component assigned to $\nu(AlFe^{3+}OH)$ appeared as a very weak band near 3570 $cm^{-1}$ in the Raman spectra and as strongly IR active band in the 3597–3566 $cm^{-1}$ interval in the IR spectra. The other components detected in the 3435–3100 $cm^{-1}$ range in both, Raman and IR spectra, have been variously attributed to the $\nu$(HOH) of water molecules (1) creating the primary hydration sphere of the exchangeable cations, (2) strongly linked through H-bonds with other water molecules, (3) weakly H-bonded to the siloxane surface, and (4) interparticle or excess surface water (see Chapter 9 of this volume).

The main Raman bands for a Sap were reported near 3680 and 3420 $cm^{-1}$ (Wang et al., 1999, 2015). The sharp intense band near 3680 $cm^{-1}$ was

**TABLE 6.2** Raman and IR Positions of the Hydroxyl Stretching Modes of Ferrian Smectites and Nontronites (in $cm^{-1}$) (Frost and Kloprogge, 2000c,d)

| | Ferruginous Smectite | | Nontronite | | | | | | | |
|---|---|---|---|---|---|---|---|---|---|---|
| | *SWa-1* | | *Hohenhagen* | | *Garfield* | | *Uley Brown* | | *Uley Green* | |
| **Assignment** | ***Raman*** | ***IR*** | ***Raman*** | ***IR*** | ***Raman*** | ***IR*** | ***Raman*** | ***IR*** | ***Raman*** | ***IR*** |
| AlFeOH stretch | 3572 | 3596 | 3574 | 3597 | 3574 | 3575 | 3574 | 3572 | 3574 | 3566 |
| HOH-stretch, $H_2O$ | 3435 | 3434 | 3437 | 3432 | 3436 | 3424 | 3436 | 3430 | 3436 | 3426 |
| HOH-stretch, $H_2O$ | 3362 | | 3356 | | 3355 | | 3355 | | 3355 | |
| Overtone of OH-bend | 3220 | 3233 | 3218 | 3232 | 3218 | 3251 | 3218 | 3262 | 3218 | 3256 |
| HOH-stretch, $H_2O$ | 3102 | 3070 | 3118 | 3047 | 3118 | 3036 | 3118 | 3043 | 3118 | |

attributed to $\nu(Mg_3OH)$ vibrations, and the broad band near 3420 $cm^{-1}$ to $\nu(HOH)$ of interlayer water. To understand the behavior of octahedral $\nu(M^{2+}{}_3OH)$ modes ($M^{2+}$ is octahedrally coordinated divalent atom) as a function of increasing tetrahedral layer charge, IR and Raman spectroscopic studies were carried out on a series of synthetic Sap samples (Pelletier et al., 2003). Both, IR and Raman spectra showed a gradual shift of the $\nu(M^{2+}{}_3OH)$ position from ~3678 to ~3686 $cm^{-1}$ with increasing layer charge. Two main bands were visible at 3683 and 3720 $cm^{-1}$ in the IR and Raman spectra of $K^+$ saponites. The intensity of the latter band increased with the layer charge, and its relative integrated intensity was roughly proportional to the layer charge, indicating that the $K^+$ was located in close proximity to the sites of isomorphic substitution in the tetrahedral sheet and influenced the structural OH. This observation also proves that the structural OH groups can form H-bonds across the ditrigonal cavity where tetrahedral substitutions exist (see Chapter 5 this volume). The band around 3720 $cm^{-1}$ was assigned to structural OH groups perturbed by the presence of $K^+$ ions above or within the ditrigonal cavity of the tetrahedral sheet (Fraser and Russell, 1994), while the band near 3683 $cm^{-1}$ was attributed to unperturbed OH groups. To assign the features obtained for hydrated $Na^+$ samples in the $\nu(OH)$ region, the Raman spectra were decomposed using four $\nu(M^{2+}{}_3OH)$ components around 3675, 3680, 3685 and 3690 $cm^{-1}$. The evolution with layer charge of the relative proportions of these four components was described by taking into account the location of hydrated interlay cations related to the distribution of tetrahedral substitutions. Such an assumption was confirmed using a simple statistical model that yielded four populations of OH groups in the octahedral layer of hydrated $Na^+$-saponites (Pelletier et al., 2003).

Rinnert et al. (2005) showed that after dehydration at 100°C at 0.1 Pa, $Na^+$ Sap still contained a significant amount of interlayer water (1 $H_2O$ molecule per 2 $Na^+$ cations). The Raman spectrum in the OH stretching range 3600–3750 $cm^{-1}$ showed two broad bands at 3675 and 3712 $cm^{-1}$. The band at 3675 $cm^{-1}$ was interpreted as corresponding to unperturbed OH groups while the second band at 3712 $cm^{-1}$ corresponded to OH groups perturbed by the presence of a $Na^+$ cation, located close to the ditrigonal cavity in accordance with the interpretation by Pelletier et al. (2003). Rinnert et al. (2005) suggested that in the dry state, each $Na^+$ ion perturbs as many as two structural OH bands, the width of the signal revealing differences in relative degrees of perturbation. This would, however, strongly depend on how well adjacent layers line up. Since in smectites the substitution is random, the perfect matching required for good crystalline stacking is not present. For partial pressures between 0.06 and 0.18, the adsorbed quantity of water increased strongly to reach an amount of 4.5 mmol/g, which corresponds to about 1.5 water molecule per cation. At the same time the basal spacing reached a value of 12.3 Å that is classically assigned to a one-layer hydrate state (1-LHS). This confirmed that the transition to the full 1-LHS was associated with the solvatation of the interlayer cation by three water molecules, as previously deduced for Ht by IR spectroscopy (Farmer and Russell, 1971).

#### 6.2.2.3 *Vermiculite*

Vermiculite (Verm) is a generally trioctahedral 2:1 clay mineral with hydrated cations occupying the interlayer space. However, the negative layer charge of Verm, arising mostly from substitution of $Al^{3+}$ for $Si^{4+}$ in tetrahedral sites, is larger than that of smectite (Brigatti et al., 2013). Haley et al. (1982) observed two sharp Raman bands at 3739 cm$^{-1}$ and 3240 cm$^{-1}$ and two broad bands near 3400 cm$^{-1}$ and 3368 cm$^{-1}$ attributed to $\nu$(HOH) of interlayer water (Fig. 6.4; Table 6.3). The band at 3739 cm$^{-1}$ was attributed to $\nu(M^{2+}{}_3OH)$ groups associated with the octahedral sites occupied by Mg. As a narrow spectral feature is typical of a well-ordered crystal site, the band at 3240 cm$^{-1}$ was attributed to OH bonded in a different lattice environment than the majority OH described above. The authors concluded from the absence of a band at 3240 cm$^{-1}$ in the Raman spectrum of talc that the band at 3240 cm$^{-1}$ was related to structural OH influenced by the minor tetrahedral $Al^{3+}$ in Verm. This conclusion, however, seems to be very questionable because the position of the 3240 cm$^{-1}$ band is too low to be connected with any stretching mode of the structural OH groups of Verm. In contrast, Wang et al. (2015) showed that Raman pattern in the OH stretching region varies widely among the samples reflecting the highly variable sites for structural OH groups and $H_2O$ molecules in the Verm structure. Three sharp bands at 3710, 3672 and 3620 cm$^{-1}$ were observed for studied Verm with ideal structural formula $(Mg,Fe^{3+},Al^{3+})(Al,Si)_4O_{10}(OH)_2.4H_2O$. Based on the IR pattern of trioctahedral biotite with similar chemical composition (see Chapter 5 of this volume) the Raman bands at 3710 cm$^{-1}$ may be related to the $\nu(Mg_3OH)$ stretching mode, at 3672 cm$^{-1}$ to the $\nu(Mg_2AlOH)$ or $\nu(Mg_2Fe^{3+}OH)$, and a peak at 3620 cm$^{-1}$ to a dioctahedral-like structural OH, *vis*, $\nu$(MgAl□OH).

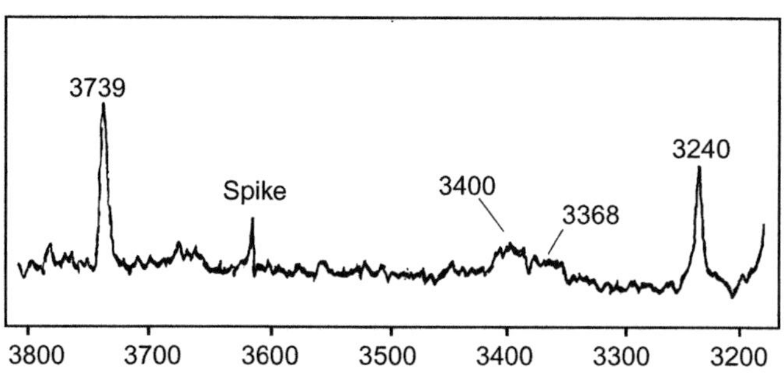


**FIG. 6.4** Raman spectrum of the hydroxyl stretching region of vermiculite, modified after Haley et al. (1982).

**TABLE 6.3** Raman positions in the hydroxyl stretching $\nu$(OH) region of vermiculite (in $cm^{-1}$)

| Haley et al. (1982) | | Wada and Kamitakahara (1991) | | | Arab et al. (2002) | | | |
|---|---|---|---|---|---|---|---|---|
| *Raman* | *IR* | *Raman* | *Raman dehydrated Na-vermiculite* | *Raman dehydrated Sr-vermiculite* | *Raman observed* | *Raman calculated* | *IR observed* | *IR calculated* |
| | | | 3707 | 3726 | | | | |
| | | | | | | | 3711 | 3880 |
| | | 3679 s | 3679vw | 3666 | | 3868 | 3561 | 3872 |
| 3739 s | | 3640 s | | | | 3660 | 3390 | 3864 |
| 3400 b | 3480–3510 | 3450 b | | | | | 3207 | 3856 |
| 3368 b | | | | | | | | |
| 3240 s | 3333 | | | | | | | |

s—sharp, b—broad, vw—very weak.

Wada and Kamitakahara (1991) also reported two relatively sharp peaks at 3640 (weak) and 3679 cm$^{-1}$ (strong) for $Na^+$-Verm in the two-layer hydrate state (2-LHS), which they assigned to $\nu(M^{2+}{}_3OH)$ bonds in the octahedral sheets. Bands with the same wavenumbers and relative intensities were also observed for $Sr^{2+}$-Verm. The difference between the wavenumbers at 3640 and 3679 cm$^{-1}$ for $\nu(M^{2+}{}_3OH)$ was similarly thought to arise from of $Al^{3+}$ for $Si^{4+}$ substitution in the tetrahedral sheets. An alternative explanation of the wavenumber difference is that it arises from the substitution of $Al^{3+}$, $Ti^{4+}$ and $Fe^{3+}$ atoms in the $Mg^{2+}$ sites of the octahedral sheets. Neither explanation, however, yielded quantitative agreement, at least on the basis of simple calculations, with regard to the observed intensity ratio of the 3640 cm$^{-1}$ Raman band to the 3679 cm$^{-1}$ band. It remains unclear what causes the large difference in the band positions of the structural OH modes. They may be associated with differences in the chemical composition (i.e. tetrahedral and octahedral substitutions) of the Verm used, with formation of domains of different composition, or with different OH-dipole orientation with respect to the *ab* plane affecting the amount of H-bonding across the ditrigonal cavity. Moreover, many of these materials may actually be interstratifications of two or three different Verm compositions.

Wada and Kamitakahara (1991) used heating experiments to dehydrate $Na^+$- and $Sr^{2+}$-Verm. At room temperature (RT), corresponding to a 2-LHS, a broad feature around 3450 cm$^{-1}$ with a full width at half maximum (FWHM) of ~300 cm$^{-1}$ was observed in Raman spectra. Upon dehydration the band became weaker and shifted to higher wavenumbers, and disappeared upon complete dehydration. As the $Na^+$-Verm sample was dehydrated, a new Raman band appeared at 3707 cm$^{-1}$, which increased in intensity at the expense of the band at 3679 cm$^{-1}$. In the 2- or 1-LHS, the interlayer cations were surrounded by water molecules, and may not have resided in the pseudo-hexagonal cavities. Upon dehydration, the $Na^+$ ions may have moved into the cavities, affecting the stretching vibrations of the hydroxyl groups located at the centres of the cavities. Since there are 1.85 $Na^+$ atoms per unit cell on average in the Verm studied by Wada and Kamitakahara (1991), almost all the cavities would have be filled with these cations. The increase in the vibration wavenumber of the $\nu(M^{2+}{}_3OH)$ mode is, therefore, probably caused by the relocation of interlayer $Na^+$ atoms to the pseudo-hexagonal cavities and the effects on the bond-strength of the OH-coordinating octahedral ions. In the 2-LHS of $Sr^{2+}$-Verm, a broad band was observed at 3450 cm$^{-1}$, originating from intercalated water molecules $\nu$(HOH) which weakens and blue-shifts with dehydration. Again, two sharp bands were evident at 3640 and 3679 cm$^{-1}$. The latter shifted to lower wavenumber as the temperature was increased, a trend that was actually also seen for $Na^+$-Verm, renormalizing to 3666 cm$^{-1}$ ($\Delta\nu = 13$ cm$^{-1}$) in the completely dehydrated sample. The persistence of this band in the anhydrous state implies that it is related to $\nu(M_3^{2+}OH)$. At the same time dehydration caused the appearance and growth of a new band at 3726 cm$^{-1}$. Since the number of interlayer cations in $Sr^{2+}$-Verm is only half that in $Na^+$-Verm, even in the completely dehydrated sample, about half of the pseudo-hexagonal cavities are not occupied with $Sr^{2+}$ cations. Consequently, the Raman band at 3679 cm$^{-1}$, which corresponds to the $\nu(M_3^{2+}OH)$ mode without a cation in the cavity, persists in the dehydrated Verm. The band at 3726 cm$^{-1}$ then corresponds to a $\nu(M_3^{2+}OH)$ vibration with a cation in the cavity. This wavenumber is higher than that found in $Na^+$-Verm (3707 cm$^{-1}$) by 19 cm$^{-1}$. The authors explained this based on the radius of $Sr^{2+}$ (1.13 Å) that is larger than that of $Na^+$ (1.02 Å), the latter could approach the OH groups more closely. The consequence would be that the greater electrostatic interactions of $Sr^{2+}$ ions produce a larger shift of the $\nu(M_3^{2+}OH)$ mode to higher wavenumbers, despite their larger ionic radii.

## 6.3 THEORY OF THE LOW WAVENUMBER VIBRATIONAL MODES

Farmer (1974c) described the pseudo-hexagonal tetrahedral sheet on the basis of the highest possible symmetry ($C_{6v}$) for $SiO_4$ as a separate unit. The vibrations can then be classified in five species: $2A_1+3B_1+1B_2+3E_1$ and $3E_2$. Of these only $A_1$ and $E_1$ are IR active, while $A_1$, $E_1$ and $E_2$ are potentially Raman active. Calculations indicate these modes ($a_1^1$), 611 ($a_1^2$), 130 ($b_2$), 915 ($b_1^1$), 779 ($b_1^2$), 319 ($b_1^3$), 1015 ($e_1^1$), 543 ($e_1^2$), 285 ($e_1^3$), 695 ($e_2^1$), 499 ($e_2^2$) and 127 ($e_2^3$) can be expected around 902 cm$^{-1}$. However, when the $SiO_4$ unit is studied as part of the tetrahedral sheet of clay minerals a reduction in symmetry is observed. Nevertheless, the $C_{6v}$ approximation allows for a reasonable starting point in interpreting the lower wavenumber region in the Raman spectra of clay minerals. Loh (1973) considered two alternative methods to the analysis of these optical vibrations: (1) to assume the lattice vibration of the entire silicate layer, and (2) to assume that each molecular species in the layer exhibits essentially its characteristic molecular vibrations and there is no strongly coupled vibrational interaction between molecules within the layer. In the first approach the phyllosilicate was treated as an assembly of nearly isolated layers with weak van der Waals bonds between them. The other extreme was to consider that the ions in the phyllosilicate tend to form several types of molecules and the vibrational coupling between the molecules is weak. These molecules are, for example, $SiO_4$ of $C_{3v}$ site symmetry, ($Mg^{2+}$, $Fe^{2+}$, $Al^{3+}$)$O_4(OH)_2$ of nearly octahedral symmetry, $S_6$ and the structural OH. Each type of molecule would exhibit its characteristic wavenumbers and thus have its own specific selection rules under these assumptions. The total number of optical vibrations would then be approximately equal to the sum of internal vibrations of each type of molecule and is of the order of 10–20 in phyllosilicates. Similar to the observation by Farmer (1974c), Loh (1973) concluded that the optical vibrational spectra of phyllosilicates seem to agree best with the molecular model (i.e. assumption (2) above). Recent advances in molecular modelling and ab initio calculations have led to new insights in the interpretation of the Raman spectra of some of the clay minerals.

### 6.3.1 The 1:1 Clay Minerals

Bougeard et al. (2000) used ab initio calculations of molecular models to compute the polarizability tensor of the kaolinite. A complete assignment of the bands observed in the experimental spectra was suggested based on the calculated IR, Raman and power spectra. In the region between 540 and 450 $cm^{-1}$ the Raman spectrum was characterized by a weak band at 511 $cm^{-1}$ and by a very strong doublet at 475 and 464 $cm^{-1}$, which was very well reproduced in the calculated spectra. Whereas the weak band at 511 $cm^{-1}$ was definitively assigned to $\delta$(O–Si–O) modes, the nature of the dominant doublet seems to be more complex. For the Kaol structure, the power spectrum of oxygen atoms shows a strong maximum at this position. Bands around 475 $cm^{-1}$ can easily be recognized in the spectra of both the $SiO_4$ and $AlO_6$ units. However, these bands were absent in the power spectrum of the tetrahedral atoms. One possible explanation is that, analogous to $SiO_2$ polymorphs, the motion may be interpreted as a motion of the bridging oxygen atoms in the plane bisecting the $T^{IV}OT^{IV}$ bridges involving $OT^{IV}O$ bending and $T^{IV}O$ stretching degrees of freedom ($T^{IV}$ = Si, Al). Three lines were observed at 790, 751 and 707 $cm^{-1}$; that were closely reproduced in the simulation. The latter two were attributed to $\nu$(Si–O) modes. In the range 1120–1000 $cm^{-1}$ the calculated IR spectrum showed only a single band around 1050 $cm^{-1}$, whereas in the experiment the absorption is clearly split into a doublet with maxima at 1031 and 1007 $cm^{-1}$ and an additional band at 1115 $cm^{-1}$. For the calculated Raman spectrum a single band was obtained at 1051 $cm^{-1}$, corresponding to a single observed band at 1112 $cm^{-1}$. This relatively large number of bands in the experimental IR spectrum was attributed to deviations of the SiO six-member rings from pure hexagonal symmetry in Kaol. The absence of the shoulder at 1115 $cm^{-1}$ in the simulated spectrum could be caused by the simplification in the used potential model where the same SiO force constant was applied for Si–O–Si and Si–O–Al bridges. Bands in this region correspond to asymmetric modes of $T^{IV}$-O-$T^{IV}$ bridges.

Prencipe et al. (2009) used molecular modelling (ab initio periodic B3LYP), to successfully calculate the vibrational modes of lizardite. The highest-wavenumber mode in which $Si^{4+}$ is involved was the one at 1053 $cm^{-1}$; this is the $A_1$ Fröhlich mode whose LO component was calculated at 1113 $cm^{-1}$, and corresponds to an $\nu$(Si–O) mode where, in addition to Si, O coordinates three $Mg^{2+}$ ions; such stretching is mixed with $\delta$(O–Si–O) involving the other three O atoms of the Si tetrahedron. The $\nu_{AS}$(Si–O) mode involving all of the four Si–O bonds was at 1043 $cm^{-1}$ but, since this is of $A_2$ symmetry, it is neither Raman nor IR-active. At lower wavenumbers, the large

IR absorption band in the 950–1020 $cm^{-1}$ range corresponds to the mixing of the TO and LO components (respectively, calculated at 927 and 1041 $cm^{-1}$) that are due to the $\nu_{AS}$(Si–O) together with $\delta$(O–Si–O) and $\delta$(Si–O–Si). Tilting of the Si-centered tetrahedra around the Si–O bond axis is at 432 $cm^{-1}$. Other bending modes involving Si were at 705 and 737 $cm^{-1}$, but they are associated with low transition moments and are not experimentally observed in IR spectra. This contrasts with the generally accepted assignment of this band in the Raman to the Mg-OH inner surface symmetric translation mode. Other possibilities that could not be ruled out by Prencipe et al. (2009), given the present data, were that the Raman signal experimentally observed at 690 or 695 $cm^{-1}$ corresponds: (1) to the calculated $A_1$ mode at 705 $cm^{-1}$ ($\delta$(Si–O–Si) and $\delta$(O–Si–O) modes, or (2) to the calculated E mode at 737 $cm^{-1}$; the latter involving the motion of H atoms so that anharmonic effects are expected to lower its wavenumber significantly. The strong Raman band observed at 233–238 $cm^{-1}$ does correspond to the LO $A_1$ mode at 234 $cm^{-1}$, which can be described as due to coupled $\delta$(Mg–O–Si) and $\delta$(O–Mg–O).

Apart from a few bending modes involving H motions, very good agreement was observed between the calculated and experimental IR and Raman spectra of lizardite. By limiting the analysis to experimentally observed modes in IR or Raman spectra, $Mg^{2+}$ was shown to be mainly involved in the following:

- 544 $cm^{-1}$, a breathing mode of the $MgO_6$ octahedra parallel to [001];
- 463 $cm^{-1}$, a deformation of the Mg-centred octahedron ($\delta$(O–Mg–O));
- 442 $cm^{-1}$, corresponding to $MgO_6$ polyhedra deformation ($\delta$(O–Mg–O)) in directions parallel to (001), coupled to $\delta$(O–Si–O);
- 413 and 361 $cm^{-1}$, are $\tau$(O–Mg–O) modes with deformation of the octahedron parallel to (001);
- 392 and 225 $cm^{-1}$, are $\tau$(O–Mg–O) modes with deformation of the octahedron along [001] (in the latter case, the LO mode at 234 $cm^{-1}$ is observed).

The strong Raman signal experimentally observed at 233–238 $cm^{-1}$ corresponds to the torsional LO $A_1$ mode at 234 $cm^{-1}$, which can be described as due to coupled $\delta$(Mg–O–Si) and $\delta$(O–Mg–O) modes. Note that this assignment strongly contradicts the assignment by Prencipe et al. (2009) as a vibrational mode of the O–H–O group. Taking into account that long-range water interactions are also present at these wavenumbers, the assignments of Prencipe et al. (2009) might be valid for water in the system.

### 6.3.2 The 2:1 Clay Minerals

In the 2:1 clay minerals an evaluation of the vibrations can be made on the basis of the highest possible symmetry $C_{6v}$ for the pseudo-hexagonal tetrahedral sheet with $A_1$, $E_1$ and $E_2$ possibly Raman active modes. In the structure of talc the symmetry of the tetrahedral sheet is reduced to $C_s$, but the approximation of $C_{6v}$ satisfactorily accounts for the vibrations above 450 $cm^{-1}$ (Farmer, 1974c). The internal vibrations of tetrahedron $SiO_4$ are known to occur in this region. They are of $\nu_1(a_1)$, $\nu_2(e)$, $\nu_3(f_2)$ and $\nu_4(f_2)$ modes, and are all Raman active (Loh, 1973). He ascribed the strong band at 678 $cm^{-1}$ as $\nu_2(a_1)$ (Si–$O_b$–Si symmetric stretching mode), a medium band at 364 $cm^{-1}$ as $\nu_5(e)$, weak bands at 1021 $cm^{-1}$ and 1054 $cm^{-1}$ as $\nu_2(a_1)$ and $\nu_4(e)$ (non-bridging Si–O stretching modes), and two bands at 435 $cm^{-1}$ and 472 $cm^{-1}$ as $\nu_3(a_1)$ and $\nu_6(e)$ modes of $SiO_4$. The $\nu_3(a_1) = 450$ $cm^{-1}$ in $SiO_4$ is due to the stretching of $SiO_3$, moving in the direction normal to the sheet, against the non-bridged O, which in turn is the nearest neighbor of octahedral cations. A strong interaction between the $\nu_3(a_1)$ mode of $SiO_4$ and the lower wavenumbers associated with the octahedrons in the middle of the layer can reasonably be expected. The absorption bands between 450 $cm^{-1}$ and 900 $cm^{-1}$ were interpreted as combination bands of such origin. Comparable assignments have been reported by other authors (e.g. Blaha and Rosasco, 1978; Šontevska et al., 2007) Finally, the band at 111 $cm^{-1}$ has been ascribed to a torsional mode of the basal $SiO_4$ vibrating about the Si–O bond parallel to the *c*-axis with $A_g$ symmetry, while neighboring tetrahedral in the pseudohexagonal ring vibrate out of phase (Holtz et al., 1993, 1994).

Arab et al. (2002) tried to come to an assignment of the observed features in the vibrational spectra of Verm based on molecular dynamics calculations. For this purpose, a force field yielding a structure and vibrational spectra in good agreement with the experimental data was developed with the help of ab initio calculations of models of vermiculite structural units. The calculated Raman spectrum showed that each important feature in the experimental spectra has its counterpart in the computed ones. In the region 700–1200 $cm^{-1}$ the power spectrum of $Mg^{2+}$ atoms has no significant intensity, in contrast to the spectra of silicon and oxygen atoms. Bands at 1075, 921 and 737 $cm^{-1}$ can clearly be attributed to Si–O bonds since Si and O atoms are the only ones for which the bands are present in this region and have a high intensity. The band at 921 $cm^{-1}$ can be attributed unambiguously to the stretching vibration of the Si–$O_{II}$ bond ($O_{II}$ = tetra-coordinated oxygen bound to one $Si^{4+}$ and to three $Mg^{2+}$ atoms) and the two bands at 1075 and 737 $cm^{-1}$ are due to the degenerate stretching coordinates $T_3$ and $T'_3$ of the $Si(O_I)_3$ pyramid ($O_I$ = oxygen linking two $Si^{4+}$ atoms). The band calculated at 1033 $cm^{-1}$ belongs to the symmetric stretching $T_2$ coordinate of the pyramid. The small shift between experimental and calculated peak positions might be attributed to differences in the chemical composition of the experimental samples and simulated structure of the clay. In addition, three bands were calculated at 240, 278 and 309 $cm^{-1}$ (observed in natural spectrum at 278 and 310 $cm^{-1}$, while the first band was missing) attributed to $\delta(O_I$–Si–$O_I)$, and two

$\delta$(Si–O$_I$–Si) modes, respectively. The band observed at 434 cm$^{-1}$ was calculated at slightly higher wavenumber at 448 cm$^{-1}$ and assigned to $\delta$(O$_I$–Si–O$_I$).

## 6.4 THE VIBRATIONAL MODES OF THE TETRAHEDRAL AND OCTAHEDRAL SHEETS IN THE LOW-WAVENUMBER REGION

### 6.4.1 The 1:1 Clay Minerals

#### *6.4.1.1 The Kaolin Group Minerals*

One of the first Raman spectra of the kaolin group minerals was published by Wiewiora et al. (1979). Michaelian (1986) gave an interpretation using the approach identified by Loh (1973) and the initial calculations by Iishi et al. (1967). Various later studies (e.g. Johnston et al., 1985, 1998; Frost et al., 1993, 1996, 1997a, 2001a; Frost, 1995, 1997; Frost and Kloprogge, 1999, 2001; Bougeard et al., 2000) using both FT-Raman and Raman microscopy provided very similar results though, as shown in Table 6.4, the exact positions varied slightly between samples and techniques. The Raman spectra of selected kaolin group minerals are shown in Fig. 6.5. The assignment of some bands has not been definitely confirmed so far, most often a combination of the vibration modes of distorted $SiO_4$ tetrahedron with $C_{3v}$ symmetry, $AlO_6$ with $S_6$ symmetry with vibrations associated with O–H···O has been considered.

Iishi et al. (1967) calculated vibrational modes at 127, 130 and 285 cm$^{-1}$ for the ideal hexagonal $C_{6v}$ $Si_20_5$ lattice; for the less symmetrical $C_{3v}$ lattice, the predicted modes were at 113, 209 and 290 cm$^{-1}$. The calculated 127 and 130 cm$^{-1}$ bands were thought to account for the observed 130 and 143 cm$^{-1}$ Raman bands, respectively, for kaolinite (API #9, Mesa Alta, New Mexico) reported by Michaelian (1986). Similar bands were reported for Keokuk Kaol at 120 and 132 cm$^{-1}$ (Wiewiora et al., 1979), while Georgia Kaol showed a single band at 141 cm$^{-1}$ (Johnston et al., 1984). One possible assignment of the bands in this region suggested by Frost (1997) is that the higher wavenumber (143 cm$^{-1}$) is the O–Al–O symmetric bending vibration and the lower wavenumber band (127 cm$^{-1}$) is that of the O–Si–O symmetric bending vibration. Michaelian (1986) assigned the 197 cm$^{-1}$ Raman band to vibrations (probably translations) of structural OH, though an alternative assignment to the 209 cm$^{-1}$ mode calculated for the $C_{3v}$ $Si_2O_5$ lattice was also thought possible. Frost et al. (1997a) attributed the band near 200 cm$^{-1}$ to $\nu_2$(e) mode of the $AlO_6$ octahedron. The 245 cm$^{-1}$ Raman band was related to a mixture of silicon-oxygen deformation and OH libration and the band at 270 cm$^{-1}$ to a combination of silicon-oxygen deformations and octahedral sheet vibrations (Michaelian, 1986).

Bands at 336, 418, 431, 463, 516 and 638 cm$^{-1}$ were attributed to the combination of Si–O deformations and octahedral sheet vibrations; the 516 cm$^{-1}$ vibration involving a Si–O–Al linkage (Michaelian, 1986).

**TABLE 6.4** Low Wavenumber Raman Band Positions in the 1200–100 $cm^{-1}$ Region of the Dioctahedral 1:1 Minerals

| Suggested Assignment | Calc. | Kaolinite | | | | | | Dickite | | | | Nacrite | | Halloysite | | | | | |
|---|---|---|---|---|---|---|---|---|---|---|---|---|---|---|---|---|---|---|---|
| *(a)* | *(b)* | *(c)* | *(d)* | *(e)* | *(f)* | *(g)* | *(h)* | *(f)* | *(g)* | *(h)* | *(i)* | *(f)* | *(j)* | *(f)* | *(g)* | *(h)* | *(a)* | *(k)* | *(l)* |
| O–Si–O sym bend | | 120 | 130 | 130 | | 129 | 120 | | 120 | 119 | | | | | | 118 | | | |
| $Si_2O_5$ out of plane | 138 | 132 | | | | | 131 | | 131 | 131 | | | | | 127 | 126 | 130 | 135 | |
| $\nu_2$(e) of $AlO_6$ octahedron | | | 141 | 143 | | 143 | 141 | | 143 | 141 | | | | | 143 | 141 | 140 | 156 | |
| | | | | | | | | | | | | | | | | | | 168 | |
| | 171 | | | | | | | | | | | | | | | | | 172 | |
| | 188 | | | | | | | | | | | | 179 | | | | | 179 | |
| $A_{1g}(\nu_1)$ of $AlO_6$ octahedron | | 200 | 202 | 197 | 205 | | 197 | | | | | | 198 | 205 | 205 | | 200 | 192 | |
| | | 215 | | | | | | | | | 223 | | | | | | | 208 | |
| $B_2(\nu_3)$ of O-H–O triangle | 244 | 248 | 244 | 245 | 244 | | 238 | 235 | 235 | | 248 | 250 | 239 | 245 | | | 245 | 238 | 245 |
| | 260 | 272 | 271 | 270 | 270 | | 265 | 275 | 275 | | 270 | 275 | 271 | 266 | | | 274 | 298 | 275 |
| $A_1(\nu_1)$ of O-H–O triangle | 302 | 338 | 334 | 336 | 335 | | 340 | 340 | 340 | | 339 | 340 | 332 | 340 | | | 334 | | 332 |
| | | | | | 360 | | | | | | | | 366 | 360 | | | 354 | | 359 |
| $\nu_2$(e) of $SiO_4$ tetrahedron | | | 396 | 394 | | | | | | | | | | | | | 355 | | |

| | | | | | | | | | | | | | | | | | | | |
|---|---|---|---|---|---|---|---|---|---|---|---|---|---|---|---|---|---|---|---|
| $\nu_2$(e) of $SiO_4$ tetrahedron | | 420 | | 418 | | 395 | | | | | | 400 | | | | | 397 | 396 | |
| $\nu_2$(e) of $SiO_4$ tetrahedron | 432 | 434 | 430 | 431 | 430 | 426 | | | 432 | | 437 | | 422 | 425 | | | 429 | | 430 |
| | 456 | | | | | | | | | | 462 | | 433 | 440 | | | | 442 | |
| $\nu_4(f_2)$ of $SiO_4$ tetrahedron | 472 | 461 | 462 | 463 | 475 | 469 | | 480 | 459 | | 484 | 480 | 461 | 465 | 459 | | 464 | | 465 |
| | | | | | | | | | | | | | 482 | 470 | | | | | |
| $\nu_4(f_2)$ of $SiO_4$ tetrahedron | 538 | | 511 | 516 | 515 | 509 | | 515 | | | 525 | 520 | | 510 | | | 511 | 510 | 503 |
| | | | | | 560 | | | 560 | | | | 550 | 543 | 550 | | | | 540 | 548 |
| Si–O–Al translation | | | 637 | 638 | 655 | 637 | 645 | 665 | | | | 660 | 646 | 635 | | | 621 | | |
| Si–O–Al translation | 709 | | | 710 | 700 | 704 | | 710 | | | | 710 | 706 | 710 | 703 | | 706 | 693 | 710 |
| | 773 | 749 | 752 | 751 | 750 | 752 | | 750 | 745 | | 750 | 720 | 748 | 750 | 745 | | 748 | 728 | 748 |
| OH translation | 790 | 791 | 790 | 790 | 790 | 785 | | 790 | 795 | | 797 | 810 | 799 | 790 | 793 | | 787 | 779 | 794 |
| | | | | | | | | | | | 847 | | 827 | | | | | 844 | 826 |
| $Al_2OH_I$ libration | 912 | 916 | 915 | 915 | 890 | 915 | | 910 | 914 | | 921 | 920 | 915 | 910 | 910 | | 912 | 910 | 910 |
| $Al_2OH_{IS}$ libration | 944 | 935 | | 938 | 940 | 936 | | | | | | | | 940 | | | | | 944 |
| | | | | | | 1020 | | 1025 | | | | 1020 | 988 | | | | | | |

*Continued*

**TABLE 6.4** Low Wavenumber Raman Band Positions in the 1200–100 $cm^{-1}$ Region of the Dioctahedral 1:1 Minerals—Cont'd

| Suggested Assignment | Calc. | Kaolinite | | | | | | Dickite | | | | Nacrite | | Halloysite | | | | | |
|---|---|---|---|---|---|---|---|---|---|---|---|---|---|---|---|---|---|---|---|
| *(a)* | *(b)* | *(c)* | *(d)* | *(e)* | *(f)* | *(g)* | *(h)* | *(f)* | *(g)* | *(h)* | *(i)* | *(f)* | *(j)* | *(f)* | *(g)* | *(h)* | *(a)* | *(k)* | *(l)* |
| $A_1(\nu_1)$ of Si–O | 1051 | 1048 | | | | 1043 | | | | | 1049 | | 1040 | | | | 1060 | | |
| $A_1(\nu_1)$ of Si–O | | 1105 | | | | 1110 | | 1080 | | | | 1085 | | 1110 | | | 1080 | 1100 | |
| | | | | | | | | 1160 | | | | 1160 | | | | | | | |
| | | | | | | | | 1190 | | | | 1175 | | | | | | | |

(a) Frost et al. (1997a).
(b) Bougeard et al (2000).
(c) Wiewiora et al. (1979).
(d) Johnston et al. (1985).
(e) Michaelian (1986).
(f) Frost et al. (1993).
(g) Frost (1995).
(h) Frost (1997).
(i) Johnston et al. (1998).
(j) Frost and Kloprogge (2000d).
(k) Frost and Shurvell (1997).
(l) Kloprogge and Frost (1999d).

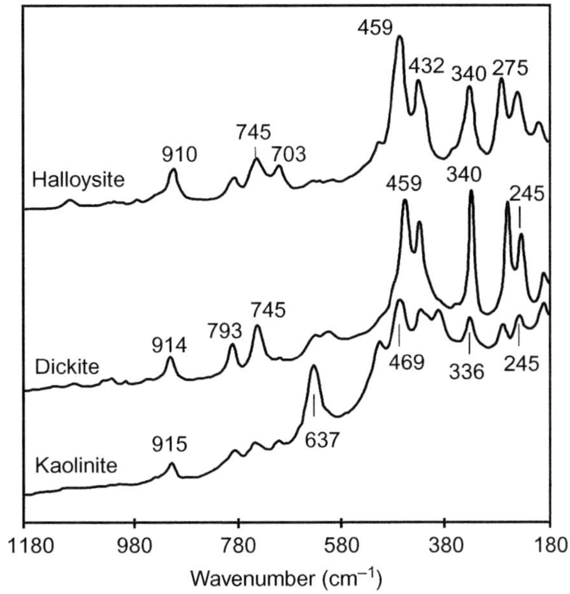


**FIG. 6.5** Raman spectra of kaolinite KGa-2, dickite (San Juan) and halloysite (New Zealand) in the low wavenumber region; modified after Frost (1995).

In contrast, four bands at 395, 426, 469 and 509 cm$^{-1}$, reported by Frost (1995) for Kaol, were attributed only to the $\delta$(O–Si–O) and $\delta$(Si–O–Si) vibrations active in both the Raman and IR spectra (Farmer, 1974c). There is a significant variation in the relative intensities of these bands as a function of order-disorder, (e.g. between KGa-1 and KGa-2). A similar variation was observed in this spectral region for wet and dry KGa-1 Kaol (Johnston et al., 1985).

The bands in the region of 700–800 cm$^{-1}$ for Kaol are associated with Al–O–Si vibrations, some of which involve structural OH (Michaelian, 1986; Frost and Kloprogge, 2001). FT-Raman spectroscopy showed mainly three bands in this region centered near 705, 756 and 789 cm$^{-1}$. Very similar Raman band positions, but without assignment, were reported for Mesa Alta and KGa-1 Kaol by Johnston et al. (1985).

The in-plain bending vibrations of the inner surface OH ($\delta(Al_2OH)_{IS}$) in Kaol occurred at 936 cm$^{-1}$ and the inner hydroxyl ($\delta(Al_2OH)_I$) at 915 cm$^{-1}$ (Farmer, 1974c). These bands are strongly IR active but are very weak in the Raman, with the inner hydroxyl showing sensitivity to the structure. The $\delta(Al_2OH)_I$ Raman band of Kaol was observed near 915 cm$^{-1}$ (Fig. 6.5; Table 6.4). Frost et al. (1993) reported a Raman spectrum of Kaol with a strong band at 900 cm$^{-1}$ and a weaker band at 940 cm$^{-1}$. Frost and Kloprogge (1999) studied the behavior of Kaol at RT (298 K) in comparison to liquid nitrogen (77 K). The 298 K Raman spectrum was fitted with four bands at 939, 923 and 916 $cm^{-1}$ with a low-intensity band at 897 $cm^{-1}$. This latter band was attributed to non-H-bonded $Al_2OH_{IS}$. Upon cooling to 77 K all bands were shifted to higher wavenumbers and the spectrum showed a distinct increase in intensity and a broadening of the spectral profile in the 935 to 945 $cm^{-1}$ region. The $\nu$(Si–O–Si) modes giving intense bands in the IR spectra are weak in the Raman, but are occasionally observable at 1020, 1043 and 1110 $cm^{-1}$ (Frost, 1997).

The other kaolin group minerals, dickite and halloysite, have similar patterns as Kaol (Fig. 6.5), though there are small shifts in positions and the intensities show significant differences. The positions and the assignments of the individual bands, reported by different authors, are given in Table 6.4. Frost (1995) observed the symmetric bending ($\delta_s$(O–Si–O)) vibration at 129, 127 and 120 $cm^{-1}$ for Kaol, halloysite and dickite, respectively. All kaolinite polymorphs were characterized by a very intense band at 143 $cm^{-1}$ that was an order of magnitude more intense than the other bands in the 200 to 1200 $cm^{-1}$ region. The dickite from San Juan shows complexity with bands at 143, 130 and 120 $cm^{-1}$, whereas the dickite from Saint Claire showed only two bands at 131 and 120 $cm^{-1}$. The halloysite from New Zealand shows peaks at 143 and 127 $cm^{-1}$ similar to those of Kaol. The spectrum of Eureka halloysite with bands at 129 and 120 $cm^{-1}$ was similar to that of dickite. For halloysite, Frost and Shurvell (1997) reported bands at 156, 168, 172 and 179 $cm^{-1}$ attributed to the $\delta_s$(O-Al–O) modes of the $AlO_6$ octahedra. These bands were assigned to $\nu_s(Al_2OH)$ of the Al–OH groups. One possibility is that the band at 156 $cm^{-1}$ is associated with the stretching vibration of the inner surface OH. The three other bands of the $\delta_s$(O-Al–O) modes of the $AlO_6$ octahedra are associated with the three inner OH. The intensities of these bands were thought to help confirm their assignment as symmetric vibrations.

Two bands occurred at 245 and 270 $cm^{-1}$ in the Raman spectrum of KGa-2 kaolinite and at 235 and 275 $cm^{-1}$ for dickite and halloysite. Their intensity increased from kaolinite through halloysite to dickite (Fig. 6.5). Johnston et al. (1998) observed similar bands for dickite at 246 and 270 $cm^{-1}$. The most intense bands in the low wavenumber region of dickite and halloysite, attributed to a combination of Si–O deformations and octahedral sheet vibrations (Michaelian, 1986) occurred at 340, 432 and 459 $cm^{-1}$. Their intensities are significantly higher than observed for corresponding bands of kaolinite (Fig. 6.5). The intense band for kaolinite was observed also at 637 $cm^{-1}$. The band is weak for dickite and almost not visible for halloysites. The bands at 703, 745 and 795 $cm^{-1}$ for halloysite and 745 and 793 $cm^{-1}$ for dickite, similar as corresponding bands of Kaol, were associated with Al–O–Si vibrations, some of which involve the hydroxyl groups (Michaelian, 1986; Frost and Kloprogge, 2001). Kloprogge and Frost (1999d) reported similar bands for halloysite from Belgium at 710, 748 and 794 $cm^{-1}$. The in-plain OH bending vibrations of the inner hydroxyl of dickite and halloysite give weak Raman bands around 914 and 910 $cm^{-1}$, respectively (Fig. 6.5).

Johnston et al. (1998) observed only the band at 919 $cm^{-1}$ in the polarized Raman spectra of the St. Claire dickite. Frost and Shurvell (1997) identified a single band at 910 $cm^{-1}$ for halloysite, but also a shoulder/low intensity band around the 936 $cm^{-1}$. In contrast, Kloprogge and Frost (1999d) clearly observed a strong band at 910 $cm^{-1}$ accompanied by a weaker band at 944 $cm^{-1}$ for a Belgian halloysite. The OH bending bands, present in the IR spectra of kaolin group minerals, have been shown to be distinguishable by selective deuteration (Wada, 1967). The band at 935 $cm^{-1}$ shifted upon deuteration to 720 $cm^{-1}$. Thus this band was assigned to the inner surface OH. The band at 915 $cm^{-1}$ remained unchanged in this process, thus confirming the assignment of this band to the deformation mode of the inner OH. Based on the IR results the Raman band observed at 935–944 $cm^{-1}$ was also supposed to correspond to the inner surface OH.

Raman spectra give very little information on the asymmetric Si–O stretching vibrations appearing in the 900–120 $cm^{-1}$ region. Frost (1995) reported bands at 1020, 1043 and 1110 $cm^{-1}$ for kaolinite (Table 6.4), which are hardly recognized in their Raman spectra (Fig. 6.5). Raman spectrum of dickite gives bands at 1025, 1080, 1160 and 1190 $cm^{-1}$ but halloysite gives only a single band at 1110 $cm^{-1}$ (Frost et al., 1993). Low sensitivity of the Raman compared to IR spectroscopy in the Si–O stretching region is attributed to the difference in the methods. Raman depends on the changes in the polarizability of the bonds and is applied to the symmetric vibrations and vibrations involving atoms of similar size. In contrast, IR depends on the changes in bond dipole and thus is useful for examination of asymmetric vibrations and vibrations that involve atoms of different size (see Chapter 2 in this volume for more details).

#### *6.4.1.2 The Serpentine Group Minerals*

In the serpentine group minerals (lizardite, chysotile, antigorite) the octahedral sheet is fully filled with central cations compared to the kaolin group where only 2/3 of the octahedral sites are filled. The three principal minerals of the serpentine group have a very similar chemical composition but significantly different structures. As such, an effect on the band positions in the low wavenumber region of the Si–O vibrations can be expected. The Raman spectra of serpentine group minerals showed similar patterns in the 1150–150 $cm^{-1}$ region (Fig. 6.6). The characteristic positions, reported by different authors, are listed in Table 6.5.

The Raman spectrum of lizardite showed mainly the bands produced by vibrations of the $SiO_4$ tetrahedra or by silicon–oxygen linkages (Rinaudo et al., 2003). The antisymmetric stretching mode ($\nu_{as}$) of the Si–$O_{nb}$ (nb = non-bridging; b = bridging) was observed as a weak band at 1096 $cm^{-1}$,

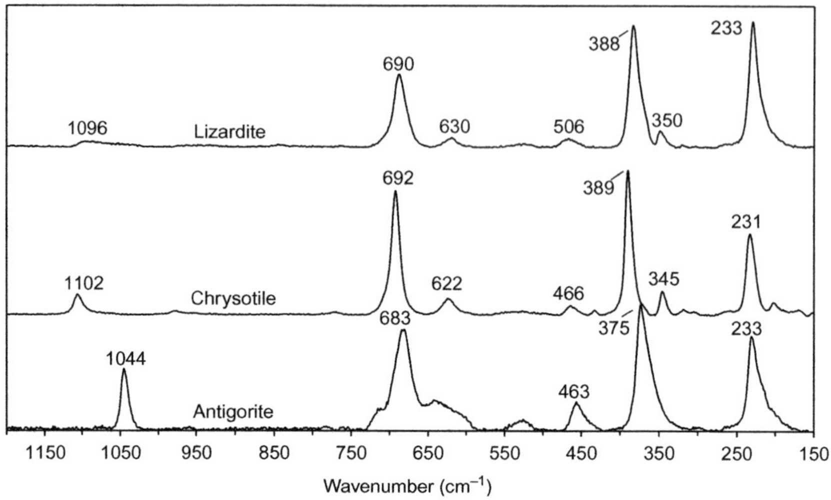


**FIG. 6.6** The Raman spectra of the serpentine minerals antigorite, chrysotile and lizardite in the low wavenumber region.

while the $\nu_s$(Si–O$_b$–Si) (b = bridging) modes together with Mg–OH inner surface translation contribute to the band at 690 cm$^{-1}$ (Fig. 6.6). Based on the polarization experiments performed by Kloprogge et al. (1999c) the band at 630 cm$^{-1}$ was assigned to antisymmetric Mg–OH translation modes. A band near 510 cm$^{-1}$ was also detected in the Raman spectrum of lizardite. This band was observed in all spectra of lizardite recorded by Rinaudo et al. (2003), and it appears to consist of two bands at 506 and 515 cm$^{-1}$. An assignment to deformation vibrations of the $SiO_4$-$AlO_4$ tetrahedra was proposed by Rinaudo et al. (2003). In the structure of lizardite, the mismatch between the tetrahedral and octahedral sheets is compensated by the deviation of the tetrahedra from an ideal hexagonal symmetry to a ditrigonal arrangement by tilting the tetrahedra about −2.8 degree and by a more significant substitution of $Al^{3+}$ for $Si^{4+}$ in the tetrahedral sheet (Mellini, 1982; Mellini and Viti, 1994). The splitting of the band at about 510 cm$^{-1}$ into two bands at 509 and 516 cm$^{-1}$ may therefore reflect deformation vibrations of the two types of tetrahedra, $AlO_4$ and $SiO_4$. Trittschack et al. (2012) reported very similar bands for lizardite powder though some bands were assigned to chrysotile (the Monte Fico sample contains up to 25% of chrysotile). The intense band at 388 cm$^{-1}$ was ascribed to the $\nu_5$(e) modes of the $SiO_4$ tetrahedra.

The Raman spectra of chrysotile was studied by Rinaudo et al. (2003), Groppo et al. (2006), Šontevska et al. (2007), Anbalagan et al. (2010), Trittschack and Grobety (2013) and Petriglieri et al. (2015). The main bands of chrysotile presented in Fig. 6.6 are assigned according to the description of the vibrational bands in the IR spectra of the clay minerals proposed by Farmer (1974c) and based on the assignment of Raman bands in tubular

## TABLE 6.5 Low Wavenumber Raman Band Positions in the 1200–100 cm$^{-1}$ Region of the Trioctahedral 1:1 Minerals

| Suggested Assignment | Lizardite | | | | | | | | Chrysotile | | | | | | | | | Antigorite | | | | | |
|---|---|---|---|---|---|---|---|---|---|---|---|---|---|---|---|---|---|---|---|---|---|---|---|
| | *(a)* | *(b)* | *(c,d)* | *(e)* | *(f)* | *(g)* | *(h)* | *(i)* | *(a)* | *(b)* | *(e)* | *(j)* | *(k)* | *(l)* | *(f)* | *(d)* | *(m)* | *(a)* | *(b)* | *(e)* | *(f)* | | *(j)* |
| | | | | | | | | | | | | | | | | | Ni-sub | | | | nat | synth | |
| O–Si–O bending | | | 129 | | | | | | | | | 132 | | | | 129 | | | | | | | 116 |
| $A_{1g}$ $Mg(O,OH)_6$ | | 195–197 | 202 | | | | | 199 | | 191–194 | | 202 | | 180 | | 202 | | | 197–202 | | | | |
| O–H···O | | 213–214 | | | | | | | | 211–213 | | | | | | | | | | | | | |
| | | | | | | | | | | | | | | | | | | | | | | | 224 |
| | 233 | 234–236 | 229 | 230 | 238 | | | 231 | 231 | 232–235 | 232 | 233 | 234 | 232 | 235 | 233 | 230 | 230 | 230–231 | 229 | 235 | 229 | 230 |
| $SiO_4$ bending | | | 305 | | | | 304 | 304 | | | | | 305 | 304 | | 302 | | | | | | | |
| | | | 320 | | | | 321 | 318 | | | | 316 | 321 | 325 | | 318 | | | | | | | |
| | 350 | | 346 | 349 | | | 345 | 345 | 345 | 348 | 345 | 346 | 348 | 348 | | 345 | 348 | | | 337 | | | |
| Mg–OH sym | | | 371 | | | | | 374 | | | | | | | | | | | | | | | |
| $\nu_5(e)$ sym $SiO_4$ | 388 | 380–388 | 384 | 386 | 393 | | 389 | 388 | 389 | 390–394 | 388 | 390 | 392 | 390 | 391 | 389 | 388 | 375 | 379 | | 377 | 378 | 372 |
| Mg–OH antisym translation | | | 430 | | | 432 | 432 | 432 | | | | | 436 | 438 | | 432 | | | | | | | |
| $\nu_3(a_1)$ $SiO_4$ | | | | | | | | 458 | | | | | | | | 455 | | | | | | | |
| Mg–OH translation $+\nu_6(e)$ $SiO_4$ | | | 465 | | | 463 | 465 | 466 | | | | | 469 | 464 | | 468 | | | 463 | | | 459 | |
| $SiO_4$-$AlO_4$ deformation | 506 | | | | | | | | | | | | | | | | | | | | | | |
| | 515 | | 518 | 532 | | | | | | | | | | | | | | 520 | | | | | |
| $Mg_3OH)_l$ libration | | | | | | 592 | | 607 | | | | | | | | | | | | | | | |
| $Mg_3OH)_{lS}$ antisym translation | | | 622 | 621 | | 622 | 623 | 629 | 620 | | 622 | 624 | 624 | 622 | | 623 | 620 | | | | | | 628 |
| | 630 | 640 | | | | | | | | 640–641 | | | | | | | | 635 | 640 | 631 | | | |
| ν sym $Si–O_b–Si$ | 690 | 688–693 | 690 | 689 | 695 | 673 | 692 | 692 | 692 | 693–695 | 691 | 692 | 694 | 692 | 694 | 691 | 690 | 683 | 683–684 | 687 | 685 | 688 | 682 |

*Continued*

**TABLE 6.5** Low Wavenumber Raman Band Positions in the 1200–100 $cm^{-1}$ Region of the Trioctahedral 1:1 Minerals—Cont'd

| Suggested Assignment | Lizardite | | | | | | | | Chrysotile | | | | | | | | | Antigorite | | | | |
|---|---|---|---|---|---|---|---|---|---|---|---|---|---|---|---|---|---|---|---|---|---|---|
| | *(a)* | *(b)* | *(c,d)* | *(e)* | *(f)* | *(g)* | *(h)* | *(i)* | *(a)* | *(b)* | *(e)* | *(j)* | *(k)* | *(l)* | *(f)* | *(d)* | *(m)* | *(a)* | *(b)* | *(e)* | *(f)* | *(j)* |
| $Mg_3OH)_{IS}$ sym translation | | | 707 | | | | | 705 | | | | | 706 | | | 707 | | | | | | |
| | | | | | | | | 709 | | | | | | | | | | | | | | |
| Si–O stretch | | | 929 | | | | | | | | | | | | | | | | | | | |
| | | | 972 | | | | | | | | | | | | | | | | | | | |
| v Antisym $Si–O_b–Si$ | | | 1050 | 1045 | | | | | | | | | | | | | | 1044 | 1045 | 1045 | 1045 | 1048 |
| v Antisym $Si–O_{nb}$ | 1096? | | 1094 | | | 1100 | 1105 | 1102 | 1105 | | | 1106 | | 1105 | | 1105 | 1089 | | | | | 1090 |
| | | | | | | | | | | | | | | | | | | | | | | 1099 |

(a) Rinaudo et al. (2003).
(b) Groppo et al. (2006).
(c) Trittschack et al. (2012).
(d) Trittschack and Grobety (2013).
(e) Petriglieri et al. (2015).
(f) Reynard and Wunder (2006).
(g) Lewis et al. (1996).
(h) Bard et al. (1997).
(i) Kloprogge et al. (1999c).
(j) Šontevska et al. (2007)).
(k) Foresti et al. (2009).
(l) Anbalagan et al. (2010).
(m) Bloise et al. (2010).

chrysotile reported by Kloprogge et al. (1999c). A band at 1102 $cm^{-1}$ was attributed to $\nu_{as}$(Si–$O_{nb}$) while an intense band near 690 $cm^{-1}$ was attributed to t $\nu_{as}$(Si–$O_b$–Si) modes. Kloprogge et al. (1999c) revealed also a weak shoulder at 705 $cm^{-1}$ under the strong 692 $cm^{-1}$ band based on a detailed study of the Raman spectrum of a single crystal of the chrysotile and its orientation dependence. In the IR spectrum of brucite, $Mg(OH)_2$, a band is also observed around 690 $cm^{-1}$. Therefore it seems reasonable for the two bands at 705 and 692 $cm^{-1}$ that the first band must be assigned to the $\nu_s$(Si–O–Si) and the second to an outer $\delta_s$(MgOH) translation mode of the OH oriented almost parallel to the *a*-axis. Lewis et al. (1996) and Bard et al. (1997) ascribed a single band at 622 $cm^{-1}$ to a $\nu_s$(Si–O–Si) mode. The band component analysis of this band revealed, in fact, two bands around 629 and 622 $cm^{-1}$ (Kloprogge et al. (1999c)). Polarization experiments showed no changes in intensity for both the 629 and 622 $cm^{-1}$ bands indicating that these represent antisymmetric modes. The band at 629 $cm^{-1}$ seems to represent the $Mg_3OH$ mode.

The double band at 466 and 458 $cm^{-1}$ for chrysotile, revealed using peak-fit analysis (Kloprogge et al. (1999c)), has been described before as a single, both Raman and IR active (Fig. 6.6). This has only been described as a metal-oxygen vibration without further details by Lewis et al. (1996) and Bard et al. (1997). Based on the comparison between talc and chrysotile and the theoretical calculations it is possible to assign the band at 458 $cm^{-1}$ as the $\nu_3(a_1)$ mode of $SiO_4$. The 466 $cm^{-1}$ band, however, is probably comparable to the OH translational mode observed for talc in the IR spectrum at 465 $cm^{-1}$, an assignment that was confirmed by deuteration experiments. It seems less likely that this band represents the $\nu_6$(e) mode of the $SiO_4$ tetrahedron although it cannot completely be ruled out. The weak band at 432 $cm^{-1}$ was ascribed to an antisymmetric Mg–OH translation mode (Kloprogge et al. (1999c)). The intense band at 389 $cm^{-1}$, similar as for lizardite, was related to the symmetric $\nu_5$(e) modes of the $SiO_4$ tetrahedra. Five Raman bands of different intensity were observed at 374, 345, 318, 304 and 231 $cm^{-1}$ by Kloprogge et al. (1999c) in the region between approximately 380 and 200 $cm^{-1}$ (Fig 6.6). These bands has never been assigned in detail for clay minerals like those of the serpentine group because of the complexity of the bands, which appear to involve mixed vibrational modes of the Si–O network, the octahedral cations and the hydroxyl groups (Farmer, 1974c). Polarization experiments proved that the 374, 318 and 304 $cm^{-1}$ bands are antisymmetric modes, while the 345 and 231 $cm^{-1}$ bands are symmetric modes. The 345 $cm^{-1}$ band shows very strong polarization. Based on the comparison between talc, brucite and chrysotile the band at 374 $cm^{-1}$ must be associated with a symmetric Mg–OH vibration. However, based on the present data and evidence from the literature a more detailed assignment of the bands at 345, 318, 304 and 231 $cm^{-1}$ is not possible. The band at 199 $cm^{-1}$ is comparable to the band around 195–197 $cm^{-1}$ observed for talc and is assigned to the $A_{1g}$ mode of a $Mg(O,OH)_6$ octahedron distorted in the direction normal to the octahedral sheet (Loh, 1973; Blaha and Rosasco, 1978; Rosasco and Blaha, 1980).

Two studies provided information on the effects of $Ni^{2+}$ and $Fe^{3+}$ doping/substitution in the structure of chrysotile (Foresti et al., 2009; Bloise et al., 2010). When the spectrum of $Ni^{2+}$-doped chrysotile was compared with the Raman spectrum for stoichiometric synthetic chrysotile, shifts in some vibrational modes were identified. The Ni- chrysotile bands at 690 and 388 $cm^{-1}$, ascribed to symmetric stretching modes of Si–$O_b$–Si linkages and to bending modes $\nu_5(e)$– of $SiO_4$ tetrahedra, respectively, appeared at 694 and 392 $cm^{-1}$ in stoichiometric synthetic chrysotile (Bloise et al., 2010). Among the bending modes of $SiO_4$ tetrahedra, detected at 348, 321 and 305 $cm^{-1}$ in stoichiometric chrysotile, only the band at 348 $cm^{-1}$ was observed for $Ni^{2+}$-substituted chrysotile, whereas the bands at 321 and 305 $cm^{-1}$ were not detected. The vibrational modes at 624 $cm^{-1}$ in Raman spectra from stoichiometric chrysotile and produced by Mg–OH translations were shifted toward lower wavenumbers, 620 $cm^{-1}$; the bands at 469 and 436 $cm^{-1}$ in the stoichiometric chrysotile, ascribed to vibrational translation of Mg–OH/bending modes of the $SiO_4$ tetrahedra and to Mg–OH translation respectively, were not detected in $Ni^{2+}$-doped chrysotile. The band at 230 $cm^{-1}$ did not show any significant shift. The observed shifts toward lower wavenumbers may be due to the high concentration of $Ni^{2+}$ in the crystallized chrysotile (Bloise et al., 2010).

Results of the Raman spectra of chrysotile with increasing iron doping amounts are similar to those shown for nickel doping. The most intense bands observed for iron-free chrysotile at 694 and 392 $cm^{-1}$ shifted to lower wavenumbers, such as 690 and 390 $cm^{-1}$, with increasing iron concentration up to 0.94% wt (Foresti et al., 2009). On the contrary, in the sample with 1.37% wt of iron, the band was observed at 695 $cm^{-1}$, similar to the iron-free sample. The shift of the bending modes $\nu_5(e)$ of the $SiO_4$ tetrahedra from 392 to 389 $cm^{-1}$, detected in highest iron doped samples was appreciable. The changes in the Si–O modes positions indicated the $Fe^{3+}$–$Si^{4+}$ substitutions in the tetrahedral sites (Foresti et al., 2005). The other bands for stoichiometric synthetic chrysotile at 469 $cm^{-1}$, assigned to vibrational translation of Mg–OH, at 436 $cm^{-1}$, and the band at 624 $cm^{-1}$, assigned to (H)O–Mg–O(H) translation mode, all shifted to lower wavenumbers with increasing Fe doping. These results indicated that the chrysotile synthesized with different Fe concentrations showed shifts of the vibrational modes of the octahedra, because of $Mg^{2+}$–$Fe^{3+}$ substitutions in the octahedral sites. This was observed when the $Fe^{3+}$ incorporation did not exceed 0.71% wt. At higher iron doping, the synthesized chrysotile crystallinity was strongly affected. The results highlighted that $Fe^{3+}$ can replace both $Mg^{2+}$ and $Si^{4+}$, differently modifying the chrysotile structure as a function of the $Fe^{3+}$ doping extent and the $Fe^{3+}$ doping process appears strongly affected by the presence of metallic $Fe^0$ in the synthetic environment. In fact, octahedral coordinated $Fe^{3+}$ has been observed in all the substitution range in $Fe^{3+}$-doped chrysotile synthesized in the absence of metallic $Fe^0$. On the contrary, tetrahedral coordinated $Fe^{3+}$ inducing a flattening of the chrysotile structure appeared prevalent in respect of octahedral coordinated $Fe^{3+}$ in highly $Fe^{3+}$-doped chrysotile synthesized when metallic $Fe^0$ is available in the synthetic environment (Foresti et al., 2009).

The Raman spectrum of antigorite shows similar bands as observed for lizardite and chrysotile; however, the positions of some bands are shifted to lower wavenumbers (Fig. 6.6). Rinaudo et al. (2003) reported (1) a band at 1044 $cm^{-1}$ corresponding to the $\nu_{as}$(Si–$O_b$–Si) modes ($E_1$) similar to the band detected and ascribed in the Raman spectrum of talc at 1049 $cm^{-1}$ by Rosasco and Blaha (1980); and (2) an intense band at 683 $cm^{-1}$ produced by $\nu_s$(Si–$O_b$–Si) linkages; and iii) the band due to $\nu_5$(e) modes of the $SiO_4$ tetrahedra at 375 $cm^{-1}$. Comparable positions of the antisymmetric and symmetric Si–$O_b$–Si stretching vibrations at 1045 and 684 $cm^{-1}$ were reported for antigorite, also by Groppo et al. (2006). The band at 530 $cm^{-1}$ was assigned to the deformation of $SiO_4$–$AlO_4$ tetrahedra (Petriglieri et al., 2015). A band was also observed at 463 $cm^{-1}$, such as in a spectral region where Mg–OH translations and $\nu_6$ $SiO_4$ vibrations are expected (Kloprogge et al. (1999c)).

The ambient condition Raman spectrum of synthetic antigorite reported by Reynard and Wunder (2006), with its low wavenumber region corresponding to the lattice and silicate layer internal vibrational modes, was similar to that obtained on a natural sample (Auzende et al., 2004). The sharp bands associated with the $SiO_4$ tetrahedra were observed at 1098, 1045, 688, 459, 378 and 132 $cm^{-1}$. Because of the complexity of the structure, many bands exhibited small splitting, shoulders, and were broader than in the higher symmetry serpentine variety lizardite (Auzende et al., 2004). The slight wavenumber differences between synthetic and natural antigorite were attributed to the chemical substitution ($Fe^{3+}$ and $Al^{3+}$ mostly) in the natural serpentine, and possibly to orientation effects that can change the relative intensity of slightly split bands (Kloprogge et al. (1999c)). With increasing pressure, all Raman bands shifted toward higher wavenumbers. Synthetic antigorite still showed well-resolved Raman spectra with no significant line broadening except at 10.9 GPa where the pressure medium solidified. This shows that amorphization of antigorite did not take place up to this pressure for hydrostatic or quasi-hydrostatic conditions. In the 1100–100 $cm^{-1}$ region the pressure-induced shifts were equal to those of natural antigorite and other serpentines, except for the band at 688 $cm^{-1}$, which had a larger pressure dependence than the same band in all natural serpentines. This mode is typically assigned to $\delta$(Si–$O_b$–Si) vibrations in clay minerals (Kloprogge et al. (1999c)). Reynard and Wunder (2006) had no clear explanation for this difference, which could be due to a difference in tetrahedral sheet geometry that increases with pressure between synthetic and natural antigorite.

The presence of a band at 1044 $cm^{-1}$ has proven to be decisive for the discrimination of antigorite from the other polymorphs lizardite and chrysotile (Fig. 6.6). A significant number of papers have used the shift in band positions in the Raman spectra of antigorite compared to lizardite and chrysotile to identify antigorite (see e.g. Bloise et al., 2014; Jourdan et al., 2012; Khedr and Arai, 2013; Levitan et al., 2009; Schwartz et al., 2013; Šontevska et al., 2007). Several studies have shown that Raman microscopy can be extremely useful to identify antigorite in Chinese jade (Wang and Li, 2011; Wang et al., 2012; Zhao et al., 2015).

## 6.4.2 The 2:1 Clay Minerals

### 6.4.2.1 *Talc and Pyrophyllite*

The 2:1 layer of a trioctahedral talc consists of a brucite-like octahedral sheet between two silicate tetrahedral sheets. The Raman spectrum of talc is shown in Fig. 6.7. Loh (1973) ascribed weak Raman bands at 1054 and 1021 $cm^{-1}$ as $\nu_4(e)$ and $\nu_2(a_1)$ non-bridging Si–O stretching modes, respectively, the strong band at 678 $cm^{-1}$ as Si–$O_b$–Si symmetric stretching mode, two weak bands at 472 $cm^{-1}$ and 435 $cm^{-1}$ as $\nu_6(e)$ and $\nu_3(a_1)$ modes of $SiO_4$ and a relatively intense band at 364 $cm^{-1}$ as $\nu_5(e)$. The brucite-like sheet is composed of octahedrally coordinated $Mg^{2+}$ ions and thus metal-oxygen vibrational modes are seen in the region below 450 $cm^{-1}$ down to near 200 $cm^{-1}$. Some of the deformation modes of the silicate network also extend into this region as well (Blaha and Rosasco, 1978). Loh (1973) reported weak spectral features at 260, 196, 162, 157 and 132 $cm^{-1}$ together with a single band at 107 $cm^{-1}$

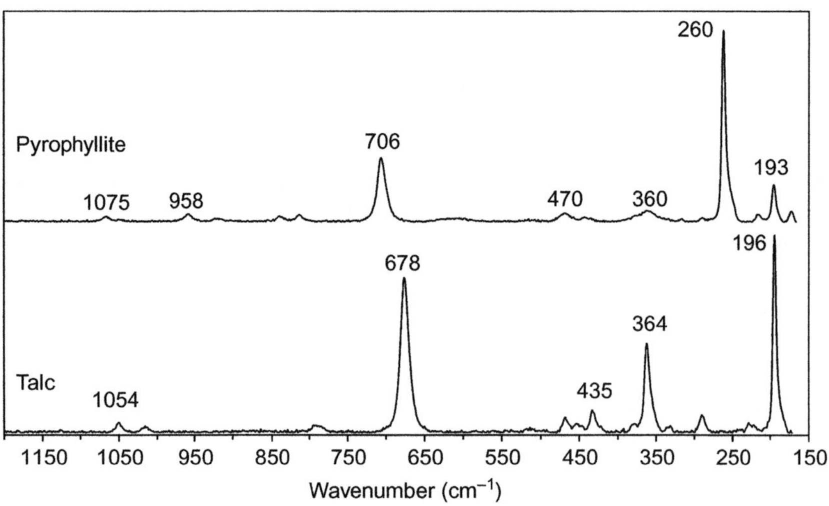


**FIG. 6.7** Raman spectra of talc and pyrophyllite in the low wavenumber region.

for talc that have not been observed in any spectrum measured by Rosasco and Blaha (1980). Loh (1973) assigned the bands at 196 $cm^{-1}$, 162 $cm^{-1}$ and 107 $cm^{-1}$ as the $A_{1g}(v_1)$, $E_g(v_2)$ and $F_{2g}(v_5)$ modes (under $O_h$ symmetry), respectively. In contrast, Holtz et al. (1993) ascribed the band near 100 $cm^{-1}$, common to all layered silicates, to the torsional mode of the $SiO_4$ tetrahedra vibrating about the Si–O bond parallel to the c-axis with $A_g$ symmetry.

For pyrophyllite, there are very few reports available on the Raman spectra, but IR spectra have been regularly reported (e.g. Farmer, 1974c). Two pyrophyllite samples from different localities showing an exact match of Raman band positions, relative intensities, and spectral patterns were reported by Wang et al. (2015). Even though the bands were not labelled or mentioned in the text an estimate of their position can be made from the spectra with weak bands at 1075 and 1045 $cm^{-1}$ as the $\nu_4(e)$ and $\nu_4(a_1)$, respectively, two broad bands around 645 and 625 $cm^{-1}$, one of which is most likely associated with the $\nu_2(a_1)$, two very weak bands around 470 and 437 $cm^{-1}$ attributed to $\nu_6(e)$ and $\nu_3(a_1)$, respectively, and finally the $\nu_5(e)$ around 360 $cm^{-1}$ (Fig. 6.7) (assignment of bands following Loh, 1973). The 958 $cm^{-1}$ should be related to the $\delta(Al_2OH)$ vibrations.

#### 6.4.2.2 Smectites

Raman spectra of Al-rich dioctahedral smectites, most notably Mt, have been reported by Frost and Rintoul (1996a,b) Bishop and Murad (2004), Kosarova et al. (2013), Wang et al. (2015) and Ritz et al. (2016). Most often the Mt standards, such as SAz-1/SAz-2, SWy-1/SWY-2, STx-1, obtained from the Source Clay Repository of the CMS, were examined and attempts to explain differences in spectra were related to the chemical composition of a mineral and/or different exchangeable cations. The FT-Raman spectrum of Mt SAz-2 reported by Ritz et al. (2016) is shown in Fig. 6.8.

The comprehensive Raman study of the low wavenumber region of montmorillonites was performed by Frost and Rintoul (1996a,b). To analyze spectra of montmorillonites these authors considered the vibrations of four structural units: (1) the $MO_6$ octahedron where M is the octahedral cation, (2) the O–H···O triangle such as occurs in H-bonding between hydroxyl groups and adjacent oxygens, (3) the $SiO_4$ tetrahedron, and (4) the librations of the OH group. It should be noted, however, that they neglected probable contributions of vibrational modes of physi-sorbed and cation hydration water molecules which might also have been present in the low wavenumber region.

Frost and Rintoul (1996a,b) observed several weak bands in the Raman spectra of montmorillonites in the 1110–750 $cm^{-1}$ region. The $\sim$1100 $cm^{-1}$ band was ascribed to the $SiO_4$ unit $\nu_3(f_2)$ vibrational mode. This band was observed at a higher wavenumber than that predicted by theory (1055 $cm^{-1}$) and this may be attributed to the distortion of $SiO_4$ tetrahedra from $T_d$ symmetry to $C_{3v}$ symmetry. The structural OH deformation modes near 915 $cm^{-1}$ ($\delta(Al_2OH)$), 885 $cm^{-1}$ ($\delta(AlFeOH)$) and 845 $cm^{-1}$ ($\delta(AlMgOH)$) can be

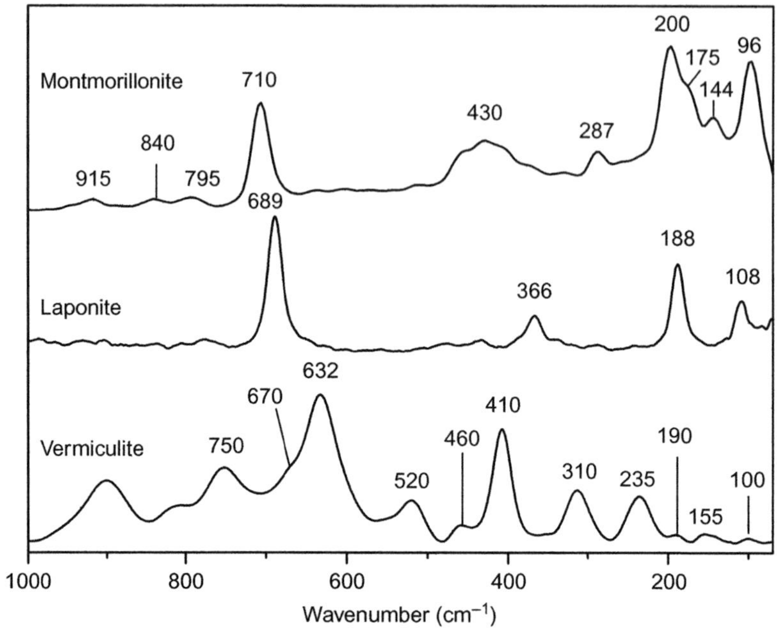


**FIG. 6.8** Raman spectra of the 2:1 clay minerals: montmorillonite, Laponite and vermiculite in the low wavenumber region.

observed in dependence on the occupancy of the octahedra by different central cations. While only 915 and 840 $cm^{-1}$ bands could be distinguished for SAz-2 (Fig. 6.8), the Raman spectrum of SWy-1 also revealed a band at 885 $cm^{-1}$ resulting from higher content of Fe in the octahedral positions (Bishop and Murad, 2004). The band near 795 $cm^{-1}$ had been assigned to wagging vibrations of OH groups attached to Al (Frost and Rintoul, 1996a), but this band was more likely the $\nu$(Si–O) vibrations of silica admixture as reported Bishop and Murad (2004). The intense band occurring near 710 $cm^{-1}$ was attributed to the $\nu_1(a_1)$ mode of the $SiO_4$ tetrahedra. The band, similar to that near 1100 $cm^{-1}$, was observed at a higher wavenumber than the predicted theoretical value due to distorted $SiO_4$ unit symmetry. The broad complex band near 430 $cm^{-1}$ (Fig. 6.8) was thought to probably arise from the $\nu_6(e)$ and $\nu_3(a_1)$ internal vibrational modes of the $SiO_4$ tetrahedral units (Frost and Rintoul, 1996a).

The band that occurred at 287 $cm^{-1}$ in the Raman spectrum of SAz-2 was probably the $\nu_1(A_1)$ mode of the OH–O triangle. The band at 200 $cm^{-1}$ was assigned to the $\nu_1(A_{1g})$ vibrational mode of the $AlO_6$ unit (Frost and Rintoul, 1996a). This was one of the most intense bands in the Raman spectra of the Mt (Frost and Rintoul, 1996a; Bishop and Murad, 2004; Wang et al., 2015) and such a band suggests a highly symmetric Raman mode. The band position of this vibration did not alter with the clay layer charge and was apparently not influenced by the counter cation. A shoulder found near 175 cm$^{-1}$ in Raman spectrum of SAz-2 (Fig. 6.8) was attributed to the $\delta_s$(O-Al–O) mode ($E_{2g}$) mode of the $AlO_6$ octahedron (Frost and Rintoul, 1996a). The band varied considerably between the different Mt. In the Wyoming Mt (SWy-1), the band was observed at 165 to 150 cm$^{-1}$. The band position is influenced by the presence of iron in the octahedral sheet, which causes the shift to lower wavenumbers. Alternatively, the influence of the cation ($Na^+$ in the case of the Wyoming Mt and $Ca^{2+}$ in SAz-2) through different cation polarizability and/or presence or absence of water molecules causes the position of this band to vary (Frost and Rintoul, 1996a).

The band between 145 and 127 is common for clay minerals and has been described as an $Si_2O_5$ ring breathing mode (Iishi et al., 1967; Farmer, 1974c). Frost and Rintoul (1996a) noticed that the band position was influenced by the layer charge. For high-charge Mt they observed this band at lower wavenumbers, such as at 135 cm$^{-1}$ for Cheto (SAz-1) (1.03 eq/unit cell) compared to the lower-charge Wyoming Mt (0.75 eq/unit cell) for which the band was observed at 141 cm$^{-1}$ (layer charge values obtained from Gates, 2005). The Raman spectrum of the higher charged SAz-2 (Ritz et al., 2016), however, shows this band at 144 cm$^{-1}$ (Fig. 6.8), for example, at the position higher than that found for the lower charged Wyoming Mt, which places doubt on the effect of the layer charge on the position of this band. The strong band at 96 cm$^{-1}$ was attributed to the vibrations of the interlayer cations (Larson et al., 1972; Fripiat, 1982).

Frost and Kloprogge (2000c,d) extensively studied the Raman spectra of ferrian smectite (SWa-1) and nontronites from various localities. The Raman spectrum of nontronite Hohen-Hagen (NG-1) is reported also in Frost and Rintoul (1996a). Weak bands at around 1092 and 1032 cm$^{-1}$ were ascribed to the O–Si–O and O–Si–O-Fe antisymmetric stretching modes. The band observed around 879 and 680 cm$^{-1}$ can be attributed to $\delta$(FeAlOH) and $\nu$(FeO) modes. Two Raman bands near 450 and 430 cm$^{-1}$ assigned to the Fe–O–Si deformation modes fit well with the IR bands at 461 and 433 cm$^{-1}$ observed for ferrian smectites and at 450 and 433 cm$^{-1}$ for the nontronite. Bands observed below 300 cm$^{-1}$ have been ascribed by Frost and Rintoul (1996a) as related to cross-vacancy H-bonding of OH–O triads, but Wang et al (2015) indicated that these may be related to the presence of hematite.

The Raman spectrum of Sap, for which $Mg^{2+}$ cations occupy the octahedral positions and substitution of tetrahedral $Si^{4+}$ with $Al^{3+}$ is characteristic, shows a rather simple pattern. The majority of the bands are vibrations belonging to distorted $(Si,Al)O_4$ tetrahedra. The most prominent bands in this region are the Si–O bands at 1082, 683, 464, 432 and 360 cm$^{-1}$. In addition, two weaker bands at 1051 and 998 cm$^{-1}$ together with 288 cm$^{-1}$ were ascribed to the vibration modes of the $SiO_4$ tetrahedra (Loh, 1973; Farmer, 1974c). At the low-wavenumber region just one mode of the distorted

$MgO_6$ at 201–202 $cm^{-1}$ was observed. Wang et al. (1999, 2015) reported some bands for a greyish green Sap occurring as clots and ovoids scattered throughout a fine-grained weathered basalt matrix. This Sap had its main Raman bands near 680, 355, 295 and 190 $cm^{-1}$ with the last band associated with $\nu_1(A_{1g})$ (Al,Mg)$O_6$. It showed only a weak feature in the 1000 $cm^{-1}$ region of the spectrum.

Kloprogge and Frost (2000) studied the effects of synthesis temperature on creation of synthetic Sap and compared the Raman spectra of synthetic with natural Sap (Fig. 6.9). The saponites synthesized at 125°C exhibited almost no Raman bands. Upon increasing the temperature, the first band to be observed was the most prominent band at 683 $cm^{-1}$ at 150°C followed by the 360 $cm^{-1}$ band at 175°C. Around 240–280°C the two bands around 430–465 $cm^{-1}$ could be observed with some difficulty (Fig. 6.9). All these bands correspond to the internal tetrahedral vibrations.

Rinaudo et al. (2004) are the only ones who reported a Raman spectrum of Ht, for which the occupancy of tetrahedra with $Si^{4+}$ and octahedra with $Mg^{2+}$ cations partly substituted with $Li^+$ is characteristic. The Raman spectrum of its synthetic counterpart, Laponite (Lap), is shown in Fig. 6.8. A weak band near 1095 $cm^{-1}$ and the most intense one at 689 $cm^{-1}$ correspondto the asymmetric and symmetric $\nu$(SiO) for other clay minerals,. Rinaudo et al. (2004) supposed that the band at 366 $cm^{-1}$, similarly as for Sap, was related to the Si–O mode. However, Arab et al. (2002) assigned the band near 355 $cm^{-1}$, often present in the Raman spectra of trioctahedral clay minerals, to the coupled $\delta(O_{III}$–Mg–$O_{III})$ and $\nu$Mg–O vibrational mode ($O_{III}$ = tetra-coordinated

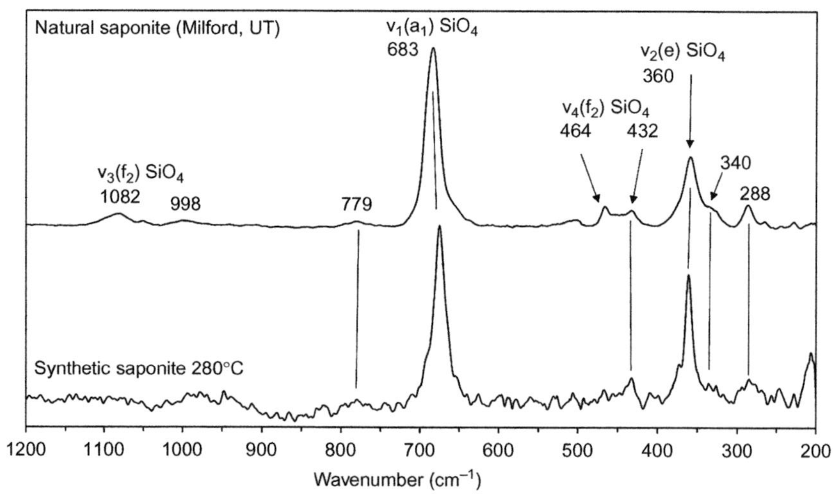


**FIG. 6.9** Raman spectra of natural (Milford, UT) and synthetic saponite prepared at 280°C in the low wavenumber region.

oxygen linked to one H atom and to three Mg atoms). The bands at 188 and 108 $cm^{-1}$ were previously assigned to the vibration of $(Mg,Li)O_6$ groups (Rinaudo et al., 2004).

In general, the Raman active symmetric stretching vibrations of the $SiO_4$ groups are characteristic for each clay mineral type. The $\nu_s$ vibrational modes of the $SiO_4$ tetrahedra appear at lower wavenumbers (680–690 $cm^{-1}$) in the trioctahedral clay minerals than in the dioctahedral clay minerals (700–710 $cm^{-1}$). A Raman band in the range 370–330 $cm^{-1}$ observed in trioctahedral clay minerals may tentatively be ascribed to $SiO_4$ bending vibrations (Kloprogge and Frost, 2000); however, the contribution of the $MgO_6$ unit (Arab et al., 2002) should also be considered.

### *6.4.2.3 Vermiculites*

The Raman spectra of Verm reported for example Rinaudo et al. (2004) and Ritz et al. (2014). The shape of a spectrum can differ significantly due to the variability in Verm chemical composition and mineralogical origin. Rinaudo et al. (2004) analyzed the experimental spectrum of Verm based on the results obtained by computer simulation using molecular dynamics calculations (Arab et al., 2002). A medium intensity band at 1039 $cm^{-1}$ was assigned to the antisymmetric stretching vibrations and the most intense band at 683 $cm^{-1}$ corresponds to symmetric stretching modes of the $SiO_4$ tetrahedra. The band at 532 $cm^{-1}$ was attributed to the deformation O–Mg–O mode. A hump at 424 $cm^{-1}$ assigned Rinaudo et al. (2004) to the vibrations of OH groups of water hydrating the interlayer cations; however, a similar band observed in experimental Raman spectrum of Verm related Arab et al. (2002) to the $\delta(O_I–Si–O_I)$ vibration. A well-defined, intense band at 354 $cm^{-1}$, observed for vermiculite (and often other trioctahedal clay minerals, e.g. some minerals of the serpentine groups) was ascribed by Rinaudo et al. (2004) to the $SiO_4$ bending vibrations. However based on the symmetry coordinate analysis, Arab et al. (2002) previously assigned the experimental band at 354 $cm^{-1}$ to the coupled $\delta(O_{III}–Mg–O_{III})$ and $\nu(Mg–O)$ vibrational mode. A broad band near 283 $cm^{-1}$ was assigned to vibrations of the O–H···O groups. The band at 186 $cm^{-1}$ was attributed to symmetric deformations of the $MgO_6$ groups. This band appeared at a lower wavenumber than the vibrations of the $AlO_6$ groups (200 $cm^{-1}$) in dioctahedral clay minerals. A band near 110 $cm^{-1}$ was assigned to vibrations of the interlayer cations (Rinaudo et al., 2004).

Ritz et al. (2014) reported the Raman spectra of three Verm samples of different mineralogical origin and chemical composition. Fig. 6.8 shows vermiculite from Bulgaria (V1) originated from the trioctahedral mica phlogopite. The positions of the bands differed from those reported by (Rinaudo et al., 2004) most probably due to the differences in chemical composition. Ritz et al. (2014) assigned the bands based on data given in Arab et al.

(2002). Raman spectrum of V1 vermiculite shows a medium band at 750 cm$^{-1}$ related to the $\nu$(Si–O) bonds and a strong band at 630 cm$^{-1}$ with a shoulder at 670 cm$^{-1}$, which was assigned to $\nu$(Mg–O) vibrations. The medium intense band at 520 cm$^{-1}$ and weak band at 460 cm$^{-1}$ were related to $\delta$(O–Mg–O) and $\delta$(O–Si–O), respectively. A strong band at 410 cm$^{-1}$ was ascribed to the $\delta$(O–Mg–O). Medium intense bands at 310 and 235 cm$^{-1}$ were assigned to $\delta$(Si–O–Si), $\delta$(O–Si–O) and/or $\delta$(Si–O–Mg) and a weak band at 190 cm$^{-1}$ to the $\delta$(Mg–O–Mg) bonds. The spectrum of vermiculite from Bulgaria shows also weak bands at 155 and 100 cm$^{-1}$ related to vibration of interlayer cations.

Wada (1987) studied the effects of water in $Na^+$-Verm with zero, one and two water layers. The Raman band at 24 cm$^{-1}$ in the anhydrous $Na^+$-Verm spectrum was assigned to the interlayer shearing mode (ILSM) and corresponds to the TO zone-centre mode. A corresponding ILSM peak was not observed in hydrous $Na^+$-Verm, most likely due to softening in the interlayer interactions upon intercalation of water. $Sr^{2+}$-Verm had the ILSM wavenumber at 21 cm$^{-1}$ in the anhydrous state and there was not a corresponding ILSM Raman band observed for the hydrated samples. The ILSM wavenumber should be quite sensitive to the interlayer cations, because the structure of the silicate layer is relatively rigid. The wavenumber difference between $Na^+$- and $Sr^{2+}$-verm may be understood in terms of the surface density of the interlayer cations. Since the charge in the verm sample was fixed, the number of $Na^+$ ions is twice that of $Sr^{2+}$ per unit area. Coulombic interactions should be similar in the two cases, but greater short-range interlamellar interactions arising from the larger number of $Na^+$ ions may slightly raise the shearing mode wavenumber in $Na^+$-Verm (the ionic radii of $Na^+$ and $Sr^{2+}$ are similar). The Raman band at 196 cm$^{-1}$ for the $Na^+$-Verm in the zero water layer hydration state showed a drastic change as a function of the hydration state. The 2-LHS sample showed a broad band at 164 cm$^{-1}$, but in the 1-LSH sample this band was split into two modes at 137 and 199 cm$^{-1}$. Further hydration (greater than 2-LSH) resulted in a shift to around 196 cm$^{-1}$ (Wada, 1987).

In subsequent work Wada et al. (1990) indicate that the narrow '*A*' torsional Raman band at 105 cm$^{-1}$ in the one water layer hydration state may have represented an ordering of the cations in the interlayer space, while in the zero and two water layer hydration states the cations were disordered between the host silicate layers causing a broadening of the signal. They also indicated that the extra band at 137 cm$^{-1}$ in the Raman spectrum from $Na^+$-Verm in the 1-LSH may be a consequence of the formation of a (2a x 2b) super-lattice. In fact, from the inelastic-neutron-scattering experiments of Wada and Kamitakahara (1991), the energy of the zone-boundary in-plane LA (longitudinal acoustic) phonon in $Mg^{2+}$ Verm in a 2-LHS may be extrapolated to be 17 meV or 136 cm$^{-1}$. It was expected that the in-plane LA mode would not be too sensitive to the interlayer cations. The 137 cm$^{-1}$ Raman peak from $Na^+$-Verm in the one-water layer hydration state;

therefore, could be assigned to be a zone-folded optical phonon from the in-plane LA branch.

### 6.4.3 Palygorskite and Sepiolite

Palygorskite and sepiolite are hydrous Mg-enriched clay minerals with fibrous-like morphologies that typically occur as fine-grained, poorly crystalline masses. Their structures are built up of ribbons of 2:1 layers that are themselves linked into modules of either double (palygorskite) or triple (sepiolite) silicate chains that sandwich (Mg,Al)-(O,OH) octahedral strips. The ribbons are linked by inversion of the $SiO_4$ tetrahedra through Si–O–Si bonds. The phases have continuous tetrahedral sheets, but with tetrahedral apices pointing in opposite directions in adjacent ribbons. In contrast to other clay minerals, they lack continuous octahedral sheets. The structures have large tunnels parallel to the ribbons that are partially occupied by water molecules (Fig. 6.10).

McKeown et al. (2002) used factor group analysis (FGA) to interpret the experimental polarized Raman spectra of palygorskite from Virginia and sepiolite from Mexico. A total of 114 fundamental optical modes have been predicted for the palygorskite structure by FGA, where 54 modes are Raman active and 60 are IR active. Variations between the $A_g$ (parallel polarized) and $B_g$ (cross-polarized) Raman spectra are subtle (Fig. 6.11). The most noticeable differences are slight intensity and band shape variations. The experimental and calculated values are compared in Table 6.6. The calculated fundamental mode wavenumbers fit very well to the wavenumbers of bands observed in the $A_g$ and $B_g$ Raman spectra allowing unambiguous assignment

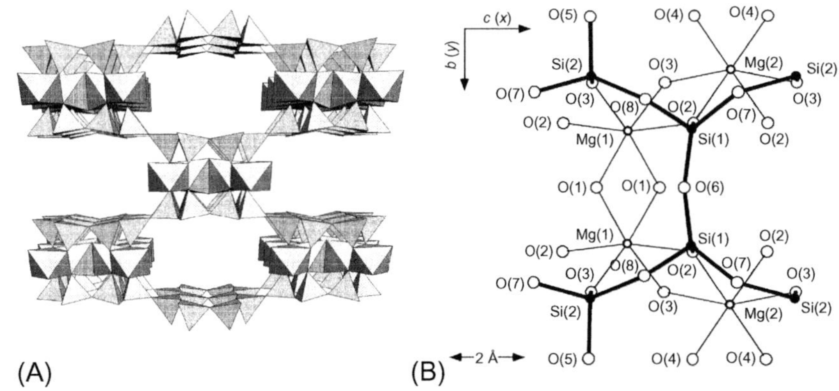


**FIG. 6.10** Polyhedral representations of the (A) palygorskite and (B) the palygorskite crystal structure cluster used in the lattice dynamics calculations, showing crystallographic *a*, *b* and *c* axes in relation to the *x*, *y* and *z* axes. *From McKeown et al. (2002). Reproduced with kind permission of The Clay Minerals Society, publisher of Clays and Clay Minerals.*

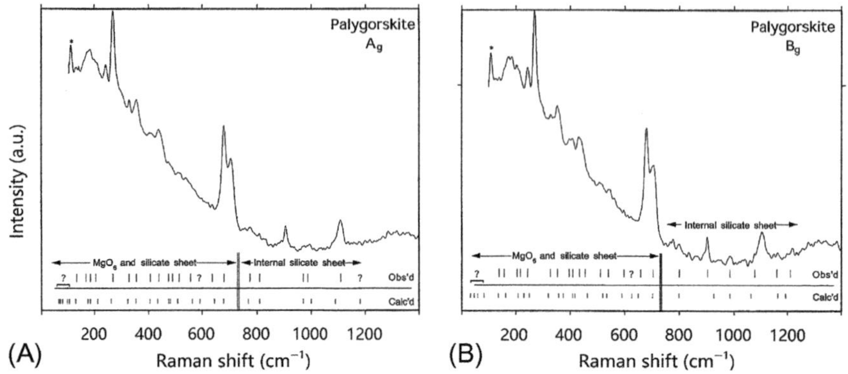


**FIG. 6.11** Raman spectra of palygorskite showing: (A) parallel polarized, Ag-dominated modes; and (B) cross-polarized, Bg dominated modes. Observed mode frequencies to the spectra and the corresponding calculated mode frequencies as well as general vibrational assignments are indicated. ? indicates a calculated mode that cannot be unambiguously assigned to an observed mode in the spectrum. * indicates an artefact from the laser line filter cut-off. *From McKeown et al. (2002). Reproduced with kind permission of The Clay Minerals Society, publisher of Clays and Clay Minerals.*

of most experimental bands. The bands that clearly stand out are the $A_g$ modes at 167, 183, 268, 327, 437, 473, 680, 971, 988 and 1109 $cm^{-1}$ and $B_g$ modes at 205, 216, 243,353, 397, 410, 704, 800, 904 and 986 $cm^{-1}$ can be readily fit and described by the calculations (Table 6.6, and see Fig. 6.10 for the atom positions used in the table).

The Raman spectra of palygorskite above $\approx$750 $cm^{-1}$ are dominated by eigenmodes related to short-range, or localized, atomic displacements within the silicate sheets. At the highest mode wavenumbers, the displacements are the most localized $\nu$(Si–O) motions, such as Si–$O_{nb}$. As mode wavenumbers become progressively lower, longer range $\delta$(Si–O) (O–Si–O) bending, silicate tetrahedral breathing and shearing within the silicate sheets dominate (Table 6.6). As mode wavenumbers decrease further, the corresponding eigenmodes evolve from more localized $\nu$(Mg–O), $\delta$(O–Mg–O) and $\delta$(O–Si–O) modes to longer-range modes such as silicate sheet deformations caused by coupled silicate tetrahedra rotation and silicate sheet shearing around the Mg octahedral sheets deforming the Mg octahedra. Below 700 $cm^{-1}$ the eigenmodes have motions within the Mg(1)- and Mg(2)-octahedra and the silicate sheets; the most prominent peaks in the Raman spectra (near 700 $cm^{-1}$) belong to this group of eigenmodes. The palygorskite structure has O(4) atoms that terminate the Mg( 2) octahedra and are not bonded to another cation. Fundamental modes with mostly Mg(2)–O(4)-stretch and O(4)–Mg(2)–O (4)-bend rotation displacements dominate the lowest frequency modes, that are generally not recorded in the Raman spectra McKeown et al. (2002).

For sepiolite, FGA has predicted 348 modes in eight vibrational species: 174 modes for the Raman-active $A_g$, $B_{1g}$, $B_{2g}$ and $B_{3g}$ species and 133 modes

**TABLE 6.6** Observed and Calculated Band Positions for Palygorksite and Their Assignments

| Raman Active Ag Fundamental Modes | | | Raman Active Fundamental Bg Fundamental Modes | | |
|---|---|---|---|---|---|
| *Experiment* | *Calculated* | *Assignment* | *Experiment* | *Calculated* | *Assignment* |
| | | | – | 32 | Shear within each sheet ± x; $MgO_6$ moving with shear: Mg(2) ± x |
| | | | – | 46 | O(4)–Mg(2)–O(4) rotation ‖ x |
| – | 63 | Mg(2) octahedra ± y; O(6) ± z | – | 60 | O(4)–Mg(2)–O(4) wagging; sheet rotation ‖ x |
| – | 68 | $MgO_6$ deformation; O(1,4) ± xz: in phase | | | |
| – | 76 | $MgO_6$ deformation; O(1,4) ± xz: out of phase | | | |
| | | | – | 83 | O(4)–Mg(2)–O(4) wagging; O(4,7) ± xz; |
| | | | | | Si(1,2)–O(2,3) ± z; sheet rotation ‖ x |
| – | 94 | O(4)–Mg(2)–O(4) bend; O(1) ± xz: out of phase | | | |
| – | 102 | O(4)–Mg(2)–O(4) bend; O(1) ± xz: in phase | | | |
| 130 | 126 | Silicate tetrahedral ring deformation: O(7,8) ± z; | 139 | 136 | Sheet shear around Mg sites: top sheet + y, bottom |
| | | O(6) ± xz; Si(1)_O(2) ± z; Si(2)_O(3) ± z | | | Sheet -y; Mg(2) ± z; O(2)–Mg(2)–O(4) bend |
| | | | 157 | 163 | $MgO_6$ def.: Mg(1) ± xz; Mg(2) ± z; O(7,8) ± z |

Continued

**TABLE 6.6** Observed and Calculated Band Positions for Palygorksite and Their Assignments—Cont'd

| Raman Active Ag Fundamental Modes | | | Raman Active Fundamental Bg Fundamental Modes | | |
|---|---|---|---|---|---|
| *Experiment* | *Calculated* | *Assignment* | *Experiment* | *Calculated* | *Assignment* |
| 167 | 174 | O(1)–Mg(1)–O(1) bend; O(6) ± xz; Si(1) + xz | | | |
| 183 | 184 | Mg(1)–O(1)–Mg(1) bend: Mg(1) ± y; O(6) ± x | | | |
| | | | 205 | 208 | Mg(1) ± xz; O(1) ± y |
| 205 | 212 | Silicate sheet deformation: SiO4 rotation \|\| y | | | |
| | | | 216 | 230 | $MgO_6$ & sheet deformation: Mg(1,2) ± xz; O(1) ± y; Si(1)–O(6) ± y |
| | | | 243 | 249 | $MgO_6$ deformation: O(nb) ± xy; Mg(1,2) ± xz; O(7) ± y |
| 268 | 262 | Mg(2) octahedral deformation, Mg(2) ± y | | | |
| | | Si(1)–O(2) ± z; O(6) ± zx | | | |
| | | | 327 | 319 | Sheet rotation \|\| x; Si(1)–O(2) + z; Si(2)O(3) -z |
| 327 | 323 | $MgO_6$ octahedral breathing (Mg stationary); | | | |
| | | O(7) + z; O(8)–z | | | |

| | | | | | |
|---|---|---|---|---|---|
| 354 | 352 | Si(2)–O(3) ± z; O(1) ± xz | 353 | 357 | Si(2)–O(3) ± z; Mg(1) ± x; O(1) ± y; O(7) ± z |
| | | | 397 | 375 | sheet deformation: Si(1) ± x; O(7) + y; O(8)-y |
| 406 | 405 | Si(2)–O(3) + z; O(7)–Si(1)–O(8) bend | | | |
| | | | 410 | 409 | O(2)–Mg(1)–O(2) rotation ∥ x; Mg(1) ± xz; O(8) ± z |
| | | | 435 | 418 | Mg(1)–O(1) stretch; Mg(1,2) + xz; O(1) ± y; O(8) + z; O(7)-z |
| 437 | 433 | Mg(1) ± y, MgO6 octahedral rotation; O(6) ± xz | | | |
| | | | 456 | 464 | Mg(1,2) ± xz, O(8) ± z; O(1) ± y |
| 473 | 474 | Mg(1) ± y; O(6) ± x; O(1) ± xz; O(6)–Si(1)–O(7) bend | | | |
| 488 | 480 | O(2)–Si(1)–O(6) bend; O(7) ± y | | | |
| 512 | 507 | Mg(2) octahedral deformation, O(6) ± z | | | |
| | | | 512 | 520 | Mg(1,2) ± xz, Mg(1)–O(2) stretch; Si(2)–O(3)Tz |
| | | | 540 | 534 | MgO(1,2)–O(2,4) stretch; Si(1)–O(6) ± y |
| 556 | 562 | O(3)–Mg(1)–O(3) bend | | | |

*Continued*

**TABLE 6.6** Observed and Calculated Band Positions for Palygorksite and Their Assignments—Cont'd

| Raman Active Ag Fundamental Modes | | | Raman Active Fundamental Bg Fundamental Modes | | |
|---|---|---|---|---|---|
| *Experiment* | *Calculated* | *Assignment* | *Experiment* | *Calculated* | *Assignment* |
| – | 591 | O(2)–Mg(2)–O(2) bend: Mg(2) ± y; | | | |
| | | O(2,3)–Si(1,2)–O(7) bend | 597 | 589 | $MgO_6$ deformation: Mg(1,2) ± zx, Mg(1)–O(2) |
| | | | | | Stretch; O(7) ± z; O(2)–Mg(2)–O(3) bend |
| | | | – | 625 | Mg(2)–O(2) stretch; O(7)–Si(2)–O(8) bend |
| 638 | 645 | O(2)–Mg(2)–O(2) bend: Mg(2) ± y; Si(2) ± y; | | | |
| | | O(3)–Si(2)–O(8) bend | | | |
| | | | 658 | 649 | Mg(2)–O(3) stretch, Mg(2) ± x, O(8) ± zx |
| 680 | 678 | O(2)–Mg(2)–O(2) bend: Mg(2) ± y; Si(2) ± y; | | | |
| | | O(2)–Si(1)–O(6) bend | | | |
| | | | 704 | 702 | Si(2) ± y, O(6) ± y alternating per sheet |

| | | | | | |
|---|---|---|---|---|---|
| 774 | 768 | Si(1)–O(6)–Si(1) bend: Si(1)+y, Si(2)–y | | | |
| | | | 800 | 799 | Silicate tetrahedral sheet shear: Si(1)+x; Si(2)-x; O(7,8)±xz |
| 809 | 809 | Si(1)–O(7,8)–Si(2) bend: Si(1)–x, Si(2)+x | | | |
| | | | 904 | 927 | Si–O(nb) stretch: Si(1)–O(2); Si(1) tetrahedral base breathing |
| 971 | 971 | Si–O(nb) stretch: Si(1,2)–O(2,3) | 986 | 986 | Si–O(nb) stretch: Si(2)–O(3) |
| 988 | 1000 | Si–O(nb) stretch: Si(1,2)–O(2,3) | | | |
| | | | 1077 | 1063 | Si–O(nb) stretch: Si(1)–O(2); Si(1) tetrahedral base breathing |
| 1109 | 1090 | O(8)–Si_(1)–O(7) bend | | | |
| | | | 1160 | 1164 | Si–O(b) stretch: Si(1,2)–O(6,8) |
| – | 1182 | Si–O(b) stretch: Si(1,2)–O(8,7) | | | |
| | | | 1211 | 1193 | Si–O(nb) stretch: Si(1,2)–O(6,7) |

From McKeown et al. (2002). Reproduced with kind permission of The Clay Minerals Society, publisher of Clays and Clay Minerals.

for the IR-active $B_{1u}$, $B_{2u}$ and $B_{3u}$ species. The large number of predicted optically active fundamental modes make it impossible to provide unambiguous assignments to all vibrational bands in the Raman spectra for sepiolite, especially considering the relatively few resolved features that could be assigned to fundamental modes in the observed $Ag$-dominated and the $Bg$-dominated Raman spectra (Fig. 6.12) (McKeown et al., 2002).

The unpolarized Raman spectrum of sepiolite in Akyuz et al. (1995) was similar (minus the pyrazine modes) to the polarized spectra. Specific eigenmode assignments made by Mckeown et al. (2002) to Raman modes did not agree particularly well with assignments made in a study of optical vibrations in sheet silicates by Loh (1973). The discrepancies are probably because Loh (1973) made vibrational assignments based on the symmetry of $MO_6$ octahedral and $TO_4$ tetrahedral atom clusters in clay minerals and observations of Raman modes from similar atom clusters in other materials; whereas the eigenmodes calculated were based specifically on the palygorskite and sepiolite structures, and thus while the arguments by Loh (1973) appear to generally agree, many discrepancies are observed in the details. According to Loh (1973), the Raman modes at wavenumbers above 300 $cm^{-1}$ are dominated by T–O-stretch motions within the tetrahedra and the $MO_6$ octahedral modes dominate at wavenumbers below 200 $cm^{-1}$. This argument is partially supported by the theoretical findings for palygorskite and sepiolite, but as described above, Raman active eigenmodes up to 700 $cm^{-1}$ can be dominated by $Mg^{2+}$ octahedral motions. The theoretical calculations showed most eigenmodes in this wavenumber range do contain $Mg^{2+}$ octahedral motions that overlap with silicate tetrahedral sheet rotations and deformations.

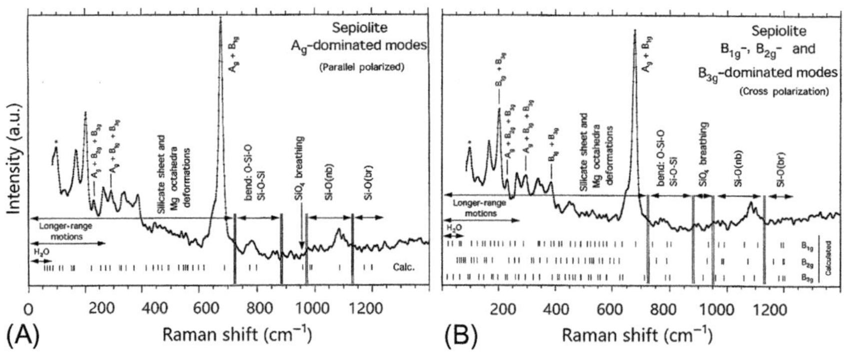


**FIG. 6.12** Raman spectra of sepiolite (A) parallel polarized Raman spectrum with some general vibrational assignments from the calculations and plausible vibrational species assignments to some peaks; and (B) cross-polarized Raman spectrum of sepiolite with some general vibrational assignments from the calculations. * Indicates an artefact from the laser line filter cut-off. *From McKeown et al. (2002). Reproduced with kind permission of The Clay Minerals Society, publisher of Clays and Clay Minerals.*

## 6.5 CONCLUDING REMARKS

Though major advances have been achieved in recent years, obtaining good quality Raman spectra of many clay minerals remains an experimental challenge with problems, especially those associated with fluorescence, sample damage and weak signals. This chapter shows, however, that a reasonable amount of literature for most common clay minerals is now available and the assignment of the different modes has gone from tentative, based on comparison with other minerals, to unambiguous, based on calculations and molecular modelling of the actual crystal structure for some minerals. With further improvements in instrumentation and in better molecular modelling software more details of the different vibrational modes in Raman spectra will become evident. Raman spectroscopy, as a complementary technique to IR spectroscopy, has grown past being mainly a technique for identifying materials to a useful tool to better understand what is happening on the molecular structure of even 'difficult' materials such as clay minerals.

Chapter 7

# Applications of NIR/MIR to Determine Site Occupancy in Smectites

W.P. Gates, S. Petit and J. Madejová

## 7.1 INTRODUCTION

This chapter will show how vibrational spectroscopy, in particular, near infrared (NIR) and mid infrared (MIR) spectra, in combination with chemical and other physical data can successfully be used to assist in the determination of site occupancies by different ions in clay minerals. Traditional approaches for determining structural chemistry of clay minerals, as emphasised herein, are still the norm. However, there is no real impediment to the implementation of multivariate and chemometric methods of analysis other than the issue of sample size.

The accurate and precise assignment of ions (cations and anions) within distinct octahedral and tetrahedral sites of the smectite structure continues to remain an important focus for some researchers where IR and Raman spectroscopies play a key role. In natural smectites, Si, often Al and sometimes $Fe^{3+}$, occupy tetrahedral sites and Al, Mg, $Fe^{3+}$, as well as $Fe^{2+}$, $Ni^{2+}$, $Zn^{2+}$ and $Li^{+}$ occupy octahedral sites within the structure. Chemical analyses, in combination with X-ray powder diffraction and other methods, has enabled smectites to be differentiated based on the occupancies of the tetrahedral and octahedral sites, but because of the small particle size and the poorly crystalline nature of smectites, only average, bulk information is available. The use of alternative techniques, including IR and Raman spectroscopies, to examine chemo-structural relationships has been essential to assist the determination of crystallographic information, such as the occupancy of octahedral and tetrahedral sites, and they have also proven extremely useful in relating chemo-structural properties to many of the macroscopic or materials properties of smectites.

Developments in Clay Science, Vol. 8. https://doi.org/10.1016/B978-0-08-100355-8.00007-2

IR spectroscopy lends itself particularly well to the question of structural chemistry and, specifically, the occupancy of various ions in the structure of smectites, because all the octahedral cations present within the structure are a constituent of at least one OH-sharing pair. Thus the IR spectra of smectites are rich in information regarding the actual octahedral cations present and how they are related to their nearest neighbours through shared OH bonds. In addition, as will be explained in more detail below, the vibrational frequencies at which OH-sharing octahedral pairs absorb are affected by the identity of next-nearest neighbours, which while adding complexity to spectral interpretations, also adds further richness of information regarding the actual occupancies with the octahedral, and to some extent, the tetrahedral sites.

In the next section, some general considerations regarding the state-of-the-science about smectite structures are briefly reviewed. Then, various methods used to interpret IR spectroscopy such as bond valence, reduced mass and ionic size are described in detail.

## 7.2 OCTAHEDRAL STRUCTURES OF SMECTITES

Based on X-ray powder diffraction analyses, enough is known about the structures of many clay minerals to enable detailed analysis of IR spectra, which in turn can assist in refining structures based on chemical analyses. The IR spectrum is sensitive to the next nearest neighbour environment of OH-sharing octahedral cations (Figs. 7.1–7.3), but this has been used to assist in quantifying the octahedral occupancies in only a few cases.

### 7.2.1 Di- and Tri-octahedral Structures of Smectites

All octahedral arrays of clay minerals have two symmetrically inequivalent octahedral sites (Fig. 7.1): M1 and M2, and there are two M2 sites for each M1 site. The M1 sites are *trans*, whereas the M2 sites are *cis*, with respect to the position of the hydroxyl groups to which at least two octahedra share

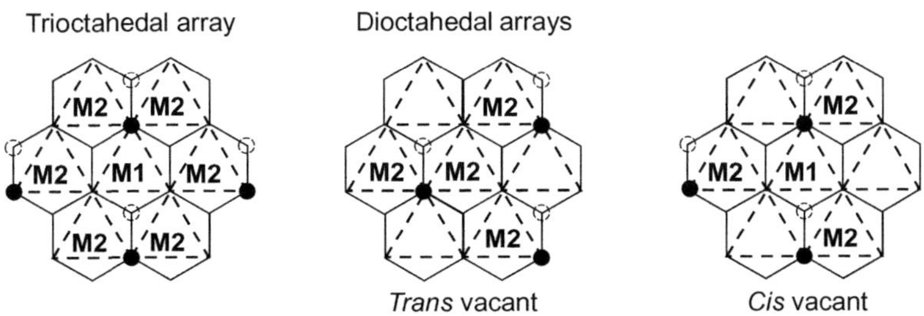


**FIG. 7.1** The octahedral arrays of smectites. In trioctahedral structures, all octahedral sites are filled with mostly divalent cations. In dioctahedral structures, only two of the three sites are filled, giving *cis*-vacant and *trans*-vacant variants. The tetrahedaral sheets are not shown for clarity. The black filled circles represent hydroxyls at the front face, while the dashed circles represent the hydroxyls at the back face of the octahedron.

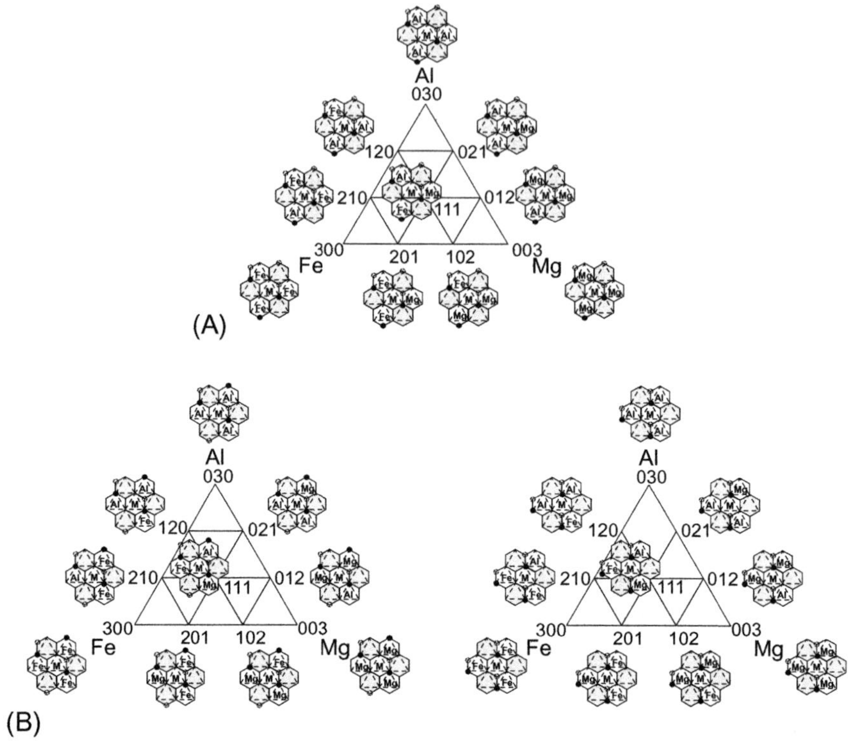


**FIG. 7.2** (A) Ternary diagrams representing 10 (where M is either Fe, Al or Mg) of the M2(*xyz*) arrays for Fe, Al, Mg occupancies of the possible 72 arrays for *trans*-vacant structures. (B) 10 M2(*xyz*) (left) and M1(*xyz*) (right) arrays each of the possible 144 arrays for *cis*-vacant dioctahedral smectites.

bonds. Within the classification scheme, there exist *dioctahedral* and *trioctahedral* structural types. In the trioctahedral structures, all three sites are filled with generally divalent cations (Fig. 7.1). Therefore each octahedral site shares an edge with six other octahedral sites as well as corners with tetrahedral sites. Each M1 cation shares one OH with two M2 cation sites and the other OH with two other M2 cation sites, and each of these pairs of M2 sites share edges. Each OH is thus linked to three neighbouring octahedral cations. In the dioctahedral structures, each octahedral cation site shares an edge with three other octahedral cation sites; the other octahedral sites being vacant (Fig. 7.1). Because only two of the three octahedral sites of the formula unit are filled with generally trivalent cations, two symmetry independent variants exist. *Cis*-vacant structural variants have cations occupying the M1 site as well as one M2 site, so each M1 site shares edges with three M2 sites, with two of which it also shares an OH. The *cis*-vacant smectites are generally represented by smectites of intermediate composition, such as montmorillonites and ferrian (referred to also as ferruginous) smectites having both $Fe^{3+}$ and Mg substitutions

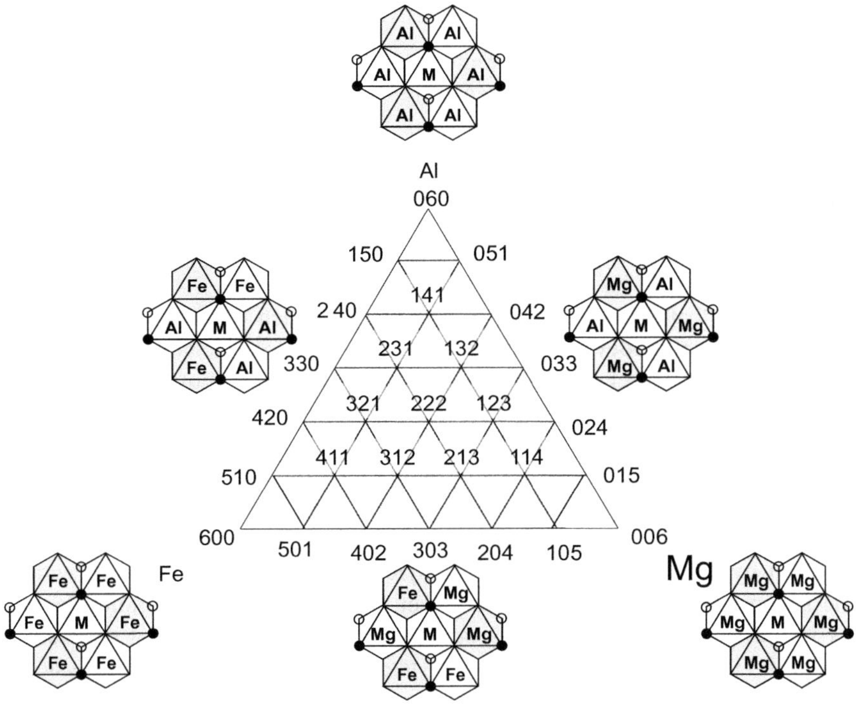


**FIG. 7.3** A ternary diagram showing some of the many possible M1(*xyz*) (where *x*, *y* and *z* are integers that sum to 6) occupancy arrays in trioctahedral smectites. In this representation, Fe is assumed to be in the ferrous, divalent ($Fe^{2+}$) oxidation state, and the presence of $Fe^{3+}$ would necessarily increase the complexity. The system can be duplicated for the M2(*xyz*) arrays, where each M2 site has 3 M2 and 3 M1 nearest (*xyz*) neighbours.

in the octahedral sheet. Only the two M2 octahedral sites are occupied in the *trans*-vacant structural variants, so each M2 cation site shares edges with three other M2 sites and both OH with only one of these. The *trans*-vacant smectites are generally represented by the end-member compositions where tetrahedral substitution is high, like nontronites and beidellites, but in ferrian smectites the probability of *trans*-vacant structures increases with octahedral $Fe^{3+}$ content (Emmerich et al., 2009; Wolters et al., 2009; Emmerich, 2013).

It follows then, that a major advantage for studying smectite structures by IR is due to the fact that hydroxyl groups are shared by at least two octahedral cations. Thus absorption bands associated with OH can be assessed quite quantitatively, as described below. However, in dioctahedral smectites, a major limitation of IR spectroscopy is that octahedral cation pairs exist in *trans*-vacant structures that do not share hydroxyl groups. In addition, corner-sharing tetrahedral cations can induce differences in the mass of the vibrating unit and/or charge that influence the positions of OH-sharing octahedral absorbencies.

### 7.2.2 Site Occupancy within a Ternary Fe-Al-Mg Field

Smectites normally have three major cation occupancies in the octahedral sites: Al, Fe, and Mg and thus depicting the chemistry by means of ternary diagrams is highly useful. Cashion et al. (2008, 2009, 2011) and Gates and Cashion (2014) developed a systematic means to identify potential nearest neighbour cation arrangements within the octahedral sheet. Originally developed to assist in the interpretation of Mössbauer spectra, it has proven to be a useful tool to assist in determining occupancies from IR spectra. It is further developed herein in its most general form.

Consider that within dioctahedral structures, each octahedral cation has three nearest neighbours (possibly Fe, Al or Mg) in the octahedral arrays of the common natural clay minerals. The details of the distribution surrounding the central cation, M, can be captured by the use of integers, which must sum to 3, representing, respectively octahedral Fe, Al and Mg atoms in nearest neighbour environments, *viz*, Ms(*xyz*), where *s* refers to the specific octahedral site (1 or 2 in smectites; 1, 2 or 3 in e.g. sepiolites). For example, in a dioctahedral smectite Al2(120) would represent an Al atom in an M2 site surrounded by 1 Fe and 2 Al atoms in adjacent edge-sharing octahedra within the trigonal array of the octahedral sheet. Note that in this example the M2 Al would share edges with 3 M1 octahedral cations in *cis*-vacant structures, but with 3 M2 octahedral cations in *trans*-vacant structures. The system, represented as ternary diagrams for the *cis*- and *trans*-vacant dioctahedral smectites, is given in Fig. 7.2.

For the dioctahedral series, there are potentially 24 symmetrically independent octahedral arrays available for each central cation M. Many of the non-redundant sites in both *cis*- and *trans*-vacant structures will only be observed as broadening or shifts in the main absorbance feature due to changes in the reduced mass of the next nearest neighbour (see Section 7.3.4). Application of the normal (random) distribution theorem (Gates and Cashion, 2014) to the chemistry of the dioctahedral smectites indicates that many of the independent arrays, depending on the specific chemistry of the considered smectite, will be of such low proportion that they can be neglected.

For the trioctahedral series (Fig. 7.3), due to (mostly complete) occupancy of all sites, there are many more possible outcomes, because Fe can be present in its ferric, trivalent ($Fe^{3+}$) as well as its ferrous, divalent ($Fe^{2+}$) state. In addition for many saponites, the octahedral charge may be quite low, or even positive, due to the presence of trivalent cations, and this is usually compensated by the presence of Al for Si substitution in the tetrahedral sheets. As for the dioctahedral arrays, many of the Ms(*xyz*) arrays in Fig. 7.3 can be neglected due to structural-chemical constraints. For example, Ms(600), Ms(060) or even Ms(330) can be neglected if M = Al or if all Fe is present as $Fe^{3+}$. Of course, if M is Mg and Fe is $Fe^{2+}$, then M1(123) is a highly probable array. Note that the sum of $x$, $y$ and $z$ is nominally 6 for trioctahedral, although the presence of some vacancies (□) can also be incorporated.

The systematic representation explored above can be expanded to include all the tetrahedral cation nearest neighbours as well (Cashion et al., 2009, 2011), yielding Ms($xyz$)Si($t$)Al($8-t$), where $t$ represents a number from 1 to 8 (for the $O_{20}(OH)_4$ unit cell). For kaolinite, a site label such as Al2(030) Si(3)Al(1) could be distinguished also by the presence of inner and inner-surface OH, or by the additional presence of nearby interlayer water in halloysite. Thus it should be appreciated that for the di- and tri-octahedral structures the number of symmetrically independent sites can become quite large.

Similar, but considerably more complex, octahedral occupancy arrays can be considered for palygorskite and sepiolite (Fig. 7.4). These non-swelling 'ribbon' clay minerals consist of discontinuous octahedral arrays linked by modulated sheets of tetrahedra (Serna et al., 1977a; Brigatti et al., 2013), resulting in three symmetrically independent octahedral sites: M1 and M2, which are the same as, respectively, the *trans* and *cis* sites in smectites, but also an M3 site, which has two coordination-bonded water molecules extending into the zeolitic channels (Fig. 7.4). Thus in sepiolite (trioctahedral structure), there would exist M1 having 4 M2 and 2 M3 ($xyz$) nearest neighbours, M2 having 2 M1, 2 M2 and 1 M3 ($xyz$) and M3 having 2 M3 and 1 M1 ($xyz$) neighbours. For idealised monoclinic palygorskite (dioctahedral structure), where the M1 site is vacant, each M2 site has two M2 and one M3 neighbours

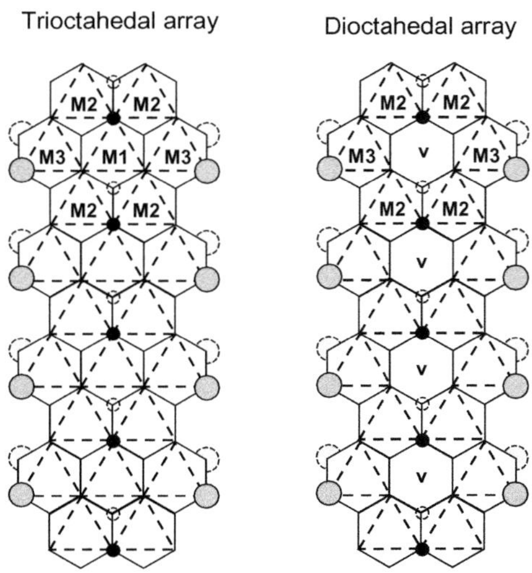


**FIG. 7.4** The idealised trioctahedral and dioctahedral arrays of, respectively, orthorhombic sepiolite and monoclinic palygorskite. As in Fig. 7.1, the small filled circles represent OH at the upper octahedron face and small dashed circles OH at the back octahedron face. Large filled and dashed circles represent coordination-bonded $H_2O$ at the edges of the octahedral extending within the zeolitic channels.

and each M3 has two M2 (*xyz*) neighbours. The introduction of the periodically inverted tetrahedral sheet, Al for Si substitution within the tetrahedral sheet, as well as the presence of coordination-bound water within the zeolitic channels, results in quite diverse possible occupancies.

As in the dioctahedral and trioctahedral arrays of smectites, there are limits to the chemical makeup of stable compositions in palygorskite and sepiolite. For example, trivalent cations such as Al or $Fe^{3+}$ are typically only found in the M2 sites, but divalent Mg or $Fe^{2+}$ can be found to be in any of the three sites (Gionis et al., 2006). The bound water is strongly associated with octahedral Mg (Serna et al., 1977a) and has discreet OH stretching, overtone and combination modes (Clark et al., 1990; Gionis et al., 2006, 2007). Thus, in addition to the sites for OH-sharing pairs, sites can be considered for water-bonding to an M3 cation. IR (and Raman) spectroscopies are well placed to assist in the continued discussion regarding the di-trioctahedral character of these minerals (Cases et al., 1991; Stathopoulou et al., 2011; Chahi et al., 2002).

## 7.3 EFFECT OF CHEMISTRY ON THE PRESENCE AND POSITION OF BANDS

Vibrational spectroscopy lends itself well to the study of site occupancy in clay minerals because collection of a single spectrum normally probes many millions of 'unit cells', each of which form part of larger (and contain many) vibrational units. Thus if our models of the crystal structure of smectites and other clay minerals hold, then vibrational spectroscopy probes the local structure, averaged over those many millions of unit cells. The importance of high quality IR spectra as well as sound chemical data on the same samples cannot be overstated. The accuracy and precision of site occupancy determination in smectites can only be as good as the chemical analysis and spectroscopic data. Assuming that the chemical data is accurate then it falls to the spectroscopist to collect the best possible spectra and conduct appropriate treatments of the spectra for analysis. For details of some of these spectral pre-treatments, see Chapter 4 in this volume. A fundamental understanding of spectroscopic selection rules also underpins our ability to quantify the structural chemistry of clay minerals.

Many earlier works were important in empirically assigning absorption bands to structural features (e.g. Stubičan and Roy, 1961; Vedder, 1964; Goodman et al., 1976; Velde, 1983; Robert and Kodama, 1988; Slonimskaya et al., 1986; Decarreau et al., 1992; Post and Noble, 1993; Madejová et al., 1994; Besson and Drits, 1997a,b; Bishop et al., 1999; Petit et al., 1999b) although some assignments are still under scrutiny (e.g. Gates, 2008; Bishop et al., 2011; Petit et al., 2015). There have been two main empirical approaches to quantifying structural chemistry from the IR spectra: band decomposition and spectral derivative intensities, both of which will be discussed in detail.

For dioctahedral smectites, starting with the initial research of Slonimskaya et al. (1986), Muller et al. (1991), Madejová et al. (1994) and Zviagina et al. (2004) have shown how the IR active OH stretching ($\upsilon(M_2OH)$) band can be used to provide quantitative information on the average occupancies of OH-sharing octahedra. Petit et al. (2004a) detailed how this information could be applied to trioctahedral talc and Vantelon et al. (2001), Gates (2008) and Petit et al. (2016) expanded this treatment to the bending ($\delta(M_2OH)$) bands, while Petit et al. (1999b, 2015, 2016) and Gates et al. (2002) expanded it for the NIR combination bands, for many different smectites. The effect of tetrahedral substitution on the OH-sharing octahedral cation bending bands has been quantified for nontronites (Gates et al., 2002; Gates, 2008; Baron et al., 2016a), but as of yet, not for other smectites due to the difficulty in assigning octahedral sites with this method alone.

### 7.3.1 Reduced Mass

As detailed in Section 7.2, most octahedral cations of smectites and other clay minerals are bonded to one or more OH, and since the O–H bond is strongly IR active, IR spectroscopy has become a valuable tool in determining octahedral site occupancy. Among the earliest explanations of the effects of chemistry on the positions of $\upsilon(M_2OH)$ or $\delta(M_2OH)$ bands relates to the reduced mass of the OH-sharing cations. The vibrational frequency of any given $\upsilon(M_2OH)$ or $\delta(M_2OH)$ (and presumably $(\upsilon+\delta)M_2OH$ combination band) is related to the force constant ($k_{OH}$) of the bond and the reduced mass ($\mu_{OH}$) of the vibrating dipole through the well-known expression

$$\upsilon(M_2OH)\,(\text{or}\,\delta(M_2OH)) = \frac{1}{2\pi c}\left(\frac{k_{OH}}{\mu_{OH}}\right)^{1/2} \tag{7.1}$$

where $c$ is the speed of light. If the force constant is considered to be constant (the veracity of this assumption will be discussed shortly) then the vibrational frequency is inversely proportional to the reduced mass. Thus, absorption frequencies for $\upsilon\left(Fe^{3+}{}_2OH\right)$ are generally lower than those for $\upsilon(AlFe^{3+}OH)$ or $\upsilon(Al_2OH)$, both of which should be lower than $\upsilon(AlMgOH)$.

Eq. (7.1) has been used empirically (e.g. Besson and Drits, 1997b; Zviagina et al., 2004; Gates, 2008) to identify the frequency of OH-sharing octahedral cation absorption features as well as to predict potential unknowns in the Al–Fe–Mg (Fe being $Fe^{3+}$ or $Fe^{2+}$) series of dioctahedral smectites (e.g. Fig. 7.5A). Unfortunately, this straightforward approach does break down, most especially in the $\upsilon(M_2OH)$ region of smectites. For example, Zviagina et al. (2004) observed in some instances that $\upsilon(AlFe^{3+}OH)$ reside at higher frequencies than $\upsilon(Al_2OH)$. Possible causes for this breakdown will be discussed further.

One useful approach to negate the effect of adsorbed $H_2O$ on interfering with the $\upsilon(M_2OH)$ region is to calculate the second derivative (Fig. 7.6).

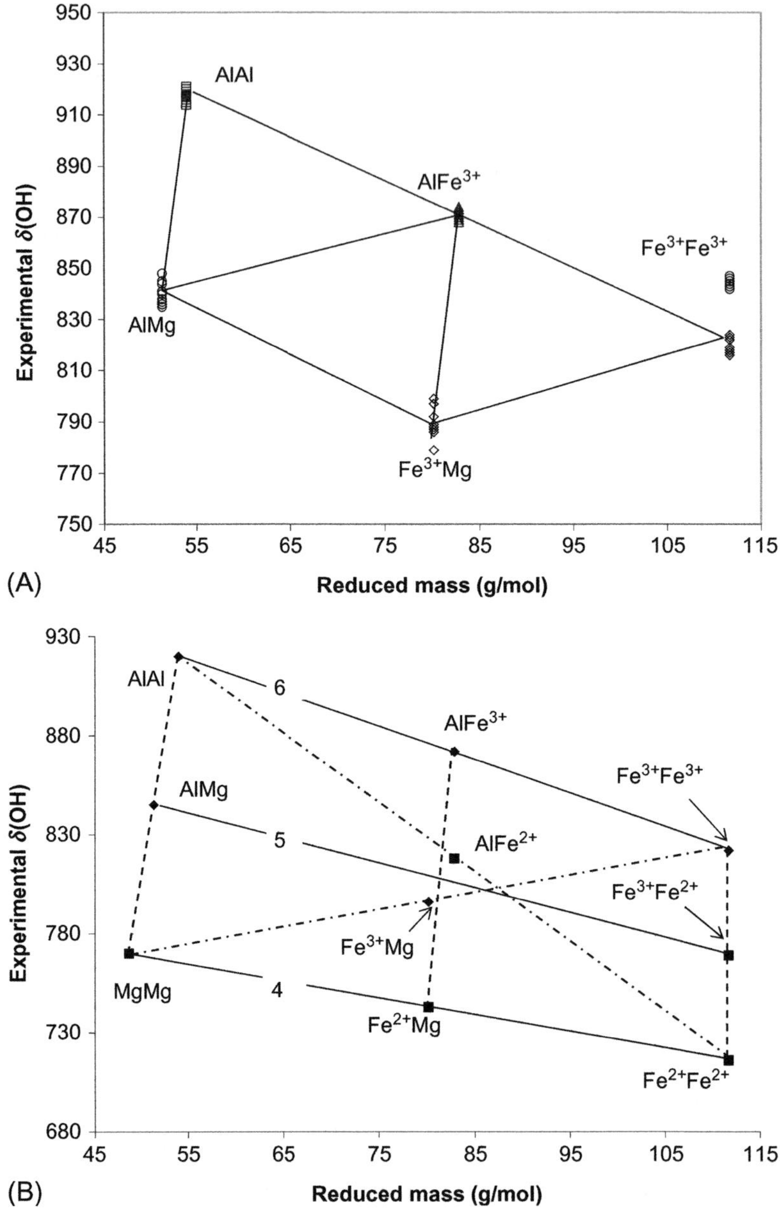


**FIG. 7.5** (A) Relationship between vibrational frequency of the $\delta(M_2OH)$ vs. the reduced mass of the octahedral cation pair in dioctahedral smectites. (B) The same relationship determined for montmorillonites, as influenced also by the valence sum. *From Gates (2008).*

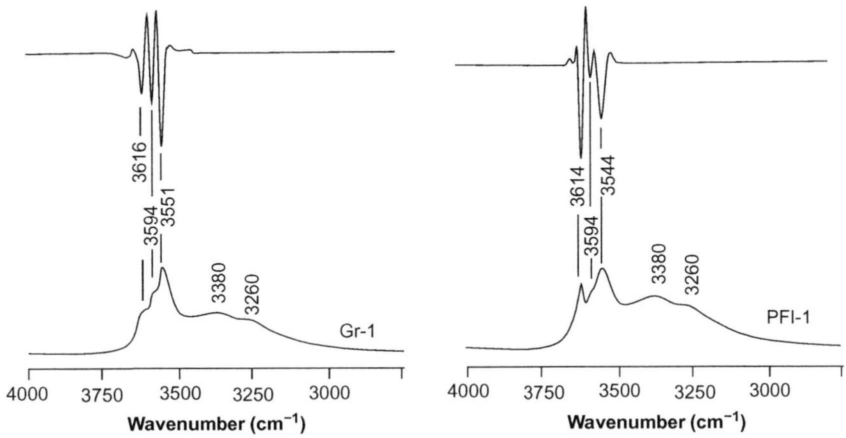


**FIG. 7.6** MIR spectra of two palygorskite samples studied by Gionis et al. (2006) depicting the $\upsilon(M_2OH)$ overlapped by $\upsilon(H_2O)$. The second derivative assists in determining the location of the $\upsilon(M_2OH)$. *Modified from Gionis et al. (2006).*

This approach has been exploited by Gionis et al. (2006, 2007) and Chryssikos et al. (2009) to make unambiguous assignments of various OH-sharing octahedral cation pairings (or triads) within the structures of palygorskites and sepiolites to better understand their structural chemistry (see Section 7.4.3)

### 7.3.2 Bond Strength (Valence)

The OH groups of hydroxyl-sharing cation pairs of the same (e.g. $Fe^{3+}$, $Fe^{2+}$) or similar (e.g. Al, Mg) mass are observed to have different vibrational frequencies that are directly related to their bond valances. The explanation is based on the bond-valence/bond-length concept originated by Pauling (1929) where decreasing bond valence of the OH-sharing octahedral cations leads to longer bond lengths with the shared bridging oxygen, and therefore allow the O–H bond strength to increase. Increased bond strength in the O–H results in an increased OH vibrational frequency. Earlier works (e.g. Vedder, 1964; Robert and Kodama, 1988) had reported trends between absorption frequencies and the bond valence. Besson and Drits (1997a,b) confirmed that when the valence sum of the OH-sharing octahedral cation pairs through the bridging oxygen is decreased, as for example $Al_2OH$ vs. AlMgOH, it resulted in an increased bond strength of the O–H and increased frequencies for $\upsilon(M_2OH)$.

The effect of bond valence has helped to explain some, but not all, of the exceptions evident in the reduced mass approach (see Section 7.3.1). For example, for $\upsilon(AlFe^{3+}OH)$ bands to reside at higher frequencies than an $\upsilon(Al_2OH)$ band, the O–H bond in the former pair must have a higher bond valence, and thus a higher bonding constant $k_{OH}$, associated with it (Zviagina et al., 2004). These authors assigned the high frequency $\upsilon(AlFe^{3+}OH)$ to OH-sharing cation pairs within a pyrophyllite-like domain where the lack of surface charge minimised the H-bonding effect of the O–H with oxygens within the vacant octahedral site, resulting in an increased valence sum of the O–H bond (see Section 7.3.4).

### 7.3.3 Reduced Mass—Valence Sum

The position of the specific $\upsilon(M_2OH)$ or $\delta(M_2OH)$ has been shown to be modified by the valence sum of the octahedral cation pair involved, which for the series concerned ($Fe^{3+}$, Al, $Fe^{2+}$, Mg) is either 4, 5 or 6. (Fig. 7.5B). The presence of divalent Mg, which is approximately the same mass as trivalent Al, decreases the resulting vibration frequency of $\upsilon(M_2OH)$ (Besson and Drits, 1997a,b; Zviagina et al., 2004) and $\delta(M_2OH)$ (Gates, 2008). This combined reduced mass—valence sum approach to interpret IR spectra of smectites brings together two of the factors considered most influential in determining vibrational frequency and has been applied previously by many researchers (e.g. Slonimskaya et al., 1986; Clark et al., 1990; Madejová et al., 1994; Petit et al., 1999b) and perhaps intuitively by many others.

For the most natural smectites (i.e. of Al–$Fe^{3+}$–Mg octahedral series, with perhaps some $Fe^{2+}$), any $M^aN^b$ cation pair (where M and N are the cation and superscript $a$ and $b$ are the valence) the position of the OH bands could be predicted from the positions of known bands of like-cation pairs through the relation:

$$M^aN^b\left(\mathrm{cm}^{-1}\right)=\frac{\left(M_2^a+N_2^b\right)}{2}\left(\mathrm{cm}^{-1}\right) \quad (7.2)$$

The following simple rules (Gates, 2008) apply:

**(i)** For OH-sharing cation pairs with the same valence sum, increased reduced mass results in decreased vibrational frequency;
**(ii)** For OH-sharing cation pairs with the same reduced mass, increased valence sum results in increased vibrational frequency;
**(iii)** Simultaneous increase in both reduced mass and valence sum of OH-sharing cation pairs results in increased vibrational frequency;

Application of Eq. (7.2) and the above rules enabled the development of the relations depicted in Fig. 7.6. These relationships hold only for the $\upsilon(M_2OH)$ or $\delta(M_2OH)$ (presumably the $(\upsilon+\delta)M_2OH$ combination bands also) of natural dioctahedral smectites within the Al–$Fe^{3+}$–Mg solid solution series. Some evidence suggests that they also apply to the synthetic dioctahedral $Ni^{2+}$ smectites. As has only very recently been shown (Petit et al., 2016), these relationships do not hold for synthetic smectites containing Ga (see Section 7.3.5).

Gates (2008) used this approach to show that Mg clustering does indeed exist in the dioctahedral smectites, which was previously suspected by others (e.g. Decarreau et al., 1987b). Cashion et al. (2008) were better able to interpret the Mössbauer spectra of some ferrian smectites only if Mg clustering, and its effect on layer charge distributions, were considered.

### 7.3.4 Effects of Next Nearest Neighbour Isomorphic Substitution

Some interesting observations have come from the application of the reduced mass-valence sum approach to well-studied samples. As discussed in detail by Besson and Drits (1997a) and by Gates (2005) isomorphic substitutions within the structure affect the strength of the O–H bond through the influence on the bridging oxygen (Robert and Kodama, 1988) and result in observed shifts in $\upsilon$ and/or $\delta$ vibrations of $M^a{}_2OH$, or $M^aN^bOH$ pairs where nearest neighbour compositions differ (Besson and Drits, 1997b; Zviagina et al., 2004). As mentioned in the previous Section 7.3.2, the presence of segregated $M^aN^bOH$ pairs within local domains in dioctahedral smectites lacking isomorphic substitution gives rise to $\upsilon(Al_2OH)$ bands at higher frequencies than those where some substitutions have occurred nearby. The mechanisms involved in changing the frequency of the O–H mode by isomorphic substitutions in the dioctahedral smectites are:

**(i)** Decreasing the valence sum of the OH-sharing octahedral cation pair directly leads to an increased valence sum of the O–H bond and a resulting increased vibrational frequency. The converse is true: the O–H bond valence sum will decrease if the valence sum of OH-sharing octahedral cations increases, resulting in a decreased vibrational frequency of the structural OH band.

**(ii)** Lower valence substitutions (e.g. Mg for Al) in any of the two (*trans*-vacant) or three (*cis*-vacant) octahedral sites (Fig. 7.7A), or in any of the 4 tetrahedral sites (e.g. Al for Si) opposite the octahedral vacancy (Fig. 7.7B) result in increased H-bonding across the ditrigonal cavity. This leads to a decrease in the O–H bond valence and a decrease in the vibrational frequency of the structural OH band.

Gates (2005) invoked these mechanisms to make some sense of the positons of $\delta(M_2OH)$ and $(\upsilon+\delta)M_2OH$ bands of 35 dioctahedral smectites with a large range in octahedral compositions. Montmorillonites with high octahedral Mg substitutions (Fig. 7.8) tended to have lower frequencies (near 915 $cm^{-1}$) associated with $\delta(Al_2OH)$ and smectites with a high degree of tetrahedral Al substitution (beidellites) had an additional band at higher wavenumber (from 935 to 960 $cm^{-1}$) attributed to $\delta(Al_2OH)$ sites that are more similar to beidellites (Kloprogge et al., 1990a; Kloprogge, 2006). The former result can be explained by a greater degree of H-bond across the ditrigonal cavity with the oxygen anionic network bonded to Mg, which, while increasing the

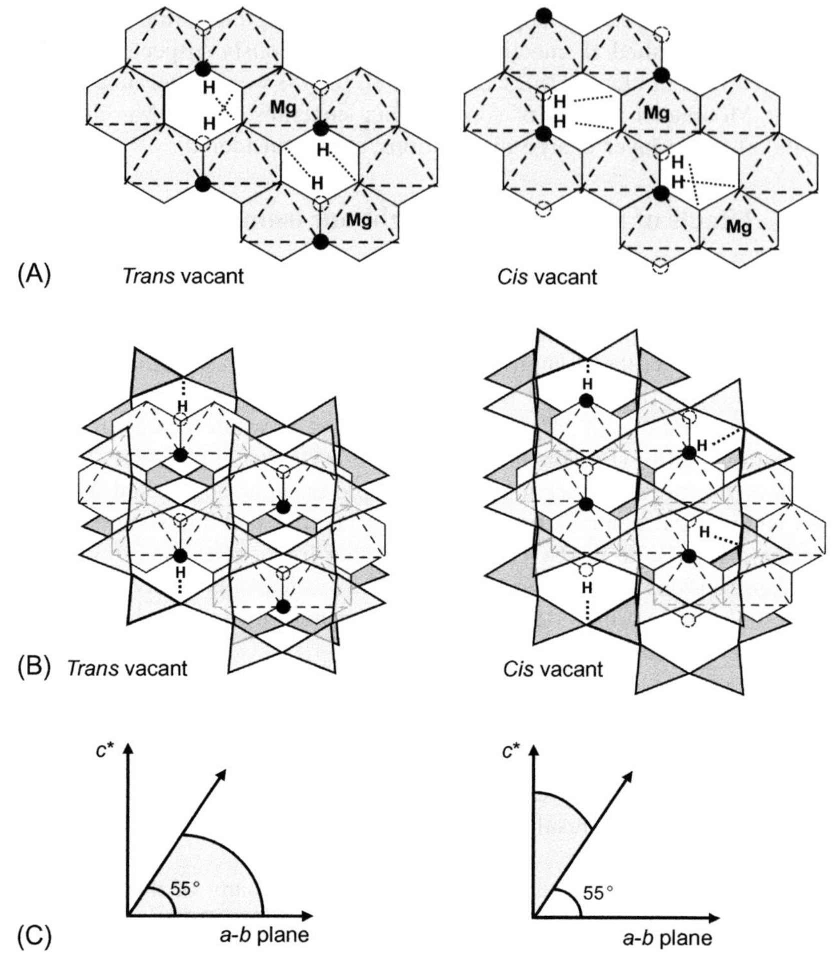


**FIG. 7.7** (A) Dioctahdral arrays depicting how Mg for Al substitutions result in increased H-bonding across the ditrigonal vacancy. (B) Al substitutions for Si in tetrahedral cause greater variation in H-bonding in *cis*-vacant compared to *trans*-vacant structures. Isomorphic substitutions resulting in layer charge depicted in (A) octahedral by 'Mg' labels and (B) in tetrahedral by thick outlines. (C) Tilt angles of the O–H vector (either stretching, bending or deformation) are <55° where octahedral Mg substitutions dominate (left) or >55° where tetrahedral substitutions dominate (right).

H-bond strength, reduces the O–H bond valence by forcing the OH vector to both bend and stretch more in line with the layer plane (Fig. 7.7C). The latter result can be explained by the stronger H-bond with the oxygen network due to Al for Si substitutions in the tetrahedral sheet, which forces the OH vector to bend and stretch at a steeper angle with respect to the layer plane (Fig. 7.7C).

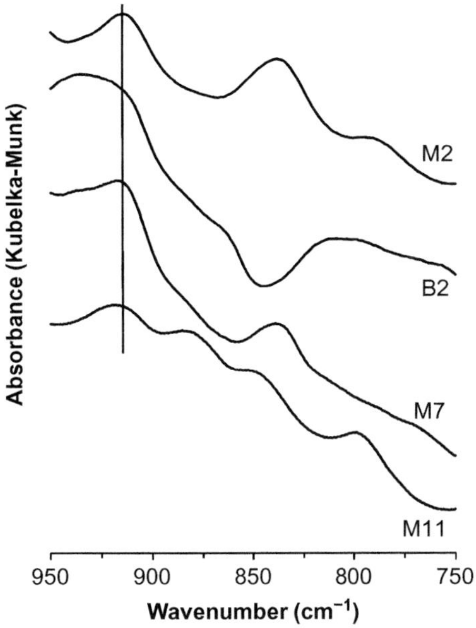


**FIG. 7.8** The effect of isomorphic substitutions of Mg for Al on the positions of the $Al_2OH$ band in montmorillonites and beidellites. Labels refer to the smectites studied by Gates (2005). Vertical line is at 915 cm$^{-1}$, typical of $\delta(Al_2OH)$ of montmorillonites. Note that for montmorillonites with less octahedral Mg (i.e. M11, M7) the position of the $\delta(Al_2OH)$ band shifts to higher frequency compared to montmorillonites with high Mg contents (i.e. M2), more similar to beidellites (i.e. B2). NB, the $\delta(M^aN^bOH)$ spectrum of M11 is broadened by the much greater concentration of octahedral $Fe^{3+}$ than is present in the other smectites.

The effect of isomorphic substitution on the orientation of the OH bending vector is clearly shown in the polarised synchrotron IR spectra displayed in Fig. 7.9, where self-supporting films were measured with the film plane at 0, 20, 35, 50 and 60° with respect to the electric field vector of the IR beam, which itself is orthonormal to the beam direction (see Manceau et al., 1998 for details of this type of experimental approach). Isotropic (i.e. invariant with measurement angle) spectral intensities would be expected for OH bending deformation vectors inclined $\approx 55°$ with respect to the film plane. For OH deformation vectors inclined at angles steeper than 55°, a positive anisotropic trend (increased spectral intensity with increasing measurement angle) would be observed, but a negative anisotropic trend in intensity would be observed for OH vectors inclined at angles $<55°$. Tetrahedral substitutions of 0.23 Al in ferrian smectite (F5) cause the $\delta(Al_2OH)$ and $\delta(AlFe^{3+}OH)$ O–H vector to be inclined at a steeper angle, but little effect is observed on the $Fe^{3+}{}_2OH$ band for the Clausthal Zellerfeld nontronite (N12) with 0.92 tetrahedral $Fe^{3+}$ (and 0.04 Al) substitutions. The strong positive anisotropy of the Fe-O out-of-plane deformation $\delta(FeO)$ band attests to the high relative amount of

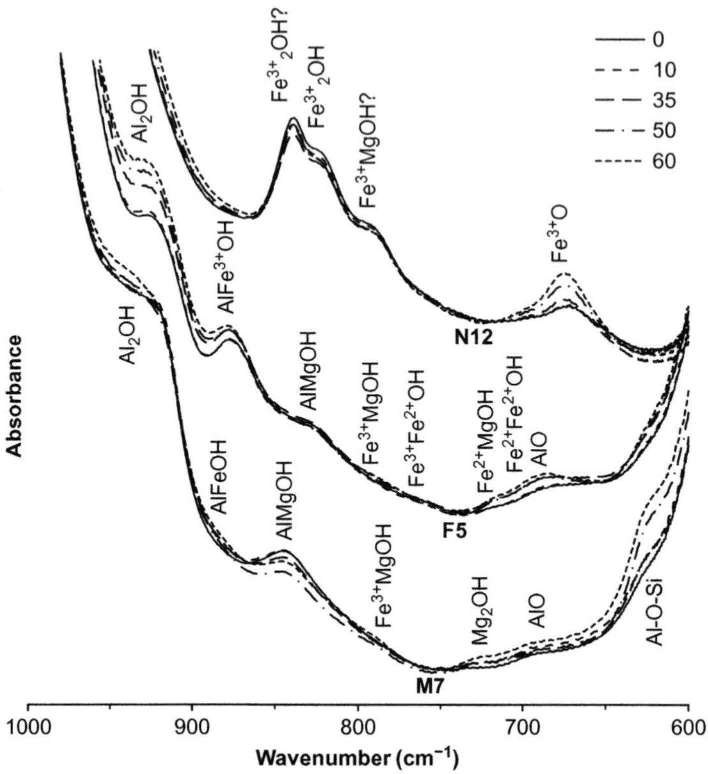


**FIG. 7.9** The effect of isomorphic substitutions in tetrahedral and octahedral sites on the orientations of the $\delta(M^aN^bOH)$ bands in montmorillonite (M7), Fe smectite (F5) and nontronite (N12) as observed from polarised Synchrotron IR spectra of oriented films. Self-supporting oriented films were inclined at the angles given relative to the electric field vector, which is >99% polarised orthonormal to the beam direction. Spectra presented are co-additions of 10 scans referenced to dry $N_2$ using a mylar beam splitter and liquid $N_2$ cooled MCT detector, operating at 0.5 $cm^{-1}$ resolution on the Far-IR high resolution beamline of the Australian Synchrotron. *Assignments based on Gates (2008).*

alignment of nontronite particles within the films. Thus, tetrahedral $Fe^{3+}$ for Si substitutions apparently do not influence the H-bond across the ditrigonal cavity to the same extent as Al (as depicted in Fig.7.7B), and this is in line with differences in the acidity of these cations.

The Mount Binjour montmorillonite (M7), which has a layer charge, *X*, of 1.23 $e^-$ per $O_{22}$ equivalents, is at the upper end of layer charge for dioctahedral smectites (Gates, 2005). This smectite has a substitution of 0.52 Al per unit cell in the tetrahedral sheet, which only marginally increased the OH bending deformation vector inclination (Fig. 7.9) for $\delta(Al_2OH)$ and $\delta(AlFe^{3+}OH)$ compared to that observed for F5. However, $Mg^{2+}$ for $Al^{3+}$

substitutions in the octahedral sheet strongly decreased the bending vector inclination for $\delta$(AlMgOH) in M7 but had little effect on that for F5. These latter effects indicate that for M7, tetrahedral Al substitutions probably occur in close proximity to sites of octahedral Mg substitutions (i.e. Al2(021)$Si_7$Al and Al2(111)$Si_7$Al) and thus the resulting H-bond is strongly influenced by octahedral substitutions. To the best of our knowledge, these data represent the first published confirmation of the effects of the location of an isomorphic substitution on the orientation of the OH deformation vectors associated with specific OH-sharing octahedral cation pairs originally postulated by Besson and Drits (1997a), Zviagina et al. (2004) and Gates (2005).

For the $\delta(Fe^{3+}{}_2OH)$ band in Fig. 7.9 marked with '?', the previous assignments remain questionable. This band grows in intensity with increasing $Fe^{3+}$ for $Si^{4+}$ substitution, up to about 1 $Fe^{3+}$ per 8 tetrahedral sites (Gates, 2008), but does not exist for synthetic nontronites having more than 1 $Fe^{3+}$ per 8 tetrahedral sites (Baron et al., 2016a). However, determination of the occupancy (see Section 7.4) appears to require this band to be associated with octahedral $Fe^{3+}$. The band indicated as $\delta(Fe^{3+}MgOH)$? for nontronites is too intense to be associated solely with octahedral Mg (Gates, 2005), as shown in the following discussion.

### 7.3.5 Ionic Radii Effects—A Generalised Approach

The reduced mass-valence sum relationship described in Section 7.3.3 breaks down when applied to $Ga^{3+}$ for $Fe^{3+}$ octahedral solid solution series of synthetic smectites (Petit et al., 2016). In this latter case, the positions of $\upsilon(M_2OH)$, $\delta(M_2OH)$ and $(\upsilon+\delta)M_2OH$ bands increase in frequency with increases in the reduced masses of $2Fe^{3+}$, $GaFe^{3+}$ and 2Ga OH-sharing octahedral pairs. As depicted in Fig. 7.10, a more generalised relationship, presumably suitable to explain all compositions and one that utilises the average ionic radii of the octahedral cations rather than the reduced mass, has been developed by Petit et al. (2016). Its generality was proven to extend to the Al-$Fe^{3+}$ series and allowed correct prediction of the position of $\upsilon(M_2OH)$, $\delta(M_2OH)$ and $(\upsilon+\delta)M_2OH$ bands of intermediate compositions.

The M-O bond distance discussed above in Section 7.3.2 represents an indirect measure of ionic radius. The valence state of the metal influences the ionic radius, bond valence and therefore, bond distance. While this explains in part the positions of some known absorbance bands, the lack of large data sets make such relationships difficult to assess. Note that in Fig. 7.10, the relationship for $\delta(M_2OH)$ and the average ionic radii of the pair differ from that for $\upsilon(M_2OH)$ (not shown) and this has consequences on the relationship for $(\upsilon+\delta)M_2OH$ (Petit et al. 2016). The $\upsilon(M_2OH)$ bands were used by Petit et al. (2016) initially in developing the respective empirical relationships described above in Section 7.3.1, because the data sets were the

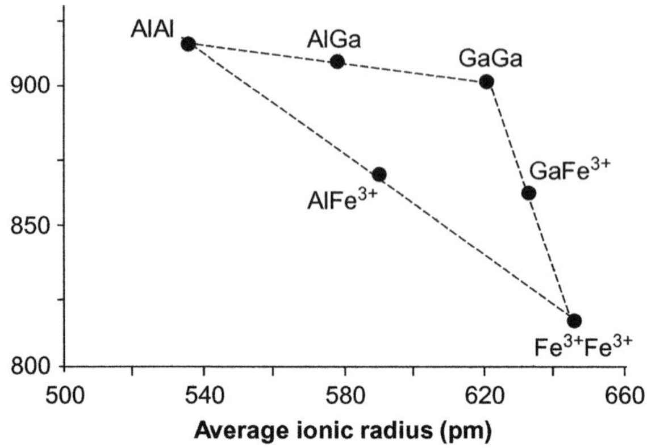


**FIG. 7.10** The relationship for the position of $\delta(M_2OH)$ bands in the Ga–Al–Fe synthetic smectite series as a function of the average ionic radius of the octahedral cation pair. Relations for $\upsilon(M_2OH)$ and $(\upsilon+\delta)M_2OH$ also exist. *Modified from Petit et al. (2016).*

most robust. However, the frequency relationships to ionic radii of the octahedral cation pairs largely explain the occurrence of apparent exceptions, or even opposite responses, of IR absorption bands as a function of reduced mass, bond valence or combined effects.

## 7.4 METHODS TO QUANTIFY OCTAHEDRAL OCCUPANCY FROM IR SPECTRA

NIR and MIR spectra provide useful probes of octahedral occupancy in clay minerals because all OH-sharing octahedral cation pairs contribute to the IR spectrum. As long as appropriate care is taken, useful and accurate complementary information can be gleaned from the IR spectra. As will be discussed further below, the main reason for widespread uptake of this method to determine site occupancy has been due to a lack of unambiguous assignments of some IR active absorbance bands.

The application of a band decomposition technique and the determination of chemistry has been the more common approach for many decades (e.g. Slonimskaya et al., 1986; Madejová et al., 1994; Frost et al., 2001b; Vantelon et al., 2001; Zviagina et al., 2004; Gates, 2005, 2008; Petit et al., 2004a, 2015, 2016). The application of spectral derivatives has been promoted more recently, largely to the more complex octahedral sheet composition of palygorskites (Gionis et al., 2007; Chryssikos et al., 2009) or $Ni^{2+}$-rich lizardites (Baron and Petit, 2016). In both of these approaches; however, successful determinations of clay mineral chemistry from IR spectra have been based on the empirical analysis of multiple samples consisting of a range of spectral characteristics that could be precisely matched to the chemistry (or other measures) of specific samples.

### 7.4.1 Band Decomposition

The band decomposition method has been used to quantify the chemistry of clay minerals. Some of the more important works have already been discussed in the context of the various models used to understand assignments. Arguably the easiest band decompositions are of the NIR spectra in the structual OH combination region (e.g. Gates et al., 2002; Petit et al., 2004a, 2015; Bishop et al., 2011; Baron and Petit, 2016; Baron et al., 2016a). Compared to MIR bands in the bending/deformation region, these bands are broad, have Gauss character, are often minimally overlapping and generally are not affected by a sloping background. Nor do the NIR combination bands have strong overlapping contributions from other vibrational features that are not related (or not of interest) to chemical quantification, such as the O–H stretching bands associated with adsorbed water in the MIR. Another feature is that the low intensity bands of interest, while often overlapped by strong tails from other bands, are sufficiently resolved to enable their quantification. An example is depicted in Fig. 7.11, where the percentage of integrated intensity based on a random distribution of cations from chemical analysis (Fig. 7.11A) and from decomposition of the NIR spectra (Fig. 7.11B) are compared to the actual percentage of integrated intensity for the occupancy assignments of synthetic talc studied by Petit et al. (2004a). Clearly the NIR spectral

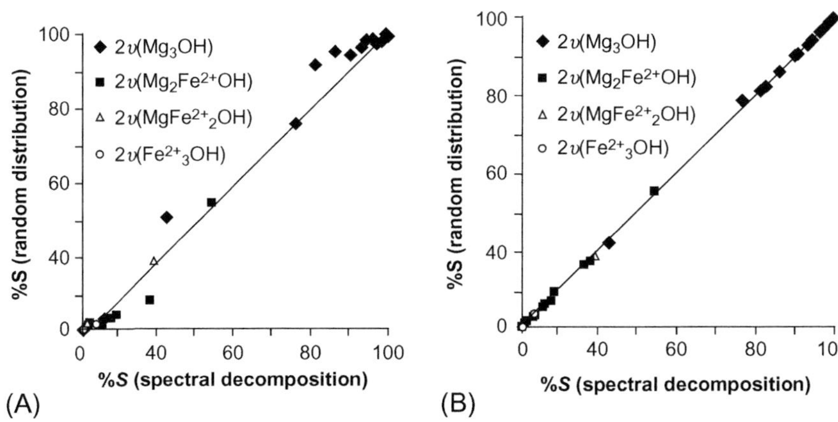


**FIG. 7.11** Results of spectral decomposition of Mg/$Fe^{2+}$ talcs depicting the determination of site occupancy from a random distribution from (A) chemical analysis and (B) decomposition of the NIR spectra as detailed in Petit et al. (2004a), vs. the integrated intensity (%S) of the NIR bands (Fig. 5.6A, Chapter 5). *Adapted from Petit et al. (2004).*

decomposition provides a more accurate determination of the actual site occupancy. A similar finding was observed using the NIR spectra to determine the site occupancy of nontronites (Gates et al., 2002) and synthetic dioctahedral ferrian smectites (Petit et al., 2015, 2016).

Absorption bands in the OH stretching region, however, are frequently more difficult to decompose because (1) they often occur as strongly overlapping doublets; (2) they are usually overlapped by bands associated with adsorbed water, and (3) the structural $\nu(M_2OH)$ bands have Gauss character, whereas the intermingling adsorbed water bands generally have Lorentz character. These effects lead to difficulties in detecting, resolving and quantitatively decomposing the less intense bands. Nonetheless, information from these bands are considered to be the most informative regarding the structural-chemistry of the layer silicates (Madejová et al., 1994; Frost et al., 2001b; Petit et al., 2004a) and require care to remove, as far as possible the effects of adsorbed water, by either heating/desiccation (see Clark et al., 1990; Cases et al., 1991; Madejová et al, 1994; Gionis et al., 2006, 2007) or by deuteration (Kuligiewicz et al., 2015a,b).

Decomposition of the bending deformation region (Vantelon et al., 2001; Gates, 2005, 2008) has proven successful for detailed occupancy assignments in dioctahedral smectites, largely because the bands are separated (non-overlapping) in this region. For the montmorillonites and ferrian smectites, most bands have been confidently assigned. A major drawback of decompositions in this region is the fact that these bands are overlain by the strong Si–O stretching band, and the position of this band itself is affected by isomorphic substitution (Madejová and Komadel, 2001) as well as by the hydration state (Yan et al., 1996a,b).

### 7.4.2 Spectral (Second) Derivative

The application of spectral derivatives for the quantification of IR absorbance bands has been limited to a few studies (Gionis et al., 2007; Chryssikos et al., 2009). The second derivative of an IR spectrum pinpoints, as a negative envelope, the location of narrow band components. It is useful because broad bands and sloping backgrounds are eliminated. The baseline for analysis then becomes the zero-line of the second derivative and the intensities of bands of interest are negative (Fig. 7.12). Within a spectrum, the negative intensities provide a quantitative measure of the chemistry for unambiguously assigned bands.

Successful application of this method requires high quality spectra because the determination of the second derivative usually results in several tens- or even a hundred-fold increase in the relative noise. Judicious use of smoothing functions (Savitzky and Golay, 1964) can minimise noise, but care must be taken to also minimise the loss of negative intensity, which impairs

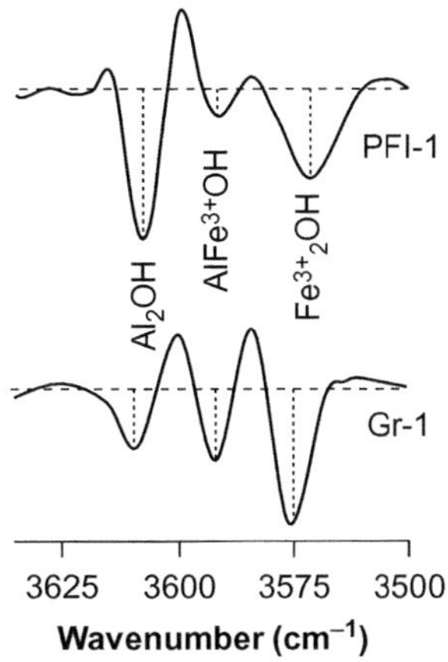


**FIG. 7.12** The second derivative of the structural υ(MOH) for the same two palyogroskite samples depicted in Fig. 7.5. While the positions of likely bands can be ascertained from the IR spectrum, their unambiguous assignment required the application of the second derivative. The dashed vertical lines represent the relative negative intensities of the υ(MOH) component bands and can be used to quantify composition (see Section 7.4.2). *Modified from Gionis et al. (2006).*

quantification. Gionis et al. (2007) report that negative intensities of the second derivative of palygorskites were reproducible with 5% error, within acceptable limits of most other methods (Gates et al., 2002). This method may also prove useful in dealing with complex mixtures where overlapping bands of different mineral phases overlap.

### 7.4.3 Assigning Occupancies

The quantitative information from band decomposition and the second derivative intensity analysis can be used to determine the octahedral composition of smectites with a high degree of accuracy. The approach requires accurate integrated intensities of all known $M^aN^bOH$ pairs within the region of interest and also of the same sort of band (i.e. for either $\upsilon(M_2OH)$ or $\delta(M_2OH)$ or $(\upsilon+\delta)M_2OH$, but not mixtures). Then for dioctahedral smectites, the number of octahedral $M^a$ sites, on the typical basis of $O_{20}(OH)_4$, can be determined from (Besson and Drits, 1997b; Gates, 2008):

$$\mathrm{Oct\,M}^a = 4x\left[2xA(\mathrm{M})^a_2\mathrm{OH} + \sum\nolimits_n A\left(\mathrm{M}^a\mathrm{N}^b\mathrm{OH}\right)\right] \tag{7.3}$$

In Eq. (7.3), $A$ is the normalised integrated peak area (or second derivative intensity), $n$ is the number of $M^aN^bOH$ bands observed in the IR spectrum For example, to determine the occupancy of Al, $M^a$, in montmorillonite, one requires information on $A$ for $Al_2OH$, $AlFe^{3+}OH$ and AlMgOH, (as well as any other $AlN^bOH$ pair) as well as $n$ and must be applied iteratively for each

$N^b$ (as $M^a$) in turn to provide a statistical octahedral assignment. The quantification is normalised to 4 for dioctahedral smectites. Similar expressions can be developed for other clay mineral species.

A major assumption that is required in order for this methodology to provide quantitative estimates of occupancy, and which has caused (and continues to cause) difficulty is the proper assignment of some of the bands in question. For example in spectra of nontronites, a low intensity band near 750 $cm^{-1}$, often only observed as a shoulder to the more intense deformation of $Fe^{3+}{}_2OH$ (cf. Fig. 7.8) has been previously assigned to $\delta(Fe^{3+}{}_2MgOH)$ (e.g. Gates, 2005) for montmorillonites and ferrian smectites based on the occurrence of $\delta(Fe^{3+}{}_2MgOH)$ in the correct frequency range on the application of Eq. (7.3). However, it is now apparent that the full intensity of this band in natural nontronites (Gates, 2008) cannot be solely associated with octahedral Mg as it occurs also in synthetic nontronites free of Mg.

### 7.4.4 Comparison to Random Distributions

Random distribution analysis is a highly useful approach to assist in determining deviations in the measured, or estimated, occupancies. Gates and Cashion (2014) have used this approach to show that most dioctahedral smectites (of the 35 studied by Gates, 2005) having chemistries close to end-member nontronite, montmorillonite or beidellite, have octahedral cation occupancies that can be largely explained by random distributions. For example, deviations from random distributions based on the chemistry of the six nontronites, were observed to indicate enrichments of tetrahedral substitution (Fe2(300) $Si_7(AlFe)_1$, Fe2(300)$Si_6(AlFe)_2$, Fe2(210)$Si_7(AlFe)_1$) and octahedral substitution (Fe2(210)$Si_8$). These sites (and Fe2(300)$Si_8$) represented more than 80 percent of all possible sites. Most of the remaining 20 percent of the sites were made up of higher levels of tetrahedral substitution, with or without one octahedral $Al^{3+}$ for $Fe^{3+}$ substitution per unit cell. For the single ferruginous smectite (SWa-1) studied, these five sites only represented 60 percent of the available sites, largely due to the increase octahedral Mg content.

Intermediate compositions, as for the ferrian smectites, or compositions with elevated Mg, as in some montmorillonites and ferrian smectites, tend to deviate from random distribution (Gates, 2008). Past research observed deviations from random distributions, but ascribed that it may have been due to the intermixing of dioctahedral (nontronites) and trioctahedral (Ni-rich) smectites (Decarreau et al., 1987b; Mano et al., 2014). In the specific example of the ferrian smectites studied by Gates (2008), excess Mg and the presence of residual $Fe^{2+}$ resulted in the clustering of sites of OH-sharing divalent cation pairs, such as $Mg_2OH$, $Fe^{2+}MgOH$ and $Fe^{2+}{}_2OH$, within a dioctahedral array.

While not surprising results in and of themselves, when compared to distributions calculated from IR spectral decompositions (e.g. Madejová et al., 1994; Vantelon et al., 2001; Gates, 2008), or from other methods, such as

X-ray absorption spectroscopy (Manceau et al., 1998) or Mössbauer spectroscopy (Cashion et al., 2009), the results do provide complementary evidence for the correctness of assignments made in the IR.

## 7.5 CONCLUSIONS AND FUTURE DIRECTIONS

The determination of the composition of the clay minerals from IR follows basic rules associated with valency, atomic or molecular mass and ionic size. These rules that have been used by mineralogists for nearly a century. Aspects of each of these parameters were initially promoted by Farmer in 1964 (Farmer and Russell, 1964) and have been used by many researchers in an attempt to unambiguously identify absorbance bands. For the most part this has been highly successful, and clay mineralogists can understand an enormous amount about a clay mineral of interest 'simply' by study of its IR spectrum. There have been some exceptions, as detailed in this chapter.

The determination of the chemical occupancy of clay minerals has also come far since Farmer (1974b) on the IR spectroscopy of the clay minerals. The applications demonstrated here mainly for Al, Fe and Mg occupancies in dioctahedral smectites may be extended to other clay minerals, as well as clay minerals containing exotic elements such as Ga and Ni. While nearly all of the advances described above have been empirical in nature, with the advent of molecular dynamics (e.g. Martínez-Alonso et al., 2002a; Cygan et al., 2004; Greathouse and Cygan, 2013; Ferrante et al., 2015; Greathouse et al., 2016) and calculations of the IR spectra of similar minerals as described in detail in previous Chapters 2 and 6 of this volume, a new generation of exciting advances in this area is about to occur.

Chapter 8

# Application of Vibrational Spectroscopy in Clay Minerals Synthesis

J.T. Kloprogge

## 8.1 INTRODUCTION

Clay minerals form an important subgroup of the phyllosilicates. The geochemical importance of clay minerals stems from their high specific surfaces, and ion-exchange properties. Consequently, because of their ubiquity in soils and sediments, clay minerals tend to dominate the surface chemistry of soils and sediments. Furthermore, these same properties have given rise to a wide range of industrial applications throughout history. Today, clay minerals are important materials with a great variety of applications in ceramics and nanocomposites, energy exploration, waste isolation and the metal and paper industries. Chemical applications of clay minerals include their use as adsorbents, decolourising agents, ion exchangers and molecular-sieve catalysts. Clay mineral deposits can be found all over the world, showing large variations in their compositions and the amount and type of impurities. While kaolins and bentonites are commercially available in large quantities, nevertheless, sufficient quantities of suitable purity of some of the more useful clay minerals are not available in nature, or they lack specific desired properties. Therefore, scientists in such fields as material science, geology, chemistry and geochemistry have taken particular interest in the laboratory syntheses of clay minerals over the past several decades to overcome these shortcomings. Much effort has been devoted to obtaining well-crystallised clay minerals with controlled composition and properties (Farmer et al., 1979; Petit et al., 1988, 2015; Petit and Decarreau, 1990; Martin et al., 1992, 1998; Grauby et al., 1993; Delineau et al., 1994; Kloprogge, 1998; Kloprogge and Frost, 1999d; Rinnert et al., 2005; Decarreau et al., 2008;

Developments in Clay Science, Vol. 8. https://doi.org/10.1016/B978-0-08-100355-8.00008-4

Lantenois et al., 2008a,b; Andrieux and Petit, 2010; Ryu et al., 2010; Baron et al, 2016a,b; Jaber et al., 2013).

A variety of characterisation methods are commonly applied to study the starting, intermediate and final products resulting from clay mineral synthesis, including X-ray diffraction (XRD), scanning electron microscopy and transmission electron microscopy. Over the past few decades vibrational spectroscopy, such as mid-infrared (MIR) and near-infrared (NIR) and Raman spectrocopies, has been applied not only to characterise these final materials, but also to gain a better understanding of the crystallisation process itself. The concept that vibrational bands, as diagnostic for particular clay mineral species, will develop, intensify and resolve during synthesis, underpins the application of vibrational spectroscopy in clay mineral synthesis. Advances in interpreting the vibrational modes of clay minerals now allow better control of the crystallisation from the early stages of nucleation to the final products, including how synthesis conditions can influence the composition and layer charge of the resulting clay minerals. In this chapter an attempt will be made to give an overview of the information available in the literature on how vibrational spectroscopy can help in this respect. Readers interested in the exact conditions of synthesis should consult the relevant papers.

## 8.2 IMOGOLITE AND ALLOPHANE

Imogolite ($Al_2SiO_3(OH)_4$) and allophane ($(Al_2O_3(SiO_2)_{1.3-2}\ ((H_2O)_{2.5-3}))$) are naturally occurring nanoparticles formed through mainly the alteration of volcanic parent material. These two aluminosilicates form under-related geochemical conditions, but have different structures related to different chemical compositions. Although these materials are both poor and para-crystalline, they do have distinct and recognisable XRD patterns (Levard et al., 2012; van der Gaast et al., 1985). Both consist of a tubular gibbsite-like outer layer linked to $Si^{4+}$ tetrahedra within: imogolites consisting of $Si^{4+}$ tetrahedral monomers and allophanes consisting of polymerised $Si^{4+}$ tetrahedra. Imogolites have a molar ratio of $Si^{4+}$ to $Al^{3+}$ (Si:Al) near 0.5, discrete XRD patterns and are single-walled aluminosilicate nanotubes of 2–3 nm in diameter and a few hundred micrometre in length (Cradwick et al., 1972; Farmer et al., 1977a,b; Levard et al., 2012). Allophanes have Si:Al between 0.65 and 1.0, and are nearly X-ray amorphous consisting of a hollow-sphere structure with a diameter of 3.5–5 nm. Because of their variable compositions, the allophane structure is much more ambiguous, but different structural types have been determined.

Basak and Ghosh (1996) reported the infrared (IR) spectra of synthetic amorphous aluminosilicate with the Si:Al ratio varying from 0.48 to 0.99 exhibiting similar features of allophane as reported by Wada et al. (1979). A single band appeared in the region of 3000–3800 $cm^{-1}$. The absorption maxima attributed to OH stretching ($\nu$(AlOH), $\nu$(SiOH) and associated adsorbed $H_2O$) shifted from 3479 to 3617 cm$^{-1}$ with increasing Si:Al in the gel. In the region between 1649 and 1663 cm$^{-1}$ a band for adsorbed water $\delta(H_2O)$ was observed. These samples also had an adsorption band maximum that narrowed with increased Si:Al within the region between 982 and 1040 cm$^{-1}$. A fourth absorption band shifted from 582 to 730 cm$^{-1}$ with increasing Si:Al. Reinert et al. (2011) hydrothermally synthesised allophane with Si:Al = 0.5 and 1.0. The FTIR spectra of all samples (Fig. 8.1) showed a broad band at about 3500 cm$^{-1}$ assigned to HOH-stretching vibrations of adsorbed water. The corresponding bending vibration of water was observed at about 1638–1639 cm$^{-1}$. The spectra of both pristine allophanes were quite similar For the Si:Al = 0.5 sample, the main band at 978 cm$^{-1}$ and its shoulder at 858 cm$^{-1}$ were assigned to the $\nu$(Si–O–M) vibrations, where M was $Si^{4+}$ or $Al^{3+}$ (Farmer, 1997). For the Si:Al = 1.0 sample, the main band was slightly shifted to higher wavenumbers (i.e. 995 cm$^{-1}$), and the shoulder was still visible (875 cm$^{-1}$). This shift towards higher wavenumbers, explained by the increase of the $Si^{4+}$ content in the framework, was

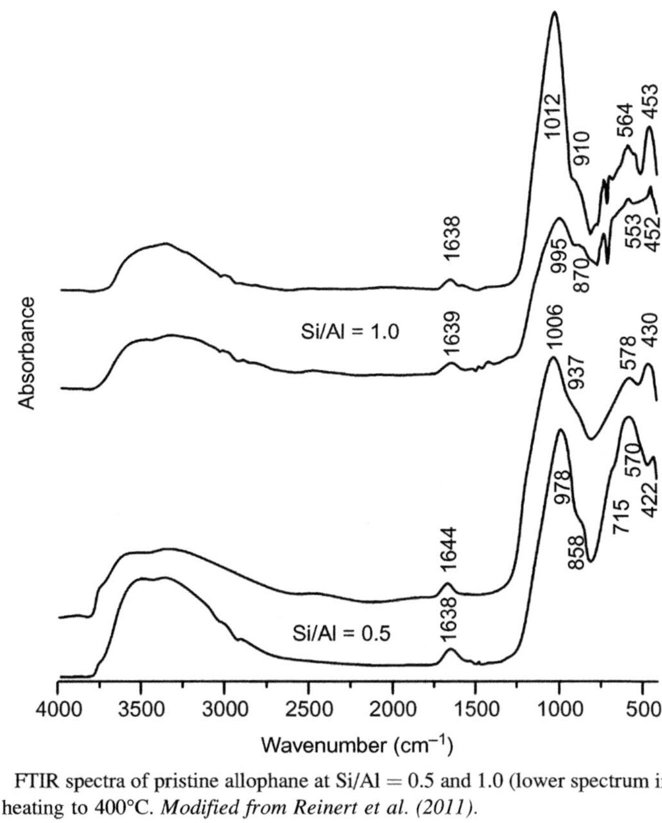


**FIG. 8.1** FTIR spectra of pristine allophane at Si/Al = 0.5 and 1.0 (lower spectrum in each pair) and after heating to 400°C. *Modified from Reinert et al. (2011).*

already reported by Lidner et al. (1998) for synthetic allophanes. The bands observed at lower wavenumbers (≈422 and 570 $cm^{-1}$ for Si:Al = 0.5 and 452 and 553 $cm^{-1}$ for Si:Al = 1.0) were typical of the allophane framework, and assigned to Al–O and AlOH bonds (Montarges-Pelletier et al., 2005). The spectra of both samples were slightly modified by heating at 400°C. The main modifications concerned the shift of the previously described characteristic bands. After heating, the *ν*(SiO) bands were shifted to higher wavenumbers (from 978 to 1006 $cm^{-1}$, and 995 to 1012 $cm^{-1}$ for Si:Al = 0.5 and Si:Al = 1.0 main bands, respectively) (Fig. 8.1). Moreover, the relative intensities of the *ν*(Si–O–Al) and *ν*(Si–O–Si) vibrations of pristine materials were changed significantly after calcination. This intensity changes by heating were attributed by Henmi et al. (1981) to the breaking of the Si–O–Al linkages, the condensation of the $SiO_4$ tetrahedra and also to the dehydroxylation of Si–OH groups resulting in the generation of new AlOH bonds as confirmed by the increased intensity and shift of the lowest wavenumber bands at ~450 $cm^{-1}$. A recent review of the IR spectra of imogolite has been provided by Thill (2016). Probably the most comprehensive overview of the MIR and NIR spectra of both imogolite and allophane has been provided by Bishop et al. (2013a) and has been combined in Table 8.1.

Farmer et al. (1979) obtained proto-imogolite forms of allophane from the acidic synthesis of various Si:Al ratios at 20°C, as indicated by a characteristic absorption band at 348 $cm^{-1}$. All silica added was incorporated into the proto-imogolite structure for Si:Al ratios of 0.5 or less, but at higher ratios not all Si were incorporated. The excess Si, recovered in the freeze-dried product (containing Al), had Si:Al ratios close to those of pure silica gel. For compositions of Si:Al > 0.5 (reacted at 96°C), the band at 348 $cm^{-1}$ had increased intensity compared with products produced at 20°C indicating development of well-ordered synthetic imogolite. Wada and Kubo (1975) synthesised aluminosilicates from solutions containing monomeric silica and $Al^{3+}$ ions at room temperature. All precipitates were non-crystalline and had resulting $SiO_2/Al_2O_3$ ratios ranging from 1.0 to 3.0 (Si:Al from 0.5 to 1.5). No para-crystalline aluminosilicate such as imogolite formed in these precipitates. Wada et al. (1979) synthesised imogolite from solutions with NaOH/Al ratios of 2.8 or less and with Si:Al ≥ 0.5 or greater. The further observed allophane precipitates from solutions with NaOH/Al ratios of 3.0, except for that with the $SiO_2/Al_2O_3$ ratio of 0.5 (Si:Al = 0.25). These precipitates formed immediately after addition of NaOH and contained only allophane with at most a trace of imogolite. The spectra also showed a shift of the *ν*(SiO) absorption maximum to higher frequency, from 970 to 1010 $cm^{-1}$, with increase of their $SiO_2/Al_2O_3$ ratio from 1.0 to 8.0 (Si:Al = 0.5 to 4).

Farmer et al. (1991a) showed that at 23°C the presence of $Mg^{2+}$ and $K^+$ had little effect on the imogolite and allophane products obtained. For starting ratios of Si:Al > 0.5, the composition of the precipitate changed progressively from that appropriate to proto-imogolite (Si:Al = 0.5) to that

**TABLE 8.1 Band Assignments for Spectral Features Observed for Imogolite and Allophane in the Near- and Mid-infrared Regions (Bishop et al., 2013a). *ν* Indicates Stretching Vibration; *d* Indicates Bending Vibration; Band Assignments are Based on the Present Study and on Previous Transmission IR Studies (e.g. Farmer et al., 1977b; Henmi et al., 1981; Parfitt and Furkert, 1980)**

| Imogolite | Allophane | Assignment |
|---|---|---|
| Wavenumber ($cm^{-1}$) | | |
| 7285–7305 | 7263 | $2\nu(OH)$ |
| 7170–7205 | 7104–7139 | $2\nu(OH)$ |
| 6930 | 6900 | $2\nu(H_2O)$ |
| 5200 | 5220 | $(\nu+\delta)H_2O$ |
| 4565 | 4560 | $(\nu+\delta)OH$ |
| 4502 | 4495 | $(\nu+\delta)OH$ |
| 3700 | | $\nu(OH)$ |
| ~3670–3755 | ~3640–3740 | $\nu(OH)$ |
| ~3500–3600 | ~3500–3600 | $\nu(H_2O)$ |
| ~3150 | ~3120–3165 | $2\delta(H_2O)$ |
| 1604–1662 | 1606–1667 | $\delta(H_2O)$ |
| ~1400–1421 | 1430–1485 | $\delta(O_3SiOH)$ |
| 1010 | 1030 | $\nu$(Si–O–Al) |
| 930 | 940 | $\nu$(Si–O–Al) |
| 825 | 825 | $\delta$(OH) |
| 595 | 610 | Out-of-plane $\delta$(OH) |
| 495 | 545 | $\delta$(Si–O–Al) |
| 415 | 420 | $\delta$(Si–O) |
| 335 | 335 | $\delta$(Si–O) |

appropriate to a Si-rich hydrous feldspathoid Si:Al = 1 as $Si^{4+}$ in solution increased from around 60 μM to > 600 μM (Farmer et al., 1991b). Boyer et al. (2014) provided a more systematic investigation into the effects of parameters such as the acid used, the Si:Al ratio and the amount of ethanol used on the formation of imogolite. The sample prepared in HCl solution showed a large but unresolved band between 3800 and 3200 $cm^{-1}$ due to the AlOH and $H_2O$ OH-stretching vibrations. A band at 1645 $cm^{-1}$ was ascribed to the bending mode of physisorbed water molecules. The nanotublar character was confirmed by the presence of two characteristic $\nu$(SiO) bands at 960 and 910 $cm^{-1}$. These lattice $\nu$(SiO) bands were shifted to lower wavenumber, due to tube enlargement, compared with unmodified Si-imogolite (Bishop et al., 2013a). No effect of different acids (HCl, $HClO_4$ and $CH_3COOH$) on the final products was observed. Both for higher or lower Si:Al ratios, turbidity increased. While IR spectra of different samples corresponded to a classical hybrid imogolite signature, several bands were affected by changes in Si:Al ratio. A broadening of the 783 $cm^{-1}$ band, an additional band at 1020 $cm^{-1}$ and additional distinct bands in the 3200–3700 $cm^{-1}$ region were observed for the lowest ratios. These features were attributed to the formation of gibbsite particles. No obvious differences in the IR nanotube spectra as a function of the ethanol content were observed. Similarly, the use of methyltriethoxysilane or methyltrimethoxysilane in the synthesis did not result in any differences in the spectra.

McBride et al. (1984) attempted to synthesise imogolite with partial $Fe^{3+}$ substitution. Analysis of the synthesised products by transmission electron microscopy, electron diffraction and IR spectroscopy indicated that while some imogolite was formed in the presence of iron, increased $Fe^{3+}$ content resulted in formation of ferrihydrite and promoted poorly organised aluminosilicates resembling proto-imogolite instead. Analysis by ESR spectroscopy suggested that little or no $Fe^{3+}$ was incorporated into the imogolite structure, although the less crystalline proto-imogolite may have accommodated some structural $Fe^{3+}$. The IR spectra of products with starting Si:Al ratios of 0.5 synthesised by Farmer et al. (1991a) all showed the general features of proto-imogolite with maxima at 969–979 $cm^{-1}$, a shoulder near 700, 565–600 $cm^{-1}$ and a shoulder near 440 and 350 $cm^{-1}$. It was uncertain whether the small displacements in band position could be ascribed to some substitution of iron into the proto-imogolite structure. The apparent increase in intensity and displacement to lower frequencies of the 600 $cm^{-1}$ band with increasing iron in the system could arise from the presence of a separate hydrous ferric oxide phase. Ookawa et al. (2006a), however, reported that for a $Fe/(Fe+Al)=0.05$, the tubular imogolite structure was preserved and the characteristic doublet in the IR was similar to that of pure imogolite, while increasing the ratio instead resulted in the disappearance of the doublet and the appearance of a very broad featureless band. With XANES and EXAFS, they confirmed that the $Fe^{3+}$ was present in octahedral coordination and that it was not adsorbed on the imogolite surface.

Imogolites in which a part or all of $Si^{4+}$ was substituted with Ge were synthesised by Wada and Wada (1982) from solutions containing $Al^{3+}$ and silicic and/or germanic acid. The products were similar to natural imogolite in their tubular morphology. However, the external diameter of the tube increased with increasing $Ge^{4+}$ substitution and decreased the curvature of the gibbsite-like sheet with which the GeO-tetrahedra are associated. As reported earlier by Farmer et al. (1977b), the synthetic and natural imogolites are not identical in terms of the diameters of the poorly ordered crystals. IR absorption bands at 995 and 930 $cm^{-1}$ assigned to $\nu$(SiO) vibrations in imogolite disappeared and new bands appeared at 910 and 810 $cm^{-1}$ on substitution of $Ge^{4+}$ for $Si^{4+}$, whereas those in the region between 700 and 300 $cm^{-1}$ associated with coupled Al–O vibrations hardly changed or remained unchanged. The ratios of the wavenumbers 995/910 (=1.09) and 930/810 (=1.15) were comparable to the ratio of the wavenumbers expected from a reduced mass effect for the corresponding $\nu$(SiO) and $\nu$(Ge–O) vibrations (1.13). The $\delta$(Al–O) vibration shifted from 570 to 550 $cm^{-1}$ with increasing $Ge^{4+}$ substitution, indicating the effect that the linking tetrahedra have on this octahedral $Al^{3+}$ band. On the other hand, a decrease in intensity of the band at 495 $cm^{-1}$ was observed with increasing Ge content and a new band appeared at 465 $cm^{-1}$ in $Ge^{4+}$-substituted imogolite. Similarly Ookawa et al. (2006b) observed two bands at 907 and 828 $cm^{-1}$ in the spectrum of the synthetic $Al^{3+}$ germanate sample. In addition, they also observed a shift of the band at 565 to 550 $cm^{-1}$, decreasing band intensity at 500 $cm^{-1}$ and an appearance of new band at 466 $cm^{-1}$.

A novel route using $F^-$ as a mineraliser was published by Chemmi et al. (2013). Small amounts of this anion, having similar size and charge as $OH^-$, have been proved to decrease synthesis times and to increase resulting crystallinity of a variety of synthetic smectites. The IR spectrum of the product was identical to that of imogolite produced conventionally with characteristic bands at 992, 953, 697, 562, 512 and 425 $cm^{-1}$. McCutcheon et al. (2005) used $^{29}Si$ labelling to study the shift in the IR bands associated with Si–O vibrations. Differences in IR absorption frequencies from conventional imogolite are small but may reflect zero-point energies. The zero-point energy of the Si–O bond of $^{29}Si$–O imogolite is different from the predominate $^{28}Si$–O in unlabelled material. Application of the stretching wavenumber equation

$$\omega = 1/2\pi c\left(\sqrt{k/\mu}\right) \tag{8.1}$$

where $k$ is a force constant and $\mu$ is reduced mass of the vibrating unite given by [M(Si)M(O)]/[M(Si)+M(O)] for $^{29}Si$ and $^{28}Si$ (and $^{16}O$), then $\omega_{29}/\omega_{28} = (\sqrt{\mu_{28}}/\sqrt{\mu_{29}}) = 0.9937$. Thus, taking the differences in the wavenumbers of the individual units ($\omega_{28} = 1082\ cm^{-1} - \omega_{29} = 1076\ cm^{-1} = 6\ cm^{-1}$) for band 1 results in a similar difference as was observed (Table 8.2). Analogous differences were expected and observed for the other bands.

**TABLE 8.2 Differences in Infrared Spectra of Unlabelled Imogolite and $^{29}Si$ Labelled Imogolite**

| | 1 | 2 | 3 | 4 |
|---|---|---|---|---|
| *Assignment* ($cm^{-1}$) | *ν(Si–O–Si)* | *ν(Si–O)* | *ν(Si–O–Al)* | *ν(O–Si–O)* |
| Unlabelled | 1082 | 990 | 942 | 573 |
| $^{29}Si$ labelled | 1076 | 986 | 937 | 572 |
| Experimental difference | −6 | −4 | −5 | −1 |
| Zero point energy difference (Eq. 8.1) | −7 | −6 | −6 | −4 |

After McCutcheon, A., Hu, J., Kannangara, G.S.K., Wilson, M.A., Reddy, N., 2005. $^{29}Si$ labelled nanoaluminosilicate imogolite. J. Non-Cryst. Solids 351, 1967–1972.

## 8.3 1:1 CLAY MINERALS

### 8.3.1 Kaolinite

Miyawaki et al. (1993) studied the formation of kaolinite (Kaol) from amorphous calcium silicate and aluminium chloride with a number of analytical techniques including IR spectroscopy. The starting material was characterised by a broad ν(SiO) band around 1000 $cm^{-1}$ with sharp bands associated with calcite at 1450, 880 and 850 $cm^{-1}$. The $\delta(Al_2OH)$ band around 900 $cm^{-1}$ became visible as a broad shoulder after 3 h synthesis time, the δ(Si–O–Al) bands typical of the 1:1 lattice occurring near 700 and 550 $cm^{-1}$, as well as the $\nu(Al_2OH)$ bands, appeared after 12 h and increased in intensity with further reaction time (Fig. 8.2). When using an amorphous mixture of silica and alumina as a starting material, Miyawaki et al. (1992) observed the first Kaol bands after 36 h hydrothermal treatment. Ryu et al. (2010) used amorphous silica gel and amorphous $Al(OH)_3{\cdot}xH_2O$ to hydrothermally synthesise Kaol and observed $\nu(Al_2OH)$ bands at 3700 and 3621 $cm^{-1}$ after 10 h. A major broad band from the starting material was still present at 3432 $cm^{-1}$, but with further reaction time it slowly diminished in intensity, while the two Kaol bands further resolved, and after 15 h the other two Kaol bands at 3670 and 3653 $cm^{-1}$ were observed but poorly developed. Huertas et al. (1993) reported on the hydrothermal synthesis of Kaol from amorphous aluminosilicate gel and assigned the broad band near 3400 $cm^{-1}$ to ν(OH) of adsorbed water with a shoulder at 3600 $cm^{-1}$ assigned to ν(OH) of structural OH-groups of condensate. With increasing reaction time and reaction temperature, the characteristic ν(OH) modes for Kaol appeared.

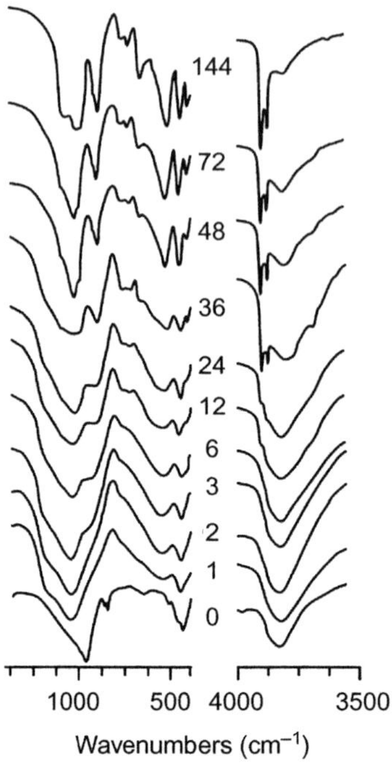


**FIG. 8.2** IR spectra of the formation of kaolinite from amorphous calcium silicate and aluminium chloride with increasing time. *Reproduced with permission from Miyawaki et al. (1993).*

Fialips et al. (2000) reported on the effects of pH on the crystallinity of synthetic Kaol prepared at 220°C under equilibrium water pressure (23.2 bar) for 14 days at an initial pH of around 5 from silico-aluminous gel. This partially crystallised material was then used as the starting material for Kaol synthesis at various pH values (at 200°C, 220°C or 240°C at equilibrium water pressure for 21 days). The four OH-stretching modes were well resolved for Kaol synthesised at pH 1.1. As the pH increased to 6.6 the intensity of the 3668 $cm^{-1}$ band decreased relative to the 3651 $cm^{-1}$ band, which was attributed to an increase in the monoclinic character of Kaol (Brindley et al., 1986). Further increasing the pH to 8.4 caused such a high density of defects that it was no longer possible to visually identify all individual bands (Fig. 8.3). In addition, at this pH a shoulder was observed at 3600 $cm^{-1}$. Very similar results were reported by Aung et al. (2015). Because the starting gel was iron free, this band could not be assigned to a $\nu(AlFe^{3+}OH)$ band as seen for $Fe^{3+}$-enriched synthetic Kaol (Petit et al., 1988; Petit and Decarreau, 1990; Delineau et al., 1994; Iriarte et al., 2005). Heating experiments proved that the band could not be associated with either the presence of pseudo-boehmite (AlO(OH)) or

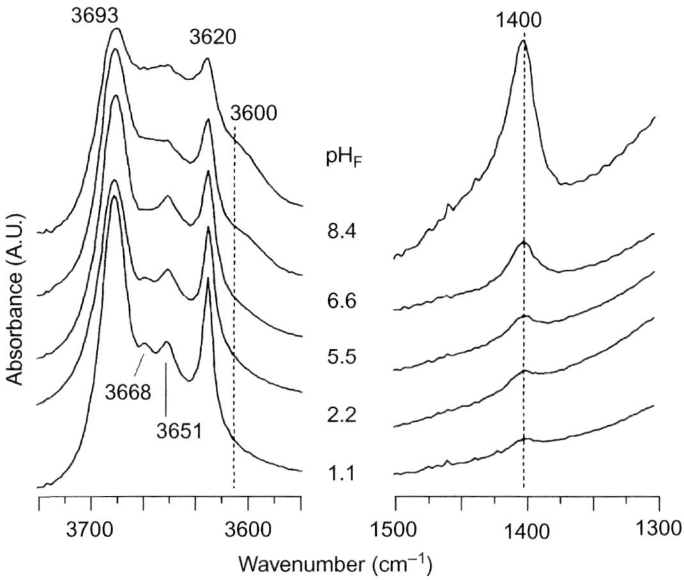


**FIG. 8.3** FTIR normalized spectra of the OH stretching region as function of pH and after exchange with $NH_4^+$. *Reproduced (modified) with permission from Fialips et al. (2000).*

intercalated water. Moreover, the intensity of this band was closely related to the reaction pH. An explanation was put forth that adsorption of cations onto the edge sites of synthetic Kaol increased with pH values (up to pH = 8.5; higher values did not result in Kaol crystallisation) induced perturbations of $\nu$(OH) (Fialips et al., 2000; Aung et al., 2015).

It is well-known that small amounts of metals like $Fe^{3+}$ can substitute for $Al^{3+}$ in the octahedral sheets of Kaol. Over the years, a number of different spectroscopic methods, such as ESR (Angel and Hall, 1972; Meads and Malden, 1975; Herbillon et al., 1976; Mestdagh et al., 1980, 1982; Brindley et al., 1986; Delineau et al., 1994), Mössbauer (Hogg et al., 1975), EXAFS (Bonnin et al., 1982), NMR (Stone and Torres-Sanchez, 1988) and IR (Mendelovici et al., 1979; Delineau et al., 1994; Hassan and Salem, 2001) have shown that small variable amounts of $Fe^{3+}$ (up to almost 3% $Fe_2O_3$ based on chemical analyses) can be accommodated into the Kaol structure. Generally, the crystals formed exhibit a decrease in crystallinity and possibly particle size, with increasing $Fe^{3+}$ substitution. Petit et al. (1988) and Petit and Decarreau (1990) hydrothermally synthesised Kaol with increasing iron content and observed a $\nu(M_2OH)$ band at 3600 $cm^{-1}$ in the in synthetic Kaols, which increased in intensity with increasing iron content. This band, which was also observed for natural iron-rich samples (Mendelovici et al., 1979), has been therefore unambiguously assigned to $\nu(AlFe^{3+}OH)$ vibrations.

Petit and Decarreau (1990) also observed a band at 875 $cm^{-1}$ for synthetic Fe-substituted Kaol, which increased in intensity with increasing iron content, and assigned it to $\delta(AlFe^{3+}OH)$. At high concentration, they observed a slight shoulder at 818 $cm^{-1}$ (Fig. 8.4), which was suggested to be due to the occurrence of $Fe^{3+}$ ions in adjacent octahedra (i.e. a $\delta(Fe^{3+}{}_2OH)$ band). Comparable results were published by Iriarte et al. (2005) with $\nu(AlFe^{3+}OH)$ at 3598 $cm^{-1}$ and $\delta(AlFe^{3+}OH)$ at 881 $cm^{-1}$. Delineau et al. (1994) pointed out that both the bands at 3600 and 875 $cm^{-1}$ were more easily observed in the diffuse reflectance spectra of iron-bearing Kaol than in transmittance spectra where the band has a much lower intensity.

Petit et al. (1995) hydrothermally synthesised $Cu^{2+}$-substituted Kaol. Their IR spectra differed from the reference Kaol sample only by changes to the respective intensities of the 3668 and 3652 $cm^{-1}$ bands, the latter being more intense for the synthetic sample. The four $Al_2OH$ bands become increasingly broadened and overlapping with increasing $Cu^{2+}$ incorporation. These features are frequently attributed to 'poor crystallinity' of samples (Barrios et al., 1977;

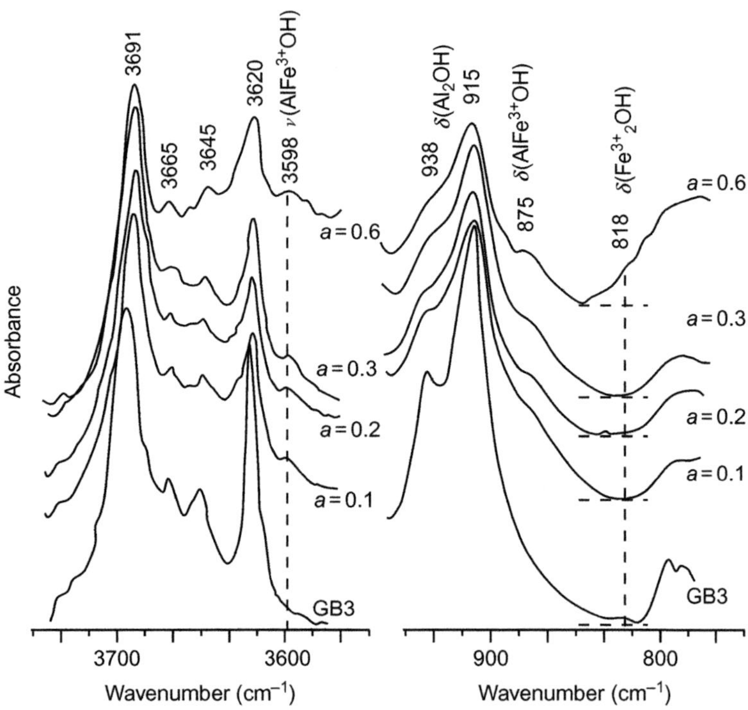


**FIG. 8.4** IR spectra in the hydroxyl stretching and bending regions of the reference product GB3 and synthetic kaolinites with various $Fe^{3+}$ content. *Reproduced (modified) with permission from Petit and Decarreau, (1990).*

Casal et al., 1982) and are more precisely empirically linked to monoclinic character of samples (Mestdagh et al., 1982). Observed in the spectra of all the samples were bands assigned to $\nu$(SiO), $\delta$(SiO), $\delta(Al_2OH)$ and $\delta(Si–O–Al^{VI})$, that broadened with increasing $Cu^{2+}$-content. The intensity of the $\delta(Al_2OH)$ band decreased with the $Cu^{2+}$ content of samples, while a small shoulder around 868 $cm^{-1}$ appeared. Another band located at 840 $cm^{-1}$ was also observed for the sample with the highest $Cu^{2+}$ content. These latter bands were tentatively attributed to vibrations of $\delta(AlCu^{2+}OH)$, and $\delta\left(Cu^{2+}{}_2OH\right)$.

Other octahedral substitutions in synthetic Kaol described include $Ga^{3+}$ (Martin et al., 1998; Bentabol et al., 2009) and $Cr^{3+}$ (Bentabol et al., 2009). The OH-stretching region spectra of the solid products confirm that these metals form a part of the Kaol structure. The NIR spectra of these products were also recorded by Petit et al. (1999b). Major NIR spectral features of Kaols are located in the 4600–4200-$cm^{-1}$ and 7000-$cm^{-1}$ regions (Table 8.3). Of importance here is the major band at 4526 $cm^{-1}$, corresponding to the $(\nu+\delta)Al_2OH$ combination band. Two bands are clearly observed at 7176 and 7065 $cm^{-1}$. These bands corresponded to the $2\nu(M_2OH)$ of fundamental modes of $Al_2OH$ groups (Hunt and Salisbury, 1970). In comparison with the pure Kaol, two additional features were observed at 4466 and 7018 $cm^{-1}$ in the spectra of Fe-rich Kaols. These bands were attributed by Delineau et al. (1994) to the combination of the bands at 3598 and 875 $cm^{-1}$, assigned to $\nu(AlFe^{3+}OH)$ and $\delta(AlFe^{3+}OH)$ vibrations, respectively, and to the $2\nu(M_2OH)$ overtone of the 3598 $cm^{-1}$ band, respectively. The 7018 $cm^{-1}$ band occurred only as a shoulder (with variable intensity) in all the spectra of this series. The very weak inflexion near 4466 $cm^{-1}$ in the spectrum of the Fe-poorest Kaol of the series increased in intensity for the intermediary samples and broadened for the Fe-richest sample. Similarly, two additional bands at 4474 and 6986 $cm^{-1}$ were observed for the two chromian Kaols in comparison with the pure Kaol. These bands were assigned to the $(\nu+\delta)M_2OH$ of $AlCr^{3+}OH$ and to $2\nu(M_2OH)$ of $\nu(AlCr^{3+}OH)$ groups, respectively. This assignment was supported by the fact that both bands were more intense for the sample higher in $Cr^{3+}$ content. In the MIR, the $\nu(AlCr^{3+}OH)$ band was observed at 3586 $cm^{-1}$, but no $\delta(AlCr^{3+}OH)$ vibration band was detected (Maksimović and White, 1973; Maksimović et al., 1981; Mosser et al., 1993). However, the wavenumber of the $\delta(AlCr^{3+}OH)$ vibration band can be calculated from the NIR data, following the equation (Petit et al., 1999b):

$$(\nu(M_2OH) + \delta(M_2OH))_{NIR} - \nu(M_2OH)_{MIR} = \delta(M_2OH) \qquad (8.2)$$

Due to anharmonicity, the calculated wavenumber of the $\delta(M_2OH)$ vibration may be lower compared with its actual value. The comparison of the observed and calculated wavenumbers of the $Al_2OH$(inner) and $AlFe^{3+}OH$ groups obtained for $Al^{3+}$- and $Fe^{3+}$-Kaols shows that approximately 10 $cm^{-1}$ should be added (Table 8.3) to the calculated $\delta(M_2OH)$ value to obtain the

**TABLE 8.3** Observed and Calculated Wavenumbers of the Vibrations of the OH Groups Linked to Various Octahedral Cations (Petit et al., 1999b)

| | Observed ($cm^{-1}$) | | | | Calculated ($cm^{-1}$) | | | |
|---|---|---|---|---|---|---|---|---|
| | *NIR* | | *MIR* | | | | | |
| *Cationic pairs* | *$2\nu(M_2OH)$ (1)* | *$(\nu+\delta)M_2OH$ (2)* | *$\nu(M_2OH)$ (3)* | *$\delta(M_2OH)$ (4)* | *From (2) and (3)* | *cor.* | *From (1) and (2)* | *cor.* |
| Al–Al* | 7065 | 4526 | 3620 | 914 | 906 | 914 | 994 | 914 |
| Al–$Fe^{3+}$ | 7018 | 4466 | 3598 | 875 | 868 | 876 | 957 | 877 |
| Al–$Cr^{3+}$ | 6986 | 4474 | 3586 | – | 888 | 896 | 981 | 901 |
| Al–$Ga^{3+}$ | 7018 | 4498 | 3599 | – | 899 | 907 | 989 | 909 |

*=Bonded to inner OH, –=not observed, cor.=corrected after comparison with the known $\delta$OH values (4).

experimental value. The $\delta(AlCr^{3+}OH)$ vibration band would be located at about 900 $cm^{-1}$ and thus would overlap with the main $\delta(Al_2OH)$ band at 915 $cm^{-1}$.

The NIR spectrum of the $Ga^{3+}$-rich synthetic Kaol exhibited additional bands at 4498 and 7018 $cm^{-1}$, assigned to $(\nu+\delta)AlGa^{3+}OH$ and $2\nu(AlGa^{3+}OH)$, respectively. The $\nu(AlGa^{3+}OH)$ vibration was observed in MIR at 3600 $cm^{-1}$, but again, no $\delta(AlGa^{3+}OH)$ vibration was detected in the MIR spectrum. Based on Eq. (8.2), the position of the $\delta(AlGa^{3+}OH)$ vibration band should be ≈ 910 $cm^{-1}$, again overlapping with the $\delta(Al_2OH)$. Assuming that in the same environment the wavenumber of the band of the OH group bonded to a cationic pair involving two different homovalent cations can be deduced from the mean of the vibration wavenumbers of the OH groups bonded to the 2 cationic pairs involving the same cations (Decarreau et al., 1992), the obtained value was 909 $cm^{-1}$. Such a wavenumber for the $\delta(AlGa^{3+}OH)$ vibration is in accordance with the results obtained from the near-IR spectra after application of the above equation. The coincidence in position of the $\delta(AlGa^{3+}OH)$ with the main $\delta(Al_2OH)$ band at 915 $cm^{-1}$ explains the absence of a separate $\delta(AlGa^{3+}OH)$ band in the MIR region.

### 8.3.2 The Serpentine Minerals

Shirozu and Ishida (1982) studied the changes in the IR spectrum of a hydrothermally synthesised aluminium lizardite (Shirozu and Momoi, 1972) on deuteration. Two bands on the OH-stretching region at 3640 and 3470 $cm^{-1}$ were partially replaced by two new bands at 2680 and 2560 $cm^{-1}$ upon deuteration. The wavenumber ratio of the original OH bands to the shifted OD bands were 1.358 and 1.355, close to the theoretical value of 1.37, indicating that no significant coupling or crystal effects were present for the different OH bands. Bentabol et al. (2010) and Bentabol and Ruiz Cruz (2013) synthesised $Al^{3+}$–lizardite and $Al^{3+}$–$Ni^{2+}$ or $Al^{3+}$–$Co^{2+}$ lizardite from Kaol. Though the lizardite had formed, a significant portion of the Kaol remained in the final product resulting in IR spectra that were heavily influenced by the Kaol bands. A very broad signal was observed, however, beneath the sharp Kaol hydroxyl bands, which was fitted with four broad overlapping bands and interpreted as the splitting of the outer $\nu(Al_2OH)$, caused by bonding to Al and vacancies in the octahedral sheet (Farmer, 1997; Serna et al., 1977b; Velde, 1980). A characteristic band at 608 $cm^{-1}$ typically for $Al^{3+}$–$Mg^{2+}$ serpentines was observed. For the $Al^{3+}$–$Co^{2+}$ and $Al^{3+}$–$Ni^{2+}$ samples, the three bands corresponding to the perpendicular vibrations in Kaol were replaced by a doublet (1033 and 1010 $cm^{-1}$) with a weak shoulder at the high wavenumber side. Baron et al. (2016b) prepared $Ni^{2+}$–$Mg^{2+}$ serpentines in the lizardite–nepouite series where the MIR spectra exhibited characteristic bands of the endmembers lizardite ($x=0$) at 436, 454, 562, 607, 634, 984 and 1084 $cm^{-1}$, and for nepouite (x = 3) at 426, 455, 674, 984 and 1075 $cm^{-1}$ (Table 8.4). The four overlapping $\nu(M_xOH$, where $x$ sums to 3) bands linked

**TABLE 8.4** Experimentally Observed IR and Raman Bands for the Serpentine Minerals

| | Baron et al. (2016b) | Shirozu and Ishida (1982) | Bentabol and Ruiz Cruz (2013) | Titulaer et al. (1993) | Falini et al. (2002) | Stubičan and Roy (1961) | | Bloise et al. (2010) | Gunnarsson et al. (2005) |
|---|---|---|---|---|---|---|---|---|---|
| | *Mg/Ni-lizardite* | *Lizardite* | *Al-lizardite* | *Chrysotile* | *Chrysotile* | *Ni chrysotile* | | *Ni chrysotile* | *Low crystal. antigorite* |
| | *IR* | *IR* | *IR* | *IR* | *IR* | *Raman* | *IR* | *Raman* | *IR* |
| M-O Lattice | | | | | | 233 | | 230 | |
| M-O Lattice | | | | | | 346 | | 348 | |
| M-O Lattice | | | | | | 390 | | 388 | |
| M-OH/Si-O bend | 436 | | | | | 432 | 429 | | 400 |
| M-OH/Si-O bend | 454 | | | | | 465 | 457 | | 450 |
| M-O perpendicular | 562 | | | | | | | | 565 |
| M-OH libration | 607 | 610 | 608 | | | 623 | 622 | 620 | 620 |
| M-O/Si-O bridge | 634 | 660 | | 655–660 | | 689 | 663 | 690 | 650 |
| | | 810 | | | | | | | |
| Si-O stretch | 984 | | | 947 | | | 971 | | 895 |

| | | | | | | | | |
|---|---|---|---|---|---|---|---|---|
| Si-O stretch | | | 1010 | 1010 | 1015 | | | 1000 |
| Si-O stretch | | | 1033 | | | | 1052 | |
| Si-O | 1084 | | | 1081 | 1082 | 1103 | | 1080 |
| *v*OH H2O | | 3470 | | | | | | |
| $\nu Mg_3OH$ | 3648 | 3640 | | 3638 | 3645 | 3685 | | |
| $\nu Ni^{2+}Mg_2OH$ | 3668 | | | | | | | |
| $\nu Ni^{2+}{}_2MgOH$ | 3686 | | | 3688 | 3692 | 3700 | | |
| $\nu Ni^{2+}{}_3OH$ | 3698 | | | | | | | |
| $2\nu Mg_3OH$ | 7124 | | | | | | | |
| $2\nu Ni^{2+}Mg_2OH$ | 7164 | | | | | | | |
| $2\nu Ni^{2+}{}_2MgOH$ | 7198 | | | | | | | |
| $2\nu Ni^{2+}{}_3OH$ | 7226 | | | | | | | |

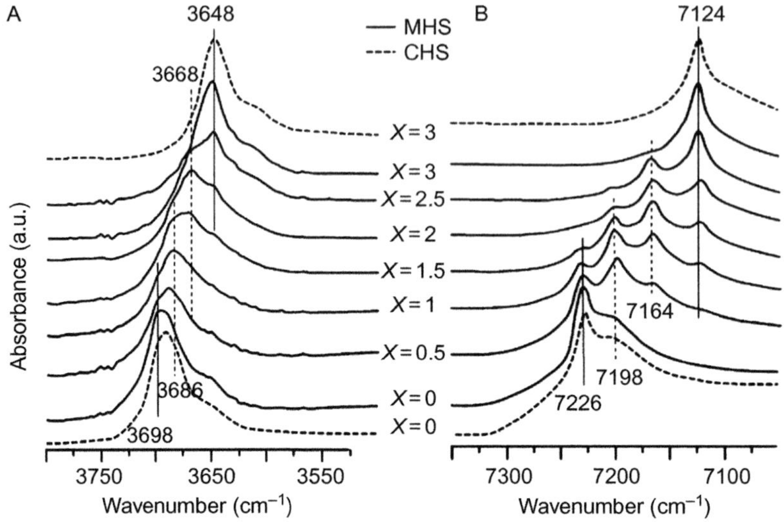


**FIG. 8.5** (A) MIR and (B) NIR spectra of the lizardite-nepouite series $Si_2Mg_{3-x}Ni_xO_5(OH)_4$. *Reproduced (modified) with permission from Baron et al. (2016).*

to four combinations of the octahedral cations (3Ni, 2Ni–Mg, Ni–2Mg and 3Mg) were observed at 3698, 3686, 3668 and 3648 $cm^{-1}$ (Fig. 8.5A). Zheng and Wang (2014) prepared $Fe^{3+}$-doped lizardite with a $\nu(Mg_3OH)$ band at 3690 $cm^{-1}$ and two broad $H_2O$ bands at 3430 ($\nu(M_3OH)$ and 1640 ($\delta(M_3OH)$ associated with the lizardite surface. Baron and Petit (2016) observed that in the NIR region, the $(\nu+\delta)M_3OH$ and $2\nu(M_3OH)$ bands for both endmembers could be attributed to two or three MIR features. The near-IR spectra exhibited the four well-resolved characteristic $2\nu(M_3OH)$ bands localised at 7226, 7198, 7164 and 7124 $cm^{-1}$ (Fig. 8.5B). For intermediate samples ($0.5<x<2.5$), the relative intensity of the $2\nu(M_3OH)$ argues for a random distribution of $Ni^{2+}$ and $Mg^{2+}$ cations in the octahedral sheet of the lizardite–nepouite series, using both NIR and MIR spectral regions (250–7500 $cm^{-1}$).

Titulaer et al. (1993) synthesised chrysotile from two different silica sources (sodium silicate solution and fumed silica). The DRIFT spectra of the chrysotile samples exhibited two characteristic $\nu(Mg_3OH)$ stretching vibrations at 3688 and 3638 $cm^{-1}$ attributed to the three external OH at one basal surface of the brucite sheet, and one internal OH at the other surface (Jolicoeur, 1981). The bands at about 1081, 1010 and 947 $cm^{-1}$ are typical for the $\nu$(Si–O–Si) modes. The 1081 $cm^{-1}$ band was attributed to the out-of-plane symmetric $\nu$(Si–O–Si) vibration of the silica sheet. The degeneracy of the single band at 1010 $cm^{-1}$ in talc to two bands at 1010 and 947 $cm^{-1}$ in chrysotile was explained by the bending of the silica sheet. The 1010 $cm^{-1}$

band was attributed to $\nu$(Si–O–Si) vibrations (mainly) parallel to the *b*-axis in the direction of the curvature, while the 947 $cm^{-1}$ was attributed to $\nu$(Si–O–Si) vibrations parallel to the *a*-axis. The $\nu$(MgO) vibration of octahedrally coordinated $Mg^{2+}$ was observed around 655–660 $cm^{-1}$. Falini et al. (2002) reported IR bands at 3692 and 3645 $cm^{-1}$ due to $\nu(M_3OH)$ and at 1082 and 1015 $cm^{-1}$ due to $\nu$(SiO), and in the Raman spectrum at 3700 and 3685 $cm^{-1}$ ($\nu(M_3OH)$), at 1103, 689 and 623 $cm^{-1}$ ($\delta$(Si–O–Si)) and at 465, 432, 390, 346 and 233 $cm^{-1}$ due to $\delta$(MgO) modes.

Several metals, such as $Fe^{2+}$, $Ni^{2+}$ and $Co^{2+}$, have been used to partially or fully substitute for Mg in chrysotile. Stubičan and Roy (1961) used a synthetic $Ni^{2+}$-chrysotile to compare with Kaol and other related silicates within the low wavenumber region of the IR spectra of these minerals. In the structures of Kaol, septechlorite and $Ni^{2+}$-chrysotile, one silica layer of the type $(Si_2O_5)^{2-}$ is combined in a unit with an octahedral layer of the type $M^{3+}{}_2(OH)_6$ (gibbsite layer) or $M^{2+}{}_3(OH)_6$ (brucite layer). The spectra obtained clearly indicated the influence of the octahedral cation: $\nu$(SiO) frequencies of Kaol at 1100, 1033 and 1010 $cm^{-1}$ moved towards lower wavenumbers for $Ni^{2+}$-chrysotile. Since the absorption bands at 1110, 1033 and 1010, as well as at 686, 645, 470 and 432 $cm^{-1}$ of Kaol correspond, respectively, to bands at 1052, 971, 663, 622, 457 and 429 $cm^{-1}$ in $Ni^{2+}$-chrysotile, it is possible to assign them to similar vibrations, impacted by the presence of $Ni^{2+}$. Bloise et al. (2010) provided a Raman spectrum of $Ni^{2+}$-chrysotile except for the band ascribed to brucite, well-defined bands produced by chrysotile were detected at 690, 620, 388, 348 and 230 $cm^{-1}$. Nevertheless, when the spectrum was compared with the Raman spectrum of stoichiometric synthetic chrysotile (Kloprogge et al., 1999c; Foresti et al., 2009), shifts in some vibrational modes were identified. The vibrational mode at 624 $cm^{-1}$ in Raman spectra from stoichiometric chrysotile and produced by $Mg_3OH$ translations shifted to 620 $cm^{-1}$; the bands at 469 and 436 $cm^{-1}$ in the stoichiometric chrysotile, ascribed to vibrational translation of Mg–OH/bending modes of the $SiO_4$ tetrahedra $\delta$(Si–O–Mg) and to MgOH translation, respectively, were not detected in the spectrum of the $Ni^{2+}$-doped chrysotile. The Raman bands produced by the vibrational modes of the $SiO_4$ tetrahedra were also affected: the bands at 690 and 388 $cm^{-1}$, ascribed to symmetric stretching modes of Si–$O_b$–Si linkages ($\nu_{AS}$(SiOSi)) and to bending modes $\nu$5(e) of $SiO_4$ tetrahedra plotted at, respectively, 694 and 392 $cm^{-1}$ in stoichiometric synthetic chrysotile. Among the bending modes of $SiO_4$ tetrahedra, detected at 348, 321 and 305 $cm^{-1}$ for stoichiometric chrysotile, only the band at 348 $cm^{-1}$ was observed in the Ni-chrysotile spectrum, whereas the bands at 321 and 305 $cm^{-1}$ were not detected. The band at 230 $cm^{-1}$ (vibrations of O–H–O groups) did not significantly shift. The observed shifts generally towards lower wavenumbers are most likely due to incorporation of $Ni^{2+}$

within the crystallised chrysotile. In contrast Foresti et al. (2009) in their work on Fe-doped chrysotile clearly observed the bands around 465, 433 and 304 $cm^{-1}$. The difference may be explained by the fact that only trace amounts of $Fe^{3+}$ were incorporated in the chrysotile structure compared with the much higher $Ni^{2+}$ amounts in the work of Bloise et al. (2010). Gunnarsson et al. (2005) prepared poorly crystalline antigorite under hydrothermal conditions with resulting IR spectra that resembled that of natural antigorite. The spectra showed a distinct trioctahedral character with two principal $\nu$(SiO) bands around 1000–1080 $cm^{-1}$, a $\delta$(SiOH) band at 895 $cm^{-1}$, a single $\delta$(Si–O–Mg) bending vibration at 450 $cm^{-1}$ (with a shoulder at 400 $cm^{-1}$) and $\delta$(MgO) out-of-plane vibrations at 565 $cm^{-1}$ (Farmer, 1997). A band at 625 $cm^{-1}$, split into two bands in spectrum after 412 h of reaction time (620 and 650 $cm^{-1}$). This was attributed to inner and surface liberation modes of the layers ($\delta$(Si–O–Mg)). The most notable difference between the spectra was a progressive narrowing of the $\nu$(SiO) split at 1000–1075 $cm^{-1}$, suggesting that the effective symmetry of the tetrahedral layer was becoming higher with time, as well as the appearance of a new band at 650 $cm^{-1}$. The $\nu$(SiO) bands were not as well defined as in a fully crystalline antigorite with two strong bands at 995 and 1075 $cm^{-1}$ with a minor band at 1200 $cm^{-1}$ and shoulders at 1120 and 960 $cm^{-1}$, probably owing to absence of $Al^{3+}$ and $Fe^{3+}$ and the pseudo-crystalline nature of the precipitate.

## 8.4 2:1 CLAY MINERALS

### 8.4.1 Trioctahedral Minerals

#### *8.4.1.1 Talc*

The primitive unit cell of talc contains two OH groups related through a centre of symmetry. Thus, two $\nu(M_3OH)$ (M = divalent cation) vibrations are possible, one active principally in the Raman spectrum in which both OH groups vibrate in phase, and the other active principally in the IR, in which the OH groups vibrate out of phase. Because coupling between the OH groups is weak, the IR and Raman frequencies are almost identical (Petit, 2005). Farmer (1974b) originally stated in his book that the wavenumber of the $\nu(Mg_3OH)$ group remains almost constant at 3677 $cm^{-1}$ whatever the octahedral environment in true talc. In natural clays, the most common changes in tetrahedral sheet are due to heterovalent $Al^{3+}$ and/or $Fe^{3+}$ for $Si^{4+}$ substitutions and are not easy to evidence through $\nu(M_3OH)$ bands. This fact is even more relevant in talcs where such substitutions are very few if any.

Divalent cations such as $Ni^{2+}$, $Fe^{2+}$, $Co^{2+}$ and $Zn^{2+}$ can easily substitute for $Mg^{2+}$, due to similar cationic size, and solid-solutions between true talc and these endmembers can be easily obtained through hydrothermal synthesis (e.g. Wilkins and Ito, 1967; Decarreau et al., 1992). Substitution by another divalent cation, due to differences between characteristics (electronegativity, atomic mass and cation size) of the substituted element from $Mg^{2+}$ induces a distinct change in the wavenumber of the corresponding $\nu(M_3OH)$. Moreover, up to four $\nu(M_3OH)$ wavenumbers can be distinguished when $Mg^{2+}$ is partially replaced by another divalent cation, corresponding to the four possible combinations of the two different cations within three octahedral sites that are directly linked to an OH group.

Decarreau et al. (1992) hydrothermally synthesised $Mg^{2+}$–$Ni^{2+}$ talc between 100°C and 240°C. The FTIR spectra revealed four $\nu(M_3OH)$ bands corresponding to the four possible cationic environments around OH ($Mg_3$, $Mg_2Ni$, $MgNi_2$ and $Ni_3$). The spectra of the synthetic $Mg^{2+}$–$Ni^{2+}$ talcs confirmed that the distribution was close to random in the series. These $\nu(OH)$ bands are among the sharpest of any purely vibrational bands in known silicates and are uniform in shape (Wilkins and Ito, 1967). The measured band width at half height was close to 3 $cm^{-1}$, reflecting the highly regular structural environment in which the hydroxyl groups reside. With increasing ionic radius of the substituting ion, two related effects became apparent: all bands broadened and all bands developed low wavenumber tails, including that at highest wavenumber ($\nu(Mg_3OH)$). The band broadening was interpreted as a second-order effect resulting from the presence of the substituting ion immediately outside the first coordination sphere of the hydroxyl ion (i.e. as a nearest neighbour effect). The most striking characteristic within the 800–400 $cm^{-1}$ region was the strong doublet at 452 and 467 $cm^{-1}$ that persisted throughout the complete series (Stubičan and Roy, 1961). Wilkins and Ito (1967) pointed out that these two bands cannot be assigned to $\delta(MgO)$ vibrations as originally favoured by Farmer (1958) and Farmer and Russell (1964), but instead to $\delta(SiO)$ vibrations. A band at 690 $cm^{-1}$, which represents a dipole oscillation normal to the sheet, did not persist even in the $Mg^{2+}{}_{84}Ni^{2+}{}_{16}$ talc. From the data, Decarreau et al. (1992) deduced (i) that $Ni^{2+}$ and $Mg^{2+}$ were randomly distributed in the talc synthesised at higher temperature and (ii) that a trend to an ordered distribution of octahedral cations with $Ni^{2+}$-clustering occurred, consistent with natural ($Ni^{2+}$–$Mg^{2+}$) clays formed at low temperature (lateritic weathering profiles), in which large $Ni^{2+}$-clusters were systematically observed (Manceau and Calas, 1985, 1986; Decarreau et al., 1987b).

Rywak and Burtlich (1996) observed that increasing the amount of $F^-$ during synthesis of talcs resulted in shift of the $\nu(Mg_3OH)$ band to lower frequencies, which was explained by an increase in hydrogen bonding (H-bonding) between the hydroxyl proton and an oxygen linked to Si, and/or by a change in charge distribution within octahedral sheet caused by the greater electronegativity of $F^-$ compared with $OH^-$. Scholtzova et al. (2003) applied a quantum–mechanical method to model talc solid-solution clusters. Changes of the geometry and of the frequency and intensities of the $\nu(M_3OH)$ bands were observed. Calculated vibrational wavenumbers correlated with experimental spectra of 2:1 trioctahedral clay minerals of the talc type. The substitution of one $Mg^{2+}$ cation by $Ni^{2+}$ resulted in formation of a $\nu(Mg_2Ni^{2+}OH)$ band approximately 14 $cm^{-1}$ lower than the $\nu(Mg_3OH)$ band and further substitution produced $\nu(MgNi^{2+}{}_2OH)$ and $\nu(Ni^{2+}{}_3OH)$ bands displaced, respectively, 12 and 18 $cm^{-1}$ to lower wavenumber. Comparatively, the largest effect of substitution was observed for $Al^{3+}$ where the displacements averaged 22 $cm^{-1}$. This effect probably related to the different formal charge of the $Al^{3+}$ atom.

Dumas et al. (2015) described in detail the NIR spectra of talc in the $Mg^{2+}$–$Ni^{2+}$ solid solution series. $Ni^{2+}$-talc showed a well-resolved band at 7080 $cm^{-1}$ corresponding to the first overtone $Ni^{2+}{}_3OH$ vibration $(2\nu(Ni^{2+}{}_3OH))$ (Fig. 8.6). The FTIR spectrum of the samples treated at 200°C and 100°C showed asymmetrical bands with maximum intensities located at 7083 and 7086 $cm^{-1}$, respectively. The broad band between 5400 and 5100 $cm^{-1}$ results from the $(\nu+\delta)H_2O$ combination (Fig. 8.7). With the increase in synthesis temperature, the intensity of the $(\nu+\delta)H_2O$ vibration decreased significantly, showing a reduction of the amount of water. Moreover, an upward shift of the $(\nu+\delta)$ $H_2O$ from 5221 to 5276 $cm^{-1}$ showed weaker H-bonds between water and the sample surface.

The NIR spectra also provide information on the growth of the octahedral sheet with the hydrothermal synthesis temperature. In synthetic $Ni^{2+}$-talc, crystallite edges are supposed to be sites where water molecules can H-bond

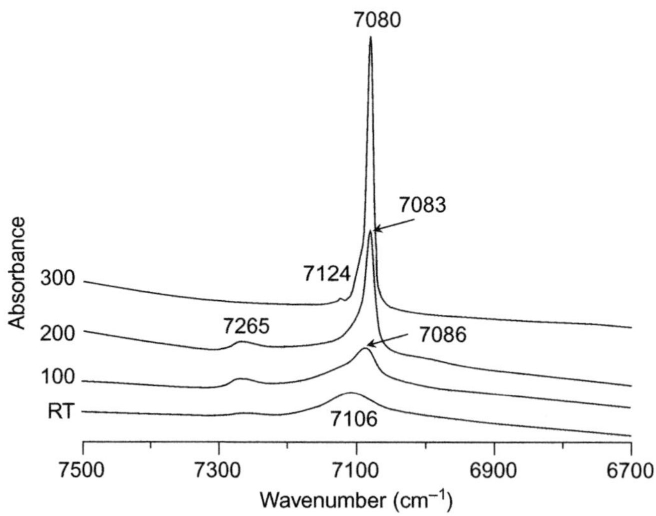


**FIG. 8.6** NIR spectra of the talc precursor (PT) and the talc precursor treated at 100°C, 200°C, and 300°C between 6700 and 7500 $cm^{-1}$. *Reproduced with permission from Dumas et al. (2015).*

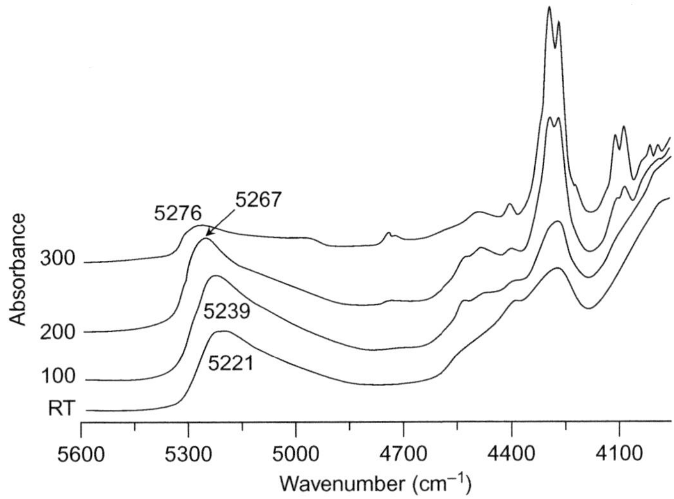


**FIG. 8.7** NIR spectra of the talc precursor (PT) and the talc precursor treated at 100°C, 200°C, and 300°C between 4000 and 5600 cm$^{-1}$. *Reproduced with permission from Dumas et al. (2015).*

to Si–OH, Mg–OH and Ni–OH groups. Increasing synthesis temperature tended to reduce the number of these sites to which water molecules are H-bonded, but, even after hydrothermal treatment at 300°C, some water remains. As shown in Fig. 8.8, the 50% $Ni^{2+}$ talc displayed four sharps bands situated at 7185, 7121, 7156 and 7080 cm$^{-1}$, characteristic, respectively, of $2\nu(Mg_3OH)$, $2\nu(Mg_2Ni^{2+}OH)$ $2\nu(MgNi^{2+}{}_2OH)$, and $2\nu(Ni^{2+}{}_3OH)$, completely in agreement with other IR and Raman studies of synthetic talcs, as described earlier.

The site symmetry of the $Mg_3OH$ and $Ni^{2+}{}_3OH$ groups in talc-willemseite endmembers is approximately $C_{3v}$, and thus one of IR active motion ($E$) forms a degenerate pair. When one or two nickel cations substitute for magnesium, the site symmetry of the $Mg_2Ni^{2+}OH$ and $MgNi^{2+}{}_2OH$ molecules becomes $C_s$ and the degeneracy of OH librational motions is lifted to be $A'+A''$ (Ishida, 1990). Russell et al. (1970) have confirmed with their deuteration experiments that the medium intensity band at 668 cm$^{-1}$ in Mg-talc and the 710 cm$^{-1}$ band in Ni-talc are due to OH libration, and that in talc and willemseite endmembers their two librational motions form a degenerate pair and produce a single IR band. In the case of partial substitution the degeneracy of these librational motions is lifted and the symmetry of the OH-molecules is lowered to $C_s$ and two OH librational bands will appear for the solid solution associated with $\delta(Mg_2Ni^{2+}OH)$ and $\delta(MgNi^{2+}{}_2OH)$ vibrations. Ishida (1990) could not distinguish individual bands but did observe a broadening of the $\delta(Mg_3OH)$ and $\delta(Ni^{2+}{}_3OH)$ between 668 and 710 cm$^{-1}$ upon partial

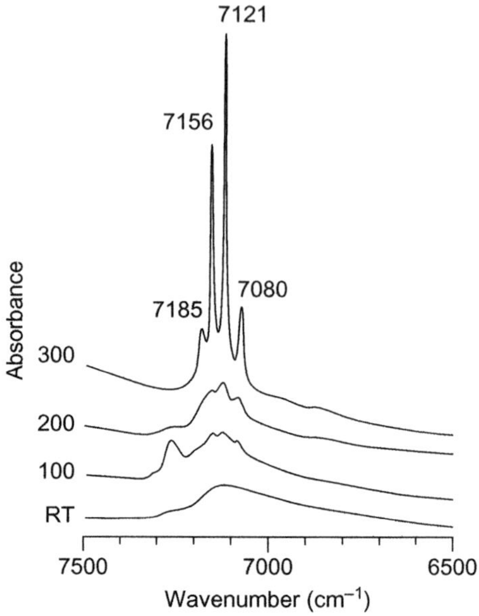


**FIG. 8.8** FTIR spectra between 6500 and 8000 cm$^{-1}$ of the 50% Ni series. *Reproduced with permission from Dumas et al. (2015).*

substitution (Fig. 8.9). In addition, changes were observed in the lower wavenumber region. The wavenumbers and intensities of the bands at 535 and 423 cm$^{-1}$, and the shoulder bands at 440 and 384 cm$^{-1}$, in $Mg^{2+}$-talc were strongly affected by ionic substitutions of the octahedral cations. After deuteration, the strong band near 465 cm$^{-1}$, which involves OH translator vibration (Russell et al., 1970) and the strong band near 450 cm$^{-1}$, which involves OH not perturbed either by $Mg^{2+}$ for $Ni^{2+}$ substitution or by deuteration, merged into the 439 and 458 cm$^{-1}$ bands of $Mg_3OD$ and $Ni_3OD$-talcs, respectively. With increasing $Ni^{2+}$ content, a rather weak band at 391 cm$^{-1}$ decreased in intensity and disappeared after deuteration.

Migration of $Zn^{2+}$ ions into vacant octahedral sites in synthetic $Zn^{2+}$ exchanged stevensite was studied by Petit et al. (2008) following the so-called Hofmann-Klemen (HK) effect. XRD analysis confirmed that the samples synthesised were smectites with site vacancies ranging from nil to 0.23 per half unit cell ($O_{10}(OH)_2$). While no simple relationship existed between temperature of HK reaction and the filling of vacancies, the samples progressed from a $Zn^{2+}$-stevensite-like structure to a $Zn^{2+}$-talc-like structure after treatment. Based on IR spectra, the synthesised samples were very similar to those of synthetic hectorite-like and saponite-like sauconites (Higashi et al., 2002). Notably, the characteristic $\nu(Zn^{2+}{}_3OH)$ band width at 3643 cm$^{-1}$ decreased with an increase in synthesis and HK temperature, and

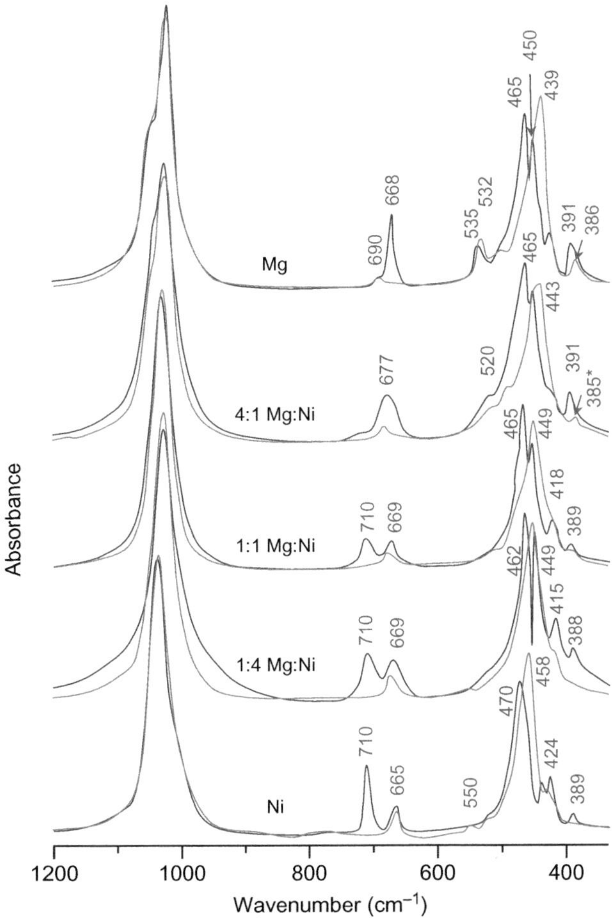


**FIG. 8.9** Infrared spectra of synthetic talc-willemseite solid solution (Blue spectra in on-line version) detailing the spectral shift (to lower wavenumber) of OH bands with deuteration (Red spectra in on-line version). *Redrawn with permission from Ishida (1990). (For interpretation of the references to colour in this figure legend, the reader is referred to the web version of this article.)*

a slight shift to lower frequency was observed with HK temperature. Another band was observed at 662 cm$^{-1}$, due to $\delta(Zn^{2+}{}_3OH)$ (Petit, 2005), perhaps overlapping with a $\nu_2$(SiO) lattice vibration which behaved similarly. In the NIR combination region, the spectra were better resolved after HK treatment. Notably, the band at 4309 cm$^{-1}$, which can be attributed to $(\nu+\delta)\,Zn^{2+}{}_3OH$, increased in intensity for the HK-treated samples.

Catalano et al. (2014) observed that the bands at 984 and 665 cm$^{-1}$ for Zn-doped and heated talc shifted to lower wavenumber and that the weak band at 796 cm$^{-1}$ split into two bands at 780 and 798 cm$^{-1}$, ascribed to $\delta(Mg_3OH)$ bands . The bands at 506, 459 and 540 cm$^{-1}$ in the Mg-talc, associated with lattice vibrations were largely unaffected. The OH stretching fundamental band at 3674 $cm^{-1}$ for Mg-talc split into four bands at 3676, 3665, 3651 and 3634 $cm^{-1}$, consistent, respectively, with $\nu(Mg_3OH)$, $\nu(Mg_2Zn^{2+}OH)$, $\nu(MgZn^{2+}{}_2OH)$ and $\nu(Zn^{2+}{}_3OH)$. The resulting splitting was dependent on the degree of substitution of the $Mg^{2+}$ in the octahedral sheet and related to the electronegativity difference between $Zn^{2+}$ and $Mg^{2+}$. Similar results were reported by Corona et al. (2015) for the substitution of $Fe^{2+}$ for $Mg^{2+}$ in talc. When a more electronegative cation ($Zn^{2+}$, $Fe^{2+}$) substitutes for $Mg^{2+}$ the vibration frequency is lowered. The additional bands formed from splitting of the main talc $\nu(Mg_3OH)$ are a result of incorporation of these cations into the talc structure (Petit et al., 2004a,b).

The influence of incorporation of $Fe^{3+}$ on the FTIR spectra of synthetic talc is largely unknown. Based on the early IR study of $Fe^{3+}$-bearing amphiboles by Bancroft and Burns (1969), octahedral $Fe^{3+}$ in an $Mg_2Fe^{3+}$ or $MgFe^{2+}Fe^{3+}$ configuration might induce a shift from the $Mg_3$ configuration of −21 and −37 $cm^{-1}$, respectively, but anticipated band positions of 3657 and 3641 $cm^{-1}$, respectively, but only one sample had a possible shoulder band at 3655 $cm^{-1}$ that might be associated with $\nu(Mg_2Fe^{3+}OH)$. However, the integrated intensity of this band corresponds to an $Fe^{3+}$ content of about 0.3% of Fe total, far less than the 15% determined from Mössbauer spectroscopy. Dumas et al. (2013) observed that the intensity of two bands around 5200 and 7150 $cm^{-1}$ due to water, which are not seen in the near IR of synthetic talc, decreased with synthesis time. All synthetic sample spectra showed a further band at 7185 $cm^{-1}$, assigned to the $2\nu(Mg_3OH)$ group, which increased in intensity and area with synthesis time. The decomposition of the synthetic talc spectra also highlighted the presence of two shoulders on the 7185 $cm^{-1}$ band near 7200 and 7230 $cm^{-1}$. These were attributed to coordination undersaturated Si–OH and Mg–OH at the edges of crystallites, and their areas decreased with growth of particles.

A few researchers (Stubičan and Roy, 1961; Martin et al., 1992, 1999a) have described hydrothermally synthesised germanium-substituted talc. The lattice symmetry group ($C\bar{1}$) was conserved in $Si_xGe_{1-x}$ talc ($0 < x < 100$) synthesised at 200°C with various octahedral (Ni, Co or Mg) cations between the two endmembers of $Si^{4+}$- and $Ge^{4+}$-talc, despite different ionic radii of $Ge^{4+}$ and $Si^{4+}$, but an increase in the cell volume and rearrangement of the crystal structure by opposed rotation of adjacent tetrahedra within the (001) plane were observed (Martin et al., 1992). The $\nu(M_3OH)$ bands of $Ge^{4+}$-talcs were shifted towards lower wavenumbers than those corresponding to $Si^{4+}$ -talc. For samples with intermediate chemistry, a broad band was observed containing two $\nu(M_3OH)$ components, influenced by tetrahedral $Si^{4+}$ and $Ge^{4+}$ environments. A shift towards the lower wavenumbers in the region 1200–200 $cm^{-1}$, as for the high wavenumber region, was observed for all the bands when the $Ge^{4+}$ content increased. A most particular effect of the tetrahedral substitution is that the wavenumbers of hydroxyl modes (bending and stretching) are strongly influenced by tetrahedral substitutions, even for co-substitutions in the octahedral sheet (Martin et al., 1992, 1999b). Table 8.5 compares some IR data for $Si^{4+}$- and $Ge^{4+}$-talcs containing $Ni^{2+}$, $Co^{2+}$ and $Mg^{2+}$ octahedral cations synthesised before and after deuteration to identify the metals associated with the OH group. Only the $\nu(M_3OH)$ and $\delta(M_3OH)$ shifted upon deuteration and the bands associated with tetrahedral cation—oxygen stretching and lattice deformation remained largely unchanged (some shifted, but by $<13\ cm^{-1}$).

In spite of very different $Si^{4+}$ and $Ge^{4+}$ cation sizes, the $Si^{4+}$-$Ge^{4+}$ talc solid-solutions were revealed as complete and continuous. A shift of almost $40\ cm^{-1}$ of the $\nu(M_3OH)$ bands towards lower wavenumbers are observed for $Ge^{4+}$-talc, revealing the strong influence of tetrahedral environment on the $\nu(OH)$ bands. The $\nu(M_3OH)$ bands are sharper for $Ge^{4+}$-talc than for $Si^{4+}$ -talc indicating a better structural organisation of the former for similar conditions of synthesis. For samples with intermediate tetrahedral chemistry $(0<x<1)$, bands broadenings were observed. Considering the talc structure, five different tetrahedral occupancies (4Ge, 3Ge1Si, 2Ge2Si, 1Ge3Si, 4Si) can interact with a given octahedral cation.

The IR spectra of the $Co^{2+}$-talc solid solutions of the $Si^{4+}$-$Ge^{4+}$ series were decomposed using five $\nu\left(Co^{2+}{}_{3-x}Ge^{4+}{}_{x}OH\right)$ bands whose wavenumbers were calculated from those of endmembers assuming a quite regular shift: 4Ge ($3587\ cm^{-1}$), 3Ge1Si ($3597\ cm^{-1}$), 2Ge2Si ($3607\ cm^{-1}$), 1Ge3Si ($3617\ cm^{-1}$) and 4Si ($3627\ cm^{-1}$). Assuming a random distribution of octahedral cations, the integrated intensity of the four structural OH stretching bands ($\nu(Mg_3OH)$, $\nu(Mg_2MOH)$, $\nu(MgM_2OH)$ and $\nu(M_3OH)$) should be in the ratio $(1-x)^3$, $3x(1-x)^2$, $3x^2(1-x)$, $x^3$, where $x$ and $(1-x)$ are the atomic proportions of cation $M^{2+}$ and $Mg^{2+}$, respectively (Wilkins and Ito, 1967).

### 8.4.1.2 Hectorite

Several studies on the synthesis and physicochemical properties of synthetic hectorite (Ht), saponite and sauconite containing $Co^{2+}$, $Ni^{2+}$, $Cu^{2+}$ and $Zn^{2+}$ in the trioctahedral sheets instead of $Mg^{2+}$ were performed (e.g. Serna et al., 1977b; Shirai et al., 1999; Higashi et al., 2002; Nakakuki et al., 2004). Torii and Iwasaki (1988) synthesised $Ni^{2+}$-Ht from a Ni–Si hydrous oxide gel. The IR band near $3630\ cm^{-1}$ was attributed to $\nu\left(Ni^{2+}{}_3OH\right)$ of synthetic $Ni^{2+}$ -talc (see previous Section 8.4.1.1), but the spectrum in the $1200–300\ cm^{-1}$ region was similar to that of Ht. The $\nu(SiO)$ bands for Ni–Ht were at higher wavenumbers than those of natural Ht. Higashi et al. (2002) observed that synthetic Zn–Ht gave two well-resolved bands at 3680 and $3638\ cm^{-1}$ in the OH-stretching region. The higher wavenumber band is comparable to that of natural Ht ($3676\ cm^{-1}$) (Farmer, 1974b) and has been unambiguously

**TABLE 8.5 Band Assignments of $M^{2+}$–$Si^{4+}$ and $M^{2+}$–$Ge^{4+}$ Talc (Martin et al., 1992)**

| | $Ni^{2+}$–$Si^{4+}$ talc | $Ni^{2+}$–$Si^{4+}$–$D_2O$ talc | $Ni^{2+}$–$Ge^{4+}$ talc | $Ni^{2+}$–$Ge^{4+}$–$D_2O$ talc |
|---|---|---|---|---|
| $\nu$(OH) | 3628 | 2682 | 3586 | 2649 |
| $\nu$(Si–O) | 1028 | 1026 | 843 | 843 |
| $\delta(Ni^{2+}{}_3OH)$ | 709 | 525 | 574 | 423 |
| $\delta$(Si–O) | 669 | 669 | 549 | 556 |
| $\delta$(Si–O) | 455 | 453 | 391 | 389 |
| | $Co^{2+}$–$Si^{4+}$ talc | $Co^{2+}$–$Si^{4+}$–$D_2O$ talc | $Co^{2+}$–$Ge^{4+}$ talc | $Co^{2+}$–$Ge^{4+}$–$D_2O$ talc |
| $\nu$(OH) | 3630 | 2682 | 3587 | alOH2647 |
| $\nu$(Si–O) | 1015 | 996 | 839 | 839 |
| $\delta(Co^{2+}{}_3OH)$ | 685 | 507 | 558 | 412 |
| $\delta$(Si–O) | 667 | 671 | 546 | 533 |
| $\delta$(Si–O) | 454 | 435 | 365 | 363 |
| | $Mg^{2+}$–$Si^{4+}$ talc | $Mg^{2+}$–$Si^{4+}$–$D_2O$ talc | $Mg^{2+}$–$Ge^{4+}$ talc | $Mg^{2+}$–$Ge^{4+}$–$D_2O$ talc |
| $\nu$(OH) | 3678 | 2724 | 3636 | 2682 |
| $\nu$(Si–O) | 1027 | 1018 | 838 | 839 |
| $\delta(Mg^{2+}{}_3OH)$ | 669 | 527 | 509 | 373 |
| $\delta$(Si–O) | 682 | 691 | 565 | 573 |
| $\delta$(Si–O) | 450 | 438 | 386 | 384 |

ascribed to $\nu(Mg_3OH)$ (Wilkins and Ito, 1967; Farmer, 1974b), whereas the band at 3638 $cm^{-1}$ was ascribed to $\nu(Mg_2Li^+OH)$. A somewhat similar feature of double bands in the OH-stretching region (3620 and 3565 $cm^{-1}$) was reported for synthetic Co-smectite by Bruce et al. (1986) and ascribed to two sites of different symmetry involving OH groups close to either octahedral Li or a vacant site. In contrast to the double OH bands of Ht and $Zn^{2+}$-Ht, sauconite (natural $Zn^{2+}$-smectite) yielded a distinctive single band at 3641 $cm^{-1}$ that was slightly higher than that of $\nu(Zn^{2+}{}_3OH)$ stretching (3634 $cm^{-1}$) for synthetic $Mg^{2+}$–$Zn^{2+}$-talc (Wilkins and Ito, 1967). In the lattice vibration region, $Zn^{2+}$–Ht and sauconite yielded distinctive bands around

1000, 660 and 450 cm$^{-1}$; the band positions were nearly the same for both $Zn^{2+}$-smectites. Hectorite also yielded comparable bands in the lattice vibration region with discernible double splitting of bands in 660 and 450 cm$^{-1}$ regions (Fig. 8.10). The IR spectra of $Zn^{2+}$–Ht showed a broad band around 3400 cm$^{-1}$ and a sharp band at 1600 cm$^{-1}$, both assigned to interlayer water (Nakakuki et al., 2004). Broad bands near 1000 cm$^{-1}$ were assigned to the tetrahedral $\nu$(SiO) whereas bands at 664 and 671 cm$^{-1}$, characteristic of $Zn^{2+}$–Ht, were assigned to $\nu(Zn^{2+}O)$ vibrations (Higashi et al., 2002).

The IR spectrum of synthetic $Fe^{3+}$–Ht (Ogawa et al., 2009) showed the characteristic bands of $\nu(Mg_3OH)$ at 3680 cm$^{-1}$, $\nu$(SiO) at 1020 cm$^{-1}$ and $\delta(Mg_3OH)$ at 655 cm$^{-1}$. The band due to the $\nu$(SiO) was at higher wavenumber compared with that of $Fe^{3+}$-free Ht (1012 cm$^{-1}$), while the $\delta(Mg_3OH)$ band remained in the same position. This was interpreted as indicating that $Fe^{3+}$ had been immobilised within or near the tetrahedral sheet of the silicate layer rather than the octahedral sheet.

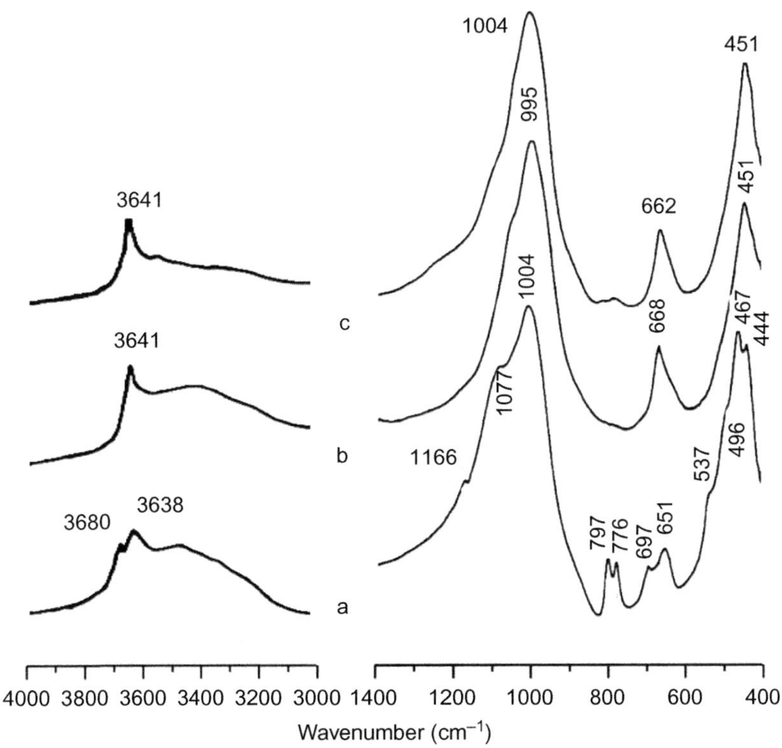


**FIG. 8.10** IR spectra of different synthetic smectites: (a) hectorite, (b) sauconite, and (c) Zn-hectorite, Q=quartz. *Reproduced with permission from Higashi et al. (2002).*

### 8.4.1.3 Saponite

In general, the crystallisation of clay minerals from dilute solution is thought to begin with the formation of octahedral sheets to which, simultaneously or shortly after the octahedral sheets are formed, the tetrahedral sheets are attached to form single layers (e.g. Kloprogge and Frost, 1999d). This model is based on the assumption that silica will never spontaneously form a two-dimensional network during polymerisation in solution, but instead a three-dimensional network is preferred. The formation of a two-dimensional silica network, as present in the tetrahedral sheets of clay minerals, therefore is thought to require some sort of template, most likely to be the initial stages of the octahedral sheet formed due to the low solubility of, for example, $Al^{3+}$ hydroxides near neutral pH values. The results the studies by Kloprogge et al. (1993) and Vogels et al. (1997) on the synthesis of saponites (Sap), however, point to a mechanism initiating as transformations within the amorphous gel (Kloprogge and Frost, 1999c,d,e, 2000; Kloprogge et al., 1999e). Their crystallisation model was based on the low solubility of silica in the pH range observed during the synthesis. The presence of amorphous silica was evident in the FTIR spectra by the bands in the 1000–1250 $cm^{-1}$ region. During crystallisation, a change in the $SiO_2$ configuration took place from a disordered, three-dimensional network in the amorphous state to a more ordered, possibly already a two-dimensional, structure as normally found in the tetrahedral sheets in clay minerals. This was clearly reflected by the disappearance of the 900–910 and 1225 $cm^{-1}$ bands and the continuous decrease in the wavenumber of the initial 1043 $cm^{-1}$ band. The fact that two overlapping bands around 660 and 690 $cm^{-1}$ were observed to appear almost simultaneously prior to the disappearance of the wide band around 1250 $cm^{-1}$ due to amorphous silica disappeared, indicated that the octahedral and tetrahedral sheets in the Sap structure formed almost simultaneously. Sharpening of the other bands associated with the various $\nu$(SiO) modes was not observed until approximately 200°C. This is in contrast to the earlier results of Kloprogge et al. (1993) and Vogels et al. (1997), where no evidence was found in the solid-state $^{27}Al$ and $^{29}Si$ MAS-NMR for the early formation of the tetrahedral layers, a difference which might be explained by the fact that IR spectroscopy is much more sensitive to these types of changes during the crystallisation than MAS-NMR spectroscopy. Tetrahedral $Al^{3+}$ substitution was only indicated in Sap synthesised at 240°C and 280°C by the new band at 837 $cm^{-1}$ band. The formation of the octahedral sheets containing both $Mg^{2+}$ and $Al^{3+}$ was demonstrated by the presence of bands related to both $Mg^{2+}$ and $Al^{3+}$ still bonded with OH at higher temperatures. This agrees well with earlier observations by Kloprogge et al. (1993) and Vogels et al. (1997) that the $Al^{3+}$ initially remained in solution until the pH dropped to values near neutral, causing the $Al^{3+}$ to be built in at a later stage of the crystallisation.

Vogels et al. (2005) reported that substitution of the octahedral cation shifted to IR bands lower wavenumbers compared with natural Sap as from

Krugersdorp (van der Marel and Beutelspacher, 1974). The $\nu(Mg_3OH)$ band at 3670 cm$^{-1}$ for natural $Mg^{2+}$-Sap shifted to 3632 cm$^{-1}$ ($Zn^{2+}$-Sap) 3620 cm$^{-1}$ ($Co^{2+}$-Sap) and 3616 cm$^{-1}$ ($Ni^{2+}$-Sap). All synthetic Sap samples exhibited sharp bands between 3600 and 3800 cm$^{-1}$. Partially or completely dehydrated natural and synthetic Sap generally displays two characteristic $\nu(M_3OH)$ and calcination at temperatures up to 500°C do not affect their positions. The position of the other $\nu(M_3OH)$ band depended on the compensating cation (Pelletier et al., 1999, 2003) and is shifted due to differences in the electronegativity of these octahedral cations (Velde, 1983). Synthetic Ni-Sap also have a small third absorption band around 3700 cm$^{-1}$, the origin of which was not clear. Discrimination of these bands is difficult when samples are hydrated due to overlap with the broad $\nu(H_2O)$ band of physisorbed $H_2O$. The Si/Al ratio affected only the $\nu_1(M_3OH)$ of a $Zn^{2+}$-Sap synthesised with excess aluminium (Si/Al = 2.3), which was at noticeably lower wavenumbers than for 'normal' $Zn^{2+}$-Sap (Table 8.6). This Sap contained a high amount of octahedral $Al^{3+}$ as observed with $^{27}Al$ MAS-NMR, indicating a structure in between Sap and montmorillonite. The position of this specific $\nu_1(M_3OH)$ band was

**TABLE 8.6 IR Strongest Absorption Maxima of M-Saponite (M = $Mg^{2+}$, $Zn^{2+}$, $Co^{2+}$, $Ni^{2+}$) (Vogels et al., 2005)**

| | $\nu$ (cm$^{-1}$) | | | Assignment |
|---|---|---|---|---|
| | 3733 | | | $\nu$(SiOH) |
| $Mg^{2+}$ | 3670 | 3731 | | $\nu(M_3OH)$, $\nu(M_2AlOH)$ |
| $Zn^{2+}$ | 3632 | 3734 | | |
| $Co^{2+}$ | 3620 | 3735 | | |
| $Ni^{2+}$ | 3616 | 3730 | 3698 | |
| $Zn^{2+a}$ | 3611 | 3730 | | |
| | ±3200 | | | Water $\nu$(OH) |
| | 1640 | | | Water $\delta$(OH) |
| | 1180h | | | $\nu$(SiO) |
| | 1000 | | | $\nu$(SiO) |
| | 880sh | | | $\delta$(MOH) |
| | 770 | | | $\delta$(Si–O–Al)/$\delta$(MOH) |
| | 460 | | | $\delta$(Si–O–M) deformation |

[a]$Zn^{2+}$-Sap with excess $Al^{3+}$ ($Si^{4+}/Al^{3+}$ = 2.33).

therefore interpreted as $\nu(AlZn^{2+}{}_2OH)$. The $\nu_1(M_3OH)$ band in $Ni^{2+}$-Sap was also significantly lower than that of pure $Ni^{2+}$-talc. As was shown in their XAFS results (Vogels et al., 2005), the $Ni^{2+}$-Sap also contained a high amount of octahedral $Al^{3+}$, presumably causing a similar shift to lower wavenumbers as in the $Zn^{2+}$-Sap with 'excess' $Al^{3+}$.

The $\nu_2(OH)$ band in this region, associated with 'free' OH groups (Pelletier et al., 1999, 2003), was present as a distinct shoulder in hydrated samples, but after some dehydration of the Sap, the band at 3732 $cm^{-1}$ was clearly visible. Farmer, 1974b initially attributed this band to $M_3$-(OH) units present in the octahedral sheets affected by the electric field associated with the interlayer cations, but Pelletier et al. (1999, 2003) observed the position of the $\nu_2(M_3OH)$ band to be insensitive to the $Si^{4+}/Al^{3+}$ ratio and the type of octahedral or interlayer cation. Thus, this band may instead originate from $OH^-$ groups coordinated to the tetrahedral sheets at the external surface of the clay particle as Si–OH. The stretching vibration of terminal Si–OH groups in zeolites is positioned at comparable wavenumbers (Borade and Clearfield, 1992).

Baldermann et al. (2014) studied the $Fe^{3+}$–$Mg^{2+}$ Sap solid solution and observed the $\nu(Mg_3OH)$ and $\nu(Mg_2AlOH)$ bands to be rather weak (3680 $cm^{-1}$) or absent (3630 $cm^{-1}$) in the synthesised ferrous Sap. Iron substitution in Mg-rich clay minerals typically moves the $\nu(Mg_3OH)$ towards a lower wavenumbers (Cuadros et al., 2008) and the synthetic clay minerals showed some, but not totally resolved modulations below 3680 $cm^{-1}$ that could correspond to such substitutions. In these precipitates a strong IR band was found around 3560 $cm^{-1}$, which was believed to correspond to dioctahedral $\nu(Fe_2{}^{3+}OH)$ or $\nu(Fe^{3+}MgOH)$ (Dill et al., 2011). Such assignment is supported by the presence of a band at 816 $cm^{-1}$. However, the $\delta(Mg_3OH)$ band at 680 $cm^{-1}$, which was observed in all samples, clearly indicated the dominance of trioctahedral structural units. The IR bands below 500 $cm^{-1}$ that could help distinguish between trioctahedral and dioctahedral structural units were not well resolved and hence could not be interpreted further.

Hydrothermally synthesised (between 125°C and 280°C) Sap were studied by FT-Raman and FTIR spectroscopy in comparison to natural Sap by Kloprogge and Frost (2000). FTIR spectra with increasing synthesis temperature exhibited information about possible crystallisation mechanisms from amorphous starting material (Fig. 8.11A). A band at 457 $cm^{-1}$ corresponding to the amorphous gel developed, at 280°C, into two separate components at 447 $cm^{-1}$ and 465 $cm^{-1}$, respectively, ascribed to $e_1^2$ translational modes of $SiO_2$ and AlOH in Sap. Above 150°C, a minor octahedral sheet band appeared around 530 $cm^{-1}$ ($\delta(M_3OH)$ perpendicular mode), followed by a band around 750 $cm^{-1}$ ($\delta$(AlMgOH) deformation mode) above 200°C. At this temperature, two bands due to the $a_1^2$ $SiO_2$ and apical Al–O bond of $AlO_4$ in the tetrahedral sheet became visible at 691 and 837 $cm^{-1}$, respectively. The strongest $SiO_2$ band exhibited a continuous shift from 1043 to 1013 $cm^{-1}$ with increasing synthesis temperature. Two overlapping bands around 660 and 690 cm$^{-1}$ appeared prior to the disappearance of the wide amorphous material ν(SiO) band around 1240 cm$^{-1}$, indicating that the octahedral and tetrahedral sheets formed almost simultaneously during crystallisation of synthetic Sap. Sharpening of other bands associated with various ν(SiO) modes was not observed until 200°C. The FT-Raman spectra indicated that upon increasing temperature, the first internal tetrahedral vibrational band observed was at 683 cm$^{-1}$ (at 150°C), followed by the 360 cm$^{-1}$ band (175°C). Around 240–280°C, two bands around 430–465 cm$^{-1}$ were observed (Fig. 8.11B).

Kloprogge and Frost (2001) used IR emission spectroscopy (IES) to study the dehydration, loss of ammonia, and dehydroxylation of synthetic

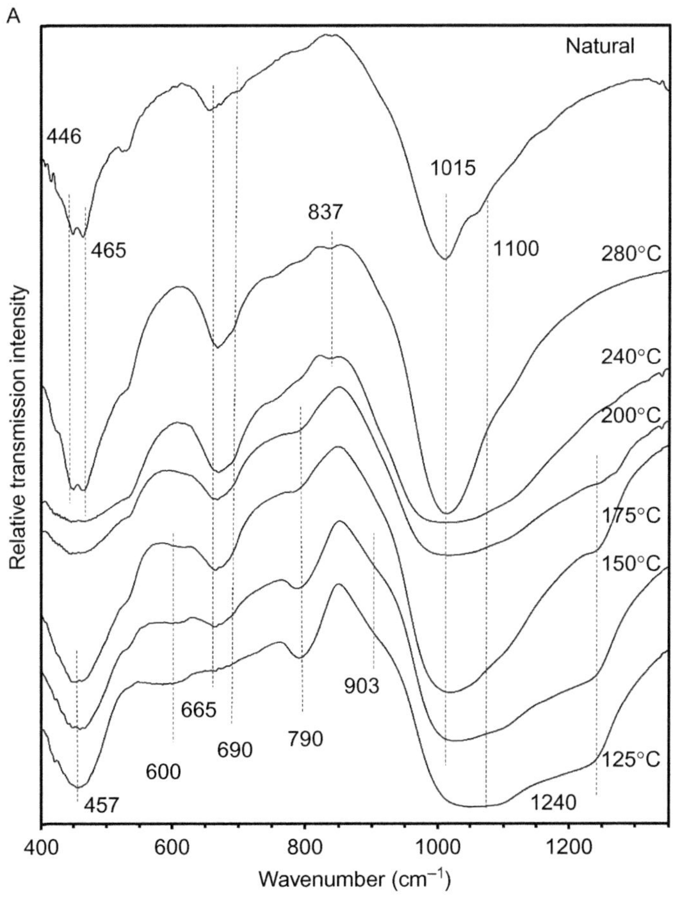


**FIG. 8.11** (A) FT-IR spectra in the region 400–1350 cm$^{-1}$ of a natural saponite (Milford, UT) and of synthetic saponites. (B) FT-Raman spectra in the region 250–1200 cm$^{-1}$ of natural saponite and synthetic saponites. *Reproduced with permission from Kloprogge and Frost (2000).*

*(Continued)*

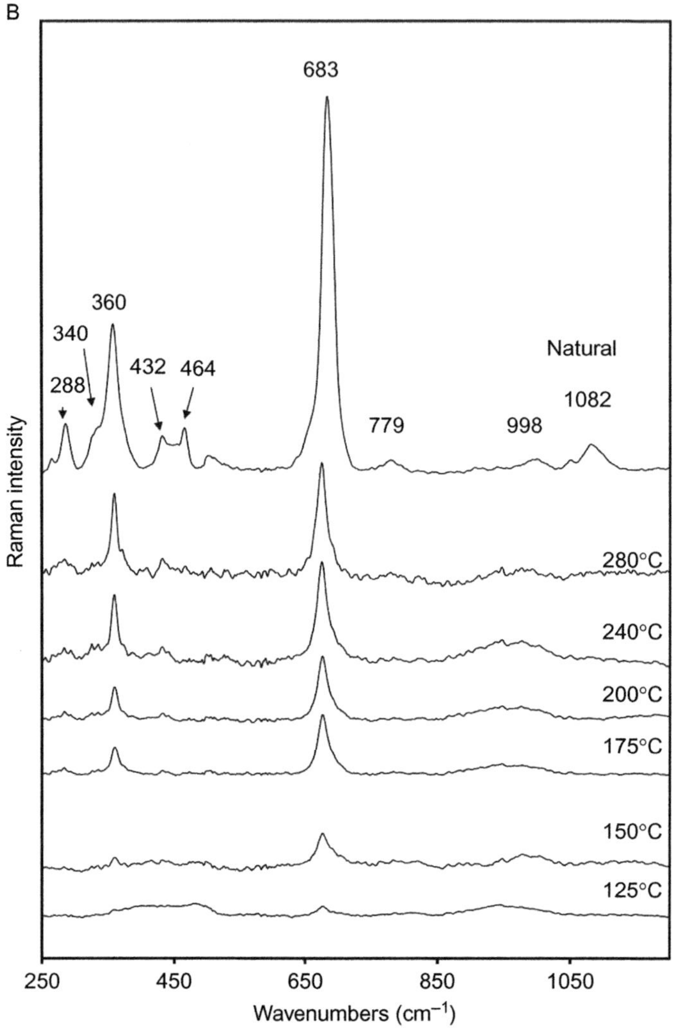


**FIG. 8 11—Cont'd**

$NH_4^+$- Sap as described in detail above. The OH-stretching region revealed for the synthetic Sap a small band around 3040–3120 $cm^{-1}$ ascribed to ammonium in the interlayer position. OH-stretching bands associated with adsorbed and interlayer water were observed around 3250 and 3425 $cm^{-1}$. The strong band around 3600–3630 $cm^{-1}$ was ascribed to the $\nu(Mg_2(Al,\square)OH)$ mode, while the band around 3670 $cm^{-1}$ was assigned to the $\nu(Mg_3OH)$ mode. Upon heating the OH-stretching region firstly showed the disappearance of the interlayer water and secondly a decrease in intensity of the $\nu(Mg_2(Al,\square)OH)$

band (Fig. 8.12A). In contrast the $\nu(Mg_3OH)$ band decreased far less in intensity. A silanol band at 3778 $cm^{-1}$ is observed over the whole temperature range from room temperature to 750°C in the spectra of all Sap. The $Mg_2(Al,\square)OH$ translation mode around 450 $cm^{-1}$ and deformation mode around 750 $cm^{-1}$ decrease in intensity and were no longer observed in the 700°C spectrum (Fig. 8.12B). Instead, a new band is observed around 730–740 $cm^{-1}$ attributed to the restructuring of the octahedral layer and the formation of a new Al–O bond after dehydroxylation. Because the dehydroxylation was not complete at 700–750°C no new bands due to the formation of new Mg–O–Mg, Mg–O–Al or Mg–O–Si bonds were observed. Only a minor decrease in the $Mg_3OH$ bands is observed. Based on the IES spectra it is possible to model the dehydroxylation of Sap. In general, dehydroxylation of clay minerals involves the liberation of OH groups from the octahedral sheet by

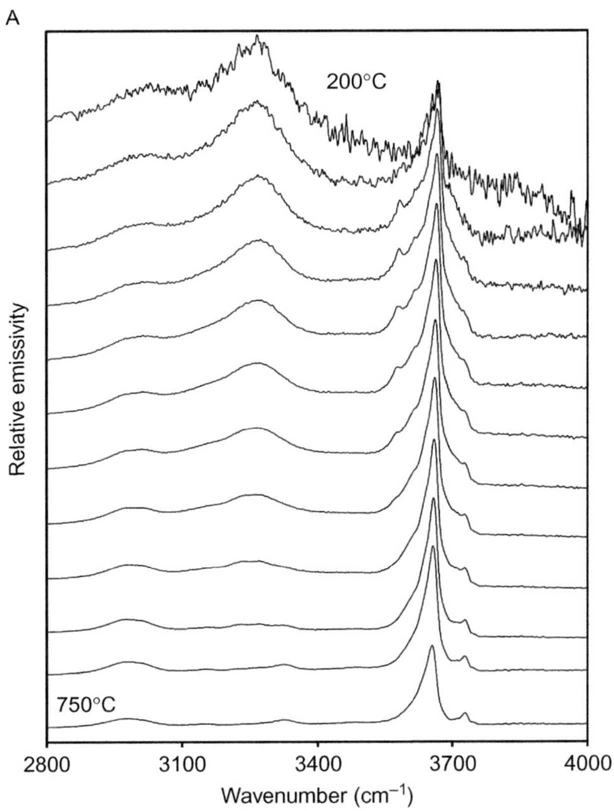


**FIG. 8.12** Infrared Emission Spectra of synthetic saponite HTSAP1 in the range (A) 2800–4000 $cm^{-1}$ and (B) 450–1800 $cm^{-1}$. *Reproduced with permission from Kloprogge and Frost (2001).*

*(Continued)*

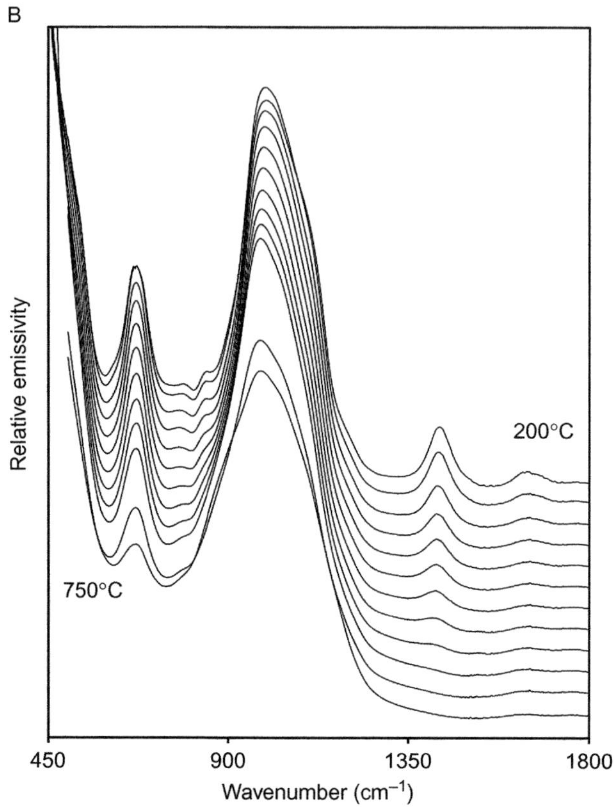


**FIG. 8 12—Cont'd**

the recombination of two OH groups into one $H_2O$ molecule leaving a newly formed metal—oxygen bond in the octahedral sheet. This process was observed in the IES spectra by the decrease and finally disappearance of the bands associated with OH-stretching and OH-bending modes. Differences in crystallinity were reflected in the thermal stability and therefore the temperature at which dehydroxylation started and ended. Kloprogge et al. (1999e) had previously shown that the dehydroxylation of beidellite proceeded via the formation of edge silanol groups above 400°C, that is, the hydroxyls and protons were transferred from the octahedral sheet to the Si(edge) of the tetrahedral sheet upon thermal treatment and were subsequently lost as water vapour. This was supported by the fact that the silanol band disappeared at the end of the dihydroxylation reaction. The IES measurements of the synthetic Sap did not provide evidence for this model as the silanol groups were already visible at room temperature and did not significantly change during the dehydroxylation process up to 700°C. However, it cannot be excluded that the silanol groups do form an active intermediate in the transfer of the hydroxyl groups and protons during the dehydroxylation. In general, it can be concluded that no significant structural changes, except for some minor rearrangements associated with the $Mg_2(Al,\square)$ OH, were observed during the loss of the OH groups from the lattice and were not expected until the dehydroxylation was completed.

Bisio et al. (2008) reported a sharp band at 3670 $cm^{-1}$ associated to the $\nu(Mg_3OH)$ (van der Marel and Beutelspacher, 1974; Kloprogge and Frost, 2000), which were virtually insensitive to the presence of water, as indicated by the well-resolved shape of the band. Nevertheless, a spectroscopic study of natural hydrated $Na^+$-Sap showed that part of the octahedral cations can also interact with water. The $\nu(M_3OH)$ stretching modes of these hydrated cations was found in the 3740–3640 $cm^{-1}$ region, the position being strongly dependent on the charge of tetrahedral sheets (Pelletier et al., 2003). The OH stretching vibration mode of Si–OH groups, that can be present both on the surface of the particles and as structural defects, also fell in the 3740–3640 $cm^{-1}$ region. Lastly, the bands between 3600 and 3000 $cm^{-1}$ were due to the OH stretching modes of water molecules, the associated bending mode falling at *ca.* 1635 $cm^{-1}$, which were bound to both Lewis, for example, $Na^+$ ions at exchange sites and Brønsted acid sites, for example, protons related to the presence of $Al^{3+}$ in the tetrahedral silica sheets. The intensity of the bands in the 3800–3000 $cm^{-1}$ range decreased, in parallel to the band at 1635 $cm^{-1}$, upon simple outgassing at room temperature. Though, it was only by increasing the temperature up to 473°K that the bands related to the stretching and bending modes of adsorbed water molecules decreased progressively until they disappeared. Dehydrated Sap showed well-defined bands at 3740 and 3670 $cm^{-1}$. Finally, the presence of silanol nests where multiple H-bonding between adjacent SiOH groups can occur, was indicated by the broad bands in the 3500–3000 $cm^{-1}$. By progressive heating up to 873°K, the intensity of the bands at 3740, 3670 and 3590 $cm^{-1}$ and the broad component in the 3500–3000 $cm^{-1}$ region strongly decreased, and after thermal treatment at 973°K only two weak bands at 3740 and 3670 $cm^{-1}$ were still present. This behaviour was associated with a partial collapse of the Sap structure, in agreement with their XRD data and the IES results of Kloprogge and Frost (2001). Very similar results were reported by Costenaro et al. (2012) in their work on the effect of the synthesis gel dilution on the physicochemical properties of Sap.

Synthesis of Sap was reported by Vicente et al. (2010) and Trujillano et al. (2010) where the appearance of one intense band around 1105 $cm^{-1}$ together with a shoulder at 1200 $cm^{-1}$ confirmed the presence of the amorphous silica phase, as observed by XRD, which was a function of the $Mg^{2+}$ source and the corresponding pH in the starting material. The observation of one band around 1027 $cm^{-1}$ corresponded to the $\nu(SiO)$ ordered within layers for the samples prepared near neutral (7–8) to slightly basic conditions using

$Mg(CH_3COO)_2$. Three other IR bands were observed at 1401, 3675 and 666 $cm^{-1}$, which are associated to $NH_4^+$ bending, vibrations related to the $\nu(Mg_3OH)$, and $\delta(Mg_3OH)$, respectively (Vicente et al., 2010). Trujillano et al. (2010) reported shoulders around 3650 and 3630 $cm^{-1}$ due to $\nu(Mg_3OH)$, and denote the trioctahedral character of the clay layers synthesised (Madejová, 2003). The band around 3440 $cm^{-1}$ was attributed to the $\nu(OH)$ of physisorbed water molecules with the corresponding bending vibration mode was observed at 1635 $cm^{-1}$. The shoulder around 3220 $cm^{-1}$ was thought to be due to the O–H stretching vibrational mode involving H-bonding. The spectra also showed the lattice vibrations around 1010, 815, 660 and 450 $cm^{-1}$ assigned to, respectively $\nu(SiO)$, apical $\delta(AlO)$ in $AlO_4$ units, $\delta(Mg_3OH)$ and $\delta(Si–O–Mg)$ bending modes.

Pelletier et al. (2003) provided a detailed study on the influence of layer charge on the OH stretching modes of Sap. The nature of the monovalent interlayer cation ($Na^+$ or $K^+$) and the hydration of the samples were shown to strongly influence spectral features. In the case of hydrated $Na^+$-Sap, the $\nu(M^{2+}{}_3OH)$ shifted from 3678 to 3686 $cm^{-1}$ with increasing layer charge. For hydrated $K^+$-Sap, two components were observed around 3720 and 3680 $cm^{-1}$, assigned to OH perturbed by compensating cation above or in the ditrigonal cavity of the tetrahedral sheet (Russell and Fraser, 1994) and free OH groups, respectively. In the case of dehydrated samples both $K^+$- and $Na^+$-Sap exhibit two components at 3677 and 3720/3612 $cm^{-1}$. The first band was assigned to unperturbed hydroxyl groups (wavenumber similar to talc, 3677 $cm^{-1}$). A linear relationship between the relative area of these two $\nu(OH)$ components and the layer charge was then evidenced. To assign the features obtained for hydrated $Na^+$-samples in the OH-stretching modes domain, the Raman spectra were decomposed using four $\nu(M^{2+}{}_3OH)$ components around 3675, 3680, 3685 and 3690 $cm^{-1}$. The bands at 3676 and 3674 $cm^{-1}$ were grouped together as they both correspond to talc-like OH groups. For low charge Sap, such a set only represented 25% of the total signal. It then gradually decreased for higher charges, and disappears completely. The component at 3681 $cm^{-1}$ exhibited a maximum for intermediate charges and decreased thereafter. The band at 3686 $cm^{-1}$ showed a wide maximum for intermediate layer charges between 1.2 and 1.6, while the band at 3691 $cm^{-1}$ appeared for layer charges around 1.4 and increased sharply for higher values.

The evolution of the free talc-like $(Si_4)Mg_3OH$ groups can easily be understood because (assuming a random distribution of $Al^{3+}$ for $Si^{4+}$ substitutions) uncharged ditrigonal cavities do not exist for layer charges $\geq$1.4 (Michot and Viliéras, 2002). For the evolution of the other three components, two parameters need to be considered for explaining the observed increase in wavenumber with layer charge: (i) the $Al^{3+}$ for $Si^{4+}$ substitutions in the tetrahedral sheet that could imply an attractive effect relative to the OH groups, or (ii) the presence of positive compensating cations in the interlayer space that could imply a repulsive effect relative to the OH groups. For one OH group linked with three octahedral $Mg^{2+}$ cations, increasing $Al^{3+}$ for $Si^{4+}$ substitutions generally decreases the stretching wavenumber. Therefore, the evolution of band positions cannot be explained by increasing charge at the tetrahedral sites. The results obtained with $K^+$-Sap and heated $Na^+$-Sap showed that cation location strongly affected the behaviour of the stretching bands. The higher component at 3715 $cm^{-1}$ corresponded to OH groups perturbed by cations located directly above the ditrigonal cavities. Such a situation is never encountered for hydrated samples since no band higher than 3700 $cm^{-1}$ is ever observed. Still, using molecular dynamics simulations Chang et al. (1995) showed that hydrated cations are always located close to the tetrahedral Al site at relative humidity less than 0.98.

A cautionary tale about the effect of sample preparation on the IR spectra was published by Pelletier et al. (1999). IR spectra in the range 3800–3600 $cm^{-1}$ of synthetic $Na^+$-Sap samples with different layer charge recorded in transmission on KBr pellets and in diffuse reflectance two bands corresponding to the OH stretching vibrations were observed at 3715 and 3682 $cm^{-1}$. In both cases, the relative area of the band at 3715 $cm^{-1}$ (ratio of the area of the 3715 $cm^{-1}$ component/sum of the areas of the bands at 3715 and around 3680 $cm^{-1}$) showed strong positive correlation with the layer charge. For transmission experiments, it was even possible to draw a linear regression. The diffuse reflectance results exhibited a similar trend. In this case, it would be incorrect to draw a linear relationship as diffuse reflectance experiments usually exhibit non-linear behaviour. As the relative area increased with layer charge, it would be logical to assign the band at 3682 $cm^{-1}$ to OH stretching, perturbed by their interaction with interlamellar water molecules, whereas the band at 3715 $cm^{-1}$ would correspond to OH groups increasingly perturbed by the $Na^+$ compensating cation associated to the $Al^{3+}$ for $Si^{4+}$ substitution in the tetrahedral sheet. Using $CaF_2$ instead of KBr as the diluting matrix the band at 3715 $cm^{-1}$ was not observed at all. This band must then be assigned to an interaction between Sap and KBr. The most likely effect is an exchange of $K^+$ for the $Na^+$ in the interlayer space. Due to their larger size, $K^+$ ions will be located more at the centre of the hexagonal cavity formed by the arrangement of silica tetrahedra and will then strongly affect the stretching vibrations of the OH groups. The spectra corresponding to $K^+$-Sap in transmission in KBr pellets and in diffuse reflectance in KBr confirmed this assignment as two bands located at 3682 and 3715 $cm^{-1}$ were found regardless of the experimental setup or diluting matrix used. The evolution of the relative area of the band at 3715 $cm^{-1}$ with the layer charge for $K^+$-Sap obtained in transmission was well described by a linear fit. The diffuse reflectance results exhibit a comparable trend. This effect, due to a $K^+$ for $Na^+$ exchange masked the true signal corresponding to the structural hydroxyls of $Na^+$-Sap. This $K^+$ for $Na^+$ exchange was clearly not limited only to the external surfaces of particles. Indeed, the external surface area of the particles is ~40 $m^2/g$, which represents ~10% of the available surface area of the layers. The relative area of the band at 3715 $cm^{-1}$ reached 0.5 for the higher charge. On the other hand, the $K^+$ for $Na^+$ exchange is certainly incomplete as the ratios obtained for $K^+$-Sap were significantly higher. Indeed, Petit et al. (1999b, 2006) showed that $K^+$ from KBr easily replaces $NH_4{}^+$ in $NH_4{}^+$-saturated Sap and other clay minerals during pellets preparation but that the extent of the exchange was closely connected to different hydration energies of exchangeable cations and to distributions of the layer charge.

Rinnert et al. (2005) studied the hydration behaviour of a synthetic $Na^+$-Sap with NIR spectroscopy. The NIR spectrum corresponding to the sample after outgassing at 100°C under a vacuum of 0.1 Pa revealed that under those conditions (basal distance around 10 Å), synthetic Sap still contained roughly for 0.5 water molecules per sodium cation. Two bands at 5260 and 5170 $cm^{-1}$, associated with $2\nu$(OH) of $H_2O$ suggested the presence of H-bonded water, under vacuum, with symmetry of $C_{2V}$ and the two H atoms similarly perturbed. The observed wavenumbers were approximately 70 $cm^{-1}$ lower compared with those for water in the vapour state and the intensity of the component at 5170 $cm^{-1}$ was relatively high, which might reveal the influence of the permittivity properties in the confined interlayer space. These remaining water molecules can then be isolated from molecules strongly linked to the interlayer $Na^+$ cations.

In the region corresponding to structural hydroxyl groups, two bands were observed at 7185 and 7260 $cm^{-1}$. The Raman spectrum in the fundamental OH stretching range also showed two bands at 3675 and 3712 $cm^{-1}$. The mode at 3675 and the corresponding overtone located at 7185 $cm^{-1}$ correspond to unperturbed OH groups whereas the second modes at 3712 and 7260 $cm^{-1}$ correspond to OH groups perturbed by the presence of a $Na^+$ cation, located close to the ditrigonal cavity. In this first hydration range, two bands grew at 5320 and 5150 $cm^{-1}$. Such a difference in wavenumber of 170 $cm^{-1}$ characterises a water molecule in which one OH bond absorbing at a high wavenumber is free of H-bonding, whereas the second one, absorbing at a lower wavenumber, is H-bonded. The appearance of those water signals was associated to a small shift of the non-perturbed structural hydroxyls. The water molecules adsorbed in this step could then be tentatively located on the edge faces of the clay mineral particles, where they would bind to edge MgOH groups. Such an interpretation was partially validated by the evolution in the high wavenumber region where the structural hydroxyls appear to be marginally affected.

In the second hydration stage the NIR displayed two additional signals at 5245 and 5100 $cm^{-1}$. In parallel, in the range corresponding to structural hydroxyls and water overtones, the main changes observed were linked to the appearance of a band around 7080 $cm^{-1}$. The observed wavenumbers in the domain corresponding to $(\nu+\delta)H_2O$ are typical of water molecules interacting with a cation. The spectral evolution around 7200 $cm^{-1}$ can then be assigned to a decrease of the perturbation of external OH groups by external $Na^+$ cations. The third hydration stage was accompanied by a strong increase of the bands at 5245 and 5100 $cm^{-1}$, that is, at the same spectral position as in the previous hydration range. In parallel, in the domain corresponding to structural hydroxyls and water overtones, two signals grew at 7080 and 6820 $cm^{-1}$, the band at 7185 $cm^{-1}$ shifted slightly towards 7190 $cm^{-1}$, and the intensity of the band at 7265 $cm^{-1}$ decreased slightly. The Raman spectra revealed modification neither on the symmetrical Si–O–Si stretching around 674 $cm^{-1}$ nor in the range of structural hydroxyls around 3710 $cm^{-1}$. In the fourth hydration stage ~1.5 water molecule per cation were present in the interlayer space. Concomitantly, the basal spacing reached a value of 12.3 Å that is classically assigned to a 'one-layer hydrate' state. Four NIR components grow at 5130, 5245, 5310 and 5550 $cm^{-1}$. In parallel, two water signals grow at 6835 and 7080 $cm^{-1}$, while the band corresponding to structural OH groups perturbed by interlayer cations disappeared; the band corresponding to unperturbed OH groups initially located at 7185 $cm^{-1}$ shifted towards 7195 $cm^{-1}$. The changes observed in the range corresponding to structural hydroxyl stretching modes can be interpreted as corresponding to the 'expulsion' of the $Na^+$ cations from their location on top of the structural hydroxyls in the middle of the ditrigonal cavities.

In the range of water combination modes, the band at 5245 $cm^{-1}$ and its companion at 5100 $cm^{-1}$ (whose existence can be inferred from the width of the IR signal between 5050 and 5200 $cm^{-1}$) can be assigned, as in the previous stages, to water molecules interacting with the $Na^+$ cation. The signals at 5310 and 5130 $cm^{-1}$ then correspond to water in which only one OH is H-bonded, as may occur for water molecules located between two other water molecules. Such an assignment is consistent with the appearance of libration modes around 5550 $cm^{-1}$. As this hydration stage involved the adsorption of approximately two water molecules per sodium cation, it could be proposed that the first of those two molecules is characterised by the signals at 5245 and 5100 $cm^{-1}$, whereas the second one could be linked to the bands at 5310 and 5130 $cm^{-1}$. A confirmation of this interpretation was obtained from Raman spectra, where the region corresponding to structural hydroxyls was strongly modified. In parallel, the wavenumber corresponding to the Si–O–Si stretching mode around 674 $cm^{-1}$ shifted strongly, which can be assigned again to the sodium cation leaving the ditrigonal cavity. In the fifth hydration stage the difference IR spectrum in the domain 4500–6000 $cm^{-1}$ displayed a very broad asymmetrical massive centred around 5160 $cm^{-1}$ and a discontinuity at 5250 $cm^{-1}$ that was interpreted as resulting from the disappearance of the component at 5245 $cm^{-1}$ observed in the previous hydration stage. In addition, it showed a significant absorption below 5000 $cm^{-1}$ that was never observed in the previous hydration ranges. In parallel, libration signals increase at 5580 $cm^{-1}$. This set of spectral results indicates the development in the interlayer of a disordered percolating H-bond network. The asymmetry towards low wavenumbers characterised rather strong H-bonds, which is coherent with the adsorption of additional water molecules in a confined space.

In the range corresponding to water and $\nu(M_3OH)$ overtones, broad bands were observed around 6900 cm$^{-1}$, whereas a narrow signal appeared at 7210 cm$^{-1}$. The existence of this latter signal suggested an increased repulsive perturbation of the structural OH compared with that of the previous hydration step, which was assigned to the presence of water molecules located close to the ditrigonal cavity. In the sixth hydration range the Sap structure showed an increase in the basal spacing to ~15.2 Å and about 4.5 water molecules per sodium cation in the interlayer space. In the spectral range corresponding to water signals, broad bands centred near 4900, 5160, 5240 and 5600 cm$^{-1}$ grew. In parallel, in the range corresponding to water overtones and structural hydroxyls broad signals centred around 6950 and 7400 cm$^{-1}$ were observed together with a sharper band at 7190 cm$^{-1}$. All the broad signals observed correspond to liquid-like water molecules. However, the full width of the massive was lower than that in bulk water suggesting that the network of interlayer water molecules is more structured than in bulk water. The band at 7190 cm$^{-1}$ could be interpreted as corresponding to structural hydroxyl groups that were less perturbed than in the previous hydration stage, which could suggest that additional water molecules added in this hydration range are located less directly above ditrigonal cavities than in the previous hydration step. In the final hydration stage, compared with those in the previous hydration range, the NIR bands observed in the range between 4500 and 6000 cm$^{-1}$ were slightly modified as broad bands were observed around 5000 and 5190 cm$^{-1}$ with shoulders at 5340 and 5650 cm$^{-1}$. In parallel, broad bands centred around 6950 and 7400 cm$^{-1}$ were found together with a sharp band at 7190 cm$^{-1}$ and a weak broader component centred around 7320 cm$^{-1}$. The shoulder at 5340 cm$^{-1}$ and the band at 7320 cm$^{-1}$ characteristic of a water molecule with one of the two OH groups free of H-bonding corresponded to a fundamental stretching vibration around 3730 cm$^{-1}$. Such a wavenumber is observed in the sum frequency generation (SFG) spectra obtained for aqueous solutions. In this case, the two components at 5340 and 7320 cm$^{-1}$ were thus assigned to water molecules located at the air/water interface in partially filled pores between clay particles.

### 8.4.2 Dioctahedral Minerals

#### *8.4.2.1 Pryophyllite*

Stubičan and Roy (1961) in their approach to assign bands in the IR for clay minerals used synthetic pyrophyllite as one of the samples. With the more symmetrical structural arrangement (one octahedral layer between two silica layers) the influence of the ions in the second coordination sphere on the

Si-0 bands mentioned is much smaller than in the 1:1 clay minerals. They assigned the pyrophyllite bands at 1120, 1047, 1010, 690, 483 and 424 $cm^{-1}$ to $\nu$(SiO) modes (Kloprogge et al., 2007). Kloprogge et al. (2000b) described the MIR spectrum of synthetic pyrophyllite. In accordance with their observations the in-plane and out-of-plane $\delta(Al_2OH)$ modes were found at 948, 879 $cm^{-1}$. Similar bands were reported by Farmer, 1974b around 915–950 $cm^{-1}$ for the $\delta(Al_2OH)$ in-plane librational modes in all dioctahedral species and around 880 $cm^{-1}$ for the out-of-plane librational mode in pyrophyllite. A band at 854 $cm^{-1}$ was initially thought to be a SiO band resulting from a trace amount of amorphous silica or beidellite present in the synthetic pyrophyllite sample (Farmer, 1974b). A band around 795 $cm^{-1}$ was interpreted as another Al-OH mode, probably as the result of local areas with high numbers of defects in the pyrophyllite structure.

Pyrophyllite exhibited a strong sharp band associated with the $\nu(Al_2OH)$ mode around 3675 $cm^{-1}$ in the MIR region and the associated FT-Raman band around 3668 $cm^{-1}$. A broad and relatively weak band associated with a minor amount of adsorbed water was observed around 3454 $cm^{-1}$ in the MIR and was associated with a FT-Raman band around 3626 $cm^{-1}$. A strong sharp band around 4615 $cm^{-1}$ in the NIR ($(\nu+\delta)Al_2OH$) was accompanied by two weak bands at both slightly higher and lower wavenumbers (Fig. 8.13). The presence of the extra band around 4548 $cm^{-1}$ may indicate the presence of a very small amount of beidellite, not observed in the XRD, corresponding also with the band around 854 $cm^{-1}$ in the MIR spectrum (Kloprogge et al., 2007). Kloprogge et al. (2000b) observed similar bands for both natural and synthetic pyrophyllites. Clark et al. (1990) observed a band at 4617 $cm^{-1}$, which agrees very well with the band observed in this study at 4615 $cm^{-1}$, and a weaker band at 4794 $cm^{-1}$ with a clear shoulder at 4850 $cm^{-1}$. This second, weaker band with its shoulder is identical to the two weak bands observed in this study around 4805 and 4844 $cm^{-1}$. For pyrophyllite a strong sharp $2\nu$(OH) band was recorded at 7172 $cm^{-1}$, together with a much broader, weaker band around 7147 $cm^{-1}$. This last band is comparable to the strongest band observed for the synthetic beidellite, supporting the presence of a minor amount of beidellite already inferred from the band around 4598 $cm^{-1}$. The band at 7172 $cm^{-1}$ agrees well with the band observed by Clark et al. (1990) at 7177 $cm^{-1}$. The synthetic pyrophyllite exhibits a weak absorption band at 8122 $cm^{-1}$, close to the value of 8110 $cm^{-1}$ reported by Clark et al. (1990) for natural pyrophyllite.

Lantenois et al. (2007) published a systematic IR spectroscopic study on the effect of iron substitution in pyrophyllite in both natural and synthetic samples. The presence of an OH-stretching vibration band at 3674 $cm^{-1}$ was observed in Al-pyrophyllite but was associated with another band at 3668 $cm^{-1}$. The intensity ratio between these two bands changed between different synthetic pyrophyllite samples (Fig. 8.14). The existence of two different dehydroxylation temperatures has been observed in some dioctahedral 2:1

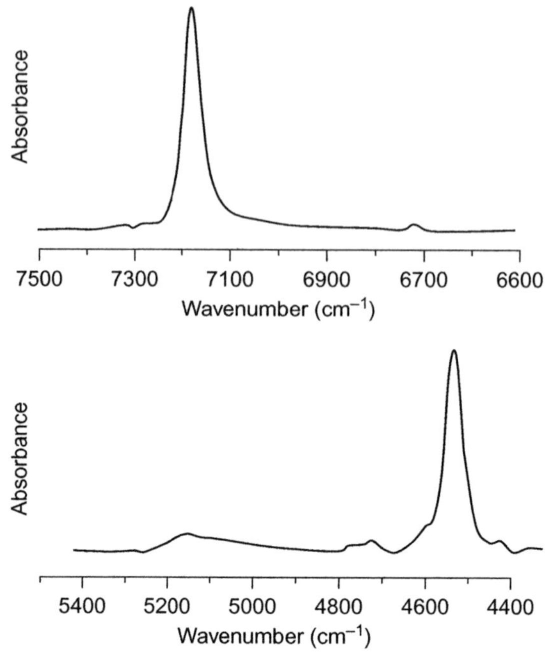


**FIG. 8.13** Near-infrared spectrum of synthetic pyrophyllite. *Reproduced with permission from Kloprogge, J.T., Duong, L.V., Martens, W.N., van der Eerden, A., Frost, R.L., 2007. Mid- and near-infrared spectroscopy of hydrothermally synthesized 2:1 phyllosilicates in the system $Na_2O$-$Al_2O_3$-$SiO_2$-$H_2O$, in: Buckley, R.W. (Ed.), Progress in Solid State Chemistry Research. Nova Science Publishers, Hauppauge, NY, pp. 285–299.*

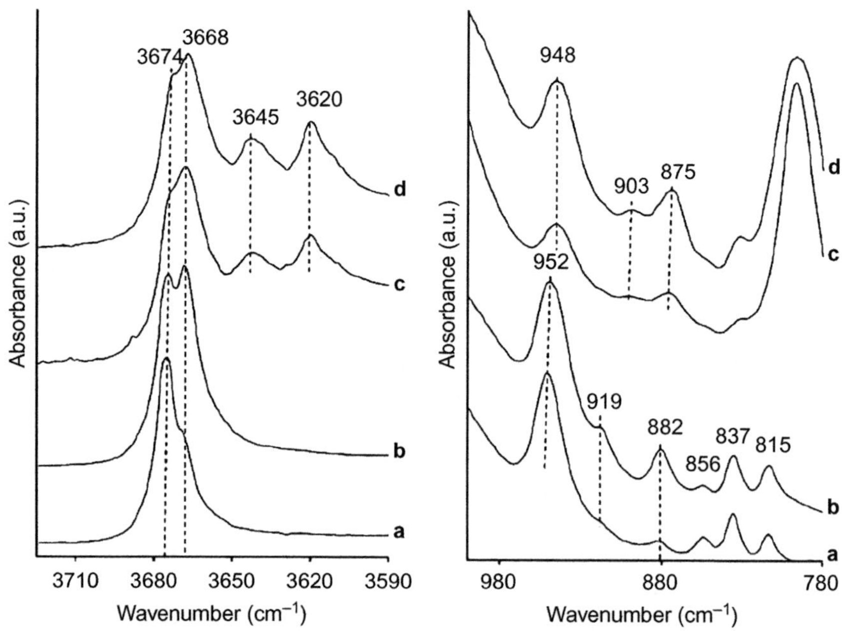


**FIG. 8.14** IR spectra of synthetic Al-pyrophyllite (a) and (b), and $Fe^{3+}$-substituted pyrophyllite (c) and (d) in the hydroxyl stretching and bending regions. *Reproduced with permission from Lantenois et al. (2007).*

clay minerals and correlated with the existence of *cis-* or *trans-*vacant layers (Drits et al., 1995; Muller et al., 2000). Drits et al. (1995) assumed that thermal energy needed for a proton to jump to the nearest OH group to form a water molecule strongly depends on the distance between the nearest OH groups. Therefore, a *cis-*vacant structure involves a greater dehydroxylation energy than *trans-*vacant structure because of the larger adjacent OH–OH distance in this dioctahedral structure.

Although pyrophyllite is classically identified as *trans-*vacant clay mineral with a dehydroxylation maximum between 550°C and 680°C, Wang and Zhang (1997a,b) identified three kinds of pyrophyllite populations: (i) a *trans-*vacant structure, (ii) a *cis-*vacant structure, and (iii) a *cis-* and *trans-*mixture of pyrophyllite phases or an interstratification between *cis* and *trans* layers. For synthetic pyrophyllite, the two structural OH losses near 550°C and 750°C could be interpreted as the presence of *trans-* and *cis-*vacant layers, respectively, and the two OH-stretching bands may be correlated with the existence of an interstratified structure. In that case, the band at 3674 $cm^{-1}$ would be characteristic of *trans-*vacant $\nu(Al_2OH)_{tv}$ bands and the band at 3668 $cm^{-1}$ of *cis-*vacant $\nu(Al_2OH)_{cv}$ bands. In the Fe-substituted samples these bands were still present, but two new ones were detected at 3645 $cm^{-1}$ as in natural Fe-containing pyrophyllite and at 3620 $cm^{-1}$. In all natural and synthetic Fe-containing samples, a band at 3645 $cm^{-1}$ was observed. Farmer and Russell (1964) originally proposed two possibilities for this band assignment: (i) it could be interpreted as the symmetric OH vibration coupled to an antisymmetric vibration at 3675 $cm^{-1}$; and (ii) the band at 3645 $cm^{-1}$ could be attributed to an $\nu(AlFe^{3+}OH)$ band. In Raman and IR spectroscopy, the symmetric and antisymmetric bands are generally inverted in intensity. The FT-Raman spectrum of the synthetic pyrophyllite sample published by Kloprogge et al. (2000b) showed that the symmetric and antisymmetric bands had similar positions and had the same intensity ratio as in the IR spectrum indicating that the first suggestion is probably incorrect. Moreover, Besson and Drits (1997a,b) developed a model that considered a $\nu(AlFe^{3+}OH)$ band at 3652 $cm^{-1}$ and a $\nu\left(Fe^{3+}{}_2OH\right)$ band at 3630 $cm^{-1}$ correlated with an increase in the Fe content in the case of a 'pyrophyllite-like' cationic environment. This was confirmed for natural smectites by Zviagina et al. (2004). In the synthetic product, the only OH-bearing phase was an $Fe^{3+}$-bearing pyrophyllite and the 3620 $cm^{-1}$ band was associated with a hydroxyl sharing bonds within an $\left(Fe^{3+}{}_2OH\right)$ environment. Furthermore, in the hydrothermal conditions used, $Fe^{3+}$ introduced in the ferric form could not change during the synthesis. Thus, if the 3645 $cm^{-1}$ vibration was correctly assigned to the $\nu(AlFe^{3+}OH)$ band, the 3620 $cm^{-1}$ band had to be associated with $\nu\left(Fe_2{}^{3+}OH\right)$.

Ferripyrophyllite is the ferric analogue of pyrophyllite with an ideal formula $Fe^{3+}{}_2Si_4O_{10}(OH)_2$. This mineral was described and characterised in its natural state by Chukhrov et al. (1979). The IR study of the two natural minerals showed two bands in the OH-stretching zone. The first had a strong intensity and is fixed near 3590 $cm^{-1}$ and the second was weaker and appeared near 3630 $cm^{-1}$. These two bands can be attributed to an $\nu(Fe^{3+}{}_2OH)$ band and to a $\nu(AlFe^{3+}OH)$ band, respectively, in accordance with the chemical composition of this mineral. These data show shifts of 15 $cm^{-1}$ and 25 $cm^{-1}$, respectively, for the $\nu(AlFe^{3+}OH)$ and $\nu(Fe^{3+}{}_2OH)$ band positions between ferripyrophyllite (Chukhrov et al., 1979) and the synthetic Fe-substituted samples. These shifts can be explained by a difference in chemical composition correlating the OH vibration band displacement with the difference in ionic radius of the octahedral cations.

### *8.4.2.2 Beidellite*

Grauby et al. (1993) synthesised smectites from gel in the beidellites (Bd) – Sap series, that is, substituting $Mg^{2+}$ for $Al^{3+}$ in the octahedral sheet changing its nature from dioctahedral to trioctahedral. The TEM chemical microanalyses indicated the $Al^{3+}$-$Mg^{2+}$ series was continuous between these two end-members at the clay particle scale. XRD indicated three different kinds of interferential coherence domains: (i) a dioctahedral essentially aluminous domains, (ii) a trioctahedral essentially magnesian domains and (iii) an intermediate case where both domains co-existed, which was the usual condition, although these were metastable. Samples of intermediate octahedral chemistry were made up of particles with stacked layers having a constant chemistry and thus, showing an apparent di-trioctahedral lattice. In the IR OH-stretching region, the prominent bands were those of $\nu(Al_2OH)$ at 3640 $cm^{-1}$ and $\nu(Mg_3OH)$ for Sap at 3675 $cm^{-1}$ (Farmer, 1974b). No clear band was detected at 3687 $cm^{-1}$, the position of the $\nu(AlMgOH)$ mode in dioctahedral smectites (Farmer, 1974b). For high aluminium contents, only one $\nu(Al_2OH)$ band was observed. As the Mg-content increased, the Sap $\nu(Mg_3OH)$ band dominated, while the intensity of the $\nu(Al_2OH)$ band decreased and disappeared. A broad band observed for intermediate samples was considered to be made up by overlapping contributions of $\nu(Al_2OH)$ and $\nu(Mg_3OH)$ bands, which changed progressively in shape. In the OH-bending region, two bands at 920 $cm^{-1}$ and near 675 $cm^{-1}$ were dominant. The first was assigned to $\delta(Al_2OH)$ vibrations (Farmer, 1974b). The second, an asymmetric band, was the considered to be an overlapping contribution of two components situated at 680 $cm^{-1}$ and 667 $cm^{-1}$, and attributed to tetrahedral $(A_1)v_1(SiO)$ and $\delta(Mg_3OH)$ vibrations, respectively (Russell et al., 1970). A third, minor band appeared at 840 $cm^{-1}$ assigned to $\delta(AlMgOH)$. IR spectra of samples with a high Al-content exhibited only the $\delta(Al_2OH)$ mode, while those with a high Mg-content only the trioctahedral $\delta(Mg_3OH)$ band. The intensities of the $\delta(Al_2OH)$ and $\delta(Mg_3OH)$ bands evolved in opposite directions based on the chemical composition of the samples within the Bd-Sap series. A minor $\delta(AlMgOH)$ vibration was present in spectra of samples with intermediate chemistry. The OH-bending region gave more accurate information than the OH-stretching region. However, IR data showed clearly that the $\delta(Al_2OH)$ and $\delta(Mg_3OH)$ bands were prevalent. Therefore, the $Al^{3+}$ and $Mg^{2+}$ cations were segregated into, respectively, dioctahedral and trioctahedral domains, and they were not randomly distributed within octahedral sheets of one phase.

Ferrian (also known as ferruginous) Bd with Al:Fe atomic ratios up to 2.36 were obtained when solutions containing Al, $Fe^{2+}$ and $H_4SiO_4$ were adjusted to pH 8.5 with $Ca(OH)_2$ and incubated at 95°C in the presence of $CaCO_3$ as a pH buffer (Farmer, 1997). Incubation took place under cyclic reducing and oxidising conditions achieved by adding 2 mM hydrazine at 14–15-d intervals over a period of 10–13 weeks. The IR spectra of the products appeared similar to that of natural Bd, but indicated little change with octahedral Al:$Fe^{3+}$ ratio in the products for starting Al:$Fe^{3+}$ ratios from 2.5 up to 3.5 (the latter being highest Al:$Fe^{3+}$ ratio at which a well-crystallised product was obtained). The positions of many of the Bd bands were sensitive to the composition of the product, showing a progressive movement to higher wavenumbers with increasing beidellitic content. The following band progressions were observed with increase in $Al^{3+}$ content:

**(1)** $\nu(M_2OH)$: 3560, 3620, 3632 $cm^{-1}$;
**(2)** $\nu(SiO)$: 1015, 1020, 1036 $cm^{-1}$;
**(3)** $\delta(Si–O–Al^{IV})$: 739, 755, 770 $cm^{-1}$ and
**(4)** $\delta(Al^{VI}–O–M^{IV})$: 507, 530, 530 $cm^{-1}$, and 460, 475, 471 $cm^{-1}$.

In addition, the $\delta(Fe^{3+}{}_2OH)$ band at 815 $cm^{-1}$ in nontronite was replaced by a $\delta(Si–O–Al^{IV})$ vibration at 816 $cm^{-1}$ in Bd with little change in intensity (Farmer, 1997), though this most likely a $\delta(Al_2OH)$ instead. The $\delta(AlFe^{3+}OH)$ mode at 870 $cm^{-1}$ decreased in intensity and moved to 882 $cm^{-1}$ in the ferrian Bd.

Kloprogge et al. (2000a) studied the effect of layer charge in synthetic Bd on the MIR spectra. The MIR spectra of the lower charge Bd (0.6 and 0.7) were very similar in the $\nu(SiO)$ region around 1000 $cm^{-1}$. The high charge Bd spectra were much more similar to that of paragonite. In the low wavenumber region the OH-libration modes associated with the octahedral $\delta(AlMOH)$ groups in the Bd structures were found at 878–882, 913–920 and 935–965 $cm^{-1}$ (Fig. 8.15). Similar bands were reported by Farmer (1974b) at 915–950 $cm^{-1}$ for the $\delta(AlMOH)$ in-plane librational modes in all dioctahedral clay minerals and around 880 $cm^{-1}$ for the out-of-plane librational mode in pyrophyllite. There is, however, evidence that points to the lower wavenumber band being the in-plane and the higher wavenumber band the out-of-plane mode. Only the last band provided a clear shift towards higher wavenumbers with increasing layer charge and was assigned as the in-plane vibration (Kloprogge et al. (2000a)).

For Bd the OH-stretching vibrations were affected by the configuration and charge distribution in the neighbouring tetrahedral layers. Early work by Vedder and McDonald (1963) has indicated that in dioctahedral clay

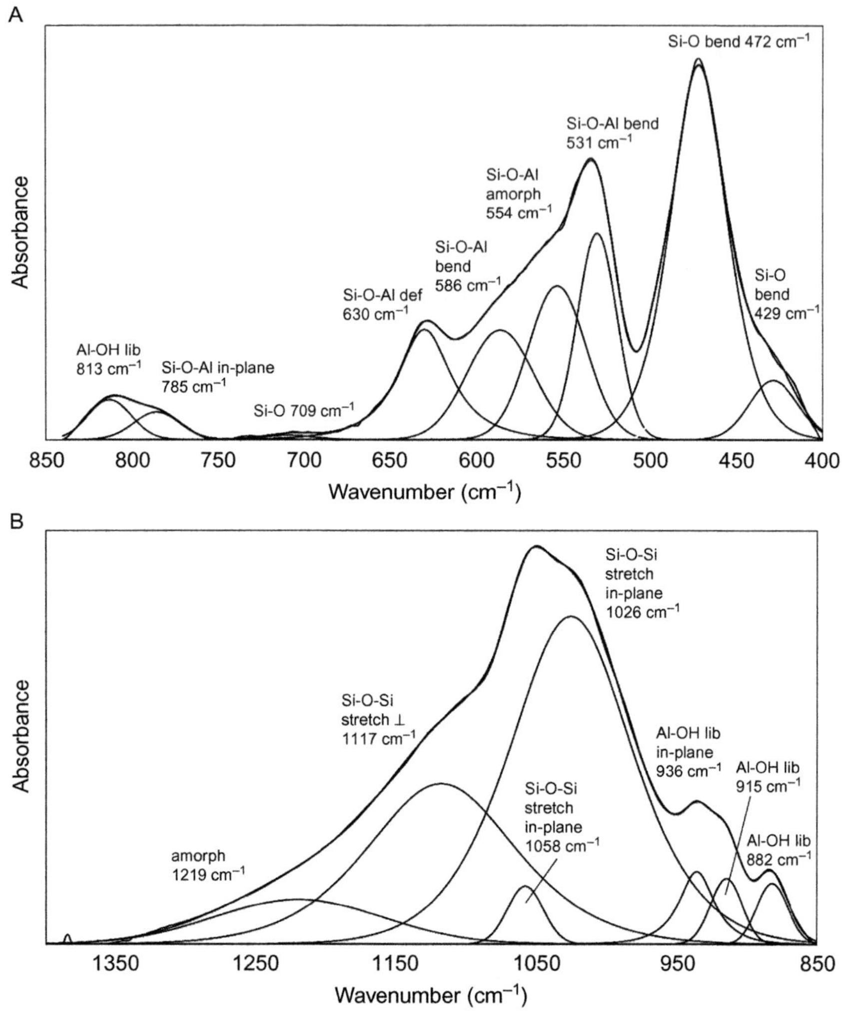


**FIG. 8.15** IR spectra of synthetic beidellite in the range of (A) 400–850 cm$^{-1}$ and (B) 850–1400 cm$^{-1}$. *Reproduced with permission Kloprogge et al. (2000).*

minerals such as Bd the OH-transition moment is tilted by a small angle out of the *001* plane, though in reality this angle is >55°. As consequence of this orientation the proton of the OH-group lies closely to two of the six oxygen atoms at the apices of the tetrahedra forming a hexagonal cavity in the tetrahedral sheet. In the relatively low-charge Bd structure this resulted in two OH-stretching modes, one associated with hydroxyl protons located near charge-bearing oxygen atoms of the $Al^{3+}$-substituted hexagonal rings and one associated with neutral oxygen atoms of non-substituted hexagonal rings.

Increasing the layer charge was consequently reflected in the intensity ratio between these OH-stretching modes. Very dominant the MIR spectra of the Bd were two bands due to interlayer water. Interlayer water in Bd was characterised by a $H_2O$ bending vibration at 1635 cm$^{-1}$ plus the associated OH-stretching vibration around 3450 cm$^{-1}$. This band is most likely associated with water molecules in the outer hydration sphere of the interlayer sodium cation (Bishop et al., 1994). The shoulder around 3240 cm$^{-1}$ was ascribed to interlayer $H_2O$ chemically bound or coordinated to the interlayer cation with an ice-like structure (van der Marel and Beutelspacher, 1974). Only one $\nu(Al_2OH)$ band can be observed around 3668 cm$^{-1}$. The small band at 3620 cm$^{-1}$ was thought to be characteristic for the asymmetric OH-stretching ($\nu_{AS}$(OH)) vibration for water associated with the inner hydration shell of the interlayer cation (Bishop et al., 1994). However, due to the presence of strong water bands overlapping the 3668 and 3620 cm$^{-1}$ bands, the positions and intensities of these bands can deviate significantly and obscure the second $\nu(Al_2OH)$ band. Therefore, no definite conclusions could be drawn from this region of the MIR spectra. Lantenois et al. (2008b) reported the IR spectra in the OH-stretching region of the three synthetic Bd showing the increase in deficit of charge was associated with an increase in the band surface near 3620 cm$^{-1}$ as compared with the band at 3665 cm$^{-1}$. The change in the *cis*/(*cis*+*trans*) ratio of these Bd with various layer-charge deficits obtained from DTG with the samples showed an increasing ratio of 70%–100% with increasing layer charge deficit for Bd accompanied an increase in the 3620 cm$^{-1}$ band relative to the 3665 cm$^{-1}$ band.

As for the MIR spectra, the NIR spectra of the low charge Bd were very similar. Band component analysis showed the presence of two bands around 4545 (weak) and 4595 (strong) cm$^{-1}$. This split disappeared in the spectra of the high charge Bd. The range around 4500 cm$^{-1}$ is characteristic for the combination ($\nu+\delta$) bands of the $\nu(Al_2OH)$-stretching modes and the $\delta(Al_2OH)$ libration modes around 3660 cm$^{-1}$ and 880 and 915 cm$^{-1}$, respectively (Madejová et al., 2000b). Three bands were observed around 5140, 5240 and 5260 cm$^{-1}$. Isolated water molecules exhibited a band near 5333 cm$^{-1}$, due to the combination mode of the stretching and bending modes of the water molecule. The presence of interlayer water in smectites will be observed as broad bands close to this position, due to the poorly ordered sites in the interlayers (Clark et al., 1990). This means that the first two bands around 5140 and 5240 cm$^{-1}$ can be ascribed to the combination modes of interlayer water modes. Similar bands were observed by Sposito et al. (1983) for Mt around 5130 and 5250 cm$^{-1}$. Band component analysis showed the relatively broad band around 5250 cm$^{-1}$ actually contained two components around 5240 and 5260 cm$^{-1}$, which can probably be explained by the fact that a part of the interlayer water was coordinated to the interlayer sodium cations, while the remaining water molecules were less tightly bound to the clay mineral structure, in agreement with the dehydration study that indicated two dehydration stages (Kloprogge et al., 1992b). The strong intensity of these bands was due to the presence of a relatively large amount of interlayer water. Clearly visible was a decrease in intensity with increasing layer charge. This increase in layer charge was compensated by the incorporation of more sodium interlayer cations leaving less interlayer space for water. The bands observed in the high wavenumber region can be attributed to the overtones and combination modes associated with fundamental stretching and bending modes observed in the MIR region. For Bd this region was relatively complex showing multiple overlapping bands with two relatively sharp bands occurred around 7100 and 7150 $cm^{-1}$ plus a much broader band around 6900 $cm^{-1}$. This broad band decreased in intensity with increasing layer charge of the Bd. Therefore, in addition to the fact that the free water molecule has an absorbance near 6878 $cm^{-1}$ (Sposito et al., 1983), this band was ascribed to the first overtone of the strongly hydrogen bonded interlayer water OH-stretching (i.e. the $2\nu(OH)$) mode (Madejová et al., 2000b). The 7100 $cm^{-1}$ band is characteristic for the first overtone of the $2\nu(Al_2OH)$ stretching modes of the Bd. The origin of the second band around 7150 $cm^{-1}$ was less clear: IES work on synthetic Bd has shown, however, that the broad band around 3660 $cm^{-1}$ consisted of two overlapping $\nu(Al_2OH)$ stretching modes around 3615 and 3650 $cm^{-1}$ (Kloprogge et al., 1999c), while for example, montmorillonite SAz showed also two bands around 3622 and 3668 $cm^{-1}$ after heating to 110°C (Madejová et al., 2000a). This means that even though the two $\nu(Al_2OH)$ modes cannot always be identified in the MIR spectra, both bands do show up in the NIR spectra. In the very high wavenumber region the low charge synthetic Bd showed a very weak band around 8670 $cm^{-1}$. For the high-charge Bd this band shifted to 8631 $cm^{-1}$. Madejová et al. (2000b) described a similar band around 8661 $cm^{-1}$ plus a band at 10550 $cm^{-1}$ for SAz Mt, assigned to $2\nu(Al_2OH)$ of structure OH and $2\nu(OH)$ of adsorbed water.

The IES spectra of synthetic Bd from 200°C to 750°C are shown in Fig. 8.16. The hydroxyl stretching region between 3000 and 4000 $cm^{-1}$ clearly shows the dehydration and dehydroxylation reactions. Although some of the interlayer water already disappeared from the interlayer at 200°C, still a large band remained visible around 3420 $cm^{-1}$, while at 450°C this band has split in two much weaker bands around 3650 and 3615 $cm^{-1}$. In addition, a band became visible at 3720 $cm^{-1}$. The band at 3720 $cm^{-1}$ was assigned to silanol groups that formed on the crystal edges during intermediate stages of dehydration and dehydroxylation as this band was not visible at lower temperatures (Kloprogge et al., 1999c). Further heating resulted in a decrease of the $Al^{3+}$-hydroxyl modes and the disappearance of the silanol band. Associated with the dehydroxylation and the changes in the $\nu(OH)$ region, were the changes in the OH-bending and -libration modes. Two bands at 788 and 820 $cm^{-1}$ decreased in intensity during heating, while two other bands at

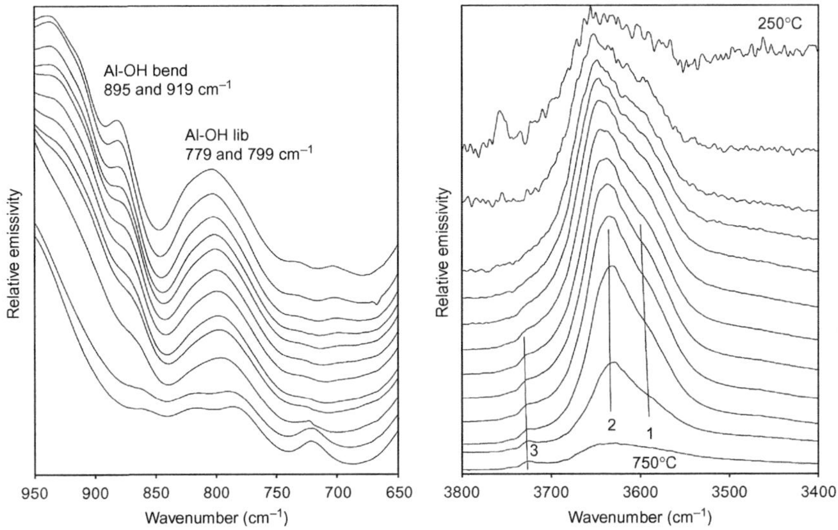


**FIG. 8.16** Infrared emission spectra of synthetic beidellite from 200 to 750°C at 50°C intervals in the OH deformation (left) and OH stretching (right) regions. 1 = 3600–3615 cm$^{-1}$ Al-OH stretching mode, 2 = 3650 cm$^{-1}$ Al-OH stretching mode, 3 = 3720 cm$^{-1}$ Si-OH stretching mode. *Reproduced with permission from Kloprogge et al. (1999c).*

876 and 929 cm$^{-1}$ completely disappeared between 650°C and 700°C, which indicates that these two bands are the $\delta(Al_2OH)$ bending vibrations. The other two bands are probably the librational modes. A new band was observed at 722 cm$^{-1}$ after the dehydroxylation associated with the formation of new octahedral Al–O or distorted tetrahedral Si–O bonds in the dehydroxylated Bd structure.

Ryu et al. (2006) hydrothermally synthesised Bd from dickite plus NaOH and amorphous $SiO_2$. The IR spectrum reaction product exhibited the following bands: $\nu(Al_2OH)$ stretching vibrations at 3672, 3650 and 3620 cm$^{-1}$; a $\nu(H_2O)$ vibration of adsorbed water at 3432 cm$^{-1}$; a $\nu$(Si–O–Si) vibration at 1030 cm$^{-1}$ and $\delta$(Si–O–Al$^{IV}$) lattice vibrations at 542 cm$^{-1}$ and 480 cm$^{-1}$. In particular, out of plane Al–O and Al–O–Si vibrations at, respectively, 818 and 770 cm$^{-1}$ were observed in pure Bd. All these bands were characteristic bands for dioctahedral smectite (Farmer, 1997; Stubičan and Roy, 1961). A synthetic Bd, produced from aluminosilicate gel at 350°C, displayed a similar spectrum (Kloprogge et al., 1990a,b). Montmorillonite has an $\nu(Al_2OH)$ band at 915 cm$^{-1}$, whereas in Bd this band was observed at 930 cm$^{-1}$.

Smectites in the Bd-Sap series were hydrothermally synthesised using silicate gels with variable $Al^{3+}/Fe^{3+}/Mg^{2+}$ ratios by Huertas et al. (2000). The characteristic OH vibration of dioctahedral smectite were identified in products from gels with $Al^{3+}/Mg^{2+}=0.8/0.15$, $Al^{3+}/Mg^{2+}=0.65/0.3$, and $Al^{3+}/Mg^{2+}=0.45/0.5$, which showed a broad OH stretching band at about $3640\ cm^{-1}$ and OH bending bands at $919\ cm^{-1}$ corresponding to $\nu(Al_2OH)$ and $848\ cm^{-1}$ from $\nu(AlMgOH)$ (Farmer, 1974b; Russell and Fraser, 1994). The OH bending region, especially the band at $848\ cm^{-1}$, had very low intensities in products from gels with increasing $Mg^{2+}$ contents. The typical trioctahedral smectite OH bending band at $663\ cm^{-1}$ (Farmer, 1974b; Russell and Fraser, 1994) was intense in products from gel with very high $Mg^{2+}$ contents and decreased in intensity in products from gels as the $Mg^{2+}$ content decreased. The lattice vibration region (about $500\ cm^{-1}$) also confirmed the transition between trioctahedral and dioctahedral smectite. Samples from the gel with high $Mg^{2+}$ content had a single band at $450\ cm^{-1}$, typical of trioctahedral smectite (Farmer, 1974b; Russell and Fraser, 1994). With increasing $Al^{3+}$, a new band developed at $530\ cm^{-1}$, typical of dioctahedral smectite (Farmer, 1974b; Russell and Fraser, 1994). The FTIR spectra always showed the presence of typical dioctahedral and trioctahedral bands without any obvious evidence for a homogeneously intermediate structure (such as a trioctahedral $AlMg_2OH$ or a dioctahedral $Mg_2OH$ arrangement), although very low intensity bands of this type could not be completely ruled out. Thus, the FTIR results did not help to discriminate between the three possible arrangements listed above, but instead set the constraint that, in case there are dioctahedral and trioctahedral domains within the layers, they should be large enough to permit a complete or nearly complete segregation of the two structures as observed in the IR spectra.

These results were similar to those of Grauby et al. (1993), but their interpretation was instead that the chemical heterogeneity of the smectite particles was due to topotactic growth of dioctahedral smectite on the trioctahedral specimen. The reason for this interpretation was related to the amount of clay mienral formed from the gels of different compositions. The amount of trioctahedral smectite produced decreased continuously from $Mg^{2+}$-rich to $Al^{3+}$-rich gels, as was expected. In contrast, the amount of dioctahedral smectite produced did not increase concomitantly in the same sequence, as would be expected, instead a minimum value existed at intermediate compositions where the concentration of $Al^{3+}$ exceeded that of $Mg^{2+}$. If trioctahedral smectite acts as a template for dioctahedral smectite, a decrease in the amount of the former would result in a retardation of the formation of the latter. This implies that trioctahedral smectite forms faster than dioctahedral smectite, at least in the early reaction stages, even for a gel with higher concentrations of $Al^{3+}$ than $Mg^{2+}$. Vogels et al. (1997) suggested that brucite and bayerite were incorporated as nucleation units during Sap growth from $Si^{4+}$-$Al^{3+}$-$Mg^{2+}$ gels of saponitic composition. Kloprogge and Frost (1999d) indicated that many authors agree the attachment of silica to octahedral units is the first step of smectite formation in smectite synthesis.

### 8.4.2.3 Nontronite and Ferrian-smectite

Syntheses of ferrian (ferruginous) smectites were initially successfully performed at low temperature (75°C) by Decarreau and Bonnin (1986). Under strictly reducing conditions only nuclei of a trioctahedral ferrous stevensite were observed and crystal growth did not take place. Spontaneous oxidisation due to contact with air resulted in a ferrian smectite. In a follow-up study by Decarreau et al. (1987a) synthetic ferric smectite was shown to contain both structural and interlayer water. In the 3000–4000 $cm^{-1}$ region, the broad band at 3430 $cm^{-1}$ was assigned to the OH-stretching vibration of the remaining interlayer water. An OH-stretching band was noted at 3565 $cm^{-1}$, in the wavenumber range of $\nu(Fe^{3+}{}_2OH)$ stretching vibrations in natural nontronite (Goodman et al., 1976). For synthetic smectite containing neither $Mg^{2+}$ nor $Al^{3+}$ as octahedral cations, only one OH-stretching absorption band was observed. Similar behaviour was shown by the $\delta(M_2OH)$ modes: the absorption at 818–819 $cm^{-1}$ was assigned to $\delta(Fe^{3+}{}_2OH)$ vibrations and bands because the $\delta(AlFe^{3+}OH)$ or $\delta(Fe^{3+}MgOH)$ vibrations were absent. An intense $\nu(SiO)$ absorption band was noted at 1020–1030 $cm^{-1}$, without any shoulder near 965 $cm^{-1}$. According to Goodman et al. (1976) these features are characteristic of ferrian smectites without $Fe^{3+}$- for $Si^{4+}$- tetrahedral substitution. In the 700–400 $cm^{-1}$ region, lattice vibration bands were observed at 681, (590 shoulder), 494 and 436 $cm^{-1}$, as in nontronite IR spectra, but these were shifted slightly towards higher wavenumbers.

A study of the octahedral $Fe^{3+}$-$Mg^{2+}$ substitution in the nontronite-Sap series was conducted by Grauby et al. (1994) where Mössbauer data revealed that only ferric iron occurred in synthesised smectites. In the OH-stretching region, the clearly observed bands were those of the $\nu(Fe^{3+}{}_2OH)$ vibrations, at 3550 $cm^{-1}$ (Goodman et al., 1976) and the $\nu(Mg_3OH)$ vibrations at 3675 $cm^{-1}$ (Farmer, 1974b). Another band was found at 3630 $cm^{-1}$. This band appeared for the samples with intermediate chemistry and was absent for the most $Fe^{3+}$- or $Mg^{2+}$-rich samples. Hence, this band was attributed to a $\nu(Fe^{3+}MgOH)$ mode. At a composition near the nontronite endmember, only the $\nu(Fe^{3+}{}_2OH)$ band was observed. As the $Mg^{2+}$-content increased, a band assigned to $\nu(Mg_3OH)$ appeared and became predominant. Simultaneously, the $\nu(Fe^{3+}{}_2OH)$ band decreased in intensity and disappeared for the most Mg-rich sample. Whatever the amount of magnesium in the clay structure, the position of the $\nu(SiO)$ vibration was constant at 1020 $cm^{-1}$, indicating only a small amount of tetrahedral iron (Goodman et al., 1976). For the most $Fe^{3+}$-rich samples, two bands were predominant. One was at 814 $cm^{-1}$ assigned to $\delta(Fe^{3+}{}_2OH)$ vibrations (Farmer, 1974b; Goodman et al., 1976; Stubičan and Roy, 1961) and the other which is classically in nontronites, at 682 $cm^{-1}$ assigned to tetrahedral $A_1\gamma_1$ SiO vibrations (Goodman et al., 1976; Stubičan and Roy, 1961) respectively. More recently it has been shown that this latter band depends on the Fe content and has been assigned to Fe-O out of plane (Gates, 2005). For the most Mg-rich samples, only one asymmetric band was present and was the sum of two components situated at 682 $cm^{-1}$ and 668 $cm^{-1}$ attributed to tetrahedral $A_1\gamma_1$ SiO vibrations and 3Mg-OH vibrations, respectively (Russell et al., 1970). For some intermediate samples, a third slight band occurred at 785 $cm^{-1}$. This band, often observed in dioctahedral smectites, was assigned to $\delta(Fe^{3+}MgOH)$ (Farmer, 1974b). Gates (2008) and Petit et al. (2015) have argued that this is now an unlikely assignment, because it is too strong for the $Mg^{2+}$ content, and is present in synthetic nontronites with no Mg and disappears with increasing octahedral $Mg^{2+}$ (Grauby et al., 1994). Thus this band is instead probably related to a $\nu$(SiO) mode. Another band located near 760 $cm^{-1}$ is observed in these same samples and was tentatively be assigned to $\delta\left(Fe^{3+}{}_2OH\right)$ or $\delta(Fe^{3+}Mg_2OH)$ groups in a trioctahedral environment. IR spectra of samples with a high $Fe^{3+}$ content present only the $\delta\left(Fe^{3+}{}_2OH\right)$ band, while those with a high Mg-content show only the $\delta(Mg_3OH)$ absorption band. The intensities of the $\delta\left(Fe^{3+}{}_2OH\right)$ and $\delta(Mg_3OH)$ bands evolved in opposite directions according to the chemical composition of the samples. For samples with intermediate chemistry, the predominant band at 682 $cm^{-1}$ is the sum of the two tetrahedral $A_1\gamma_1$ SiO vibrations influenced by $Fe^{3+}$ or $Mg^{2+}$. The IR spectra showed clearly that the $\delta\left(Fe^{3+}{}_2OH\right)$ and $\delta(Mg_3OH)$ bands were prevalent. Consequently, the $Fe^{3+}$ and $Mg^{2+}$ cations are not randomly distributed in the octahedral sheets, they are segregated into clusters, similar to what was observed for the Bd-Sap series for $Al^{3+}$ and $Mg^{2+}$ (Grauby et al., 1993). However, the occurrence of $Fe^{3+}$-OH-Mg bands is evidence for some $Fe^{3+}$-Mg pairs. Thus $Fe^{3+}$ and $Mg^{2+}$ cations were segregated, at least partly, within the same octahedral sheet.

The $Al^{3+}$-$Fe^{3+}$ chemical series using stoichiometric co-precipitated hydrogels investigated by Andrieux and Petit (2010) indicated the formation of nontronites at lower temperatures (below 150°C) and higher pH (around 12), while higher temperatures (200°C) and lower pH (down to 7) were suitable for the more $Al^{3+}$-rich smectites ($Fe^{3+}$-Bd). For intermediary compositions, the pH and temperature of synthesis leading to pure smectite crystallisation could be concomitantly adjusted. In the $Al^{3+}$-$Fe^{3+}$ smectitic chemical series, three octahedral environments may be encountered: $2Fe^{3+}$, $AlFe^{3+}$ and 2Al. Up to three vibrational bands due to the structural OH groups were effectively observed in the synthesised smectites. The relative intensities of bands associated with the presence of $Fe^{3+}{}_2OH$, $AlFe^{3+}OH$ and $Al_2OH$ pairings varied for all the chemical series depending on the starting composition. Smectites which contained only $Fe^{3+}$ in octahedra exhibited the $Fe^{3+}{}_2OH$ bands at 3566 $cm^{-1}$ $\left(\nu\left(Fe^{3+}{}_2OH\right)\right)$, 815 $cm^{-1}$ $\left(\delta\left(Fe^{3+}{}_2OH\right)\right)$, and 4378 $cm^{-1}$ $\left((\nu+\delta)\left(Fe^{3+}{}_2OH\right)\right)$ originating from stretching, bending, and combination vibrations, respectively, like those observed for natural nontronites (e.g. Gates, 2005, 2008). The synthetic nontronites corresponding to such spectra were not necessarily $Al^{3+}$-free: some octahedral $Al^{3+}$ occurrence, in too small amounts to be revealed via OH bands, is possible, and tetrahedral $Al^{3+}$ may also exist. Subtle changes in the MIR spectrum may reflect these crystal-chemical variations. The amount of $Al^{3+}$ atoms occupying the octahedral sites increased in parallel with the $Al^{3+}$ content of the synthetic nontronite and hence, the relative intensities of the $AlFe^{3+}OH$ bands at 869 $cm^{-1}$ $\delta(Fe^{3+}{}_2OH)$, and 4462 $cm^{-1}$ $(\nu+\delta)Fe^{3+}{}_2OH$ for bending and combination modes, respectively, increased. The bands observed in the OH first overtones and in the stretching region shifted towards higher wavenumbers compared with less $Al^{3+}$-rich nontronites, in agreement with the results obtained for natural ferrian smectites (Gates, 2005). When the $Al^{3+}$ content of the system was high, $Al_2OH$ bands were observed at 3629 $cm^{-1}$ ($\nu(Al_2OH)$), 915 $cm^{-1}$ ($\delta(Al_2OH)$), and 4567 $cm^{-1}$ ($(\nu+\delta)Al_2OH$), in parallel with those observed for natural Bd (e.g. Kloprogge et al., 1998; Gates, 2005), clearly reflecting the effective $Al^{3+}$ enrichment of the synthesised smectites.

A subsequent paper (Petit et al., 2015) provided a more in-debt study of the crystal chemistry of these synthetic smectites. The OH stretching bands ranged from 3620 $cm^{-1}$ for Al-rich smectites to 3562 $cm^{-1}$ for $Fe^{3+}$-rich smectites assigned to $\nu(Al_2OH)$ and $\nu(Fe^{3+}{}_2OH)$, respectively, and were broad and were largely overlapped by $H_2O$ OH-stretching bands. For octahedral $Fe^{3+}$ $(Fe^{3+}{}_{oct})$ contents $< 1.4$ the band position remained at about 3620 $cm^{-1}$. Three OH-bending bands at 920, 877 and 814 $cm^{-1}$ were observed in the synthetic smectites and assigned to $\delta(Al_2OH)$, $\delta(AlFe^{3+}OH)$ and $\delta(Fe^{3+}{}_2OH)$ (Farmer, 1974b). The $\delta(Fe^{3+}{}_2OH)$ band was relatively strong and well observed for $1.2 < Fe^{3+}{}_{oct} < 2$. The $\delta(Al_2OH)$ band appeared for octahedral $Al^{3+}$ ($Al_{oct}$) greater than 1.4 per $O_{11}$ and was observed as a shoulder probably linked to the poor crystallisation of the $Al^{3+}$-rich smectites (Andrieux and Petit, 2010). The $\delta(AlFe^{3+}OH)$ band was found between 0.36 $0.36 < Fe^{3}{}_{+oct} <$ 1.9 per $O_{11}$ and ranged from 870 to 877 $cm^{-1}_{,}$ consistent with those reported by Gates (2005) for natural smectites, where $\delta(AlFe^{3+}OH)$ band ranged from 868 to 870 $cm^{-1}$ for nontronites and from 872 to 875 $cm^{-1}$ for Bd. However, different, also to the results of Craciun (1984) and Vantelon et al. (2003), no linear correlation was found between the $\delta(AlFe^{3+}OH)$ wavenumber and $Fe^{3+}{}_{oct}$. OH-bending bands were not sufficiently resolved to see possible splitting of the $\delta(Al_2OH)$ and $\delta AlFe^{3+}OH$ bands. A strong $\nu(SiO)$ stretching band from 1036 to 973 $cm^{-1}$ was observed (Fig. 8.17). Higher wavenumbers indicated the presence of significant amounts of unreacted gel. The wavenumber of the strong $\nu(SiO)$ band decreased with total $Fe^{3+}{}_{oct}$ content of the synthesised smectites. A linear relationship was

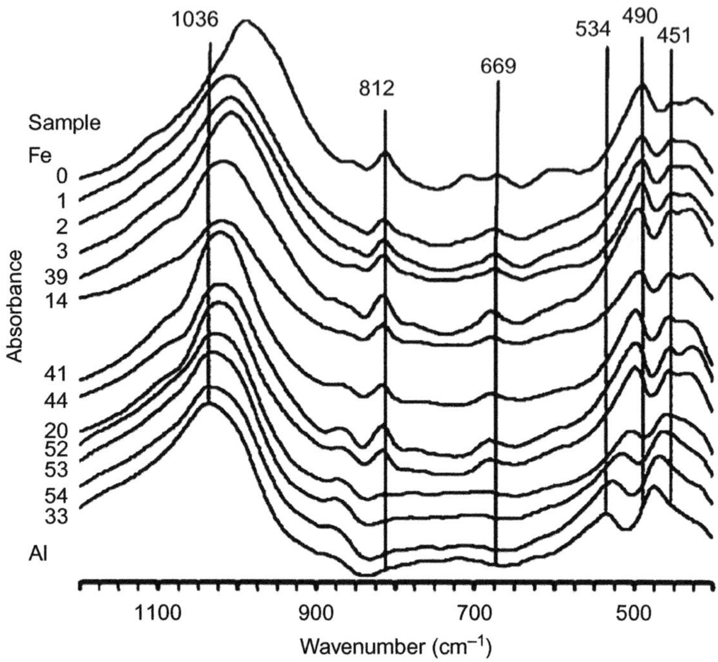


**FIG. 8.17** Mid-IR spectra of synthetic ferrian smectites and nontronites in the 1200–400 cm$^{-1}$ region. Decreasing Fe/(Fe+Al) with sample number. *Reproduced with permission from Petit et al. (2015).*

shown to exist between the wavenumber of this band and the octahedral $Fe^{3+}$ following the regression:

$$\nu(\text{SiO}) = -9.5^*\text{Fe}^{3+}{}_{\text{oct}} + 1040\,\text{cm}^{-1}$$

This indicates that the wavenumber of the $\nu$(SiO) for a pure aluminous ($Na_{0.4}(Si_{3.6}Al_{0.4})Al_2O_{10}(OH)_2$) Bd would be at 1040 cm$^{-1}$, and at 1021 cm$^{-1}$ for a $\left(Na_{0.4}(Si_{3.6}Al_{0.4})Fe^{3+}{}_2O_{10}(OH)_2\right)$ nontronite. Most natural nontronites having $Fe^{3+}{}_{oct} \approx 1.8$ were observed to have a $\nu$(SiO) near 1020 cm$^{-1}$ (Gates, 2005). For $Fe^{3+}$ contents of the synthetic smectites $<2$, the decrease of the wavenumber was amplified and the following linear regression was obtained:

$$\nu(\text{SiO}) = -32^*\text{Fe}^{3+}{}_{\text{oct}} + 1019\,\text{cm}^{-1}$$

This regression line was obtained with only four data points and as such the equation is only indicative. Even so, comparing the two relations for $\nu$(SiO), tetrahedral Fe has three times more influence than $Fe^{3+}{}_{oct}$ in lowering the $\nu$(SiO) wavenumber. From the latter regression, the wavenumber of the $\nu$(SiO) band would be at 1006 $cm^{-1}$ for a pure $Na_{0.4}(Si_{3.6}Fe^{3+}{}_{0.4})Fe^{3+}{}_2O_{10}(OH_2)$ nontronite. These results were also in agreement with the position of this band observed for essentially endmember natural nontronites (e.g. samples N11 and N12 in Gates (2005). Three other bands have their position obviously linked to $Fe^{3+}{}_{oct}$ concentration: the $M^{3+}{}_{oct} - O_{ap}$ out-of-plane deformation band (669–698 $cm^{-1}$) and two Si-O-$M^{3+}{}_{oct}$ bands (490–535 $cm^{-1}$ and 449–475 $cm^{-1}$, respectively).

Lastly, two small bands corresponding to $M^{3+}{}_{oct} - O_{ap} - Si$ were found at 611 $cm^{-1}$ for Al and 588–598 $cm^{-1}$ for $Fe^{3+}$. A low intensity band at 851–856 $cm^{-1}$ was observed for samples with the highest $Fe^{3+}{}_{oct}$ values as previously observed by Decarreau et al. (2008). This band is similar to the 841–850 $cm^{-1}$ band observed in IR spectra of natural nontronites (e.g. Gates, 2005) and whose attribution is still debated (Decarreau et al., 2008; Gates, 2008). As the band was observed for smectites that contain no or very few octahedral Al, this band cannot be attributed to a $\delta(AlFe^{3+}OH)$ mode as earlier hypothesised by Komadel et al. (1995). Gates (2005, 2008) observed that the band at 842–850 $cm^{-1}$ was only present for natural nontronites with $Fe_{oct} \geq 1.48$ and attributed this to a second $\delta(Fe^{3+}{}_2OH)$ bending band. Gates (2008) observed a strong correlation of the ratio of the integrated intensities of this band and the $\delta(Fe^{3+}{}_2OH)$ band near 818 $cm^{-1}$ with tetrahedral layer charge, which for nontronites is related to tetrahedral $Fe^{3+}$ content. Nevertheless, as has been demonstrated for synthetic nontronites (Petit et al., 2015), this band is very weak, if present at all, and the NIR combination region obviously showed the existence of only one band corresponding to $\delta(Fe^{3+}{}_2OH)$ vibrations. Therefore, assignment of this 841–850 $cm^{-1}$ band in synthetic nontronite to a second $\delta(Fe^{3+}{}_2OH)$ bending mode (Goodman et al., 1976; Gates, 2005) is therefore highly unlikely. Three other small lattice bands were observed around 775, 710 and 414–430 $cm^{-1}$, but were not assigned due to insufficient information as to their origin.

As for many natural ferrian smectites, the $(\nu+\delta)OH$ combination bands were broad and poorly resolved (Fig. 8.18). The $(\nu+\delta)OH$ combination bands shifted gradually from the combination bands observed for natural nontronites (4370 $cm^{-1}$) (e.g. Madejová et al., 2000b; Bishop et al., 1999; Gates, 2005) to the combination bands observed for natural Mt-Bd samples (4550 $cm^{-1}$). Up to five bands were needed to fit all spectra. The band near 4375 $cm^{-1}$ is the major band of sample containing only $Fe^{3+}$ as octahedral cation and was then attributed to $(\nu+\delta)Fe^{3+}{}_2OH$ (Gates, 2005; Decarreau et al., 2008; Andrieux and Petit, 2010; Neumann et al., 2011). This band resulted evidently from the combination of $\nu(Fe^{3+}{}_2OH)$ (stretching: 3565 $cm^{-1}$) and $\delta(Fe^{3+}{}_2OH)$ (bending: 813 $cm^{-1}$) observed in the MIR, and occurred in all spectra except those of samples that had the lowest amounts of $Fe^{3+}$. For samples that were the highest in iron content, a small isolated band occurred at 4510–4529 $cm^{-1}$. As one

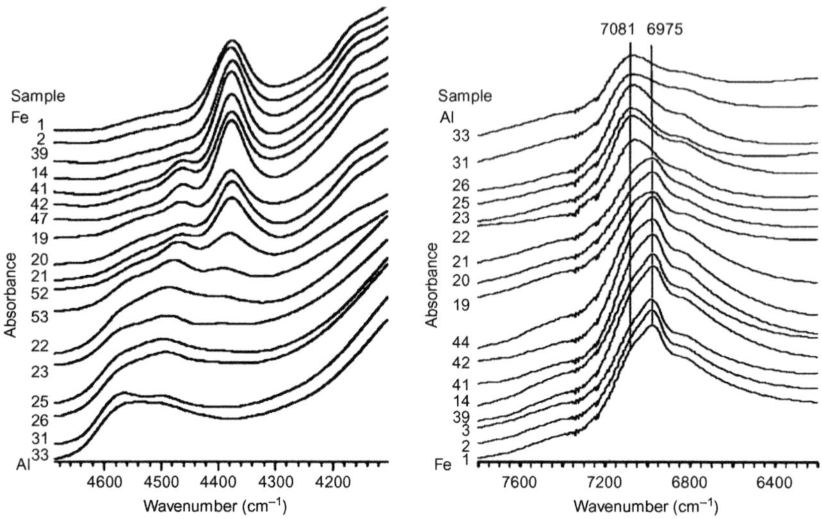


**FIG. 8.18** NIR combination bands (left) and overtone (right) of synthetic ferrian smectites and nontronites. Decreasing Fe/(Fe+Al) with sample number. *Reproduced with permission from Petit et al. (2015).*

sample did not contain any Al at all, this band cannot be attributed to $(\nu+\delta)$ $Al_2OH$ or $(\nu+\delta)AlFe^{3+}OH$ and was probably due to another type of combination. The next block of samples were satisfactorily fitted with three bands located around 4379, 4471 and at 4530 $cm^{-1}$ assigned to $(\nu+\delta)Fe^{3+}{}_2OH$, $(\nu+\delta)AlFe^{3+}OH$, and $(\nu+\delta)Al_2OH$ bands, respectively. The band at 4379 $cm^{-1}$ was due to $Fe^{3+}{}_2OH$. The band at 4530 $cm^{-1}$ was attributed to $(\nu+\delta)Al_2OH$ combination of $\nu(Al_2OH)$ (3620 $cm^{-1}$) and $\delta(Al_2OH)$ (912 $cm^{-1}$) bands observed in the MIR. Finally, the 4471 $cm^{-1}$ band was assigned to $(\nu+\delta)AlFe^{3+}OH$ ($\nu(AlFe^{3+}OH)$ (3598 $cm^{-1}$) and $\delta(AlFe^{3+}OH)$ (875 $cm^{-1}$)). Similar assignments were made by Gates (2005) for natural samples. The intensities of the $(\nu+\delta)AlFe^{3+}OH$ and $(\nu+\delta)Al_2OH$ bands increased, while the $(\nu+\delta)Fe^{3+}{}_2OH$ band intensity simultaneously decreased, with the increase of Al content of the synthetic smectites. A band at 4573 $cm^{-1}$ formed and reached its maximum intensity only for the most Al enriched samples. This band was therefore attributed to a second $(\nu+\delta)Al_2OH$ combination. By analogy, the 4419 $cm^{-1}$ band was tentatively attributed to a second $(\nu+\delta)AlFe^{3+}OH$ combination. While the two $(\nu+\delta)Al_2OH$ bands (4576 and 4536 $cm^{-1}$) had similar intensities, this was not the case for the two $(\nu+\delta)$ $AlFe^{3+}OH$ bands: the 4419 $cm^{-1}$ band always had a lower contribution than the 4471 $cm^{-1}$ band. Similar bands to those described for the synthetic smectites were observed previously for natural smectites (Gates, 2005), with the exception of the 4419 $cm^{-1}$ band. While Gates (2005) did not separate $(\nu+\delta)$ $AlFe^{3+}OH$ into two distinct bands, both a second $(\nu+\delta)Al_2OH$ and the $(\nu+\delta)$ $AlFe^{3+}OH$ bands assigned in this study matched the positions he observed for natural Al-rich smectites with sufficient octahedral Fe content. Both the low intensity of the second $(\nu+\delta)AlFe^{3+}OH$ as assigned by Petit et al. (2015), and a possible overlap with another band such as $(\nu+\delta)$AlMgOH, could account for the fact that this second $(\nu+\delta)AlFe^{3+}OH$ has never been distinctly observed previously in natural Fe–Al smectites. NIR $2\nu$(OH) first overtone bands were broad, with notably a large band centred around 6850 $cm^{-1}$ due to water partially overlapping (Fig. 8.18). Maxima were around 6975 $cm^{-1}$ for iron-rich smectites, corresponding mainly to the first overtone of the $Fe^{3+}{}_2OH$ vibrations, or around 7081 $cm^{-1}$ for $Al^{3+}$-rich smectites, corresponding mainly to the first overtone of the $Al_2OH$ vibrations. Following Petit et al. (2004 a,b)), the relation between the $\nu$(OH) wavenumber in the MIR and the $2\nu$OH wavenumbers in the NIR is:

$$W_{\nu(OH)} = \left(0.5 \times W_{2\nu(OH)}\right) + 85.6.$$

Consequently, the MIR $\nu\left(Fe^{3+}{}_2OH\right)$ vibrations would be expected around 3573 $cm^{-1}$ and those of $\nu(Al_2OH)$ around 3626 $cm^{-1}$, which agrees well with literature (e.g. Farmer, 1974b; Petit et al., 2015). Baron et al. (2016a) synthesised at 150°C a chemical series of Fe(III)-nontronites with variable tetrahedral $^{[4]}$Fe(III) content (x) $([Si_{4-x}Fe(III)_x]Fe(III)_2O_{10}(OH)_2Na_x)$, across a range of initial aqueous pH values between 11 and 14. The permanent layer charge, due to Fe(III)-for-Si(IV) tetrahedral substitutions only, ranged from 0.43 to as high as 1.54 per half-unit cell. The OH stretching band was always at 3650 $cm^{-1}$, whatever the $^{[4]}$Fe(III) content. The level of Fe for Si tetrahedral substitution was quantified using the position of the main Si–O band: $\nu(SiO)\ (cm^{-1}) = -54.5 \times {}^{[4]}Fe(III) + 1037$.

### 8.4.2.4 Montmorillonite

The difficulties in synthesising pure Mt are due to the presence of substitutions in both octahedral (e.g. ferrous and ferric Fe replacing $Mg^{2+}$ and $Al^{3+}$, respectively) and tetrahedral sheets, where some $Al^{3+}$ can replace $Si^{4+}$. Mt synthesis needed four conditions: i) basic to neutral pH conditions; ii) a low silica concentration (under-saturated with respect to amorphous silica); iii) a more or less stoichiometric composition of the initial precipitate; and iv) at least 6% $Mg^{2+}$ content (Kloprogge and Frost (1999d)). These conditions agree quite well with the Mt-forming conditions observed in nature. For example, volcanic material (ash, glass) can react with seawater, or with fresh water under basic conditions, to form smectite minerals (Tomita et al., 1993). The IR spectrum showed a band at about 3600 $cm^{-1}$ due to OH stretching vibration that was not well defined. The broad bands at 3400 $cm^{-1}$ and

1640 cm$^{-1}$ were attributed to interlayer water and adsorbed water. A broad band around 1000 cm$^{-1}$ was visible in their spectrum together with a number of other bands in the low wavenumber region that were not assigned (around 725–750, 650 and 475 cm$^{-1}$). Similar spectra were also reported for Mt (though always accompanied by analcime) prepared from sericite and pyrophyllite in the presence of NaOH and $MgCO_3$ with a broad band at about 3300–3500 cm$^{-1}$, ascribed to the existence of interlayer water, while a band due to OH stretching vibration was observed around 3625 cm$^{-1}$. An absorption at 1640 cm$^{-1}$ due to adsorbed water was visible (Tomita, 1970).

A more recent work was performed to synthesise Mt at different pressures and with various layer charges to obtain a well-characterised reference smectite series (Lantenois et al., 2008a). Pressure was found to be an important factor in this study. At 320°C and 120 MPa for 30 days, only low-charged Mt (Mg content $x=0.33$) were synthesised, with impurities and residual amorphous phases of less than 3%. On the IR spectra, a vibration band was observed for all samples between 3648 and 3660 cm$^{-1}$ in the OH bond stretching range. This band was not present in the starting gel phase spectrum and corresponds to the smectite structural OH vibration. In the reacted products, some characteristic smectite bands appeared in the OH bending range between 800 and 950 cm$^{-1}$. Three bands at 805, 885 and 918 cm$^{-1}$ were identified. A band at 850 cm$^{-1}$ was observed in some samples and at 830 cm$^{-1}$ in other samples. Some differences between samples spectra were observed in the OH bending region. The differentiation between Mt, Bd and Sap was difficult in the OH stretching region of the IR spectra. However, some differences could be observed in the (cation, cation)-OH bending region. In dioctahedral smectites, two bands are classically associated with $\delta(Al_2OH)$ and $\delta$(AlMgOH) bending vibrations (915 and 845 cm$^{-1}$, respectively, Farmer, 1974b; Madejová et al., 2000a; Gates 2004). In the study by Lantenois et al. (2008a) two bands, $\delta$(AlMgOH) at 850 cm$^{-1}$ and $\delta(Al_2OH)$ at 918 cm$^{-1}$, were identified, but other bands observed in the OH bending region were not easily assigned. The di-trioctahedral sample can undiscernibly correspond to two segregated smectite phases or to a mixture between trioctahedral and dioctahedral layers in the same smectite phase. After lithium exchange and heating (Greene-Kelly test), a second band appeared at 3668 cm$^{-1}$ and was assigned to a $\nu(AlMgLi^+OH)$ mode (Calvet and Prost, 1971).

## 8.5 VERMICULITE

IR absorption spectra of synthetic dioctahedral vermiculite (Kitagawa and Watanabe, 1970) showed the bands at 3450 cm$^{-1}$ and at 1650 cm$^{-1}$, enhanced by treatment of muscovite with molten $LiN0_3$ and boiling $AlCl_3$, were correlated with the enhancement of an endothermic peak near 120°C. The bands at

760, 540 and 490 $cm^{-1}$, originating, respectively, from $\delta$(Si–O–Al), $\delta$(Si–O–$Al^{VI}$) and $\delta$(SiO) modes, were not changed by the treatment, but those at 840 ($\delta$(AlMgOH)) and 810 $cm^{-1}$ ($\delta$(AlO)) almost disappeared. A much more extensive IR study on vermiculite formed from muscovite was provided by Sayin and von Reichenbach (1979). The 5–20 μm fraction of the sample was vermiculitised by repetitive treatments of 200 mg portions by 1 litre of 0–2 N $BaCl_2$ solution in an autoclave at 120°C and the products were treated with solutions of chloride salts of $Li^+$, $Na^+$, $K^+$, $Rb^+$, $Cs^+$, $Be^{2+}$, $Mg^{2+}$, $Ca^{2+}$, $Sr^{2+}$ and $Ba^{2+}$. Virtually no shift is observed in the OH stretching band at 3638 $cm^{-1}$, although satellites at high (near 3660 $cm^{-1}$) and low wavenumber (near 3630 $cm^{-1}$) were observed. The intensity of the OH stretching band was strongest with $Na^+$-saturation as the $Na^+$ ion dehydrated readily, leading to the formation of a well-ordered paragonite-like structure. On heating to 400°C, the intensity of the $\nu(Al_2OH)$ decreased for the $Ca^{2+}$- and $Mg^{2+}$-saturated samples, not for the samples treated with $Na^+$, $Rb^+$, $Cs^+$, $Sr^{2+}$ and $Ba^{2+}$. Thus, water molecules contributed to the OH stretch under hydrated conditions for the $Mg^{2+}$ and $Ca^{2+}$ forms. The region on the low wavenumber side of the OH stretching vibration showed variations depending on the type of the cation as well as on the amount of adsorbed water (Fig. 8.19). The intensity of this band on $Li^+$-saturation was very strong and it moved from 3450 $cm^{-1}$ to higher wavenumbers with hydration and split into two maxima at 3530 and 3485 $cm^{-1}$. On heating to 300°C these bands lost some intensity, but remained in the same position. The strong band following $Li^+$-saturation appeared to be connected with the presence of a stable monohydrate phase as found by XRD.

In the lattice vibrational region the spectrum of the parent mica was essentially returned with $K^+$ re-saturation, but important changes were observed with $Rb^+$- and $Cs^+$-saturation (Fig. 8.20). The most consistent changes were the high wavenumber shift of the band at 1065 $cm^{-1}$ and the low wavenumber shift of the OH deformation band at 930 $cm^{-1}$. These were a direct consequence of the fact that $Cs^+$, and to a lesser extent $Rb^+$, invoked a change in tetrahedral rotation angle. The $Cs^+$ ion brought about shifts even in the $\delta$(SiO) vibrations below 600 $cm^{-1}$. The $Mg^{2+}$ ion caused variable effects on the lattice vibrations depending upon the amount of water. When the sample was heated to 110°C and pressed in a KBr disk without further heating, it still contained a large amount of water. The excess water changed the spectral contours. On heating the sample to 220°C in the KBr disk, the spectral features of the $Mg^{2+}$-saturated sample became like those of the parent mica. X-ray data indicated that at this temperature the sample dehydrated completely, that is, $Mg^{2+}$ ions were positioned at the mid-plane of the interlayer space. On preheating the sample to 220°C the tetrahedra-related bands at 900–700 $cm^{-1}$ underwent an intensity decrease, particularly the band at 826 $cm^{-1}$. Under this dehydration condition, part of the hydration water was removed and the $Mg^{2+}$ ions moved to the ditrigonal holes to perturb the lattice

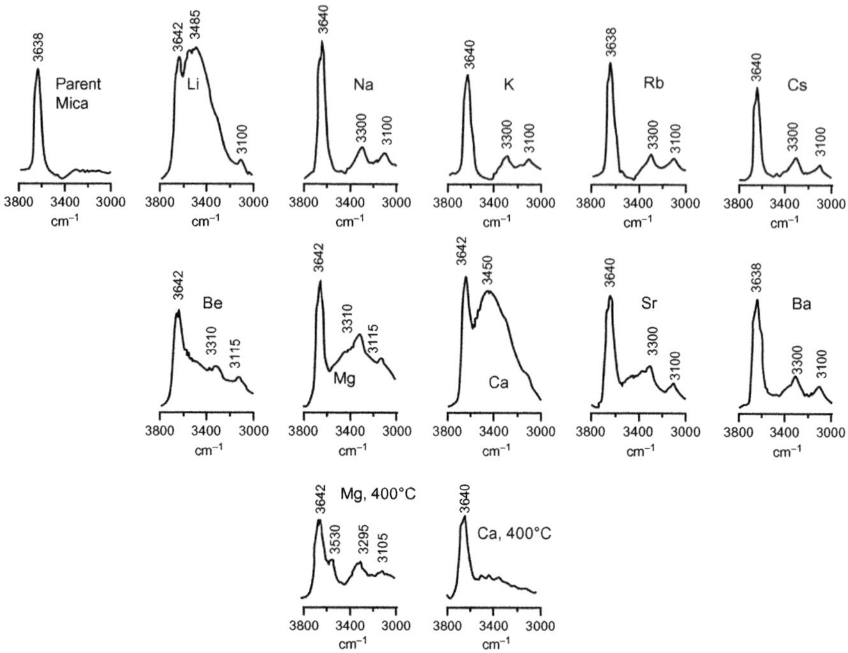


**FIG. 8.19** OH stretching region of Spittal muscovite and its vermiculitized product saturated with alkali and alkali earth cations. For Mg and Ca forms, the spectra after heating to 400°C are also shown. *Reproduced from Sayin and von Reichenbach (1979) with the permission of the Mineralogical Society of Great Britain & Ireland.*

vibrations. On pre-heating to 400°C the band at 342 $cm^{-1}$ almost disappeared and the one at 406 $cm^{-1}$ moved up to 427 $cm^{-1}$. Furthermore, all the main bands underwent a high wavenumber shift except for $\nu$(SiO) band at 1030 and the $\delta$(SiO) band at 530 $cm^{-1}$, which slightly decreased in wavenumber. The changes of the vibrations below 450 $cm^{-1}$ were attributed to the migration of small $Mg^{2+}$ ions into the vacant octahedral sites. The librational motions at 342 and 406 $cm^{-1}$ involved octahedral cations and were therefore affected by this migration. The $Ca^{2+}$-form exhibited some similarities to the $Mg^{2+}$-form. Upon dehydration the lattice vibrations between 700 and 950 $cm^{-1}$ were affected, but no important change occurred in low lying $\delta$(Si–O–Si) or $\delta$(Si–O–Al) modes. The $Ca^{2+}$ ion was too large to move into the octahedral sites and it caused only limited rotation of the surface oxygen triads. Saturation of vermiculite with $Sr^{2+}$ and $Ba^{2+}$ resulted in similar spectra. The $Sr^{2+}$ ion is smaller than the ditrigonal hole so that it can approach the silicon-oxygen network. Since the $Ba^{2+}$ ion is as large as the $K^{+}$ ion, its perturbing effect was unexpected at first sight. However, the higher valency of this ion caused a stronger electrostatic attraction towards structural OH. Both $Sr^{2+}$ and $Ba^{2+}$ ions perturbed the OH deformation band at 930 $cm^{-1}$ as well as the band at 830 $cm^{-1}$.

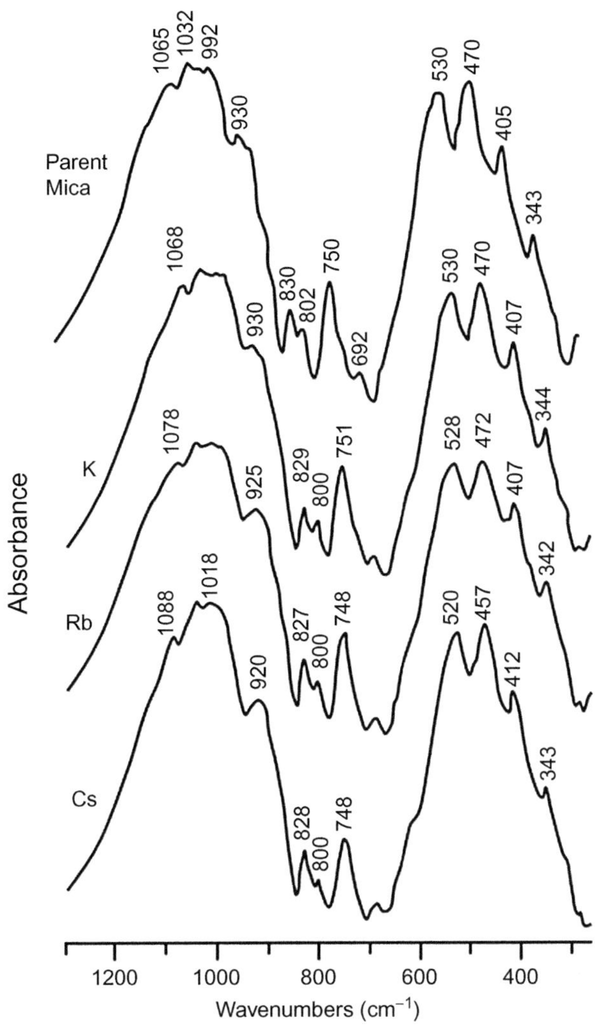


**FIG. 8.20** Lattice vibrations of Spittal muscovite and its vermiculitized product saturated with K, Rb and Cs. *Reproduced from Sayin and von Reichenbach (1979) with the permission of the Mineralogical Society of Great Britain & Ireland.*

The $Na^+$ ion was exceptional among the exchange cations studied because it dehydrated readily and completely and permitted maximum rotation of the surface oxygen triads. Its effect on the lattice vibrations was expected to occur in the reverse direction to that obtained with $Rb^+$ and $Cs^+$ ions. In fact, the band at 1065 cm$^{-1}$ disappeared and the main $\nu$(SiO) mode at 1032 cm$^{-1}$ moved to 1005 cm$^{-1}$. In addition, the $\delta$(SiO) modes at 530 and 470 cm$^{-1}$ and the $\delta(Al_2OH)$ mode at 930 cm$^{-1}$ shifted towards higher wavenumbers.

The band at 830 $cm^{-1}$ lost its intensity. The sharpness of the bands indicated the formation of a well-ordered paragonite-like structure.

Important changes were observed with $Li^+$- and $Be^{2+}$-saturated samples, particularly below 450 $cm^{-1}$. With $Li^+$-muscovite the band at 343 $cm^{-1}$ became a shoulder and a new band developed at 370 $cm^{-1}$. The band at 405 $cm^{-1}$ disappeared completely with $Be^{2+}$-saturation. The changes below 450 $cm^{-1}$ were like those obtained with the $Mg^{2+}$-saturated sample at 400°C and suggest that the small $Li^+$ and $Be^{2+}$ ions migrated into the empty octahedral sites. The migration seemed to occur at lower temperatures with $Li^+$ and $Be^{2+}$ than with $Mg^{2+}$. Perturbations of the lattice vibrations above 450 $cm^{-1}$ were the combined effect of the cations which remained in the interlayer and those which moved into the octahedral sites.

## 8.6 CHLORITE

Gupta and Malik (1969b) showed that a magnesium hydroxide-montmorillonite complex prepared at an $OH^-/Mg^{2+}$ molar ratio of 2 (16 meq $Mg^{2+}$/g clay) and immediately cleaned of any free $Mg(OH)_2$ behaved like magnesium-chlorite. Magnesium hydroxide completely precipitated within the interlayer space of Mt resulting in a 14.7 Å basal spacing. The Mt spectrum showed bands at 3703, 3450, 1655, and 1042 $cm^{-1}$ (with some minor shoulders at 982 and 945 $cm^{-1}$). The $\delta(Al_2OH)$ band, between 900–950 $cm^{-1}$, though not completely missing, was very broad. Additional absorption bands appeared in the spectrum of acid-washed $Mg(OH)_2$-montmorillonite complex at about 3780, 3600, 2750, 2445, and 910 $cm^{-1}$, indicating transformation of the original montmorillonite structure. The band at 3703 $cm^{-1}$ of Mt shifted to a much sharper band at 3780 $cm^{-1}$ on chloritisation. Ageing of the specimen did not change the spectra in any way. The spectra for the acid washed material (both fresh and 6 months aged) after heating (200°C, 2 h, or after another acid leaching, 0.5 N HCl) did not change, pointing towards the absence of any metal hydroxide external to Mt layers.

In subsequent work, Gupta and Malik (1969a) prepared $Ni^{2+}$-chlorite by the co-precipitation of nickel hydrous oxide and Mt at an $OH^-/Ni^{2+}$ ratio of 2.0 with a resulting basal spacing of 14.6 Å. Extra absorption bands appeared in the $Ni^{2+}$ hydrousoxide-clay complex (prewashed with 0.05 N HCl, System II) spectrum at about 2750, 2445 and 910 $cm^{-1}$ (with a shoulder band at 1310 $cm^{-1}$) pointing towards the development of new atomic groupings and possibly the transformation of the original montmorillonite structure. The spectrum of $Ni^{2+}$ hydrousoxide was clear of bands other than the $\delta(OH)$ vibrations and remained unchanged on ageing these samples for 6 months. Spectra obtained after heating or further acid leaching did not change, confirming the absence of any metal hydrous oxide external to Mt layers. In neither paper an attempt was made to interpret the spectra and give a detailed band assignment, in particular the three bands around 2750, 2445 and 1310 $cm^{-1}$, which are not normally observed in any of the clay minerals.

Shirozu and Momoi (1972) showed there were close relations between the IR patterns and chemical compositions. In the lattice vibration region, the main $\nu$(SiO) bands in the region between 950 and 1100 $cm^{-1}$ and the strong band between 660 and 680 $cm^{-1}$ ($\delta$(Si–O–Si) and $\delta$(Si–O–Al)) showed the most significant change with chemical composition. In the OH stretching region, with increasing aluminium content the two broad $\nu(M_3OH)$ bands between 3400 and 3650 $cm^{-1}$, which were attributed to structural OH of brucite, shifted in the direction of lower wavenumbers, and simultaneously the relative intensity of the band with lower wavenumber increased. The wavenumbers of the synthetic chlorites bands, however, were higher than those of the corresponding natural $Mg^{2+}$-chlorites. It was thought that it could indicate longer H-bond distances of the synthetic chlorites between the basal tetrahedral oxygens and the brucite OH. It is noticeable that the intensity ratios of the two bands are almost identical in both the synthetic and natural $Mg^{2+}$ -chlorites: the relative intensities of the OH bands can be correlated with the aluminium contents in the tetrahedral layers.

Kleppe et al. (2003) described the Raman spectrum of synthetic endmember clinochlore $(Mg_5Al)(Si_3Al)O_{10}(OH)_8$. The observed ambient Raman wavenumbers of synthetic endmember clinochlore agreed well with the previous ambient Raman study of natural Mg-chlorite by Prieto et al. (1991) (Table 8.7). The Raman spectrum was characterised by several strong modes in the range from 100 to 1100 $cm^{-1}$ and by OH modes between 3470 and 3700 $cm^{-1}$. The OH-stretching bands between 3470 and 3650 $cm^{-1}$ were attributed to H-bonded interlayer OH (Prieto et al., 1991; Shirozu, 1980; Welch et al., 1995). The 3477 $cm^{-1}$ mode results from $\nu(Mg_2AlOH)$ that are H-bonded to the interlayer surface oxygen (OSi,Al) ($\nu(Mg_2AlOH–OSiAl)$) and the 3605 and 3647 $cm^{-1}$ bands are due similar H-bonding with O(SiSi), that is, $\nu(Mg_2AlOH–OSiSi)$. The highest-wavenumber and narrow OH band at 3679 $cm^{-1}$ was attributed to the non H-bonded hydroxyls of the talc-like 2:1 layer. Its position was identical to the position of the talc OH mode. The IR spectrum of the same sample revealed only three OH bands (Welch et al., 1995) but at lower resolution than the present Raman experiment. Deconvolution of the 3630 $cm^{-1}$ feature into two overlapping bands suggests at least two Raman-active vibrations contributed to this band rather than the one expected for space group *C2/m*. As the 3477 and 3630 $cm^{-1}$ bands are both due to structurally similar H-bonded OH groups. Corresponding additional components in the 3477 $cm^{-1}$ band were expected, although these were not resolved in the spectrum. To explore this possibility further the ambient Raman spectrum of a synthetic aluminous clinochlore, $(Mg_{4.5}Al_{1.5})(Si_{2.5}Al_{1.5})O_{10}(OH)_8$, also studied by Welch et al. (1995), and related compositionally to endmember clinochlore by $0.5[^6Al\ ^4Al^6Mg^4_{-1}Si_{-1}]$, were collected. The composition of the brucite like layer is not affected by the

**TABLE 8.7 Raman Wavenumbers of the Most Intense Lattice Modes and OH-stretching Modes of Clinochlore; for Comparison the Raman Wavenumbers of Aluminous Clinochlore and of a Previous Study by Prieto et al. (1991) are Given (Kleppe et al., 2003)**

| Prieto et al. (1991) | | | |
|---|---|---|---|
| *$(Mg_5Al)(Si_3Al)O_{10}(OH)_8$* | *$(Mg_{4.5}Al_{1.5})(Si_{2.5}Al_{1.5})O_{10}(OH)_8$* | *Natural $Mg^{2+}$-rich chlorite* | |
| ***$\nu/(cm^{-1})$*** | ***$\nu/(cm^{-1})$*** | ***$\nu/(cm^{-1})$*** | ***Assignment*** |
| 104 | 105 | 105 | $\nu_5(F_{2g})$, $MO_6$ |
| 198 | | 197 | |
| 358 | 360 | 356 | $MO_4(OH)_2$, 2:1 |
| 548 | 552 | 543 | $E_1^2$ $\nu(T_2O_5)$ |
| 679 | 676 | 682 | $A_1^2$, $E_2^1$, $\nu(T_2O_5)$ |
| 3477 | 3455 | ~3480 [⊥(*001*)] | $\nu$(OH), $\nu$(Si,Al) O···OH |
| 3605 | 3541 | | |
| 3647 | 3587 | 3626 [‖ (*001*)] | $\nu$(OH), $\nu$(Si,Si) O···OH |
| | | 3642 [⊥(*001*)] | |
| | 3641 | | $\nu$(OH), 2:1 |
| 3679 | 3675 | 3681 | |

$Mg^{2+}$-Tschermak's exchange, and so the same octahedral configuration surrounds the H-bonded protons in this chlorite as in endmember clinochlore. The only significant spectral difference from endmember clinochlore was the presence of a $\nu(Mg_2AlOH)$ contribution from the 2:1 layer at 3641 $cm^{-1}$ and a corresponding reduced intensity for the $\nu(Mg_3O–H)$ band at 3675 $cm^{-1}$. The latter band was more clearly separated from the highest-wavenumber $\nu(Mg_2AlOH···OSiSi)$ band in aluminous clinochlore than it was in $Mg^{2+}$-rich clinochlore. The association of the 3641 $cm^{-1}$ mode with $\nu(Mg_2AlOH)$ was supported by its pressure dependence, which was identical to that of the 3675 $cm^{-1}$ 2:1 layer mode, and different from the neighbouring 3541 and 3587 $cm^{-1}$ bands that are associated with $\nu(Mg_2AlOH···OSiSi)$. At 2.5 GPa both the 3641 and 3675 $cm^{-1}$ mode were only 2 $cm^{-1}$ higher than at 1 bar, effectively constant within error. On the other hand, the 3541 and 3587 $cm^{-1}$ bands showed clear positive pressure shifts, their band positions being 13 $cm^{-1}$ higher at 2.5 GPa than at ambient conditions. The increased intensity of the $\nu(Mg_2AlOH\cdots OSiAl)$ Raman band at 3455 $cm^{-1}$ in aluminous clinochlore agrees with the increased $^4Al$ content. Additional components in this band could not be resolved, but could be present as is indicated by its large full-width at half-maximum [FWHM was 130 $cm^{-1}$ compared with 54 $cm^{-1}$ of the $\nu(Mg_2AlOH\cdots OSiSi)$ bands]. The two-band nature of the $\nu(Mg_2AlOH\cdots OSiSi)$ hydroxyl vibration in Mg endmember clinochlore was confirmed by a similar doublet structure in the spectra of the aluminous phase. It was suggested that the fine structure of the 3630 $cm^{-1}$ band reflected a splitting of the corresponding H site in the structure.

## 8.7 CONCLUDING REMARKS

Clay mineral synthesis deservedly is an important field of study as it assists us in understanding the various physical and chemical conditions necessary for these minerals species to form in natural environments, including those on other planets or extra-terrestrial bodies other than Earth. Studies of synthesis also have provided insight into how clay minerals nucleate and grow, and these have led to increased complexity in chemical composition of the synthetic clay minerals in the past decade alone.

The combination of vibrational spectroscopic techniques with the controlled synthesis of various species of clay minerals, largely only attainable within the past two–three decades, has provided a solid basis for clarifying the assignments of many IR and Raman active bands that previously were either unassigned or misassigned. Lack of compositional series largely was responsible for past misassignments (or inability of making assignments) of various vibrational modes. IR and Raman spectroscopy can be expected to play a continued important role in characterisation and understanding of conditions of clay mineral synthesis.

Chapter 9

# Infrared Studies of Clay Mineral-Water Interactions

C.T. Johnston

## 9.1 INTRODUCTION

The nanoscale dimensions of clay minerals (along at least one dimension) translate to a group of naturally occurring and synthetic nanomaterials, which have expansive surface areas that can exceed 800 $m^2/g$. These surfaces are further characterized by high reactivity due to charge defects and broken edges that impart reactive site densities of 1–2 sites/$nm^2$ (Schoonheydt and Johnston, 2013). In terms of abundance, clay minerals are among the most abundant nanomaterials on both Earth and Mars, with densities as high as 1022 particles/$m^3$ (Hochella, 2008; Hochella et al., 2008; Bridges, 2014). Water is comprised of the first and third most abundant elements in the Universe, is one of its most abundant molecular constituents, and is the most abundant molecule found on Earth to a depth of several thousand meters (Fraser et al., 2002; Mottl et al., 2007). Water is the defining molecule of the Blue Planet and its continual interaction with exposed minerals at the rock-water interface has produced a broad array of weathering products, including clay minerals (Hazen and Ferry, 2010). Thus the interaction of water with 'high area to volume' clay minerals is critically important to many different disciplines including all aspects of geology, soil and water science, geotechnical engineering, geomorphology, seismology, geophysics, petroleum engineering, atmospheric science, and astrobiology. As a class of functional industrial materials, the interaction of water with clay minerals is crucial to our understanding of catalysts, clay-polymer-nanocomposites, drilling fluids, ceramics, rheological additives, water treatment additives, cosmetics, nuclear waste storage, and drug delivery (Bergaya and Lagaly, 2013b).

Several examples of clay mineral-water complexes, where water is an integral component of the clay mineral structure, are shown in Fig. 9.1. Smectite (A) and vermiculite (B) are common 2:1 clay minerals that contain varying


Developments in Clay Science, Vol. 8. https://doi.org/10.1016/B978-0-08-100355-8.00009-6

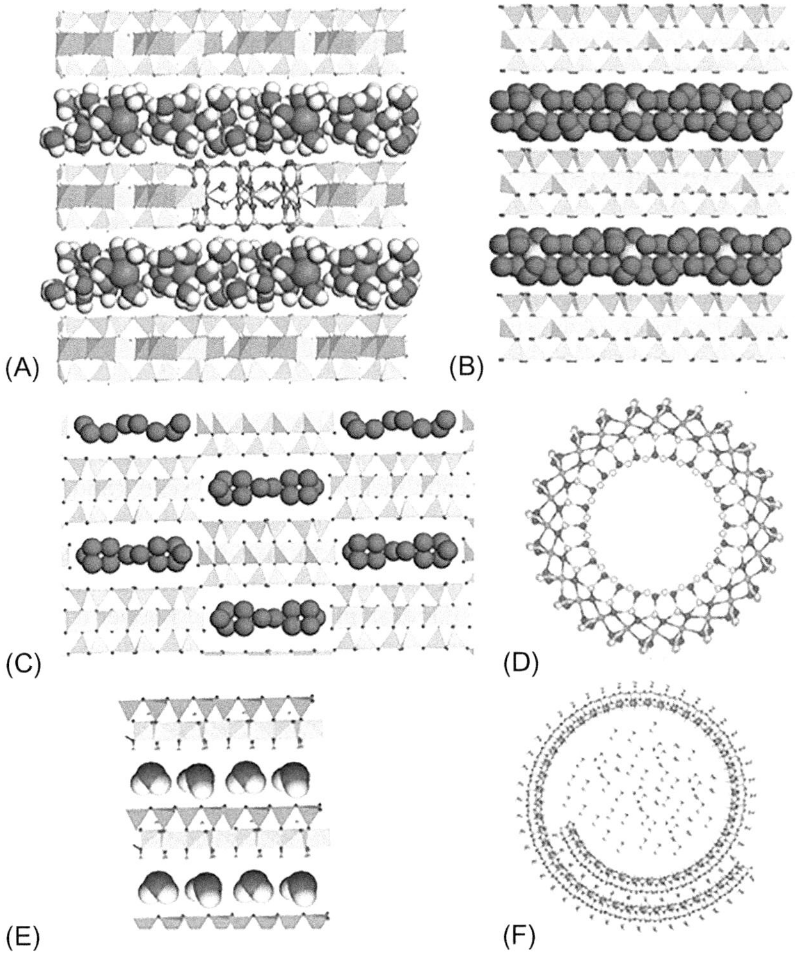


**FIG. 9.1** Examples of clay mineral structures that are closely associated with $H_2O$. Smectite (A) where $H_2O$ molecules are clustered around exchangeable cations in the interlamellar region (3 $H_2O$ layer structure is shown in A). Vermiculite (B) is a higher-layer charge analog of smectite and has two layers of $H_2O$ in the interlamellar space. Sepiolite (C), which contains 'zeolitic $H_2O$' in channels and tunnels of its structure. Imogolite (D) and tubular halloysite (E, F) have tubular structures where $H_2O$ resides internally in the lumen. In the case of halloysite, a single layer of $H_2O$ resides between the fundamental 1:1 layers (E).

amounts of water depending on their layer charge and type of exchangeable cation (Brigatti et al., 2013). For smectites (A), $H_2O$ molecules are loosely coordinated around interlayer cations through ion-dipole and hydrogen-bonding interaction. Additional $H_2O$ adsorption occurs through capillary condensation and the layers can then expand from 1 to >150 layers of $H_2O$. The higher layer charge of vermiculites compared to smectites restricts the interlayer spacing to two layers of $H_2O$. Sepiolite (C) and palygorskite (not shown) are low charge modulated structures that are characterized by hydrated channels and tunnels that extend through the structure (Bergaya and Lagaly, 2013a; Brigatti et al., 2013a). The $H_2O$ molecules in these structures are constrained mainly through hydrogen bonding (H-bond) interactions with the sepiolite/palygorskite surfaces. Imogolite (D) and tubular halloysite (E-F) have $H_2O$ residing within the lumen. In the case of halloysite, an additional 'type' of $H_2O$ is intercalated between the fundamental 1:1 layers (E). A common feature to all of the structures shown in Fig. 9.1 is that attempts to remove all of the $H_2O$ from these structures results in irreversible changes and partial collapse of the structure. This is particularly acute for halloysite, where even a slight lowering of water activity (e.g. to $\approx p/p_o$ of 0.5), results in the irreversible loss of the interlayer $H_2O$ molecules shown in Fig. 9.1E (Churchman et al., 1972; Joussein et al., 2005).

Clay mineral-water interactions have been the subject of intensive study using a diverse array of experimental methods for many years (Sposito and Prost, 1982; Newman, 1987; Johnston, 2010; Schoonheydt and Johnston, 2011, 2013). The structure of water bound to clay minerals is timescale dependent and each experimental method has an intrinsic timescale associated with it (Sposito and Prost, 1982). For example, Raman and Fourier transform infrared (FTIR) spectroscopy measure transitions that occur in the ($10^{-12}$ to $10^{-15}$ s) time domain whereas nuclear magnetic resonance (NMR) measurements are over longer time domains ($10^{-3}$ to $10^{-10}$ s). In addition to vibrational spectroscopy, many other types of experimental methods have been used to study clay mineral-water interactions including NMR (Bank et al., 1989: Weiss et al., 1990; Bowers et al., 2011), electron spin resonance (ESR) (Miwa et ,al., 2006) and dielectric relaxation spectroscopy (Salles et al., 2015), inelastic neutron scattering (Edge et al., 2014; Cygan et al., 2015; Michot et al., 2016), neutron and X-ray diffraction (Post et al., 2007), along with thermodynamic studies (Blanc et al., 2013). Furthermore, computational methods have greatly extended our understanding of clay mineral-water interactions (Greathouse and Cygan, 2013; Cygan et al., 2015; Greathouse et al., 2016). The nature of clay mineral-water interactions are complex. A more complete understanding can be achieved using multiple methods and timescales.

The use of IR spectroscopy to study water in minerals predates the discovery and application of X-ray diffraction (XRD) in 1912 (Coblentz, 1910, 1911). XRD and IR spectroscopy are surprisingly complementary methods. Structural details about the atomic structure and composition of minerals have been steadily revealed over the past eight decades (Hendricks, 1929; Pauling, 1930). Modern X-ray studies, including synchrotron based methods, have played a prominent role in revealing detailed information about the atomic structures of clay minerals and the local structures of adsorbed species with sub-angstrom resolution (Brown et al., 1999; 2005). One limitation of XRD, however, is that the proton has an exceptionally small cross section (meaning that hydrogen atoms do not diffract incident X-rays). This limits the ability of XRD methods to determine crystal structures that include the positions of hydrogen atoms. IR spectroscopy, on the other hand, is particularly sensitive to the vibrational mode associated with both structural OH groups and adsorbed $H_2O$ (Farmer, 1974a; McMillan, 1985; Johnston and Aochi, 1996a; Johnston, 2010; Petit and Madejová, 2013a). Because the O–H bond has a large IR-cross section, IR methods can be combined with XRD methods to provide enhanced structural insight (Bish and Johnston, 1993; Johnston et al., 1998, 2002a, 2008). Similarly, Raman methods have provided valuable insight about structural OH groups in clay minerals (Wiewiora et al., 1979; Johnston et al., 1985, 1998; Murad, 1997; Ritz et al., 2016).

## 9.2 MOLECULAR PROBES AND REPORTER GROUPS

Information gained from spectroscopic studies of clay minerals can generally be grouped into one or both of the following categories: (1) reporter group and/or (2) molecular probe studies (Johnston et al., 1993; Zecchina and Arean, 1996; Johnston, 2010). A common feature is that both the reporter group and the molecular probe have one or more diagnostic properties that are sensitive to changes in their local environment. The diagnostic properties of interest in this chapter are changes in band position, intensity, polarization/orientation, or isotopic shift of vibrational bands of the clay mineral itself, or of $H_2O$ molecules associated with the clay mineral surface. Reporter group studies of clay minerals include the extensive use of structural OH groups whose stretching and bending vibrations are sensitive to small changes in their local environment (Johnston, 2010). Correlation of vibrational band positions with accurate structural information for kaolin group minerals, for example, has shown that a change in interlayer O⋯(H–O) distance of 0.001 nm, or 0.01 Å, produces a shift in band position of 4 $cm^{-1}$. In this context, IR and Raman methods have afforded significant structural insight about the structural OH groups of kaolin group minerals, enabling a distinction between polytypes and the determination of structural disorder (Prost et al., 1987; Johnston et al., 2008; Kogure et al., 2010). In a broader context, the change in position of the $\nu$(OH) bands of structural OH groups of minerals can be evaluated on the basis of distance-frequency relationships (Beran and Libowitzky, 1999; Libowitzky, 1999; Johnston et al., 2008; Johnston, 2010). The study of structural OH groups with IR and Raman spectroscopies has been used to follow changes that occur during cooling, heating and dehydroxylation, intercalation with guest molecules, under pressure, perturbations induced by interlayer cations, oxidation-reduction processes, molecular orientation, and hydration-dehydration cycles.

Molecular probe studies rely on similar diagnostic properties (e.g. vibrational bands) but are associated with molecules external to the layer structure. In this chapter, $H_2O$ (and $D_2O$), are used as examples, but many different types of molecules have been used as molecular probes to study clay minerals using IR and Raman spectroscopies (Johnston, 2010). The vibrational bands of $H_2O$ include three 'fundamental vibrations' corresponding to two types of H–O–H stretching vibrations (both symmetric and asymmetric, designated $\nu$(HOH), which will collectively be referred to in this chapter as $\nu$) and the O-H bending vibration designated as $\delta$(HOH). To illustrate these bands, representative Raman and IR spectra of water and of a semi-hydrated smectite are shown in Fig. 9.2A–C, respectively, showing the overall spectral assignments of the fundamental $\nu$(HOH) and $\delta$(HOH) bands of adsorbed water. In addition, some of the vibrational features present in these spectra are overtone bands such as the 3250 cm$^{-1}$ band corresponding to the $H_2O$ 2$\delta$(HOH) overtone, and a combination band at 2127 cm$^{-1}$ of the $H_2O$ $\delta$(HOH) band and a librational (lattice deformation) band $\tau$ at 490 cm$^{-1}$ (Walrafen et al., 1988; Venyaminov and Prenderast, 1997; Johnston and Premachandra, 2001). In the near-IR (NIR) region (not shown), additional combination and overtone bands are present. A band in the 5000–5300 cm$^{-1}$ region corresponds to the combination band of the $\nu$ and $\delta$ modes and an overtone band of 2$\nu$ occurs near 7000 cm$^{-1}$.

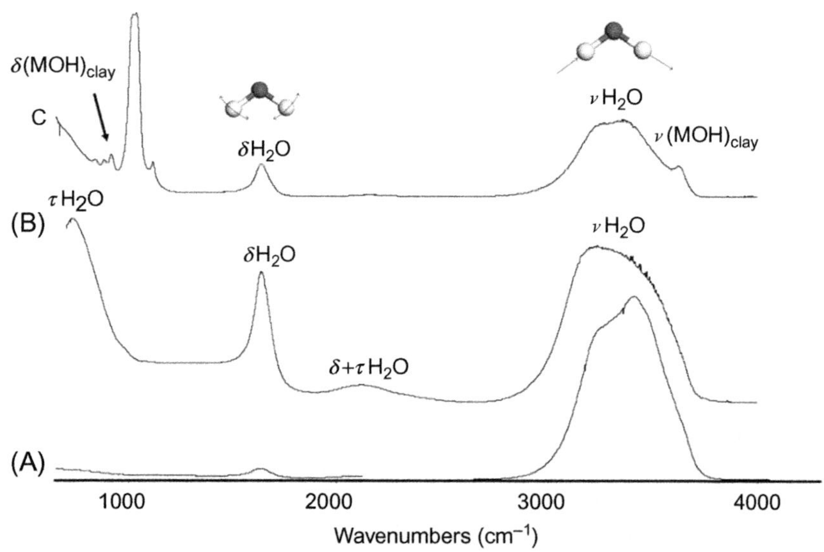


**FIG. 9.2** Raman and ATR-FTIR spectra of water in the 700–4000 cm$^{-1}$ region showing the regions where the three fundamental vibrational mode of $H_2O$ absorb (IR) and scatter (Raman) incident radiation. These include the smectite structural OH ($\nu$(OH) and adsorbed $H_2O$ stretching ($\nu$(HOH)) regions from 3000 to 3600 cm$^{-1}$ and $\delta$(HOH) bending of $H_2O$ at 1640 cm$^{-1}$. In addition, the band at 2127 cm$^{-1}$ represent the combination of the $\delta$(HOH) and librational bands and lower frequency librational and torsional modes of $H_2O$ occur at <800 cm$^{-1}$. For comparison, the ATR-FTIR spectrum of a hydrated Na-smectite is shown that includes spectral contributions of both sorbed $H_2O$ and from the clay mineral itself.

The vibrational modes associated with $H_2O$, $D_2O$ and structural OH groups are summarized in Fig. 9.3. The bottom rectangular regions shaded in black correspond to the vibrational bands associated with structural OH (both ν(MOH) and δ(MOH)) groups (Farmer, 1974b; Pálková et al., 2015; Bukas et al., 2013). The boxes shaded in darker gray in the middle of the figure correspond to vibrational modes of $H_2O$ and the top boxes shaded in light gray correspond to $D_2O$ (and HOD). The generic cutoff for the mid-IR (MIR) bands is 4000 $cm^{-1}$ or 2.5 μm with vibrations <4000 $cm^{-1}$ considered MIR bands and those >4000 $cm^{-1}$ are near-IR (NIR) modes. Fundamental vibrations generally occur <4000 $cm^{-1}$ and combination bands are >4000 $cm^{-1}$, however, some combination bands are present in the MIR region, which include the 3250 $cm^{-1}$ (2$\delta$) and the 2127 $cm^{-1}$ ($\delta+\tau$) bands. One of the experimental challenges in vibrational studies of clay mineral-water interactions can be illustrated in Fig. 9.3 where the position of the $\nu$(MOH) bands overlap with the $\nu$(HOH) bands represented by the dashed-box. In general, the vibrational bands in the MIR region are more diagnostic and 'easier' to interpret because they are not resulting from combination bands and do not suffer from anharmonicity effects. On the other hand, NIR methods have some significant advantages in that water has lower absorptivity

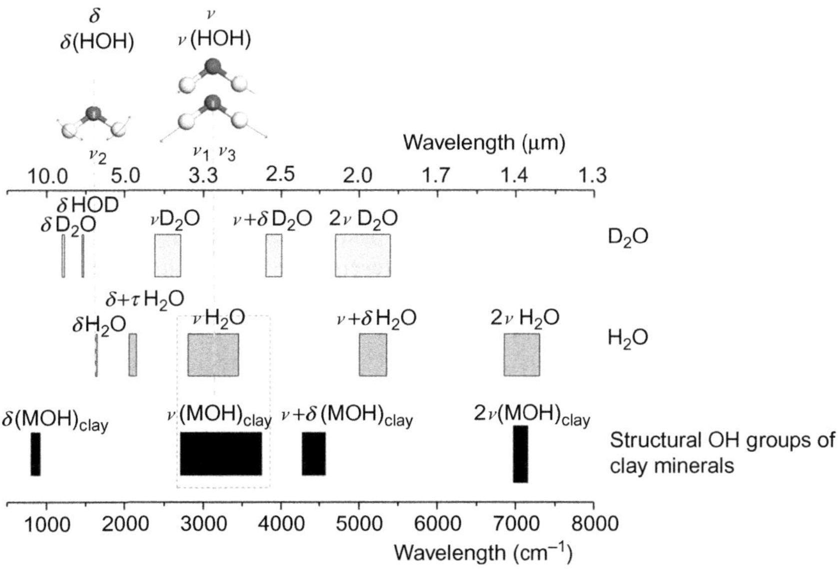


**FIG. 9.3** Spectral regions and assignment of $H_2O$ bands (middle row) that occur in the mid- (400–4000 $cm^{-1}$; 25–2.5 μm) and near-infrared regions (>4000 $cm^{-1}$; <2.5 μm). $D_2O$ bands are represented by the top row and the vibrational bands associated with structural OH groups of clay minerals are shown by the lower row. The energy scale of wavenumber ($cm^{-1}$) is shown on the bottom and the corresponding wavelength scale in microns (μm) is shown by the upper scale.

in this region that allows a wider range of sampling options, including non-invasive sampling options.

## 9.3 WATER CONFINED IN CLAY MINERAL INTERLAYER SPACES

From the standpoint of IR and Raman spectroscopies, the most interesting types of $H_2O$ molecules associated with clay minerals are those found in the interlayer spaces of both 1:1 and 2:1 clay minerals (Fig. 9.1). For over 100 years, the IR spectra of clay mineral-water complexes has attracted interest (Coblentz, 1910, 1911; Buswell et al., 1937) where IR methods were used to detect and study 'structural $H_2O$' even before crystal structures were known. The length scales that define clay mineral-water interactions range from short-range H-bonding and ion-dipole interactions (<0.2 nm) to intermediate-scale interactions that include electrical double layer and capillary condensation interactions associated with clay mineral swelling (>10 nm). These short and intermediate range interactions (<10 nm) are manifest at much larger macroscopic scales and are central to many important large-scale processes including shrink-swell behavior in soils, soil liquefaction, geomorphology, landslides, and deep focus earthquakes. With regard to the application of vibrational methods to study clay mineral-water interactions, most studies focus on the 'short range' interactions that included H-bonding and ion-dipole interactions. The remainder of this chapter examines the distinct $H_2O$ environments that occur in clay minerals (Fig. 9.1).

### 9.3.1 Smectites and Vermiculites (Ion Dipole)

Smectites and vermiculites are groups of 2:1 clay minerals that possess intermediate levels of layer charge ($\chi$) ranging from 0.2 to 0.9 charges per formula unit (Fig. 9.1A and B). The $\chi$ value represents the net negative charge per layer resulting from isomorphous heterovalent substitution and is expressed as a positive number (Brigatti et al., 2013a). Examples of isomorphous heterovalent substitution commonly found in clay minerals include $Al^{3+}$ for $Si^{4+}$ substitution in the tetrahedral sheet or $Mg^{2+}$ for $Al^{3+}$ substitution in the octahedral sheet and are shown in Fig. 9.4 (Schoonheydt and Johnston, 2013). Substitution in the octahedral sheet results in a more delocalized charge (i.e. zone of influence) that is present on both the 'top' and 'bottom' of the structure (Fig. 9.4A). Tetrahedral substitution, in contrast, results in a more localized charge present only on one surface (Fig. 9.4B). In order to develop an improved understanding of how water is associated with smectites and vermiculites, consider the four siloxane sheet surfaces shown in Fig. 9.5. In the top left portion (Fig. 9.5A), a 'neutral' siloxane surface is shown, representative of the types of siloxane surfaces found on talc or pyrophyllite which have no appreciable layer charge (i.e. $\chi=0$). These neutral siloxane surfaces are

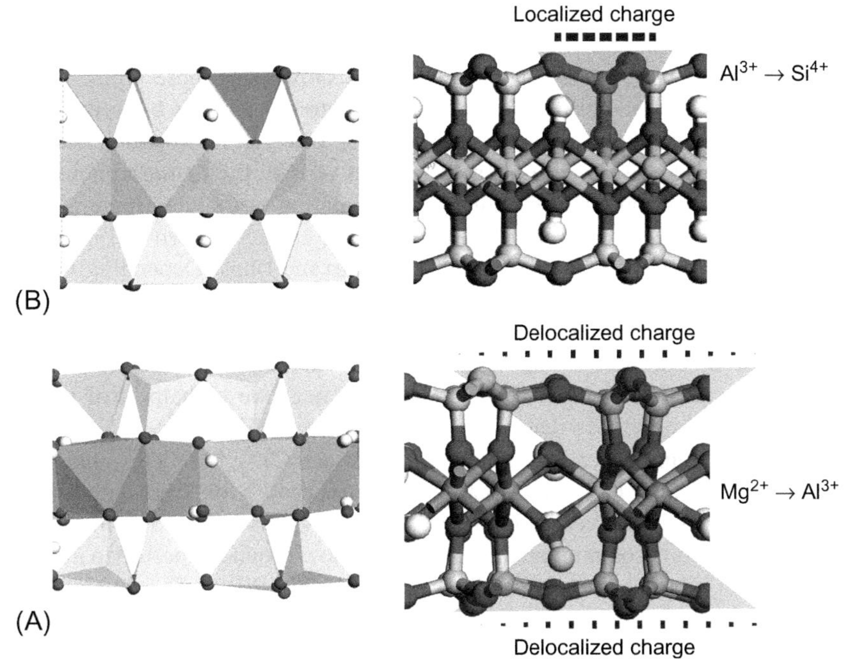


**FIG. 9.4** Common isomorphous heterovalent substitutions in 2:1 phyllosilicate. Substitution of $Mg^{2+}$ for $Al^{3+}$ is shown in the octahedral sheet in polyhedral and ball and stick representation (A). In this case, the charge is more delocalized and is manifest on both the 'top' and 'bottom' of clay structure. Substitution of $Al^{3+}$ for $Si^{4+}$ is shown in the upper figure resulting in a more localized charge that is mainly manifest on one surface (B).

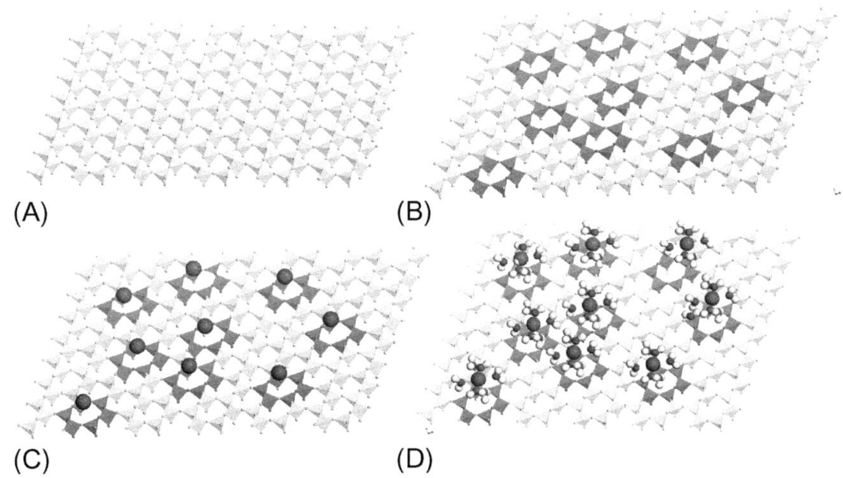


**FIG. 9.5** The top siloxane surface of a 4 nm × 4.2 nm 2:1 phyllosilicate particle is shown. Surface (A) shows a neutral siloxane surface without any isomorphous substitution. Surface (B) illustrates the type of charge pattern present in a smectite with octahedral isomorphous heterovalent substitution where the darker polyhedral structures mark the location of charge substitution. Surface (C) illustrates the presence of exchangeable cations that compensate the negative charge created from isomorphous heterovalent substitution. Surface (D) depicts the $H_2O$ molecules surrounding the exchangeable cations.

mainly hydrophobic, having little affinity for water (Jaynes and Boyd, 1991; Michot et al., 1994). Molecular dynamics has shown that these types of surfaces do not have the ability to overcome the attractive forces between water molecules, resulting in a large contact angle and overall hydrophobic surface (Solc et al., 2011). In the case of smectites and vermiculites, isomorphous heterovalent substitution occurred during the formation of the clay mineral, and this is schematically represented by the siloxane surface shown in Fig. 9.5B, where the charge deficit is shown by dark siloxane rings. Depending on the $\chi$ value, these sites are separated by distances of 0.6–1.2 nm. In nature, charge neutrality is maintained by the presence of exchangeable cations shown in Fig. 9.5C. In addition to the exchangeable cations, water molecules are strongly attracted to these cations due to their negative enthalpies of hydration, resulting in surfaces that are hydrophilic (Fig. 9.5D).

The most common cations found in nature are $Ca^{2+}$, $Mg^{2+}$, $K^+$, and $Na^+$; however, $Al^{3+}$ and $Fe^{3+}$ are often important because of their large ionic potentials. In addition, weakly hydrated cations, such as $Cs^+$ or $Sr^{2+}$ have high importance in soils contaminated by above or below ground exposure to nuclear waste/fallout (Zachara et al., 2007; Yasunari et al., 2011). The bubble plot shown in Fig. 9.6 illustrates the relationship between the size (ionic radii) of the common exchangeable ions, enthalpy of hydration (kJ/mol), and ionic potential. The enthalpy of hydration ranges from −264 kJ/mol for $Cs^+$ to −4680 kJ/mol for $Al^{3+}$ (House, 2013). It follows that the enthalpy of

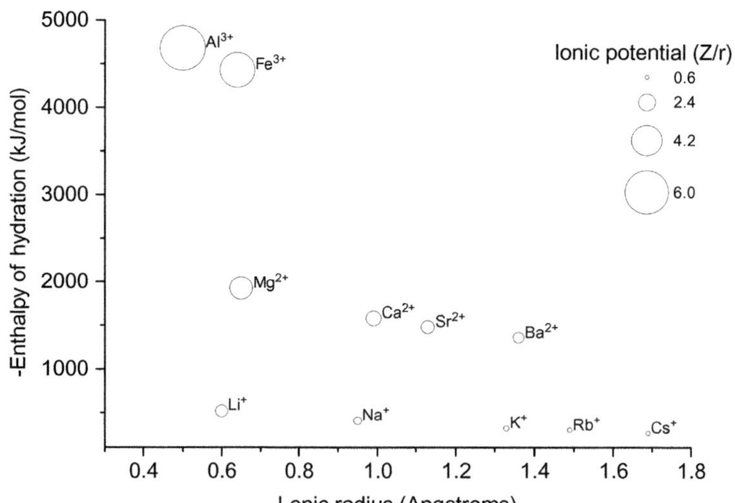


**FIG. 9.6** Ion properties of cations commonly associated with clay minerals. The enthalpy of hydration (kJ/mol) of the cation is plotted versus ionic radius. The size of bubble corresponds to the ionic potential of the cation defined as the valence of ion $Z$ divided by the ionic radius (House, 2013).

hydration of the exchangeable cation(s) present on the clay mineral surface plays a significant role in governing clay mineral-water interactions (Fig. 9.5D). For clay minerals exchanged with weakly hydrated cations (e.g. $Cs^+$ and $K^+$), collapse of the interlayer space is possible. For the more strongly hydrated monovalent cations (e.g. $Li^+$ and $Na^+$), these ions will be more resistant to dehydration and further accumulation of interlayer $H_2O$ molecules can occur due to capillary condensation (Prost et al., 1998). In the case of the divalent cations $Ca^{2+}$ and $Mg^{2+}$, these ions have much larger enthalpies of hydration (−1600 to −2000 kJ/mol) and will be surrounded by $H_2O$ molecules (Cancela et al., 1997). However, the clay mineral surface is also in competition' for the charge compensation provided by these exchangeable cations that can limit 'free expansion'. In the case of $Ca^{2+}$- and $Mg^{2+}$-exchanged vermiculites (i.e. high charge), expansion is limited to 2 $H_2O$ layers, compared to >6 nm, or potentially room for 18 $H_2O$ layers, for low charge $Na^+$-smectites layers. The trivalent ions, $Al^{3+}$ and $Fe^{3+}$, have large enthalpies of hydration (>−4000 kJ/mol) and ionic potential values (Z/r) which translates to their ability to promote hydrolysis reactions. Hydrolysis (i.e. splitting) of $H_2O$ molecules in the interlayer space results in high surface acidity, which can lead to the transformation of both inorganic and organic molecules in the clay mineral interlayer spaces (Mortland and Raman, 1968; Helsen, 1982; Dontsova et al., 2005). Clay minerals have been used as catalysts in organic chemical synthesis for many years, ever since they started being used to 'crack' petroleum products (Thomas et al., 1950; McCabe and Adams, 2013; Schoonheydt, 2016).

Only one surface is shown in Fig. 9.5, but this type of clay mineral-water interaction generally occur between opposing charged siloxane surfaces. A structure containing ≈3 layers of water molecules is shown in Fig. 9.7. From the standpoint of surface chemistry, the $H_2O$ molecules surrounding the exchangeable cations in Figs. 9.5 and 9.7 are attracted to the interlayer cations through ion-dipole interactions and the clay mineral surface itself functions as a rigid axial anion (Maes et al., 1980). The cations are attracted to the clay mineral surface through electrostatic attraction, which can limit their hydration behavior depending on the charge density and location of isomorphous heterovalent substitution (Fig. 9.4) of the clay mineral, and on the size, hydration enthalpy, and ionic potential of the cation (Fig. 9.6). Smectites and vermiculites with isomorphous heterovalent substitution in the tetrahedral sheet (Fig. 9.4B) exchanged with cations that have lower enthalpies of hydration (Table 9.1) will be less hydrated than their octahedral counterparts (Fig. 9.4A). In terms of the composition of the interlayer aqueous solution, the molar concentration of cations in the interlayer space can be quite large. The interlayer molar concentration of $Na^+$ for the structure shown in Fig. 9.6 is 4 mol/L (M) and dehydration of the clay mineral to a one-layer hydrate (i.e. *d*-value of 1.26 nm) can produce 'concentrations' in excess of 12 M.

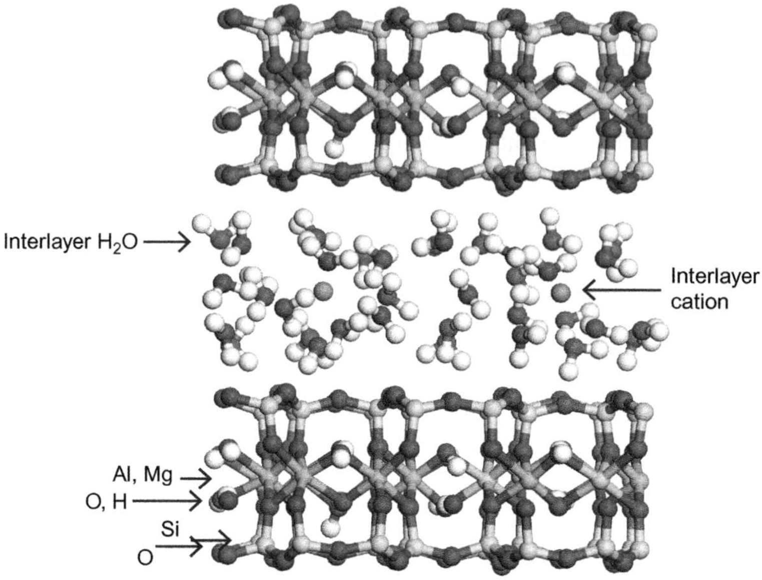


**FIG. 9.7** Side view of a Na-exchanged dioctahedral smectite with isomorphous heterovalent substitution in the octahedral sheet with three layers of $H_2O$ molecules in the interlamellar region along surround exchangeable cations. The molar concentration of $Na^+$ in the interlamellar region is 4 M.

#### *9.3.1.1 Studies of Smectites and Vermiculites in the $H_2O$ Bending Region*

Upon inspection of the spectral regions shown in Figs. 9.2 and 9.3, the $\delta$(HOH) band does not overlap with the absorption bands of the structural OH groups, or with other bands of the clay mineral itself (Xu et al., 2000; Tang et al., 2016). Thus the $\delta$(HOH) band can be used to probe interlayer $H_2O$ because it occurs in a region free of spectral overlap from vibrational bands of the clay mineral itself (Fig. 9.3). Researchers have investigated the change in band position and molar absorptivity of this band as a molecular probe to study the clay mineral-water interactions. Earlier work established that the spectral properties of $\delta$(HOH) were sensitive to changes in water content (Russell and Farmer, 1964; Poinsignon et al., 1978). More recent studies have provided quantitative insights using methods that measure $H_2O$ adsorption isotherms and FTIR spectra simultaneously (Johnston et al., 1992a; Tipton et al., 1993; Xu et al., 2000; Schuttlefield et al., 2007b). These studies have shown that the vibrational properties of interlayer $H_2O$ molecules change as a function of $H_2O$ content and that the $H_2O$ molecules are strongly polarized by their proximity to exchangeable cations. In a gravimetric-FTIR study of $H_2O$ adsorbed to $Na^+$- and $Li^+$-smectites (Xu et al., 2000), the position of the $\delta$(HOH) at $\sim$1635 $cm^{-1}$ is plotted as a function of $H_2O$ content (Fig. 9.8A). At high water content (>12 $H_2O$ molecules/$Na^+$ ion, $\sim$3 $H_2O$ layers), the position of the $\delta$(HOH) band is the same as bulk $H_2O$

**TABLE 9.1** Properties of Common Cations

| Ion | Atomic Number | Effective Ionic Radius (nm)[a] | Ionic Potential[b] | Single Ion Hydration Enthalpies (kJ/mol)[c] | Hydrolysis Constants (pKh)[d] |
|---|---|---|---|---|---|
| $Li^+$ | 3 | 0.076 | 0.7 | −515 | 13.8 |
| $Na^+$ | 11 | 0.102 | 0.5 | −405 | 14.5 |
| $K^+$ | 19 | 0.138 | 0.45 | −321 | |
| $Cs^+$ | 55 | 0.167 | 0.4 | −263 | |
| $Mg^{2+}$ | 12 | 0.072 | 2.8 | −1922 | 11.4 |
| $Ca^{2+}$ | 20 | 0.100 | 2.15 | −1592 | 12.7 |
| $Sr^{2+}$ | 38 | 0.118 | | −1443 | |
| $Ba^{2+}$ | 56 | 0.135 | | −1302 | |
| $Al^{3+}$ | 13 | 0.053 | 6.9 | −4660 | 5.1 |
| $Fe^{2+}$ | 26 | 0.078 | 2.5 | −1920 | |
| $Fe^{3+}$ | 26 | 0.055 | 6.3 | −4376 | 2.2 |
| $Co^{2+}$ | 27 | 0.065 | 2.6 | −2054 | 9.6 |
| $Cu^{2+}$ | 29 | 0.073 | 2.6 | −2100 | 7.5 |
| $Zn^{2+}$ | 30 | 0.074 | 2.6 | −2044 | 9.6 |
| $Cd^{2+}$ | 48 | 0.95 | 2.3 | −1806 | 11.7 |
| $Pb^{2+}$ | 82 | 0.119 | 2 | −1480 | 7.8 |

[a] *Shannon (1976).*
[b] *Huheey et al. (1997).*
[c] *Burgess (1978).*
[d] *Yatsimirksii and Vasil'ev (1960).*

(Venyaminov and Prenderast, 1997). Upon lowering the $H_2O$ content, the position of the $\delta$(HOH) band increases slightly for $H_2O$ contents of ~6 $H_2O$ molecules/cation followed by a sharp drop in position at lower $H_2O$ content. Similar results have been reported for other cations ($Ca^{2+}$, $Mg^{2+}$, $Cu^{2+}$) with both specimen clay minerals and clay minerals fractionated from soils (Johnston et al., 1992a; Dontsova et al., 2004). Fundamental studies of water using traditional IR and non-linear optical methods, such as sum-frequency generation (SFG) have shown that the position of the $\delta$(HOH) band of water is anti-correlated with the position of the $\nu$(HOH) band. Thus a shift in position of the $\delta$(HOH) band to lower energy (red-shift) corresponds to a

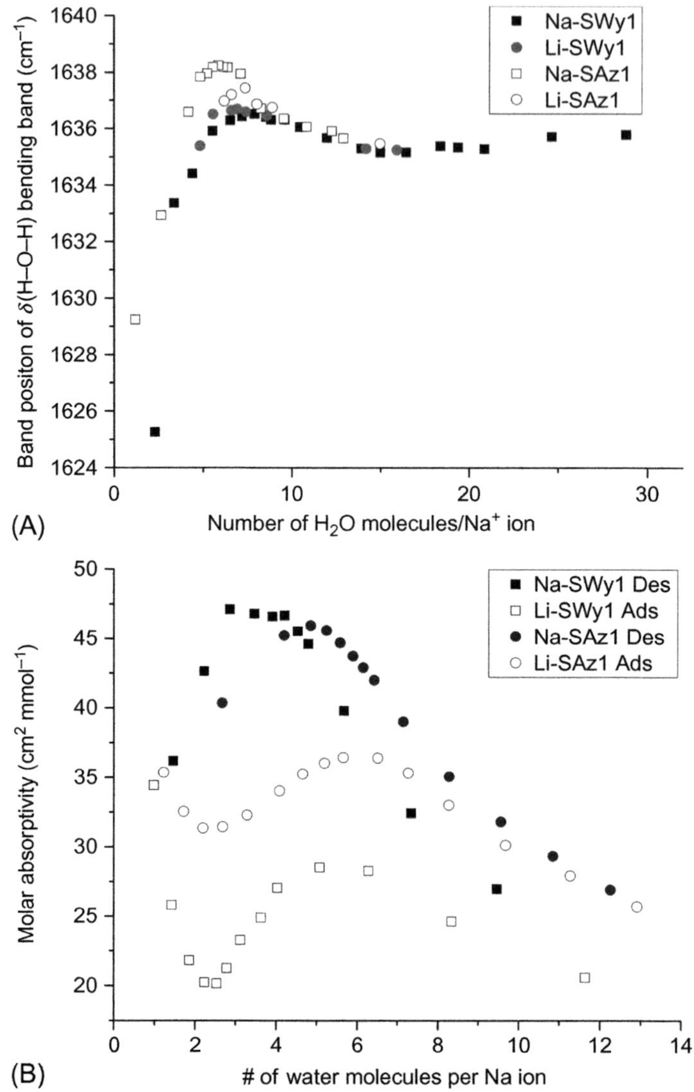


**FIG. 9.8** Band positon of the $\delta$(HOH) bending band of $H_2O$ sorbed to Na- and Li- exchanged low charge (SWy-1) and high charge (SAz1) smectites as a function of water content (A). Corresponding change in molar absorptivity of the $\delta$(HOH) bending band of $H_2O$ sorbed to Na-exchanged SWy-1 and SAz-1 smectite as function of water content. *(Adapted from Xu, W., Johnston, C.T., Parker, P., Agnew, S.F., 2000. Infrared study of water sorption on Na-, Li-, Ca- and Mg-exchanged (SWy-1 and SAz-1) montmorillonite. Clays Clay Miner. 48, 120–131).*

blue-shift of the $\nu$(HOH) band (Ni and Skinner, 2015). The position of the $\delta$(HOH) band of $H_2O$ adsorbed to clay minerals at low water contents is consistently shifted to lower energy (Johnston et al., 1992a; Xu et al., 2000; Dontsova et al., 2004; Schuttlefield et al., 2007b; Schnetzer et al., 2016;

Tang et al., 2016). These results show that $H_2O$ bound to smectites at low $H_2O$ content is less H-bonded that at high $H_2O$ content.

Additional insight can be obtained from plotting the molar absorptivity of the $\delta$(HOH) band as a function $H_2O$ content (Fig. 9.8B) (Xu et al., 2000). The molar absorptivity of $H_2O$ adsorbed to $Na^+$-smectite at high $H_2O$ content is similar to that of bulk $H_2O$ with a value of 20 $cm^2$/mmol (Venyaminov and Prenderast, 1997). Upon lowering the $H_2O$ content, the molar absorptivity increases to a value of 47 ($cm^2$/mmol $H_2O$) (solid symbols shown in Fig. 9.8B), followed by a reduction to 35 $cm^2$/mmol. Strong hysteresis is observed upon rehydration of the smectite (open symbols in Fig. 9.8B), where the molar absorptivity decreases further to values representative of bulk $H_2O$. Upon rehydration, the initial stages of $H_2O$ adsorption occur on external surfaces as reflected by the low molar absorptivity values. These results indicate that changes in molar absorptivity and position of the $\delta$(HOH) band can be used to study the water adsorption process.

These results reflect the somewhat surprising result that the water molecules adsorbed to smectites and vermiculites at low water contents are *less* H-bonded than at high water contents or in bulk water. For a fully hydrated smectite, two populations of $H_2O$ molecules are present: $H_2O$ molecules bound to cations through ion-dipole interactions (blue), and 'intervening' $H_2O$ molecules (pink) (Fig. 9.9A), which are less-strongly bound. Lowering the water content results in the preferential loss of $H_2O$ molecules not directly bound to these interlayer cations (Fig. 9.9B and C) (Gates et al., 2012). This loss of the intervening $H_2O$ molecules has the net effect of lowering the extent H-bonding for the few remaining $H_2O$ molecules (Fig. 9.9C). These residual $H_2O$ molecules are associated with the interlayer cations through its oxygen.

### 9.3.1.2 Studies of $H_2O$ in the $H_2O$ Stretching Region for Smectites and Vermiculites

Studies in the $\nu$(HOH) of adsorbed $H_2O$ have also been used to gain insight into -clay-mineral water interactions. The general result here is that the position of both the 3420 and 3250 $cm^{-1}$ bands shift to higher frequency upon dehydration of the clay mineral (Xu et al., 2000; Madejova et al., 2002; Kuligiewicz et al., 2015b). The results can be attributed to the model shown in Fig. 9.9A–C. Dehydration of the clay mineral results in the preferential loss of the latter population of $H_2O$ molecules, sometimes called 'intervening' $H_2O$ molecules, leaving the remaining ion-dipole-bound $H_2O$ molecules less H-bonded (Fig. 9.9).

Unambiguous spectral interpretations of adsorbed $H_2O$ in the stretching region is somewhat complicated by the spectral overlap with the $\nu$(MOH) bands of structural OH groups (Figs. 9.2 and 9.3). As an alternative method, deuteration methods can be used to isotopically shift the bands of adsorbed $D_2O$ away from spectral overlap with the structural $\nu$(OH) bands. These methods have been used for over fifty years by clay researchers to study structural and surface OH groups (Ledoux and White, 1964a; Wada, 1967; Russell et al., 1974; Parfitt et al., 1976; Annabi-Bergaya et al., 1980). The central

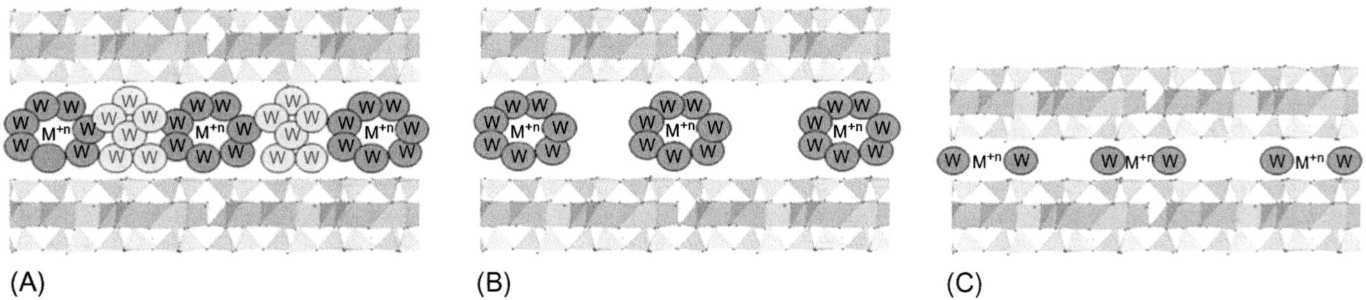


**FIG. 9.9** Schematic illustration of $H_2O$ retained by a smectite at high $H_2O$ content (A), intermediate (B), and low $H_2O$ content (C). The $H_2O$ molecules surrounding the exchangeable cation represented by the blue spheres are strongly retained by ion-dipole interactions and 'difficult' to remove. The $H_2O$ molecules 'between' cations shown in pink more are preferentially lost at lower $H_2O$ activities. At low $H_2O$ content, some $H_2O$ molecules can be strongly retained by the exchangeable cation.

concept here is that exposed OH groups (both structural OH groups and surface exposed $H_2O$ molecules) rapidly exchange with deuterium oxide to form O–D. The molecular motion of the $\nu$(HOH) band can be considered as a simple coupled oscillator originating from atomic displacements of both the hydrogen and oxygen atoms. Because the mass of the oxygen atom is 16 times that of the proton, the atomic displacement of the proton is dominant. When exchanged with deuterium, the deuteron has twice the mass of the proton and the position of the O-D oscillator (i.e. the $\nu$(DOD) band) is isotopically shifted by $\sim \frac{1}{\sqrt{2}}$ the frequency of the $\nu$(HOH) (or corresponding $\nu$(HOH)) band. The structural OH groups within the clay mineral structure (for smectites and vermiculites) are inaccessible to $D_2O$ and do not exchange with $D_2O$ under these conditions. Thus the $\nu$(DOD) groups associated with adsorbed $D_2O$ are isotopically shifted into a region free of spectral interference. When this method is applied to smectites, a 'high-energy' $\nu$(DOD) band appears at frequencies ranging from 2694 to 2684 $cm^{-1}$. This unusual band was assigned by Kuligiewicz et al. (2015a,b) as an 'oriented dangling O-D' band. Although this shift is modest and not uniform for all cations, the position of this band is strongly correlated with a layer charge and supports the utility of spectroscopic measurements to derived parameters that are difficult and time consuming using traditional methods (Farmer and Russell, 1971; Kuligiewicz et al., 2015a,b). Earlier work examining $D_2O$ bands adsorbed to a suite of smectites demonstrated a strong correlation with specific surface area (Mulla et al., 1985).

### *9.3.1.3 Vibrational Bands of the Clay Mineral that are Influenced by $H_2O$*

Many studies have shown that the internal vibrational modes of the clay mineral layer itself (e.g. $\nu$(HOH), $\delta$(MOH), and $\nu$(SiO)) are influenced by the difference in $H_2O$ content and the nature of the exchangeable cation. Sposito et al. (1983) were the first to show that the intensity of the structural OH bending band of the MgAlOH group was decreased at low water content. In a later study using a gravimetric-FTIR cell, all of the structural $\delta$(OH) bands of SWy-2 and SAz-1 smectites exchanged with $Na^+$, $Li^+$, $Ca^{2+}$, and $Mg^{2+}$ were found to be sensitive to changes in $H_2O$ content (Xu et al., 2000). More specifically, the molar absorptivity (i.e. a quantitative measure of intensity) of the structural $\delta$(OH) bands of a both a high charge (SAz-1 smectite) and a lower charge (SWy-1 smectite) were significantly reduced at low water content. The positions of these bands were all influenced by water content as well. Because these bands change with water content, it is important to control water content and relative humidity when studying the structural chemistry aspects of clay minerals. Clearly, vibrational bands of the clay mineral itself are influenced by changes in water content as some studies have reported that the $\nu$(SiO) bands are 'coupled' to changes in water content (Yan et al., 1996a,b). Some caution is needed in evaluating these studies, however, as the $\nu$(SiO) bands are strongly polarized and consist of both in-plane and out-of-plane components (Farmer, 1974b; Johnston and Premachandra, 2001; Ras et al., 2007). Changes in water content are often associated with changes in particle orientation during dehydration so some of the changes are simply due to orientation effects. In fact, even the vibrational modes of adsorbed $H_2O$ can be influenced by preferred orientation of $H_2O$ molecules at low $H_2O$ content. Recent structural studies of clay mineral-water complexes have shown that the b-dimension of the clay mineral lattice is compressed at low water content (Schnetzer et al., 2016). Because the exchangeable cations have different sizes and enthalpies of hydration, it is difficult to distinguish between changes in $H_2O$ content with cation-specific effects. Although some of the changes reported in the literature that have been assigned to surface-$H_2O$ coupling mechanisms can better be attributed to particle orientation effects, some current evidence does point to surface-$H_2O$ interactions (Michot et al., 1994).

### 9.3.2 Nanoconfined $H_2O$: Sepiolite and Palygorskite

The modulated crystal structures of sepiolite and palygorskite, which are important industrial minerals, are characterized by tunnels that run along the length of the *c* axis (Ahlrichs et al., 1975; Serna et al., 1977a; Gionis et al., 2006; Post et al., 2007; Bukas et al., 2013) shown in Fig. 9.1C for sepiolite. The $H_2O$ molecules residing in these tunnels are a type of 'zeolitic' water (Serna et al., 1977a; Gionis et al., 2006; Bukas et al., 2013). These nano-confined $H_2O$ molecules are distinct from the molecules located in the interlayer space of smectites because, ideally, charge-compensation cations are not present in the tunnels. In a recent study, Bukas et al. (2013) combined MIR and NIR methods to study structural OH groups, $H_2O$, and the silicate framework vibrations of sepiolite under different stages of hydration using both $H_2O$ and $D_2O$. This study distinguished the spectra of $H_2O$ on the external surface of the particles (which, contrary to smectite, bears a high concentration of Si-OH terminal defects) and inside the tunnels. Regarding the occupancy of the latter, three different arrangements of H-bonded tunnel $H_2O$ were identified as well-defined phases upon drying. The transition between these phases has a pronounced effect of the spectra of the 2:1 framework (structural OH and Si–O framework). They reported on essentially all of the spectral regions shown in Fig. 9.3 and use the observed spectral changes in both the MIR and NIR regions to follow the hydration characteristics of sepiolite.

### 9.3.3 Nanoconfined $H_2O$: Halloysite and Imogolite

Two related examples of nanoconfined $H_2O$ in clay minerals are halloysite and imogolite, (Fig. 9.1D–F). A more detailed representation of tubular halloysite is shown in Fig. 9.10. The SEM image shown in the upper left region reveals the tubular structure. For this particular halloysite sample from southern Indiana (US), the tubes were 100–200 nm in length, had an outer diameter

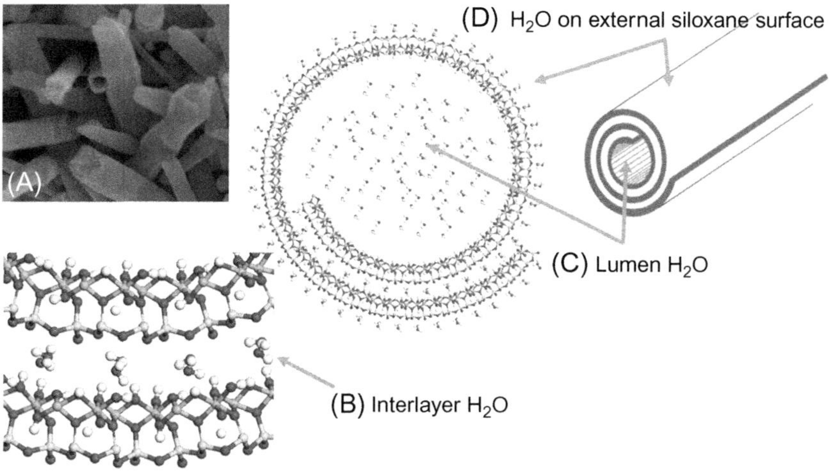


**FIG. 9.10** Cryo-SEM micrograph and conceptual structure of tubular halloysite. The cryo-SEM image (A) was obtained from a hydrated halloysite from the Gardner Mine Ridge deposit in Southern Indiana, US (Ambers and Murray, 1995). TEM analysis revealed that the average inner diameter was $15 \pm 4$ nm and the outer diameter was $45 \pm 20$ nm. A schematic of the tubular structure is shown in the upper right portion. An expanded view of the halloysite crystal structure is adapted from a recent modelling study (Ferrante et al., 2015) and a view of the $H_2O$ molecules in the interlayer are shown in the lower left portion. Three types of $H_2O$ molecules are present consisting of interlayer $H_2O$ (B), $H_2O$ present in the tubular structure or lumen (C), and $H_2O$ external to the halloysite structure (D). *(Data were adapted from Santagata, M.C., Premachandra, G.S., Johnston, C.T., 2009. Nanoconfined water in clay minerals. Book of Abstracts for XIV International Clay Conference "Micro et Nano Scientiae Mare Magnum", vol. 1. Castellaneta Marina (TA), Italy. AIPEA, p. 428.)*

of 45 nm and an inner diameter of 14 nm (Santagata et al., 2009). The particle-crystal structure is shown at three different scales. Halloysite can be considered as a type of 'hydrated kaolinite' where a single layer of $H_2O$ molecules are intercalated in the structure resulting in a *d*-value of 1.0 nm (Figs. 9.1E and 9.10B) (Bailey, 1989; Giese, 1990). In addition to the intercalated $H_2O$ molecules, water is found in the lumen (Fig. 9.10C) and external to the crystal structure (Fig. 9.10D). Dehydration of halloysite leads to irreversible collapse of the structure and loss of the interlayer $H_2O$ (Fig. 9.10B). There are relatively few studies that have examined fully hydrated specimens (Churchman et al., 1972; Churchman and Carr, 1972) and much of the IR data published for halloysite has been obtained from desiccated samples (Yariv and Shoval, 1975; Joussein et al., 2005). Powder XRD and environmental SEM studies of hydrated halloysite reveal the irreversible loss of $H_2O$ occurs 'easily' (i.e. at relatively high relative humidity values) ($p/p_o < 0.5$) (Churchman et al., 1972; Kohyama et al., 1978)). The interlayer environment of halloysite is unique in that the interlayer water molecules reside in a 0.3 nm space juxtaposed between a siloxane surface and an aluminum-hydroxy surface resulting in a material with a d-spacing of 1 nm. ATR-FTIR spectra of

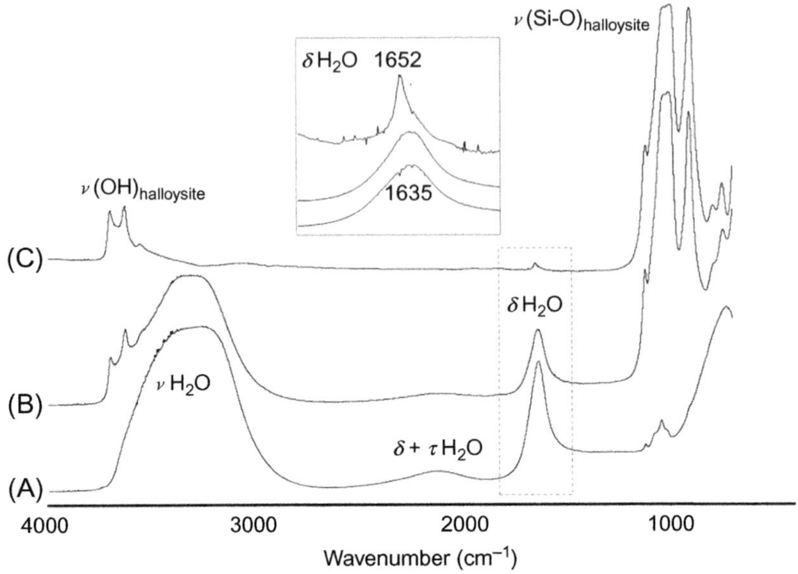


**FIG. 9.11** ATR-FTIR spectra in the 4000–700 $cm^{-1}$ region of Gardner halloysite (A) at high $H_2O$ content ($p/p_o \approx 1$), intermediate $H_2O$ content (B) and low $H_2O$ content (C). *(Data were adapted from Santagata, M.C., Premachandra, G.S., Johnston, C.T., 2009. Nanoconfined water in clay minerals. Book of Abstracts for XIV International Clay Conference "Micro et Nano Scientiae Mare Magnum", vol. 1. Castellaneta Marina (TA), Italy. AIPEA, p. 428.)*

a fully hydrated halloysite that was allowed to dehydrate are shown in Fig. 9.11 (Santagata et al., 2009). Initially, the spectra are dominated by the presence of $H_2O$ features in the $\nu$(HOH), $(\delta+\tau)$ combination band, $\delta$(HOH), and low frequency librational bands of $H_2O$ (Fig. 9.13A). The $\nu$(SiO) bands of halloysite can also be observed in the fully hydrated sample at 1000–1100 $cm^{-1}$ (Fig. 9.11A). Upon dehydration, the $H_2O$ features are strongly diminished and both the structural $\nu$(OH) and $\nu$(SiO) bands of halloysite become prominent (Fig. 9.11B and C). In the fully desiccated sample, the $\nu$(HOH) bands are clearly resolved and the position of the $\delta$(HOH) band is present at 1652 $cm^{-1}$ (Fig. 9.11C). These results are consistent with published IR spectra of halloysite where the position of the $\delta$(OH) band between 1635 and 1666 $cm^{-1}$ and $\nu$(HOH) bands of intercalated $H_2O$ between 3100 and 3450 $cm^{-1}$ (Tarasevi and Gribina, 1972; Yariv and Shoval, 1975). The main point to emphasize here is that $H_2O$ is an integral part of the halloysite structure and vibrational spectroscopy can be used to probe the nature of $H_2O$ associated with this important industrial mineral.

Imogolite and allophane (Fig. 9.1D) are commonly found mineral phases in soils derived from volcanic ash. From a structural perspective, imogolite is roughly analogous to an 'inverted kaolinite/halloysite'. In this case, the siloxane surface is found on the 'internal surface' and the aluminum-hydroxy surface is presented on the external surface. This is a unique 'nano-environment' where the $H_2O$ molecules reside in the lumen of imogolite tubes. The position of the $\delta$(HOH) band of these water molecules is 1650 cm$^{-1}$ and reflects a more strongly H-bonded environment than in bulk water (Venyaminov and Prenderast, 1997).

## 9.3.4 Physisorbed $H_2O$

For some hydrous clay minerals, water adsorption only occurs on the external surfaces of the particle (e.g. kaolinite). For nano-sized clay minerals, the amount of $H_2O$ adsorption can be quite high and measuring the specific surface area of these materials using traditional methods, like $N_2$ Brunauer-Emmett-Teller (BET) theory, is not reliable because the use of a strong vacuum can result in the irreversible collapse of the mineral structure. In this regard, gravimetric-FTIR methods (Xu et al., 2000; Wang et al., 2003) have been developed (Fig. 9.12) where the sample, suspended from a micro-balance within an FTIR spectrometer (Fig. 9.12B and C), is held under controlled water vapor activity (Fig. 9.12A). Correlation of the intensity of the $\delta$(HOH) band with gravimetric water content (Fig. 9.13) can be used to determine the dry mass of the sample through extrapolation. The apparatus is comprised of a computer-controlled humidity system (Fig. 9.12A) that delivers known water activity vapor to the sample that is both mounted in the FTIR spectrometer

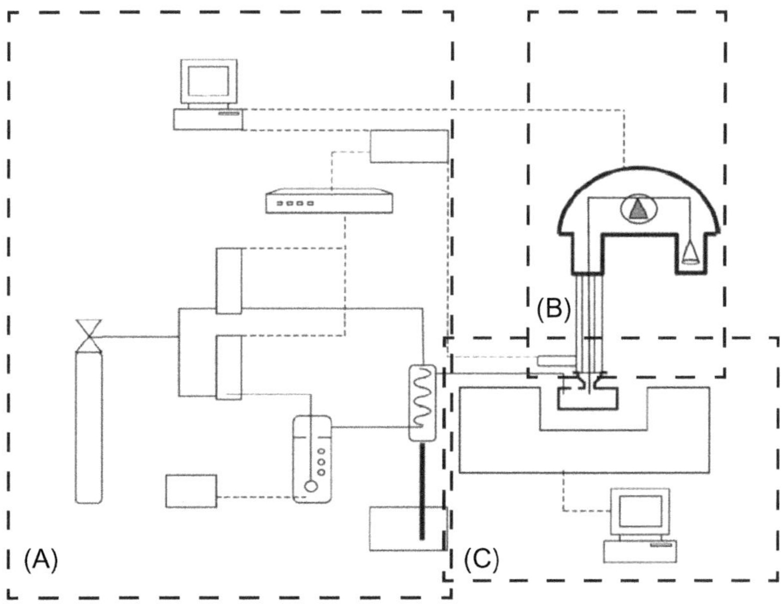


**FIG. 9.12** Schematic of the gravimetric FTIR apparatus showing a flowing gas manifold that mixes dry and 'wet' streams of $N_2$ gas (A), delivers the vapor stream to a specially designed gravimetric cell (B) that permits simultaneous recording of the FTIR spectrum (C). *(Adapted from Wang, S.L., Johnston, C.T., Bish, D.L., White, J.L., Hem, S.L., 2003. Water-vapor adsorption and surface area measurement of poorly crystalline boehmite. J. Colloid Interface Sci. 260, 26–35.)*

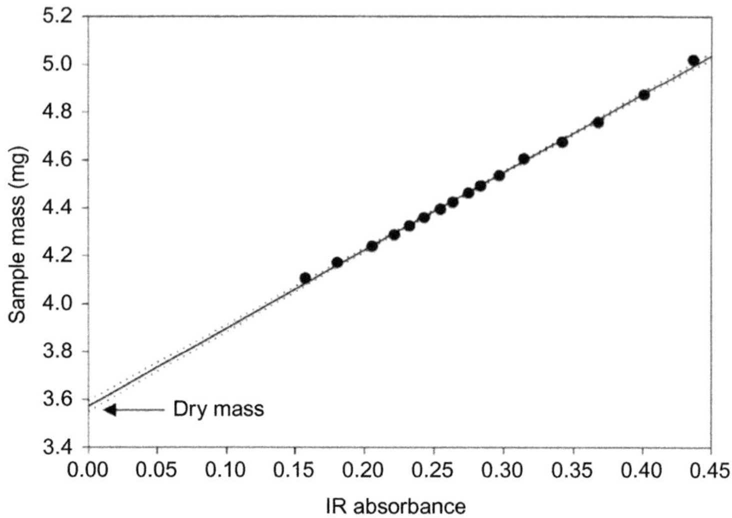


**FIG. 9.13** Plot of the sample mass of poorly crystalline boehmite as a function of IR absorbance of the $\delta$(HOH) band. The coupled IR-gravimetric method permitted the dry mass of the sample to be determined without taking the sample to dryness. If the sample had been completely dried by heating >100°C, the sample would have been irreversibly altered. The accurate IR-derived dry mass then permitted application of the BET theory to determine a IR-$H_2O$ derived surface area of $514 \pm 36$ $m^2/g$. *(Data adapted from Wang, S.L., Johnston, C.T., Bish, D.L., White, J.L., Hem, S.L., 2003. Water-vapor adsorption and surface area measurement of poorly crystalline boehmite. J. Colloid Interface Sci. 260, 26–35.)*

and suspended from a gravimetric balance (Fig. 9.12B and C). The advantage of this method is that the dry mass of a sample can be determined by extrapolating the intensity of the $\delta$(HOH) band to zero absorbance without having to take the sample to dryness, which avoids potentially irreversible collapse of the structure.

## 9.4 CLAY MINERAL-WATER INTERACTIONS AS DIRECTORS OF CLAY MINERAL-ORGANIC ADSORPTION PROCESSES

Mortland and his colleagues were among the first to show that clay mineral-water interactions played a significant role in governing the interaction of organic molecules with clay mineral surfaces (Doner and Mortland, 1969; Mortland and Pinnavaia, 1971; Pinnavaia and Mortland, 1971). In the case of aromatic hydrocarbons, minimal adsorption of aromatic hydrocarbons occurred at high $H_2O$ content. However, at low $H_2O$ contents adsorption was greatly enhanced; depending on the nature of the exchangeable cation, single electron transfer reactions as well as chemisorption reactions could occur. FTIR methods were included in these studies to observe both the adsorbed $H_2O$ and the organic solutes (Johnston et al., 1991, 1992b; Stevens and Anderson, 1996). The view that began to emerge from these studies is that the siloxane surface of smectites has partial hydrophobic character (Jaynes and Boyd, 1991). In seminal studies, Laird et al., (1992, 1994) and Laird and Fleming (1999) showed that semi-polar organic compounds (e.g. atrazine) could be described on the basis of layer charge and cation hydration. Our current understanding is that clay mineral-water interactions play a significant role in determining how organic compounds are adsorbed (and sometimes transformed) in clay mineral interlayer spaces (Boyd et al., 2011). Conversely, the presence of organic compounds can significantly alter and even block the clay mineral from rehydration (Tsiantos et al., 2012). FTIR spectroscopy has played a prominent role in these types of studies to determine the mechanisms of clay mineral-$H_2O$-organic interactions. Because water is ubiquitous in nature ($\sim$55 M), understanding clay-mineral water interactions is critical to determine how organic compounds (pollutants, pesticides, proteins, and polymers) will interact with the clay mineral surface.

## 9.5 CONCLUSIONS

Historically, the presence of water was often viewed as a constituent that needed to be eliminated in IR studies of clay minerals before 'good' spectral data could be obtained. Despite the early bias towards studying fully dehydrated clay minerals, the $H_2O$ molecule itself has been found to serve as a useful molecular probe of clay minerals in a wide range of spectral regions. The diagnostic properties of $H_2O$ include the fundamental $\nu$(HOH) stretching and $\delta$(HOH) bending vibrations, librational, combination and overtone bands that extend from a few hundred $cm^{-1}$ well into the NIR region (10,000 $cm^{-1}$). The structure of many clay minerals is intimately associated with $H_2O$ molecules, to the extent that $H_2O$ in some cases should be considered as a fundamental part of the clay mineral structure. Vibrational studies of these $H_2O$ molecules reveals valuable insight that is complementary to data obtained from other methods, including XRD and NMR spectroscopy. Essentially all physicochemical properties of clay minerals are influenced by $H_2O$-surface interactions that extend from short-range to intermediate- and long-range interactions. IR and Raman spectroscopies are particularly sensitive to short-range interactions, although the impact of these short-range interactions is often manifest at much larger scales. For clay minerals characterized by isomorphous heterovalent substitution, IR studies have provided key insights about the role of interlayer cations in directing clay mineral-$H_2O$ interactions. One interesting aspect here is that clay mineral-$H_2O$ interactions can be geochemically controlled through cation exchange. Clay mineral organic interactions are also critically linked to clay mineral-$H_2O$ interactions. In order for organic solutes to associate with clay minerals, they must 'compete' with $H_2O$ molecules present at orders of magnitude greater concentration. Because multiple constituents are interacting with the clay mineral surface (e.g. $H_2O$, organic solutes, exchangeable cations), IR and Raman methods have provided insight about the surface chemistry of clay mineral-organic interactions.

Chapter 10

# Analysis of Organoclays and Organic Adsorption by Clay Minerals

H.P. He and J. Zhu

## 10.1 ORGANOCLAY

The term 'organoclay' mainly denotes a family of hydrophobic materials that are obtained by modifying clays and clay minerals with various organic compounds through an intercalation process (Lagaly, 1981; Favre and Lagaly, 1991; Bergaya and Lagaly, 2001; Paiva et al., 2008; Bergaya et al., 2011a; He et al., 2013; Lagaly et al., 2013). Organoclays, usually abbreviated as OC, have been found to have some important practical applications, notably as adsorbents of organic contaminants (Stockmeyer, 1991; Zhu et al., 2000; Theng et al., 2008; Bergaya et al., 2013; Zhu et al., 2015) and as components in the formation of clay-polymer nanocomposites (Ray and Okamoto, 2003; Theng, 2012; Bergaya et al., 2011a, 2013). Over the last two decades, a large volume of literature has accumulated where various aspects of OC have been studied, including: (1) synthesis and characterization (e.g. Lagaly, 1981; Zhu et al., 2003; Lagaly et al., 2013); (2) surface properties and stability (e.g. He et al., 2005; Theng et al., 2008; Zhu et al., 2008); (3) synthesis of clay-based nanocomposites (e.g. Ray and Okamoto, 2003; Theng, 2012; Lambert and Bergaya, 2013; Bergaya et al., 2013); and (4) synthesis of novel materials using OC as precursors (e.g. Ishii et al., 2005; Bergaya and Lagaly, 2011).

In these studies, various natural and synthetic clay minerals have been used to prepare OC, including montmorillonite, saponite, hectorite, vermiculite and mica. Among them, smectites have held the limelight in the preparation of OC because of their moderate charge densities and cation exchange capacity (CEC), interlayer swelling ability, and propensity for intercalating simple and polymeric organic compounds (Swartzen-Allen and Matijevi, 1974; Theng, 1974, 2012; Murray, 2000; Bergaya et al., 2011b, 2012;


Developments in Clay Science, Vol. 8. https://doi.org/10.1016/B978-0-08-100355-8.00010-2

Lagaly et al., 2013), not to mention their wide availability and low cost as constituents of bentonites.

The most common organic molecules used for modification of clay minerals are the quaternary ammonium surfactants (Sarkar et al., 2012; He et al., 2014), which are cationic in nature. Commercially available quaternary ammonium surfactants used in OC preparation generally contain short (5–10 carbons) aliphatic chains, benzyl, and sometimes hydroxyl groups that are incorporated along with at least one long (16–24 carbons) aliphatic chain resulting in a high interlayer expansion (Kim et al., 2003; Lee and Kim, 2004; Slade and Gates, 2004a; Zha et al., 2005). Other organic modifiers include alkyl amines (Li and Ishida, 2003a; Akcay, 2006; Akcay et al., 2006; Xi et al., 2007), alkyl carbazol (Chigwada et al., 2005), polydimethylsiloxane (Li and Ishida, 2005), alkyl carboxy (Sheng et al., 1999), quinolinium, or pyridinium (Andrades et al., 2004; Chigwada et al., 2005), propylene carbonate (Onikata et al., 1999), glycerol carbonate (Fehervari et al., 2016) and derivatives (Gates et al., 2016), oligomers (Chen and Vyazovkin, 2006), or more complex molecules (Yao et al., 2002; Yei et al., 2005) and other reactive molecules (Zhang and Wilkie, 2004; Ding et al., 2005).

Since the behaviors and properties of OC mainly depend on their interlayer structure and the molecular environments of the intercalated organic compounds, this chapter first focuses on the interlayer structure of OC as evidenced from X-ray diffraction (XRD) studies. Then, a review of recent advances in the applications of the Fourier transform infrared (FTIR) spectroscopy in better elucidating the structures of OC intercalates is provided.

## 10.2 BASAL SPACING OF ORGANOCLAY

Smectites are 2:1 type swelling clay minerals with a relatively low charge density (0.2–0.6 equivalence per formula unit (eq/f.u.)) (Bergaya and Lagaly, 2013b). Their unit layers consist of an octahedral sheet sandwiched between two opposing tetrahedral sheets. Due to isomorphous substitution, that is, a replacement of structural cations by other cations of lower valency (e.g. $Al^{3+}$ by $Mg^{2+}$, or $Mg^{2+}$ by $Li^{+}$, in the octahedral sheet; and $Si^{4+}$ by $Al^{3+}$ in the tetrahedral sheet), the layers are negatively charged, which is balanced by inorganic cationic counterions in the interlayer space (Likos et al., 2010; Brigatti et al., 2013). These compensator cations are exchangeable, and the sum of their charges represents a cation exchange capacity (CEC), which is conventionally expressed in milliequivalents per 100 grams clay (meq/100 g). The magnitude of the CEC determines the amount of quaternary ammonium cation (QAC) that can enter into the interlayer space by cation exchange. Fig. 10.1 presents a structure schematic for a common group of quaternary alkylammonium cations with different chain substitutions.

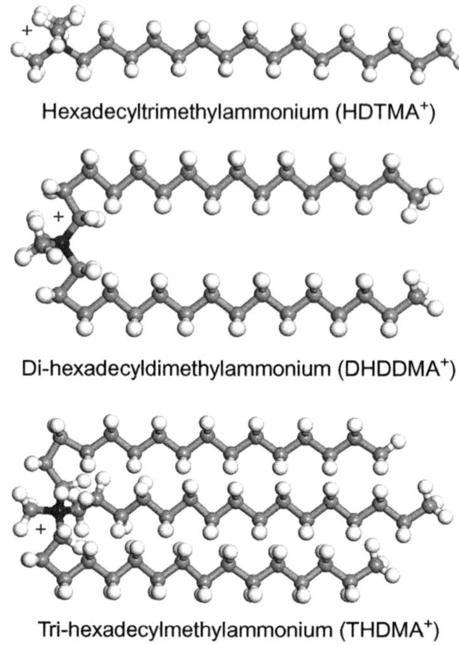


**FIG. 10.1** Structure schematics of $HDTMA^+$, $DHDDMA^+$ and $THDMA^+$.

A number of studies have been conducted regarding the effects of clay mineral properties, such as CEC, alkyl chain length and chain number, the nature of the polar head, and the concentration of QAC (added QAC/CEC ratio), on the structures and properties of resulting OC, including basal spacing, alkyl chain conformation, thermal stability, and surface affinity (Lagaly, 1981; Vaia et al., 1994; Xie et al., 2002; Mandalia and Bergaya, 2006; He et al., 2010, 2014). Basal spacing values represent the thickness of the structural unit of clay mineral and the intercalate (Bailey, 1980b), such as the total assemblage of the layer plus the organo-modified interlayer. Generally, the basal spacing of OC increases with alkyl chain length (expressed as the number of carbon atoms, $n_c$) and concentration of QAC. The intercalated QAC may adopt different configurations, ranging from a flat monolayer to a paraffin bilayer through a lateral bilayer and a paraffin monolayer (Lagaly, 1981; Zhu et al., 2003). When the concentration of the organocation exceeds the CEC of the smectite, the QAC may intercalate as an ion pair, which is accompanied by the corresponding (halide) anion (Tahani et al., 1999; Klapyta et al., 2001; Lee and Kim, 2002; Janek and Lagaly, 2003). At concentrations well exceeding the CEC, the QAC may enter into the interparticle pore space or attach to the outer surfaces of the smectite, decreasing the thermal stability of the resultant OC (He et al., 2005). Occupancy of interparticle pores by the QAC also leads to a dramatic decline in pore volume and (external) specific surface area (SSA), while the propensity of OC for adsorbing organic contaminants in solution is significantly increased (He et al., 2006b). These observations provide an insight into the microstructure and properties of OC that are highly important and relevant to their industrial applications.

The structures and properties of OC are mainly controlled by the nature of the clay minerals and the characteristics of the intercalated organic materials that are used. As for the former, the CEC of the clay mineral used appears to be the key factor. This is evidenced by structurally different OC, which were prepared using the same QAC, and under identical experimental conditions (Lagaly, 1981; Lee and Kim, 2002; Xi et al., 2005; He et al., 2006a,b; Zidelkheir and Abdelgoad, 2008).

Table 10.1 lists some experimental basal spacings ($d_{001}$) of OC prepared from montmorillonites (Mt) with different CECs and varying concentrations (0.5–2.0 CEC) of hexadecyl trimethylammonium bromide (HDTMAB), as occurring in the literature. The relationship of the CEC of Mt and the basal spacing of the resultant OC is plotted in Fig. 10.2. Although the basal spacing is dependent on the layer charge density and its distribution in clay mineral layers (Slade and Gates, 2004b; He et al., 2010; Su et al., 2016), most organo-Mt (OMt) intercalates prepared using HDTMAB at a concentration of 0.5 CEC, give very similar basal spacings (1.4–1.6 nm). Since the amount of added HDTMAB, in this instance, is less than the CEC of Mt, essentially all $HDTMA^+$ are intercalated, as expected through a cation exchange mechanism, giving rise to a lateral (monolayer) arrangement within the interlayer space (Lagaly, 1981; Zhu et al., 2003). This proposed arrangement model is based on an assumption that the alkyl chains of the intercalated quaternary ammonium cations are fully elongated (*all-trans* conformation). When the $HDTMA^+$ lies flat and parallel to the clay mineral surface, the height of alkyl-chain is 0.4 nm and that of the 'nailhead' (positively charged N) holding three methyls is 0.51 nm. Thus the structure unit's thickness (the smectite layer thickness plus the height of alkyl chain) is about 1.47 nm, which is consistent with the experimental $d_{001}$ value (1.48 nm) (Zhu et al., 2003). An exception was reported by Carrizosa et al. (2004) who measured a basal spacing of 2.2 nm for an OMt prepared from Arizona Mt, which with a CEC of 120 meq/100 g, is among the highest CEC value reported for Mt. In this case, according to the calculation method reported by e.g. Zhu et al. (2003), the intercalated $HDTMA^+$ adopted a paraffin monolayer (instead of a lateral layer) arrangement, probably because the absolute $HDTMA^+$ concentration was appreciably larger than that used in preparing OC from Mt with lower CEC (Slade and Gates, 2004b).

As shown in Fig. 10.2, the CEC of Mt has little effect on the basal spacing of the resultant OC when the concentration of added $HDTMA^+$ is 0.5 CEC. On the other hand, when the concentration equals the CEC of Mt, the basal spacing of the intercalate generally increases with the increment of CEC,

**TABLE 10.1 Experimentally Observed Basal Spacings of Organo-Montmorillonite Intercalates Prepared at Different HDTMAB Concentrations from Solution**

| CEC[a] (meq/100g) | Layer Charge[b] | $d_{001}$ (nm) *0.5 CEC* | *1.0 CEC* | *1.5 CEC* | *2.0 CEC*[c] | References |
|---|---|---|---|---|---|---|
| 19 | 0.14 | n.d. | 1.36 | n.d. | n.d. | Slade and Gates (2004b) |
| 40.8 | 0.30 | n.d. | 1.37 | n.d. | n.d. | Slade and Gates (2004b) |
| 45 | 0.33 | n.d. | n.d. | 2.51 | n.d. | Koswojo et al. (2010) |
| 52 | 0.38 | 1.44 | 1.84 | n.d. | n.d. | Marsal et al. (2009) |
| 54 | 0.40 | n.d. | 2.1 | n.d. | n.d. | Xin et al. (2011) |
| 57 | 0.42 | 1.67 | 2.02 | n.d. | n.d. | Li et al. (2009) |
| 57.9 | 0.43 | 1.48 | 1.98 | n.d. | n.d. | He et al. (2006a) |
| 63 | 0.46 | 1.52 | 2.0 | n.d. | 2.0 | Mojović et al. (2011) |
| 63.9 | 0.47 | 1.41 | 2.00 | n.d. | n.d. | Lee et al. (2002) |
| 69 | 0.51 | 1.86 | 1.90 | n.d. | 2.02 | Fatimah and Huda (2013) |
| 70 | 0.52 | 1.44 | 1.97 | n.d. | n.d. | Feng et al. (2009) |
| 76 | 0.56 | n.d. | 1.54 | n.d. | n.d. | Slade and Gates (2004b) |
| 76 | 0.55 | 1.7 | 1.8 | n.d. | n.d. | Carrizosa et al. (2004) |
| 76 | 0.56 | n.d. | n.d. | n.d. | 3.9 | Ma (2010) |
| 78 | 0.57 | n.d. | 2.06 | n.d. | n.d. | Jovic-Jovicic et al. (2010) |
| 80 | 0.59 | 1.46 | 1.99 | 2.15 | 2.39 | Gammoudi et al. (2012) |
| 80 | 0.59 | 1.69 | 2.15 | n.d. | n.d. | Sanchez-Camazano and Sanchez-Martin (1994) |

**TABLE 10.1 Experimentally Observed Basal Spacings of Organo-Montmorillonite Intercalates Prepared at Different HDTMAB Concentrations from Solution—cont'd**

| CEC (meq/100g) | Layer Charge | $d_{001}$ (nm) 0.5 CEC | 1.0 CEC | 1.5 CEC | 2.0 CEC | References |
|---|---|---|---|---|---|---|
| 80 | 0.59 | n.d. | 2.06 | n.d. | n.d. | Vianna et al. (2005) |
| 82 | 0.60 | 1.68 | 1.90 | n.d. | 3.65 | Volzone et al. (2006) |
| 87 | 0.64 | n.d. | 1.81 | n.d. | n.d. | Zhao and Vance (1998) |
| 88.8 | 0.65 | n.d. | n.d. | n.d. | 3.7 | Ma (2010) |
| 90.8 | 0.67 | 1.48 | 1.95 | 2.23 | 3.61 | He et al. (2006b) |
| 90.8 | 0.67 | 1.48 | n.d. | 2.71 | n.d. | Zhou et al. (2007b) |
| 90.8 | 0.67 | 1.44 | 2.25 | n.d. | 2.95 | Chen (2008) |
| 91 | 0.67 | 1.58 | 1.80 | n.d. | n.d. | Yilmaz and Yapar (2004) |
| 92 | 0.68 | 1.45 | 1.82 | 3.26 | 4.05 | Bergaya et al. (2005) |
| 92 | 0.68 | n.d. | 1.80 | n.d. | n.d. | Mahadevaiah et al. (2011) |
| 93 | 0.68 | n.d. | 2.03 | n.d. | n.d. | Dentel et al. (1998) |
| 97 | 0.71 | n.d. | 2.1 | n.d. | n.d. | Zaghouane-Boudiaf and Boutahala (2011) |
| 98 | 0.72 | 1.76 | 2.57 | n.d. | 3.94 | Volzone et al. (2006) |
| 100 | 0.74 | n.d. | 2.55 | 3.84 | n.d. | Wang et al. (2012a) |
| 100 | 0.74 | n.d. | 1.89 | n.d. | n.d. | Yan et al. (2007) |
| 100 | 0.74 | n.d. | 1.95 | n.d. | n.d. | Slade and Gates (2004b) |

*Continued*

**TABLE 10.1 Experimentally Observed Basal Spacings of Organo-Montmorillonite Intercalates Prepared at Different HDTMAB Concentrations from Solution—cont'd**

| CEC (meq/100g) | Layer Charge | $d_{001}$ (nm) *0.5 CEC* | *1.0 CEC* | *1.5 CEC* | *2.0 CEC* | References |
|---|---|---|---|---|---|---|
| 104 | 0.77 | 1.51 | 1.95 | n.d. | n.d. | Lee et al. (2004) |
| 106 | 0.78 | 1.46 | 2.01 | n.d. | n.d. | Zhou et al. (2010) |
| 106.5 | 0.78 | 1.7 | 2.3 | 3.4 | 3.9 | He et al. (2010) |
| 108 | 0.79 | n.d. | 2.39 | n.d. | n.d. | Zhu et al. (2009) |
| 110.6 | 0.81 | 1.43 | 2.21 | 3.47 | 3.73 | He et al. (2014) |
| 120 | 1.08 | 2.2 | 2.3 | n.d. | n.d. | Carrizosa et al. (2004) |
| 130 | 0.96 | n.d. | 2.2 | n.d. | n.d. | Zhao and Vance (1998) |
| 138 | 1.02 | n.d. | 2.45 | n.d. | n.d. | Liu et al. (2012) |

n.d. = Not determined.
[a]*Cation exchange capacity (CEC) of the untreated (parent) montmorillonite.*
[b]*Layer charge refers to electrons ($e^-$) per unit cell $[O_{20}(OH)_4]$. For montmorillonite whose structural formula was not provided in the literature, the formula weight of a typical Wyoming montmorillonite, $[Na_{0.75}(Al_{3.5}Mg_{0.5})(Si_{7.75}Al_{0.25})O_{20}(OH)_4]$, is used in the calculation of its layer charge.*
[c]*Surfactant concentration in solution: 1 CEC denotes a concentration equivalent to the CEC of montmorillonite.*

and it follows that the larger the CEC of Mt, the greater is the concentration of intercalated $HDTMA^+$. Basal spacings of OC prepared at HDTMAB concentrations of 1.5 CEC and 2.0 CEC generally show a marked step-increase for Mt at CEC ≥ 80 meq/100 g, from 2.0–2.5 nm to 3.2–3.8 nm at 1.5 CEC and to 3.5–4.0 nm at 2.0 CEC (Fig. 10.2). The CEC of Mt thus has a profound effect on the basal spacing of OC prepared at HDTMAB concentrations of 1.5 to 2.0 CEC. When the concentration of added HDTMAB exceeds 2.5 CEC, the OC apparently reach a 'saturated' state, giving a (maximum) basal spacing of ~4.0 nm, where the CEC of OC has little influence on the maximum basal spacing (He et al., 2010). The observed increase in basal spacings

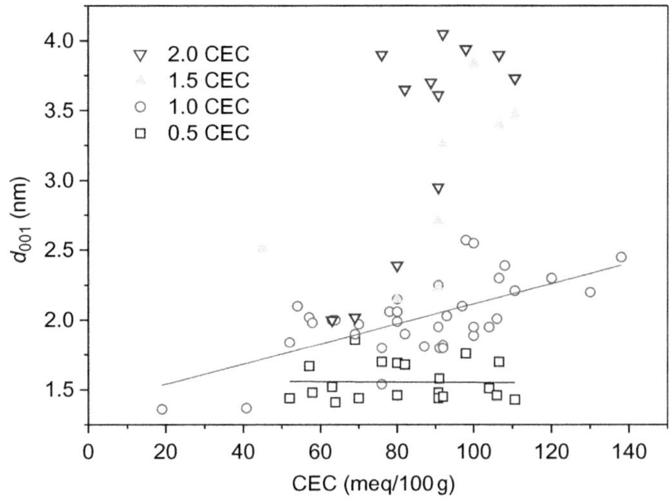


**FIG. 10.2** Relationship between the CEC of montmorillonite and the basal spacing ($d_{001}$) of the resultant organoclay intercalates formed at different HDTMAB concentrations (0.5–2.0 CEC). *Modified from He et al. (2014).*

of OC may be explained in terms of the intercalation of $HDTMA^+$ as lateral monolayer, lateral bilayer, pseudo-trilayer, paraffin monolayer, or paraffin bilayer as depicted in Fig. 10.3 (Lagaly, 1981; Choy et al., 1997; Zhu et al., 2003; Theng et al., 2008; Lambert and Bergaya, 2013).

Alkyl chain length, surfactant loading, and the arrangement of intercalated surfactants, in addition to their chemical properties (e.g. polarity, valence, composition), are also key factors controlling the basal spacing of OC, which have been the subject of many investigations (Lagaly, 1981; Heinz et al., 2007; He et al., 2010; Ma et al., 2016). While the basal spacing of OC intercalates prepared from $n_c \geq 16$ QAC increases through configuration rearrangement as described above, intercalates formed from short alkyl chain ($n_c < 14$) QAC are composed of lateral (flat) layers rather than in a paraffin arrangement over a wide range of QAC concentrations (Lagaly, 1981; He et al., 2010). The basal spacing of the resultant OC strongly depends on the size of the QAC used, but is only slightly affected by the CEC of clay minerals. For example, OC prepared from octyl trimethylammonium chloride (C8-S) and dodecyl trimethylammonium bromide (C12-S) at concentrations lower than 1 CEC have a basal spacing of approximately 1.4 nm, indicating a lateral monolayer arrangement (He et al., 2010). When the surfactant concentration is increased to 1.0–4.0 CEC, XRD patterns of OC formed from QAC with relatively short alkyl chain length ($n_c < 14$) display broad *001* reflections at 1.8–2.0 nm, corresponding to a lateral bilayer arrangement. Here, it is worthwhile noting that with an increase in Mt CEC, the *001* reflection is obviously broadened, which is accompanied by a slight basal spacing increase (from 1.8

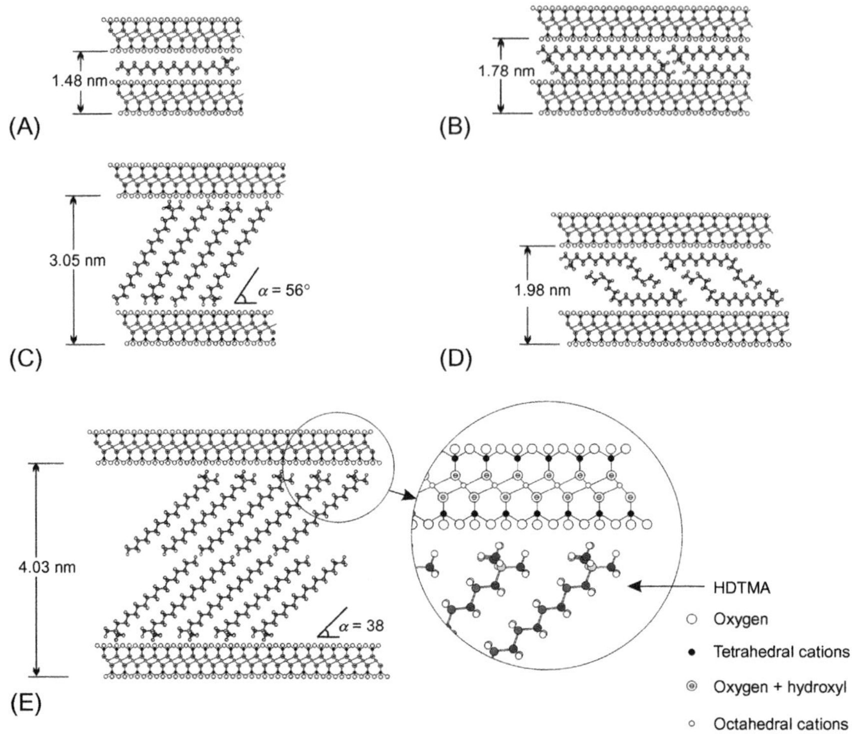


**FIG. 10.3** HDTMA$^+$ aggregation in the interlayer space of montrmoillonite. (A) Lateral monolayer, (B) lateral bilayer, (C) paraffin monolayer, (D) pseudotrilayer, (E) paraffin bilayer. *Reproduced with permission from Zhu et al. (2003).*

to 2.0 nm) and an occurrence of a weak reflection at ~4.0 nm (He et al., 2010). This implies a layer charge heterogeneity occurred in the studied clay minerals resulting in rearrangement of the organic cation into paraffinlike layers within the interlayer space (Su et al., 2016).

A variation occurs in maximum basal spacing of OMt prepared from $Ca^{2+}$-Mt (CEC = 106.5 meq/100g) and single alkyl chain surfactants with different chain lengths ($n_c$ = 8–18) (Fig. 10.4). The basal spacings of OMt derived from surfactants with relatively short single chain lengths (C8-S, C10-S and C12-S) were markedly distinct from those of their counterparts with long chain length (C14-S, C16-S and C18-S). An abrupt increase of the maximum basal spacing was observed from OC prepared from C12-S to that from C14-S, where the intercalated QAC in the former adopts a lateral model while the latter adopts a paraffin model. This suggests that the size of the intercalated QAC has a profound effect on their arrangement as well as the resulting (maximum) basal spacing. In other words, for QAC with short single chain length (i.e. $n_c \leq 12$), there is plenty of room for them to lie flat within the interlayer space; whereas

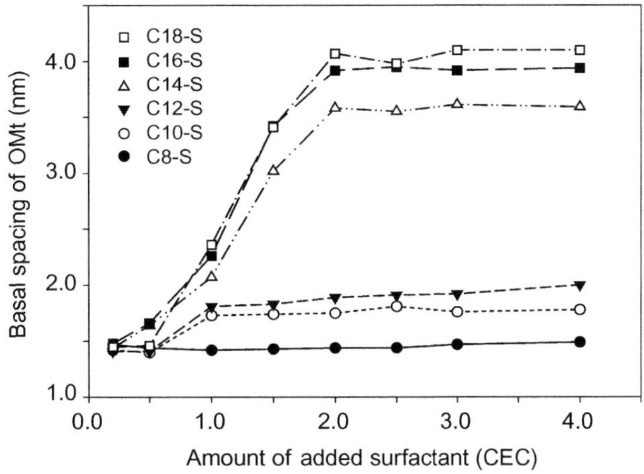


**FIG. 10.4** Basal spacings of organoclays prepared from Ca-montmorillonite and different surfactants. *Modified from He et al. (2010).*

for QAC having longer chain lengths (i.e. $n_c \geq 14$), the interlayer space cannot accommodate the intercalated QAC, when laying parallel to the clay mineral surface (i.e. lateral monolayer), and when the loaded QAC ≥ 1.0 CEC (Fig. 10.4). Thus the alkyl chain orientation orients away from the surface resulting in a paraffin-like arrangement (He et al., 2004a; Slade and Gates, 2004a).

The alkyl chain number of QAC used also has a significant effect on the basal spacing of the resultant OC. The maximum basal spacings of OC prepared from single alkyl chain surfactants are markedly lower than those modified with double alkyl chain surfactants. As shown in Table 10.2, the maximum basal spacing of OMt with C18-S (single alkyl chain) at 4xCEC is 4.1 nm, while that of OMt with C18-D (double alkyl chains) is about 5.5 nm. In the case of C16 surfactant, an increase of the alkyl chain number from single to double can increase the maximum basal spacing from 3.9 to 5.2 nm. The corresponding values for C12 range from 2.0 to 4.8 nm, while those for C8 are from 1.5 to 3.9 nm. However, for a surfactant with three alkyl chains, the change in maximum basal spacing is very complex. For example, the basal spacing of Mt(NM)-C12-T (triple chain) is 3.9 nm while that of Mt(NM)-C12-D (double chain) is 4.8 nm, both at 4xCEC (Table 10.2). Therefore, the OMt modified with a triple alkyl chain surfactant is less expanded than its double-chain counterpart (He et al., 2010). On the other hand, OMt prepared from C18-T at 4xCEC have a larger basal spacing than those intercalated with C18-D at the same loading (Zhou et al., 2007b).

Vermiculites are of the trioctahedral type in which an octahedrally coordinated sheet of (Mg, Al, Fe) ions is sandwiched between two inward-pointing sheets of linked (Si, Al)$O_4$ tetrahedra. The amount of isomorphous

**TABLE 10.2 Maximum Basal Spacings of Organo-Montmorillonites When the Added Surfactants are 4.0 Times of the CEC (He et al., 2010)**

| Montmorillonite | Mt(AB) | Mt(HN) | Mt(NM) |
|---|---|---|---|
| *CEC (meq/100g)* | *72.4* | *88.8* | *106.5* |
| $d_{001}$ (nm) | | | |
| C18-S | 4.1 | 4.1 | 4.1 |
| C18-D | 5.5 | 5.6 | 5.5 |
| C16-S | 3.9 | 3.7 | 3.9 |
| C16-D | 5.2 | 5.3 | 5.2 |
| C12-S | 1.8 | 1.9 | 2.0 |
| C12-D | 4.7 | 4.4 | 4.8 |
| C12-T | 4.0 | 3.9 | 3.9 |
| C8-S | 1.4 | 1.5 | 1.5 |
| C8-D | 4.0 | 3.9 | 3.9 |

S: single alkyl chain; D: double alkyl chains; T: triple alkyl chains.

substitution in vermiculite, however, is generally greater than that in smectite group minerals and much of this occurs in the tetrahedral sheet ($Al^{3+}$ for $Si^{4+}$) (Brigatti et al., 2013b; Su et al., 2016). Accordingly, the layer charge density of vermiculite is higher than that of smectite group minerals, and the electrostatic interaction between vermiculite and interlayer cations is stronger than that of smectite (de la Calle and Suquet, 1988; Vahedi-Faridi and Guggenheim, 1997). Thus intercalation of organic modifiers into the interlayer space of vermiculite is difficult. For example, $Mg^{2+}$-saturated vermiculites fail to expand beyond a basal spacing of ~1.45 nm with glycerol whereas $Mg^{2+}$-Mt give a basal spacing at ~1.78 nm (MacEwan, 1961; Walker, 1961; Theng, 1974; Meunier, 2005). By the same token, the replacement of interlayer cations by alkylammonium ions in vermiculite is more difficult than that for smectite group minerals. A complete replacement of interlayer inorganic cations by alkylammonium ions in Mt can be reached after bringing the clay mineral into contact with an aqueous solution of appropriate alkylammonium halides for two hours at room temperature (Theng et al., 1967), whereas much longer periods of contact and elevated temperatures are usually required to obtain a similar interchange in vermiculite crystals ( John and Sen Gupta, 1967; Walker, 1967). Meanwhile, some mechanical methods (e.g. ball milling method) are also helpful for the preparation of organo-vermiculites. The organo-vermiculites prepared by hot solution and ball milling methods have larger basal spacing than those obtained in solution at room temperature. For example, HDTMAB modified vermiculite prepared at room temperature displays maximum basal spacings at 2.8–4.0 nm (Plachá et al., 2008; Wu et al., 2015) while those prepared by hot solution and ball milling methods have basal spacings larger than 5.0 nm (Wang et al., 2011), where the thermal energy and mechanical energy can accelerate the intercalation of surfactant into the interlayer space via exchanging with pre-existing inorganic cations. The large basal spacing obtained in organo-vermiculites is mainly due to the high charge density of vermiculite, for example, more cationic surfactant can be adopted in vermiculite interlayer space (Mittal, 2012). However, the different preparation methods required may have significant influences on the interlayer structure and thermal stability of the resulting organo-vermiculite intercalates.

Wang et al. (2011) conducted a comparison study of organo-vermiculites prepared by hot solution and ball milling methods, respectively. The organo-vermiculite prepared by the ball milling method displays a slight larger basal spacing (~5.24 nm) than that obtained in a hot solution (~5.01 nm), where a paraffin-bilayer arrangement model is adopted. In the case of organo-vermiculite prepared in a hot solution, the intercalated components are mainly surfactant cations whereas both surfactant cations and molecules (ion-pair) enter into the vermiculite interlayer space in the case of ball milling. This leads to very high loadings of surfactant and a low thermal stability of organo-vermiculite prepared by ball milling in comparison to that prepared from a solution. Certainly, the maximum basal spacing of organo-vermiculite is significantly affected by the chain length of surfactant, the CEC of the vermiculite, as well as the exposure to solvent (Lagaly, 1982; Ghabru et al., 1989; Mermut, 1994; Mermut and Lagaly, 2001b; Slade and Gates, 2004a; Lagaly et al., 2013). When surfactants with longer alkyl chains and vermiculites with larger CECs are used, greater basal spacings can be obtained in the resultant organo-vermiculites (Lagaly et al., 2013).

Generally, the basal spacing of organo-vermiculite is larger than that of OMt when they are prepared under similar experimental conditions (Slade and Gates, 2004a,b; Osman, 2006). However, the intercalation process of surfactant into vermiculite by cationic surfactants is very different from that of smectite. In the case of vermiculite, one-stepped increase of the basal spacing of cationic surfactant modified vermiculite was observed (Slade and Gates, 2004a,b; Plachá et al., 2008; Su et al., 2016), instead of a gradual stepwise increase of the basal spacing for cationic surfactant modified smectite (Lagaly, 1981; Zhu et al., 2003).

In the case of OMt, with an increase of the loaded surfactant amount, the intercalated surfactant may adopt different arrangement models, from lateral monolayer to lateral bilayer, then to paraffin monolayer and last to paraffin bilayer. Different from OMt, however, only two kinds of arrangement models (i.e. paraffin monolayer and paraffin bilayer) were reported in QAC modified vermiculites. This is evidenced by an abrupt increase of the basal spacing from 1.2 to 1.5 nm for pristine vermiculite (e.g. ~1.2 nm for $Na^+$-vermiculite and ~1.5 nm for $Ca^{2+}$-vermiculite) to 2.6–5.2 nm, depending on the preparation method, concentration, and alkyl chain length of the surfactant (Slade and Gates, 2004a; Wang et al., 2011; Su et al., 2016). More interestingly, the intercalation of vermiculite by QAC is inhomogeneous, for example, parts of the interlayer spaces are intercalated at low surfactant concentration and reach a large basal spacing close to the maximum with a paraffin-type arrangement while the others remain unchanged (without intercalation). With an increase in the surfactant loading amount, more and more interlayer spaces are occupied by the surfactant and finally reach a 'saturation' status (Su et al., 2016). This may be due to the high layer charge density and charge distribution on Si-O tetrahedral sheet in vermiculite, which results from the substitution of $Al^{3+}$ for $Si^{4+}$. During the exchange of interlayer cations by cationic surfactants, both repulsive forces (one is between the hydrophilic vermiculite surface and hydrophobic alkyl chain, and the other is between the two adjacent vermiculite layers with high negative charge density) may drive the vermiculite layers to be apart, resulting in a prominent increase in the interlayer height. Meanwhile, the positively charged amine group, the 'head' of the surfactant molecule, is fixed at the vermiculite layer surface through a strong electrostatic interaction between the positively charged 'head' and the negatively charged layer, and the hydrophobic alkyl chain orientation orients away from the surface, resulting in a paraffin arrangement (He et al., 2004a; Slade and Gates, 2004a), which through van der Waals interactions among alkyl chains, results in a strongly ordered all-*trans* conformation of the intercalated surfactant (He et al., 2004a). In the all-*trans* conformation, methyl groups are furthest apart, this configuration has the lowest energy, which was well evidenced by the FTIR spectra of organo-vermiculite showing $CH_2$ stretching vibrations at certain wavenumbers (see Section 10.3 for further details).

Natural mica is typical 2:1 type non-swelling clay minerals with higher layer charge density (1–2 eq/f.u.). The layer charge originates mainly from the substitution of $Al^{3+}$ for $Si^{4+}$ in tetrahedral sheets, which is balanced mostly by interlayer $K^+$ cation (Yariv, 1992a). These $K^+$ ions are all tightly confined in coordination cages above surface hexagons of Si-O tetrahedral sheets, leading to an inhibition of mica swelling (Liu and Lu, 2006). However, after exchanging natural mica (e.g. muscovite and sericite) with metal ions such as $Li^+$ (Yu et al., 2006) and $Na^+$ (Liang et al., 2013; Ding et al., 2014), alkylammonium cations can be successfully intercalated into the interlayer space of the resultant Li- and Na-mica. The basal spacing of the obtained organo-mica with ordered structure can reach 4.5–5.0 nm, controlled by the surfactant concentration, reaction temperature, and nature of preparation solutions.

Preparation of organo-mica by using synthetic micas has attracted great interest due to their swelling capacities similar to that of smectite or vermiculite, but with tuned layer charge values between 2.0 and 4.0 per unit cell (Kodama and Komarneni, 1999; Kodama et al., 2001; Park et al., 2002; Alba et al., 2006). These synthetic micas have a higher crystallinity, a controllable composition, and fewer impurities than natural clay minerals, giving them more advantages for being used as host materials than natural clay minerals (Pazos et al., 2012, 2015). Organo-micas were successfully synthesized by using synthetic micas with different layer charge densities and QAC with different chain lengths (Alba et al., 2011; Pazos et al., 2012, 2015). The basal spacing and arrangement of the intercalated QAC depend on both the layer charge density and the alkyl chain length. The structural arrangement of the QAC with a relatively short alkyl chain (e.g. dodecyl ammonium, C-12 or tetradecyl ammonium, C-14) in the interlayer space of high-charge micas is more sensitive to the effect of the mica layer charge than the QAC with a relatively long alkyl chain (e.g. octadecyl ammonium, C-18). For example, in organo-micas formed from dodecyl ammonium and tetradecyl ammonium, a transition from a parallel layer ($d_{001}=1.34$–$1.43$ nm) to a paraffin bilayer arrangement ($d_{001}=3.50$–$4.05$ nm) can be observed when the layer charge of the mica increases. However, when octadecyl ammonium is the interlayer cation, the molecular arrangement of the surfactant is found to follow the paraffin bilayer model ($d_{001}=4.48$–$4.60$ nm) for all values of layer charge (Pazos et al., 2012, 2015).

## 10.3 FTIR OF ORGANOCLAY INTERCALATES

FTIR has been a powerful technique to probe the structure of OC and molecular environments of the intercalated organic modifiers in the interlayer space, which have significant effects on behavior and properties of the hybrids (Vaia et al., 1994). The FTIR spectra of untreated Mt, neat HDTMAB and HDTMAB-Mt intercalates prepared at different HDTMAB concentrations, are shown in Fig. 10.5, and the respective band assignments are listed in Table 10.3. The major bands of the clay mineral backbone are in the range of 470–1120 $cm^{-1}$. These bands are associated with bending, deformation, and stretching of Si-O-Si, structural OH and adsorbed water. As these vibrations are within the crystal structure, the adsorbed HDTMAB or exchanged $HDTMA^+$ has little effect on the linkage of structural OH and Si-O bonds (He et al., 2004a; Li et al., 2008). However, the intercalation of $HDTMA^+$ had a profound effect on the presence and position of H-O-H stretching and bending of adsorbed water, due to changes in surface affinity of the Mt from predominantly hydrophilic to hydrophobic character. This is evidenced most significantly by intensity change and frequency shift of the bands in the H-O-H bending region (1600–1700 $cm^{-1}$) and stretching region of adsorbed water molecules (3100–3700 $cm^{-1}$) (He et al., 2004a; Li et al., 2008).

The frequency, width, and integrated intensity of the $CH_2$ and $CH_3$ stretching modes at 2800–3000 $cm^{-1}$, the scissoring modes at 1420–1475 $cm^{-1}$, and the rocking modes at 715–735 $cm^{-1}$ are all sensitive to the *gauche*/*trans* conformer ratio of alkyl chains as well as intermolecular interactions between the chains (Scheuing and Weers, 1990; Kung and Hayes, 1993; He et al.,

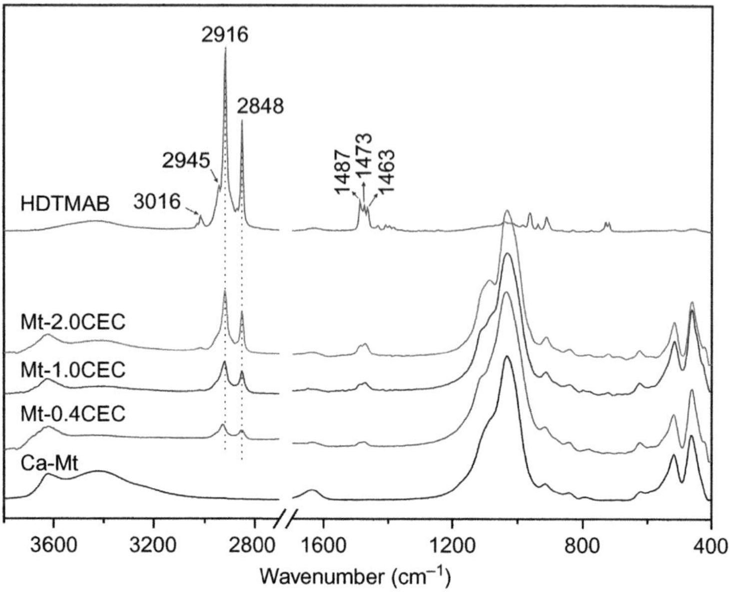


**FIG. 10.5** FTIR spectra of montmorillonite, HDTMAB and OMt prepared at different HDTMAB concentrations.

2004a; Ma et al., 2010). In an all-*trans* conformation the alkyl chain backbone is linear and the protons of linking $H_2C$-$CH_2$ groups alternate so that they are separated by the greatest distance possible. In the *gauche* conformation, the alkyl chain backbone is not linear; the alternation of $CH_2$ groups is interrupted such that the protons of linking $H_2C$-$CH_2$ groups are in a *cis* configuration (Lagaly et al., 2013). Thus the spectral parameters associated with e.g. the $CH_2$ stretching vibrations can be used to monitor structural rearrangements and molecular environments of the intercalated QAC, resulting from changes in packing density of surfactant, temperature, pressure and mechanical stress.

## 10.3.1 FTIR Spectrum of Surfactant in Organoclay

Generally, three regions of interest are particularly useful in the FTIR spectra of OC to investigate molecular arrangements and environments of the intercalated QAC within the interlayer space of clay minerals. These include C-H stretching (3100–2800 cm$^{-1}$), scissoring (wagging) (1500–1450 cm$^{-1}$), and rocking (deformation) (735–715 cm$^{-1}$) vibrational regions (Table 10.3). Their frequencies and intensities have been extensively used to evaluate the *gauche/trans* conformation ratio in alkyl chains, intermolecular interactions among alkyl chains, as well as the packing density of the intercalated QAC (Weers and Scheuing, 1990; Kung and Hayes, 1993; Venkataraman and Vasudevan, 2002; Barman et al., 2003).

**TABLE 10.3 Infrared Frequencies and Band Assignments for Crystalline Hexadecyltrimethylammonium Bromide (Weers and Scheuing, 1990; Li et al., 2008)**

| Wavenumber ($cm^{-1}$) | Assignment |
|---|---|
| 3040 | Asymmetric stretching of $CH_3$ ($\nu_{as}(CH_3)$) coupled to N-(CH3)$_3$ |
| 3016 | Symmetric stretching of $CH_3$ ($\nu_s(CH_3)$) coupled to N-$(CH_3)_3$ |
| 2945 | Asymmetric stretching of $CH_3$ ($\nu_{as}(CH_3)$) coupled to R[a]-$CH_3$ |
| 2916 | Asymmetric stretching of $CH_2$ ($\nu_{as}(CH_2)$) (uncoupled) |
| 2870 | Symmetric stretching of $CH_3$ ($\nu_s(CH_3)$) coupled to R-$CH_3$ |
| 2848 | Symmetric stretching of $CH_2$ ($\nu_s(CH_2)$) (uncoupled) |
| 1487, 1480 | Asymmetric deformation of $CH_3$ ($\delta_{as}(CH_3)$) coupled to N-$(CH_3)_3$ |
| 1472 (singlet) | $CH_2$ bending ($\delta(CH_2)$) in triclinic subcell |
| 1473, 1463 (doublet) | $CH_2$ bending ($\delta(CH_2)$) in orthorhombic subcell |
| 1468 (singlet) | $CH_2$ bending ($\delta(CH_2)$) in hexagonal or rotator phase |
| 1468–1466 | $CH_2$ bending ($\delta(CH_2)$) in 'liquid-like' disorder |
| 1465–1456 | Asymmetric deformation of $CH_3$ ($\delta_{as}(CH_3)$) coupled to R-$CH_3$ |
| 1420 | α-$CH_2$ bending (deformation) ($\delta(\alpha CH_2)$) |
| 1405 | Asymmetric deformation of $CH_3$ ($\delta_{as}(CH_3)$) coupled to N-$(CH_3)_3$ |
| 1380 | Symmetric deformation of $CH_3$ ($\delta_{as}(CH_3)$) coupled to R-$CH_3$ |
| 912 | C–N stretching ($\nu(CN)$) |
| 729 | $(CH_2)_n$ bending (rocking mode) ($\tau(CH_2)_n$) |
| 719 | |

[a]*R: alkyl chain.*

Vaia et al. (1994) was among the first to apply FTIR spectroscopy in conjunction with XRD to investigate the interlayer structure and phase state of intercalated alkylammonium silicates. By monitoring frequency shifts of the asymmetric $CH_2$ stretching and bending vibrations, they found that the intercalated chains exist in states of varying degrees of order. In general, with a decrease in the interlayer packing density or the chain length or an increase in the temperature, the intercalated chains adopt a more disordered, liquid-like structure, resulting from an increase in the *gauche/trans* conformation ratio.

As the available surface area within the interlayer space per alkylammonium molecule approaches the size of the molecule, which strongly depends on the packing density and the configuration of the intercalated alkylammonium molecules, the alkyl chains are not completely disordered, but retain preferred orientation order similar to that in a liquid crystalline state. Since the chains are free to move around and reorient in the liquid-like state, they attain an equilibrium configuration much faster compared to those in a liquid crystalline or solid-like state. These findings differ from the proposed conventional models based only on XRD analysis, which imply or assume all-*trans* conformations as described above, and related re-orientations with increasing loadings (e.g. Lagaly, 1981; Zhu et al., 2003; Slade and Gates, 2004a) but the extensive recent studies on the interlayer structure of OC with FTIR (e.g. Vaia et al., 1994; He et al., 2004a; Ma et al., 2010) indicate that changes to *gauche* conformation is important to consider.

The antisymmetric and symmetric $CH_2$ stretching vibrations are the most common modes used to investigate the configuration of the intercalated QAC. In this regard, the antisymmetric $CH_2$ stretching mode is more sensitive to the conformational ordering than the symmetric stretching mode (He et al., 2004a; Ma et al., 2010).

The frequency and bandwidth of both the antisymmetric and symmetric $CH_2$ stretching modes of OC intercalates strongly depend on their packing density (Vaia et al., 1994; Li and Ishida, 2003a). He et al. (2004a) studied the FTIR spectra of $HDTMA^+$- Mt intercalates in the range of 2800–3000 $cm^{-1}$ (Fig. 10.6). The intensities of the antisymmetric ($\nu_{as}$) and symmetric ($\nu_s$) C–H stretching vibrations in both -$CH_2$ (at ~2916 and 2848 $cm^{-1}$, respectively) and –$CH_3$ (at 2952 and 2867 $cm^{-1}$, respectively) groups increase gradually as a function of the packing density of the intercalated amine chains. This suggests that the packing density of amine chains also contributes to changes in the spectra. The frequencies of the $CH_2$ stretching vibrations of alkyl chains are extremely sensitive to the conformation of the chains (Mendelsohn et al., 1995; Venkataraman and Vasudevan, 2002; Barman et al., 2003; Li and Ishida, 2003a). Only when the chains are highly ordered (i.e. in an all-*trans* conformation), do the narrow absorption bands of antisymmetric $\nu_{as}(CH_2)$ (near 2916 $cm^{-1}$) and symmetric $\nu_s(CH_2)$ (near 2848 $cm^{-1}$) appear (Li and Ishida, 2003a) (Fig. 10.6). The frequency and bandwidth of both $\nu_{as}(CH_2)$ and $\nu_s(CH_2)$ modes of the confined alkyl chains are therefore dependent on the QAC packing density. In the case of high loading levels ($>2 \times$ CEC), both the frequencies of $\nu_{as}(CH_2)$ and $\nu_s(CH_2)$ are close to those for pure QAC (respectively, 2916 and 2848 $cm^{-1}$) with a very narrow bandwidth. At relatively high packing densities (e.g. $>1 \times$ CEC), the confined alkyl chains adopt an essentially all-*trans* conformation. However, at relatively low packing density (e.g. $<1 \times$ CEC), the frequency shifts significantly to high wavenumber and the vibration band is obviously broadened, reflecting that a large number of *gauche* conformers introduced into the alkyl chain

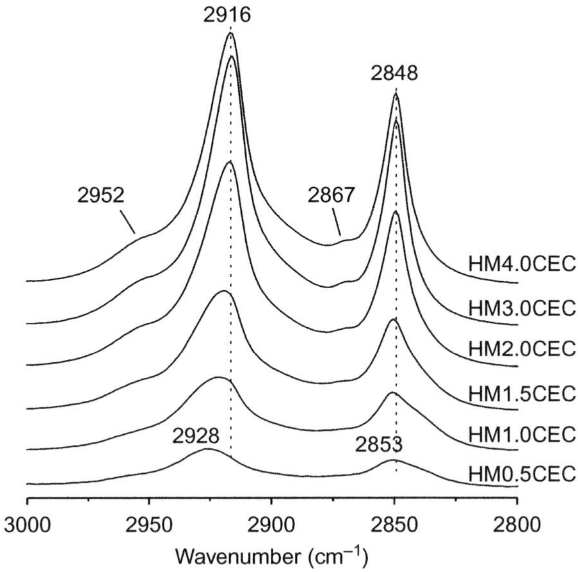


**FIG. 10.6** FTIR spectra of the HDTMA$^+$ intercalated montmorillonites prepared at 0.5–4.0 CEC (2800–3000 cm$^{-1}$). *Modified from He et al. (2004).*

(Yu et al., 2006). Consequently, both the frequency and bandwidth of antisymmetric and symmetric $CH_2$ stretching vibration bands can reflect the change of *gauche*/*trans* conformer ratio in the intercalated QAC (Vaia et al., 1994; Yu et al., 2006).

He et al. (2004a) found that the wavenumber variation of the $\nu_{as}(CH_2)$ mode was as much as 12 cm$^{-1}$ (from 2928 cm$^{-1}$ for OMt prepared at 0.5 CEC of HDTMAB (HM0.5CEC) to 2916 cm$^{-1}$ for the one prepared at 4.0 CEC of HDTMAB (HM4.0CEC)). However, the $\nu_s(CH_2)$ mode shifted by 5 cm$^{-1}$ (from 2853 cm$^{-1}$ for HM0.5 CEC to 2848 cm$^{-1}$ for HM4.0 CEC). This suggests that the symmetric $CH_2$ stretching mode is less sensitive to the alkyl chain conformation than the antisymmetric $CH_2$ stretching mode (He et al., 2004a; Ma et al., 2010).

Methyl scissoring vibrations at 1480 and 1440 cm$^{-1}$ and rocking vibrations at 750–700 cm$^{-1}$ are also sensitive to local environments of intercalated surfactants. For example, the spectrum of pure C16-S displays doublets at 1473 and 1463 cm$^{-1}$ corresponding to the scissoring modes, and those at 730 and 720 cm$^{-1}$ to the rocking modes (Fig. 10.7), respectively. The splitting with 10 cm$^{-1}$ of the -$CH_2$ scissoring and rocking vibrations is due to the intermolecular interaction between two adjacent hydrocarbon chains in a perpendicular orthorhombic sub-cell (Snyder, 1979; Casal et al., 1982), and further requires an all-*trans* conformation for its detection (Flach et al., 1997;

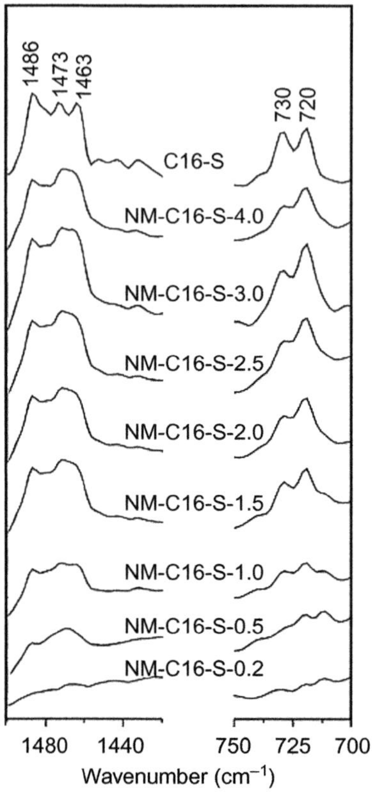


**FIG. 10.7** The scissoring and rocking vibrations of $CH_2$ in organo-montmorillonites prepared from C16-S. *Modified from Ma et al. (2010).*

Li and Ishida, 2003a). For OC prepared at a low surfactant concentration (i.e. low packing density), the doublet at 1473 and 1463 cm$^{-1}$ disappears, and is substituted by a single broad band at 1468 cm$^{-1}$. The broad single absorption band is related either to a liquid-like structure or to the existence of relatively disordered hexagonal sub-cell packing, where the hydrocarbon chain freely rotates around its long axis (Li and Ishida, 2003a). With an increase in surfactant loading, the intensity of this vibration increases with an occurrence of band splitting (Zhu et al., 2005; Ma et al., 2010). A similar phenomenon is also observed for the -$CH_2$ rocking modes. When OC are prepared at a low surfactant concentration (e.g. <1 CEC), the FTIR spectra only display a broad singlet at ~723 cm$^{-1}$. However, with a further increase in surfactant loading, two well-resolved absorption bands at 730 and 720 cm$^{-1}$ can be observed (Li and Ishida, 2003a; Ma et al., 2010).

Whether or not the –$CH_2$ rocking and scissoring mode doublets can be detected might strongly depend on the used techniques. He et al. (2004a) pointed out that two well-resolved absorption bands at 730 and 720, and 1473 and 1463 $cm^{-1}$ were observed in the FTIR spectra with the KBr pressed disk technique, while the spectra obtained using the attenuated total reflection (ATR) technique only displayed singlets at 1468 and 720 $cm^{-1}$. This suggests that the KBr pressed disk technique may be more suitable for probing conformational ordering of the confined alkyl chains within the interlayer space of clay minerals than the ATR technique.

The splitting of methylene scissoring and rocking modes not only relates to the investigation technique used (and surfactant loading), but it also relates to the configuration of surfactant intercalate. For OC prepared from surfactants with more than one alkyl chain, their FTIR spectra rarely display doublets of methylene scissoring and rocking modes (Ma et al., 2010). The frequency shifts of both $\nu_{as}(CH_2)$ and $\nu_s(CH_2)$ modes in OC, formed from surfactants with more than one chain, also differ from those from surfactants containing only single alkyl chains. For OC prepared from HDTMAB, with an increase in surfactant loading, $\nu_{as}(CH_2)$ shifts from 2926 to 2919 $cm^{-1}$ and $\nu_s(CH_2)$ shifts slightly from 2855 to 2851 $cm^{-1}$. Furthermore, for OC prepared from dihexadecyl dimethylammonium bromide (double chain), the shifts of $\nu_{as}(CH_2)$ and $\nu_s(CH_2)$ modes are, respectively, 4 $cm^{-1}$ (2923–2919 $cm^{-1}$) and 1 $cm^{-1}$ (2852–2851 $cm^{-1}$). This suggests that the double chain surfactant prefers to adopt a highly ordered conformation in comparison to single chain surfactant, probably due to higher packing density from the additional alkyl chains.

The chain length also has a significant effect on the configuration of the intercalated QAC. Vaia et al. (1994) depicted a relationship between $\nu_{as}(CH_2)$, alkyl chain length, and basal spacing of OMt with a similar packing density of intercalated surfactant (Fig. 10.8). With an increase in the chain length, the $\nu_{as}(CH_2)$ mode shifts in a stepwise fashion from 2932 $cm^{-1}$, for OC prepared from hexyl trimethylammonium (HTMA, C6), to 2921 $cm^{-1}$ for the one prepared from octadecyl trimethylammonium (ODTMA, C18). The frequency shifts of $\nu_{as}(CH_2)$ are mirrored by a similar step decrease in the interlayer height (Fig. 10.8) suggesting that the two are interrelated. The large variation in $\nu_{as}(CH_2)$ for OC prepared from alkyl amines with varying chain length suggests that the chains exist in a variety of molecular environments. The shifts in the $\nu(CH_2)$ vibrations largely reflect changes in the chain conformation, with the frequency shifting from lower frequencies for ordered, all-*trans* chains, to higher frequencies as the chain disorder (*gauche*/*trans* conformer ratio) increases, as described above. Conversely, inter-chain interactions dominate the $\delta(CH_2)$ mode, which is less sensitive to the alkyl chain conformation in comparison to the $\nu_{as}(CH_2)$ mode (He et al., 2004a; Ma et al., 2010).

In general, with an increase in temperature, both the $\nu_{as}(CH_2)$ and $\nu_s(CH_2)$ modes shift from a lower frequency (characteristic of the liquid crystalline state) to higher frequencies before finally plateauing at a frequency approaching that of a liquid-like state (Vaia et al., 1994; He et al., 2004b). The more

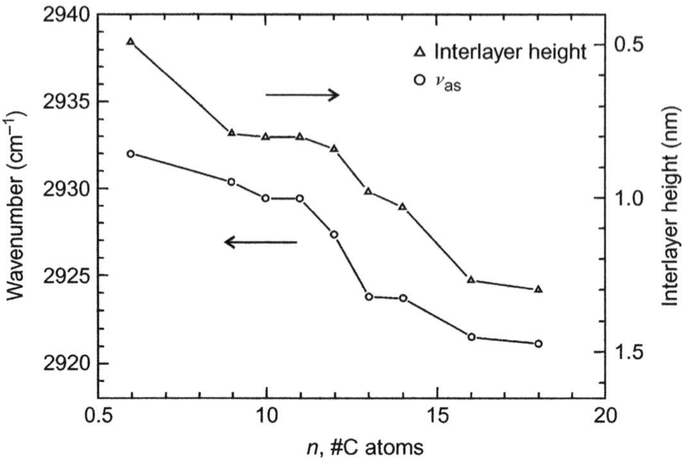


**FIG. 10.8** Antisymmetric $CH_2$ stretching ($\nu_{as}(CH_2)$) and interlayer height as a function of chain length ($C_n$) for surfactants measured at room temperature. *From Vaia et al. (1994).*

disordered environment at elevated temperatures is further supported by band broadening and decreasing intensity. The chain disordering on heating is consistent with molecular dynamics (MD) simulations where the chains are shown to pass gradually from a low-temperature solid-like state to a high-temperature liquid-like state over a broad range of temperatures between 200 and 300 K (Klatte and Beck, 1993).

## 10.3.2 FTIR of Clay Mineral in Organoclay

A typical IR spectrum of clay mineral shows stretching vibrations of structural OH, H-O-H stretching and bending of physi-sorbed, hydration and hydrogen bonded water, Si-O stretching, and deformation related to structural OH. The IR spectrum of Mt shows a sharp intense band at ~3630 cm$^{-1}$ assigned to the asymmetric stretching vibration of structural hydroxyl ($\nu$(OH)) and a broad band at ~3430 cm$^{-1}$ ascribed to stretching modes of physi-sorbed water molecules adsorbed within the interlayer space of the clay mineral, and bands at 1640 cm$^{-1}$ attributed to in-plane bending of water in the hydration sphere of the interlayer cation. A broad band peaking near 1030 cm$^{-1}$ is ascribed to the stretching vibration of Si–O groups. This complex band generally consists of four poorly resolved components. Three of the components (near 1000, 1035 and 1120 cm$^{-1}$) correspond to 'in-plane' $\nu$(SiO) vibrations of Si with basal oxygen atoms ($O_{bas}$), while the other, near 1080 cm$^{-1}$, is due to an 'out-of-plane' $\nu$(SiO) (or perpendicular) vibration between Si and apical oxygen atoms ($O_{ap}$). The bands at 520 and 455 cm$^{-1}$ correspond to Al–O–Si and Si–O–Si bending vibrations, respectively (Madejová et al., 2011; Zhuang et al., 2015). The character of these interlayer water molecules is greatly dependent on the moisture level and the interlayer cation (Bishop et al., 1994; Ma et al., 2010).

The introduction of organic surfactants by exchange reaction and/or adsorption mainly results in the presence of some of the vibrations associated with the surfactant, accompanied with frequency shifts of the bands associated with the clay mineral host (He et al., 2004a; Chen et al., 2005; Ma et al., 2010; Tomić et al., 2012b). This occurs most notably for physi-sorbed and cation hydration water, but it also influences some of the absorbance bands associated with the clay mineral structure.

A moderately intense band at ~3620 $cm^{-1}$ is attributed to structural $\nu$(OH) while broad bands at 3100-3550 and 1600-1700 $cm^{-1}$ are ascribed to, respectively, stretching and bending modes of adsorbed water molecules (Fig. 10.9). The frequency of the maxima for these bands change with an increase in the surfactant loading, while structural $\nu$(OH) are largely independent of the surfactant loading. For example, with an increase in surfactant loading, the frequency of $\delta$(H–O–H) vibrations corresponding to water molecules hydrating the interlayer cation, gradually shifts from 1636 $cm^{-1}$ (typical for $Na^+$-smectite) to 1647 $cm^{-1}$ (He et al., 2004a). Simultaneously, the intensity of this band decreases significantly. The decrease of the band intensity reflects that the amount of hydrogen bonded $H_2O$ molecules present in the OC with

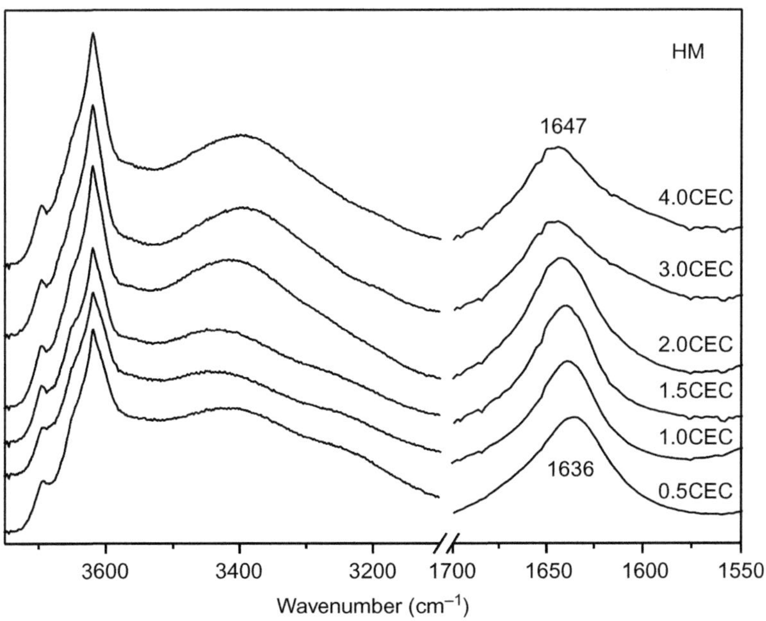


**FIG. 10.9** FTIR spectra of organo-montmorillonites at 3100-3550 and 1600-1700 $cm^{-1}$. *Modified from He et al. (2004).*

higher surfactant loading is less than those with lower surfactant loading because the hydrophobic surfactant cations displace the interlayer $Na^+$ and its hydrated water molecules. Hence, $H_2O$ molecules are not easily adsorbed by OMt and are restricted in their ability to form hydrogen bonds with adjacent $H_2O$ molecules (Xu et al., 2000). Similarly, the broad band at 3100–3550 $cm^{-1}$, ascribed to physi-sorbed water, is also highly dependent on the surfactant loading. At a relatively low surfactant loading, FTIR spectra of OC display a broad absorption band at approximately 3440 $cm^{-1}$ and a shoulder at around 3250 $cm^{-1}$, corresponding to the overlapping antisymmetric $\nu_3$ and symmetric $\nu_1$(HOH) stretching vibrations of H-bonded water (Madejová et al., 2002a) and an overtone ($2\nu_2$) of the bending mode of cation hydration water (Farmer, 1974b), respectively. At relatively high surfactant loading, FTIR spectra only show one broad absorption band due to stretching vibrations of H-bonded water and the band shifts to lower frequency. It is noteworthy that various IR techniques show different sensitivities to adsorbed water. Ma et al. (2010) proposed that ATR and diffuse reflectance infrared Fourier transform (DRIFT) techniques are sensitive to the content and state of the water molecule physi-sorbed on clay minerals in comparison to the KBr pressed disk technique. Since the KBr matrix is hygroscopic, absorption bands of water molecules adsorbed on KBr and those present on clay minerals are overlapped. Therefore it is difficult to distinguish individual water vibrations in unheated samples and assess the water content in OC (Madejová, 2003; Ma et al., 2010).

The intercalation of surfactant into the interlayer spaces of clay minerals has a significant influence on the arrangement of $SiO_4$ tetrahedral layers, due to the interaction between the intercalated surfactant and silicate surfaces. The band shape and frequencies of these vibrations strongly depend on the surfactant loading and alkyl chain length in OC (Ma et al., 2010; Tomić et al., 2012b). In the FTIR spectrum of OMt, the bands in the region of 950–1100 $cm^{-1}$, corresponding to $\nu$(SiO), split into a sharp band at 1033 $cm^{-1}$ with a shoulder near 1088 $cm^{-1}$ attributed to perpendicular Si–O stretching (Slosiariková et al., 1992). For OC with a low surfactant loading (e.g. 0.2 and 0.5 CEC), the intensity of the shoulder at approximately 1088 $cm^{-1}$ is markedly reduced. With an increase in surfactant loading, this shoulder is recorded again and shifts to a higher frequency. When the surfactant loading increases further, the frequency remains essentially unchanged but the band intensity increases (Ma et al., 2010). The number of surfactant alkyl chains also affects the Si–O vibrations of clay mineral host. When the intercalated surfactant chain number increases from single to double, the band of perpendicular $\nu$(SiO) is more prominent. It is noteworthy that the splitting of the $\nu$(SiO) is also affected by the basal spacing of OC; therefore, the larger the basal spacing, the more obvious the splitting of the $\nu$(SiO) observed (Ma et al., 2010). The intercalation of surfactant into the interlayer space is accompanied by a marked interlayer swelling, during which a perpendicular adsorbate orientation is reached (Lerot and Low, 1976; Slosiariková et al., 1992). In this way, the interlayer distance is markedly increased (similarly as in Mt slurries), and the orientation of the $SiO_4$ tetrahedral sheets is changed, most probably toward a better-ordered arrangement and more marked manifestation of the perpendicular ν(SiO) mode (Slosiariková et al., 1992; Ma et al., 2010).

## 10.4 IN SITU XRD AND FTIR OF ORGANOCLAY

As a family of important adsorbents for the cleanup of organic pollutants in the environment, OC are often used in media with bulk water. Although the structure and properties of OC in a dry state have been well documented, these results may not represent completely the real in-field conditions experienced by OC in water (Xu and Boyd, 1995; Kung and Hayes, 1993). The structural information of OC in a wet state is, therefore, of high importance for understanding the interaction between OC and pollutants, and potential industrial applications. To date, in situ XRD and FTIR are the main characterization techniques to investigate the structure of OC in a wet state (Ma et al., 2010; Zhu et al., 2008).

Zhu et al. (2008) investigated the interlayer structure of OC in a wet state by using wet paste samples rather than dry powder, which contain excess water. Generally, OC in wet state have a larger basal spacing than the corresponding ones in a dry state. Fig. 10.10 displays the interlayer height of OMt in both wet and dry states, prepared from HDTMAB, and the difference between the corresponding samples. The interlayer height difference between the wet and dry samples increases from 0.47 nm for OMt prepared at 0.2 CEC to 1.5 nm for the ones prepared at 0.4–1.2 CEC, and then

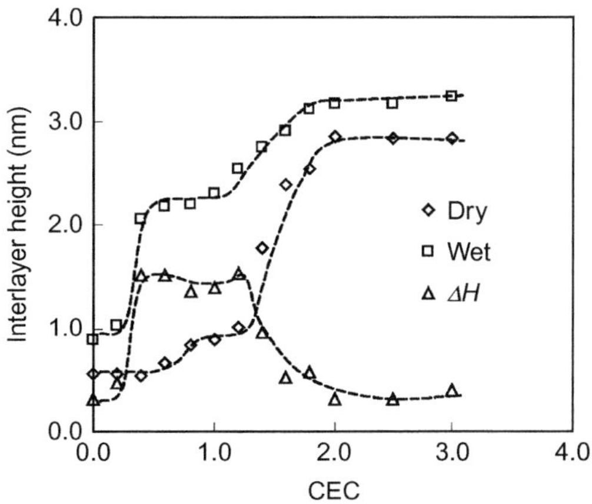


**FIG. 10.10** Interlayer height of OMt in dry and wet states. $\Delta H$ is the interlayer height difference between OMt in dry and wet states. *From Zhu et al. (2008).*

decreases to 0.31 nm for the one prepared at 2.0 CEC (Zhu et al., 2008). Such variation strongly suggests that the microstructure of OC in a wet state differs from that in a dry state. The structure information obtained from wet paste samples is more representative for the real state of clay minerals and their OC derivatives in dispersion in comparison to those obtained from dry samples or hydrated ones in equilibrium with water vapor under different relative humidity conditions (Gates, 2004; Ferrage et al., 2005, 2010; Fehervari et al., 2016; Gates et al., 2016). This finding suggests that water molecules can in fact enter the interlayer space of OC, despite their increased hydrophobicity, and therefore play an important role in regulating the microenvironments of intercalated surfactants (Zhu et al., 2008).

Ma (2010) conducted a similar study on OMt prepared from a series of surfactants with different alkyl chain lengths, which showed that, in comparison to the corresponding samples in dry state, the basal spacing of OMt in a wet state strongly depended on the surfactant loading and the chain length of the surfactant used. For OMt prepared from surfactants with relatively long alkyl chains and at high surfactant concentrations (e.g. C16 and 2.0–5.0 CEC), all OMt in both dry and wet states have a similar basal spacing at ~4.0 nm and a series of highly ordered (00l) reflections. For OMt prepared at surfactant concentrations lower than 2.0 CEC, a prominent increase of the basal spacing from dry state to wet state was observed. However, for OMt prepared from surfactants with relatively short alkyl chains (e.g. C8 and C12) and at surfactant concentrations of 0.2–5.0 CEC, the basal spacing of OMt in a wet state was obviously larger than that of the corresponding samples in dry state (Ma, 2010).

In order to elucidate the microenvironment of the intercalated surfactants in OC under different conditions, Ma et al. (2010) further conducted a comparison study on OMt in dry and wet states by using FTIR, where KBr pressed disk, ATR, and DRIFT techniques were applied. Similar to that for OC in a dry state, the intercalated alkyl chains within the interlayer space of OMt in a wet state also experience an evolution from liquid-like state (disordered) to solid-like state (ordered) with an increase in surfactant loading. Fig. 10.11 displays FTIR spectra of OMt in dry and wet states, prepared from HDTMAB (C16) at concentrations of 0.2–4.0 CEC. For an OC sample (NM-C16-0.2, prepared at 0.2 CEC), the $\nu_{AS}(CH_2)$ occurs at 2926 $cm^{-1}$ in dry state whereas that of the corresponding sample in wet state moves to 2930 $cm^{-1}$. This shift suggests that a greater proportion of *gauche* conformers were introduced into the alkyl chains in the wet state compared to the dry state. Such difference between the samples in dry and wet states gradually diminishes with an increase of surfactant loading. For OMt prepared at surfactant concentrations $\geq$ 2.0 CEC, FTIR spectra of OMt in dry and wet states are identical, displaying $\nu_{AS}(CH_2)$ and $\nu_S(CH_2)$ at, respectively, 2919 and 2851 $cm^{-1}$ (Figs. 10.11 and 10.12). However, for OMt prepared from surfactants with relatively short alkyl chains (e.g. C8 and C12), the frequencies of the $\nu_{AS}(CH_2)$ mode of wet samples always are higher than those in the corresponding dry

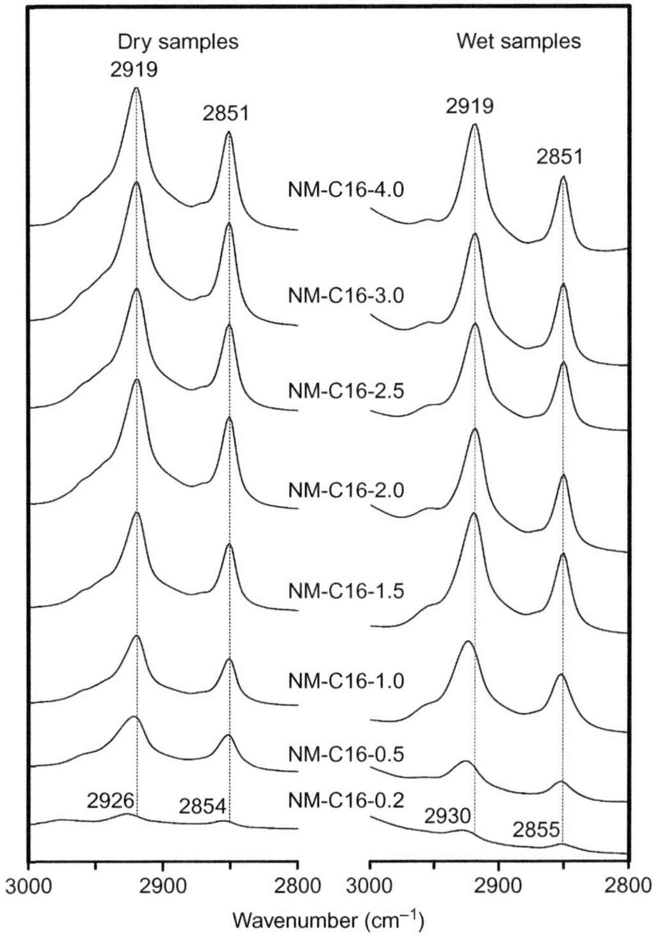


**FIG. 10.11** FTIR spectra of organo-montmorillonites prepared from HDTMA (C16) at 0.2-4.0 CEC in dry (left) and wet (right) states. *From Ma et al. (2010).*

ones, even in the cases of OMt prepared at high surfactant concentrations (2.0–4.0 CEC) (Fig. 10.12) (Ma et al., 2010). For example, the FTIR spectrum of OMt prepared from octyl trimethylammonium chloride (C8) at 2.0 CEC displays $\nu_{AS}(CH_2)$ at 2934 cm$^{-1}$ while that of the corresponding dry sample displays at 2928 cm$^{-1}$ (Fig. 10.12).

As indicated by Fig. 10.12, the $\nu_{as}(CH_2)$ difference between the samples in dry and wet states strongly depends on the alkyl chain length and surfactant loading, i.e. such difference decreases with the increase of alkyl chain length and increase of surfactant loading (Ma et al., 2010). Lagaly and Dekany (2005) proposed that in addition to liquid molecules strongly organized

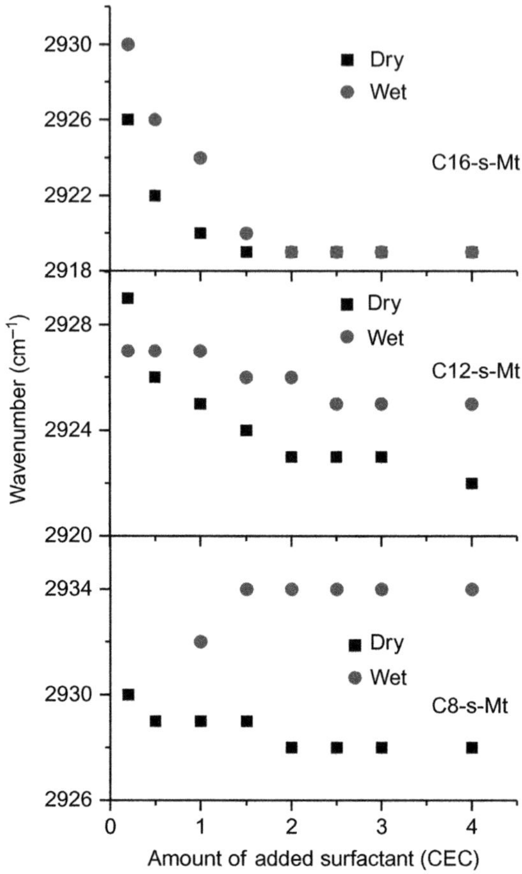


**FIG. 10.12** Comparison of $\nu_{as}(CH_2)$ in organo-montmorillonites prepared from surfactants with different alkyl chain lengths at dry and wet states. *From Ma et al. (2010).*

around the alkyl chains and on the bare silicate surface (composing the adsorption phase), other molecules are taken up that do not belong to the adsorption phase (i.e. they have bulk-like properties). Further these molecules diminish the entropy loss of the system because they increase the conformational freedom of the alkyl chains (which are probably not in the all-*trans* conformation). They also increase the mobility of the liquid molecules. At relatively high surfactant loading, most hydration cations can be exchanged by surfactant cations, resulting in a decrease in hydration water content and a surface property change from hydrophilic to hydrophobic. In this case, water molecules are difficult to introduce into the interlayer space and the interlayer cannot be further expanded even if the Mt layers are surrounded by an abundance of water in a wet condition. This is strongly evidenced by the similar basal spacings of OC with high surfactant loading in wet and dry states (Ma et al., 2010).

On the basis of the in situ XRD measurements, for the first time, Zhu et al. (2008) discussed the interlayer structure of OMt in water and its effect on the adsorption toward organic contaminants. For OMt with low surfactant loading (Fig. 10.13A), the height of the interlayer is relatively low, in which the intercalated surfactant cations lie parallel to the clay mineral surface; thus the available interlayer space for accommodating a contaminant is limited. On the other hand, as part of the inorganic cations remains unexchanged, the interlayer spaces still show relatively strong hydrophilicity, causing low affinity toward organic contaminants. When the exchangeable sites are completely occupied by surfactant cations in the case of 1.0 CEC sample (Fig. 10.13B) the intercalated surfactants adopt an arrangement perpendicular to the silicate

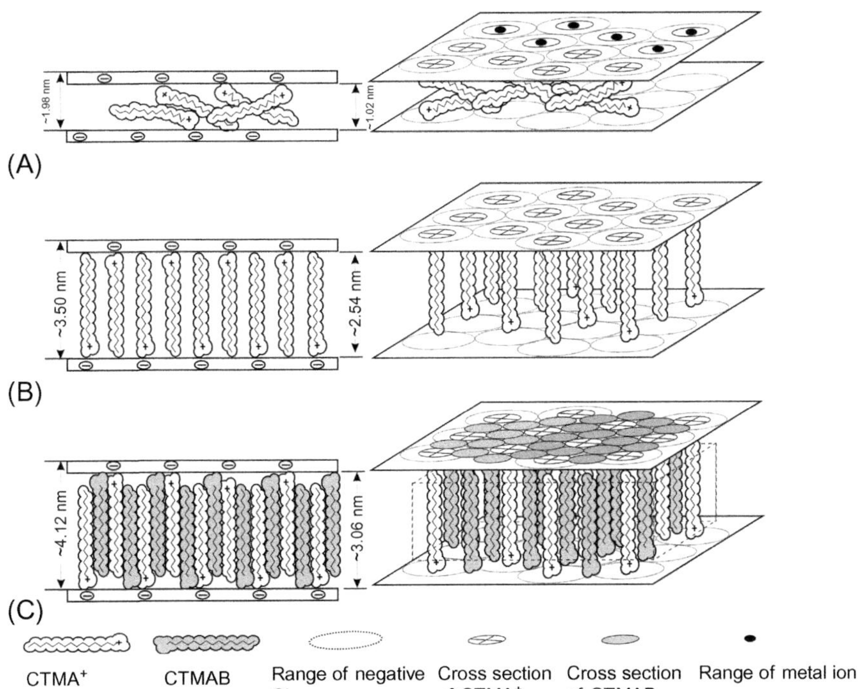


**FIG. 10.13** A schematic drawing of three-dimensional structure of surfactant molecules or/and ion's location on clay layers (including top view). (A) Low surfactant loadings (0.2–0.8 CEC); (B) Middle surfactant loadings (1.0 CEC); (C) High surfactant loadings (1.2–3.0 CEC). The dash-line cube is donated as a partition phase, and the available space confined by hydrophobic siloxane surface is limited gradually with the increase in surfactant loadings. *Modified from Zhu et al. (2008).*

layer surface. The space between the surfactant cationic 'pillars' can provide sufficient room for organic contaminants. Thus these samples have the highest adsorption capability. For OMt with the high surfactant loading, the accessible room is occupied by the surfactant through cation exchange and adsorption of surfactant molecules (ion pair) (Fig. 10.13C) (Tahani et al., 1999; Klapyta et al., 2001; Lee and Kim, 2002; Janek and Lagaly, 2003). Consequently, the adsorption capacity of OC toward organic contaminants prominently decreases. According to the structure characterization of OC in a wet state and the adsorption results of OC toward model organic contaminants (nitrobenzene, phenol, and aniline), a relationship between surfactant loading, structure of OC, and the adsorption capability toward the organic contaminants was established (Zhu et al., 2008).

## 10.5 FTIR OF ORGANOCLAY WITH ADSORBED ORGANIC CONTAMINANTS

The effectiveness of different applications of OC mainly depend on their physical and chemical affinity toward the organic contaminants. With the addition of organic modifiers, the surface of clay minerals shows better affinity toward the given organic contaminants. As the application of OC for environmental remediation is gaining increasing attention, some studies have been conducted to explore how the FTIR technique can be applied toward investigating the interaction mechanism between OC and organic contaminants, and the structural change of OC before and after adsorption (Singh et al., 2003, 2004; Zhou et al., 2007a,c, 2008; Vidal and Volzone, 2012; Liu et al., 2013; Yan et al., 2014).

Fig. 10.14 shows the FTIR spectra of para-nitrophenol (PNP), Mt, OMt, and OMt with adsorbed PNP. The FTIR spectrum of PNP displays OH stretching vibrations at ~3400 and ~3318 $cm^{-1}$, attributed to non-hydrogen bonded phenol OH units and hydrogen bonded ones, respectively (Zhou et al., 2007a). The intensity of the vibration at ~3318 $cm^{-1}$ is much stronger than that at ~3400 $cm^{-1}$. Two types of hydrogen bonding are proposed: (1) hydrogen bonding between PNP molecules and (2) hydrogen bonding between PNP and water molecules. The latter bonding is proposed as a mechanism for the adsorption of PNP on clay minerals and OC with low organic modifier loading. The IR spectrum of $Na^+$-Mt displays two vibrations at ~3612 and ~3394 $cm^{-1}$, which correspond to structural hydroxyls within the clay mineral structure and adsorbed water, respectively. A less intense band observed at ~3235 $cm^{-1}$ is assigned to water hydrogen bonded to other water molecules within the interlayer space of Mt. These water molecules are involved in the structure of the hydration sphere of the cation within the Mt interlayer space. The IR spectrum of $Na^+$-Mt with adsorbed PNP shows vibrations at the same frequencies as those for $Na^+$-Mt. While the intensity of

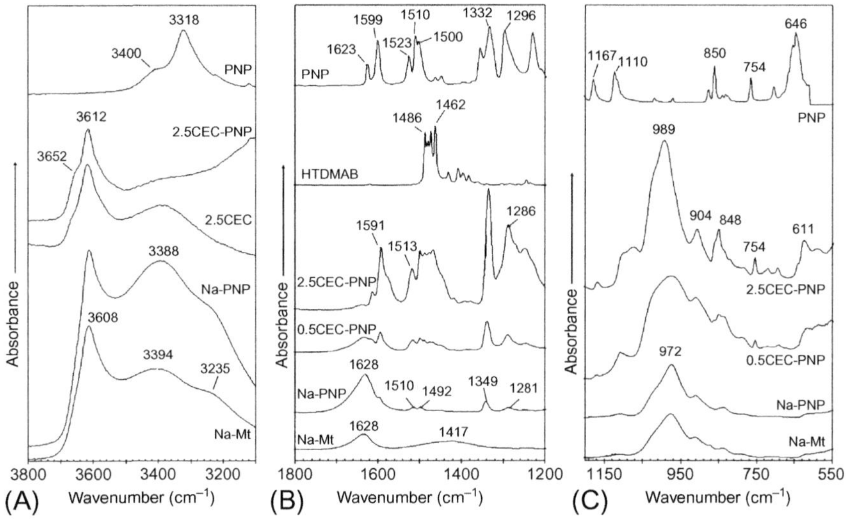


**FIG. 10.14** Infrared absorption spectra of PNP, $Na^{+}$-Mt, OMt and the corresponding samples with adsorbed PNP. (A) 3100–3800 $cm^{-1}$, (B) 1200–1800 $cm^{-1}$ and (C) 550–1150 $cm^{-1}$. *Modified from Zhou et al. (2007).*

vibrations at ~3388 and ~3235 $cm^{-1}$ obviously increase, indicating that PNP is adsorbed through hydrogen bonding to the water molecules in the cation hydration sphere.

For OMt, the intensity of the vibrations attributed to water molecules decreases, which can be attributed to an increase in the surface hydrophobicity and a subsequent decrease in adsorbed water content (He et al., 2005). A further decrease in vibration intensities of these water molecules is observed upon the adsorption of PNP (Fig. 10.14A and B), reflecting that the adsorption of organic contaminants can also contribute to an increase in surface hydrophobicity. It is noteworthy that in the FTIR spectra of both OC (e.g. 2.5CEC-Mt) and corresponding OC with adsorbed PNP (e.g. 2.5CEC-PNP), an additional band is observed at approximately 3652 $cm^{-1}$ (Fig. 10.14A). Zhou et al. (2007a,b) proposed that this vibration corresponded to an interaction between the intercalated surfactants/adsorbed organic contaminants and the structural hydroxyls of Mt.

Table 10.4 illustrates a summary of the characteristic vibrations of para-nitrophenol and their assignments (El-Eswed et al., 1997; Abkowicz-Bienko et al., 1999; Zhou et al., 2007c). When PNP is adsorbed on Mt, obvious frequency shifts of these vibrations can be observed. For instance, the $\nu(C=C)_{arom}$ of PNP shifts from 1599 to 1591 $cm^{-1}$ while that of $\nu_s(NO_2)$ shifts from 1332 $cm^{-1}$ to 1340 $cm^{-1}$. The cation–π interaction happens between the electron-rich π system (benzene ring in PNP) and the cation of the surfactant. Cations behave as electron withdrawing groups that withdraw electron

**TABLE 10.4 Characteristic Vibrations of Para-Nitrophenol and Their Assignments**

| Wavenumber ($cm^{-1}$) | Assignment |
|---|---|
| 3400 | Free $\nu$(OH) |
| 3318 | Bonded $\nu$(OH) |
| 1623 | Aromatic C=C stretching ($\nu$(C=C)) |
| 1599 | |
| 1510 | Antisymmetric $NO_2$ stretching ($\nu_{AS}$(ONO)) |
| 1500 | In-plane CH bending ($\delta$(CH)) |
| 1332 | Symmetric $NO_2$ stretching ($\nu_S$(ONO)) |
| 1286 | C–O stretching ($\nu$(CO)) |
| 1225 | In-plane OH bending ($\delta$(OH)) |
| 1167 | In-plane CH bending ($\delta$(CH)) |
| 1110 | |
| 865 | In-plane CH bending ($\delta$(CH)) |
| 850 | CH wagging ($\tau$(CH)) |
| 754 | $NO_2$ wagging ($\tau$(ONO)) |

density of the π system in PNP and thus reduce its energy level (Ringwald and Pemberton, 2000; Qu et al., 2008; Keiluweit and Kleber, 2009). This is accompanied by a red-shift of the $\nu(C{=}C)_{arom}$. In addition to the intermolecular hydrogen bonds between the OH and para-substituted $NO_2$ groups in PNP (Zhou et al., 2007c), hydrogen bonds can also form between the surface hydroxyl groups of clay minerals and $NO_2$ groups (Ringwald and Pemberton, 2000; Qu et al., 2008). Due to the formation of hydrogen bonds, the kinetic energy of molecules is increased, and blue-shift of symmetric $NO_2$ stretching vibration can be observed. These frequency shifts of characteristic vibrations of PNP provide convincible evidence for the interaction between adsorbed PNP and Mt. Meanwhile, a prominent decrease in the vibration intensity at 1628 $cm^{-1}$, corresponding to the bending mode of water molecules in the hydration sphere within the clay interlayer space, can be observed with an increase in surfactant loading and adsorption of PNP. This reflects an increase in surface hydrophobicity of the OC and the corresponding ones with adsorbed PNP. It is noticeable that a significant intensity increase of Si–O stretching vibrations in the region of

950–1100 $cm^{-1}$ can be seen with an increase in the amount of intercalated surfactant and adsorbed PNP (Fig. 10.14C), which is due to the interaction between loaded organics (i.e. surfactant and PNP) and silicate surfaces (see Section 10.3.2).

Modification of clay minerals with organic modifiers changes the surface of clay minerals from hydrophilicity to hydrophobicity, which is a key step for the application of OC to remove organic contaminants from aqueous solutions. However, the interaction between OC and organic contaminants, as well as the influences of adsorbed organic matter on the structure of OC, are very complicated (Frost et al., 2008). So far, studies about adsorption of OC toward organic contaminants have demonstrated that organic contaminants can penetrate into the interlayer space of OC accompanied by a replacement of pre-intercalated surfactant, which has a significant influence on the clay interlayer structure, i.e. the arrangement of intercalated surfactants.

## 10.6 CONCLUDING COMMENTS AND A FUTURE OUTLOOK

FTIR has proven to be a powerful technique to investigate the microstructure of OC, and the interaction between organic modifiers and clay minerals. So far, most FTIR studies of OC focus on the molecular environment of intercalated surfactants with special attention on their configurations and arrangements. On the basis of FTIR spectral changes obtained from OC powders, such as frequency shifts, and changes in band intensity and width, the interaction between organic matter and clay minerals are discussed with due consideration of the complexities of the interactions. Many of these results are based on a basic assumption that organic modifiers are loaded within the interlayer space of the clay minerals. In fact, the modifiers enter not only into the interlayer space, but also attach to the outer surface of clay minerals and within interparticle pores, most notably for OC with loaded organic modifiers > CEC of the host clay mineral. The modifiers in the latter environment have significant effects on the structure and property of OC, which cannot be distinguished from the intercalated ones by FTIR spectra alone. On the other hand, some other characterization techniques can provide complementary structural information about OC. For example, $^{13}C$ MAS NMR spectra can provide relative ratios of ordered (all-*trans*) and disordered (*gauche*) conformers (Wang et al., 2000; He et al., 2004c; Zhu et al., 2005), and thermal analysis can determine where the surfactant is loaded (He et al., 2005). Hence, a combination of various characterization techniques with FTIR can provide the best options for complete understanding of structural aspects of OC and the molecular environment of the loaded organic modifiers.

Recently, molecular simulation and quantum calculation have been demonstrated as powerful techniques to investigate the structure and property of OC, which can provide visualized images and dynamic course of the formation of OC and interactions between organic matter and clay mineral in the OC (Heinz et al., 2007; Scholtzová et al., 2013, 2014). A combination of characterization and modeling techniques is of high importance for a comprehensive understanding the structure and property of OC. For example, Scholtzová et al. (2013, 2014) have investigated the structural and vibrational features of Mt intercalated with tetra alkylammonium cations and tetra methylammonium cations, respectively. A theoretical study was performed using density functional theory and compared with IR spectroscopy. The calculated vibrational spectra showed very similar features with the experimental FTIR spectra. However, some important effects can only be observed and analyzed from the calculated spectra.

Organoclays have been commonly considered for use as adsorbents of organic contaminants in aqueous systems. The characterization results obtained on OC powder in a dry state are very different from those obtained on media with water, and the results from dry state cannot reflect the real state of OC in dispersion. Thus development and application of in situ FTIR characterization methods are critical for a better understanding of the structures and properties of OC in the presence of bulk water, as well as the interaction between OC and contaminants under these circumstances.

Chapter 11

# Raman and Infrared Spectroscopies of Intercalated Kaolinite Groups Minerals

J.T. Kloprogge

## 11.1 INTRODUCTION

Kaolins (a group of 1:1 clay minerals including kaolinite (Kaol), dickite, nacrite and halloysite or hydrated, high-defect Kaol) have often been described as non-swelling clay minerals in contrast to smectites (swelling 2:1 clay minerals) (Brigatti et al., 2013). The structure of kaolins can be described as a repetition of units consisting of an $Al(O_2OH)$ octahedral sheet coupled to a $SiO_4$ tetrahedral sheet. Within this structure a differentiation can be made between four symmetrically independent hydroxyl (OH) groups. Three of these extend from the layers and form hydrogen bonds (H bonds) to adjacent layers and are referred to as the inner surface OH groups. The other OH group is located within the vacancies of the octahedral sheets and is known as the inner OH group. The neutral siloxane surface of the Si-tetrahedral side of kaolins is regarded as among the least reactive surface found on clay minerals under ambient conditions (Yariv, 1992b; Giese and van Oss, 1993). In the case of Kaol, the two basal surfaces have very different structures and corresponding surface chemistries. The external oxygen atoms on the siloxane surface are relatively weak electron donors (weak Lewis bases) and therefore do not interact with water molecules and are thus considered hydrophobic (Schrader and Yariv, 1990). The surface dominated by the inner surface OH groups, however, is strongly hydrophyllic. These OH group differentiations were supported recently by an ab initio molecular dynamics study of water interacting with both types of surfaces (Tunega et al., 2002).

The infrared (IR) and Raman spectroscopies of the kaolins are complex, especially to those not familiar with these techniques and/or materials. Because the selection rules for vibrational transitions are different in Raman and IR

Developments in Clay Science, Vol. 8. https://doi.org/10.1016/B978-0-08-100355-8.00011-4

spectroscopy, the use of both the techniques can gain complementary information on the interactions of intercalants with the surfaces of Kaol. Whereas IR tends to provide greater sensitivity to surface reactions, Raman spectroscopy is less affected by the presence of water (Johnston et al., 1985; Johnston and Stone, 1990). Wiewiora et al. (1979) first used dispersive Raman spectroscopy to determine the Raman spectrum of Kaol. As expected from the application of normal mode calculations, all the OH groups described above can be discerned from various bands in both the IR and Raman spectra. The bands that are observed in the IR spectrum around 3695 ($\nu_1$), 3670 ($\nu_2$) and 3650 ($\nu_3$) wavenumbers ($cm^{-1}$) have been ascribed to vibrational modes of the inner surface OH groups, whereas the band around 3620 $cm^{-1}$ ($\nu_5$) has been attributed to the inner OH group (Johnston et al., 1985, 1998; Johnston and Stone 1990; Frost et al., 1993; 1996; Frost, 1997; Frost and Van der Gaast, 1997; Frost and Vassallo, 1997; Farmer, 1998, 2000; Frost and Kloprogge, 2000a, 2001). Similar bands can be observed in the Raman spectrum, where an additional band ($\nu_4$) can also be observed generally as a component of an unresolved doublet of medium intensity at 3695 and 3685 $cm^{-1}$. Michaelian (1986) proposed likely origins of this band in terms of uncoupled surface OH stretching and transverse-longitudinal splitting.

Wada (1961) initiated a new field of research when he proved that Kaol can be expanded during washing of the clay with solutions of organic salts such as potassium acetate. Nowadays, many molecules are known to be intercalated into Kaol, the most well known being: hydrazine ($N_2H_4$), urea ($N_2H_4CO$) and formamide ($NH_3CO$). Intercalation of these molecules occurs via a number of mechanisms depending on the specific interaction between the organic molecule and the Kaol surface (e.g. Lagaly, 1984), but require that the reactive guest molecule intrude between adjacent Kaol layers, thereby breaking the H bonds occurring between the OH groups on one interlayer surface and the oxygen atoms of the siloxane layer on the other surface. The organic molecule then forms new bonds, generally with the more hydrophilic hydroxyl groups (van Oss et al., 1992; Giese and van Oss, 1993), but also sometimes with the more hydrophobic siloxane layer (e.g. Servagent-Noinville et al., 2000), depending on the specific chemistry of the guest molecule. Intercalation reactions have also been reported for nacrite, dickite and halloysite (Weiss and Orth, 1973; Weiss et al., 1973; Adams and Jefferson, 1976; Adams, 1979; Ben Haj Amara et al., 2000; Churchman et al., 1984; Franco and Ruiz Cruz, 2003; Zamama et al., 1995).

Intercalation complexes are usually prepared by reacting Kaol with guest molecules in the form of melts or concentrated solutions, usually at about 60–80°C and often over several days. Intercalation of Kaol with organic molecules results in an expansion in the *c* axis direction, which is clearly observed by X-ray diffraction (XRD) where the original basal spacing (*001*) increases from approximately 0.72 nm to values around 1.0–1.2 nm. Because the OH groups in the Kaol structure are very sensitive to the presence of molecules within the interlayer, interactions such as the formation of new H bonds can be followed in detail by spectroscopic techniques such as IR and Raman spectroscopies. In addition, changes in the guest molecules themselves caused by the interaction with the clay mineral layers can be followed by these spectroscopic techniques. As will be discussed in detail below, IR and Raman spectroscopies have indeed been regularly used to demonstrate these surface solvation reactions.

The reactive and intercalating organic molecules can be divided into three major groups, Group A, Group B and Group C. Group A such as hydrazine, urea, formamide and acetamide, consists of molecules that can form strong H bonds with the Kaol siloxane layer. To break the H bonds between the layers, the guest molecules must contain two separate functional groups able to either accept or donate H bonds, such as carbonic acid and amide: The carbonyl group accepts and the amide functional groups donate H bonds. In this case, alcohol functional groups are unable to intercalate directly because the OH group itself combines both H bond donor and acceptor properties. Group B includes molecules with high dielectric permittivity such as dimethylsulphoxide $(CH_3)_2SO$ (DMSO) or dimethylselenoxide (DMSeO), the strong dipole interaction of which can interact directly with the siloxane layer of the Kaol. Group C contains alkali salts of short chain fatty acids (acetates, propionates, butyrates and isovalerates), especially acetic and propionic acid. The reaction mechanism for group C molecules is still unclear (e.g. Wada, 1961; Ledoux and White, 1966; Weiss et al., 1966; Thompson and Cuff, 1985; Raupach et al., 1987), but whereas compounds with bulky substituents fail to intercalate on their own, many can be intercalated by displacement or entraining reactions.

Most intercalated guest compounds in kaolins can be easily desorbed by washing with water or by heating, because of the relative weakness of H bond or van der Waals interactions compared with electrostatic or covalent interactions. As the state with the maximum number of intercalated molecules is only achieved in the presence of an excess of guest molecule's external to the interlayer space, the analytical composition of many intercalated complexes can only be approximately determined. Removing excess guest molecules (by washing or heating) is usually accompanied by desorption of a certain amount of intercalated molecules.

In this chapter, an overview is given on how the intercalation of a variety of organic molecules in Kaol takes place, and how IR and Raman spectroscopies provide evidence for how these molecules bond or otherwise interact with the clay mineral surface structures. It would go beyond the scope of this article to discuss in detail the experimental background, but this information can easily be retrieved from the literature cited in the following sections.

## 11.2 GROUP A MOLECULES

### 11.2.1 Hydrazine

Hydrazine ($N_2H_4$ also written as $NH_2$–$NH_2$ or $H_2N$–$NH_2$) is a colourless, flammable and highly alkaline inorganic liquid that is miscible with water in all proportions. Hydrazine is highly toxic and dangerously unstable unless handled in solution. The N–N single bond distance is 145 picometer (pm), and the molecule adopts a *gauche* conformation. A relatively high temperature (250°C) is required, in the absence of catalysts, for significant decomposition to occur (Schirmann and Bourdauducq, 2000). Durig and Zheng (2002) have reported the IR and Raman spectra of pure hydrazine hydrate (Table 11.1).

#### *11.2.1.1 Intercalation of Hydrazine in Kaol*

Intercalation of hydrazine in Kaol, as described by various groups (Barrios et al., 1977; Barron et al., 1983; Johnston, 1989; Johnston and Stone, 1990; Frost et al., 1998b,c; Kloprogge et al., 1999b; Kristof et al., 1999; Johnston et al., 2000; Deng et al., 2002; Martens et al., 2002b; Zhou et al., 2013) results in an expansion of the *001* basal spacing of Kaol from ≈0.72 to ≈1.04 nm. Frost et al. (1999f) and Kristof et al. (1999) observed that a number of different structures developed during the intercalation process. Expanded phases with *d*-values of 1.03 and 0.95 nm were related to the conformation of the hydrazine in the interlayer space between adjacent Kaol layers. These conformations were shown to be dependent on the degree of stacking of the Kaol layers (i.e. crystallinity), duration of the intercalation and exposure to air or inert gases.

The IR spectrum exhibited two new bands around 3600 and 3555 $cm^{-1}$, on intercalation of Kaol by hydrazine in water, which can only be attributed to $\nu$(HOH) of $H_2O$ as both Kaol and hydrazine have no bands at these positions (Fig. 11.1). However, in the Raman spectra no new OH bands were detected, the OH stretching ($\nu(Al_2OH)$) bands ascribed to the inner surface OH groups disappeared. The ≈1.03 nm phase of the hydrazine hydrate Kaol intercalate complex was formed presumably by the intercalation of partially hydrolysed hydrazine $(NH_2\text{-}NH_3)^{\delta+}OH^{\delta-}$ units. Removal of water brought about a reorientation of the hydrazine molecules in the interlayer spaces of the Kaol and resulted in the appearance of new $\nu(Al_2OH)$ bands at 3626 $cm^{-1}$ and at 3611 $cm^{-1}$ in both IR and Raman spectra. These two bands are most likely associated with the 0.95 or 0.88 nm phases. The band around 3626 $cm^{-1}$ may be inferred as being due to the inner surface OH ($\nu$(MOH)) forming a relatively weak H bond with the free electron pair of the nitrogen in hydrazine. Other authors (Johnston, 1989; Johnston and Stone, 1990; Johnston et al., 2000) proposed that this band is caused by a shift of the inner OH group.

In the OH deformation ($\delta$(OH)) region, for the hydrazine-intercalated Kaol an additional band was observed at 957 $cm^{-1}$ in addition to the two IR bands at 915 and 940 $cm^{-1}$ associated with the inner-surface and inner OH groups of Kaol. The new band is assigned to the H-N-H rocking motion of the $NH_2$ group ($\nu_{12}$ in Table 11.1) (Johnston and Stone, 1990). A distinct blue shift of the IR band has been reported for the inner OH group from the normal

**TABLE 11.1** Observed IR and Raman Vibrations of Hydrazine (Durig and Zheng, 2002). Values are in Wavenumbers ($cm^{-1}$)

| Species | | Description | Raman Gas | IR Gas | IR Liquid | Raman Liquid | IR Solid | Raman Solid |
|---|---|---|---|---|---|---|---|---|
| Hydrazine | $\nu_1$ | HNH antisymmetric stretch of $NH_2$ | 3398 | 3325 | 3352 | 3277 | 3310 | 3301 |
| | $\nu_2$ | HNH symmetric stretch of $NH_2$ | 3329 | | 3189 | 3189 | 3200 | 3169 |
| | $\nu_3$ | HNH deformation of $NH_2$ | 1642 | 1493 | 1806 | 1628 | 1603 | 1609 |
| | $\nu_4$ | HNH wag of $NH_2$ | | 1098 | 1283 | 1295 | 1304 | 1335 |
| | $\nu_5$ | N–N stretch | 1076 | | 1098 | 1111 | 1126 | 1114 |
| | $\nu_6$ | HNH rocking mode of $NH_2$ | | 780 | 871 | 882 | 884 | 884 |
| | $\nu_7$ | Torsion | | | | | 627 | 534/510 |
| Hydrazine-d$_4$ | $\nu_8$ | HNH antisymmetric stretch of $NH_2$ | – | 3350 | 3332 | 3336 | 3310 | 3194 |
| | $\nu_9$ | HNH symmetric stretch of $NH_2$ | – | 3297 | 3332 | 3336 | 3310 | 3188 |
| | $\nu_{10}$ | HNH deformation of $NH_2$ | – | 1608 | 1608 | 1628 | 1655 | 1662 |
| | $\nu_{11}$ | HNH wag of $NH_2$ | – | 1275 | 1324 | 1295 | 1350 | 1318 |
| | $\nu_{12}$ | HNH rocking mode of $NH_2$ | – | 950 | 1042 | 1000 | 1066 | 994 |

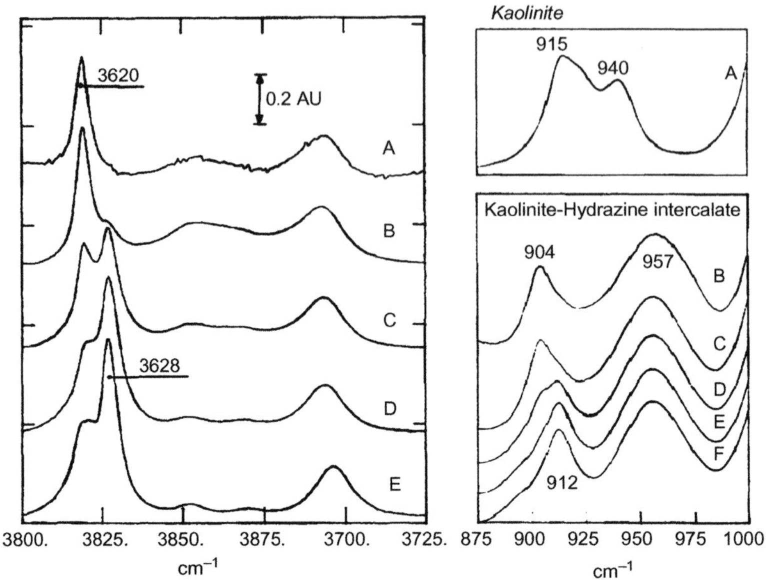


**FIG. 11.1** Controlled-environment FTIR spectra of kaolinite-hydrazine intercalate as function of pressure at 298 K. Left: OH stretching region at: a, 760; b, 1.0; c, 0.1; d, 0.001; and $e < 1.0 \times 10^{-4}$ Torr. Right: OH-deformation region of: a kaolinite; and kaolinite-hydrazine intercalate at: b, 760; c, 1.0; d, 0.1; e, 0.001; and $f < 1.0 \times 10^{-4}$ Torr. *From Johnston and Stone (1990). Reproduced with kind permission of The Clay Minerals Society, publisher of Clays and Clay Minerals.*

3620 $cm^{-1}$ to 3628 $cm^{-1}$ under reduced pressure, indicating a slight electrostatic repulsion between the $NH_2$ moiety and the OH group. A similar observation can be made in the OH deformation region with a shift from 904 to 912 $cm^{-1}$ compared with 915 $cm^{-1}$ for Kaol (Fig. 11.1).

For the 1.03 nm intercalate, two new bands in the IR spectrum at 3465 and 3568 $cm^{-1}$ with a difference of 103 $cm^{-1}$ indicate the formation of H bonds with two distinct N–O bond lengths for the 1.03-nm hydrazine Kaol intercalate. Johnston et al. (2000) have indicated that this difference corresponds to the N–O (where the O is $Al_2OH$ of Kaol) interactions. On partial collapse to 0.95 nm these two bands disappeared. The reduction in the interlayer spacing forces the hydrazine molecules closer to the Kaol layer surfaces resulting in a strong red-shift of these two bands to two very broad bands in the 3100–3400 $cm^{-1}$ region. Well-defined $\nu$(NH) bands were observed at 3368, 3363, 3312 and 3303 $cm^{-1}$ in the IR spectrum of the 1.03 nm complex, whereas only two bands at 3363 and 3303 $cm^{-1}$ were apparent in the 0.95 nm complex. The line widths and positions of the well-resolved $\nu$(NH) bands of hydrazine Kaol intercalate complexes are unexpectedly comparable with those of matrix-isolated hydrazine molecules. Observation of these narrow, well-resolved bands is very rare in clay–organic complexes and shows that the $NH_2$ groups corresponding to these bands are in a well-ordered, highly shielded, neat-phase like environment.

The spectral features of hydrazine Kaol intercalate complexes can be interpreted in terms of two types of $\nu$(NH) bands in both the 0.95 and 1.03 nm forms, corresponding to, respectively, non H-bonded and H-bonded $NH_2$ groups. The presence of both types of bands implies the existence of two different environments for the $NH_2$ groups, assuming all hydrazine present is intercalated. The narrow $\nu$(NH) bands in the 3300–3370 $cm^{-1}$ region are close in wavenumber to those of matrix-isolated hydrazine (non-interacting single molecules in the gas phase) and thus show that some of the associated $NH_2$ groups have minimal interaction with the Kaol layer surface, or with other hydrazine molecules. Because of the limited distances (0.32 or 0.24 nm) between the adjacent 1:1 layers in the hydrazine Kaol complexes, $NH_2$ groups interact with both sides of the Kaol interlayer surface probably by reorienting so the N–N molecular axis is parallel with the interlayer surface, but isolated so as not to interact with other hydrazine molecules.

The $\nu$(NH) bands, corresponding to H bonded $NH_2$ groups, are represented in the IR spectrum of the 1.03 nm hydrazine Kaol complex by a strong, broad band around 2975 $cm^{-1}$. This band is red-shifted by 275 $cm^{-1}$ relative to the average position of the $\nu$(NH) bands of liquid hydrazine (Table 11.1), pointing to $NH_2$ groups significantly more strongly H bonded than those of liquid hydrazine molecules. There are two possible proton acceptors that can account for the strongly H-bonded $NH_2$ groups. The first are O atoms at the siloxane surface of the Kaol layer, and the second is the lone-pair of electrons of adjacent hydrazine molecules. Formation of H bonds with the siloxane O atoms would imply an N–H···O distance of 0.285 nm. In the latter case, the band at 2975 $cm^{-1}$ corresponds to an N–N distance of approximately 0.297 nm. On partial collapse to 0.95 nm, the broad band at 2975 $cm^{-1}$ disappeared and the dominant feature in the spectral region is an equally broad, complex band centred at 3270 $cm^{-1}$. This apparent increase in the frequency of the $\nu$(NH) bands in the 0.95-nm intercalate indicates that these $NH_2$ groups are less strongly H bonded than those present in the 1.03-nm intercalate. Formation of the 0.95-nm hydrazine Kaol intercalate is accompanied by a reduction in height of the interlayer spacing from 0.32 to 0.24 nm, and the intercalated hydrazine molecules have to reside closer to the Kaol surface. The blue-shift of the $\nu$(NH) band from 2975 to 3250 $cm^{-1}$ on formation of the 0.95-nm complex supports the assignment of the 2975 $cm^{-1}$ band to intermolecular H bonding between adjacent hydrazine molecules. The cross-sectional surface of hydrazine is 0.135 $nm^2$ and the monolayer surface available on Kaol is 0.458 $nm^2$. Thus, three or four hydrazine molecules per unit cell can be accommodated in the interlayer space of Kaol to result in a monolayer surface coverage, depending on the arrangement of hydrazine molecules in the interlayer space. The surface loading of hydrazine for the 0.95-nm complex was determined by Johnston et al. (2000) to be 0.5 hydrazine molecules per unit cell. This lower surface loading (compared with the 1.03 nm phase) is responsible for the loss of the 2975 $cm^{-1}$ band as the intermolecular H bonding between hydrazine molecules is reduced. The large, broad band at 3270 $cm^{-1}$ corresponds to spatially separated hydrazine molecules on the Kaol surface with one end of the hydrazine molecule keyed into the siloxane ditrigonal cavity and the other $NH_2$ group interacting with the inner surface OH groups of the Kaol surface.

### *11.2.1.2 Deintercalation of Hydrazine from Kaol*

Complete deintercalation of hydrazine from the hydrazine Kaol intercalate complexes by exposure to moist air can be achieved. During deintercalation, Kaol exhibited a degree of disorder along the *c*-axis, as the Kaol layers initially realigned incompletely. After a prolonged period of time (e.g. four days), Kaol returns to its original basal spacing.

Frost et al. (2002a) used controlled rate thermal analysis (CRTA) in combination with diffuse reflectance IR spectroscopy (DRIFT) to determine different steps in deintercalation reactions of hydrazine Kaol intercalates. The thermoanalytical curves of the intercalate dried under flowing nitrogen for two hours prior to heating are given in Fig. 11.2. Three mass loss events occurred at 59, 77 and 115°C, implying that hydrazine is bound within Kaol

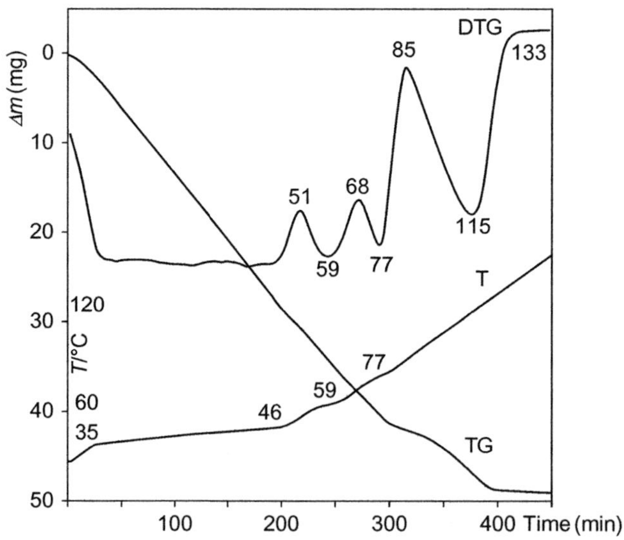


**FIG. 11.2** Controlled rate thermal analysis (CRTA) of treated hydrazine–Kaol intercalate complexes dried under flowing N2 before CRTA analysis. *Reproduced with permission from Frost et al. (2002a).*

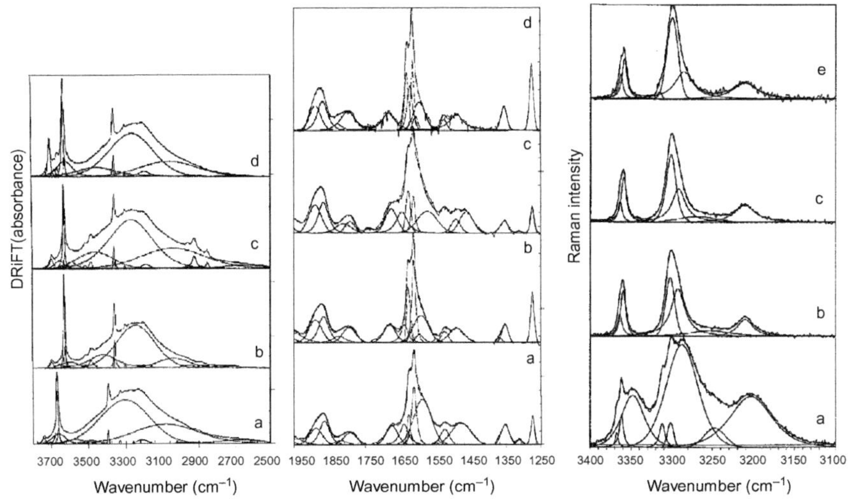


**FIG. 11.3** DRIFT (Left) spectra of 3800–2500 cm$^{-1}$ region and 1950–1250 cm$^{-1}$ and Raman (Right) of Controlled Rate Thermal Analysis (CRTA)-treated hydrazine Kaol intercalate complexes at conditions of: a dry nitrogen; b CRTA treated at 51°C; c 70°C; and d 85°C. *DRiFT spectra reproduced with permission from Frost et al. (2002a). Raman spectra reproduced with permission from Martens et al. (2002b).*

by three different mechanisms. The DRIFT and Raman spectra of the NH and OH stretching region of these same hydrazine Kaol intercalates are shown in Fig. 11.3. The DRIFT spectra of untreated Kaol display four bands in the OH stretching region at 3695 ($\nu_1$), 3668 ($\nu_2$), 3652 ($\nu_3$) and 3619 ($\nu_5$) cm$^{-1}$. In addition, a band may be resolved at 3684 ($\nu_4$)cm$^{-1}$ for the low-defect Kaol. On intercalation of the low-defect Kaol with hydrazine, the $\nu_1$ band decreased in intensity and the $\nu_4$ and $\nu_2$ modes were not observed in spectra a–c. In spectrum d, the $\nu_2$ band reappeared. In addition, a new band at 3626 cm$^{-1}$ attributed to the inner surface OH of Kaol affected by the intercalated hydrazine was observed. Furthermore, for the spectra a–c, a band at 3604 cm$^{-1}$ was observed, which was ascribed to the $\nu$(HOH) of interlayer water. The band at 3695 cm$^{-1}$ was of low intensity and represents the inner surface OH, which are not H bonded to the hydrazine.

In the spectra (b) and (c) of hydrazine Kaol intercalates, two sharp bands were detected at 3514 and 3481 cm$^{-1}$, not visible in spectra (a) of hydrazine Kaol intercalate exposed to dry $N_2$ for 1 h and neither in spectra (d) when the hydrazine Kaol intercalate was CRTA-treated at 85°C (Fig. 11.3). These two bands were assigned to the $\nu$(HOH) of water molecules involved in the hydrazine Kaol intercalate. When the hydrazine Kaol intercalate was CRTA treated at 70°C, the hydrazine Kaol intercalate collapsed to the 0.96-nm phase. In the DRIFT spectra of this complex, two additional bands were observed at 2885 and 2921 $cm^{-1}$. These bands were assigned to H-N-H vibrations of hydrazine involved in strong H bonding with the Kaol surface. These bands were not observed in the CRTA-treated intercalate at 85°C. Some low-intensity broad bands were visible around these positions for the $N_2$-dried intercalate prior to heating. In the spectra a–c (Fig. 11.3), a broad band at 2976 $cm^{-1}$, similar to that observed at decreased pressure, was observed for the dry $N_2$-treated hydrazine Kaol intercalate and at 2959 and 2957 $cm^{-1}$ for the 51 and 70°C CRTA-treated intercalates. The band had very low intensity and was not visible in the spectrum of the 85°C treated intercalate. The Raman spectrum showed a set of bands at 2890 and 2948 $cm^{-1}$, which corresponded with the bands observed at 2921 and 2962 $cm^{-1}$ for pure hydrazine (Martens et al., 2002b). Johnston et al. (2000) reported on partial collapse of the intercalate to 0.95 nm that the broad band at 2975 $cm^{-1}$ was replaced by a band at 3270 $cm^{-1}$. A similar broad band at 3270 $cm^{-1}$ was also observed for the dry $N_2$-treated intercalate. The band appeared to shift to lower wavenumbers on CRTA treatment and was observed at 3244, 3266 and 3260 $cm^{-1}$. However, this broad band was visible in the spectra for all the CRTA-treated intercalates. The 70°C CRTA-treated intercalate had a partially collapsed basal spacing of 0.96 nm, and at the same time two new bands were observed at 2885 and 2921 $cm^{-1}$. These two bands are assigned to the H-N-H vibrations of very strongly H-bonded hydrazine molecules. For such increased H bonding of hydrazine, reduction in the interlayer space by 0.08 nm would be expected. Two bands were observed at 3362 and 3356 $cm^{-1}$ and can be assigned to the antisymmetric H-N-H stretching $\nu_{as}$(NH) vibrations. In the corresponding Raman spectrum, two bands were observed at 3367 and 3362 $cm^{-1}$ (Martens et al., 2002b). These two bands were assigned to two different $NH_2$ units and their intensities varied with the conditions and thermal treatment of the hydrazine intercalated complex. Two bands observed at 3301 and 3292 $cm^{-1}$ are assigned to the H-N-H symmetric stretching vibrations. Therefore, two different types of $NH_2$ units were observed with slight differences in the bonding strengths of these units similar to what was observed in the DRIFT spectra. A band was also observed at 3210 $cm^{-1}$. This band may be due to $NH_2$ affecting the inner surface hydroxyl group. On obtaining the spectra at 85°C, the $\nu_{as}$(NH) and $\nu_s$(NH) bands were observed at 3363, 3358 $cm^{-1}$ and 3300, 3291 $cm^{-1}$. The intensity ratio is different from that at 50°C, suggesting that the hydrazine intercalation at 85°C is significantly different to that at 50°C. At 85°C, the two $\nu_{as}$(NH) bands were observed at 3362 and 3358 $cm^{-1}$. The two $\nu_s$(NH)are observed at 3315 and 3299 $cm^{-1}$. There was a significant shift of the bands to higher wavenumbers at the elevated temperature (Martens et al., 2002b).

The 1250 to 1950 $cm^{-1}$ region of the DRIFT spectra of hydrazine Kaol intercalates is significant as this region contains the H-N-H twist and scissor modes together with the water deformation $\delta$(HOH) modes (Fig. 11.3). One of the problems is that the vibrational modes overlap and this makes the exact assignment of bands around 1630 $cm^{-1}$ difficult. In this range, based on the presence of hydrated hydrazine, bands were observed around 1598, 1612 and 1627 $cm^{-1}$, which can be assigned to: (i) free or non H-bonded water in the interlayer spaces, (ii) water around hydrazine and (iii) water adsorbed on the Kaol surface. In addition, the two bands around 1650 and 1679 $cm^{-1}$ were ascribed to water oriented on the siloxane surface (Frost and Kloprogge, 2012). As a result of the presence of water, one has to consider the possibility of overlapping bands with the twist and scissor modes of hydrazine. Johnston et al. (2000) reported, assuming the presence of dry hydrazine, the H-N-H twisting mode ($\nu_{11}$ or $\nu_4$ wagging mode) of hydrazine at 1305 $cm^{-1}$ in the spectrum of the 1.03 nm complex, which shifted to 1276 $cm^{-1}$ in the 0.95-nm complex, corresponding to the reduction of intermolecular H-bonded isolated hydrazine molecules. The scissor mode ($\nu_{10}$ mode) of hydrazine appeared at 1616 $cm^{-1}$ in the 1.03 nm complex and at 1613 $cm^{-1}$ in the 0.95-nm complex, compared with 1595 $cm^{-1}$ for matrix-isolated hydrazine. Two bands with narrow bandwidths were found at 1628 and 1613 $cm^{-1}$ for hydrazine intercalated Kaol treated for 1 h with dry $N_2$ (Frost et al., 2002a,c). Similar bands are observed at 1608 and 1606 $cm^{-1}$ for liquid hydrazine and at 1655 and 1603 $cm^{-1}$ for solid hydrazine (Table 11.1). These two bands can be assigned to the antisymmetric and symmetric H-N-H scissor modes of the $NH_2$ units. A band was observed at 1354 $cm^{-1}$ for all the treated hydrazine intercalated Kaol complexes and a low-intensity band at 1314 $cm^{-1}$ for the dry nitrogen-treated hydrazine-intercalated Kaol and for the 70°C CRTA-treated complex. Frost et al. (2002a) observed a similar band at 1276 $cm^{-1}$ in all the spectra for all the hydrazine-intercalated Kaols. Bands observed at around 1596, 1620 and 1644 $cm^{-1}$ may be attributed to the presence of water: The broad band at 1596 $cm^{-1}$ corresponds to the wavenumber of non H-bonded water molecules. In the dry $N_2$ and 51°C CRTA-heated intercalated Kaols bands at 1641 and 1681 $cm^{-1}$ are attributed to the hydroxyl deformation modes of water strongly bonded to the hydrazine. A strong hydroxyl deformation mode of water is observed at 1592 $cm^{-1}$ corresponding to the position of non H-bonded water and to the hydroxyl-stretching vibration at 3605 $cm^{-1}$.

The position of the $\nu_s$(N–N) band of hydrazine ranges from 1086 $cm^{-1}$ for matrix-isolated hydrazine to 1126 $cm^{-1}$ for bulk solid. This mode appears at comparable wavenumbers in the IR spectra of both hydrazine Kaol complexes, at 1095 and 1091 $cm^{-1}$, and in the Raman spectrum at 1113 $cm^{-1}$ for liquid hydrazine. These data point to the fact that the N–N bond length is not strongly affected by the partial collapse of the complex from 1.03 to 0.95 nm. The $\nu_{12}$ mode of hydrazine (asymmetric wag) occurs in the 937–1066 $cm^{-1}$ region, depending on the state of hydrazine (liquid, vapour, matrix isolated or solid). The most intense bands in the entire IR spectrum of Kaol and its hydrazine intercalates however are the $\nu$(SiO) bands at 1010, 1040 and 1116 $cm^{-1}$. The positions of these bands are influenced by the intercalation of hydrazine and by the partial collapse of the 1.03-nm complex to 0.95 nm (Table 11.2).

**TABLE 11.2** Observed FTIR and Raman Bands for the 1.03 nm and 0.95 nm Kaolinite–Hydrazine (KH) Complexes in the 900–4000 $cm^{-1}$ Region (Johnston et al., 2000). Values are in Wavenumbers ($cm^{-1}$)

| FTIR | | | Raman | | | |
|---|---|---|---|---|---|---|
| *1.03-nm KH* | *0.95-nm KH* | *Kaolinite* | *kaolinite* | *1.03 nm KH* | *hydrazine* | Assignment |
| 904 | | | | | | $\delta$(Al–OH) inner OH perturbed by hydrazine |
| 940 | 940 | 941 | 940 | 948 | | $\delta$(Al–OH) inner surface OH–kaolinite |
| 1010 | 1017 | 1011 | 1018 | | | $\nu$(Si–O) kaolinite |
| 1040 | 1044 | 1033 | 1040 | | | $\nu$(Si–O) kaolinite |
| 1091 | 1095 | | | 1088 | 1114 | $\nu_s$(N–N) hydrazine |
| 1125 | 1124 | 1102 | | 1130 | | $\nu$(Si–O) kaolinite |
| 1305 | 1276 | | | 1313 | 1335 | HNH twist $\nu_4$ or $\nu_{11}$ |
| 1355 | 1366 | | | | | HNH twist $\nu_4$ or $\nu_1$1 |
| 1473 | 1471 | | | | | |
| 1593 | 1580 | | | | | HNH scissor $\nu_3$ or $\nu_{10}$ |
| 1616 | 1613 | | | 1618 | | HNH scissor $\nu_3$ or $\nu_{10}$ |
| 1628 | | | | | | HNH scissor $\nu_3$ or $\nu_{10}$ |
| 2975 | 2964 | | | | | $\nu$(NH) hydrazine–hydrazine |
| 3198 | 3214 | | | | 3200 | $\nu$(NH) |
| 3303 | 3302 | | | 3302 | | $\nu$(NH) (s–s HNH stretch) |
| 3312 | | | | 3311 | | $\nu$(NH) (s–a HNH stretch) |

| | | | | | | |
|---|---|---|---|---|---|---|
| 3363 | 3363 | | | 3362 | | $\nu$(NH) (s–s HNH stretch) |
| 3368 | | | | | | $\nu$(NH) (s–a HNH stretch) |
| 3465 | | | | | | $\nu$(OH) inner surface H-bonded to hydrazine |
| 3568 | | | | | | $\nu$(OH) inner surface H-bonded to hydrazine |
| 3620 | 3620 | 3620 | 3621 | 3620 | | $\nu$(OH) inner OH |
| 3654 | 3653 | 3652 | 3652 | | | $\nu$(OH) inner surface OH |
| 3695 | 3695 | 3695 | 3696 | 3692 | | $\nu$(OH) inner surface OH |
| | | | | 871 | 899 | |
| | 913 | 918 | 915 | | | $\delta$(Al–OH) inner OH–kaolinite |
| | | | | 1273 | 1283 | HNH twist $\nu$4 or $\nu$11 |
| | | | | 3196 | | $\nu$(NH) |
| | 3270 | | | 3283 | 3283 | $\nu$(NH) |
| | | | | | 3347 | $\nu$(NH) liquid hydrazine |
| | 3402 | | | | | $\nu$(OH) inner surface H-bonded to hydrazine |
| | 3268 | | | | | $\nu$(OH) inner OH perturbed by hydrazine |
| | 3670 | 3668 | 3668 | | | $\nu$(OH) inner surface OH |
| | | | 3688 | | | $\nu$(OH) inner surface OH |

Kaol has two prominent $\nu$(SiO) bands at 1010 and 1034 cm$^{-1}$ that shift to higher wavenumbers intercalation of the hydrazine to 1010 and 1040 cm$^{-1}$ for the 1.03 nm complex, and 1017 and 1044 cm$^{-1}$ for the partially collapsed 0.95-nm complex, respectively. Changes to the framework vibrational modes of Kaol can be attributed to minor structural changes induced by the presence of hydrazine and by dielectric effects. On intercalation of hydrazine, the layers are separated by distances of 0.24 or 0.32 nm. In non-intercalated Kaol, the inner-surface OH groups have predominantly H bonding and van der Waals interactions with the opposing siloxane surface. Deng et al. (2002) observed that the IR spectra of the hydrazine monohydrate–Kaol intercalate had a sharp band at 1124 cm$^{-1}$ that arose from the apical $\nu$(SiO) vibration. Compared with the same vibration in the starting Kaol, this band shifted 9 cm$^{-1}$ from 1115 cm$^{-1}$ and varied less in intensity than the starting Kaol when the incidence angles were changed from 0 to 45°. The pleochroism change of the apical $\nu$(SiO) was more obvious in the IR spectra during slow deintercalation. On deintercalation, the new band at 1113 cm$^{1}$ resulted from the apical $\nu$(SiO) of the deintercalated Kaol and was strongly positively pleochroic as shown by the almost doubled intensity when the incidence angle was rotated from 0 to 45°. The intensity changes of the 1113 and 1124 cm$^{-1}$ bands relative to each other confirmed that the apical $\nu$(SiO) became less-strongly pleochroic when Kaol was intercalated. Intercalation-induced perturbations associated with the $\nu$(SiO) band shifts and pleochroism are considered to be ascribed to increased tetrahedral tilting in the intercalate-reduced pleochroism of the apical $\nu$(SiO) compared with the starting Kaol (Johnston et al., 2000; Deng et al., 2002). When the hydrazine molecules moved out of the interlayer space, the tetrahedra returned to its original configuration as shown by the recovered pleochroism and band position. The pleochroism changes of the apical $\nu$(SiO) also points to the fact that the tilting of the tetrahedra during intercalation and deintercalation cycles is reversible.

Frequency shifts and changes in the intensity of the structural $\delta$(OH) bands on deintercalation of the hydrazine Kaol intercalate complexes reflect the decrease in surface loading from 2 to 4 hydrazine molecules per unit cell to ~0.5 hydrazine molecules per unit cell (Johnston and Stone, 1990). The symmetric wag, $\nu_6$, was observed by Raman spectroscopy in the bulk solid at 780 cm$^{-1}$. This band is split into two components for both complexes, 770–791 cm$^{-1}$ (for the 1.03 nm complex) and 765–805 cm$^{-1}$ (for the 0.95 nm complex). This splitting may reflect the fact that the interactions of the two ends of the hydrazine molecule with the host differ appreciably. Although an isolated hydrazine molecule has only one torsional mode ($\nu_7$), the intercalated molecule will have both an out-of-phase and in-phase torsion where the two ends of the molecules rotate in the opposite ($\nu_I$) and same directions ($\nu_E$), respectively. Both of these torsional modes are subject to an internal barrier to rotation and an external barrier at each end of the molecule. If the external barriers at the two ends of the molecule are the same, the two torsional modes $\nu_I$ and $\nu_E$ are related to $\nu_7$ as $\nu_I^2 - \nu_E^2 = \nu_7^2$.

### 11.2.2 Urea

The urea molecule ($N_2H_4CO$ or $(NH_2)_2-C{=}O$) has eight atoms, which leads to 24 degrees of freedom. Taking into account the six (3 translations + 3 rotations) motions, 18 internal modes of vibration are considered. In the solid state urea has $C_{2v}$ point group symmetry, giving rise to 7 $A_1$, 2 $A_2$, 3 $B_1$ and 6 $B_2$ modes. The two $A_2$ vibrations are IR inactive, but all the vibrational modes are Raman active. In the gas phase, urea has $C_2$ point group symmetry resulting in 9 A and 9 B modes. All these vibrations are both IR and Raman active.

An overview of the IR and Raman bands is given in Table 11.3. The antisymmetric $NH_2$ $A_1$ and $B_2$ vibrations appear at 3450 and 3444 $cm^{-1}$ and the symmetric analogues at 3349 and 3341 $cm^{-1}$, respectively. The band at 3264 $cm^{-1}$ is a combination band of the symmetric $NH_2$ deformation ($\delta_s$(NH)) and the C=O stretching vibration observed at 1683 and 1601 $cm^{-1}$, respectively. The bands at 1625 and 1466 $cm^{-1}$ are assigned to the $\delta_{as}$(NH) and to the $\nu_{as}$(CN) vibrations, respectively of urea. The band at 1683 $cm^{-1}$ is assigned to $\delta_s$(NH) and the band at 1601 $cm^{-1}$ to the $\nu$(C=O) vibration, but force field calculations on the solid-state show that these two vibrations are strongly coupled. In the 1700–1400 $cm^{-1}$ region the problem is that not four but six bands are clearly visible in this part of the Raman spectrum. Keuleers et al. (1999) observed the same six absorptions, although two of them only appeared as shoulders. The $\delta_{as}$(NH) and $\nu_{as}$(CN) vibrations are relatively intense bands at 1625 and 1466 $cm^{-1}$, respectively, in both the Raman and IR spectra of urea. The band at 1683 $cm^{-1}$ appears as a very strong band in the IR spectrum, but a very weak band in the Raman spectrum of urea. In contrast, the band at 1648 $cm^{-1}$ is very strong in the Raman, but only appears as a shoulder in the IR. The same effect is observed with the bands appearing at 1601 and 1540 $cm^{-1}$.

In the case of urea the absorptions in both IR and Raman occur at the same wavelengths, and because of the relative intensity change in the IR and Raman spectra as described above, one cannot assign two of these four bands to combination bands or overtones. Based on the crystal structure of urea, a splitting of the $A_1$ and $A_2$ vibrations is expected when the factor group symmetry ($D^3_{2d}$) of urea is considered instead of its site symmetry ($C_{2v}$). The $B_1$ and $B_2$ vibrations are not affected. Looking at the 1700–1400 $cm^{-1}$ region, six bands are indeed expected: splitting of the two $A_1$ ($C_{2v}$) vibrations in two $A_1$ ($D^3_{2d}$) and two $B_2$ ($D^3_{2d}$) vibrations and the two $B_2$ ($C_{2v}$) vibrations that become two E ($D^3_{2d}$) vibrations. All these $D^3_{2d}$ vibrations are Raman active, resulting in the six clearly observable Raman bands. The $A_1$ ($D^3_{2d}$) modes are IR inactive and consequently only appear as shoulders, and only four absorptions are clearly visible in the IR spectrum. A very strong band at 1153 $cm^{-1}$ and a weak band at 1057 $cm^{-1}$ were shown to disappear on deuteration. As a consequence, these bands can be assigned to the H-N-H rocking vibrations. The weak band at 1003 $cm^{-1}$ is assigned to the symmetric C-N stretching vibration.

**TABLE 11.3** Observed IR and Raman and Calculated Vibrations of Solid Urea (Keuleers et al., 1999). Values are in Wavenumbers ($cm^{-1}$)

| | | IR 20°C | IR-196°C | Raman | Calculated | |
|---|---|---|---|---|---|---|
| $\nu_{as}$(NH) of $NH_2$ | ($A_1$) | 3450 | 3448 | | | |
| $\nu_{as}$(NH) of $NH_2$ | ($B_2$) | 3444 | 3435 | | | |
| $\nu_s$(NH) of $NH_2$ | ($A_1$) | 3349 | 3345 | | | |
| $\nu_s$(NH) of $NH_2$ | ($B_2$) | 3341 | 3330 | | | |
| Combination band | | 3264 | | | | |
| $\delta_s$(NH) of $NH_2$ | ($A_1$) | 1683 | | 1647 | 1663 | 51% $\delta_s$(NH)+21% $\nu$(CO)+19% $\nu_s$(CN) |
| $\delta_{as}$(NH) of $NH_2$ | ($B_2$) | 1625 | | 1623 | 1636 | 72% $\delta_{as}$(NH)+24% $\nu_{as}$(CN) |
| $\nu$(CO) | ($A_1$) | 1601 | | 1540 | 1602 | 41% $\nu$(CO)+33% $\delta_s$(NH)+10% $\rho_s$(NH) |
| $\nu_{as}$(CN) | ($B_2$) | 1466 | | 1466 | 1473 | 54% $\nu_{as}$(CN)+21% $\delta_{as}$(NH)+15% $\delta$(CO) |
| HNH rocking | | 1153 | | | 1177 | 70% $\rho_s$(NH)+28% $\nu$(CO) |
| HNH rocking | | 1057 | | | 1056 | 81% $\rho_{as}$(NH)+19% $\nu_{as}$(CN) |
| $\nu_s$(CN) | | 1003 | | | 1010 | 84% $\nu_s$(CN) |
| $\pi$CO | | 789 | | | | |
| Antisym. HNH torsional | | 727 | | | | |
| CO deformation | | 569 | | | | |
| CN deformation | | 530 | | | | |
| Antisym. HNH wagging of $NH_2$ | | 508 | | | | |

Force field calculations on crystalline urea showed no coupling between the symmetric H-N-H rocking and the symmetric C-N stretching vibrations. The two weak bands at 1177 and 1056 $cm^{-1}$ and the intense band at 1010 $cm^{-1}$ in the Raman spectrum of urea were converted on deuteration into two intense bands and one weak band, indicating that a coupling between the $\nu_s$(CN) and the $\rho_s$(DND) vibrations must exist. The band at 789 $cm^{-1}$ showed a small shift to a higher wavenumber on lowering the temperature and clearly appeared at a lower wavenumber for the $^{13}$C and the $^{18}$O isotopomers. This band can be assigned to the πC=O vibration. As this band also appears at a lower wavenumber on deuteration, a coupling with the $\tau_{as}$(NH) modes of $NH_2$ must occur. On the basis of force field calculation, the bands at 569 and 530 $cm^{-1}$ have been attributed to the C=O and C-N deformation vibrations, respectively. The antisymmetric H-N-H torsional and wagging modes occur at 727 and 508 $cm^{-1}$, respectively, and appear at 550 and 376 $cm^{-1}$ on deuteration. The symmetric modes are IR inactive and the Raman bands are too weak so that they cannot be experimentally observed (Keuleers et al., 1999).

### 11.2.2.1 Intercalation of Urea in Kaolinite

Ledoux and White (1966) observed an increase in the basal spacing from 0.715 nm to 1.077 nm on intercalation of urea in Kaol. After intercalation, these authors removed excess urea outside the Kaol by heating at 110°C for 15 min which resulted in a striking reduction in the intensity of the 3695, 3670 and 3650 $cm^{-1}$ bands of Kaol. Heating the complex for 1 h at 110°C produced a large decrease in the intensity of v(NH) bands and shifted the 3440 and 3345 $cm^{-1}$ bands to higher wavenumbers and split them into two doublets at 3520 and 3500 $cm^{-1}$, and 3415 and 3380 $cm^{-1}$, respectively. The expanded complex was shown to be stable at 110°C up to 5 h. The prominent features of the IR spectrum of the Kaol urea intercalate compared with that of pure urea and Kaol were a better resolution of the urea absorption bands between 1800 and 1500 $cm^{-1}$ (Fig. 11.4), a decrease in the intensity of the 935 $cm^{-1}$ band, and an additional band at 895 $cm^{-1}$. Heating the complex at 110°C for 1, 2, 3, 4 and 5 h caused a steady decrease in the intensity of the 1680, 1620 and 1470 $cm^{-1}$ bands and the occurrence of a shoulder at 1725 $cm^{-1}$. The band at 1600 $cm^{-1}$ seemed to have shifted to 1590 $cm^{-1}$.

In order to investigate the possible orientations of the transition moments in urea molecules adsorbed on the hydroxyl or oxygen surfaces of Kaol, Ledoux and White (1966) studied the IR pleochroism of the Kaol urea complex which was heated at 110°C for 5 h. The pleochroism of the 3415 $cm^{-1}$ absorption band reveals that the direction of the transition moment change of the corresponding $NH_2$ group is in a plane perpendicular or making a large angle with the 'ab' cleavage plane of Kaol. The band at 1680 $cm^{-1}$ results from a contribution of $\delta$(NH) bending and v(C=O) stretching modes of urea; the pleochroic behaviour of this absorption indicates that the direction of

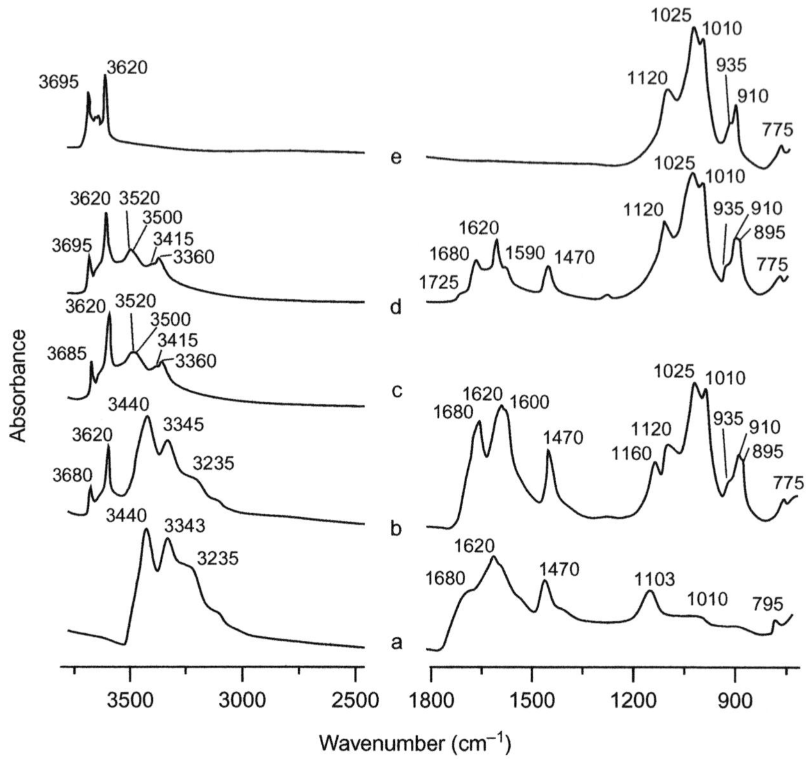


**FIG. 11.4** Infrared spectra of: a urea crystals heated 15 min at 110°C; and urea kaolinite complex heated at 110°C for: b, 15 min; c, 1 h; d, 5 h and e, 48 h. *Redrawn from Ledoux and White (1966).*

the net transition moment is normal or nearly normal to the '*ab*' cleavage plane of Kaol. One-dimensional Fourier-projections of the urea Kaol intercalate indicated a configuration for the interlayer urea molecules where one $NH_2$ group is located near the oxygens of the basal tetrahedral sheet and the remaining $NH_2$ and C=O groups are opposite to the inner-surface hydroxyls. Thus, the Kaol urea intercalate contains two $NH_2$ groups, each having a different environment; one $NH_2$ group is in the vicinity of the OH surface, whereas the other is adjacent to the oxygen surface. Intercalation of urea molecules resulted in the shift of the $\nu_{as}$(NH) and $\nu_{as}$(NH) band frequencies of pure urea, 3440 and 3345 to 3500 and 3380 cm$^{-1}$, respectively. The shift of the $\nu$(NH) of pure urea to higher frequencies after intercalation implies that H bond between the $NH_2$ of urea and the siloxane surface of Kaol is longer and weaker than those of pure urea crystals. The C=O groups of the intercalated urea are located in the vicinity of the inner-surface OH and are susceptible to the formation of H bonds (Frost and Kristof, 1997; Frost et al., 1997b).

As a result, in the Raman spectrum new bands were observed around 3387, 3408 and 3500 $cm^{-1}$.

Small shifts of 2–4 $cm^{-1}$ were observed between urea-intercalated high-defect Kaol and urea-intercalated low-defect Kaol by Frost et al. (2000e), whereas an additional band was observed at 3373 $cm^{-1}$ for the low-defect Kaol at −196°C. The bands ascribed to the inner surface OH groups strongly decreased in intensity but not as strong as for hydrazine. The band at 3500 $cm^{1}$ in the Raman spectrum has been interpreted as being due to the formation of an H bond between the $NH_2$ groups of urea and the siloxane layer, whereas in the IR spectrum a strong band was observed around 3504 $cm^{-1}$. The two other bands around 3387 and 3408 $cm^{-1}$ form the symmetric and antisymmetric vibrations of the H-bonded $NH_2$ groups. Not all the intensities of the various Kaol $\nu$(OH) bands diminish to the same degree; particularly, the 3684 $cm^{-1}$ band decreased less strongly in intensity, perhaps due to the formation of H bonds between the C=O group of the urea with the inner surface OH groups of the Kaol. Evidence for this can be found in the strong shift of the C=O symmetric stretching vibration and the remaining intensity in the inner surface OH stretching vibrations after intercalation. Cooling to −196°C of the urea-intercalated Kaol resulted in strong changes in the Raman spectrum (Frost et al., 2000e) with a split of the 3620 $cm^{-1}$ inner OH stretching mode into two bands around 3615 and 3621 $cm^{-1}$. Thus, urea forms an H bond between the C=O group and the inner surface OH groups as well as between one of the $NH_2$ group of urea with the siloxane layer, but the other $NH_2$ group remains free. It is possible that the single band around 3620 $cm^{-1}$ at room temperature (RT) in reality involves two strongly overlapping bands of the inner OH group and of the H-bonded C=O group with the inner surface OH group. Alternatively, the urea may be bonded solely to the siloxane layer at RT and, due to the strong cooling to -196°C, the C=O group of the urea and the OH groups are brought so closely together that they can then H bond.

Additional support for this model can be found in the C–N stretching region. A single band at 1009 $cm^{-1}$ at RT split in two evenly intense bands at 1014 and 1004 $cm^{-1}$ on cooling to −196°C, which points to two non-equivalent $NH_2$ groups in the urea Kaol intercalate. Such an effect is reinforced by the shift in band position of the carbonyl unit during cooling from RT to −196°C. In pure urea, this band is observed around 1649 $cm^{-1}$ and does not shift in the intercalate complexes, indicating that there is no interaction between the C=O group and the hydroxyl groups at RT. However, on cooling to −196°C, a strong shift to 1616 $cm^{-1}$ was observed due to the formation of a chemical interaction between the C=O group and the Kaol surface.

#### 11.2.2.2 *Intercalation of Urea in Halloysite*

Intercalation of urea in dehydrated halloysite showed some significant differences on thermal deintercalation. Whereas the initial intercalation resulted in

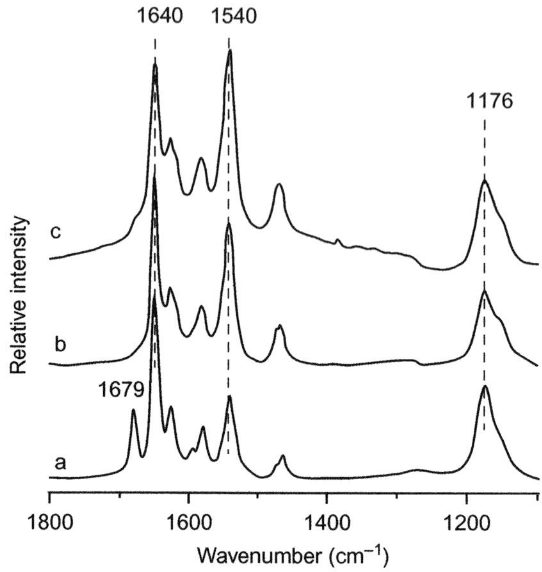


**FIG. 11.5** FT-Raman spectra of the 1100–1800 cm$^{-1}$ region for: a, urea; b, urea halloysite intercalate and c, urea kaolinite intercalate complexes. *Redrawn from Frost et al. (1997a).*

a similar increase in the basal spacing from 0.74 nm to 1.07 nm and the appearance of new bands in the IR spectrum at 3503 and 3390 cm$^{-1}$, the $\delta_s$(NH) was found at 1676 cm$^{-1}$, $\delta_{as}$(NH) at 1624 cm$^{-1}$, $\nu$(CO) at 1591 cm$^{-1}$ and $\nu_{as}$(CN) at 1476 cm$^{-1}$ (Nicolini et al., 2009). Differences in the Raman spectra of urea and the spectrum of the halloysite urea intercalate were also observed (Frost et al., 1997a). The band at 1679 cm$^{-1}$ attributed to the $v_1(A_1)$ stretching vibration of the C=O group was not observed in the intercalate (Fig. 11.5). A shift in the $v_1(B_1)$ C-N asymmetric stretching band from 1464 to 1468 cm$^{-1}$ was detected on intercalation. A noticeable increase in the intensity of the 1540 cm$^{-1}$ band was attributed to an increase in the symmetry of the H-N-H bend as the $NH_2$ groups were H bonded to the silica layer. Decreases in the relative intensities of the bands assigned to urea at 1176 and 1640 cm$^{-1}$ were also observed on intercalation.

### *11.2.2.3 Deintercalation of Urea from Halloysite and from Kaol*

Thermal analysis of the intercalated halloysite showed three mass loss steps with endothermic peaks at 70°C associated with the loss of physisorbed water, at 238°C due to volatilization/decomposition of the intercalated urea and at 480°C dehydroxylation of the halloysite structure. The urea halloysite intercalate was stable up to 150°C (Fig. 11.6). The first basal reflection of the intercalated urea broadened (onset temperature of 142°C in TGA). After the partial

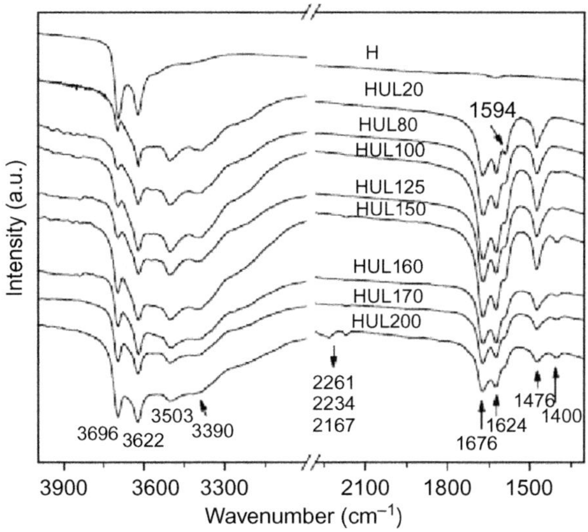


**FIG. 11.6** FTIR spectra of halloysite (H), urea intercalated halloysite washed with isopropanol (HUL20), and HUL20 subjected to different heat treatments (HULT). *Reproduced with permission from Nicolini et al. (2009).*

removal of urea, the disorder returned to the same level as that observed in raw halloysite. Above 100°C, in the IR spectrum a new band, which was likely associated with biuret ($C_2H_5N_3O_2$ compound obtained by condensation of two urea molecules with elimination of ammonia ($NH_3$)), began to appear at 1400 $cm^{-1}$ and persisted up to 200°C. Schaber et al. (2004) reported that biuret appeared at 160°C in an open reaction vessel. Above 125°C, a small band appeared at 2160 $cm^{-1}$, splitting into two bands at 170°C (2223 and 2167 $cm^{-1}$) and three bands at 200°C (2261, 2234 and 2167 $cm^{-1}$). The three new bands can be assigned to cyanic acid (HNCO) and/or its salts ($NH_4NCO$ and $H_3ONCO$) (Schaber et al., 2004). Intercalated urea was partially retained even after heat treatment at 200°C, unlike the urea-intercalated Kaol (Fukamachi et al., 2007). These authors reported the new bands of the urea-intercalated Kaol at 3500 and 3389 $cm^{-1}$, associated with urea molecules binding to the inner surface OH (Frost et al., 2000e), persisted up to a heating temperature of 160°C. At 170°C, it also showed small stretching bands of the urea carbonyl groups at 1673 and 1625 $cm^{-1}$. The FTIR spectrum of the sample heated to 200°C was very similar to that of raw Kaol, confirming the thermal decomposition of urea. These FTIR results of the intercalated sample heated at different temperatures are confirmed by XRD results, confirming that at 170°C all the urea was eliminated from the interlayer spaces of Kaol.

A much more complex thermal deintercalation profile of urea from Kaol was reported by Liu et al. (2014). Heating the complex to 250°C resulted in a prominent reduction in intensities of the 3500, 3380, 1675, 1595, 1623 and 1474 $cm^{-1}$ bands, suggesting that the decomposition of the complex was incomplete. Consequently, the urea intercalated Kaol maintained its structure up to 200°C. At temperatures of ≥300°C, all the characteristic bands of the intercalated urea disappeared (Fig. 11.7). Interestingly, the spectra of the deintercalated Kaol at a higher wavenumber are not the same as the spectra of the untreated Kaol, due to the structural disorder caused by the intercalation and deintercalation processes, which was also observed in the XRD patterns. The spectra show no evidence to support the complex thermal behaviour observed with TGA and DSC.

The evolved gases and volatile products during the thermal decomposition of Kaol–urea intercalation complex was also studied using FTIR. The

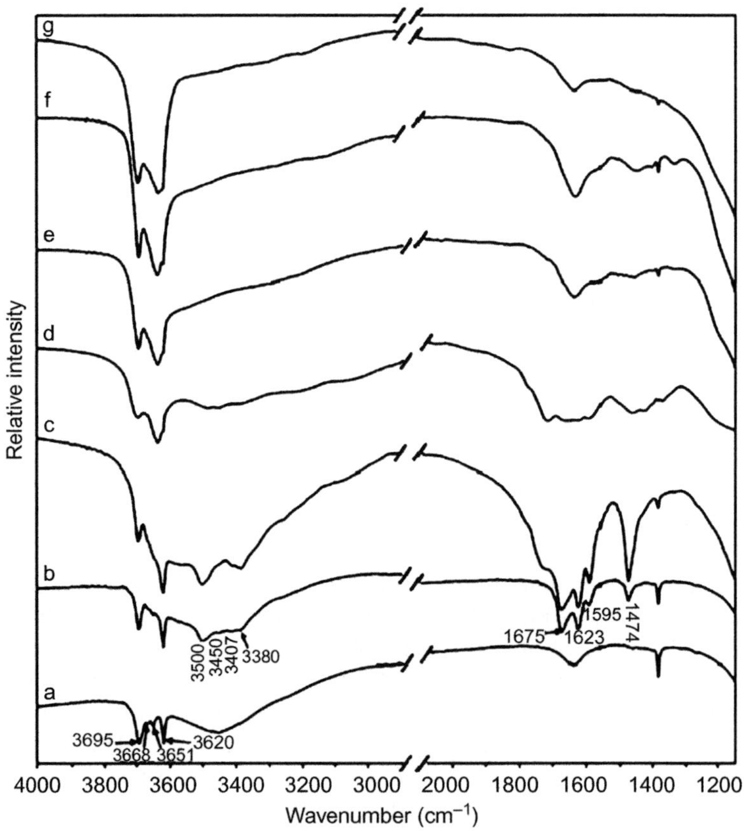


**FIG. 11.7** FTIR spectra of: a kaolinite; and the kaolinite–urea intercalation complex heated at: b room temperature; c 200°C; d 250°C; e 300°C; f 350°C and g 400°C. *From Raupach et al. (1987). Reproduced with kind permission of The Clay Minerals Society, publisher of Clays and Clay Minerals.*

intensities IR bands associated with $H_2O$, $CO_2$, $NH_3$ and HNCO were very high, and the evolved products obtained were mainly divided into two processes: (i) the major evolved products $CO_2$, $H_2O$, $NH_3$ and HNCO were mainly released at the temperature between 200 and 450°C with a maximum at 355°C due to thermal decomposition of urea; (ii) up to 600°C, the main evolved products were $H_2O$ and $CO_2$ with a maximum at 575°C due to thermal decomposition of residual nitric acid and the dehydroxylation of Kaol (Fig. 11.8). The release of $NH_3$ was confirmed by the appearance of absorption bands at 3335, 1632 and 966 $cm^{-1}$. The cyanic acid (HNCO) has two bands with equal intensity at 1095 and 1720 $cm^{-1}$ corresponding to C–N stretching vibration and C=O stretching vibration, together with a band at 3474 $cm^{-1}$. Water was generated mainly by the dehydroxylation of Kaol and confirmed by the disappearance of structural υ(OH) bands at 3500–3640 $cm^{-1}$. The presence of $CO_2$ was confirmed by the appearance of bands at 2290 and 2253 $cm^{-1}$.

Liu et al. (2014) attempted to model the intercalation structure through molecular dynamics simulation. Three main distribution types of the urea molecules can be observed according to the angle between the dipole moment vectors of the urea molecules and the basal surface of Kaol. Most of the dipole moment vectors point to the silica tetrahedron sheet, with the angle between them and the basal surface of Kaol ranging approximately from 20 to 40°, which is consistent with the study of Rutkai et al. (2009). In addition, a few urea molecules are nearly parallel to the basal surface of Kaol, with the urea molecules situated closer to the silica tetrahedron sheet. Finally, some urea molecules are almost perpendicular to the basal surface of Kaol.

The preceding observation may serve as an indicator of the three endothermic peaks at 218°C, 250°C and 261°C observed in the DSC curve of Kaol–urea intercalation complex. These peaks are attributed to the sequential decomposition of the three types of urea molecules distributed between the Kaol layers. Unfortunately, there is no spectroscopic evidence in either the IR or Raman spectrum for the presence of three different possible orientations of urea in the interlayer space. On top of this, the snapshot shows that for a diffuse distribution a distance between the layers of 0.602 nm is necessary, which is much larger than what is observed with XRD for urea-intercalated Kaol, where the separation between the layers is no more than about 0.35 nm.

### 11.2.3 Formamide

Formamide ($NCH_3O$ or HC=O–$NH_2$) is an amide compound with two H bond donor sites in the molecule, has a melting point of 2°C and crystallizes at atmospheric pressure to a monoclinic unit cell containing four molecules forming H bonds of adjacent C=O and N–H units. Crystalline formamide is the simplest molecular solid with $C_s$ symmetry on general sites in a centrosymmetric unit cell belonging to the point group $C_h$. The four molecules per unit cell lead to

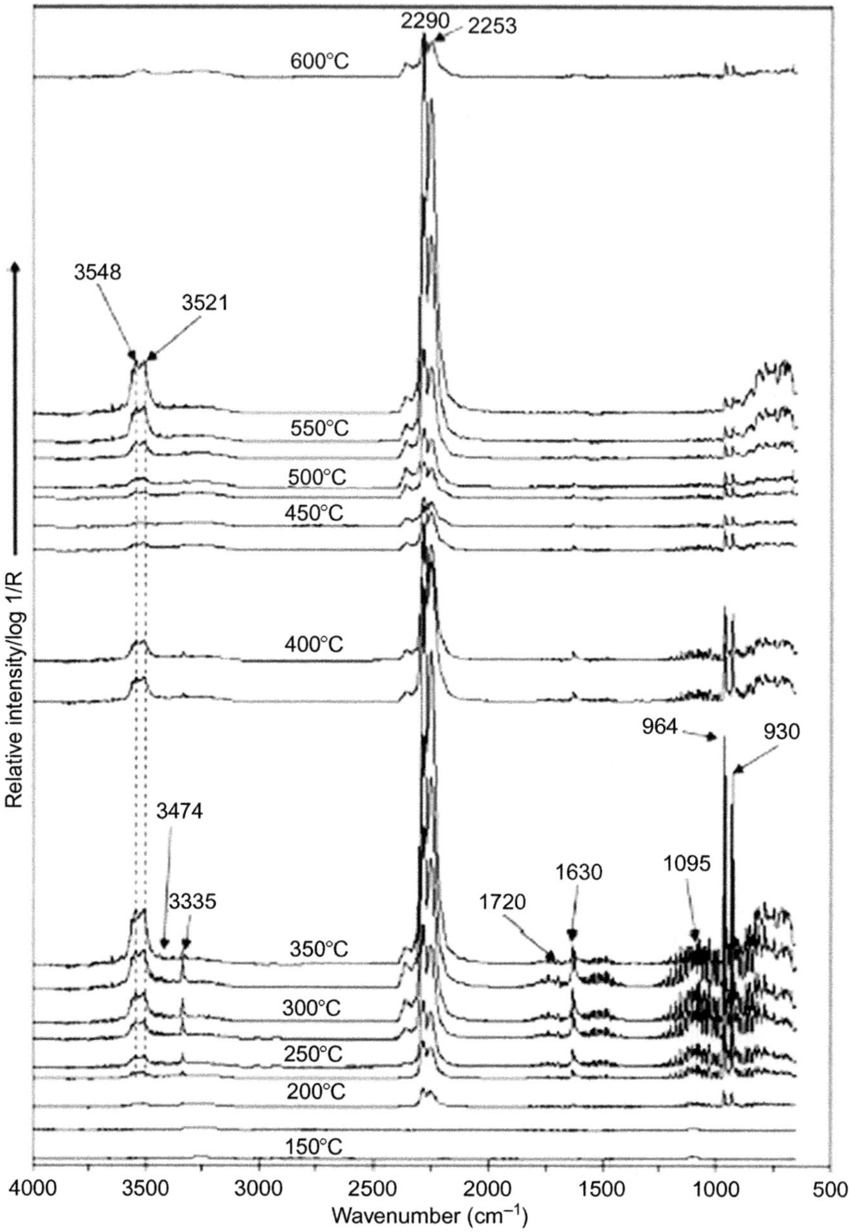


**FIG. 11.8** Infrared spectroscopy analysis of the evolved gases for kaolinite urea intercalation complex. *Reproduced with permission from Cheng et al. (2014).*

12 Raman and 9 H-active lattice modes and an expected splitting of the internal modes to two modes in both the Raman and IR spectra (Torrie and Brown, 1994). In its liquid state, formamide constitutes the smallest unit in a peptide chain which forms extended H bond networks. Because the protons of the $NH_2$ group are oriented *cis* (*cis*-NH H) to the C=O group, the formamide molecule is capable of forming a cyclic H-bonded dime, but H bonding with the $NH_2$ having protons oriented *trans* (*trans*-NH H) to the C=O group gives rise to formation of a linear (zigzag) chain structure. A regular two-dimensional H bond network is formed in the crystalline state (Torii and Tasumi, 1998).

The IR spectrum of formamide in the liquid state in the region of the $\nu$(NH) modes shows three strong bands at 3404, 3329 and 3190 $cm^{-1}$ with a weak band at 3248 $cm^{-1}$ (Table 11.4). The highest band has been attributed to the $\nu_{as}$(NH) and the lower two to the Fermi doublet due to $\nu_{as}$(NH) and the overtone of $\delta$(NH) of the $NH_2$ group (Bukowska, 1979), although the four bands are due to a distorted cyclic dimer. The C=O stretching band was observed at 1690 $cm^{-1}$ with a shoulder at 1712 $cm^{-1}$. The former and the latter are assigned to C=O stretching modes attributed to H-bonded molecules and monomers or molecules at the end of a chain, respectively (Mortensen et al., 1995b), although an asymmetric IR band profile has been calculated on the basis of a transition–dipole coupling (TDC) mechanism and liquid structures derived from molecular dynamics simulations (Torii and Tasumi, 1998). In the isotropic Raman spectrum, the C=O stretching is observed at 1668 $cm^{-1}$ (Mortensen et al., 1995b), which is weak in the IR spectrum (Kanesaka and Mitsuhashi, 2008). In solution, effects of intermolecular interactions can be seen not only in the low-wavenumber region but also in the amide I region (1750–1600 $cm^{-1}$). Mortensen et al. (1994, 1995b; 1995a) have shown that the isotropic Raman band in the amide I region is significantly lower in wavenumber than both that in the IR and anisotropic Raman bands (only a minor wavenumber difference between the latter two).

The IR spectrum of solid formamide obtained at −243°C, exhibited a broad band from 3800 to 2944 $cm^{-1}$ with two bands positioned at 3372 $cm^{-1}$ and 3179 $cm^{-1}$ corresponding to the asymmetric and symmetric $\nu$(NH) modes, respectively of the $NH_2$ group. Five other fundamental bands were observed at 2895 $cm^{-1}$, 1698 $cm^{-1}$, 1628 $cm^{-1}$, 1386 $cm^{-1}$ and 1334 $cm^{-1}$ corresponding to the C-H stretching, C=O stretching, in plane H-N-H scissor, in plane H-C-H bending, and C-N stretching modes, respectively (Table 11.4) (Sivaraman et al., 2013). All these bands agree well with those reported by Torrie and Brown (1994), but four more bands were observed at 1172 $cm^{-1}$, 1117 $cm^{-1}$ (H-N-H rocking), 836 $cm^{-1}$ (H-N-H wagging) and 710 $cm^{-1}$ (H-N-H twisting), which were also observed by Torrie and Brown (1994) but in solid formamide at −253°C. In contrast to the IR spectrum reported by Sivaraman et al. (2013), Torrie and Brown (1994) showed Raman and IR spectra with much more details and splitting of bands with the single band at 710 $cm^{-1}$ split into four bands, the 1334 $cm^{-1}$ in three bands and the 1628 and 1698 $cm^{-1}$ band into two bands each

**TABLE 11.4** Observed IR and Raman and Calculated Vibrations of Liquid and Solid Formamide. Values are in Wavenumbers ($cm^{-1}$)

| Liquid | | | | | Solid | | | | | Assignment | |
|---|---|---|---|---|---|---|---|---|---|---|---|
| *IR*[a] | *IR*[b] | *IR*[c] | *Raman*[a] | *IR*[a] | $IR_b$ | *IR*[d] | | *a,b* | c | d | |
| | | | *−253°C* | *−253°C* | *−65°C* | *−165°C* | *−243°C* | *Model*[c] | | | |
| | | | 619 | 636 | 633 | 635 | | | | | |
| 608 | 608 | 608 | 641 | 655 | 649 | 657 | 710 | 602 | NCO deformation | NCO deformation | HNH twist |
| | | | 646 | 664 | | 675 | | | HNH twist | | |
| | | | 676 | 685 | | | | | | | |
| | 750 | | | 841 | 825 | 843 | 836 | | HNH wag | | HNH wag |
| | | | | | 1041 | 1047 | 1022 | | CH out of plane bend | | CH out of plane bend |
| | 1056 | | 1061 | 1068 | 1057 | 1063 | 1056 | | | | |
| 1090 | 1090 | 1090 | 1143 | 1134 | 1133 | 1133 | 1117 | 1094 | HNH rocking | HNH rocking | HNH rocking |
| | | | 1154 | 1142 | | 1140 | 1172 | | | | |
| | | | | | | | 1226 | | | | |
| 1309 | | 1309 | | 1315 | | 1310 | 1334 | | CH deformation | CN stretch | CN stretch |
| | | | | 1333 | | | | | | | |
| | | | 1358 | 1388 | | | | | | | |

| | | | | | | | | |
|---|---|---|---|---|---|---|---|---|
| 1391 | 1391 | 1373 | | 1386 | 1392 | HNH bend | CH bend | In plane CH bend |
| | | 1396 | | | | | | |
| 1608 | 1608 | | 1602 | 1628 | 1597 | CN stretch | HNH bend | In plane HNH scissor |
| | | | 1661 | | | | | |
| 1690 | 1690 | 1692 | 1704 | 1698 | 1690 | CO stretch | CO stretch | CO stretch |
| | | | 1748 | | | | | |
| 2882 | 2882 | 2893 | 2891 | 2895 | 2887 | CH stretch | CH stretch | CH stretch |
| 3190 | 3190 | 3124 | 3217 | 3179 | 3239 | HNH symm. stretch | HNH symm. stretch | HNH symm. stretch |
| | 3324 | | | | | | | |
| 3330 | 3404 | 3265 | | 3372 | 3399 | HNH antisymm. stretch | HNH asymm. stretch | HNH antisymm. Stretch |

[a]*Torrie and Brown (1994)*[c], *Kanesaka and Mitsuhashi (2008)*[b] *Itoh and Shimanouchi (1972)* [d] *Sivaraman et al. (2013).*

(Table 11.4). The 710 $cm^{-1}$ band has also been assigned to N-C=O deformation by Itoh and Shimanouchi (1972), Torrie and Brown (1994) and Kanesaka and Mitsuhashi (2008). Three bands at 1022 $cm^{-1}$, 1056 $cm^{-1}$ and 1226 $cm^{-1}$ were observed for the first time at this low temperature of −243°C. The two bands observed in the region 1065–1040 $cm^{-1}$ in solid formamide at −165°C, by Itoh and Shimanouchi (1972), were assigned to the out of plane $\delta$(CH); therefore, the bands observed at 1056 $cm^{-1}$ and 1022 $cm^{-1}$ in the spectra at 30 K were also assigned to the out of plane $\delta$(CH) by Sivaraman et al. (2013).

### *11.2.3.1 Raman Non-coincidence*

The wavenumber difference between the isotropic and anisotropic Raman bands is known as the 'Raman non-coincidence'. The IR, isotropic Raman and anisotropic Raman bands in the amide I region of liquid formamide originate from modes with different vibrational patterns. The vibrational coupling between molecules in the liquid results in the difference in the wavenumbers of these bands. The observed bands in the low-wavenumber region are well reproduced by the spectra of the antiparallel hexamer, but not by those of the cyclic and linear hexamers. In most cases, Raman non-coincidence is observed for a vibrational mode with a strong IR intensity and is taken as arising from interactions between the transition dipoles of molecular vibrations in the liquid. The IR and Raman spectra in the amide I region calculated on the basis of the TDC mechanism and the liquid structures obtained from molecular dynamics simulations show the isotropic Raman band at a significantly lower wavenumber than the IR and anisotropic Raman bands (Torii and Tasumi, 1998), in agreement with the experimental results (Mortensen et al., 1994, 1995b).

Disorder in H bonded chains of formamide (or any molecule) is an important factor for the Raman non-coincidence ($\Delta\nu_{aniso\text{-}iso}$) in the amide I region. The intermolecular interactions within linear H-bonded chains can explain about two thirds of the observed value of $\Delta\nu_{aniso\text{-}iso}$. However, this is not true for the value of $\Delta\nu_{IR\text{-}iso}$ in the amide I region, if the disorder in the chain structure is taken into account. Interchain interactions by the TDC mechanism are significant for elucidation of this quantity. Therefore, it can be concluded that there are a large number of H bonds formed with the *cis*-oriented $NH_2$, whereas cyclic hexamers, which consist of H bonds formed only with the *trans*-oriented $NH_2$, are not dominant in liquid formamide (Torii and Tasumi, 1998).

### *11.2.3.2 Intercalation of Formamide in Kaol*

Intercalation of Kaol with formamide results in an expansion of the basal spacing from approximately 0.72 to 1.01 nm (Ledoux and White, 1966; Frost et al., 1998e, 2000c). In the IR spectra, new bands appear around 3629 and 3606 $cm^{-1}$. The first band can be ascribed to the $\nu$(OH) mode of the inner surface OH ($\nu(Al_2OH_{IS})$) of the Kaol H bonded to the C=O group of the formamide. The band around 3606 $cm^{-1}$ has been interpreted as the $\nu$(HOH) mode of non-bonded $H_2O$ within the interlayer space. In an earlier study, a similar band was observed around 3590 $cm^{-1}$, which was then interpreted as the H bond perturbed $\nu(Al_2OH_{IS})$ by interaction with formamide C=O. This band, however, was very broad, and an interpretation as the υ(HOH) for water seems much more plausible. In addition, to these bands also the C–H stretching modes around 2772 and 2884 $cm^{-1}$ and the N–H symmetric ($\nu_s$(NH)) and antisymmetric ($\nu_{as}$(NH)) stretching modes around 3167, 3257, 3340 and 3448 $cm^{-1}$ were observed in the IR spectrum (Fig. 11.9) (Frost et al., 2000c).

The normally observed $\nu(Al_2OH_{IS})$ bands are no longer observed in the Raman spectra of Kaol formamide intercalates. However, two bands are visible around 3627 and 3620 $cm^{-1}$, which are due to the formation of the H bonds between the Kaol $Al_2OH_{IS}$ and the formamide C=O group and the original, unchanged, inner OH group ($Al_2OH_I$), respectively (Frost et al., 1998e, 1999d). These bands shifted to 3631 and 3616 $cm^{-1}$ on cooling to 77 K (Fig. 11.9) (Frost et al., 1998e). The band at 3463 $cm^{-1}$ is assigned to the H bond perturbed symmetric stretching vibration of the O–Si unit ($\nu_s$(Si–O) via interaction with formamide N-H. If the formamide NH group was not bonded to the siloxane layer, a band at 3374 $cm^{-1}$ would be expected. In the region of $\delta$(OH) vibrations, a similar picture emerges: The inner surface OH bands ($\delta(Al_2OH_{IS})$) strongly decreased in intensity after intercalation with formamide, and a new band was observed around 905 $cm^{-1}$, which can be ascribed to H-bond perturbation of the $Al_2OH_{IS}$ via interaction with formamide C=O.

To show the presence of water, one normally studies the IR spectral region around 1650 $cm^{-1}$. In this case, the situation is slightly more complicated because there is an overlap with the so-called amide bands of formamide in this region. Three bands were observed around 1715, 1695 and 1674 $cm^{-1}$ for formamide Kaol intercalate. The first band has been ascribed to amide 1 (actually $\nu$(C=O)), whereas the other two have been ascribed to amide 2 ($\delta$(NH) of the primary amide). The presence of two amide bands indicates that only one $NH_2$ group H bonded with the Kaol siloxane layer, whereas the other group remains free. Water gave a single band around 1597 $cm^{-1}$, which points to water functioning to fill the space between the formamide molecules in the interlayer space of Kaol without the formation of any bonds.

The DRIFT spectra of the untreated Kaol and formamide and 50/50 wt% formamide–water intercalated Kaol showed the changes in vibrational modes of N-H and C-H of the inserting molecule (Fig. 11.10). The intensity of these bands was greater for the formamide water Kaol intercalate than in the formamide Kaol intercalate. In the DRIFT spectra of the formamide intercalated Kaol (Fig. 11.10) bands were observed at 1700, and

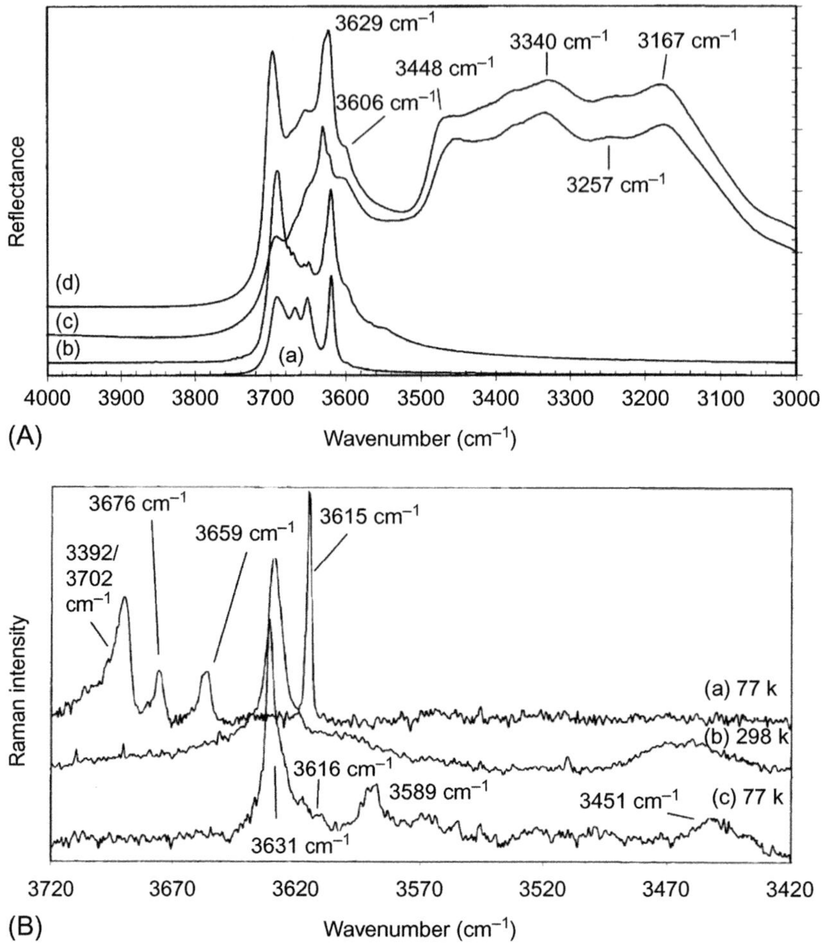


**FIG. 11.9** (A) DRIFT and (B) Raman spectra of the hydroxyl-stretching region of formamide kaolinite intercalation complexes. In (A) spectra of: a, low-defect kaolinite; b, high-defect kaolinite; c, formamide-intercalated low-defect kaolinite and d, formamide intercalated high-defect kaolinite. In (B) spectra of: a, low-defect kaolinite at 77 K; b, low-defect kaolinite intercalated with formamide at 298 K and c, low-defect kaolinite intercalated with formamide at 77 K. *Part figure (A) reproduced with permission from Frost et al. (2000c). Part figure (B) reproduced with permission from Frost et al. (1998e).*

1671 cm$^{-1}$ and can be attributed to the amide 1 and amide 2 vibrations, respectively. The 1700 cm$^{-1}$ band was attributed to $\nu$(C=O), the normal position of which is between 1640 and 1680 cm$^{-1}$. A substantial shift in this vibration was observed on intercalation of the formamide into Kaol. The bands at 1695 and 1674 cm$^{-1}$ can be attributed to formamide $\delta$(NH). The remaining bands at 1630 and 1595 cm$^{-1}$ can attribute to the $\delta$(HOH)

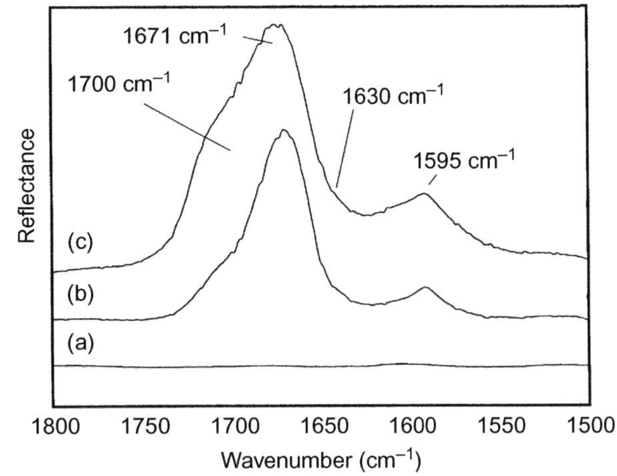


**FIG. 11.10** DRIFT spectra of the water bending modes for: a, kaolinite; b, formamide intercalated kaolinite and c, formamide–water intercalated kaolinite. *Reproduced with permission from Frost et al. (2000a).*

mode of water. It was concluded that water is not strongly H-bonded and simply functions as a space-filling molecule in the expanded interlayer space. The relative intensity of the 1630 $cm^{-1}$ band was 5.6% for the formamide intercalated low-defect Kaol.

This observation agreed with the TGA data where intercalation of the Kaol with water increased the amount of formamide loaded into the interlayer spaces of Kaol (Frost et al., 2001d; Fig. 11.11A). On intercalation of the Kaol with the 50/50 wt% formamide water mixture, Raman bands were observed at 3647, 3630 and 3605 $cm^{-1}$. Bands were observed in the same positions and with approximately the same intensity irrespective of whether Kaol was intercalated with 100% formamide or with 50/50wt% formamide water, but a significant difference in the intensity of the band at 3605 $cm^{-1}$ was observed, being 17% for the formamide-intercalated Kaol and 26% for the formamide water intercalate. The TGA (Frost et al., 2001d) confirms the earlier assignment of this band to the $\nu$(HOH) of $H_2O$ within the interlayer space. The 3648 $cm^{-1}$ and 3630 $cm^{-1}$ bands, which are both Raman and IR active are the H-bond perturbed $\nu(Al_2OH)_{IS}$. The difference in the two band positions can be accounted for by the strength of the H bonds formed between formamide with other formamide molecules or with hydrated formamide. A weaker H bonding results in the observation of the structural $\upsilon$(HOH) bands at high wavenumbers (3648 $cm^{-1}$ compared with 3630 $cm^{-1}$). Significant differences in the intensities of the $\nu$(NH) bands were observed between the two intercalates. The 3460 $cm^{-1}$ band intensity decreased from 33.7 to 5.4%. The intensity of the 3340 $cm^{-1}$ band

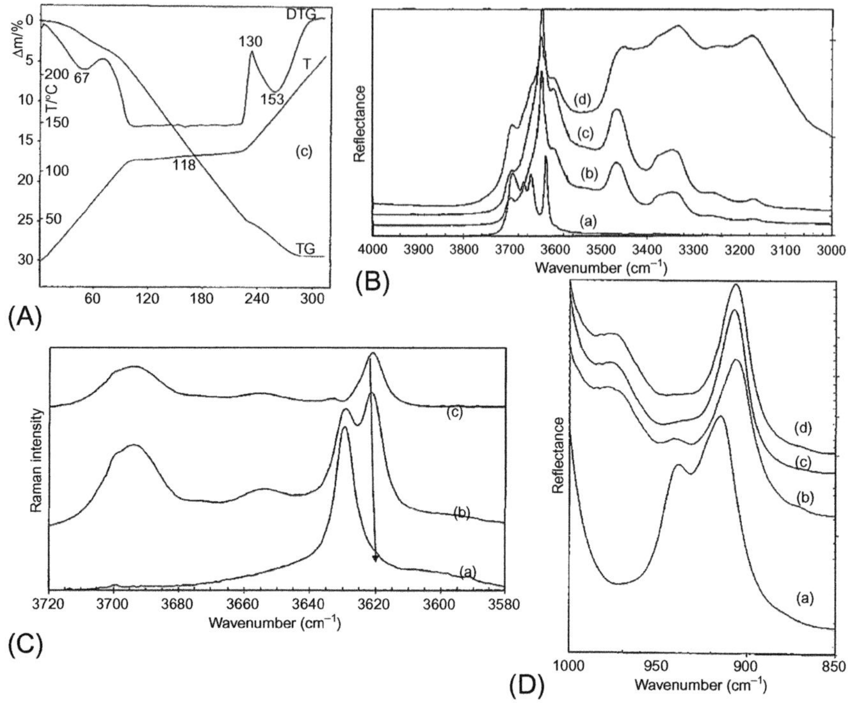


**FIG. 11.11** (A) Controlled rate thermal analysis of formamide kaolinite intercalate complexes. (B) DRIFT spectroscopy of the NH and OH stretching region of: a, low-defect kaolinite; b, CRTA-treated intercalated with 100% formamide; c, CRTA-treated intercalated with 50% formamide/50% water; and d, intercalated with 100% formamide. (C) Raman spectrum of the hydroxyl stretching region of: a, formamide-intercalated kaolinite; b, CRTA-treated formamide-intercalated kaolinite; and c, the difference spectrum. (D) DRIFT spectroscopy of the OH deformation region of: a, low-defect kaolinite; b, intercalated with 100% formamide; c, CRTA-treated intercalated with 100% formamide; and d, CRTA-treated intercalated with 50% formamide/50% water. *Reproduced with permission from Frost et al. (2001d).*

remains unchanged, but significant increases in the 3242, 3177 and 3106 $cm^{-1}$ bands were observed. It was suggested that a high-energy conformation is minimised and significant $NH_2$ groups are not H bonded in the formamide water intercalate. The additional 3106 $cm^{-1}$ band, which may correspond with the very broad 3028 $cm^{-1}$ band of formamide, was attributed to $\nu$(NH) H bonded to $H_2O$.

Dawley et al. (2012) and Michálková Scott et al. (2012) calculated the IR spectra of all formamide Kaol complexes studied using the density functional theory indicating that formamide binds with the four Kaol octahedral OH through the formation of four H bonds formamide oxygen atom and the $NH_2$ group and with the Kaol siloxane surface via formation of three H bonds of formamide CH and $NH_2$ groups. The adsorption of formamide to the octahedral

OH groups was found to be stronger than that to the siloxane surface as demonstrated by adsorption energies of −14.8 kcal/mol (octahedral OH) and −13.7 kcal/mol (siloxane surface). A decrease in positions of 40–70 $cm^{-1}$ was determined for the $\nu_s$(NH) (3112 $cm^{-1}$), $\delta$(NH) (1585 and 1597 $cm^{-1}$) and $\nu$(CN) (1273 $cm^{-1}$) modes when H bonding with the octahedral OH groups when compared with experimental results. A small comparative increase (~30 $cm^{-1}$) in the $\nu$(CH) was observed. However, a much larger increase (200 $cm^{-1}$) in the theoretical wavenumber was found for the $\nu_{as}$(NH) mode (from 3297 in the 110 K experiment to 3490 $cm^{-1}$ in the calculation). Two theoretical $\nu$(NH) bands at 3112 and 3490 $cm^{-1}$ were calculated considered due to H-bonded formamide NH with the surface oxygen atom (3112 $cm^{-1}$) and non H-bonded formamide NH (3490 $cm^{-1}$). However, experimentally, such splitting and shifts were not observed. In contrast, a good consistency in theoretical and experimental values for the C═O stretching band at 110 K (calculated at 1706 $cm^{-1}$, difference only 6 $cm^{-1}$) and only a slight decrease for the CH scissors mode (calculated at 1371 $cm^{-1}$, difference about 18 $cm^{-1}$) were found. The theoretical bands seen in the region of 3150–3350 $cm^{-1}$ corresponded to the stretching vibrations of the terminated OH groups of the siloxane sheet that were added to the Kaol cluster model to saturate the dangling bonds.

The calculated spectrum of the the interaction of formamide with Kaol octahedral OH compared with the calculated spectrum of formamide showed significant shifts to lower wavenumbers for the $\nu$(NH). Among other shifts was an increase in the wavenumber of the $\nu$(CN) mode from that of pure formamide (shift of ~50 $cm^{-1}$). A similar increase in wavenumber of this vibration was observed by Frost et al. (1998e) experimentally for formamide-intercalated Kaol (from 1300 to 1320 $cm^{-1}$). The in-plane $\delta$(CH) (1373 $cm^{-1}$), $\nu$(C═O) (1718 $cm^{-1}$), and $\nu$(CN) (1275 $cm^{-1}$) modes were close to the experimental results (difference < 15 $cm^{-1}$). In contrast, the bands above 2900 $cm^{-1}$ assigned to $\nu$(CH) (2934 $cm^{-1}$) and $\nu$(NH) (3546 and 3414 $cm^{-1}$) were found at much higher wavenumbers than those in Kaol octahedral-site-formamide.

The theoretical spectrum of the Kaol siloxane surface formamide interaction also exhibited good agreement with the experiment for the $\delta$(CH) and $\nu$(C═O) modes (difference < 20 $cm^{-1}$). However, the $\nu$(NH) and $\delta$(NH) vibrations were shifted from experimental results more compared with the shifts observed in the Kaol octahedral OH formamide complexes (~250 and ~80 $cm^{-1}$, respectively). Prior work using DRIFT spectroscopy proposed that a band at 1667 $cm^{-1}$ is C═O H-bonded to the Kaol inner surface hydroxyl (Wang et al., 2007), whereas Frost et al. (1998e, 2000a, 2000c, 2001f) attributed the band at 1700 $cm^{-1}$ to this vibration and the bands at 1695 and 1674 $cm^{-1}$ to the $\delta$(NH) modes of a primary amide.

Experimental studies (e.g. Olejnik et al., 1971; Frost et al., 2001d) have also shown that formamide and other amides are adsorbed predominantly onto the Kaol surfaces by H bonding to $Al_2OH_{IS}$. The theoretical IR spectra results and adsorption energy values for the interactions between formamide and the octahedral OH and siloxane surface of Kaol confirmed that the octahedral surface of Kaol is the favourable adsorption surface while interacting with formamide.

#### *11.2.3.3 Deintercalation of Formamide from Kaol*

Analysis by CRTA allows differentiation of the adsorbed formamide from the intercalated formamide in formamide–Kaol complexes (Frost et al., 2001c) (Fig. 11.11A). Under quasi-isothermal conditions, two mass losses were observed: (i) at 118°C under quasi-isothermal conditions and (ii) at 153°C, but under dynamic conditions only a single mass loss was observed. TG-MS studies indicated that the mass loss at 67°C is due to removal of water (3.27 wt%). Bonded formamide was released in two stages under quasi-isothermal conditions at 118°C (21.57 wt%) and under non-isothermal conditions at 153°C (4.24 wt%). Significant differences were observed in the thermal behaviour of differently bonded formamide molecules. Whereas the first type of bonded formamide was lost under isothermal conditions at 118°C, the second type of formamide molecule exhibited a drastically different behaviour on heating. The XRD pattern of the CRTA-treated Kaol formamide intercalate was identical to that of the untreated intercalate, demonstrating that the CRTA-treated intercalate was still fully expanded at 130°C. On heating the formamide Kaol intercalate to 300°C, formamide was lost and the structure collapsed to its original *d*-value of ≈7.2 Å (Frost et al., 2001d).

The differences in the DRIFT spectra between the formamide and CRTA-treated Kaol formamide intercalates were quite noteworthy. The spectra of the CRTA-treated samples are shown in Fig. 11.11B showed intensity in the 3300–3500 $cm^{-1}$ region, whereas the untreated intercalate showed intensity in the 3100–3300 $cm^{-1}$ region. As discussed above, on intercalation of formamide into Kaol, additional bands are observed at 3606 and 3629 $cm^{-1}$. Only two $\nu(Al_2OH)$ bands are resolved at 3695 and 3629 $cm^{-1}$ of the CRTA-treated formamide-intercalated Kaols. The band at 3629$cm^{-1}$ can be attributed to H-bond perturbed $\nu(Al_2OH_{IS})$. The band at 3695 $cm^{-1}$ can be assigned to the non-bonded $\nu(Al_2OH_{IS})$. In the $\nu$(NH) region (Fig. 11.11B), bands at 3460 and 3365 $cm^{-1}$ are attributed to the intercalated formamide, whereas bands 3344, 3227 and 3148 $cm^{-1}$ can be attributed to adsorbed formamide and the bands (Frost et al., 2001d). The Raman spectra (Fig. 11.11C) showed a single intense band ($\nu_6$) at 3629 $cm^{-1}$ attributed to formamide H-bond perturbed $\nu(Al_2OH_{IS})$. No intensity was observed for non H-bond perturbed $\nu(Al_2OH_{IS})$ or $\nu(Al_2OH_I)$, the latter could be hidden by the 3629 $cm^{-1}$ band. A broad band was observed at around 3603 $cm^{-1}$ and may be attributed to the $\nu$(HOH) mode of water, which is in the interlayer space. On subtraction of the formamide intercalated Kaol from the CRTA-treated formamide-intercalated Kaol, the Raman spectrum of the only formamide adsorbed Kaol is obtained (Fig. 11.11C). Significant intensity remains in the band ($\nu_1$) attributed to the inner surface OH stretching vibration in both spectra. In the formamide-intercalated Kaol, the percent relative intensity is 10% and in the water formamide-intercalated Kaol the value is 29%. These values indicate that a significant number of OH groups are involved with adsorbed formamide molecules as opposed to intercalated formamide molecules (Frost et al., 2001f).

Bands were also observed at 905, 915, 938 and 974 cm$^{-1}$ (Fig. 11.11D). Significantly, as with the $\nu(Al_2OH_{IS})$ modes, the intensity of the bands attributed to the $\delta(Al_2OH_{IS})$ at 939 and 924 cm$^{-1}$, which can be attributed to formamide H-bond perturbed $\delta(Al_2OH_{IS})$ modes, were reduced to half the total intensity. The band at 974 cm$^{-1}$ can be assigned to an out-of-plane CH wag. The 1715 cm$^{-1}$ band of formamide Kaol intercalates has been attributed to the C=O stretching vibrations. The bands at 1695 and 1674 cm$^{-1}$ were attributed to $\delta$(NH) modes of a primary amide. These bands are amide-2 bands and are often described as a mix of the $\beta NH_2$ and $\gamma$CN vibrations in a 60/40 ratio and indicates that one NH is bonded to the siloxane layer, while the second NH of the amide group remains unbonded. The band at 1630 cm$^{-1}$ for the adsorbed formamide in the formamide-intercalated Kaol corresponds to $\delta$(HOH) of H bonded water. In this case, the water is H bonded to the formamide. In the CRTA-treated Kaol formamide intercalates (Fig. 11.12), no band at around 1630 cm$^{-1}$ is observed, thus confirming the TGA that water was removed. No band was observed at 1709 cm$^{-1}$ for the CRTA-treated intercalates as well, indicating that this vibration involved water in the vibrating unit (Table 11.5). Therefore, it can be envisaged that the 1709 cm$^{-1}$ band arises from an interaction between the C=O and the protons of water. In the absence

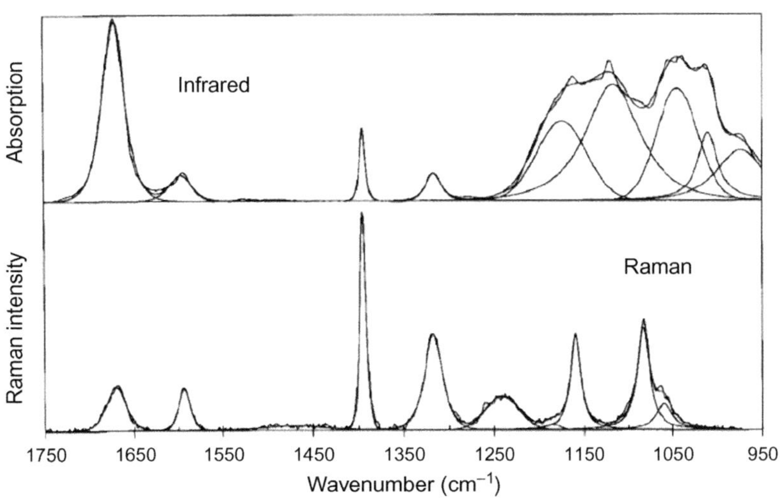


**FIG. 11.12** Raman and infrared spectra of the 950–1750 cm$^{-1}$ region of the CRTA-treated formamide-intercalated kaolinite (Frost et al., 2001c).

**TABLE 11.5** Results of the Band Component Analysis of the DRIFT Spectra of Formamide–Kaol Intercalate Complexes in the NH-Stretching Region, 1800–1000 $cm^{-1}$ Region and OH Deformation Region (Frost et al., 2001d). Values are in Wavenumbers ($cm^{-1}$)

| | $\nu_1$ | $\nu_2$ | $\nu_3$ | $\nu_4$ | $\nu_5$ | $\nu_6$ | Form C=O stretch | NH-def | $\nu_{H2O}$ | $\nu_1$ | $\nu_2$ | $\nu_3$ | $\nu_4$ | Form | $\nu_1$-def | $\nu_2$-def | $\nu_3$-def | $\nu_4$-def |
|---|---|---|---|---|---|---|---|---|---|---|---|---|---|---|---|---|---|---|
| Water | | | | | | | | | ≈1630 | | | | | | | | | |
| Kaolinite | | | | | | | | | | | | | | | 939 | 924 | 913 | |
| Form-intercalated | 3460 | | 3365 | 3344 | 3227 | 3184 | 1709 | 1675 | 1630 | 1594 | 1392 | 1314 | 1290 | 971 | | | 913 | 905 |
| Form–$H_2O$–intercalated | 3460 | | 3365 | 3344 | 3248 | 3167 | 1710 | 1675 | 1630 | 1595 | 1392 | 1314 | 1290 | 971 | | | 913 | 905 |
| Form-intercalated CRTA | 3462 | | 3365 | 3332 | | | | 1671 | | 1595 | 1394 | 1317 | | 972 | | 926 | | 907 |
| Form–$H_2O$–intercalated CRTA | 3462 | | 3371 | 3335 | | | | 1671 | | 1595 | 1395 | 1317 | | 972 | | 922 | | 907 |

Form = formamide, Form-$H_2O$ = mixture 50/50 wt% formamide/water, CRTA = after controlled rate thermal analysis at 130°C.

of water, this interaction no longer exists as is observed in the CRTA-treated intercalates.

Bands at 1290 $cm^{-1}$ and at 1317 $cm^{-1}$ of the formamide Kaol intercalate can be assigned to $\delta$(CH) of the intercalated, and the adsorbed formamide, respectively. The 1709 $cm^{-1}$ vibration, which has been assigned as an amide band, composed of the combination of the C=O (80%) and the C–N vibration (20%) is only associated with adsorbed formamide. Thus, the assignment of the 1709 $cm^{-1}$ band is questionable, as it seems only to be observed in the presence of water. There is also no signal at this position in the IR spectrum (Fig. 11.12), proving that the CRTA technique applied removed all adsorbed and intercalated water. Based on the thermal analysis, the amount of intercalated formamide was 4.24 wt%. This means that only 1 in 4 of the inner surface OH were potentially bonded to, or interacted with, the formamide. Therefore, a likely model is more likely to be based on formamide–water units space filling the interlayer spaces and adsorbed on the internal surfaces of the Kaol (Frost et al., 2001d). The band observed at 1670 $cm^{-1}$ was therefore attributed to the $\nu$(C–O) of the intercalated formamide molecule. The band at 1594 $cm^{-1}$ was assigned to $\delta$(NH) of a primary amide. Two bands were observed at 1394 and 1319 $cm^{-1}$, which are intense in the Raman spectrum compared with the IR spectrum. These bands were assigned to the H-C-H deformation and rocking vibrations. After the CRTA treatment, an additional band was observed at 1370 $cm^{-1}$ attributed to a second H-C-H rocking vibration, which appeared only after the CRTA treatment (Frost et al., 2001c).

## 11.2.4 Acetamide

Amides are the simplest model compounds for peptide linkage in polypeptides and proteins. Acetamide ($NC_2H_5O$, also written $CH_3$–C=O–$NH_2$) is a representative compound for the amide linkage in peptides and proteins. The FTIR gas spectrum of acetamide qualitatively resembles that of formamide, with the exception that the lowest transition is split due to interaction with the methyl-torsion, which is absent in the formamide case.

At high resolution (0.7 $cm^{-1}$), the bands at 256, 348 and 378 $cm^{-1}$ show an additional small splitting, and the band at 259 $cm^{-1}$ seemed to consist of several overlapping bands. The bands at 256, 348 and 378 $cm^{-1}$ represent the $0 \rightarrow 1$, the $1 \rightarrow 2$ and the $2 \rightarrow 3$ transitions of the amino-inversion, respectively, and assignment confirmed by MIR (Hansen et al., 1980). The same transitions were seen at 184, 250 and 285 $cm^{-1}$, respectively, in $CH_3COND_2$. The stronger bands of each split pair were assigned to the pure inversion transitions. The weaker bands were then thought to arise from the same inversion transitions occurring in the excited state of some other low-energy vibration. The only very low-energy motion expected in acetamide is the internal rotation of the methyl group, a mode seen at 143 $cm^{-1}$ in acetaldehyde. The C-type band at 507 $cm^{-1}$ was interpreted as the H-N-H torsional motion of the $NH_2$ group (this vibration has been located near 600 $cm^{-1}$ in formamide). The lower wavenumber of this motion in acetamide is consistent with the expectation that the resonance structure in which the carbon and nitrogen atoms are joined by a double bond is more important in formamide than in acetamide. The weaker C-type bands at 492 and 480 $cm^{-1}$ probably represent the same torsional transition occurring in the excited state of some other vibrations. The A-type band at 548 $cm^{-1}$ was interpreted as a vibration in the plane of the heavy atoms, and is probably best described as a deformation of the angles around the carbonyl carbon atom.

Ganeshsrinivas et al. (1996) reported the results of a study of the vibrational spectra of acetamide by an extended molecular mechanics method. The bands at 1726 and 1585 $cm^{-1}$ in the matrix are identified, respectively, at 1665 and 1640 $cm^{-1}$ in the solid and at 1680 and 1594 $cm^{-1}$ in solution. The shifts were larger for the out-of-plane vibrations. The bands at 628, 508 and 269 $cm^{-1}$ were identified in the solid at 810, 700 and 582 $cm^{-1}$, respectively (Table 11.6).

In the Raman spectrum of acetamide, the in-plane vibrations of the amide I band can be observed at 1662 $cm^{-1}$ originating primarily from $\nu$(C=O) and was most evident with 560-nm excitation. For acetamide, an amide II-like band occurs at 1616 $cm^{-1}$ and is associated with $\delta$(NH) of the $NH_2$ group and contains a small contribution from $\nu$(C=O). The amide III-like band of acetamide, located at 1404 $cm^{-1}$, contains a large contribution from $\nu$(CN), as well as minor contributions from $\nu$(C–C), $\delta$(NH) and symmetric deformations of the N-C=O group and $CH_3$ groups. Other Raman active vibrations of acetamide include a band at 871 $cm^{-1}$, which has a large contribution from $\nu$(C–C) and a minor contribution from $\nu$(C–N). The 1005-$cm^{-1}$ vibration results from rocking and asymmetric deformation of the $CH_3$ group. The band at 1131 $cm^{-1}$ derives from $\delta$(NH) and $\nu$(C=O), and the bands at 1360 and 1457 $cm^{-1}$ result from symmetric and asymmetric deformations of the $CH_3$ group (Dudik et al., 1985). Usually, the OH stretching band around 3500 $cm^{-1}$ is used for the characterization of the H-bond state. However, as acetamide does not have the OH group, the Raman spectra were obtained in the region between 1300 and 1700 $cm^{-1}$, where there is a variety of Raman bands related to the H-bond formation.

At $\approx$230 nm the $\pi$–$\pi^*$ transition of the peptide group started to be significant and, in parallel, the Raman band position of the C-N stretching mode shifted from 1403 to 1405 $cm^{-1}$. The Raman spectrum obtained below 210 nm was significantly affected by the resonant Raman effect as confirmed by the difference from the 532 nm spectrum. It should also be noted that the band position moving back at ca. 210 nm cannot be explained by the vibrational mixing phenomenon known in the C=O vibration mode, considering the simple Raman intensity trends of acetamide bands on the excitation wavelength. The Raman band position of the H-N-H mode obtained between 210 and 230 nm (1613 $cm^{-1}$) can be related to a stronger H-bond configuration at the $NH_2$ group compared with the other H-bond states observed above

**TABLE 11.6** Experimental Infrared and Raman Spectra of Acetamide and the Calculated Spectral Positions and Intensities (Dudik et al., 1985; Ganeshsrinivas et al., 1996). Values are in Wavenumbers ($cm^{-1}$)

| | Wavenumber ($cm^{-1}$) | | | | | | Intensity (km/mol) | | | |
|---|---|---|---|---|---|---|---|---|---|---|
| | Experimental IR | | | | | | Experimental | | | |
| | *Solid* | *Solution* | *Vapour* | *Matrix* | *Raman* | *Calcd.* | *Vapour* | *Matrix* | *Calcd.* | *Assignment* |
| A′ | 3380 | 3360 | 3550 | 3554 | | 3572 | m | m | 216.4 | $\nu_{as}$(NH) of $NH_2$ |
| | 3190 | 3161 | 3450 | 3433 | | 3438 | m | m | 140.4 | $\nu_s$(NH) of $NH_2$ |
| | 3005 | 2967 | 2976 | | | | m | vw | 1.5 | $\nu_{as}$(CH) of $NH_2$ |
| | 2930 | 2860 | 2943 | 2903 | | | m | vw | 2.9 | $\nu_s$(CH) of $NH_2$ |
| | 1665 | 1680 | 1733 | 1726 | 1662 | 1720 | vs | | 749.4 | $\nu$(C=O) |
| | 1640 | 1594 | 1600 | 1585 | 1616 | 1613 | m | | 243.3 | $\delta$(NH) of $NH_2$ |
| | 1470 | 1440 | 1432 | 1436 | 1457 | | w | | 11.5 | $\delta_{as}$(CH) of $CH_3$ |
| | 1398 | 1378 | 1385 | 1368 | 1360 | 1381 | m | | 8.3 | $\delta_s$(CH) of $CH_3$ |
| | 1358 | 1336 | 1319 | 1315 | 1401 | 1315 | | | 101.8 | $\nu$(CN) + $\delta$(NH) |
| | 1154 | 1118 | 1134 | 1100 | 1131 | 1130 | w | vw | 13.5 | ρ(NH) of $NH_2$ |
| | 1010 | 981 | 965 | 968 | 1005 | 985 | w | wm | 0.5 | ρ(CH) |
| | 878 | 872 | 858 | 838 | 871 | 833 | w | w | 5 | $\nu$(CC) |
| | 582 | 557 | 548 | 547 | | 554 | w | wm | 8.4 | $\delta$(N-C=O) |
| | 465 | 443 | 427 | 431 | | | w | | 6.4 | $\delta$(CC) |

*Continued*

**TABLE 11.6** Experimental Infrared and Raman Spectra of Acetamide and the Calculated Spectral Positions and Intensities (Dudik et al., 1985; Ganeshsrinivas et al., 1996). Values are in Wavenumbers ($cm^{-1}$)—Cont'd

| | Wavenumber ($cm^{-1}$) | | | | | | Intensity (km/mol) | | | |
|---|---|---|---|---|---|---|---|---|---|---|
| | Experimental IR | | | | | | Experimental | | | |
| | *Solid* | *Solution* | *Vapour* | *Matrix* | *Raman* | *Calcd.* | *Vapour* | *Matrix* | *Calcd.* | *Assignment* |
| A″ | 3000 | 2900 | 2989 | 2973 | | | m | w | 0.9 | $\nu_{as}$(CH) of $CH_3$ |
| | 1457 | 1433 | 1445 | | | | w | | 4.9 | $\delta_{as}$(CH) of $CH_3$ |
| | 1046 | 1041 | 1040 | 1035 | | 1022 | w | w | 34.5 | $\rho$(CH) of $CH_3$ |
| | 810 | 625 | 628 | 621 | | | w,sh | | 50.4 | $\tau$(NH) of $NH_2$ |
| | 700 | 507 | 508 | 508 | | | w | wm | 89.4 | $\pi$(C=O) |
| | 582 | 527 | 269 | 269 | | | | | 421.5 | $\omega$(NH) |
| | 155 | 160 | | | | | | | 1.2 | $\tau$(CH) of $CH_3$ |

230 nm (the Raman band position of which is 1610 $cm^{-1}$). Contrary to the $NH_2$ case, the positive shift of the Raman band position of $\nu$(C=O) is known to be caused by a weaker H bond formation at the site. Mennucci and Martínez (2005) showed the absence of H bond formation at the C=O site caused a further red shift of UV absorption. This is consistent with the result of Saito et al. (2015) because the weaker H bond state at the C=O site was selectively excited at the preresonance regime by the red shift. Based on the idea of H-bond selectivity, the detected tendency in the band position can be explained as follows: Generally, peptides are in many different H-bond states in aqueous solutions and, in the wavelength region far from the π-π* resonance, an average of these H-bond states (mainly major H-bond state) is observed in Raman spectrum. When the incident wavelength decreases to about 230 nm, acetamide in the stronger (and minor) H-bond state at the $NH_2$ group (with a longer excitation wavelength) starts to be selectively excited due to the resonance Raman effect. With decreasing the excitation wavelength further, the major H-bond state starts to be additionally excited and at ca. 210 nm intensity of the resonance Raman signal from the major state starts to exceed that from the minor state.

#### *11.2.4.1 Intercalation of Acetamide in Kaol*

Although acetamide is a molecule comparable with formamide, its intercalation into Kaol follows a slightly different mechanism. Intercalation results in a larger expansion of the Kaol basal spacing from 0.72 to 1.09 nm (Frost et al., 1999e), due to the extra methyl group in acetamide in comparison with formamide. In this case, H bonds are formed between the $NH_2$ groups of the acetamide and the oxygen atoms of the Kaol siloxane layer. This can be seen in the IR spectrum by the formation of two new bands around 3400 and 3509 $cm^{-1}$ (Fig. 11.13A). In addition, a new band appears in both the IR and Raman spectra around 3596 $cm^{-1}$, which have been ascribed to H bonding between the acetamide C=O and the Kaol $Al_2OH_{IS}$. Similar bands were earlier reported by Olejnik et al. (1971) at 3394, 3415 and 3501 $cm^{-1}$, though they ascribed the last band to the H-bonded $\nu_{as}(NH_2)$ (Table 11.7). Two types of intercalation of acetamide were proposed depending on the type of interaction: (i) through the H bond formation of the amine H to the oxygens of the siloxane layer and (ii) by H bond formation of the C=O group of the acetamide to the $Al_2OH_{IS}$. The Raman spectra of the $Al_2OH_{IS}$ show diminished intensity, which supports the latter assignment. This may be attributed to several factors: (i) all the inner surface OH groups are involved in intercalation with the acetamide. This is unlikely as it would be expected that the DRIFT spectrum would also show such diminished intensity in these bands and (ii) a second consideration is that the polarisability of the bands has altered to such an extent that the bands are no longer Raman active. A common feature to the Raman spectra is the existence of the additional band at 3596 $cm^{-1}$, which is also observed in the DRIFT spectra of the acetamide-intercalated Kaol.

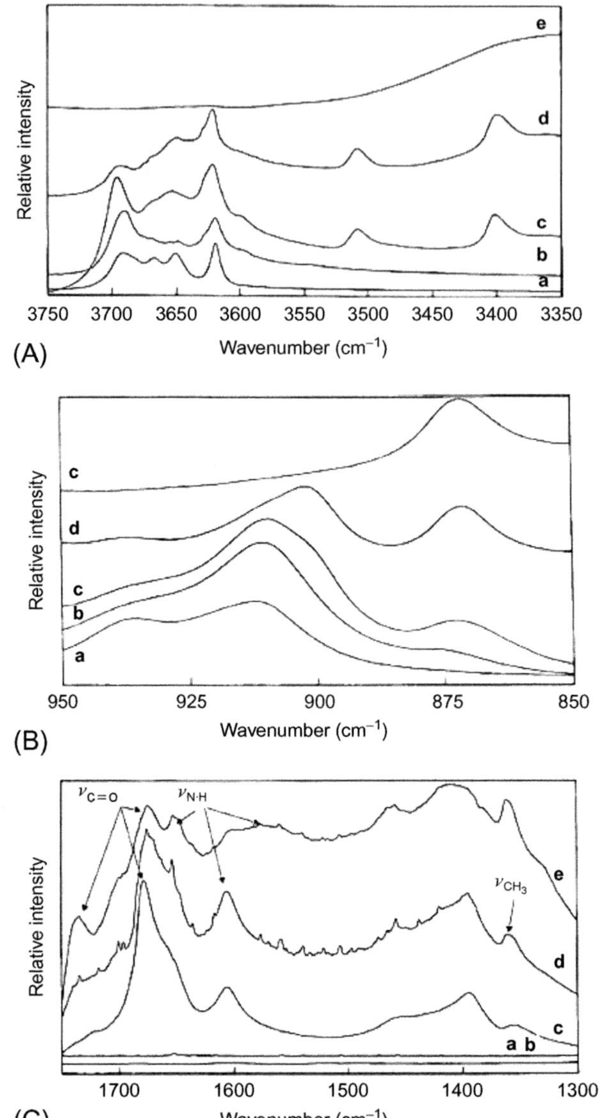


**FIG. 11.13** DRIFT spectra in the (A) OH stretching region, (B) OH deformation region and (C) CO and NH stretching regions of acetamide Kaol intercalate complexes. Spectra are: a, ordered kaolinite; b, disordered kaolinite; c, acetamide intercalated ordered kaolinite; d, acetamide intercalated disordered kaolinite and e, acetamide. *Reproduced with permission from Frost et al. (1999e).*

**TABLE 11.7 Infrared and Raman Band Positions for Acetamide Intercalated Kaolinites (Frost et al., 1999e; Olejnik et al., 1971). Values are in Wavenumbers ($cm^{-1}$)**

| Olejnik et al. (1971) | Frost et al. (1999e) | | | |
|---|---|---|---|---|
| | *DRIFT* | | *Raman* | |
| *IR* | *Ordered* | *Disordered* | *Ordered* | *Disordered* |
| 3690 | 3696 | 3693 | Absent | 3697 |
| | Absent | Absent | 3682 | 3686 |
| | 3673 | 3673 | Absent | Absent |
| 3648 | 3651 | 3650 | 3655 | 3652 |
| 3617 | 3621 | 3621 | 3620 | 3620 |
| 3585–3520bsh* | 3599 | 3599 | 3596 | 3596 |
| 3501 | 3509 | | | |
| 3475sh | | | | |
| 3415 | 3400 | | | |
| 3394 | | | | |
| 3292sh | | | | |
| 2850 | | | | |
| | 1740sh | | | |
| 1685sh | 1680 | | | |
| 1665 | 1650 | | | |
| 1604 | 1607 | | | |
| 1558 | 1570 | | | |
| 1455w | 1450b | | | |
| 1434vw | | | | |
| 1416 | 1400b | | | |
| 1386 | | | | |
| 1376 | | | | |
| 1293 | | | | |
| | 941 | Absent | | |
| | 931 | 938 | | |
| | 913 | 903 | | |
| | 874 | 874 | | |

*b=broad, sh=shoulder, w=weak, vw=very weak.*

The DRIFT spectra of the δ(OH) regions showed considerable changes in the intensity of the OH deformation modes on intercalation of acetamide in Kaol. The band component analyses of the DRIFT spectra show three δ(OH) bands around 940, 924 and 913 $cm^{-1}$ (Fig. 11.13B). The ordered Kaol exhibited about 30% of the total normalised band intensity for the 940 $cm^{-1}$ band, whereas the disordered Kaol had only 20%. This difference can be attributed to the differences in the H bonding of the inner surface OH groups. The more ordered Kaol showed a higher degree of H bonding. For the disordered Kaol, a band was observed at 876 $cm^{-1}$ attributed to the δ(OH) mode of non H-bonded OH groups. The disordered Kaol shows little intensity in the 925 $cm^{-1}$ band, which was also attributed to the lack of H bonding. The acetamide was thought to have broken the H bonding between adjacent Kaol layers through the formation of H bonds between the inner surface OH and the C=O group of the acetamide. For the disordered Kaol, the changes in the δ(OH) modes were even more pronounced. No intensity remained in the 940 $cm^{-1}$ band. An additional band was instead observed on intercalation at 903 $cm^{-1}$. This band now makes up about 25% of the total band intensity and was attributed to weakly H-bonded inner surface OH groups. The band at 874 $cm^{-1}$ in the DRIFT spectrum of acetamide also appeared in the spectra of the acetamide-intercalated Kaols. This band, attributed to δ(NH) showed an increase in bandwidth on intercalation due to the H bonding of the acetamide with the siloxane groups.

The acetamide molecule itself also shows some changes in its structure due to the intercalation in Kaol and the subsequent formation of new H bonds with the Kaol. The amide II bands around 1650 and 1570 $cm^{-1}$ disappear completely, and instead, a well-defined HNH scissor mode is observed around 1607 $cm^{-1}$ due to the formation of an H bond between the amide group and the oxygen atoms of the Kaol siloxane layer (Fig. 11.13C). Thus, the acetamide molecules was thought to have lost their internal H bonds. However, Olejnik et al. (1971) reported two bands at 1604 and 1558 $cm^{-1}$. The C=O bands (amide I) around 1680 and 1740 $cm^{-1}$ were replaced by a single band around 1680 $cm^{-1}$ in the study by Frost et al. (1999e), whereas Olejnik et al. (1971) described a band at 1665 $cm^{-1}$ with a shoulder at 1685 $cm^{-1}$ for the amide I. The two amide III bands around 1400 and 1450 $cm^{-1}$ remain nearly unaltered on intercalation. The very low intensity of the 1740 $cm^{-1}$ band can be explained by the formation of H bonds between the inner surface OH groups of the Kaol and the carbonyl group of the acetamide.

## 11.3 GROUP B MOLECULES

### 11.3.1 Dimethylsulphoxide, $(CH_3)_2SO$ (DMSO) and Dimethylselenoxide, $(CH_3)_2SeO$ (DMSeO)

DMSO and DMSeO are dipolar aprotic solvents with high dielectric constants. The large dipole moment (3.9 D) and polarisability ($n_d^{20} = 1.478$)

enable DMSO to stabilize molecules and ions through dipole and induced-dipole interactions. DMSO strongly solvates protons and cations by its high electron density on the unscreened oxygen atom. The physicochemical properties of DMSO are determined mostly by the S=O group, which contains two potential basic centres. Pure liquid DMSO has a cluster-like structure caused by strong dipole–dipole interaction. On adding water, the cluster structure gradually breaks up. The DMSO–water and water–water interactions play a significant role: but no protonation of the S=O group has been found. In the presence of weak organic acids, some protonation occurs.

For DMSO at 298 K, two bands are observed in the Raman spectrum at 2994 and 2913 $cm^{-1}$ which have been assigned to the antisymmetric ($\nu_{as}$(CH)) and symmetric ($\nu_s$(CH)) CH stretching modes, respectively (Table 11.8). On cooling the DMSO to 77 K, the $\nu_{as}$(CH) mode split into three bands at 3001, 2995 and 2988 $cm^{-1}$, whereas two weak $\nu_s$(CH) bands were observed at 2830 and 2809 $cm^{-1}$. For DMSO at 298 K, a complex SO stretching band profile is centred at 1050 $cm^{-1}$. Three bands were resolved at 1058, 1042 and 1026 $cm^{-1}$, attributed to the unassociated monomer and the out-of-phase and the in-phase vibrations of the dimer, respectively. A minor component was observed at 1015 $cm^{-1}$ normally associated with polymeric DMSO. A band was also observed at 951 $cm^{-1}$ assigned to the symmetric methyl rocking vibration. On cooling to 77 K, three bands were observed at 1057, 1038 and 1019 $cm^{-1}$. The latter predominant band is ascribed to the DMSO polymer. It is apparent that on cooling to 77 K, the molecules are brought closer together and hence more H bonding occurs between adjacent molecules suggesting that DMSO at 77 K is polymeric. The concept of perturbed and unperturbed DMSO molecules is supported by the observation of three $\nu_s$(S=O) bands which may be ascribed to, respectively, the 'free' or non-interacting DMSO molecules, the perturbed DMSO vibrations to monomeric DMSO and the highly perturbed vibrations to polymeric DMSO molecules. The antisymmetric $\nu_{as}$(CS) and symmetric $\nu_s$(CS) modes of the C-S group of DMSO were observed at 698 and 667 $cm^{-1}$, respectively. A shift of the bands to higher wavenumbers on cooling to 77 K (705 and 672 $cm^{-1}$, respectively) suggests a reduction in the influence of adjacent units on the $\nu$(CS). A complex spectral profile was observed for the HCH bending modes centred around 1426 $cm^{-1}$. This band was broad suggesting that more than one component was in the spectral profile. It could be resolved into two bands at 1426 and 1417 $cm^{-1}$. This observation of two bending modes suggests two different types of structures. One description of these modes might be in terms of H-bonded and non H-bonded $CH_3$ units. The H atoms of the $CH_3$ unit may interact with the S=O units of an adjacent DMSO molecule. On cooling to 77 K, these bands are separated into four components at 1455, 1450, 1422 and 1410 $cm^{-1}$. The solid DMSO now has four different HCH bending modes, two of which were now found at high wavenumbers (1457 and 1450 $cm^{-1}$). It has been proposed that these two modes belong to $CH_3$ units which are now strongly H bonded to the

**TABLE 11.8** Raman and IR Band Positions of DMSO and DMSO-$d_6$ at 298 and 77 K (Bertoluzza et al., 1981; Martens et al., 2002a). Values are in Wavenumbers ($cm^{-1}$)

| Bertoluzza et al. (1981) | | Martens et al. (2002a) | | | | | Bertoluzza et al. (1981) | | Martens et al. (2002a) | | | | |
|---|---|---|---|---|---|---|---|---|---|---|---|---|---|
| *DMSO-d₆* | | *DMSO* | | *DMSO-d₆* | | | *DMSO-d₆* | | *DMSO* | | *DMSO-d₆* | | |
| *Raman* | *IR* | *Raman 298 K* | *77K* | *298 K* | *77 K* | *Assignment* | *Raman* | *IR* | *Raman 298 K* | *77K* | *298 K* | *77 K* | *Assignment* |
| | | 3010 | 3001 | 2256 | 2253 | CH antisymmetric stretch | | | | | | 1032 | SO symmetric stretch of dimer |
| | | 3000 | 2995 | 2250 | 2246 | CH antisymmetric stretch | | | 1026 | 1019 | | 1026 | SO symmetric stretch of dimer |
| | | 2995 | 2988 | 2244 | 3343 | CH antisymmetric stretch | 1010 | 1009 | | | | 1011 | SO symmetric stretch of polymer |
| | | | 2923 | | | CH symmetric stretch | | | | | | 1003 | |
| | | | 2921 | | 2131 | CH symmetric stretch | | | | | | 977 | |
| | | | 2909 | 2125 | 2129 | CH symmetric stretch | 826 | 821 | | | 819 | 832 | |
| | | 2913 | 2908 | 2118 | 2120 | CH symmetric stretch | | | | | | 784 | HCH rocking |
| | | | 2885 | 2117 | 2113 | CH symmetric stretch | 778 | 773 | | | 784 | 768 | HCH rocking |
| | | 2876 | 2877 | | 2099 | CH symmetric stretch | 764 | 758 | | | 768 | 705 | HCH rocking |
| | | | 2856 | 2072 | 2065 | CH symmetric stretch | | | 953 | 965 | 757 | 686 | HCH rocking |
| | | | 2845 | | 2045 | CH symmetric stretch | | | | 720 | | 672 | HCH rocking |

| | | | | | | | | | | | | | |
|---|---|---|---|---|---|---|---|---|---|---|---|---|---|
| | | 2830 | 2831 | | 2005 | CH symmetric stretch | 682 | 679 | 698 | 705 | 624 | 631 | CS antisymmetric stretch |
| | | 2809 | 2808 | 1995 | 1989 | CH symmetric stretch | | | 667 | 672 | 612 | 617 | CS symmetric stretch |
| | | | 1455 | | | HCH deformation | 628 | 626 | | | | | CSC antisymmetric stretch |
| | | | 1450 | | | HCH deformation | 617 | 613 | | | | | CSC symmetric stretch |
| | | | 1448 | | | HCH deformation | | | | | 346 | | CSO in-plane rock |
| | | 1426 | 1422 | 1007 | 994 | HCH deformation | | | 382 | 387 | 340 | 345 | CSO in-plane rock |
| | | 1417 | 1410 | | | HCH deformation | 340 | 339 | 333 | 337 | 307 | 307 | CSO out-of-plane bend |
| | | 1307 | 1320 | | | HCH deformation | | | | 323 | | | CSO out-of-plane bend |
| 1058 | 1059 | 1058 | 1057 | 1057 | 1053 | SO symmetric stretch of monomer | | | | 315 | | | CSO out-of-plane bend |
| 1034 | 1030 | 1042 | 1038 | 1034 | 1045 | SO antisymmetric stretch of dimer | 307 | 306 | 309 | 304 | | | CSO out-of-plane bend |

S=O units. In addition, the band observed at 1307 cm$^{-1}$ shifted to 1320 cm$^{-1}$ on cooling to 77 K. Another low-intensity band is observed in the 77 K spectra at 1371 cm$^{-1}$. These bands are also attributed to methyl deformation modes. The Raman spectral region between 300 and 400 cm$^{-1}$ includes the $\delta$(CSO) and $\delta$(CSC) units with two bands in the spectrum of DMSO at 298 K at 333 and 382 cm$^{-1}$. These bands were ascribed to the in-plane and out-of-plane $\delta$(CSO). The band around 304 cm$^{-1}$ was attributed to the $\delta_s$(CSC). On cooling to 77 K, bands were observed at 387, 337, 323 and 315 cm$^{-1}$. The in-plane band increased in wavenumber from 382 to 387 cm$^{-1}$. The next three bands were assigned to out of-plane bending vibrations. Such an observation suggests at least three different molecular arrangements of the DMSO at 77 K. Bands were observed for pure DMSO at 302 and 309 cm$^{-1}$ in the Raman spectrum measured at 298 K and were ascribed to $\delta_s$(CSC) and $\delta_{as}$(CSC) vibrations (Martens et al., 2002a).

### 11.3.1.1 Intercalation of DMSO and DMSeO in Kaol

Intercalation of DMSO into Kaol results in the expansion of the basal spacing from approximately 0.72 to 1.12 nm, which is greater in comparison with Kaols intercalated with the group A molecules. Intercalation of Kaol with DMSO is generally used to distinguish Kaol from chlorite in XRD. In addition, it is used as a precursor for the intercalation of Kaol with inorganic alkali salts, where DMSO functions as a molecular 'crowbar' to separate the Kaol layers so that the salt molecules can freely enter the interlayer space (Lahav, 1990; Lapides et al., 1997). The reason why molecules such as DMSO are used to expand the Kaol layers is that these inserting molecules form stronger H bonds with the siloxane or gibbsite-like Kaol surfaces than do the adjacent (siloxane and gibbsite-like) Kaol surfaces (Martens et al., 2002a).

Earlier research has shown that the intercalation of DMSO in Kaol results in 3-D ordering in the interlayer spaces. The S=O group of the DMSO forms an H bond with the inner surface OH groups, whereby the sulphur atom orients itself in relation to the oxygen atoms in the siloxane layer of the Kaol (Thompson and Cuff, 1985; Raupach et al., 1987). As a result of this ordering, the molecular freedom of movement of the methyl groups of the DMSO also changes ( Hayashi and Akiba, 1994; Hayashi, 1995). In the original model, it was assumed that each DMSO molecule formed three H bonds with the OH groups around the vacant position in the octahedral layer of the Kaol. One methyl group would be oriented roughly into the ditrigonal cavity in the siloxane layer of the Kaol with the other C–S bond parallel to the layers. To get the vacant position and ditrigonal cavity in the subsequent layers in the correct positions to form the bonds with the DMSO molecule requires a significantly forced horizontal displacement of the layers (Thompson and Cuff, 1985). Further research has changed this model significantly (Frost et al., 1998d, 1999b,c; Kristof et al., 1999).

Intercalation of Kaol with DMSO resulted in three new Raman bands around 3660, 3536 and 3501 $cm^{-1}$ (Fig. 11.14A), whereas the $\nu(Al_2OH_{IS})$ bands almost completely disappeared (Frost et al., 1998d). The IR spectra (Fig. 11.14B) exhibited comparable bands around 3660, 3538 and 3502 $cm^{-1}$ (Johnston et al., 1984). The corresponding IR spectra for DMSO intercalated dickite showed similar behaviour, with the 3622 $cm^{-1}$ band assigned to the inner OH not changing, whereas the other OH groups were sensitive to the presence of DMSO giving two bands at 3537 and 3505 $cm^{-1}$ (Zamama et al., 1995). For the interpretation of these spectra, it is important to know that even if dewatered DMSO is used, there is still a small amount of water present in the DMSO. Thermogravimetric analysis has shown that DMSO intercalated Kaol contains about 7 wt% water. The bands around 3536 and 3501 $cm^{-1}$ were broad compared with the original Kaol hydroxyl bands. Therefore, these bands have been interpreted as water bands associated with the DMSO in the interlayer spaces of the Kaol (Frost et al., 1998d; Yariv and Lapides, 2008; Zhang et al., 2012). The band around 3660 $cm^{-1}$ is probably the result of the formation of H bonds between the Kaol inner surface hydroxyl groups and the S=O group of the DMSO. A comparable image shows up from the hydroxyl bending spectral region, where a band around 905 $cm^{-1}$ has been interpreted as the bending equivalent of the 3660 $cm^{-1}$ stretching mode.

The DMSeO molecule has a similar structure to that of dimethylsulfoxide (DMSO), but unlike DMSO, DMSeO is solid at RT, the melting point being 94°C, and it is not amenable to intercalation from a saturated vapour as used by, for example, Thompson and Cuff (1985) in their synthesis of a DMSO Kaol intercalate. A 1.10 nm intercalate was prepared by repeated leaching of DMSeO from either of the other two intercalates with chloroform. The 1.14 nm intercalate was obtained by reacting oven-collapsed hydrazine Kaol intercalate ($d_{001}$-value = 0.7–0.80 nm, as a result of incomplete expulsion of hole water) with a 50% excess of DMSeO in aqueous solution at 60°C until all the water had evaporated. A 1.13-nm intercalate was also obtained from an untreated Kaol (#2 of Thompson and Cuff, 1985). The DMSeO Kaol intercalates gave IR spectra showing perturbation of the 3689 $cm^{-1}$ Kaol $\nu(Al_2OH)$ bands by about 6 $cm^{-1}$ due to the close approach of the Se=O linkage of the DMSeO molecule. The relative intensity of this Kaol band was reduced because of H bonding. Other OH stretching bands of Kaol were modified by DMSeO intercalation. For the 1.10- and 1.14-nm DMSeO Kaol intercalates the inner OH Kaol band at 3621 $cm^{-1}$ gave two bands centred at 3628 and 3593 $cm^{-1}$. In contrast, the 1.13 nm intercalate had two adjacent and relatively intense bands at 3623.1 and 3616 $cm^{-1}$, together with an H bond wavenumber centred at 3388 $cm^{-1}$. The 1.10 nm intercalate probably contained unoccupied ditrigonal cavities, which gave rise to the inner-OH band at 3593 $cm^{-1}$ (Raupach et al., 1987).

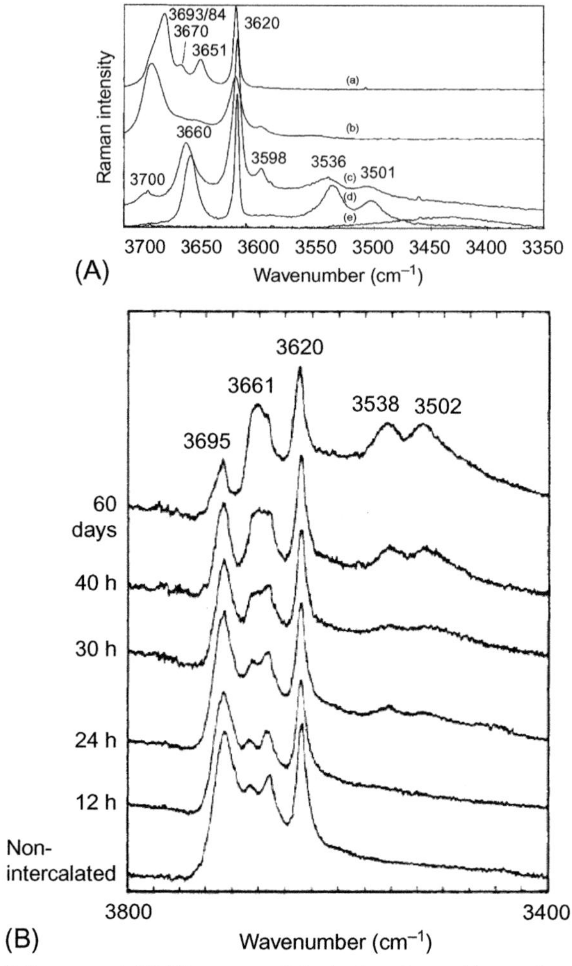


**FIG. 11.14** (A) Raman and (B) IR spectra of the hydroxyl stretching region of DMSO Kaol intercate complexes. In (A) spectra a, low-defect kaolinite; b, high-defect kaolinite; c, DMSO intercalated high-defect kaolinite; d, DMSO intercalated low-defect kaolinite and e DMSO. In (B) spectra a, kaolinite; and b, kaolinite-DMSO complexes at reaction times indicated. *Part figure (A) reproduced with permission from Frost et al., (1998). Part figure (B) reproduced with permission from Johnston et al. (1984).*

Much more important for the model, however, are the changes observed in the DMSO molecule itself after the intercalation in Kaol. Pure DMSO reveals two broad bands around 2911 and 2996 cm$^{-1}$, being ascribed to the symmetric and antisymmetric $\nu$(CH) modes. After intercalation, the 2911 cm$^{-1}$ band splits up in two bands around 2917 and 2935 cm$^{-1}$. At the same time, the other band splits up in four new bands around 2998, 3015, 3021 and 3029 cm$^{-1}$, indicating the presence of different $\nu_{as}$(CH) bands (Fig. 11.15). Comparable results were obtained from the IR spectra. In both the Raman and IR spectra, the bands of the intercalated DMSO are sharper than for pure DMSO. This suggests that the methyl groups are positioned in more rigid positions in the interlayer spaces of Kaol. In the 1.13 nm DMSeO intercalate, a medium intensity $\nu_{as}$(CH) band at 3012 cm$^{-1}$ was found, along with two equally intense, strong bands at 3041 and 3026 cm$^{-1}$; the equivalent symmetric bands were observed at 2946 and 2920 cm$^{-1}$. The bands of the first set were interpreted as perturbed upwards from 3000 cm$^{-1}$ for DMSeO (Raupach et al., 1987).

The methyl H-C-H bending vibrations of the Raman spectra around 1419 cm$^{-1}$ for DMSO also showed a split in two much sharper bands around 1411 and 1430 cm$^{-1}$ (Fig.11.16A). The IR spectrum, however, is much more complex and shows the splitting of two DMSO bands around 1414 and 1445 cm$^{-1}$ in seven new bands around 1438, 1428, 1408, 1404, 1392, 1374 and 1318 cm$^{-1}$, indicating the presence of seven different $\delta$(CH) vibrations. Similar bands were reported by Raupach et al. (1987) at 1318, 1393, 1408, 1428, 1433 and 1437 cm$^{-1}$. This complexity is a clear indication that multiple molecular configurations are possible for the DMSO molecules in the interlayer spaces. Pure DMSO shows three $\nu_s$(S=O) vibrations in the Raman spectrum around 1058, 1042 and 1026 cm$^{-1}$ associated with the DMSO monomer

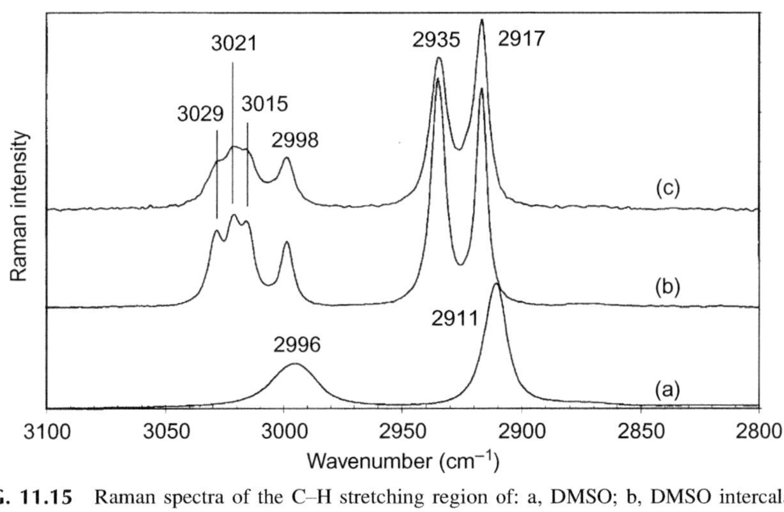


**FIG. 11.15** Raman spectra of the C–H stretching region of: a, DMSO; b, DMSO intercalated low-defect kaolinite and c, DMSO intercalated high-defect kaolinite. *Reproduced with permission from Frost et al. (1998).*

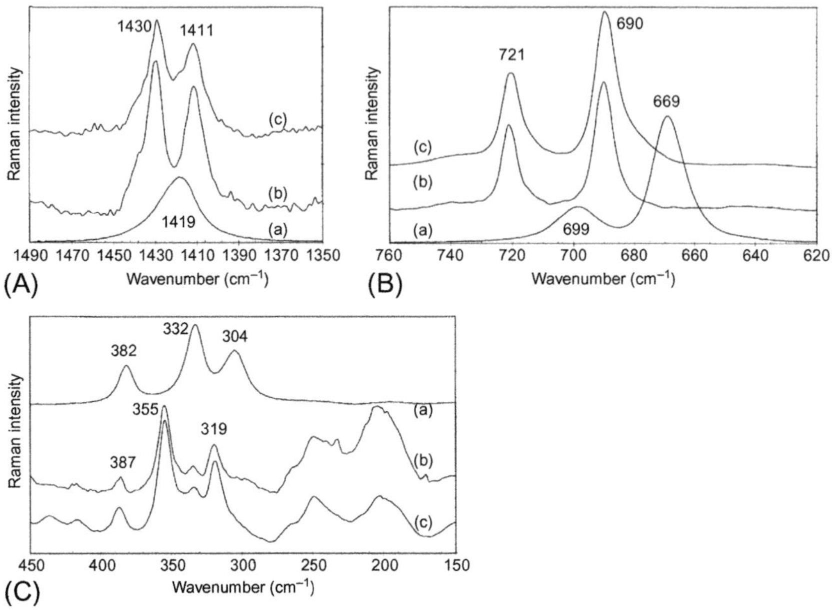


**FIG. 11.16** Raman spectra of the (A) in-plane methyl bending region, (B) the C–S stretching region and (C) C–S–C and S–O bending region of DMSO kaolinite intercalate complexes. Spectra are of: a, DMSO; b, DMSO intercalated low-defect kaolinite and c, DMSO intercalated high-defect kaolinite. *Modified after Frost et al. (1998).*

and the in- and out-of-phase vibrations of the dimer, respectively. A very small band around 1015 cm$^{-1}$ indicates the presence of a very minor amount of a DMSO polymer (Martens et al., 2002a). After intercalation, three new bands were observed around 1066, 1023 and 1010 cm$^{-1}$. The last two are vibrations of two different S=O units of DMSO, whereas the first band has been ascribed to the S=O of a free DMSO monomer.

The bands in the low-wavenumber region can be ascribed to the $\nu$(CS), $\delta$(CSO) and $\delta$(CSC) vibrations. All these vibrations exhibited a shift and sharpening due to the intercalation of the DMSO in the Kaol. Significant is that the intensity of the C–S=O bands increased much more strongly than those of the C–S–C bands due to a change in symmetry of the S=O unit. All these observations lead to a much more complicated model than sketched at the beginning of this paragraph. It has become clear that water plays an important role in the intercalation process. Water can form H bonds with the free electron pair of the S atom and act as a sort of bridge between two adjacent DMSO molecules, resulting in some oligomerisation. The presence of two S=O bands around 1023 and 1010 cm$^{-1}$ indicates that possibly two types of oligomers can exist. A second type is formed by the intercalation of single DMSO monomers, which form H bonds with the inner surface OH groups of the Kaol. Again, two types can be distinguished: DMSO molecules without He bonds with water and DMSO molecules with H bonds with water via one or, simultaneously both OH groups, of the water. The presence of multiple types of intercalated DMSO has consequently the presence of a large number of different methyl vibrations in the IR spectra. Here, the $\delta_s$(CH) and $\delta_{as}$(CH) vibrations have split into, respectively, two and seven new vibrations due to the intercalation in line with the observations for the corresponding bending vibrations (Frost et al., 1998d, 1999b,c, 2000d). The $\nu$(Se=O) band at 820 cm$^{-1}$ in the IR spectrum of DMSeO was perturbed to 831 cm$^{-1}$ in the 1.13 nm Kaol–DMSeO intercalate. This vibration consisted of two bands at 834 and 827 cm$^{-1}$ for the 1.01-nm intercalate, and at 847 and 811 cm$^{-1}$ for the 1.14 nm intercalate, showing that the Se=O bond lengths were different in the three intercalates (Raupach et al., 1987).

Bands are observed at 669 and 699 cm$^{-1}$ for pure DMSO and have been assigned to the symmetric and antisymmetric $\nu$(CS) vibrations (Table 11.9). The $\delta_s$(CS) and $\delta_{as}$(CS) vibrations were found at 690 and 721 cm$^{-1}$ in the DMSO-intercalated low-defect Kaols (Fig. 11.16B). The $\delta_s$(CS) and $\delta_{as}$(CS) bands are substantially narrower for DMSO intercalated low-defect Kaol. There is about 21 cm$^{-1}$ difference between the unperturbed C–S stretching wavenumbers and the perturbed stretching wavenumbers. For the high-defect Kaol, four bands were observed at 680, 690, 720 and 712 cm$^{-1}$. The first two bands were attributed to the symmetric stretching modes, and the second to the antisymmetric stretching modes. No differences in the $\nu$(CS) between the high-defect Kaol and the low-defect Kaol were observed. However, a second set of bands was observed for the DMSO-intercalated high-defect Kaol at 680 and 712 cm$^{-1}$. The two sets of values for both the $\delta_s$(CS) and $\delta_{as}$(CS) vibrations indicates that two different types of molecular arrangements of the DMSO molecules are present in the Kaol intercalate. The ratio of the two types of molecules is $\approx$1:3. Multiple bands not yet assigned were found by Johnston et al. (1984) at 637, 673, 688, 705 and 720 cm$^{-1}$.

The bands in the Raman spectrum of DMSO at 382 and 333 cm$^{-1}$ have been assigned to the in-plane and out-of-plane $\delta$(CSO) (Fig. 11.16C). The band at 304 cm$^{-1}$ has been attributed to the C–S–C symmetric bend. The bands assigned to the $\delta$(CSO) and $\delta$(CSC) modes moved to higher wavenumbers on intercalation of the Kaol with DMSO. The $\delta$(CSC) band was observed at 319 cm$^{-1}$. The two $\delta$(CSO) bands were observed at 355 and 387 cm$^{-1}$. There was a very large increase in intensity of the in-plane and out-of-plane $\delta$(CSO) relative to the intensity of the $\delta$(CSC) band. This apparent increase could be the result of changes in symmetry of the S=O unit in the DMSO intercalation complex.

**TABLE 11.9** Observed Raman and Infrared Bands for DMSO and DMSeO Intercalated Kaolinites. Values are in Wavenumbers ($cm^{-1}$)

| Frost et al. (1998d) | | Johnston et al. (1984) | | | | Raupach et al. (1987) | | | |
|---|---|---|---|---|---|---|---|---|---|
| | | | | | | *DMSO* | *DMSeO 1.14 nm* | *DMSeO 1.13 nm* | *DMSeO 1.10 nm* |
| *Raman* | *DRIFT* | *Raman* | *IR* | *ATR-IR* | | | | | |
| *Low def.** | *High def.* | *Low def.* | *High def.* | | | | | | |
| | 3700 | 3697 | 3700 | | 3695 | 3695 | 3695 | 3604 | 3695 |
| 3660 | 3664 | 3660 | 3663 | 3661 | 3661 | 3661 | | | |
| | | | | | sh | | | | |
| 3620 | 3620 | 3620 | 3621 | 3620 | 3620 | 3620 | 3624 | 3623sh | 3629 |
| | | | | | | | | | 3622 |
| 3536 | 3598 | 3538 | 3599 | | | | | 3616 | 3593 |
| | 3543 | | 3558 | | | | | 3551 | |
| | | | 3539 | | 3538 | 3536 | | | |
| 3501 | 3509 | 3502 | 3504 | | 3502 | 3504 | | | |
| | | 3492 | 3468 | | | | | 3455 | |
| | | 3413 | | | | | | 3338 | |
| | | | | 3116 | | | | | |
| 3029 | 3028 | 3029 | 3020 | | | 3022 | 3033 | 3041 | 3042 |
| 3021 | 3021 | 3021 | 3021 | | | | | 3026 | |
| 3015 | 3015 | 3016 | 3016 | 3006 | | | | 3012 | |

| | | | | | | | | | |
|---|---|---|---|---|---|---|---|---|---|
| 2998 | 2999 | 2998 | 2998 | 3001 | | | | | |
| 2935 | 2934 | 2935 | 2935 | 2936 | | 2936 | 2944 | 2946 | 2948 |
| 2917 | 2917 | 2918 | 2918 | 2919 | | | 2923 | 2920 | 2932 |
| 1430 | 1430 | 1438 | 1438 | | | 1433 | | | |
| | | 1429 | 1427 | | | 1428 | 1421 | 1418 | 1421 |
| | | 1427 | | | | | | | |
| 1411 | 1411 | 1408 | 1408 | | | 1408 | 1404 | 1402 | 1408 |
| | | | 1404 | | | sh | | | |
| | | 1392 | 1392 | | | 1393 | 1394 | 1386 | |
| | | 1373 | 1374 | | | | | | |
| | | | | | 1320 | 1318 | 1284 | 1285 | 1284 |
| | | | | | | 1304 | 1266 | 1270 | 1261 |
| | | 1083 | 1094 | | 1110 | | | | |
| 1066 | 1058 | 1058 | | | | | | | |
| | 1040 | | 1040 | | | | | | |
| 1023 | 1026 | 1028 | 1015 | 1024 | | | | | |
| 1010 | 1010 | 1004 | | | | | | | |
| | | 987 | 989 | | | | | | |
| 959 | 959 | 958 | 958 | 955 | 960 | | | | |
| 941 | 942 | 940 | 943 | | 940 | | | | |
| 923 | 923 | | | Sh | | | | | |

*Continued*

**TABLE 11.9** Observed Raman and Infrared Bands for DMSO and DMSeO Intercalated Kaolinites. Values are in Wavenumbers ($cm^{-1}$)—Cont'd

| Frost et al. (1998d) | | Johnston et al. (1984) | | | Raupach et al. (1987) | | | |
|---|---|---|---|---|---|---|---|---|
| | | | | | *DMSO* | *DMSeO 1.14 nm* | *DMSeO 1.13 nm* | *DMSeO 1.10 nm* |
| *Raman* | *DRIFT* | *Raman* | *IR* | *ATR-IR* | | | | |
| *Low def.* | *High def.* | *Low def.* | *High def.* | | | | | |
| 913 | 913 | | | | | | | |
| 905 | 905 | 906 | 904 | | | | | |
| | | | 869 | | | 847 | 831 | 834 |
| | | | | | | 812 | | |
| | | | | | | 787 | 784 | 787 |
| | | | | | | | | 764 |
| 721 | 720 | | | 720 | | 750 | 736 | sh |
| | 712 | | | 708 | | | | |
| 690 | 690 | | | 688 | | | | |
| | 680 | | | 673 | | | | |
| | | | | 637 | | | | |
| 387 | 387 | | | 385 | | | | |
| 355 | 354 | | | 335 | | | | |
| 319 | 318 | | | 305 | | | | |

**Low def. and high def. are DMSO intercalated low-defect and high-defect kaolinite.*

The spectra in this region for the DMSO intercalated high-defect Kaol were substantially different. The relative intensity of the Raman spectrum of the $\delta_s$(CSC) mode was 34% compared with the 16% for the intercalated low-defect Kaol. Frost et al. (1998d) concluded the DMSO molecule was more perturbed for the low-defect Kaol-DMSO intercalate. For the high-defect Kaol, the $\delta_s$(CSC) was not as perturbed and so its symmetry was retained and consequently the relative intensity of this band was higher. Far-IR bands for the $\delta$(CSeO) were observed at 298, 270, 246,216 and 207 $cm^{-1}$ in the spectrum of the 1.13 nm intercalate. Untreated Kaol had bands at 278, 248 and 195 $cm^{-1}$, the latter two bands being poorly resolved (similar bands were observed in the Raman spectra of DMSO-Kaol, Fig. 11.16C).

Because of the structural and spectroscopic similarities between the DMSO and DMSeO Kaol intercalate, the structures appear to be analogous. The structural description previously given for the 1.13 nm DMSO Kaol intercalate was thought to apply equally to the 1.14 nm DMSeO intercalate, in which the seleninyl oxygen is triply H bonded to the inner-surface hydroxyls, and one methyl group is keyed into the ditrigonal hole in the silicate sheet above the seleninyl oxygen. The structure of the 1.14 nm DMSeO intercalate, from consideration of the XRD and spectroscopic evidence, is closely related to the structure of the 1.13 nm DMSeO Kaol intercalate. The main differences observed for the 1.14 nm relative to the 1.13-nm intercalate are the multiplicity of DMSeO orientations from the introduction of a mirror image DMSeO site, incomplete occupancy of the ditrigonal cavity and slightly less distortion of the DMSeO bond angles. The 1.10 nm intercalate, however, on the basis of IR (and also NMR) evidence has equivalent DMSeO methyl groups in contact with the silicate sheet, but which do not penetrate the ditrigonal cavities. Adjacent cavities cannot be fully occupied because of the space taken by each DMSeO molecule in this orientation. The disorder parallel to *110* can be accounted for by the lack of keying of one of DMSeO methyl groups into the ditrigonal cavities. Assuming that the seleninyl oxygen is triply H bonded to the octahedral sheet above the octahedral vacancy, as for the other models, and that the DMSeO dipoles tend to align at a low angle to the *ab* plane in the absence of keying as indicated by the IR evidence. In the 1.13 nm DMSeO and DMSO Kaol intercalates, the two methyl groups are structurally and spectroscopically resolved. The $^{13}$C methyl NMR resonances suggested that one methyl group is more sterically hindered than the other. This hindrance is clearly seen from the splitting of the far-IR bands, the different polarisations of the deformation vibrations, and the two symmetric CH abending vibrations at 1386 and 1402 $cm^{-1}$ (see Table 11.9) in the 1.13 nm intercalate (Fig. 11.17) (Raupach et al., 1987).

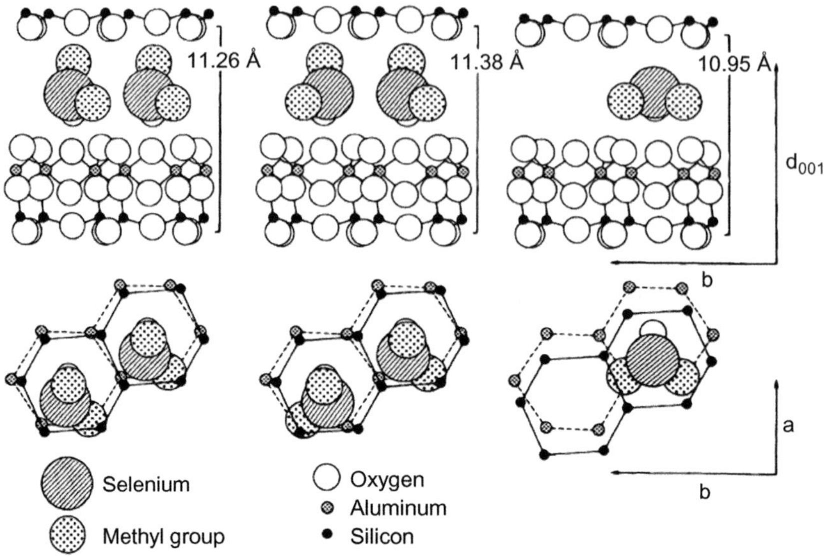


**FIG. 11.17** Comparison of the three kaolinite:dimethylselenoxide (DMSeO) intercalate models projected along [100] (upper) and projected onto *001* with kaolinite oxygens omitted (lower). Selenium and oxygen atoms and methyl groups have been set to half their van der Waals radius. In the 1.14 nm (11.4 Å) intercalate each intercalant site may be occupied by a DMSeO molecule in either orientation. *Reproduced with permission from Raupach et al. (1987).*

### *11.3.1.2 Deintercalation of DMSO from Kaol*

The thermal deintercalation of DMSO from Kaol has been reported by several groups. Frost et al. (1999c) observed with TGA-DTA, a two steps water loss at 77°C and in a minor amount at 173°C, amounting to ≈7wt% water. On heating, the DMSO-intercalated low-defect Kaol, two stages of loss of DMSO were observed by mass spectrometry at 117°C and 173°C. DMSO was released from the sample in two different stages representing a temperature difference of 56°C. This implies DMSO is present in the Kaol intercalate in two completely different forms. At 50°C, decomposition started as is evidenced by an increase in the intensity of the Raman bands at 3687 $cm^{-1}$ and 3702 $cm^{-1}$. As the sample was heated through the deintercalation temperatures, the intensity of the band at 3660 $cm^{-1}$ decreased and the intensity of the Kaol $Al_2OH$ bands increased in intensity. Both the Raman and DRIFT spectra of the Kaol at 173°C showed the Kaol to be completely deintercalated. The spectrum of the deintercalated Kaol closely resembled that of the non-intercalated Kaol at 130°C. Significant intensity remained in four bands observed at 2998, 3015, 3021 and 3029 $cm^{-1}$ even at 110°C. Some of the DMSO was retained in the intercalated complex at 110°C, but at 130°C all the DMSO was lost. Similar observations were made by Zhang et al. (2012) though at slightly higher temperature with the main DMSO mass loss around 210°C. On increasing the temperature, the intensities of the bands at 3021 and 2935 $cm^{-1}$ decreased and disappeared at 175°C, and the bands of water *ν*(H-O-H) modes persisted up to 150°C. The intensities of the bands at 1428 and 1318 $cm^{-1}$ gradually decreased and disappeared at 175°C. Meanwhile, the intensities of the Al–O and Si–O bands of Kaol gradually recovered and the original Kaol spectrum was observed when the temperature reached 200°C.

Yariv and Lapides (2008) observed on thermal analysis that the absorbance of the band at 3996 $cm^{-1}$ gradually increased, whereas that of the 3663 $cm^{-1}$ band decreased. The latter was not observed after heating the complexes at 160°C. The band shifted with the rise in temperature from 3663 $cm^{-1}$ before the thermal treatment to 3656 $cm^{-1}$ after heating at 140°C, which they interpreted as the thermal strengthening of the O–H···O=S$(CH_3)_2$ H bond, probably due to water escape. Thermo-XRD-analysis showed that type II complex was obtained by the thermal treatment of type I with a $d_{001}$-value of 1.12 nm complex. It seems that the band at 3663 $cm^{-1}$ represents type I complex, whereas the band at 3656 $cm^{-1}$ represents the type II complex. The spectra of DMSO Kaol intercalate complexes showed very weak H-C-H stretching bands of $CH_3$ at 3022, 2937, 2919, 1431, 1412, 1397 and 1320 $cm^{-1}$, respectively. The intensities of these bands significantly decreased with the temperature, but they disappeared only at 160°C. At 130°C, three small shoulders appeared attached to the 3022 band at 3032, 3017 (vw) and 2995 (vvw) $cm^{-1}$. The appearance of the υ(CH) bands in spectra recorded after heating the samples at temperatures between 130 and 150°C points to DMSO being present in the interlayer space as a component of type II complex.

Franco and Ruiz Cruz (2003) described the thermal behaviour of DMSO-intercalated dickite. Heating of the complex at a continuous rate caused an intense mass loss between RT and 300°C, which was accompanied by three endothermic effects: Two weak effects were centred at 57 and 98°C; a third (more intense) occurred at about 212°C. The first represented a mass loss about 8% of the total loss below 300°C. The mass loss associated with the second endothermic effect was about 6% of the loss below 300°C. The third endothermic effect correlates with a faster mass loss, which represents about 86% of the total below 300°C. Assuming that the first mass loss (below 75°C) corresponds to the loss of adsorbed molecules as deduced from ATR-FTIR spectra, the TG curve permits an estimate of the ratio $[Si_2Al_2O_5(OH)_4]:[(CH_3)_2SO]$ in interlayer positions of about 1:0.88, at 75°C. This ratio becomes 1:0.82 at the beginning of the third mass loss stage (at 125°C). The low-temperature endothermic effects were almost immediately followed by a weak exothermic effect at 318°C, which probably corresponds to a rearrangement of the dickite layers. Heating to 75°C caused a slight decrease in the intensity of the methyl deformation bands, whereas the OH stretching bands of the complex remained practically unaffected at this temperature. This is in contradiction with the TGA-MS results of DMSO intercalated

Kaol where only water was lost at this temperature on and no DMSO (Frost et al., 1999c). However, at 115°C, the intensities of the methyl deformation bands decreased, whereas band 1 intensifies and shifts from 3705 to 3702 $cm^{-1}$. Concomitantly, the maximum of the 3658 $cm^{-1}$ band shifted to lower wavenumbers as a consequence of the partial regeneration of the 3644 $cm^{-1}$ band of dickite, and the bands at 3531 and 3499 $cm^{-1}$ decreased in intensity. The behaviour of these bands was thought to suggest that the elimination of intercalated molecules began after heating above 75°C. A new decrease in the intensities of the methyl deformation bands was evident in the spectrum obtained at 175°C. This decrease was accompanied by an increase in the intensity of the bands at 3702 and 3644 $cm^{-1}$, and the drastic decrease of the bands at 3656, 3530 and 3498 $cm^{-1}$. The complete loss of the basal reflection of the dickite–DMSO complex, above 150°C, without restoration of the 7 Å (0.7 nm) dickite reflection indicated that the dickite formed a single layer complex.

## 11.4 GROUP C MOLECULES

### 11.4.1 Potassium Acetate

Vibrational spectroscopic studies of acetate salts in both aqueous media and solid state have been comprehensive and have been studied for decades. Raman spectra of aqueous acetate is more easily obtained than IR spectra. Solid phases like hydrated magnesium acetate (MgAc), sodium acetate (NaAc) and anhydrous potassium acetate (KAc) differ in their crystal structures which should be reflected in the vibrational spectra. For example, in NaAc the $CH_3$ group will show free rotation resulting in a higher symmetry than in MgAc or Ni-acetates, where this rotation is partly or completely lost resulting in a lowering of the Ac space group symmetry from $C_{2v}$ to $C_s$ (Bickley et al., 1990). However, others assumed that even though rotation is limited the space group symmetry $C_{2v}$ is preserved (Raghuvanshi et al., 1983, 1985). Vibrational spectra of the Ac ion, $CH_3CO_2^-$, in sodium acetate have been assigned on the basis of $C_{2v}$ molecular symmetry, assuming free rotation of the $CH_3$ group about the C–C bond. This formula of Ac leads to a theoretical 15 degrees of vibrational freedom with the following vibrational spectroscopic activity:

$$\Gamma_{\text{vib}} = 5A_1 + A_2 + 5B_1 + 4B_2$$

in which the ($5A_1+5B_1+4B_2$) normal modes are all Raman and IR active; the $A_2$ mode is a methyl torsion and is active in the Raman but inactive in the IR. K-Ac does not contain any crystal-water, which will result in a different crystal structure in which $K^+$ is only coordinated by Ac groups. This has a significant impact on especially the low wavenumber region where various metal–oxygen vibrations can be expected.

When the salts of carboxylic acids such as acetic acid are formed, the C=O and C–O bonds are replaced by two equivalent C$\cdots$O bonds with a bond order of 1.5. These two C$\cdots$O bonds interact as in-phase and out-of-phase vibrations to give two bands: the antisymmetric stretching vibration ($\nu_{as}$(C$\cdots$O)) which is intense in IR, but weak in Raman spectra; and a symmetric stretching ($\nu_s$(C$\cdots$O)) band which is weak IR, but strong and polarised in Raman spectra. These two vibrations are normally found in the 1540–1650 and 1360–1450 cm$^{-1}$ regions, respectively. The Raman spectra of KAc showed a $\nu_s$(C$\cdots$O) band at 1437 cm$^{-1}$ (Table 11.10). The Ac methyl HCH deformation vibration ($\nu_2$) of KAc showed a very low intensity, broad complex $\nu_{as}$(C$\cdots$O) with a maximum at 1564 cm$^{-1}$. The Raman spectrum of KAc shows complexity with three bands at 1452, 1437 and 1407 cm$^{-1}$. The band at 1452 cm$^{-1}$ was assigned to the $\nu_{13}$ mode attributed to the methyl deformation vibration. The band at 1437 cm$^{-1}$ is attributed to $\nu_s$(C$\cdots$O) vibration

**TABLE 11.10 Vibrational Assignments for Raman Spectra of the Acetate Group in Solid Potassium Acetate at Room Temperature and at Liquid Nitrogen Temperature (Frost and Kloprogge, 2000). Values are in Wavenumbers (cm$^{-1}$)**

| Symmetry | Class | Assignment | 298 K | 77 K |
|---|---|---|---|---|
| $A_1$ | $\nu_1$ | C–H stretch | 2930 | 2930 |
| | $\nu_2$ | Me deformation | 1344 | 1336 |
| | $\nu_3$ | C–O stretch | 1437 | 1438 |
| | $\nu_4$ | C–C stretch | 923 | 917 |
| | $\nu_5$ | COO deformation | 650 | 648 |
| $A_2$ | $\nu_6$ | Torsion | – | – |
| $B_1$ | $\nu_7$ | C–H stretch | 2983 | 2981 |
| | $\nu_8$ | C–O stretch | 1564 | 1584 |
| | $\nu_9$ | Me deformation | 1407 | 1417 |
| | $\nu_{10}$ | Me rocking | 1018 | 1015 |
| | $\nu_{11}$ | COO rocking (in-plane) | 471 | 463 |
| $B_2$ | $\nu_{12}$ | C–H stretch | 3001 | 3006 |
| | $\nu_{13}$ | Me deformation | 1452 | 1453 |
| | $\nu_{14}$ | Me rocking | 1052vw | 1045 |
| | $\nu_{15}$ | COO rocking | 617 | 622 |

of the Ac group and the band at 1407 $cm^{-1}$ to the methyl deformation mode $\nu_9$. At 77 K, three bands were observed at 1452, 1437 and 1407 $cm^{-1}$. The highest wavenumber is probably the $\nu_{13}$ mode. The band at 1407 $cm^{-1}$ may be the $\nu_9$ mode. This would then leave the band at 1437 $cm^{-1}$ as the $\nu_s$(C$\cdots$O) band $\nu_3$. The band 1439 $cm^{-1}$ is in a position similar to that observed for the C$\cdots$O stretching vibration of nickel acetate (Raghuvanshi et al., 1983; Bickley et al., 1990). Such a band has been defined as the Ac coordinating to the cation. It was therefore concluded that the 1437 $cm^{-1}$ band of KAc is the C$\cdots$O stretching vibration of the coordinating acetate. The HCH bending vibrations was observed in the 298 K spectra at 1347 $cm^{-1}$, accompanied by a low-intensity band at 1344 $cm^{-1}$. The fact that the band is so broad suggests the possible existence of more than one band in the spectral profile, particularly when compared with the double band observed for NaAc.

On cooling to 77 K, the KAc showed a single methyl $\delta$(HCH) band, which remained invariant with temperature, indicating that the methyl groups are all identical in the structure. At 298 K, a very weak band was observed at 1018 $cm^{-1}$. On cooling to 77 K, this band shifted in the opposite direction compared with NaAc to 1015 $cm^{-1}$. This difference in either blue or red shift may be explained by the fact that both NaAc and MgAc contain crystal-water influencing the freedom in bending motion of the H–C–H bond in the methyl group, whereas KAc is anhydrous. The band observed around 930 $cm^{-1}$ was assigned to the $\nu_s$(C–C) vibration ($\nu_4$). This band is observed at 892 $cm^{-1}$ for acetic acid. The C–C stretching vibration for KAc also showed a shift from 923 to 917 $cm^{-1}$ on cooling to 77 K. Raman bands in the 600–680 $cm^{-1}$ region have been ascribed to the $\delta$(OCO). The 298 K Raman spectrum of KAc had two bands at 650 and 617 $cm^{-1}$. At 77 K, KAc has two bands at 648 and 622 $cm^{-1}$. Additional information can also be obtained from the in-plane O-C-O rocking vibration. These bands are normally found in the 480–510 $cm^{-1}$ region. A band was observed at 471 $cm^{-1}$ in the 298 K spectra, which shifted on cooling to 77 K to 463 $cm^{-1}$. The band at 471 $cm^{-1}$ for KAc has been assigned to the O-C-O in-plane rocking mode $\nu_{11}$ of the free $CH_3COO_2$ (Frost and Kloprogge, 2000).

#### *11.4.1.1 Intercalation of KAc in Kaol*

Some of the earliest papers on the intercalation of KAc in Kaol date back more than 50 years (Ledoux and White, 1964b, 1966). Intercalation of Kaol in KAc results in an expansion of the basal spacing to approximately 1.42 nm, whereas for halloysite an expansion to approximately 1.39 nm has been observed (Frost and Kristof, 1997; Frost et al., 1997a, 1998f). These values agree well with, e.g. the results published by Brindley and Brown (1980) for KAc-intercalated Kaol. Intercalation with KAc resulted in a new band around 3605 $cm^{-1}$ in the Raman spectrum, whereas the band belonging to the inner OH group did not change in intensity but showed significant broadening compared with the untreated Kaol. It is possible that $K^+$ exactly fits into the ditrigonal hole of the siloxane layer of the Kaol, thereby influencing the polarisation of the inner OH groups. The bands ascribed to the inner surface OH groups disappear almost completely on intercalation, which indicates that these groups may form H bonds with the carbonate group of the acetate ion (Frost et al., 1997b, 1998g). A comparable but much weaker and broader band was observed in the IR spectrum. This indicates the formation of an H bond between the OH group and the centre between the two oxygen atoms in the Ac ion. Dickite (Fig. 11.18A) and halloysite, in contrast to Kaol, displayed two new Raman active bands in their spectra around 3609 and 3599 $cm^{-1}$ and around 3603 and 3585 $cm^{-1}$, respectively, after intercalation with potassium acetate. The second band is only Raman active, whereas the first one is both Raman and IR active (Frost and Kristof, 1997).

Kaol showed two overlapping bands around 916 and 936 $cm^{-1}$, whereas the KAc-intercalated Kaol showed two bands around 916 and 926 $cm^{-1}$ with a 10 times higher intensity and higher polarisation. The IR active band around 916 $cm^{-1}$ has been ascribed to deformation vibration of the inner surface OH group (Farmer, 1974b). Considering the orientational effects, these changes indicate that the inner surface OH group is no longer oriented at an angle of 34° to the clay layers but more or less perpendicular. Comparable changes were observed in the Raman spectra of halloysite and KAc-intercalated halloysite (Frost and Kristof, 1997; Frost et al., 1997a). The carboxyl Raman active bands of the Ac ion also show changes on intercalation (Fig. 11.18B). The $\nu_s$(OCO) band shifted from 1420 to 1412 $cm^{-1}$. The sharp deformation band of the methyl group around 1352 $cm^{-1}$ (Kaklhana et al., 1983) split bands at 1355 and 1345 $cm^{-1}$ due to the loss of symmetry caused by the interaction of the methyl group with the siloxane surface of

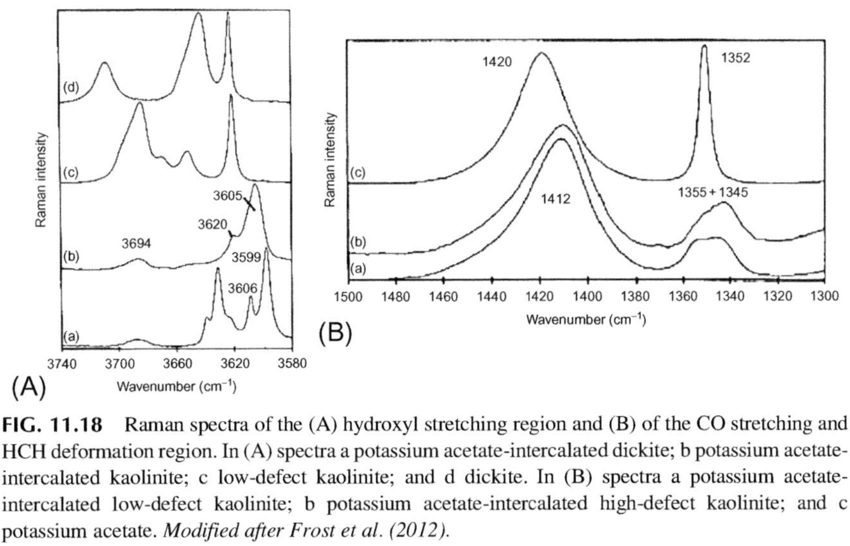


**FIG. 11.18** Raman spectra of the (A) hydroxyl stretching region and (B) of the CO stretching and HCH deformation region. In (A) spectra a potassium acetate-intercalated dickite; b potassium acetate-intercalated kaolinite; c low-defect kaolinite; and d dickite. In (B) spectra a potassium acetate-intercalated low-defect kaolinite; b potassium acetate-intercalated high-defect kaolinite; and c potassium acetate. *Modified after Frost et al. (2012).*

the Kaol. Apparently, Ac forms H bond with $Al_2OH_{IS}$ and siloxane surfaces of adjacent Kaol layers.

Intercalation of Kaol with KAc under pressure (about 20 bar) and increased temperature (up to 300°C) resulted in a number of interesting differences compared with the RT intercalation (Frost et al., 1998a, 1998c, 2000b, 2001e). At 220°C and 20 bar, only partial intercalation was observed compared with complete intercalation at RT. Instead of one, now three new bands were observed in the Raman spectrum in the OH stretching region around 3590, 3604 and 3609 $cm^{-1}$. It was also much more difficult to obtain a good-quality Raman spectrum with the Raman microscope. However, IR spectroscopy is not influenced by this crystallinity effect. At 120°C, intercalation with potassium acetate resulted in two new bands around 3595 and 3605 $cm^{-1}$ and at 220°C to three new bands around 3592, 3600 and 3606 $cm^{-1}$ comparable with the Raman spectra. A similar effect of splitting up of the OH stretching band was observed on the intercalation of halloysite at RT (Frost and Kristof, 1997). A possible explanation can be that under increased pressure and temperature in the presence of KAc, the stacking of the layers decreased (increased stacking disorder) causing a differentiation of the OH groups. As soon as this occurs, the H bonds between adjacent layers are broken, creating free space for the intercalating molecule. This model is supported by the observation of two new OH deformation bands around 895 and 906 $cm^{-1}$ due to the formation of H bonds with the Ac ions.

The NIR spectra in the 4700–4000 $cm^{-1}$ region for Kaol and the KAc Kaol intercalate are shown in Fig. 11.19A. An intense band at 4531 $cm^{-1}$ appears in the spectrum of Kaol and corresponds to the OH combination mode $(\nu+\delta)Al_2OH$ in the structure of Kaol (Madejová et al., 2009). After intercalation, the four bands at 4319, 4296, 4246 and 4200 $cm^{-1}$ became two bands at 4319 and 4200 $cm^{-1}$. The disappearance of these two bands after intercalation was attributed to KAc H bonding with $Al_2OH_{IS}$. The band at 4412 $cm^{-1}$ is common to the Ac and to the Ac Kaol intercalate complex. This band is due to the combination of the $\nu(CH)$ (2994 $cm^{-1}$) and the $\delta_s(CH)$ modes of the $CH_3$ groups (1346 $cm^{-1}$). The band around 4508 $cm^{-1}$ for the KAc–Kaol intercalation complex is not observed in the NIR spectrum of Kaol. This band was assigned to the combination $(\nu+\delta)$ of the H bond perturbed $\nu(Al_2OH_{IS})$ (3604 $cm^{-1}$) and $\delta(Al_2OH_{IS})$ (908 $cm^{-1}$) modes. Four bands at 4636, 4601, 4559 and 4531 $cm^{-1}$ appeared in similar positions but with different intensities for the Kaol and its intercalation complex. These bands are attributed to the non-perturbed $(\nu+\delta)Al_2OH_{IS}$ mode. After intercalation, the intensity of these bands appeared decreased, in line with the observations for the corresponding modes in the Raman and MIR spectra.

The water $(\nu+\delta)HOH$ region for the intercalated Kaol is shown in Fig. 11.19B. The relatively low-intensity bands observed at 5249, 5230, 5203 and 5160 $cm^{-1}$ in the spectrum of Kaol were assigned to the $(\nu+\delta)HOH$ modes of a small quantity of water adsorbed on the $Al_2OH_{IS}$ of Kaol. The band

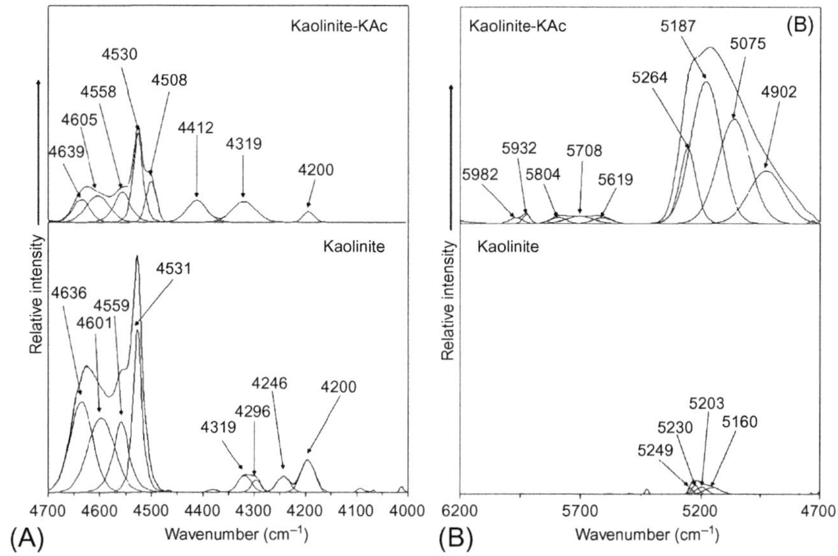


**FIG. 11.19** (A) The 4700–4000 cm$^{-1}$ region and (B) The 6200–4700 cm$^{-1}$ region of NIR spectra of kaolinite (bottom) and kaolinite-KAc intercalation complex (top). *Reproduced with permission from Zhang et al. (2011).*

observed at 5264 cm$^{-1}$ for the intercalated Kaol was assigned to the H bond perturbed ($\nu(Al_2OH_{IS})$ (3604 cm$^{-1}$) and water $\delta$(HOH) at 1643 cm$^{-1}$. However, three broad bands were observed at 5187, 5075 and 4902 cm$^{-1}$ for the KAc–Kaol intercalation complex. The combination mode of water molecules in the AlOH···$OH_2$ complexes has been reported at 5270 cm$^{-1}$. Therefore, the bands observed in the 5500–4700 cm$^{-1}$ region were ascribed to the combination of the water $\nu$(HOH) modes (3475 and 3255 cm$^{-1}$) with the water $\delta$(HOH) mode (1643 cm$^{-1}$). The KAc–Kaol intercalation complex was formed from the expansion of Kaol with both KAc and water molecules. Therefore, these new bands are attributed to the water coordinated to KAc in the interlayer of Kaol.

The NIR 5500–6000 cm$^{-1}$ region exhibits the first fundamental overtones of Ac ion. The relatively low-intensity bands were observed at 5982, 5932, 5804, 5708 and 5619 cm$^{-1}$ in the spectrum of KAc–Kaol intercalation complex and were not observed in this region for Kaol. The bands at 5982 and 5932 cm$^{-1}$ are the first overtones of the $CH_3$ stretching modes observed at 2994 cm$^{-1}$. The bands of 5804 and 5619 cm$^{-1}$ were almost the same in intensity. These two bands were assigned to the overtone of the $CH_3$ stretching bands observed at 2994 cm$^{-1}$ and twice the wavenumber of the symmetric deformation band of the $CH_3$ (1346 cm$^{-1}$) in the Mid-IR spectrum. The bands in the 7200–7050 cm$^{-1}$ region arise from the first overtone of the OH stretching bands. The bands were observed for Kaol at 7247, 7208, 7170 and 7123 cm$^{-1}$ and were assigned to the overtones of the OH stretching vibrations. The bands observed at 6895 and 6530 $cm^{-1}$ were due to the overtone of water H-O-H stretching vibration (3475 and 3255 $cm^{-1}$). The bands at 7039 and 7027 $cm^{-1}$ were assigned to the overtone of the band observed at 3604 $cm^{-1}$. This is due to the inner surface hydroxyl of Kaol H bonding with the acetate ion (Zhang et al., 2011).

### *11.4.1.2 Deintercalation of KAc from Kaol*

In addition to the intercalation process, research has been performed on the stability of the intercalated Kaol, e.g. during washing and during heating. Deintercalation has received only minimal attention so far (Wada, 1965; Churchman and Carr, 1973). Churchman and Carr (1973) showed the deintercalation of halloysite with $K^+$ acetate resulted in a basal reflection around 0.84 nm together with the associated *002* and *003* reflections. This is in contrast to hydrated nacrite as observed by Wada (1961) and Deeds *et al.* (1966). Frost et al. (1998f) observed for a well-ordered Kaol intercalated with KAc, after I min of washing, 75% of the KAc had disappeared; after 10 min 95%, had deintercalated and after one hour deintercalation was nearly complete at 99.9%. Significant is that it appeared to stop here, even after 24 h there was still something visible of the 1.42 nm phase in the XRD pattern. Simultaneous with the disappearance of the 1.42 nm reflection in the XRD disappeared the 3605 $cm^{-1}$ band in the Raman spectrum, whereas the bands ascribed to the inner surface hydroxyl groups reappeared and increase in intensity with time. In the end, deintercalation resulted in an increase in the number of stacking faults of the Kaol layers and, therefore, in a lower order. The intercalation of the so-called high-defect (low-order) Kaol resulted in a much more complicated intercalation pattern than the highly ordered Kaols. Washing with water for 30 min results in a full deintercalation, as can be seen in XRD results, where the basal reflection at 1.39 nm disappears and was replaced by a new reflection around 0.72 nm accompanied by a small reflection around 1.03 nm This was an indication that not only deintercalation had taken place but also on a small scale a form of hydration to a sort of halloysite-like material. Because the material was dried first before the XRD measurements, it cannot be excluded that directly after deintercalation all the material was initially hydrated. Heating of the deintercalated material resulted in the full recovery of the stacking order and the disappearance of the hydrated phase.

Intercalation has a clear effect on the stability of the Kaol and its dehydroxylation behaviour. IES at increased temperatures can provide a clear picture of this process (Frost et al., 2000f). Directly after the dehydration step, dehydroxylation started, which was about 175°C lower than that for untreated Kaol. All emission spectra showed a high concentration of water, and it has been proposed that the intercalation of KAc in Kaol cannot proceed without water present. This water is supposed to hydrate $K^+$ before and during the intercalation (Cheng et al., 2011b; Frost and Kloprogge, 2012).

### 11.4.2 Caesium Acetate

Intercalation of Kaol with caesium acetate (CsAc) is more difficult than the intercalation with KAc. Under similar conditions where KAc is intercalated for at least 95%, CsAc is only intercalated for about 65% (Frost et al., 1999a). Intercalation of the Kaol with CsAc resulted in the appearance of three new bands in the ν(OH) region of the Raman spectra around 3606, 3598 and 3580 $cm^{-1}$ (Fig. 11.20A; Frost et al., 2001g). In the IR spectrum, only a single new band is observed around 3605 $cm^{-1}$ (Fig. 11.20B). Associated with the changes in the ν(OH) region of the IR spectrum is the formation of two new

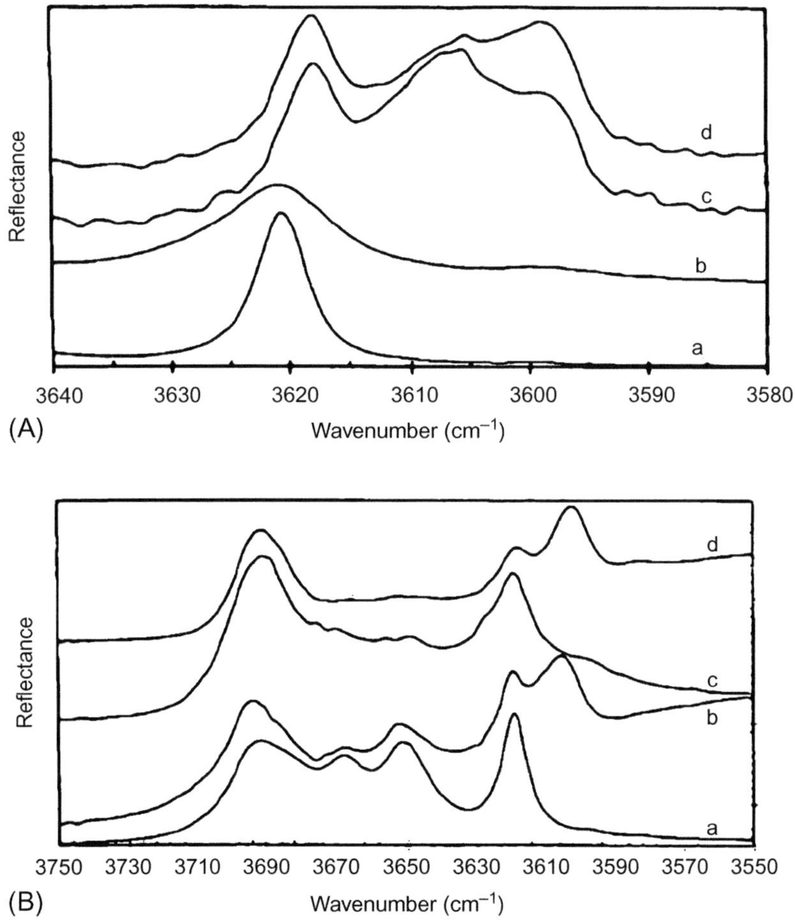


**FIG. 11.20** Raman (A) and infrared (B) spectra of part of the hydroxyl stretching region of kaolinite and caeseium-acetate kaolinite intercalate complexes. Spectra a, low-defect kaolinite; b, high-defect kaolinite; c, caesium acetate-intercalated low-defect kaolinite; and d, caesium acetate intercalated high-defect kaolinite. *Reproduced with permission from Frost et al. (2001g).*

bands in the hydroxyl deformation region around 898 and 910 $cm^{-1}$ (Frost et al., 1999a). This indicates that, as with KAc, H bonds were formed between the Ac group and the inner surface OH groups of the Kaol. In addition, it showed the inner OH stretching band shifted in both the IR and the Raman spectra from 3620 $cm^{-1}$ to 3617 $cm^{-1}$, while this band did not change in bandwidth as it did with the intercalation with KAc. This may be because $Cs^+$, in contrast to the $K^+$ cation, is too large to fit into the ditrigonal hole of the siloxane layer of Kaol and, hence, will have no effect on the bandwidth and the polarisation of the inner OH group. The inner OH group can be oriented in two directions within the Kaol structure: (i) in the direction of the vacant octahedral site or (ii) in the direction of the ditrigonal hole of the tetrahedral or siloxane layer. Due to the slightly larger ionic radius of the $Cs^+$ compared with the $K^+$, it is more plausible that $Cs^+$ is located on top of the vacant octahedral site, thus forcing the inner OH group into the direction of the ditrigonal hole causing the shift of the $\nu$(OH) band in both the IR and the Raman spectra (Kristóf et al., 1997).

Intercalation experiments of CsAc at increased temperature and pressure (120°C and approximately 20 bar) did not result in any expansion of Kaol to *d*-values observed for the intercalation of potassium acetate. The XRD patterns, however, showed a strong asymmetry of the basal reflection towards higher basal spacings. This may indicate that some water had been intercalated during the experiments at increased temperature and pressure, supported by some changes observed in the IR spectra. Here two broad bands were observed around 3582 and 3550 $cm^{-1}$, which were interpreted as water $\nu$(HOH). In addition, new bands were observed in the $\delta$(HOH) region around 900 $cm^{-1}$, and the water $\delta$(HOH) region around 1640 $cm^{-1}$ also support this interpretation.

## 11.5 CONCLUDING REMARKS

The study of the intercalation complexes of the kaolinite group minerals has a long history in which vibration spectroscopy has played an important role. Interpretation of the vibrational spectra of the various intercalate complexes of kaolinite and other kaolin minerals has underpinned our understanding of the mechanisms by which these intercalate are formed. Advances in the use of coupled instrumentation, such as controlled rate thermal analysis with Raman or DRiFT has provided further insight into the stabilities of the complexes, and has led to confirmation of many assignments.

Chapter 12

# Infrared and Raman Spectroscopies of Pillared Clays

J.T. Kloprogge

## 12.1 INTRODUCTION

Extensive research into the synthesis of suitable cracking catalysts for heavy petroleum, having larger pore sizes and greater accessible pore volumes than available from zeolites, was stimulated by the 'oil crisis' that adversely effected the world economy in 1973. Synthetic zeolite catalysts have limited micropore sizes, and their surfaces are rapidly poisoned by metals present in the heavy oil fractions. Clay minerals intercalated with certain inorganic species retain their micro- (and meso-) porosity after heating at $\geq$300°C. These so-called pillared clays (abbreviated PILC for pillared interlayered clay minerals) have a long and interesting history in clay science research, where their surface properties have been explored for their catalytic activity, most importantly in petroleum processing. A PILC is considered to be similar to a two-dimensional zeolite-like material, but with a larger pore size. These materials began to compete with zeolites in the 1970s as oil cracking catalysts.

The term 'pillaring' has often been linked to the synthesis of catalytically active heteroporous materials using swelling 2:1 clay minerals, where the expandable interlayer space is 'propped' into a partially opened position by thermal reaction with a hydrolyzed interlayer metal ion (the so-called polymeric hydroxyaluminium cation, such as $[AlO_4Al_{12}(OH)_{24}(H_2O)_{12}]^{7+}$ the $Al_{13}$ keggin cation) that chemically bonds with the interlayer siloxane surfaces on heating. Polymeric (hydr)oxymetal cations have been used as pillaring agents for clay minerals since the late 1960s (Sawhney, 1968). However, the paper by Brindley and Sempels (1977) can be regarded as the first work detailing the preparation of an inorganic pillared clay mineral. The pillaring process consisted of exchanging the $Na^+$ cations in the smectite with $Al_{13}$ cations, which were then converted by a calcination step into Al-oxides. By propping the smectite layers permanently apart, these Al-oxide species acted as 'pillars' in the interlayer

Developments in Clay Science, Vol. 8. https://doi.org/10.1016/B978-0-08-100355-8.00012-6

space. In a technical report of the IUPAC, Schoonheydt et al. (1999) gave a definition of pillared clays (enlarged to pillared layered solids): "*…pillaring is a process by which a layered compound is transformed in a thermally stable micro- and/or mesoporous material with retention of the layer structure*". The report adds that "*a pillaring agent is any compound which maintains the spacing between adjacent layers upon removal of the solvent and which induces an experimentally observable pore structure between the layers called 'interlayer region' accessible to molecules at least as large as $N_2$*". Thus the definition of a pillaring agent includes inorganic pillars but also organic surfactants.

The choice of clay mineral host for the production of PILC is very important. Natural smectites such as montmorillonite (Mt), hectorite (Ht), beidellite (Bd), saponite (Sap), and synthetic smectites such as Laponite (Lap) are generally used as PILC hosts because the first step in pillaring is the intercalation of the pillaring agent by cation exchange. Vicente et al. (2013) presented the different metal ions that have been studied as pillaring agents in the form of a "Periodic Table of Elements," From all the different metals, $Al^{3+}$ is the most commonly used and studied. Other PILC include $Si^{4+}$, $Fe^{3+}$ and $Ti^{4+}$. The intercalation and pillaring of clay minerals with solutions containing two different metals has been investigated since the 80s. One of the pillaring metals is typically $Al^{3+}$. The second inorganic metal is then added in various molar fractions to expand the thermal, adsorptive and catalytic properties of the resulting PILC products (Gil et al., 2000). Incorporation of $Zr^{4+}$, $Ga^{3+}$, transition metal elements like $Cr^{3+}$, $Fe^{3+}$, $Cu^{2+}$, $Mo^{2+}$, $Ru^{2+}$ as well as metals of the lanthanide series into Al-pillaring solutions have been published (Vicente et al., 2013).

The most commonly defined characteristic of PILC is its accessible porosity, which is determined from a BET adsorption method or from measured *d*-values of the PILC and the initial layer charge characteristics of the host mineral. PILC have a permanent porosity and a higher $N_2$ accessible specific surface area (SSA) than the starting clay mineral. A smectite with completely exfoliated layers is assumed to have an SSA of more than 800 $m^2/g$ but the highest SSA recorded for PILC is 600 $m^2/g$. A significant proportion of this SSA of PILC corresponds to the remaining free basal surface of the smectite and the other proportion includes the lateral surfaces of the pillars. The highest reported total pore volume value of 0.6 $cm^3/g$ (Vaughan et al., 1979) corresponds to pores of different size, although micropores (<2 nm) are usually obtained by Al pillaring. In some cases, such as the mixed Al–Fe PILC, mesopores (2–50 nm) have been observed (Mandalia et al., 1998). Macropores (>50 nm) have been also obtained with delaminated pillared Lap (Butruille and Pinnavaia, 1996).

The structure of the polymeric $Al_{13}$ Keggin cation, the most studied intercalant for PILC, is well known. However the structures of some other polymeric metal intercalants are less understood. For example, the structure of the polymeric $Ga_{13}$ cations is assumed to be similar to the $Al_{13}$ structure (Bradley et al., 1990a,b); hypothetical structures have been proposed for Zr and Si pillars (Figueras, 1988; Kloprogge, 1998; Ding et al., 2001) while the structures of others are still unknown.

Despite the knowledge accumulated on PILC and particularly on $Al_{13}$, two main questions remain (1) the real structure of some intercalated species before and after heating with the pending question whether the structure of the intercalated polymerical cations is maintained after heating, or whether the intercalant species are transformed into oxide pillars, as suggested by Vaughan et al. (1979), and (2) the type of bonds formed between the clay mineral host and the pillar guest that maintains the stability of the pillared clay minerals. Even though some hypotheses have been proposed such as cross-linking between pillars and silica tetrahedra that become inverted after calcination (Plée et al., 1985; Fripiat, 1988; Lambert et al., 1994; Lambert and Poncelet, 1997), a universally recognized mechanism explaining the thermally stable linkage that exist between pillars and the clay mineral surfaces is still missing.

So far most of the characterization of PILC has been based on a combination of XRD to determine the basal spacing and $N_2$ adsorption to determine the SSA, pore volumes and pore sizes. While solid state NMR has been applied to assess the molecular structure of the pillars and their interactions with the clay mineral, relatively little attention has been paid to techniques such as infrared (IR) or Raman spectroscopy. The major aims of this chapter are to provide an overview of the existing results from IR and Raman spectroscopies and what can be inferred about the structures in question.

## 12.2 OLIGOMERS SALTS

### 12.2.1 $Al_{13}$-Sulfate and $Al_{13}$ Nitrate

In order to gain a better understanding of PILC based on the $Al_{13}$ keggin structure, sulfate and nitrate salts of pure $Al_{13}$ are studied. The Keggin structure $[AlO_4Al_{12}(OH)_{24}(H_2O)_{12}]^{7+}$ is preserved in the crystals as shown by $^{27}$Al solid-state MAS-NMR, XRD and by XPS (Kloprogge et al., 1992a, 1995; Duong et al., 2005). Kloprogge et al. (1992a; 2001a) and Kloprogge and Frost (1998) described the MIR and NIR spectrum of the monoclinic basic aluminium sulfate. The spectrum of basic aluminium sulfate was dominated by two strong water H–O–H stretching and bending modes at 3247 and 1640 $cm^{-1}$, and a strong Al–OH stretching band at 3440 $cm^{-1}$. The IR spectrum was further dominated by the $\nu_1$ and $\nu_3$ bands at 981 and 1051 $cm^{-1}$ of the sulfate group in the $Al_{13}$ sulfate structure. Furthermore, the band at 724 $cm^{-1}$ was assigned to an $(Al–O)_{Td}$ (Td = tetrahedral) mode of the polymerized Al–O–Al bonds in the $Al_{13}$ Keggin structure. In addition the other Al–O vibrations were assigned as 627, $(Al–OH)_{Oh}$ (Oh = octahedral); 546,

$(Al\text{–}O)_{Oh}$, and 492 $cm^{-1}$, $(Al\text{–}OH_2)_{Oh}$ stretch. In addition, a trace amount of nitrate was included in the crystal structure and was readily observed in the IR and Raman spectra. The Raman spectrum showed the $\nu_2$ and $\nu_4$ $SO_4^{2-}$ triplets at 446, 459 and 496 $cm^{-1}$ and 572, 614 and 630 $cm^{-1}$ (Fig. 12.1). The $\nu_1$ was observed as a single band at 990 $cm^{-1}$, partly overlapped by the $\nu_3$ triplet at 979, 1009 and 1053 $cm^{-1}$ (Kloprogge and Frost, 1999a,b). Even though the NIR spectra of the $Al_{13}$ sulfate and nitrate are very similar indicating comparable crystal structures, there are minor differences related to the strength with which the crystal water molecules are bonded to the sulfate and nitrate groups (Fig. 12.2). The interaction between crystal water and nitrate is stronger than with the sulfate as shown by the shift of the crystal water band positions from 6213, 4874 and 4553 $cm^{-1}$ for the $Al_{13}$ sulfate

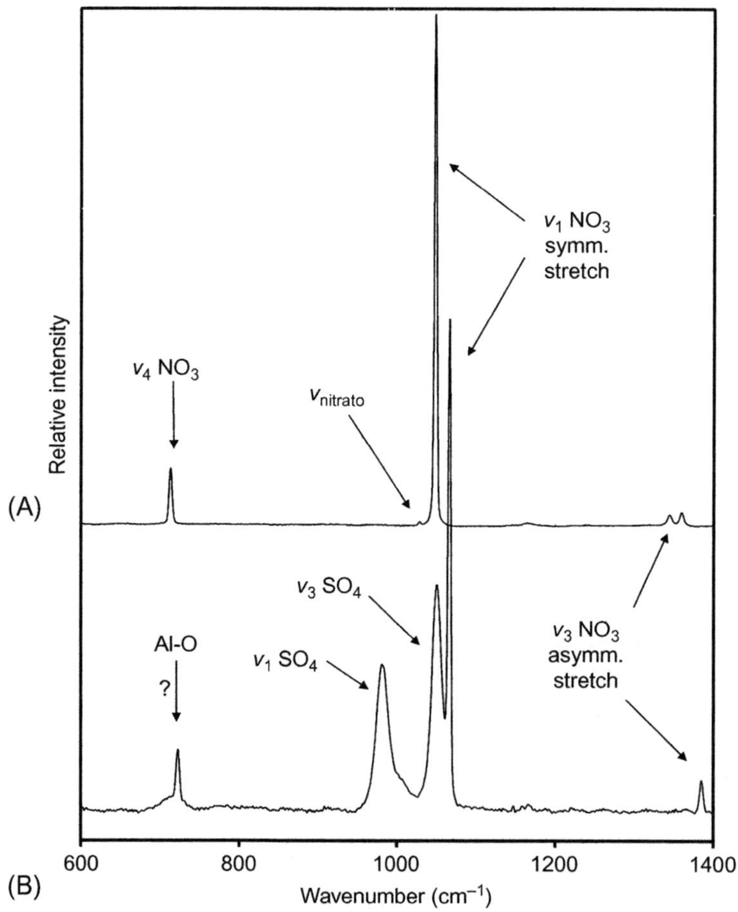


**FIG. 12.1** Raman spectra of $Al_{13}$ nitrate (A) and sulfate (B). Reproduced (modified) with permission from Kloprogge and Frost (1999b).

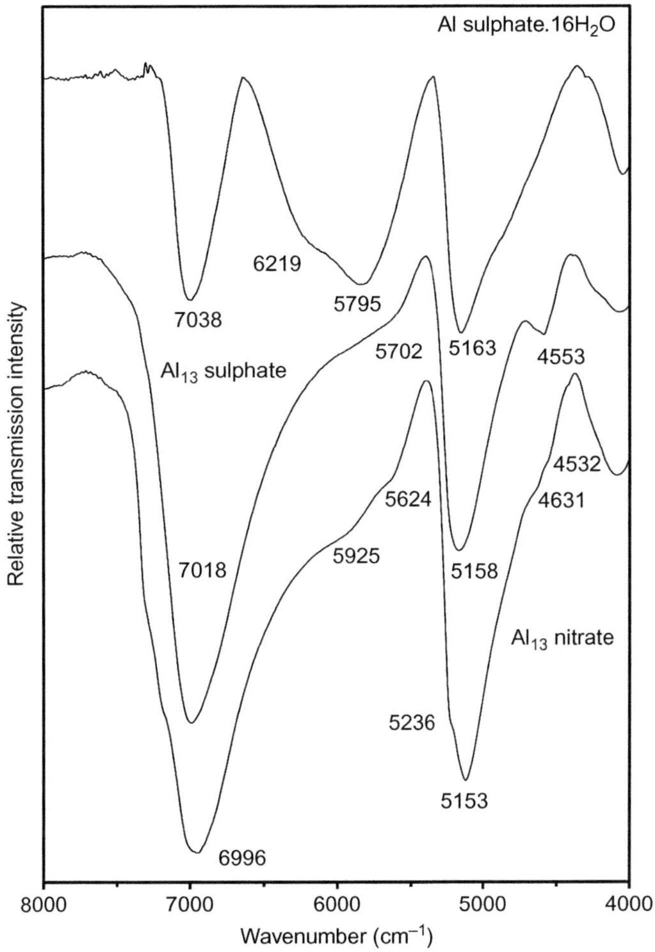


**FIG. 12.2** Near-infrared spectra of $Al_2(SO_4)_3.16H_2O$, $Al_{13}$-sulfate and $Al_{13}$-nitrate. Reproduced (modified) with permission from Kloprogge et al (2001a).

towards 5925, 4848 and 4532 $cm^{-1}$ for the nitrate. A reverse shift from 5079 and 5037 $cm^{-1}$ for the sulfate towards 5238, and 5040 $cm^{-1}$ for the nitrate for the water molecules in the $Al_{13}$, shows that the nitrate-$Al_{13}$ bond is weakened due to the influence of the crystal water on the nitrate. The Al–OH bond in the $Al_{13}$ complex is unaffected by changing the salt group due to the shielding by the water molecules of the $Al_{13}$ complex.

### 12.2.2 $Ga_{13}$-Sulfate and $Fe_{13}$-Sulfate

Bradley and Kydd (1993) indicated the relative positions of the IR bands would be expected to vary significantly between an $Al_{13}$ keggin ion and an isostructural $[GaO_4Ga_{12}(OH)_{24}(H_2O)_{12}]^{7+}$ ion, primarily due to the different masses of aluminium and gallium. Comparisons between the spectra of γ-$Al_2O_3$ and γ-$Ga_2O_3$ found the M-O stretching vibrations to occur approximately 100 $cm^{-1}$ lower for the gallium spinel and on this basis the bands for $Al_{13}$ and $Ga_{13}$, which can be considered to be sub-units of the γ-$M_2O_3$ structure, would probably occur at correspondingly lower wavenumbers for the $Ga_{13}$ ion. There were two strong bands observed at wavenumbers that would be appropriate for (Ga–O) and $(Ga–OH)_{Oh}$ bands in a $Ga_{13}$ ion (605 and 471 $cm^{-1}$), and in addition there appeared to be a broad underlying band. The band at ca. 605 $cm^{-1}$ could not be ascribed to the $\nu_4$ vibration of $SO_4$ as it did not shift its position when the counter ion was changed to selenate. This spectrum did not, however, exhibit the other (weaker) stretching bands that were observed for the $Al_{13}$ sulfate and selenate salts (i.e. the $(Al–O)_{Oh}$ and the $(Al–OH_2)_{Oh}$ bands). This was thought to be due to the gallium sulfate being very poorly crystalline and amorphous material also being present, while the $Al_{13}$ sulfates and selenates were highly crystalline. The very broad band underneath the two relatively sharp bands might be due to a poorly crystalline gallium alunite phase.

When an $Fe^{3+}$ solution was hydrolyzed, and a sulfate salt thereof was prepared instantly, the IR spectrum was practically indistinguishable from that of the gallium sulfate salt with bands assigned as tetrahedrally $(Fe–O)_{Td}$ and octahedrally coordinated $(Fe–OH)_{Oh}$, at 606 and 465 $cm^{-1}$ (Fig. 12.3). In addition, there was a very broad underlying band, like that attributed to gallium alunite in the spectra obtained for the $Ga_{13}$ sulfate salt. As Fe is also known to form a poorly crystalline alunite type of phase (jarosite), $[H_3O][Fe_3(OH)_6(SO_4)_2]$, this result was not surprising. The overall appearance of the two spectra show that $Ga^{3+}$ and $Fe^{3+}$ behaved in a very similar manner. The kinetics in the case of $Fe^{3+}$ was much faster; the $(Fe–O)_{Td}$ band vanished if the hydrolyzed solution was allowed to age for more than just a few seconds before the sulfate solution was added. This reaction occurred even when all of the solutions were cooled to 0°C, suggesting that a tridecamer was formed upon the hydrolysis of iron(III) salt solutions, although the tridecamer was extremely unstable. The major difference between the spectra of the $Ga^{3+}$ and $Fe^{3+}$ sulfate salts was the degree of splitting of the $\nu_3(SO_4^{2-})$ bands, indicating differences in structural arrangements.

### 12.2.3 Mixed $(Al–Fe)_{13}$-Sulfate, $(Al–Cr)_{13}$-Sulfate and $(Al–Mn)_{13}$ Sulfate

In the IR spectra of the co-hydrolyzed $Fe^{3+}$ and $Al^{3+}$ ions, only bands due to $(Al–O)_{Td}$, $(Al–OH)_{Oh}$, $(Al–O)_{Oh}$ and $(Al–OH_2)_{Oh}$ were observed (Table 12.1, Kiricsi et al., 1995; Palinko et al., 1997). The bands retained the positions observed in the $Al_{13}$-Keggin sulfate. At low stoichiometric ratios (Fe: Al = 1:12 and 2:11) no band typical of either tetrahedrally or octahedrally

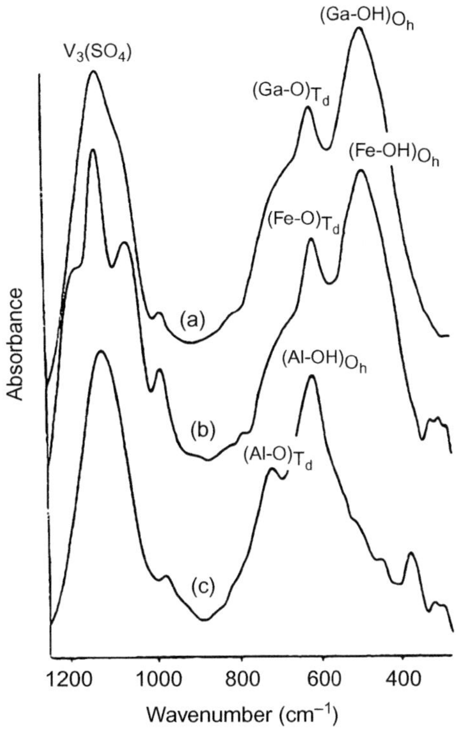


**FIG. 12.3** Infrared spectra of CsI pellets of sulfate salts of hydrolyzed solutions of $Ga^{3+}$ (a), $Fe^{3+}$ (b) and $Al^{3+}$ (c). Reproduced (modified) with permission from Bradley and Kydd (1993).

coordinated iron could be detected. When the proportion of iron was increased at the low wavenumber end of the spectrum (near 465 $cm^{-1}$) a new band started to develop. It was believed to belong to octahedrally coordinated iron in the coprecipitated hydrous iron oxide. Octahedrally coordinated iron (actually in three different octahedral positions in the as-prepared sample) was detected by $^{57}Fe$ Mössbauer spectroscopy at a low stoichiometric ratio (Fe: Al = 2:11) when IR spectroscopy was not sensitive enough to indicate its presence.

The IR characteristics of the Cr–Al sulfates were different. Even at the lowest Cr–Al ratio two bands at 764 and 525 $cm^{-1}$ were observed. The latter was so strong (and its spectral area increased with chromium content) that it suppressed every other weaker band. Only the band typical for $(Al–OH)_{Oh}$ remained unchanged, indicating that octahedral Al was not significantly substituted by chromium. Although the $(Al–O)_{Td}$ band remained undetected in any of the Cr–Al sulfates, tetrahedral substitution did not seem probable as a band typical for $Cr^{3+}$, $(Cr–O)_{Td}$, tetrahedral was not found. Like other transition metals $Cr^{3+}$ tends to favor octahedral rather than tetrahedral coordination (Palinko et al., 1997).

**TABLE 12.1** Location of Absorption Bands in the 800–400 $cm^{-1}$ Range of co-hydrolyzed Keggin-Type Ions with Various Metal/ Al Ratios in Co-Crystallized $Al_{13}$-Keggin Sulphate. Data from Palinko et al. (1997).

| | M/Al | Location of Adsorption Bands in the 800–400 $cm^{-1}$ Region | | | |
|---|---|---|---|---|---|
| *Ga/Al* | *Ratio* | *(Al–O)$_{Td}$* | *(Al–OH)$_{Oh}$* | *(Al–OH)$_{Oh}$* | *(Al–OH$_2$)$_{Oh}$* |
| | 0/13 | 720 | 611 | 548 | 494 |
| | 1/12 | 698 | 610 | 543 | 491 |
| Fe/Al | | | | | |
| | 1/12 | 720 | 614 | 551 | 494 |
| | 2/11 | 723 | 616 | 551 | 492 |
| | 3/10 | 720 | 609 | 545 | 465 |
| | 4/9 | 722 | 615 | 552 | 465 |
| Cr/Al | | | | | |
| | 1/12 | 764 | 610 | 525 | |
| | 2/11 | 764 | 610 | 526 | |
| | 3/10 | 764 | 611 | 525 | |
| | 4/9 | 764 | 610 | 525 | |

A 9 GHz electron paramagnetic resonance (EPR) study of manganese ($Mn^{2+}$) substituted $Al_{13}$ sulfate revealed that the number of $Mn^{2+}$ ions embedded in the axial crystal field had a characteristic maximum at the stoichiometric molar concentration equal to 7.69%. In confirming the EPR results, Kawano et al. (2002) showed that the FT-IR measurements for $Al_{13}$ sulfate salts with various concentrations of $Mn^{2+}$ indicated bands at 555 $cm^{-1}$ (Mn–O), 614 $cm^{-1}$ (Al–OH), 722 $cm^{-1}$ (Al–O) and 1120 $cm^{-1}$ ($SO_4$). The intensity ratio of bands at 555 $cm^{-1}$ and 722 $cm^{-1}$ indicated that substitutions of $Mn^{2+}$ for $Al^{3+}$ in tetrahedral environments were stoichiometric over the range of 0.5–7 mol% Mn.

## 12.3 Al PILC

Mechanistic models of how the many different Al-pillared smectites form depend largely on the degree of substitution in the tetrahedral sheet of the smectite host. The most comprehensive study on the changes in the structures of $Al_{13}$ exchanged clay minerals upon calcination has been performed by Kloprogge et al. (1999d) who used IR emission spectroscopy (IES) to study the in situ changes upon heating from room temperature to 750°C. Based on $^{27}Al$ and $^{29}Si$ solid-state MAS-NMR for Bd a mechanism was proposed where linkages are formed between the pillar and the tetrahedral clay mineral layer by inversion of a Si tetrahedron or a Si–O–Al bond in a tetrahedron (Plée et al., 1985). A similar mechanism has been proposed for Al-pillared Sap (Li et al., 1993; Lambert et al., 1994; Bergaoui et al., 1995).

## 12.3.1 $Al_{13}$ Pillared Smectites with Tetrahedral Substitutions

### *12.3.1.1 $Al_{13}$-Pillared Beidellite*

Kloprogge and Frost (1999c) studied the structural changes of synthetic Bd and $Al_{13}$-pillared analogues during dehydroxylation with IES. The OH-stretching region of Bd was characterized by two OH-stretching modes around 3605 and 3650 $cm^{-1}$. These bands strongly decreased in intensity upon dehydroxylation of the Bd up to 700–750°C. Compared to unpillared Bd, IR spectra of $Al_{13}$-pillared Bd showed a much stronger absorption band at 3456 $cm^{-1}$, which was assigned to silanol groups formed by the breaking of Si–O–Al bonds in the tetrahedral layer. A reaction between the Al pillar and the protonated Si–OH–$Al_{13}$ linkage would yield either $^{IV}$Si–O–$Al^{pillar}$ or $^{IV}$Al–O–$Al^{pillar}$ linkages. The observation of the $^{IV}$Si silanol groups in the IR spectra of pillared Bd seemed to favor the second assignment (Kloprogge and Frost, 1999c). It should be noticed that no reactions between the pillar and the tetrahedral layer were observed for smectites without tetrahedral substitutions, which supports the assignment.

In the $Al_{13}$-pillared high charge Bd, strong hydroxyl stretching bands were observed compared to the low charge pillared Bd, but in contrast, this sample showed a minor silanol band at 3733 $cm^{-1}$ when the temperature increased above approximately 400°C (Fig. 12.4A). The new band at 3450 $cm^{-1}$ observed for the low charge $Al_{13}$ pillared Bd shifted slightly to higher wavenumbers around 3485 $cm^{-1}$ indicating a stronger bonding between the non-calcined (intercalated) pillar and the Bd structure and is thought to be due to the higher charge on the tetrahedral sheets of the Bd. However, this band, also observed by Schutz et al. (1987) as a shoulder after heating to 100°C on a broad band around 3640 $cm^{-1}$, is interpreted as a silanol group created by the breaking of Si–O–$Al^{IV}$ bonds. A very broad band present around 3200 $cm^{-1}$ was assigned to water molecules in the $Al_{13}$ pillar. This band quickly disappeared upon heating to 300°C while the HOH stretching band at 3485 $cm^{-1}$ weakened. Band component analysis of the HOH bending region revealed the presence of three different water bands at 1710, 1623 and 1564 $cm^{-1}$ for the $Al_{13}$-pillared low charge Bd and at 1705, 1645 and 1574 $cm^{-1}$ for the $Al_{13}$-pillared high charge Bd. This complexity was not observed for the synthetic Bd itself. The first two bands disappeared at temperatures around 350°C, whereas the band at 1564/1574 $cm^{-1}$ disappears around 500–550°C. These bands are ascribed to, in the same order, strongly surface adsorbed

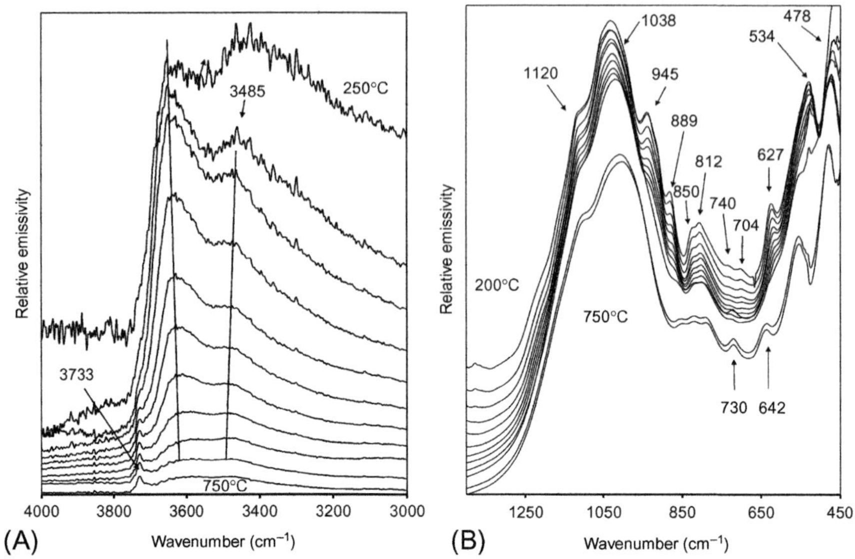


**FIG. 12.4** (A) Infrared emission spectra (IES) of $Al_{13}$-pillared synthetic beidellite in the OH-stretching region. Reproduced (modified) with permission from Kloprogge and Frost (1999c). (B) Infrared emission spectra (IES) of $Al_{13}$-pillared synthetic beidellite in the low wavenumber region (1350–450 $cm^{-1}$). Reproduced (modified) with permission from Kloprogge and Frost (1999c).

water, interlayer water and water associated with the $Al_{13}$ structure. The position of the interlayer water band at 1623/1645 $cm^{-1}$ agrees with the observation of one water band in the IR spectrum at 1624 $cm^{-1}$ for the unpillared synthetic Bd. Since each $Al_{13}$ complex contains 12 water molecules, it is reasonable to assume that the large band at 1564/1574 $cm^{-1}$ has to be assigned to these water molecules.

The presence of two different OH-groups in Bd was also reflected in the structural $\delta$(OH) modes around 880 and 916 $cm^{-1}$ and in the $\tau$(OH) modes around 770 and 804 $cm^{-1}$ in the IR absorption spectra (Fig. 12.4B). These bands showed an intensity decrease with heating related to dehydroxylation of the clay mineral. Significant differences between the Bd and its pillared analogue were the splitting of the single band at 814 $cm^{-1}$ into two bands at 812 and 830 $cm^{-1}$ and the bands associated with $\delta$(AlOH) modes showed increased intensities for the pillared Bd. These results showed that upon calcination the pillars dehydroxylate between 300°C and 400°C (Schutz et al., 1987), thereby releasing protons that could diffuse through the tetrahedral sheet into the octahedral sheet where they interacted with the structural OH groups and formed water during dehydroxylation of the octahedral sheet which occurs between 500°C and 700°C (Heller-Kallai, 2013). In addition, the formation of Si-OH groups was observed in the IES spectra of the

$Al_{13}$-pillared Bd indicating that these protons could also interact with Si–O–Al bonds upon calcination and form new Si–OH–Al bonds. A reaction between the $Al_{13}$ pillar and the protonated Si–OH–Al linkage then possibly yielded $Al^{IV}$–O–$Al^{pillar}$ linkages. Still, the IES data do not give conclusive evidence for the inversion of the apical oxygen of the $AlO_4$ in the tetrahedral sheet from being oriented towards the octahedral sheet to being oriented to the interlayer space as was assumed by Plée et al. (1985) in their model for the pillaring of Bd with $Al_{13}$.

### 12.3.1.2 $Al_{13}$-Pillared Sap

Li et al. (1993) reported the IR spectra of $Al_{13}$-pillared synthetic Sap. Exchange of the Sap with $Al_{13}$ resulted in a shift towards higher wavenumbers for the 1054 and 988 $cm^{-1}$ bands assigned to the Si(Al)–O stretching modes and towards lower wavenumbers for the 738 and 652 $cm^{-1}$ bands assigned to Si–O–$Al^{pillar}$ modes. Further, a reduction in intensity of the 820 and 691 $cm^{-1}$ bands was observed. Calcination of the $Al_{13}$-exchanged Sap decreased the intensity of the Si–O–Al bending mode at 646 $cm^{-1}$. This decrease was interpreted to be caused by a reaction between the $Al_{13}$ pillar and the tetrahedral clay mineral sheets probably on Si–O–Al linkages. Chevallier et al. (1994) reported a new band in the OH-stretching region around 3594–3597 $cm^{-1}$, which they interpreted as a $\nu$(OH) band created by the pillaring process and is comparable to the band observed by Ocelli and Finseth (1986) for H-bonded OH groups in pillared Ht. Sap displayed in the $\nu$(OH) region a sharp band at 3669 $cm^{-1}$ and a broad band around 3490 $cm^{-1}$ assigned to the $Mg_3OH$ and the interlayer $H_2O$ $\nu$(OH) modes, respectively. Exchange with $Al_{13}$ resulted in new bands at 3626 and 3244 $cm^{-1}$ assigned to Al–$H_2O$ and Al–OH $\nu$(OH) modes, respectively (Kloprogge et al., 1999d). This is probably caused by the different interlayer environment. Heating resulted in a strongly diminishing intensity of the interlayer water band. The Al–OH band broadened and shifted to higher wavenumbers and strongly diminished in intensity above 500°C. At 700°C this band and the Al–$H_2O$ band completely disappeared, indicating that in Sap the transition to aluminium oxide was complete. The $Mg_3OH$ band of Sap was hardly influenced by the calcination, indicating the pillaring mechanism in Sap does not involve or affect the octahedral sheet. Trombetta et al. (2000) observed a sharp and weak band near 3740 $cm^{-1}$, due to silanol groups (Si–OH) exposed in defect sites of the external layer surface, while the stronger peak at 3675 $cm^{-1}$ is due to the octahedral $Mg_3OH$ groups of the lattice. The spectra in the OH stretching region of the two pillared Sap were similar to those of the parent material, but a clear increase of the absorption in the region 3750–3600 $cm^{-1}$ was observed after calcination at 400°C. In particular, the appearance of some absorption between the two main bands and of a tail towards lower wavenumbers were interpreted as likely to be due to OH groups at the surface of the aluminium pillars.

$Al_{13}$-exchanged Sap exhibited in the low wavenumber region new bands at 1701, 1644 and 1525 $cm^{-1}$. Calcination of the $Al_{13}$-exchanged Sap resulted in a disappearance of these bands around 600°C, while the unassigned band at 1312 $cm^{-1}$ disappeared around 450°C. In the low wavenumber region, the band at 963 $cm^{-1}$ strongly decreased in intensity but was still observed at 750°C with a similar, but less strong decrease for the 793 $cm^{-1}$, band.

## 12.3.2 $Al_{13}$ Pillared Smectites with Octahedral Substitutions

### *12.3.2.1 $Al_{13}$-Pillared Ht*

Pinnavaia et al. (1985) observed a structural transformation upon calcination of intercalated synthetic fluoro–Ht that has no tetrahedral charge. These authors concluded that the structural F partly replacing the OH groups in the structure causes the Si–O bands to become more labile. These labile Si–O bands then promote coupling to the $Al_{13}$ pillar during the thermal transformation. More likely the strong electronegativity of F withdrew sufficient electrons from the Si-O (via the octahedra) to provide points of electrostatic attraction.

Based on IES results, Kloprogge et al. (1999d) showed in Ht the $Al_{13}$ complex completely converted upon calcination. In the OH-stretching region the Ht was characterized by two bands at 3671 and 3453 $cm^{-1}$ assigned to the M-OH (M = octahedral cation) and interlayer $H_2O$ OH-stretching modes. Exchange with the $Al_{13}$ complex resulted in two new bands at 3641 and 3334 $cm^{-1}$, which were considered to represent the Al–$H_2O$ $\nu(AlOH_2)$ and Al–OH $\nu(AlOH)$ modes of the $Al_{13}$ complex while the original $\nu(MOH)$ bands for Ht shift to 3680 and 3548 $cm^{-1}$. The interlayer water band strongly reduced in intensity after the exchange with $Al_{13}$. Calcination resulted in the rapid disappearance of the remaining interlayer water. The transformation of the $Al_{13}$ complex resulted in the disappearance of the $\nu(AlOH_2)$ mode between 400 and 500°C and of the $\nu(AlOH)$ mode at 600°C, which suggests that the pillar structure had completely converted at this temperature. The $\nu(MOH)$ band remained visible in the entire temperature range without significant changes in position but with decreasing intensity when the dehydroxylation of the PILC proceeded. The PILC showed no complete dehydroxylation up to 750°C. Butruille et al. (1993) observed for synthetic fluoro Ht that the OH-stretching region was much broader and the signal quickly disappeared except for a band at 3680 $cm^{-1}$, which persisted even after calcination at 500°C.

In the low wavenumber region, the $Al_{13}$-exchanged Ht displays new bands at 1227, 1315, 1395 and 1510 $cm^{-1}$ in comparison to the original clay mineral. The bands at 1510, 1395 and 1315 $cm^{-1}$ strongly diminished in intensity but remained visible up to 800°C. It is assumed that these bands represent a differentiation of the initial water band of the $Al_{13}$ complex around 1640 $cm^{-1}$ due to the geometrical constraints of the clay mineral interlayer space, causing different interactions between the various water molecules on the outside of the $Al_{13}$ complex and the clay mineral layers. The initial water band observed for the Ht was strongly diminished in the $Al_{13}$-exchanged Ht.

No changes in the low wavenumber region pointed to the formation of some sort of linkage between an aluminium-oxide pillar and the Ht structure. In contrast, Butruille et al. (1993) reported a new band at 756 $cm^{-1}$ tentatively assigned to an Al–O stretching vibration of the $Al_{13}$ group, which was replaced by two bands at 863 and 801 $cm^{-1}$ between 350 and 500°C. These changes might indicate a possible modification of the pillar, though there was no change in the basal spacing between the pillared Ht calcined at 350 and 500°C. The Si–O–Si band was observed to shift from 1001 to 1012 $cm^{-1}$ after calcination at 500°C.

#### *12.3.2.2 $Al_{13}$-Pillared Mt and $Al_{13}$-Pillared Acid Activated Mt*

Kloprogge et al. (1999d) showed the changes In the $\nu$(OH) region of IES spectra of the $Al_{13}$-Mt. In the OH-stretching region, exchange of SWy-1 Mt with $Al_{13}$ complex resulted in two new bands at 3682 and 3538 $cm^{-1}$, assigned to the $\nu$(AlOH) and $\nu(AlOH_2)$ modes of the $Al_{13}$–complex, in addition, to the $\nu(Al_2OH)$ band at 3628 and the $\nu$(OH) of physi-sorbed water at 3363 $cm^{-1}$. The $Al_{13}$–complex $\nu$(AlOH) band was previously observed around 3695 and 3700 $cm^{-1}$ in IR absorption spectra of Mt (Brydon and Kodama, 1966; Gupta and Malik, 1969a) while $Al_{13}$-complex $\nu(AlOH_2)$ band was observed at 3570 $cm^{-1}$ (Goh and Huang, 1986). The interlayer water band of SWy-1 strongly decreased in intensity after exchange with $Al_{13}$ and calcination resulted in rapid disappearance of any remaining interlayer water. The transformation of the $Al_{13}$ complex caused the disappearance of the $Al_{13}$-complex $(AlOH_2)$ and $\nu$(AlOH) bands between 400 and 500°C and their replacement by two new bands around 3574 and 3505 $cm^{-1}$, indicating a structural rearrangement within the $Al_{13}$ complex. Although these latter bands strongly decreased in intensity at temperatures above 500°C, they were still visible at 800°C, suggesting that the pillar structure retained structural $\nu$(OH) and had not completely converted to aluminium oxide as generally assumed. These bands may represent OH bridges between adjacent partly dehydroxylated $Al_{13}$ molecules. The smectite $\nu(Al_2OH)$ of the Al–PILC was observed over the entire temperature range without changes in position but with decreasing intensity when the dehydroxylation of the Mt proceeded. Since the untreated Mt showed complete dehydroxylation above 700°C, thermal stability was increased on pillaring.

In the low wavenumber region of the IES, the $Al_{13}$-exchanged Mt displays new bands at 642, 1321, 1402 and 1512 $cm^{-1}$ in comparison to the pristine Mt. The first two bands corresponded well with the bands observed for $Al_{13}$-sulfate (Kloprogge et al., 1992a). The other bands possibly signify a differentiation of the initial water band of the $Al_{13}$ complex around 1640 $cm^{-1}$

due to the geometrical constraints of the clay mineral interlayer space causing different interactions between the various water molecules on the outside of the $Al_{13}$ complex and the Mt tetrahedral sheets. Molecular simulation studies have shown that the ideal $Al_{13}$ molecule will be oriented in the interlayer space with its threefold axis perpendicular and the aluminium and oxygen plains parallel to the Mt layers (Čapková et al., 1998). This results in water molecules oriented toward either the Mt layer above or below and toward the interlayer space. The band at 1512 $cm^{-1}$ disappeared around 500°C, followed by the bands at 1402 and 1321 $cm^{-1}$ above 600°C. The band at 642 $cm^{-1}$ diminished in intensity but was still observed at 800°C. At 700°C a new band was observed at 722 $cm^{-1}$, indicating the formation of an Al–O bond, like the very strong band observed for corundum, within the pillar structure. Tichit et al. (1988) observed ion exchange did not affect the bands at 1125 and 1035 $cm^{-1}$, and only a shift to 735 $cm^{-1}$ of the band initially at 710 $cm^{-1}$ was noticed. After calcination, the IR spectrum showed the dehydroxylation process through the disappearance of the band at 935 $cm^{-1}$ at calcination temperature of 300°C. For higher temperatures, the shoulder at 1125 $cm^{-1}$ decreased progressively, accompanied by a broadening of the band centered at 1035 $cm^{-1}$. The bands initially at 800 and 735 $cm^{-1}$ shifted to 775 and 700 $cm^{-1}$, respectively, at 300°C and were no longer detectable at about 100°C. In contrast, Xu and Chen (2011) claimed the disappearance of the antisymmetric Si–O stretching mode at 1087 $cm^{-1}$ upon calcination at 500°C as proof that the $SiO_4$ tetrahedron had changed its coordination and connection changes, which they used as an argument for the inversion of the tetrahedra to form Si–O–Al or Al–O–Al bridges with the pillars. However, close examination of their IR spectra shows that the band at 1087 $cm^{-1}$, though not visible as a separate band is still present in the much broader asymmetric band with its maximum at 1054 $cm^{-1}$. The model of Tennakoon et al. (1986), in which condensation of the terminal OH groups of the $Al_{13}$ complex with the lattice OH groups in the clay mineral octahedral sheets results in a direct linkage via oxygen to the cations (Mg or Al) in the octahedral sheet, must be considered improbable based on the above observations and on the fact that this will require the formation of a very long metal(clay mineral)–oxygen–metal(pillar) bond through the hexagonal holes in the tetrahedral sheet. This seems to be very unlikely based on both the geometrical constraints and the energies involved in the formation of such a bond. Kloprogge et al. (2002) noted that the low wavenumber regions of the IR spectra of the untreated Mt and the $Al_{13}$-intercalated Mt (in air and at 60°C) were very similar but the latter show a very small low intensity band at around 667 $cm^{-1}$. This is not surprising, as the $Al_{13}$ polyoxocation has not formed any permanent bonds with the adjacent Mt layers at 60°C. The lattice vibration at 667 $cm^{-1}$ was ascribed to the Al—O bond of tetrahedral coordinated Al cations in the center of the $Al_{13}$ pillars similar to the band around 642 $cm^{-1}$ observed in the IES experiment. In addition, the intensity of the water bending vibration band around 1635 $cm^{-1}$ increased after expansion with $Al_{13}$ polyoxocations due to the water molecules in the $Al_{13}$ structure. Over the last few decades a number of papers have briefly described the IR spectra of the $Al_{13}$ intercalated and pillared Mt showing no significant differences other than minor shifts in band positions and intensities (e.g. Tennakoon et al., 1987; Zheng et al., 1992; Bodoardo et al., 1998; Katdare et al., 1999; Salerno et al., 2001; Manohar et al., 2006; Molu and Yurdakoc, 2009; Okoye and Obi, 2011b; Timofeeva et al, 2011; Gashti and Eslami, 2012). These small shifts are an indication that upon calcination and the transformation of the $Al_{13}$ complex to an Al oxide pillar, these pillars have some effect on the local structure of in particular the tetrahedral sheet, but there is no evidence for any tetrahedral inversion or the formation of new $^{IV}Si–O–Al^{pillar}$ or $^{IV}Al–O–Al^{pillar}$ bonds between the tetrahedral sheet of the Mt and the pillar.

Faghihian and Mohammadi (2012, 2013) described the pillaring of acid activated (2 and 4M $H_2SO_4$ at 80–100°C) Mt. In the FTIR spectra of the acid-activated sample the position and the shape of the bands were slightly changed. The band at 1045 $cm^{-1}$ is shifted to higher wavenumbers ($\Delta\nu = 5$ $cm^{-1}$). The intensity of the $\delta$(OH) vibration at 840 $cm^{-1}$ (AlMgOH) and 875 $cm^{-1}$ ($AlFe^{3+}OH$) was significantly decreased. This was related to a decrease in $Mg^{2+}$ and $Fe^{3+}$ content in the octahedral sheet. In the pillared Mt, the band at 3438 $cm^{-1}$ was broadened, indicating the introduction of the OH groups of the pillars and was interpreted as the effect of pillaring. The increase in the broadness and intensity of the band at 3438 $cm^{-1}$ was related to the presence of $Al_{13}$ oxocations in the samples that increased the positive surface charge on oxygen involved in Si–O–Si linkages. The introduction of more positively charged aluminium groups causes a change in the electric field around Si groups and consequently a change in the symmetry of the tetrahedral sheet surface Si–O–Si vibration could happen and the related stretching vibration band is shifted from 1050 $cm^{-1}$ to a higher position at 1060 $cm^{-1}$.

## 12.4 MIXED $(Al–METAL)_{13}$ PILC

### 12.4.1 $(Al–Fe)_{13}$–PILC

Vaughan (1987) has indicated that one or more of $Fe^{3+}$ ions may substitute the $Al^{3+}$ in the $Al_{13}$. However, $AlO_2^-$ and $Al^{3+}$ ions may occur in the pillaring solution with $Fe/Al \geq 0.5$. This is further supported by the fact that $Fe_{13}$ sulfate can be precipitated. IR spectra for the PILC in the $\nu$(OH) region exhibited a prominent OH band centered between 3629 and 3635 $cm^{-1}$. After calcination at 400°C, a weaker shoulder near 3737 $cm^{-1}$ was visible in all the samples. This band is probably due to silanol groups formed in the tetrahedral sheet upon acidification. After calcination at 500°C, the spectra for the PILC in the $\nu$(OH) region depended on the Fe:Al ratio. Their intensity decreased strongly with increasing Fe:Al ratio. The changes appeared to correspond to the surface acidity changes with the Fe: Al ratio and hence related to the surface OH of the PILC. Several bands around 3635, 3650 and 3709 $cm^{-1}$ were observed in the spectra for hydroxy-Fe-Al PILC products with Fe:Al $\geq$ 0.5, but were barely observed for hydroxy-Fe PILC after calcination above 400°C (Zhao et al., 1993). In the MIR and FIR spectra of the ion-exchanged or pillared Mt, bands at 465 $cm^{-1}$, typical for Fe–$OH_{Oh}$, were seen for the ion-exchanged and the Fe-pillared samples. These bands were not observed for Al–Fe-pillared Mt. Here, the bands of Mt and bands typical of $(Al–O)_{Td}$ (729$cm^{-1}$), $(Al–O)_{Oh}$ (627 $cm^{-1}$), $(Al–O)_{Oh}$ (546 $cm^{-1}$) and $(Al–O)_{Oh}$ (492 $cm^{-1}$) were found. Controlled hydrolysis of Fe together with Al can only be accomplished if it is slow and the resulting species survive ageing without precipitation. The hydrolysis was fast when Fe(III) salt was applied and ageing caused the precipitation of hydrous iron(III) oxide mainly on the outer surface of Mt. Even the intercalated species migrated onto the surface when calcined. Thus slow hydrolysis is essential for optimizing the probability of isomorphous substitution of Fe for Al in preparing Al–Fe–PILC. It is known, however, that in the presence of $Al^{3+}$, $Fe^{3+}$ salts undergo hydrolysis instantly forming hydrous iron oxide with a highly ordered structure (Singh and Kodama, 1994). The rate of hydrolysis can be slowed down using a $Fe^{2+}$ salt, which in acidic solution is slowly oxidized by air to $Fe^{3+}$. Then the formation rate of $Fe^{3+}$ is sufficiently low to allow $Fe^{3+}$ incorporation without hydrolysis. However, MIR and FIR spectroscopic measurements on the Keggin salts (Kiricsi et al., 1995) and the intercalated materials, and the facile reducibility of $Fe^{3+}$ in Al–Fe–PMt revealed by $^{57}Fe$ Mössbauer spectroscopy all indicated that, notwithstanding earlier statements obtained from indirect measurements (e.g. Vaughan, 1987; Bergaya et al., 1993; Zhao et al., 1993), isomorphous substitution of Fe for Al did not occur here. Instead, separate phases of polynuclear hydrous $Fe^{2+}$, $Fe^{3+}$ (oxy)hydroxides were formed in the solution and were intercalated together with the $Al_{13}$ Keggin ion (Palinko et al., 1996). Caglar et al. (2015) reported only minor shifts and intensity changes in the IR bands associated with the Mt Si–O and Al–O vibrations indicating that the pillaring did affect the clay mineral structure but no new bonds were formed. Even the addition of $Ni^{2+}$ to the Fe/Al system did not result in any major changes in the IR spectrum (Gao et al., 2014).

### 12.4.2 $(Al–Cr)_{13}$–PILC

Kiricsi et al. (1997) reported that the IR spectra of the Al and Al–Cr-pillared Mt as well as the ion-exchanged Mt were basically identical. According to Tomul and Balci (2009) the FTIR spectra of the Mt band at 1045 $cm^{-1}$ characteristic of the Si–O–Si stretching vibration remained unchanged in the PILC. The intensity of $\delta$(SiO) and $\delta$(AlO) modes occurring at 535–580 $cm^{-1}$ decreased by pillaring. The IR OH stretching modes at 3628 $cm^{-1}$ and

3450 $cm^{-1}$ were attributed to structural OH groups in the Mt layer and water molecules in the interlayer space. The location and the intensity of both bands were preserved during pillaring. A small increase in the intensity of water bending vibration caused by surface hydroxyl groups at 1635 $cm^{-1}$ occurred by intercalation and this increase became dominant for mixed PILC, compared to Al–PILC and Cr–PILC. In contrast, De León et al. (2014) observed the bands at 910 and 842 $cm^{-1}$ assigned to the bending vibrations of Al–OH and Al/Mg–OH groups, respectively, (which were also observed in the PILC samples) especially for those containing chromium, shifted to higher wavenumber values of 934 and 877 $cm^{-1}$. These authors thought that this fact suggested some interaction of the pillaring species with Al–OH and Al/Mg–OH groups of the octahedral sheet of the host clay mineral, but strangely enough had no effect whatsoever on the tetrahedral sheet. Contrary to Vaughan (1987), $^{27}$Al MAS NMR and FTIR spectra on the polyhydroxy-ion salts as well as the pillared structures convincingly proved that isomorphous substitution of Cr for Al did not occur, but rather co-hydrolysis and co-pillaring of the Mt. Substitution of the tetrahedral Al should not even be anticipated as $Cr^{3+}$ strongly prefers octahedral to tetrahedral coordination. The lack of substitution in any site may be explained on one hand, by the widely different conditions needed for the partial hydrolysis of the two salts. Whereas $AlCl_3$ polymerizes around pH = 6, $CrCl_3$ precipitates as a hydroxide at this pH. On the other hand, there is a significant difference in size between the two cations and the Keggin unit is too small to release the strain, which would arise from the substitution. Upon calcination, the loss of outer-sphere water and dehydroxylation at higher temperatures caused the formation of alumina pillars decorated with chromia species.

### 12.4.3 $(Al–Zr)_{13}$–PILC

The structural changes after modification of $Na^+$-Mt by Al–Zr-intercalating solution by FTIR showed a band at 1041 $cm^{-1}$ assigned to the stretching vibrations of silica-oxygen tetrahedrons $\nu_{as}$(SiOSi) and band at 468 $cm^{-1}$ assigned to the bending $\delta$(Si–O) and stretching $\nu$(M–O) vibrations shifted to low wavenumber with increasing of Zr content in Al–Zr Mt. The intensities of the bands at 842 $cm^{-1}$ ($\nu_s$(OSiO)) and 628 $cm^{-1}$ ($\delta$(SiO)) also decreased with increasing Zr content in Al–Zr Mt. Moreover, the increase in Zr content led to the shift of the band position for the Si–O–Al mode (octahedral Al, 520–535 $cm^{-1}$) and $Al_2OH$ mode (914–924 $cm^{-1}$). These changes can be an indirect evidence of the incorporation of Zr ions into the pillar structure of the Mt. $^{27}$Al NMR(MAS) data were in accordance with these results. The insertion of Zr into the $Al_2O_3$ structure of Al–Zr PILC resulted in the shift of band from 258 to 266 nm. The bands in the 205–212 nm and 270–400 nm regions were observed in all the studied samples.

### 12.4.4 (Al–Co)-PILC

Bertella and Pergher (2015) reported on the preparation of Al–Co-pillared Mt. Only for low Co to Al ratios were stable pillared products obtained with basal spacings in the range of 18.4–19.3 Å, $Al_{13}$ PILC give smaller basal spacings between 17 and 18 Å. It was therefore believed that $Al_{90}Co_{10}$- and $Al_{75}Co_{25}$-PILC could be explained by the presence of mixed Al–Co pillars. These structures were thought to be possible due to the similar sizes of the cations. These samples exhibited the same IR bands related to structural features as the natural clay mineral indicating that the insertion of Al–Co pillars did not significantly impact the clay mineral structure.

### 12.4.5 (Al–REE)-PILC

Though several papers have reported the combination of rare earth elements (REE) with Al only a few have published IR results (Flego et al., 1998; Valverde et al., 2000; Zuo et al., 2011). McCauley (1988) described the preparation of PILC with large basal spacings of ~28 Å, followed by other publications (Sterte, 1991a,b; Booij et al., 1996a,b; Kloprogge et al., 2001b; Valverde et al., 2001). These authors used hydrothermally treated or refluxed solutions containing the Al polyoxocation and REE metals in the form of their chloride salts. The presence of Ce or La seems to promote polymerization of the Al polyoxocation. Until now, however, nobody has been able to determine the structure of the newly formed pillaring molecules. Although the interlayer spacing is about twice that of the initial Al-PILC in some cases, chemical analyses indicate smaller amounts of $Ce^{4+}$ and $La^{3+}$ than expected from polymerization of the Al polymer alone. Several other studies on the combination of Al pillars with $Ce^{4+}$ or $La^{3+}$ resulted in the formation of PILC with basal spacings characteristic of Al-PILC. However, the catalytic activity was enhanced owing to Ce or La (e.g. Gonzalez et al., 1992; Trillo et al., 1993). Zuo et al. (2011) reported the spectra of Al- and Al/REE-PILC were very similar and only an increase of the 3440 $cm^{-1}$/3640 $cm^{-1}$ ratio for the Al/REE-PILC was observed. The calcination at 500°C of PILC modified with either Ce or La caused an intensity reduction of the main bands, especially in the stretching mode of the hydroxyl groups at 3627 $cm^{-1}$ (Valverde et al., 2000). Also, there was an intensity reduction of Al–OH–Al and Al–OH–Mg vibrational modes at 910 and 842 $cm^{-1}$, respectively, accompanied by a broadening of the band around 1047 $cm^{-1}$. However, when PILC modified with $Ce^{4+}$ or $La^{3+}$ were calcined at 800°C, no bands at 910, 842, 523 and 462 $cm^{-1}$ were present. The resulting materials behaved similarly to Al-PILC that was calcined at 500°C. Comparable results were published by Flego et al. (1998) for Al/Ce- and Al/Ce/Mg-PILC.

## 12.5 Ti PILC AND MIXED (Ti-METAL) PILC

### 12.5.1 Ti-PMt

Ti-PILC have been of interest for some time, not only because of their relatively high basal spacing compared to other metal-PILC, but also for the photocatalytic activity of the anatase $TiO_2$ phase, which can form as the potential pillar inside the interlayer space. To understand the changes upon exchange and calcination of the Ti-PILC the MIR provides only limited information. In most cases intercalation is observed to increase the intensities of the bands in the $\nu$(OH) and $\delta$(OH) regions. Calcination results in a strong decrease in intensity, while bands in the low wavenumber region associated with different Si–O and Al–O vibrations may show some minor shifts as a result of the interaction with the pillars (Xu et al., 2004; Binitha and Sugunan, 2006; Nikolopoulou et al., 2009; Okoye and Obi, 2011a; Okoye et al., 2012; Zhao et al., 2012; Tuo et al., 2014). Some papers have indicated the formation of anatase by XRD (Yang et al., 2008; Bineesh et al., 2010a; Tahir and Amin, 2013) but in most of these cases there is no direct evidence that it is also present in the interlayer space as a clear 001 reflection of the clay mineral was either not observed at all, or was extremely broad and weak. It is therefore likely that the anatase is actually situated on the outside surface of the clay mineral particles instead as shown by SEM and TEM ( Yang et al., 2008; Tahir and Amin, 2013). Sun et al. (2006) reported on the FIR spectra of Ti-PILC. The incorrectly assigned bands at 526 and 468 $cm^{-1}$ instead correspond to $\delta$(AlOS)i and $\delta$(SiOS)i modes in Mt. $TiO_2$ sol dried at 80°C showed two bands at 461 and 318 $cm^{-1}$, which were characteristic of stretching vibrations of Ti-O. However, the band at 318 $cm^{-1}$ blue-shifted to 325 $cm^{-1}$ in microwave calcined Ti PMt and to 338 $cm^{-1}$ in normal calcined Ti PMt. This blue-shift may be the result from quantum size effects generally arising for particle smaller than 10 nm. Yang et al. (2008) showed the Raman spectra in the low wavenumber region below 700 $cm^{-1}$, which was characterized mainly by the vibrations of anatase $TiO_2$. The main band around 142 $cm^{-1}$ is the $E_g$ mode of Ti–O. The blue-shift and the asymmetrically broadening of the peaks at 142 and 630 $cm^{-1}$ are both associated with the same quantum effects as observed in the FIR.

### 12.5.2 Mixed (Ti-Metal) PILC

Bahranowski et al. (2015) prepared mixed Ti–Zr PILC with a basal spacing of 21.6 Å. In all intercalated clay minerals the pillar component remained amorphous after calcination at 400°C. In order to get more insight into the nature of mixed Ti–Zr pillars, the calcined Ti–Zr PILC was subjected to additional calcination at 800°C and 1000°C, in an attempt to trigger crystallization of pillars. Crystallization of pillars in the Ti–Zr PILC sample led to a new broad reflection, which appeared after heating at 800°C, attributed to the most intense peak of an orthorhombic $ZrTiO_4$ phase ($d$=2.93 Å), especially since further heating at 1000°C leaves no doubt that the phase nucleating in Ti–Zr PILC is indeed $ZrTiO_4$. The spectrum of Ti–Zr PILC was much less intense than the corresponding Ti PILC, nevertheless a clear set of Raman bands was observed. The bands positions showed that all bands could be assigned to the vibrational modes of zirconium titanates: $ZrTiO_4$ or related $Zr_5Ti_7O_{24}$ structure. The result provides a direct evidence that pillars in Ti–Zr PILC are indeed Ti–Zr mixed oxide nanoclusters.

A number of publications have described the introduction of other metals to Ti PILC such as Cu, Ag, Fe, Y, Ce and Mn, but the IR spectra did not show any changes when compared to the Ti PILC without the introduction of these metals (e.g. Valverde et al., 2005; Liu et al., 2007; Shen et al., 2012; Tomul et al., 2016; Han et al., 2015).

## 12.5.3 Impregnated Ti-Metal-PILC

### *12.5.3.1 Palladium (Pd)- and Chromium (Cr)-Impregnated Ti–Zr PILC*

Michalik-Zym et al. (2015) prepared Ti–Zr PILC supports for Cr and Pd catalyst by wet impregnation of the calcined PILC with the appropriate amounts of the nitrate or chloride solutions. The impregnation did not change the basal spacing of the Ti–Zr PILC. The Raman spectra of Cr-doped catalysts exhibited a band at 865–869 $cm^{-1}$ (Fig. 12.5). The appearance of a band around 865 $cm^{-1}$ is characteristic of hydrated monochromates (Weckhuysen et al., 1996 and references therein), showing that such species dominate at the surface of the Ti–Zr PILC impregnated with 1 wt.% of chromium. The asymmetry of this band was probably caused by slightly different locations of chromate species within the PILC structure, while the increased background intensity in the range 900–1100 $cm^{-1}$ might point to the contribution from vibrational modes of polymeric chromate species.

In the case of palladium ($Pd^{2+}$) supported on Ti–Zr PILC, a narrow band at 644 $cm^{-1}$ (Fig. 12.6) may be attributed to the Raman active $B_{1g}$ phonon mode of PdO (McBride et al., 1991). In nanocrystals of PdO the $B_{1g}$ mode becomes red-shifted with respect to bulk PdO where it is observed at 651 $cm^{-1}$ (Bardhan et al., 2013). The effect is due to confinement of phonons within nanocrystal grain boundaries, and its magnitude increases with the decreasing size of PdO nanocrystals. Thus, the red-shift of the PdO $B_{1g}$ bands with respect to the value characteristic of bulk PdO confirmed the nanocrystalline character of PdO species supported on PILC carriers.

### *12.5.3.2 Vanadium-Impregnated Ti PILC*

Chae et al. (2004) and Bineesh et al. (2011) described the liquid phase impregnation of vanadium pentoxide ($V_2O_5$) in Ti PILC. According to

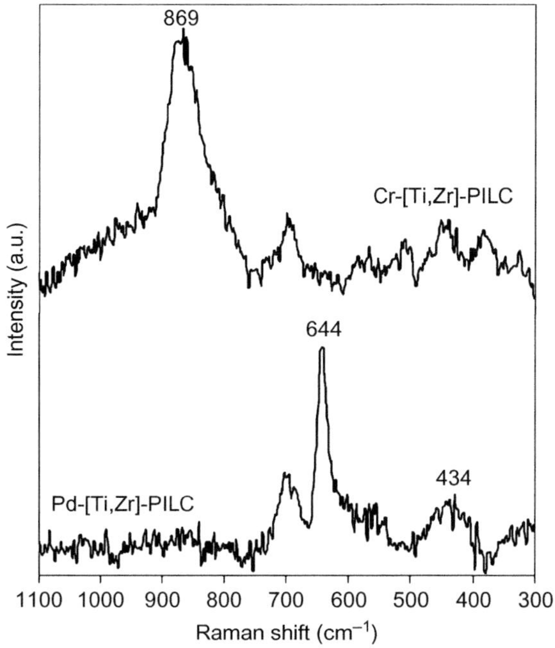


**FIG. 12.5** Raman spectra of Cr-[Ti-Zr]-PILC, and Pd-Zr PILC. Reproduced with permission from Michalik-Zym et al. (2015).

Bineesh et al. (2011) after doping the Ti PILC with $V_2O_5$, the FTIR band positions remained nearly the same. However, at a high $V_2O_5$ loading of 10 wt.%, a new band was observed at around 820 $cm^{-1}$, which was assigned to the $\nu$(VO) vibrations arising from the formation of crystalline $V_2O_5$ species (Reddy et al., 1999; Shylesh and Singh, 2004; Bineesh et al., 2010b). Chae et al. (2004) used Raman spectroscopy as the main technique to probe the structure of $V_2O_5$ (Fig. 12.6A). The narrow band at 1040–1020 $cm^{-1}$ was assigned to a $\nu$(V=O) mode of the isolated mono-vanadate species. The broad bands in the region of 1000–700 $cm^{-1}$ can be ascribed to $\nu$(V=O) (1000–800 $cm^{-1}$) and $\nu$(V–O–V) (800–600 $cm^{-1}$) modes of poly-vanadate species. Bulk crystalline vanadate was identified by the bands around 283, 305, 405, 485, 527, 700 and 997 $cm^{-1}$. The strongest band for the formation of crystalline $V_2O_5$ at 995 $cm^{-1}$ was due to the symmetric stretching of V=O groups existing in a form of bulk $V_2O_5$ on the Ti PILC surface. $V_2O_5$ impregnated on Ti PILC mainly contained poly-vanadate, although a weak isolated $V_2O_5$ band was detected at 1024 $cm^{-1}$. The shift in the position of the band for the surface $V_2O_5$ species was mainly attributed to the slight alteration of the length of the terminal V=O bond, approximately 0.02Å for the surface $O{=}V(–O–S)_3$ species, where S represents the cation in the support. The

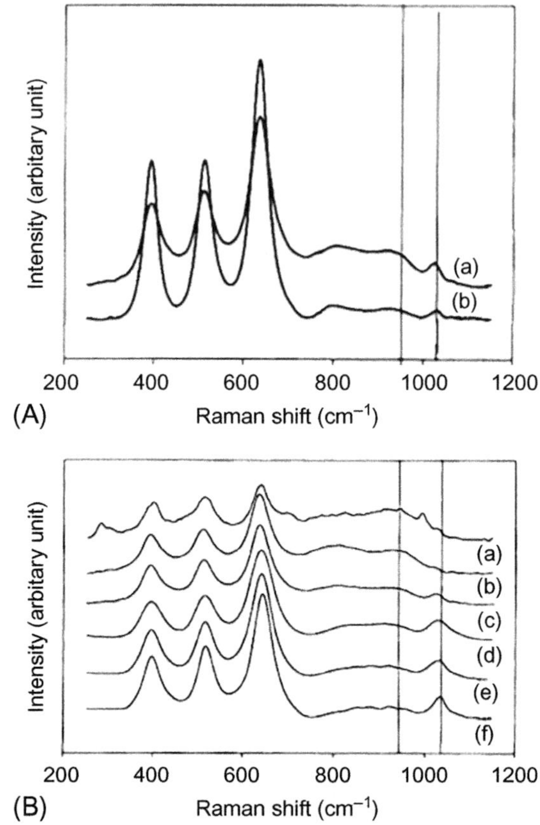


**FIG. 12.6** (A) Raman spectra of supported vanadia catalysts with similar $VO_x$ surface density with respect to the type of support: (a) $10VO_x$/Ti-PILC4 and (b) $3VO_x/TiO_2$. Reproduced with permission from Chae et al. (2004). (B) Raman spectra of $VO_x$/Ti-PILC catalysts with respect to the content of vanadia: (a) $20VO_x$/Ti-PILC4, (b) $15VO_x$/Ti-PILC4, (c) $10VO_x$/Ti-PILC4, (d) $5VO_x$/Ti-PILC4, (e) $3VO_x$/Ti-PILC4, (f) $2VO_x$/Ti-PILC4. Reproduced with permission from Chae et al. (2004).

wavenumber of the V=O bond for isolated $V_2O_5$ species shifted from 1040 to 1020 cm$^{-1}$ with respect to the Ti-PILC support. It suggests that the V–O bond on the support became stronger as the terminal bond strength of V=O became weaker. In addition, the maximum intensity of the broad band for poly-vanadate shifted to higher wavenumber, containing no polymeric vanadate. It revealed the increase of the formation of V–O–V species on the Ti PILC surface compared to V=O bonds.

The Raman spectra of the dehydrated vanadia catalysts supported on Ti PILC with respect to the content and the surface density of $VO_x$ can be observed in Fig. 12.6B. The strong bands at 400, 515 and 640 cm$^{-1}$ mainly represent the anatase type of $TiO_2$. The intensity of these bands decreased with $VO_x$ loading due to the absorption of the Raman laser light by the development of a yellow-orange-colored surface as the loading of $V_2O_5$ increased on the Ti PILC surface. As the content of the surface $V_2O_5$ increased up to the monolayer coverage of $V_2O_5$, the position of the broad bands indicating the existence of poly-vanadate shifted toward higher wavenumbers, while the sharp band for the isolated surface $V_2O_5$ species slightly shifted toward lower wavenumbers. The ratio of the polymerized vanadate species to the isolated one increased with respect to the loading of $V_2O_5$ up to the monolayer coverage of the Ti-PILC surface. In addition, the formation of polymerized vanadate seems to be more favorable on the surface of Ti-PILC than that of $TiO_2$ at similar surface density of $VO_x$. At high surface density beyond the monolayer coverage, the Ti PILC contained the microcrystalline structure of $V_2O_5$ on the surfaces, which was observed as a distinctive phase from the monomeric and polymeric vanadates revealing a unique Raman band around 995 $cm^{-1}$. The formation of crystalline $V_2O_5$ was observed at the similar surface density of $VO_x$ for both $VO_x$-$TiO_2$ and $VO_x$-Ti-PILC. However, at a given $VO_x$ surface density, the amount of highly dispersed $V_2O_5$ on Ti PILC may be three to four times greater than that on $TiO_2$ due to higher surface area of Ti PILC than that of $TiO_2$. Chae et al. (2004) observed that $V_2O_5$ species on bentonite exist in a crystalline phase as indicated by a moderately intense and narrow Raman band at 950 $cm^{-1}$. This is due to the weak interaction between $V_2O_5$ and bentonite.

## 12.6 Fe-PILC AND MIXED (Fe-METAL) PILC

### 12.6.1 Fe-PILC

In catalysis there is a great interest in $Fe_2O_3$ PILC in particular as a Fischer-Tropsch catalyst. This type of PILC can be synthesized in two ways: (1) the intercalation of hydrolyzed $Fe^{3+}$ salts and (2) the intercalation of the acetate complex. Hydrolyzed Fe salts give complex species of different size and shape. Maes et al. (1994) and Maes and Vansant (1995) have described the MIR and FIR spectra of a $Fe^{3+}$-PILC prepared via the acetate complex route. The IR spectrum of the complex reflected four COO vibrations, two of them caused by bridging acetato groups (1590 and 1448 $cm^{-1}$; $\Delta = 142$ $cm^{-1}$) and two others attributed to mono-dentate acetato groups (1685 and 1294 $cm^{-1}$; $\Delta = 391$ $cm^{-1}$). The FIR region gave information about the skeletal vibrations of the complex, especially the $M_3O$ and $MO_4$ groups (Table 12.2). The presence of a strong vibration at 229 $cm^{-1}$ showed clearly the presence of a triangular $M_3O$ unit. The hydrolysis was monitored under two conditions: in pure water and in aqueous clay mineral dispersion. Both resulted in the disappearance of the mono-dentate acetyl group and the formation of acetic acid (typical bands at 1712, 1284 and 1409 $cm^{-1}$). Hydrolysis in water left the bridged coordination unchanged (1589 and 1444 $cm^{-1}$); however, in the presence of

**TABLE 12.2 Position of the Most Interesting Mid- and Far-IR Vibration Bands of the Tri-Nuclear Fe(III)-Acetato Complex (Maes and Vansant, 1995; Maes et al., 1994)**

| Observed | Literature | Assignment |
|---|---|---|
| 1685 | | $\nu COO_{asym}$, monodentate |
| 1590 | 1585 | $\nu COO_{asym}$, bridged |
| 1448(vs)/1421(sh) | 1444(vs)/1420(sh) | $\nu COO_{sym}$, bridged |
| 1384 | 1384 | $NO_3$ |
| 1346 | 1348 | $\delta CH$ |
| 1294 | – | $\nu COO_{sym}$, monodentate |
| 1049 | 1044 (sh) | $\delta CH_3$ |
| 1037 | 1030(m) | ? |
| 950 | 948 | 7C–C |
| 676(sh)/661 (s) | 663 (sh)/657 (s) | $\delta OCO$ |
| 615 | 613 | $\pi COO$ |
| 526 | 523 | $\rho_{\Gamma} COO$ |
| $M_3O$ Unit | | |
| 607 | 609 | $\nu_{asym}$ $M_3O$ |
| 229 | 234 | $\delta_{sym}$ |
| $MO^4$ unit | | |
| 370 | 368 | $\nu_d$ (MO) |
| 304 | 302 | $\nu_s$ (MO) |
| 202 | 200 | $\delta_d$ (OMO) |
| 150 | 146 | $\pi$ |

clay mineral the intensity of the 1589 $cm^{-1}$ band decreased, while a new band appeared at 1542 $cm^{-1}$, leading to a lower Δ value of 98 $cm^{-1}$. The formation of acetic acid was promoted by the clay mineral, showing a much faster hydrolysis. The low Δ value of 98 $cm^{-1}$ could indicate the formation of a bidentate acetyl configuration; however, only the asymmetric COO vibration was influenced. If there is a change in configuration, the symmetric COO vibration should also change, which is not the case. So, the observed shift must be due to the perturbation of the asymmetric COO vibration probably caused by the ion-exchange process of the clay mineral. After washing and drying the intercalate, two COO vibration bands (at 1533 and 1440 $cm^{-1}$) are observed at the same wavenumbers as those found when the complex was hydrolyzed in dispersion. The $Fe^{3+}$-acetyl configuration became perturbed during the ion-exchange process. No traces of COO vibration bands at a higher wavenumber, indicating a non-exchanged $Fe^{3+}$-acetato, were observed, meaning that all excess complex was removed during the washing step. In the FIR no traces of the typical triangular structure were observed. Further identification of the $Fe^{3+}$-oxide pillars after calcination by IR and XRD failed. No $Fe^{3+}$-O vibrations specific for a certain $Fe^{3+}$-oxide phase were detected.

Chen et al. (2009) prepared Fe-PILC through the hydrolysis of $Fe^{3+}$-nitrate with sodium carbonate. Comparing the FTIR spectrum of pure Mt with the spectrum of $Fe^{3+}$-exchanged Mt and dried at 80°C, two new bands emerged at 1384 $cm^{-1}$ and 699 $cm^{-1}$. The former was the $NO_3^-$ stretching mode, which showed that there were excess positively-charged $Fe^{3+}$ aggregates outside the interlayer space of $Fe^{3+}$-exchanged Mt, while the latter was a bending vibration of FeOOH, which also suggested that hydrolyzed $Fe^{3+}$ intercalated into interlayers prior to pillaring. No spectra were provided of the calcined product. Very similar results were reported by Hou et al. (2011).

## 12.6.2 Mixed (Fe-Metal) PILC and Modified Fe-PILC

Organo–inorgano modified clay minerals (Zhu et al., 2007b) by surfactant were reported to exhibit much larger SSA, adsorptive capacity and thermal stability than organoclay or inorganic clay minerals. Lin et al. (2001) and Wang and Pinnavaia (2003) reported polyetheramine D2000 could expand the interlamellar space of Mt by ca. 9 nm. Yu and Yang (2010) used polyetheramine D2000 (O,O′-Bis(2-aminopropyl)polypropyleneglycol) as a structural regulator to modify the intercalated $Fe^{3+}$-Mt, and treated it with acetic acid vapor prior to calcination to improve the magnetization of the resulting $Fe^{3+}$ PILC. The *d*-value of the 001 reflection of $Fe^{3+}$-exchanged Mt increased from 1.26 nm to 1.42 nm due to the intercalation of poly-hydrated $Fe^{3+}$. After addition of polyetheramine D2000 (abbreviated D), the $d_{001}$ value of Fe-D Mt increased to 1.79 nm. There was no obvious change in $d_{001}$ after it was exposed to acetic acid vapor. Calcination at 400°C, Fe-D-$A_CA$ Mt was transformed to Fe Mt. But, the $d_{001}$ of the product collapsed to 1.22 nm similar to the $d_{001}$ (1.26 nm) of the original Mt, and the corresponding reflection became smaller and broader compared to the original Mt. Therefore, the magnetic species did not exist in the interlayer space of the Mt. There were few differences between the IR spectra of Mt and $Fe^{3+}$-Mt, which implied that the final modified clay mineral kept its original structural configuration after intercalation. However, $Fe^{3+}$-$A_CA$ Mt gave out two distinct bands at 1726 and 1451 $cm^{-1}$ assigned to $\nu_{as}(COO^-)$ and $\nu_s(COO^-)$ of the acetate anion. The intercalated Mt was considered to have transformed to $Fe^{3+}$-$A_CA$ Mt due to the interaction of $Fe^{3+}$ within the interlayer space of Mt with the $A_CA$ vapor. Additionally, the stretching vibrations of the combined water molecules from 3620 to 3630 $cm^{-1}$ became weaker due to water loss after polyetheramine D2000 was introduced. The symmetric and asymmetric stretching vibrations of C–H bonds appear in the region 2857–2985 $cm^{-1}$ in the modified Mt. The Raman active deformation vibration of $NH_2$ at 1578 $cm^{-1}$ and those of C–O–C at 936 and 1100 $cm^{-1}$ indicated that polyetheramine D2000 was introduced into the modified Mt.

Li and Wu (2010) prepared $Fe^{3+}$-PILC in combination with sodium dodecyl nitrate and studied the effect on the order of addition of the organic molecule and hydrolyzed $Fe^{3+}$ on the final product. The modified Mt samples exhibited broad bands near 3434 $cm^{-1}$. A band around 1385 $cm^{-1}$, attributed to the $NO_3$ stretching mode, was observed in the FTIR spectra of modified clay minerals. Shoulders characteristic of symmetric stretching of the $CH_3$ groups on the alkyl chains at 2958 $cm^{-1}$ and bands due to the $CH_2$ groups of the alkyl chains at 2922 $cm^{-1}$ and 2852 $cm^{-1}$ (antisymmetric and symmetric stretching), 1468 $cm^{-1}$ (bending) were present for the modified Mt. All of these demonstrated the existence of -$CH_n$ and can be evidence for binding of the polar segment of the surfactant. The results of FTIR reveal that the organic ion has been intercalated into the clay mineral layers but no differences were found between the different preparation methods (hydrolyzed Fe first, sodium dodecyl nitrate first or both simultaneous).

## 12.7 Si-PILC AND DERIVED MATERIALS

### 12.7.1 Hybrid Mesostructured Si-PILC

A new class of organic–inorganic hybrid mesostructured silica-PILC was prepared by Mao et al. (2011) through the surfactant directed assembly of organosilica in the interlayer spaces of Mt. The surfactant templates were removed from the pores by solvent-extraction of the calcined products. All products had *d*-values corresponding to the basal spacing from approximately 4.2 to 4.8 nm. Evidence for the removal of the templates was observed on calcination and extraction of Si-PILC and hybrid Si PILC precursors. While a band at 1630 $cm^{-1}$, associated with the OH bending mode of water, and a sharp band at 1050 $cm^{-1}$, associated with $\nu_{as}$(Si–O) modes of the Mt layers, the synthesized organic intercalated Mt showed distinct bands in the 2930–2856 $cm^{-1}$ region and a band at 1380 $cm^{-1}$, assigned to the different C-H deformation vibrations of the templates. Additionally, the bands at 1596 $cm^{-1}$, 1489 $cm^{-1}$ and 1431 $cm^{-1}$ suggested the presence of a phenyl group. After extraction, these bands almost disappeared indicating that most of the templates were successfully removed. The shoulder at 1134 $cm^{-1}$ was associated by the authors with the Si–C vibration of the sample.

### 12.7.2 Phospho-Tungstate Functionalized Si-PILC

In a follow up paper Mao et al. (2015) used the pore-expanded Si-PILC as the catalytic supports for phospho (P)–tungstate (W) based $H_3PW_{12}O_{40}$ (termed PW herein) loading as high as 26.9%, 35.8% and 48.2%. For pure PW, the Keggin unit has four main IR bands at 1080 $cm^{-1}$ ($P–O_a$), 980 $cm^{-1}$ ($W–O_d$), 890 $cm^{-1}$ ($W–O_b–W$) and 800 $cm^{-1}$ ($W–O_c–W$). The strong, broad bands around 1200–1000 $cm^{-1}$ were associated with the asymmetric Si–O stretching modes of the Mt layers and the intercalated silica and the spectrum of Si-PILC also showed a band at 803 $cm^{-1}$, which was attributed to the symmetric stretching frequency of Si–O–Si. In FTIR spectrum of PW-Si-PILC, the peak was shifted to 816 $cm^{-1}$, which was interpreted to be due to the chemical interactions between PW and Si-PILC. As reported by Hu et al. (2007), the band at 816 $cm^{-1}$ was attributed to the formation of Si–O–W bonds. Moreover, with the increase of PW loading, a band at 980 $cm^{-1}$ also increased slowly. These results indicated that the Keggin unit of PW had been loaded into the Si-PILC porous channels. Li et al. (2012) reported similar incorporation of this Keggin structure in Si PILC. The FTIR spectra of the Si-PILC, pure PW, and PW–Si-PILC samples in the region of 700–1350 $cm^{-1}$ are shown in Fig. 12.7A. The spectrum of PW (spectrum e, Fig. 12.7A) showed a broadened band at 802 $cm^{-1}$, which was assigned to $\nu_{as}(W\text{-}O_c\text{-}W)$ in edge shared octahedral (Rocchiccioli-Deltcheff et al., 1983). In the spectrum of Si-PILC sample (spectrum a, Fig. 12.7A), the band at 795 $cm^{-1}$ was attributed to the symmetric stretching frequency of Si–O–Si. In the FTIR spectra of the PW-Si-PILC samples (spectra b-d, Fig. 12.7A), the band at 816 $cm^{-1}$ was attributed to the formation of Si–O–W bonds (Hu et al., 2007), which was due to the chemical interactions between the *in situ* formed PW and Si-PILC. With the increase of the P and W content, more PW was in situ formed in the interlayer of the PW-Si PILC samples, and the peak at 816 $cm^{-1}$ was evidently enhanced. Therefore, the phenomenon led to the conclusion that PW has been in situ formed in the mesoporous silica frameworks and formed new Si–O–W bonds within the silica wall of the Si-PILC material. The characteristic bands of Keggin-type PW were also observed at 1080, 980 and 890 $cm^{-1}$, which were usually assigned to $\nu_{as}(P–O)$, $\nu_{as}(W–O)$ and $\nu_{as}(W–O_b–W)$ in corner shared octahedra (Rocchiccioli-Deltcheff et al., 1983), respectively. However, in the FTIR spectra of the PW-Si-PILC samples, the band at 1080 $cm^{-1}$ moved to a higher wavenumber. They attributed this to possible interactions of the $Na_2HPO_4$ and the silica wall of the Si PILC during the formation of PW but ignored the effects of overlapping bands with the Si-O stretching modes that occur here as well. In addition, the bands of the 25% PW-Si PILC sample were evidently strengthened compared to that of Si PILC sample, which further indicated the formation of PW and the incorporation of PW into the Si PILC frameworks.

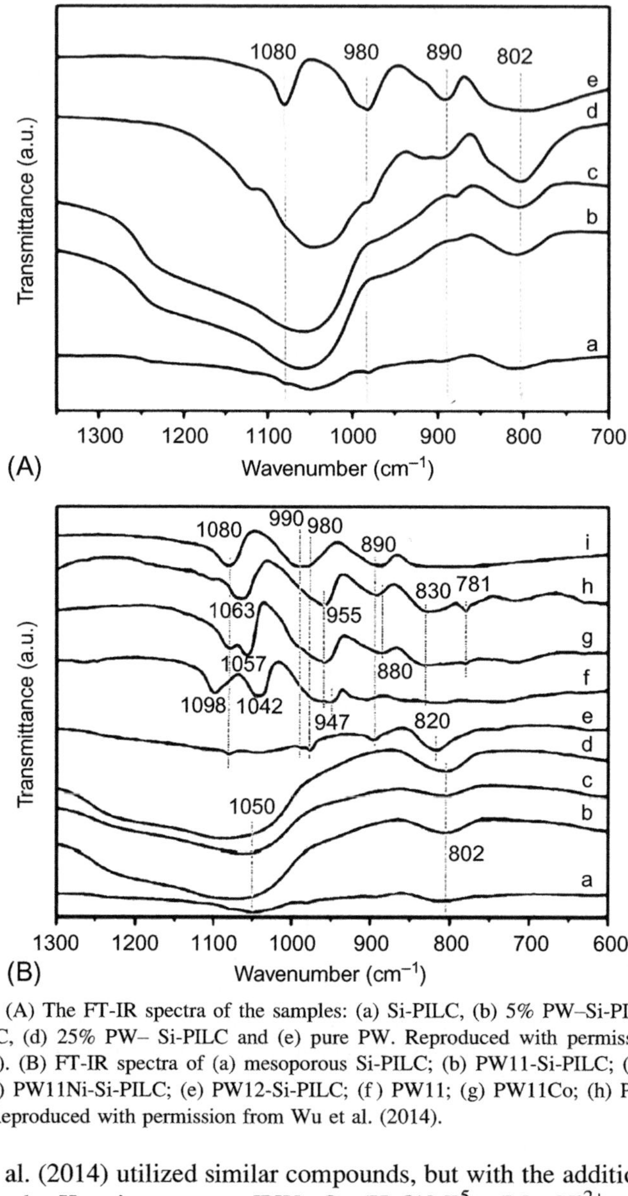


**FIG. 12.7** (A) The FT-IR spectra of the samples: (a) Si-PILC, (b) 5% PW–Si-PILC, (c) 15% PW–Si-PILC, (d) 25% PW– Si-PILC and (e) pure PW. Reproduced with permission from Li et al. (2012). (B) FT-IR spectra of (a) mesoporous Si-PILC; (b) PW11-Si-PILC; (c) PW11Co-Si-PILC; (d) PW11Ni-Si-PILC; (e) PW12-Si-PILC; (f) PW11; (g) PW11Co; (h) PW11Ni; and (i) PW12. Reproduced with permission from Wu et al. (2014).

Wu et al. (2014) utilized similar compounds, but with the addition of $Co^{2+}$ or $Ni^{2+}$ in the Keggin structure $[PW_{11}O_{39}(H_2O)M]^{5-}$ ($M = Ni^{2+}$ or $Co^{2+}$) to prepare mesoporous Si PILC. Herein, two materials are compared, one with the PW complex (PW11), and one with the PW complex doped with either $Ni^{2+}$ or $Co^{2+}$ (PW11Ni or PW11Co). The broad bands around 802 and 1050 $cm^{-1}$ associated with PW11-Si PILC, PW11Ni-Si PILC and PW11Co-Si PILC (spectra b-d, Fig. 12.7B) overlapped with the broad Si–O–Si bands associated with the Mt support, but indicated increased intensity associated with formation of mesoporous Si PILC (spectrum a, Fig. 12.7B). Hu et al. (2007) attributed these bands to the contribution of $\nu_{as}$(W-$O_c$-W or W-$O_c$-M) in edge shared octahedra. The spectra of PW11-Si PILC and PW11M-Si PILC were very different from the spectrum of PW12-Si PILC (spectrum e, Fig. 12.7B), in which all characteristic bands of PW12 (spectrum i, Fig. 12.7B) and mesoporous Si PILC were present, consistent with higher symmetry of PW12 compared with PW11 (spectrum f, Fig. 12.10B) and its metal doped counterparts (spectra g and h, Fig. 12.7B). The band of 830 $cm^{-1}$ in the spectra of PW11 and PW11M shifted to lower wavenumber in the spectra of PW11-Si-PILC and PW11M-Si PILC, which was assigned to the formation of Si–O–W bonds between PW11 or PW11M and the mesoporous Si PILC in the interlayer space of mesoporous Si PILC (Hu et al., 2007), indicating that poly(oxo)tungstates had been encapsulated into the mesoporous Si PILC framework.

### 12.7.3 Titanium Functionalized Si PILC

Titanium-functionalized Si PILC has been synthesized through a post synthetic route by Mao et al. (2014a). The FTIR band at 960 $cm^{-1}$, which is often used as evidence for the vibration of Ti–O–Si confirming the silica-titania inorganic network (Haskouri et al., 2001), was initially rather weak and could not be observed in pure Si PILC. With the titanium functionalization, the intensity of 960 $cm^{-1}$ band increased with the increase of tetrabutyl titanate during the preparation process. The FTIR patterns confirmed that silica-titania inorganic network formed in the interlayer spaces and the Ti O Si content was strongly correlated with the added amount of organic Ti source during post processing.

### 12.7.4 Iron Functionalized Si PILC

A similar synthesis method was used by Mao et al. (2014b) prepare magnetic Si PILC. These samples with a basal spacing of about 3.85–4.21 nm were formed irrespective of synthetic conditions. TEM images of magnetic Si PILC showed (001) lattice fringes of magnetic Si PILC of 3.8–4.1 nm, which agree well with the XRD results. Almost all $Fe_3O_4$ domains were homogeneously dispersed on the clay mineral surfaces and the mesoporous Mt interlayer spaces, with a few larger particles aggregated together. $C_{18}$DMBAB molecules were rearranged accompanying the hydrolysis and polymerization of the intercalated TEOS. This was strongly supported by FTIR results. In comparison with $C_{18}$DMBAB-Mt, the IR bands of the C-H bonds of the methylene groups in uncalcined magnetic Si-PILC clearly shifted to higher wavenumbers, indicative of the formation of a liquid-crystal $C_{18}$DMBAB micelle. In a first step, a viscous gel mixture of Mt, surfactant, TEOS, $Fe^{2+}$, $Fe^{3+}$

and PEG-400 allowed simultaneous intercalation of these compounds in the interlayer space of the Mt. In a second step, the addition of ammonia solution to the gels led to the rapid hydrolysis of TEOS in the interlayer spaces, resulting in Si PILC. Moreover, the presence of PEG-400 would decrease the crystal growth rate of $Fe_3O_4$, resulting in highly dispersed magnetic nanoparticles. With increasing iron oxides content, a slight red shift was observed for the stretching vibration of the tetrahedral $SiO_4$ and $FeO_4$ unit (ca. 1080 and 803 $cm^{-1}$), indicating that heteroatoms were incorporated into the interlayer space of Si PILC and M–O–Si bonds formed. This was explained by the fact that Fe has a higher atomic mass than Si and the bond length of Fe-O is greater than that of Si-O, thus leading to a decrease in the wavenumber. A new band around 570 $cm^{-1}$ was ascribed to the stretching mode of Fe-O in $Fe_3O_4$ for the samples with higher iron oxides amounts.

### 12.7.5 Nickel and Cobalt Doped Si PILC

Mao et al. (2009) used a similar method to prepare Ni- and Co-doped Si PILC. The Mt had a wide band around 1200–1000 $cm^{-1}$. However, the other three Si PILC samples showed intense bands around 1050 and 495 $cm^{-1}$ (Fig. 12.8), suggesting that the crystallinity of the Mt increased because of the silica framework formed in the interlayer spaces. The characteristic bands of Si–O–Si at 803 $cm^{-1}$ could be observed for all the samples. The intensity of the band at 803 $cm^{-1}$ of Si PILC was stronger than the other samples, and suggested that in Co–Si PILC and Ni–Si PILC some Si–O–Si bond had been

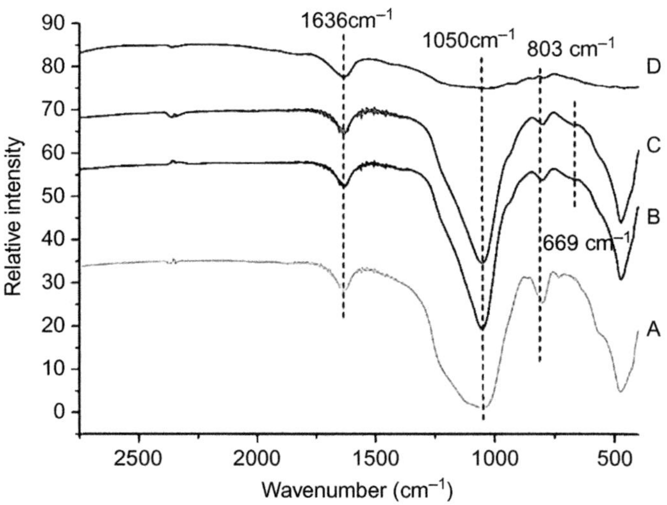


**FIG. 12.8** FTIR patterns for (A) Si-PILC, (B) Co–Si-PILC, (C) Ni– Si-PILC and (D) Mt. Reproduced with permission from Mao et al. (2009).

replaced by Si–O–X (X = Ni, Co). A shoulder was observed at 669 cm$^{-1}$ in the Ni- and Co- containing samples, which was assigned to the Si–O–X (X = Ni, Co) in the interlayer space of the pillared structure.

### 12.7.6 Si-Zr-Porous Clay Heterostructures

Porous clay heterostructures with increasing amounts of $Zr^{3+}$ substituting for $Si^{4+}$ in the pillars were prepared from Mt by Cecilia (2013). In the OH-stretching region Mt presented a band with a maximum located at 3646 cm$^{-1}$ ($\nu$(MOH)) that represents the sum of several bands in the range 3756 and 3525 cm$^{-1}$, and two wide bands centered at 3371 and 3265 cm$^{-1}$ ($\nu$(HOH) of H-bonded physi-sorbed surface water). The FTIR spectra of Si-BG Mt and Si-Zr BG Mt samples showed a well-defined band at 3742 cm$^{-1}$, a series of overlapping and poorly defined bands between 3714–3533 cm$^{-1}$, and a wide band in the range 3533–2921 cm$^{-1}$. Zhao et al. (1997) assigned the band at 3742 cm$^{-1}$ to a stretching vibration of the Si-OH groups located in the pillars that hold the 2:1 layers. The overlapping bands between 3714 and 3533 cm$^{-1}$ of Mt $\nu$(MOH) shifted to higher wavenumbers due to interactions with the inserted pillars. From Si-BG Mt to Si-Zr-BG Mt, the band at 3742 cm$^{-1}$ lost intensity and broadened, and the band in the range 3533–2921 cm$^{-1}$ slightly increased intensity as the Zr content increased. This suggests that the presence of Zr influenced the orientation of the OH groups (Zhao et al., 1997).

### 12.7.7 Tungsten Impregnated Mixed Si-Zr-PILC

Saboya et al. (2016) used similar Si-Zr PILC to wet-impregnate with increasing concentrations of $WO_3$ (5–30 wt%). The Raman spectra revealed an increase of the intensity of bands at 807 and 717 cm$^{-1}$ when the Zr content decreased, which were attributed to W–O stretching mode and W–O bending mode, respectively, while the bands located at 272 and 326 cm$^{-1}$ were assigned to W–O–W deformation modes (Ross-Medgaarden and Wachs, 2007). The decreased intensity of the bands suggested the presence of isolated tungstate species ($WO_x$) diluted into the tetragonal $ZrO_2$, which agreed with the XRD data.

## 12.8 Zr-PILC

### 12.8.1 Humic Acid Impregnated Zr PILC

Anirudhan et al. (2010) used humic acid (HA)-immobilized Zr-PILC to study the adsorption of uranium ($U^{VI}$). Zirconium oxychloride solution ($ZrOCl_2$. $8H_2O$) was refluxed at 40°C for 2 h and added dropwise to a $Na^+$-Mt dispersion and aged at 90°C. After ageing, the Zr PILC was separated, washed, dried, ground and heated at 150°C. This powdered sample was then used for the preparation of HA-immobilized Zr PILC (HA-Zr PILC). HA was impregnated on Zr PILC at pH 3.5 using the batch adsorption techniques. In the case of Zr PILC, the OH stretching modes of Zr PILC were observed in the region 3600–3700 $cm^{-1}$. On calcination, dehydration of interlayer water species occurred and shifting and broadening of some bands were observed in the region 900–1100 $cm^{-1}$, which the authors ascribed to the formation of protonic sites associated with Si-OH-Zr bridges appearing due to the calcination process. The intensity reduction may also be due to the sintering of the pillars. In HA-Zr PILC, bands similar to that of the Mt structure was seen in the region 3350–3420 $cm^{-1}$ together with a number of vibrations associated with organic groups (NH, $CH_3$, etc.) within the HA in the region between roughly 2800 and 3400 $cm^{-1}$. The characteristic bands for –COOH group were observed at 1720 and 1453 $cm^{-1}$. Another difference from the Zr PILC was the appearance of four bands at 900, 845, 775 and 600 $cm^{-1}$ which were ascribed by the authors to the presence of polyphenyls in the HA. A medium intensity band at 918 $cm^{-1}$ was assigned to the -OH out of plane bending of the -COOH group which is hydrogen bonded to another carboxylic group. In the case of U-HA-Zr PILC, since the hydrogen ions get replaced from the –COOH group, a band was found at 1380 $cm^{-1}$. This is due to the conversion of the acid group into its inorganic salt. Also, the characteristic –COOH peak at 1720 $cm^{-1}$ shifted to the 1670 $cm^{-1}$ region, indicating the replacement of $H^+$ ion with metal ions. An absorption band of uranyl (VI) ion was found at 696 $cm^{-1}$ that was absent in the spectrum of HA-Zr-PILC.

### 12.8.2 Organo-Sulfonated Zr-PILC

A series of sulfonated polyvinyl alcohol (PVA)-Zr PILC nanocomposites were synthesized by Kar et al. (2016). A low molar mass PVA was sulfonated using chlorosulfonic acid. The sulfonated PVA was intercalated inside the parent Mt and Zr-pillared Mt (Zr PMt) matrix. In an alternative approach, the PVA was initially intercalated into the clay mineral and Zr PMt structure followed by in-situ sulfonation inside the clay mineral matrix (sulfonated-Zr PMt). For the parent clay mineral, a sharp intense band was observed at 3636 $cm^{-1}$ along with a broad less intense band at 3450 $cm^{-1}$ in the OH stretching region of the FTIR spectrum. These bands were assigned to, respectively, $\nu$(MOH) and $H_2O$ $\nu$(OH) modes of the clay minerals. In addition to the $H_2O$ $\nu$(OH), the clay-polymer nanocomposites showed two less intense bands at 2850 and 2920 $cm^{-1}$ ascribed to the symmetric and asymmetric stretching vibrations of $CH_2$ group present in the structural skeleton of the polymeric species as well as in the cetyl group of CTAB (cetyl trimethylammonium bromide). The band at 1630 $cm^{-1}$ was assigned by the authors as the bending vibration mode of OH groups of sulfonated-Zr PILC and PVA-sulfonated Zr PILC (PVA dispersed in Zr PILC followed by sulfonation). Its intensity suggested that pillaring with Zr-polycation increased the surface acidity. In addition to the discrete peaks between 700 and 950 $cm^{-1}$ related to the octahedral $\delta$(MOH) modes of Mt, two additional bands were observed at 1155 and 1030 $cm^{-1}$ for all clay-polymer nanocomposites. These bands were ascribed to the S=O and O-S-O vibrations of the sulfonate (-$SO_3H$) group present in the polymer skeleton (Cristovan et al., 2006; Zhang et al., 2010). The FTIR study clearly indicated the presence of sulfonated polymeric species in the interlayer spaces and that the clay mineral retained its structural integrity in the final clay polymer nanocomposite.

The Raman spectra of PVA intercalated Mt followed by sulfonation inside the clay mineral matrix and PVA sulfonated PMt nanocomposite materials are presented in Fig. 12.9. Both materials showed an intense band around 2920 $cm^{-1}$ corresponding to the stretching vibration of the $CH_2$ group present in the polymeric skeleton. A series of Raman bands were also observed at 1492, 1105 and 1055 $cm^{-1}$. The Raman band at 1492 $cm^{-1}$ was assigned to the shear mode of vibration of -$CH_2$ group of the sulfonated polymeric species

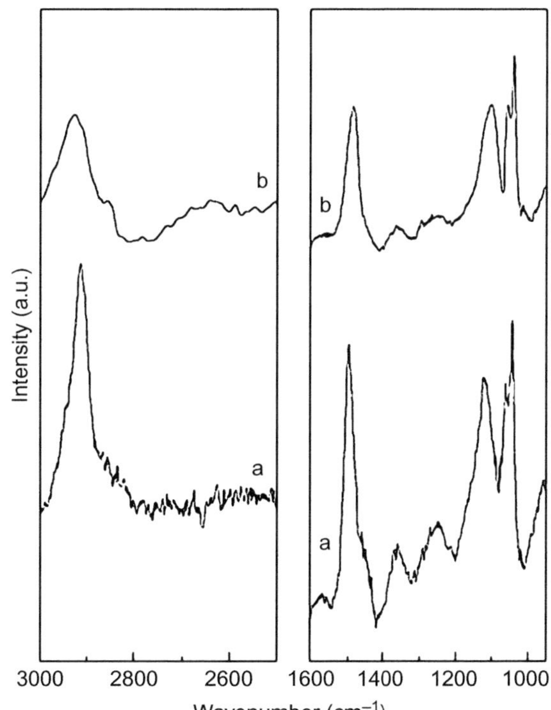


**FIG. 12.9** Raman spectra of (a) Clay-SP and (b) Clay-PIS. Reproduced with permission from Kar et al. (2016).

(Prosanov and Matvienko, 2010; Kozai et al., 2012). The Raman band at 1105 $cm^{-1}$ was due to an OH deformation vibration (Prosanov and Matvienko, 2010). This band occurred in the spectrum due to presence of residual -OH groups that were not sulfonated during the sulfonation step. The band at 1055 $cm^{-1}$ in the Raman spectra of the clay polymer nanocomposites was due to an S-O vibration present in the sulfonated PVA. Such vibration has been observed earlier for nafion polymer and has been ascribed to the S-O vibration of the sulfonate functional group (Hara et al., 2011). Another notable feature was the occurrence of broad and less intense Raman bands for PVA sulfonated PMt nanocomposites as compared to the sulfonated PVA intercalated Mt. This observation indicated that the sulfonated PVA polymeric species were present in a well dispersed state in the PILC. However, agglomeration of the polymeric species occurs in the sulfonated PVA intercalated Mt with consequent development of partial crystalline character that was responsible for the sharp and intense Raman bands (Zhen et al., 2012). This observation corroborated XRD results indicating incomplete intercalation of sulfonated PVA species in the sulfonated PVA intercalated materials.

## 12.9 OTHER-METAL PILC

### 12.9.1 Macrocyclic Transition Metals

The transition metal macrocyclic complex cations M (trans-14-diene)$^{2+}$ (M = $Co^{2+}$, $Ni^{2+}$, $Cu^{2+}$ and $Zn^{2+}$) were introduced into the interlayers of a sodium Mt by Liu et al. (1995). The resulting clay minerals were calcined at various temperatures. All samples were thermally stable up to 300°C with interlayer space distances of 0.49 to 0.51 nm and SSA of 57–86 $m^2\ g^{-1}$. The IR spectra of the metal complex [M(trans-14-diene)]$(ClO_4)_2$ indicated that the band at 1670 $cm^{-1}$ present in all spectra was due to the C=N vibration, while the bands at 1300–1470 $cm^{-1}$ were attributed to the C-H bending mode and that at 2960 $cm^{-1}$ to the C-H stretching mode of methyl group (Bellamy, 1975). The broad band around 3400 $cm^{-1}$ was ascribed to the hydrogen bonded water molecules and the band at 1100 $cm^{-1}$ to the vibration of a perchlorate group. Typical IR spectra of the original clay mineral and of $Cu^{2+}$ exchanged clay mineral calcined at various temperatures are shown in Fig. 12.10. For the latter sample dried at 25°C, the bands appearing at 2960 $cm^{-1}$ and 1300–1470 $cm^{-1}$ implied the existence of pillaring species [Cu(trans-14-diene)]$^{2+}$ in the intercalated clay mineral. These bands became broad and less intense when calcined at 300°C and finally disappeared at 400°C, showing the collapse of the pillar structure at this temperature in accordance with the XRD results that showed a complete collapse of the interlayer space to less than 0.1 nm.

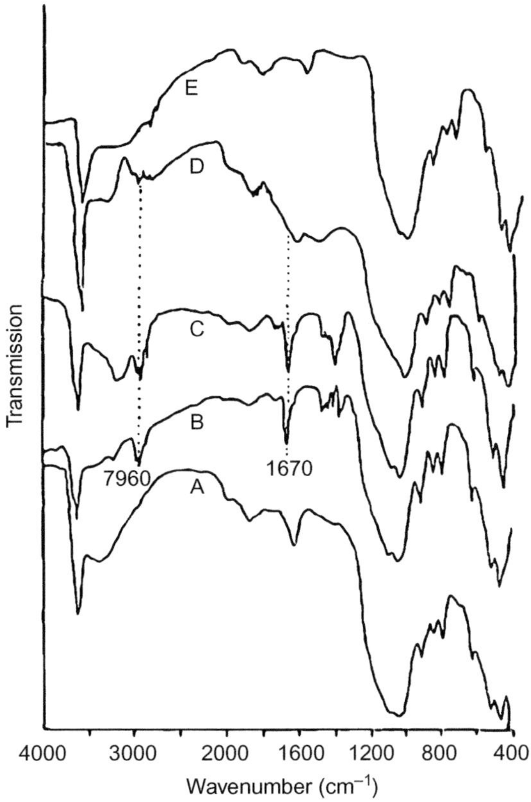


**FIG. 12.10** IR spectra of the original clay (A) and the copper intercalated clay calcined at (B) 25°C, (C) 110°C, (D) 300°C and (E) 400°C. Reproduced with permission from Liu et al. (1995).

## 12.9.2 PVMO-Bentonite Heterogeneous Catalysts

A heterogeneous catalyst (PVMo/bentonite), consisting of $V^{3+}$ substituted polyphosphomolybdate with Keggin structure $H_4[PVMo_{11}O_{40}].13H_2O$ (PVMo) supported on acid activated bentonite, was prepared by wet impregnation by Boudjema et al. (2015). As for other systems, spectra showed a band around 3600 $cm^{-1}$ for all the samples (octahedral $\nu$(MOH)) and one band at 1620–1640 $cm^{-1}$ ($H_2O$ $\delta$(OH)) associated with the host clay mineral. The spectrum of a pure PVMo sample exhibited bands at 1063, 961, 867 and 780 and 596 $cm^{-1}$ that were attributed to the stretching vibrations of $\nu_{as}(P–O_a)$, $\nu_{as}(Mo–O_d)$, $\nu_{as}(Mo–O_b–Mo)$, $\nu_{as}(M–O_c–M)$ (M=Mo, V) and $\delta(P–O_a)$ vibrations, respectively. In a Keggin-type unit, $O_a$ refers to the oxygen atom common to $PO_4$ tetrahedron and one trimetallic group $Mo_3O_{13}$, $O_b$ connects two trimetallic groups, $O_c$ binds two octahedral

$MoO_6$ units inside a trimetallic group and $O_d$ is the terminal oxygen atom. After the PVMo were intercalated in the HMt, some of the characteristic Keggin bands were observed at 982 and 893 $cm^{-1}$ while others overlapped with the HMt bands. A decrease in the intensities of these bands was detected as the PVMo loading increased.

## 12.10 CONCLUDING REMARKS

It has been more than 50 years now since the earliest research on PILC. Over these decades enormous progress has been made in the development of these materials suitable for a variety of applications in catalysis and adsorption. A reasonable understanding of some of the pillaring agents in solution has been gained, in particular on the Keggin type structures such as the $Al_{13}$. What is less clear, though, after all this time are the exact structures of the pillars following calcination. Specifically, how these pillars are bonded to the clay mineral El-Korso interlayer surfaces. A range of analytical techniques such as XRD and solid-state NMR in combination with vibrational spectroscopy (IR and Raman), as well as UV-Vis, has allowed some insight in the structures of some PILC. But up to this day there is no definitive answer to how the pillars are exactly connected to the clay mineral layers and models about the different behavior between dioctahedral and trioctahedral smectites and the possibility of tetrahedral inversion in the tetrahedral sheet to form permanent $^{IV}Si–O–M^{pillar}$ or $–^{IV}Al–O–M^{pillar}$ remain merely hypothetical.

Chapter 13

# NIR Contribution to The Study of Modified Clay Minerals

J. Madejová and H. Pálková

## 13.1 INTRODUCTION

Near infrared (NIR) spectroscopy is a useful analytical technique with increasing popularity in both science and industrial applications due to its low running costs, and none or only minimal sample preparation. Moreover, it is a non-destructive, environmental friendly, rapid technique with the ability to analyze a sample through glass or other packaging materials that are not opaque to NIR. The NIR spectra, covering the region between approximately 12,000 and 4000 $cm^{-1}$, contains bands corresponding to overtones and combinations of fundamental O–H, C–H and N–H vibrations that are related to absorption frequencies of these bonds observed in the middle infrared infrared (MIR) region.

In the clay minerals studies the NIR region can provide useful information because the observed bands related to the OH overtones and combination vibrations are sensitively affected by variations in the clay mineral structure itself. Most often, the NIR spectra in the 8000–4000 $cm^{-1}$ range contain bands due to the first overtone of the OH stretching vibrations $2\nu(OH)$ and the combination of OH stretching and bending fundamental modes $(\nu+\delta)OH$ (see Chapter 5 in this Volume and the references therein).

A NIR database of silicate minerals was established by Hunt and Salisbury (1970) and Hunt et al. (1973). The assignments of NIR absorption bands of water molecules and lattice-hydroxyl groups of montmorillonites (Mt) were reported by Cariati et al. (1981), and the effect of exchangeable cations and layer charge on the water bands positions was reported by Cariati et al. (1983a,b). The history and the development of NIR reflectance spectroscopy was summarized by Clark et al. (1990) including the NIR spectra of selected minerals. Bishop et al. (1994) analyzed the NIR spectral features attributed to interlayer water molecules in smectites containing different interlayer cations. In the last twenty years the NIR spectroscopy was successfully used in

**Developments in Clay Science, Vol. 8. https://doi.org/10.1016/B978-0-08-100355-8.00013-8**

different fields of clay minerals research, such as for identification and basic characterization of clay minerals, for crystal-chemical studies, clay-water interactions and also for identification and characterization of clay minerals occurring on Mars (all aspects are developed in this volume). The NIR spectroscopy can also offer useful information about the modification of clay minerals' structure following different physical and chemical treatments, such as grinding, heating, dissolution in acids or interaction with organic species, providing materials with tailored properties for specific industrial applications. The application of NIR spectroscopy to these types of problems, however, is not as common as the more traditional MIR spectroscopy. Therefore the main purpose of this chapter is to introduce selected examples documenting the potential of the NIR spectroscopy (1) to monitor different types of clay minerals modifications and (2) to assess new information often not available from traditional MIR spectroscopy.

## 13.2 MECHANO-CHEMICAL TREATMENT

Mechano-chemical treatment (dry grinding) of clay minerals is among the oldest industrial treatments for modifying their SSA and bulk properties. Depending on the conditions applied, dry grinding leads to the alteration and complete or partial amorphization (loss of crystallinity) of a clay mineral structure. Moreover, the process is accompanied by reduction of the particle size and changes in specific surface area (SSA). This treatment was also applied to enhance clay mineral surface activity in reaction in the solid state, for example in intercalation of some reactive molecules such as urea (Makó et al., 2013), preparation of fertilizers systems (Solihin et al., 2011), or in preparation of clay polymer nanocomposites (CPN) (Vertuccio et al., 2009). Among a wide variety of instrumental techniques, MIR spectroscopy has been employed to study the effect of grinding on the structure of kaolins (e.g., Frost et al., 2002c; Horváth et al., 2003; Breen et al., 2007; Makó et al., 2013; Borges et al., 2015, 2016), vermiculite (Hongo et al., 2012), pyrophyllite (Mohammadnejad et al., 2014; Zhang et al., 2015), talc (Čavajda et al., 2015) and smectites (Christidis et al., 2004; Hrachová et al., 2007a,b; Petra et al., 2015). Only the paper of Frost et al. (2002c), however, has reported on observed changes in NIR spectra of clay minerals subjected to grinding so far.

Upon grinding, chemical bonds in clay minerals are modified, directly affecting the shape of the bands in the MIR spectra. MIR and NIR spectra of two smectites are reported in Fig. 13.1: a Fe-rich Mt Blšany from Czech Republic (BLS), representing dioctahedral smectite and a synthetic saponite Sumecton from Japan (SUM), representing trioctahedral smectite. Spectra were measured before and after dry grinding in a high-energy planetary mill for 6 min. The distinct band of structural OH mode found at 3621 $cm^{-1}$ for BLS lost intensity and merged with the $\nu(H_2O)$ (also termed $\nu(HOH)$) band near 3400 $cm^{-1}$ into a broad featureless band (BLS-6 min). The most

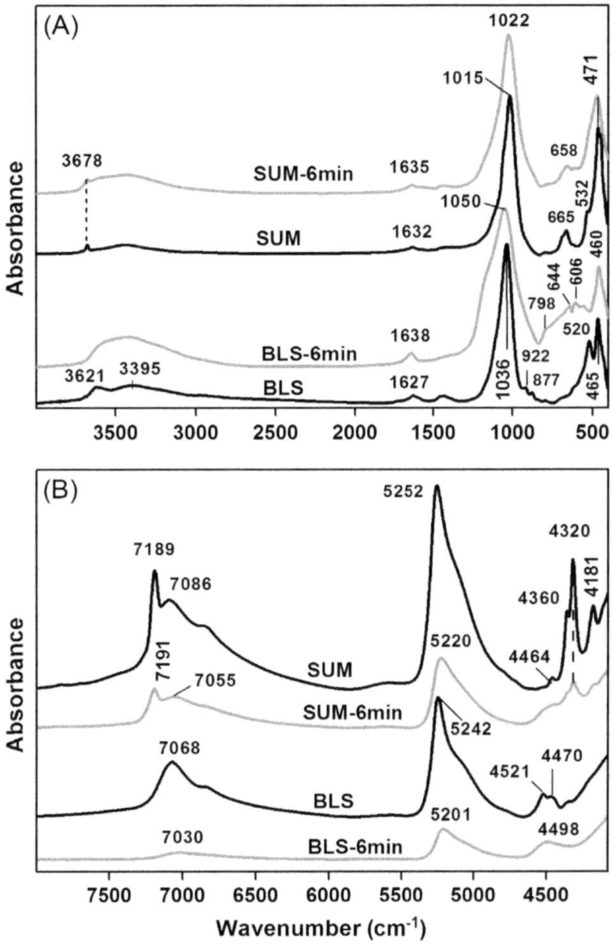


**FIG. 13.1** (A) MIR and (B) NIR spectra of smectites Blšany and Sumecton before and after grinding for 6 min.

pronounced changes resulting from dry milling of Mt occurred in the region below 1300 cm$^{-1}$, where the $\nu$(SiO) absorption band shifted from 1036 to 1050 cm$^{-1}$. Moreover, the band broadened, not only due to the reduced crystallinity of the smectite crystals, but also as a result of the formation of an amorphous Si–O(H) phase as revealed by a broad shoulder near 1200 cm$^{-1}$. Well-resolved $\delta(Al_2OH)$ and $\delta$(AlFeOH) bands at 922 and 877 cm$^{-1}$ were observed in the BLS spectrum, but were subsequently lost after grinding because of an extensive broadening of the 2$\nu$(SiO) band and a partial release of both structural OH groups and octahedral cations. Disappearance of the band at 520 cm$^{-1}$, assigned to bending Al–O–Si vibrations, confirmed a breakdown of the linkages between tetrahedral and octahedral sheets of ground BLS Mt. A broad complex band in the 800–500 $cm^{-1}$ region of the MIR spectrum was observed for BLS-6 min, with resolved peaks at 644 and 606 $cm^{-1}$. Upon grinding the Mt layers were supposed to create amorphous Al–Fe-silicate network containing [$SiO_4$], [$AlO_4$] and [$FeO_4$] units absorbing in the 1200–400 $cm^{-1}$ region (Hrachová et al., 2007b).

Different trends could be observed in the MIR spectra of trioctahedral smectite SUM (Fig. 13.1A), and for the most part, the changes were less pronounced compared to dioctahedral smectite BLS. The $Mg_3OH$ stretching mode at 3678 $cm^{-1}$ was still clearly visible in the SUM-6 min spectrum, although its intensity decreased and the width of the band increased. The position of the Si–O stretching vibration shifted by only 7 $cm^{-1}$ and broadening of this complex band was less pronounced than for BLS-6 min, indicating that the integrity of the smectite SUM structure was better preserved. No features due to amorphous silica, such as a shoulder near 1200 $cm^{-1}$ or a band near 800 $cm^{-1}$, indicated lower extent of SUM destructions compared to BLS after 6 min of grinding. The complex band consisting of perpendicular Si–O (692 $cm^{-1}$) and bending $Mg_3OH$ (665 $cm^{-1}$) vibrations slightly decreased in intensity after SUM grinding confirming a partial release of the OH groups. A broad band with components at 532 $cm^{-1}$ (Mg-O perpendicular), 464 $cm^{-1}$ (Mg-OH translation) and 448 $cm^{-1}$ (Si–O–Si bending) became merged into one band with a maximum at 471 $cm^{-1}$. All of the features observed in the MIR spectra of the studied smectites prove that the trioctahedral saponite structure is more resistant to mechanical treatment compared to the dioctahedral Mt structure. The XRD analysis (not shown) confirmed the IR results. While the $d_{001}$ diffraction of the initial (not-ground) BLS was intense and sharp, the $d_{001}$ of SUM appeared as a broad peak indicating a microcrystalline character of synthetic saponite. After 6 minutes of grinding $d_{001}$ can not be recognized in the XRD pattern of BLS-6 min in contrast to SUM-6 min showing weak but still identifiable basal reflection.

NIR spectra of initial and 6 min ground smectites were also analyzed in order to provide complementary information to MIR region (Fig. 13.1B). The spectrum of BLS has all bands characteristic of dioctahedral smectite. The complex band at 7068 $cm^{-1}$ corresponds to the first O–H overtone ($2\nu$) of the structural OH groups and water molecules that are weakly H-bonded to the Mt siloxane surface, while a shoulder near 6830 $cm^{-1}$ is related to the $2\nu(H_2O)$ of water molecules strongly linked through H-bonds with other water molecules, including also those directly bonded to the exchangeable cations. A broad band near 5242 $cm^{-1}$ involves the combination modes ($\nu+\delta$) of water molecules engaged in H-bonds of different strengths with other water and surfaces. The bands at 4521 and 4470 $cm^{-1}$ are due to the $(\nu+\delta)Al_2OH$ and $(\nu+\delta)AlFeOH$ combination modes, respectively. After BLS grinding a significant decrease in the intensities of all bands can be detected. Both the 2$\nu$OH overtone and $(\nu+\delta)H_2O$ band are shifted to lower wavenumbers indicating a decrease of the amount of $H_2O$ weakly bonded to siloxane surface of the Mt and a simultaneous increase of water molecules that are strongly H-bonded to each other. The combination bands of the structural OH groups are merged into one broad complex band at 4498 $cm^{-1}$ as a result of the modification of the octahedral sheet.

The NIR spectrum of trioctahedral saponite (SUM) shows the 2$\nu$(OH) at 7189 $cm^{-1}$, 2$\nu(H_2O)$ at 7086 $cm^{-1}$, $(\nu+\delta)H_2O$ at 5252 $cm^{-1}$, and $(\nu+\delta)$OH at 4360 and 4320 $cm^{-1}$. The 4181 $cm^{-1}$ is a combination of 3678 and 532 $cm^{-1}$ band. Similarly as in the MIR region (Fig. 13.1A) a clear difference between the modification of a saponite structure upon grinding can be observed in NIR region (Fig. 13.1B). Though the intensities of all bands decreased, the shape of Sum-6 min spectrum resembles that of an un-ground sample. The downward shift of the water bands indicates the formation of a more extensive network of strong H-bonds between water molecules. The preservation of all bands related to the octahedral Mg, such as at 7191, 4320 and 4181 $cm^{-1}$ proves that the decomposition of saponite layers from dry grinding is not as pronounced as in Mt despite the same treatment conditions used.

The results presented above illustrate well the complementary information gleaned from both MIR and NIR regions. Both regions provide evidence of the modifications of the octahedral sheets of smectites that occur upon grinding. The MIR spectra, in particular, allows observation of the changes in the arrangement of the $SiO_4$ tetrahedra. The NIR region, on the other hand, provides information regarding the character of water molecules adsorbed within the interlayer spaces. In addition, a release of the Mg from the octahedral sheets of SUM-6 min sample can be better recognized in the OH overtone and combination band regions than in the OH stretching and bending band regions.

The only paper reporting the utilization of NIR spectroscopy for examination of the structural changes of clay minerals upon grinding so far has been that of Frost et al. (2002c). These authors applied both MIR and NIR to analyze the changes in kaolinite structures upon grinding. They observed a decrease in the intensity of the 2$\nu$OH band in the region of 7400–7065 $cm^{-1}$ with time of milling. Concomitantly, the intensities of the overtones assigned to water molecules increased. Decomposition analysis of the $(\nu+\delta)H_2O$ band revealed the presence of several components due to heterogeneous H-bonding of water molecules within sample. The intensity of the particular components varied as a function of grinding time, shifting the maxima of this band to the lower wavenumbers and high frequency components continuously disappeared as the time of the treatment progressed. The shift to lower wavenumbers (energy) indicates that grinding promoted the formation of stronger H-bonds between water molecules or between surfaces of kaolinite disturbed by grinding.

## 13.3 LAYER CHARGE REDUCTION

The modification of the layer charge of clay minerals, mainly dioctahedral smectites, has been a frequently studied topic because it significantly affects the properties of these minerals. The negative charge of the layers, arising from the partial non-equivalent substitution of the central atoms in tetrahedra and/or octahedra can be reduced by a thermal treatment of octahedrally charged smectites saturated with small exchangeable cations, notably $Li^+$, $Cu^{2+}$, $Mg^{2+}$ or $Ni^{2+}$. Upon heating to 200 – 300°C or even less, the cations are dehydrated and migrate into the smectite layers where they become trapped. If these cations are not exchangeable by the methods normally used for determining cation exchange capacity (CEC) of smectites, the charge on the layers is effectively reduced, irrespective of whether the migrating cations are embedded within the ditrigonal cavities or occupy octahedral vacancies (Komadel et al., 2005, and the references therein).

MIR spectroscopy has been used to answer questions concerning the location of migrating cations (e.g., Calvet and Prost, 1971; Sposito et al., 1983; Srasra et al., 1994; Mosser et al., 1997; Madejová et al., 1996, 2000a; Hrobáriková et al., 2001; Petit et al., 2002; Madejová, 2005; Skoubris et al., 2013; Steudel et al., 2015). If small interlayer cations penetrate deeper into the smectite layers, any change in the local environment and/or orientation of the hydroxyl groups should be reflected in their OH vibrations. To compensate the negative layer charge, $Li^+$ are attracted to sites close to octahedra where Mg(II) exists as the central atom. Occupation of the previously vacant octahedral position (in dioctahedral smectites) results in creation of local trioctahedral domains, such as AlMgLiOH or $Al_2LiOH$, identified by new bands at 3670 and 3640 $cm^{-1}$, respectively, appearing in the MIR spectra of heated $Li^+$-Mt (e.g., Calvet and Prost, 1971).

Madejová et al. (2000a) used MIR spectroscopy to follow the structural changes occurring upon heating of five $Li^+$-saturated dioctahedral smectites of different layer charge size and location. Fig. 13.2A1, A2 show the OH stretching region of selected smectites from this study, namely SAz-1 and HD Mt, both with prevailing octahedral layer charge, and ferruginous smectite SWa-1 (Fig. 13.2.A3) with dominant tetrahedral layer charge. The coefficients obtained for $Mg^{VI}$ and $Al^{IV}$ from the structural formula (per $O_{20}(OH)_4$) and CEC of unheated and heated smectites, as the main parameters affecting the layer charge of clay minerals, are given in Table 13.1.

The spectra of heated Mt show a gradual shift of the OH stretching band to higher wavenumbers and the appearance of a new band near 3670 $cm^{-1}$. Fixation of $Li^+$ in the previously vacant octahedra causes the creation of AlMgLiOH groups absorbing near 3670 $cm^{-1}$. The 3638 $cm^{-1}$ band is most likely related to the $Al_2LiOH$ domains in the octahedra, probably where $Li^+$ occupies previously vacant *cis* sites. However, the effects of $Li^+$, penetrating upon heating deep into the hexagonal cavities, on the vibrations of the OH

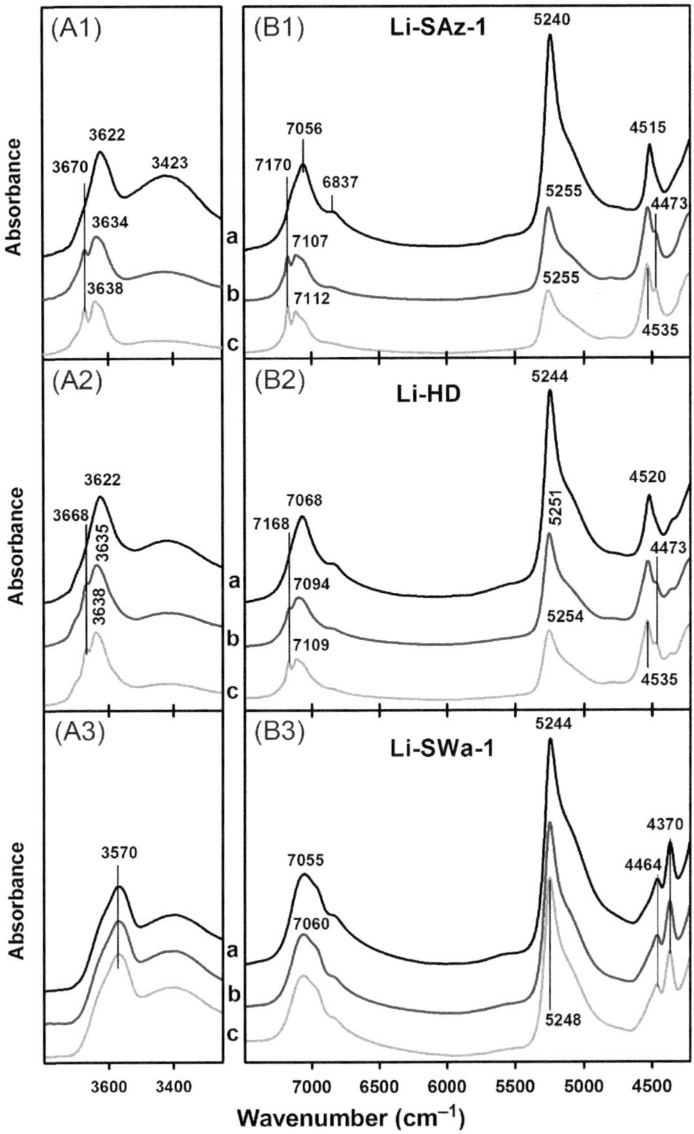


**FIG. 13.2** (A1, A2, A3) MIR and (B1, B2, B3) NIR spectra of Li-SAz-1, Li-HD and Li-SWa-1 smectites: (a) unheated and heated at (b) 150 and (c) 200°C for 24 h.

groups need to be taken into consideration as well. Higher intensity of the $3670\ cm^{-1}$ band of heated SAz-1 compared to HD results from its higher octahedral charge deficit attracting more $Li^+$ to form AlMgLiOH groups. In contrast to Mt, no new band indicating the presence of $Li^+$ in the octahedral positions and/or the shift of the OH stretching band is observed in the spectra

**TABLE 13.1 The Content of Octahedral Magnesium ($Mg^{VI}$), Tetrahedral Aluminium ($Al^{IV}$) and CEC of Studied Smectites (Madejová et al., 2000a)**

| Samples | SAz-1 | HD | SWa-1 |
|---|---|---|---|
| $Mg^{VI}$ | 1.2 | 0.85 | 0.27 |
| $Al^{IV}$ | 0 | 0.11 | 0.68 |
| *CEC in mmol/g* | | | |
| Unheated | 1.202 | 1.004 | 1.025 |
| 150°C/24 h | 0.33 | 0.392 | 0.896 |
| 200°C/24 h | 0.154 | 0.217 | 0.85 |

of heated SWa-1, such as an $Fe^{3+}$ enriched trans vacant smectite with a prevailing tetrahedral charge (Fig. 13.2A3). Because fixation of $Li^+$ upon heating occurs most readily for dioctahedral smectites with dominating octahedral charge, the IR results correlate very well with CEC values (Table 13.1) determined for the individual samples.

Though NIR spectroscopy provides information on the local environment of the OH groups, few papers were published dealing with NIR spectra of reduced-charge Mt. Sposito et al. (1983) compared the NIR spectra of the $Na^+$-Mt (100% CEC) with $Na^+$/$Li^+$-Mt (58% of original CEC) but did not find any noticeable shift of the principal absorption bands with charge reduction. The first study showing the advantages of the NIR region for the identification of structural changes upon $Li^+$-fixation was that of Madejová et al. (2000b) studying the same series of samples as reported in Madejová et al. (2000a). The NIR spectra of selected representative smectites from this study, for example, SAz-1, HD and SWa-1 are given in Fig. 13.2B1–B3. The spectra of heated Mt show an upward shift and a splitting of the OH overtone band into two components near 7170 and 7100 $cm^{-1}$. The presence of a band near 7170 $cm^{-1}$, assigned to the $2\nu$(AlMgLiOH), implies that local trioctahedral domains have been created upon $Li^+$-fixation within the previously vacant *cis* octahedra of Mt. The 7100 $cm^{-1}$ band is the first overtone of the OH stretching vibrations appearing near 3640 $cm^{-1}$. The OH overtone region shows lower intensity of the 7170 $cm^{-1}$ band for heated HD than for SAz-1, a result which infers a lower content of AlMgLiOH groups in the former arising from lower charge deficit resulting in migration of less $Li^+$ into the vacant octahedral positions. No new band and only a slight shift to higher wavenumber (Fig. 13.2B3) due to partial dehydration confirms that the amount of fixed $Li^+$ in SWa-1 upon heating at 200°C is very low (Madejová et al., 2000a).

The presence of $Li^+$ in the previously vacant octahedra and a gradual collapse of the interlayer space of Mt (Komadel et al., 1996a) evokes significant changes in the intensity and position of the combination water band observed in the 5240–5255 $cm^{-1}$ region (Fig. 13.2B1, B2). Reducing content of water in a heated Mt decreases the intensity of the $(\nu+\delta)H_2O$ band. Moreover, most of $Li^+$ is trapped in the octahedra and only a very low amount of exchangeable $Li^+$ cations coordinated with water molecules involved in strong H-bonds with other water molecules is present in the interlayer spaces. The majority of the residual water molecules are weakly H-bonded to neutral siloxane surface of the Mt layers, which shifted the $(\nu+\delta)H_2O$ band to higher wavenumbers. Only a minimal shift and decrease in the intensity of the $(\nu+\delta)H_2O$ band was observed for SWa-1 (Fig. 13.2B3), and this correlates with the very low decrease of the CEC of smectite with dominant tetrahedral charge (Table 13.1).

Migration of $Li^+$ into the SAz-1 and HD Mt layers influences also the spectral features in the structural OH combination region near 4500 $cm^{-1}$. A gradual shift of the 4515 $cm^{-1}$ band to higher frequencies (up to 4535 $cm^{-1}$) and a new band at 4473 $cm^{-1}$ is observed in the spectra of Mt heated at 150 and 200°C (Fig. 13.2B1-b,c and B2-b,c). The modification of the combination bands upon heating of $Li^+$-Mt reflects the perturbation of the structural OH groups vibrations by $Li^+$ fixed in the previously vacant octahedral positions. As expected, the OH combination region of ferruginous smectite SWa-1 shows no features indicating the presence of $Li^+$ in the octahedral positions and/or within the hexagonal cavities. The intensity of the first AlMgLiOH overtone band was clearly higher than stretching the AlMgLiOH band for both Mt and both heating temperatures (i.e. 150 and 200°C) which benefits the NIR region for $Li^+$-fixation studies.

The NIR region, together with X-ray diffraction (XRD), MIR spectroscopy, and analytical electron microscopy was used to analyze the distribution of the layer charge in the Fe-rich smectite from Ölberg, Germany (Petit et al., 2002). A $Li^+$-saturated sample was heated at different temperatures (105–300°C) for 24 h to obtain a set of samples with a gradually decreasing layer charge. A broad band near 7050 $cm^{-1}$, present in the spectrum of the unheated Li-Ölberg sample, was assigned to the first overtones of both the structural OH groups and water molecules weakly H-bonded to the surface oxygen atoms of the tetrahedral sheets. Upon heating of the $Li^+$-Ölberg sample, the shape of the $2\nu$OH band was significantly modified due to the presence of fixed $Li^+$ in previous vacant octahedral positions. The $2\nu$(OH) region exhibited the bands at 7175, 7118 and 7053 $cm^{-1}$ assigned to the $2\nu(AlFe^{3+}LiOH)$, $2\nu(MgFe^{3+}LiOH)$ and $2\nu(Fe^{3+}{}_2LiOH)$, respectively. A less resolved broad band centered near 6995 $cm^{-1}$ was supposed to correspond to the local dioctahedral environment that was less affected by $Li^+$ migration. A detailed analysis of the smectite from Ölberg showed that the sample was heterogeneous, consisting of two populations. One population, which comprises a minor portion of the sample, was a '$Fe^{3+}$-Mt', a ferrian smectite with little or no tetrahedral charge and $Fe^{3+}$ as the major octahedral cation. The other population, a major constituent of the sample, contained less Mg and more Al than the first one and exhibits some tetrahedral charge. This fraction may be considered to be an inter-grade between nontronite (dominant tetrahedral charge) and $Fe^{3+}$-Mt (dominant octahedral charge). These two populations may occur as separate particles but also as interstratified layers (Petit et al., 2002).

Skoubris et al. (2013) investigated the structural changes of two $Li^+$-saturated Mt with a different layer charge upon heating at 80–300°C. The migration of $Li^+$ cations into the Mt layers was accompanied by a decrease in the layer charge, CEC, and the amount of adsorbed water, whereas the dehydroxylation temperature remained unaffected. Two new sharp OH-stretching bands at ~3640 and 3670 $cm^{-1}$ and their overtones at ~7115, 7170 $cm^{-1}$ were observed in the spectra of both heated $Li^+$-smectites. Based on the analysis of various models of $Li^+$-fixation capable of removing the degeneracy of the original AlMgOH species, the authors assigned the changes observed in the FTIR spectra to migration of $Li^+$ to the vacant octahedral sites in both smectites.

The results reported above are related to the utilization of the NIR spectroscopy for the monitoring the $Li^+$-fixation in dioctahedral smectites. This technique, however, has also been used to follow the migration of other small exchangeable cations such as $Ni^{2+}$ and $Cu^{2+}$ into smectite structure. Pálková et al. (2003) compared the structural changes of the SAz-1 Mt saturated with $Li^+$ or $Ni^{2+}$ upon heating at different temperatures. The MIR spectra of heated $Li^+$-SAz showed new OH stretching and overtone bands near 3670 $cm^{-1}$ and 7170 $cm^{-1}$, respectively, indicating the creation of local trioctahedral domains containing Li(I) in the previously vacant octahedra. These authors found no evidence of OH groups in trioctahedral coordination in the spectra of heated $Ni^{2+}$-SAz. Although $Li^+$ and $Ni^{2+}$ have similar ionic radii of 0.68 and 0.69 Å, respectively, the lower charge of $Li^+$ facilitates its penetration into the previously vacant octahedra, while $Ni^{2+}$ appears to be trapped in the ditrigonal holes of the tetrahedral sheet.

IR spectroscopy in the NIR together with MIR and FIR regions was used to study the structural changes of a thermally treated Mt saturated with $Li^+$ or $Cu^{2+}$ cations (Madejová et al., 2006). The OH stretching and OH overtone regions proved that the $Li^+$ migrated into the octahedral vacancies, thus creating local trioctahedral domains, such as AlMgLiOH. Though $Cu^{2+}$ has similar ionic radius as $Li^+$, no spectral features indicating its presence in the octahedral positions were found, even in the sample heated at 300°C. Fixed $Cu^{2+}$ was supposed to be located deep in the ditrigonal cavities of the tetrahedral sheets of a $Cu^{2+}$-Mt (Jelšový Potok). The IR spectra confirmed that the charge of exchangeable cations significantly affected their final position after fixation upon heating.

The ability of NIR spectroscopy to follow the solvation of $Cu^{2+}$ cations in two series of reduced-charge $Cu^{2+}$-Mt differing in their chemical composition was presented by Brtáňová et al. (2014). Structural changes occurring upon heating of $Cu^{2+}$-Mt significantly modified the shapes and the positions of the bands in the NIR spectrum. The $2\nu(OH)$ band was shifted upon heating from 7053 $cm^{-1}$ to 7095 $cm^{-1}$ as a result of the migration of $Cu^{2+}$ cations into the ditrigonal cavities of the tetrahedral sheets. A new component at 7049 $cm^{-1}$ was related to the isolated $Cu^{2+}$-$H_2O$ clusters tightly bound close to ditrigonal cavities. Changes in the 7200–7000 $cm^{-1}$ region upon acetonitrile and DMSO adsorption on reduced charge samples indicated that fixation of $Cu^{2+}$ cations in the Mt layers was a partially reversible process. Both solvents were able to coordinate with the $Cu^{2+}$ cations located deep in the ditrigonal cavities of the tetrahedral sheets of Mt layers and moved them toward the interlayer space.

While the migration of small exchangeable cations into the layers of dioctahedral smectites has been studied quite often (Komadel et al., 2005) few studies reported the migration of cations into the structure of trioctahedral smectites, mainly hectorite (Jaynes and Bigham, 1987; Jaynes et al., 1992). The possible migration of the exchangeable $Li^+$ and $Zn^{2+}$ cations within synthetic trioctahedral smectites ($Zn^{2+}$-stevensite, with $Zn^{2+}$ in the octahedral positions) under heating was studied by Petit et al. (2008). The permanent layer charge of $Zn^{2+}$-stevensite came from octahedral vacancies only. The MIR spectra showed the characteristic $Zn_3OH$ stretching and bending bands at 3643 and 662 $cm^{-1}$, respectively, but no $Zn_2OH$ vibration bands were detected even in the sample with the most octahedral vacancies, possibly because $Zn_2OH$ stretching and bending bands occurred too close to the $Zn_3OH$ bands. To confirm the presence of vacant octahedral sites in $Zn^{2+}$- stevensite, the samples were $Li^+$ and $Zn^{2+}$ saturated and heated overnight at 300°C. Their MIR spectra varied only slightly. However, in the OH-combination region the spectra were better resolved with narrower bands after heat treatment. Moreover, the increase of the band at 4309 $cm^{-1}$ attributed to $(\nu+\delta)Zn_3OH$ was clearly observed in a $Zn^{2+}$-saturated sample. While the migration of $Li^+$ into the previously vacant octahedral sites upon heating resulted only in partial layer charge decrease, the structure of the $Zn^{2+}$-saturated sample changed from $Zn^{2+}$-stevensitelike to $Zn^{2+}$-talclike. These results indicated that both $Zn^{2+}$ and $Li^+$ could migrate into the vacant sites in smectites of appropriate crystallochemical properties.

## 13.4 ACID TREATMENT

Acid activation is a conventional chemical modification of clays used for scientific and manufacturing purposes. Typically it includes a treatment of clay ore with a solution of hydrochloric or sulfuric acid to gain partly dissolved clay mineral of increased SSA, porosity and surface acidity (Carrado and Komadel, 2009; Komadel, 2003, 2016; Komadel and Madejová, 2013). Acid-activated clays are used as solid sources of protons in technologically important reactions. Upon acid-treatment, protons penetrate into the clay mineral layers and attack the structural OH groups. Resulting dehydroxylation is connected with successive release of the central atoms from the octahedra and with removal of Al from the tetrahedral sheets. Simultaneously, a gradual transformation of the tetrahedral sheets into a three-dimensional framework proceeds. The modification of the clay mineral structure has been often followed by FTIR spectroscopy in the MIR region (Kooli and Jones, 1997; Madejová et al., 1998; Belver et al., 2002; Hassan and Baioumy, 2006; San Cristóbal et al., 2009; Pentrák et al., 2009, 2012a,b; Alver and Alver, 2013; Alves et al., 2014). The MIR spectra of trioctahedral hectorite (with mainly $Mg^{2+}$ in the octahedral positions) dissolving in 0.5 M HCl at 30°C are presented in Fig. 13.3A. The most intensive band observed for acid-untreated sample at 1016 $cm^{-1}$ is attributed to Si–O stretching vibrations of the tetrahedral sheets. The complex band with maxima at 704 and 665 $cm^{-1}$ involves the contributions of the perpendicular Si–O and bending $Mg_3OH$ vibrations, respectively. The absorption near 535 $cm^{-1}$ is due to the perpendicular Mg–O vibration, and the 469 $cm^{-1}$ band corresponds to overlapping Si–O–Si and Si–O–Mg bending bands. Pronounced degradation of the hectorite structure was observed even after 1 h of treatment. The intensity of the $Mg_3OH$ bending band decreased due to the release of the octahedral cations. The Si–O stretching band shifted to higher position and a new band corresponding to the $\nu$(SiO) vibrations of amorphous silica with a three-dimensional framework appeared at 1072 $cm^{-1}$. Another characteristic band of silica near 805 $cm^{-1}$ increased in intensity. With prolonged time of treatment the characteristic bands of hectorite gradually disappeared and only the bands related

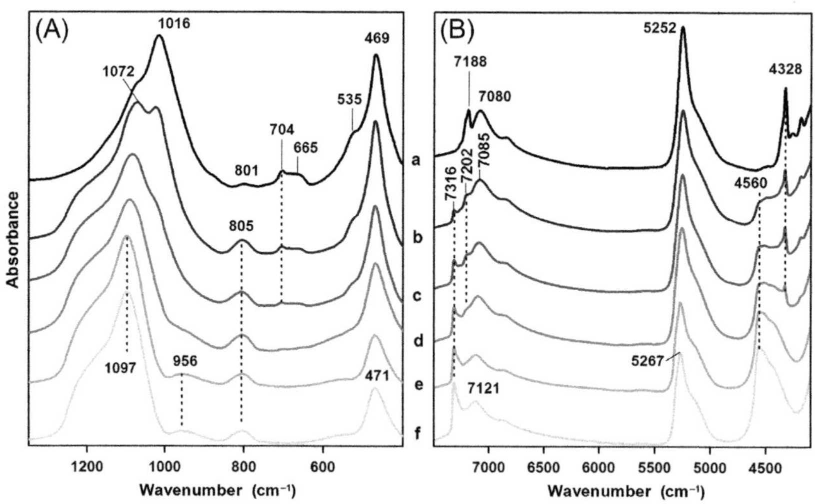


**FIG. 13.3** (A) MIR and (B) NIR spectra of (a) untreated hectorite and hectorite treated with 0.5 M HCl at 30°C for (b) 1 h, (c) 2 h, (d) 4 h, (e) 6 h and (f) 8 h.

to the Si–O vibrations of silica were present in the spectra. The MIR spectra of hectorite dissolved for 4 and 6 hours show also a new absorption band near 950 $cm^{-1}$ assigned to Si–O stretching vibrations in Si–OH groups (Madejová et al., 1998).

In addition to MIR spectroscopy the NIR region can also provide information on the structural changes of clay minerals upon acid treatment (Pálková et al., 2003; Madejová et al., 2007, 2009). Gradual release of the octahedral atoms can be followed in the regions where the bands related to the overtones (~7500–6500 $cm^{-1}$) and combinations (~5000–4000 $cm^{-1}$) of the structural OH groups occur (Fig. 13.3B). The OH overtone region of acid-untreated hectorite shows well-resolved bands at 7188 and 7080 $cm^{-1}$ corresponding to the first OH stretching overtones of the $Mg_3OH$ groups and water molecules. An intensive band at 5252 $cm^{-1}$ is due to $(\nu+\delta)H_2O$ and the combination of the stretching and bending vibrations of $Mg_3OH$ groups appear at 4328 $cm^{-1}$.

The NIR spectra of acid-treated hectorite show a reducing intensity of the $2\nu(OH)$ band near 7200 $cm^{-1}$ reflecting a decrease of the octahedral cations content. Moreover, a new band at 7316 $cm^{-1}$ corresponds to the first OH overtone of the Si–OH groups formed as the octahedral sheets in the hectorite layers become depleted and amorphous silica phase is created. Another feature that can be clearly observed in the NIR spectra of acid-treated hectorite is the decreasing intensity and gradual shift of both bands related to water molecules, such as $2\nu(H_2O)$ and $(\nu+\delta)H_2O$, to higher wavenumbers. Reduced intensities reflect a diminishing amount of water molecules due to the decomposition of the hectorite layers. An upward shift, on the other hand, provides evidence that $H_2O$ molecules adsorbed on the surface of destroyed layers and/or amorphous silica are less strongly H-bonded than those in untreated clay minerals. The dissolution of hectorite also changes the spectra in the 5000–4000 $cm^{-1}$ region. With progressive leaching of the octahedral cations, the $(\nu+\delta)Mg_3OH$ band at 4328 $cm^{-1}$ decreases in intensity and a broad band emerges at 4550–4560 $cm^{-1}$. This complex band corresponds to the $(\nu+\delta)$Si–OH combination modes of different SiOH groups, free or H-bonded to water molecules, present in the reaction product.

Madejová et al. (2009) presented a comprehensive paper on the advantages of the NIR spectroscopy in research of acid-treated clay minerals. The effects of the mineral structure, chemical composition and the availability of the interlayer space on the extent of clay mineral dissolution were examined. The changes of the NIR spectra of acid-treated Mt, ferruginous smectite, nontronite, illite/smectite and kaolinite were analyzed in detail. The NIR spectra confirmed that the mechanism of dissolution of the clay mineral in inorganic acids, as well as the final reaction product is the same regardless of the structural arrangement and chemical composition of the mineral. The extent of the dissolution reaction, however, depends on both the clay mineral type and reaction conditions used. Trioctahedral hectorite was much more susceptible to dissolution in HCl than dioctahedral nontronite. The presence of non-swelling interlayer spaces in illite/smectite substantially restricted the access of protons to the layers and the extent of the acid attack decreased.

The effect of non-swelling layers on clay mineral dissolution in HCl was confirmed also by Pentrák et al. (2010). The MIR and NIR regions showed that Mt with swelling interlayer spaces was completely dissolved within 18 hours in 6 M HCl at 95°C. In contrast, after 36 hours of treatment in HCl, residual illite was distinguished in the reaction products of mixed-layer illite/smectite and illite samples. The negative influence of the non-swelling interlayer spaces (interlayers) on the dissolution rate of reduced-charge smectites was observed also by Pálková et al. (2003). Heating of $Li^+$ and $Ni^{2+}$ saturated SAz-1 Mt at 150°C resulted in interstratified non-swelling and swelling interlayer spaces in LiSAz150 in contrast to fully expandable interlayers in NiSAz150. Heating at 300°C caused collapse of the interlayer spaces in LiSAz300 but some expandable interlayers were still identified in NiSAz300. The NIR spectra of the unheated $Li^+$- and $Ni^{2+}$-SAz showed a substantial degradation of their structure after acid dissolution in 6 M HCl. As the treatment time progresses (2–8 h), the Si–OH overtone increased in intensity, while a pronounced decrease was observed in the intensity of the OH overtone near 7060 $cm^{-1}$, related to the vibrations of octahedral cations. The changes in the spectra of both samples were comparable indicating that the character of the interlayer cation did not influence dissolution rate of unheated SAz in hydrochloric acid. The slower dissolution of LiSAz150, LiSAz300 and NiSAz300 was related to the development of non-swelling interlayer spaces due to the fixation of interlayer cations in the structure. The Si–OH overtone at 7314 $cm^{-1}$ permitted the extent of acid attack of smectites to be followed by IR spectroscopy, even in the case when no differences were observed in the 1300–400 $cm^{-1}$ region, which is traditionally used to monitor this process.

NIR spectroscopy was also successfully used for the characterization of products obtained by acid leaching of $Fe^{3+}$-bentonite from Slovakia (Madejová et al., 2007). Dissolution of bentonite in HCl resulted in damage of the $Fe^{3+}$-smectite structure and in the formation of a protonated amorphous silica phase. Kaolinite, present as admixture in bentonite, was more resistant to acid attack than smectite. A more detailed study of kaolinite dissolution is presented in Pentrák et al. (2009). Two kaolins containing kaolinites of different crystallinity were treated in HCl at 95°C. Due to the higher resistivity of these non-swelling 1:1 clay minerals, the time of treatment was set to 36 hours. The samples were characterized by several methods including NIR spectroscopy. No changes were observed in the OH overtone and combination regions of well-ordered Gold Field (GF) kaolinite. The same intensities and positions of the first overtones of the inner-surface (7166 and 7116 $cm^{-1}$) and inner (7064 $cm^{-1}$) OH groups indicated sufficient stability of GF kaolinite against acid attack. In contrast, partial decomposition of disordered KGa-2 kaolinite layers in HCl was confirmed. The intensities of the OH overtones near 7171 and 7064 $cm^{-1}$ as well as both OH combination bands near 4620 and 4525 $cm^{-1}$ decreased after acid treatment, and a new band at 7315 $cm^{-1}$ due to the first Si–OH overtone demonstrated partial protonization of Si–O bonds in KGa-2 kaolinite.

Acid-activation of smectites results in an increase in their SSA and pore volume. Both parameters significantly affect the hydration properties of smectites, which are of a great importance for their industrial applications. Pentrák et al. (2012a) investigated the effect of acid degradation of Mt structure on its hydration properties using a special UpDRIFT accessory enabling the measurement of NIR spectra of dried and hydrated samples directly in closed glass vials. The hydration of the samples at 52% relative humidity increased the intensity of the $2\nu(OH)$ band and caused its shift to lower wavenumbers due to the presence of more strongly H-bonded water molecules. The disappearance of the $2\nu(SiOH)$ band at 7315 $cm^{-1}$ in hydrated acid-treated Mt revealed the creation of H-bonds between the water molecules and Si–OH groups. The OH combination region also confirms the presence of H-bonded silanol groups. The variability of the H-bonds affected the position of the $(\nu+\delta)H_2O$ band near 5250 $cm^{-1}$. The second derivative analysis of this band distinguished water molecules bonded to Mt inner or outer surfaces from water H-bonded to different types of Si–OH groups.

The results presented above show that the NIR region provides not only information on the release of octahedral cations from clay mineral layers (available also from the MIR spectra), but also on the amount of water molecules adsorbed on the samples, the strength of the water H-bond interaction with clay mineral and/or silica surfaces, and also on the creation of Si–OH groups, the amount of which also reflects a degree of the mineral decomposition. Unfortunately the utilization of this technique for the investigation of acid-treated clay minerals is rather rare despite its benefits such as the speed and simplicity of sample preparation and the non-destructive nature of the technique.

## 13.5 ORGANO-MODIFIED CLAY MINERALS

### 13.5.1 NIR Spectra of Organoclays

The modification of clay minerals with organic compounds is currently an active research area targeting the preparation of progressive materials with desired properties required for novel industrial and environmental applications. A detailed review of the clay mineral-organic species interactions is given for example in Theng (1974), Paiva et al. (2008), Lagaly et al. (2013), He et al. (2014). Among the spectroscopic techniques, MIR spectroscopy has been most often used to characterize the interactions of organic substances with clay minerals (Yariv, 2002, and also Chapter 10 in this volume).

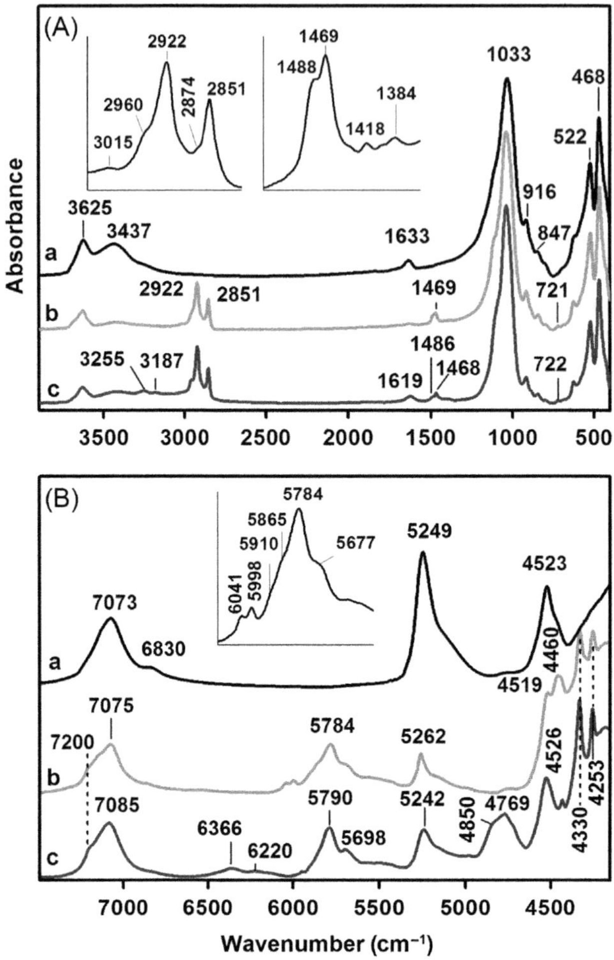


**FIG. 13.4** (A) MIR and (B) NIR spectra JP montmorillonite saturated with (a) Na (b) ODTMA and (c) ODA cations. The inserts in (A) are zoomed CH stretching and bending regions of ODTMA. The insert in (B) is zoomed CH overtone region of ODTMA.

The MIR spectrum of $Na^+$-Mt from the Jelšový Potok (JP) deposit (Slovakia) shows characteristic stretching and bending vibrations of OH (3625, 916 and 847 $cm^{-1}$) and Si–O (1033, 522 and 468 $cm^{-1}$) groups of Mt and water molecules (3437, 1633 $cm^{-1}$) (Fig. 13.4A,a). After the exchange of $Na^+$ with octadecyl trimethylammonium ($ODTMA^+$) cations, the spectrum shows also the absorption bands related to vibrations of $CH_3$ and $CH_2$ groups (Fig. 13.4A,b) details of which are zoomed in inserts. A weak band at 3015 $cm^{-1}$ belongs to stretching vibration of methyl groups bonded directly to nitrogen. The bands at 2922 $cm^{-1}$ and 2851 $cm^{-1}$ correspond to asymmetric

($\nu_{as}$) and symmetric ($\nu_s$) stretching vibrations of $CH_2$ groups, respectively, a shoulder near 2960 $cm^{-1}$ is due to $\nu_{as}(CH_3)$. The position of the $\nu_s(CH_3)$ band can be identified using only the second derivative (2874 $cm^{-1}$) because the band is fully overlapped by the stronger $\nu_s(CH_2)$ band. The deformation ($\delta$) vibrations of the $CH_2$ groups gives the band at 1469 $cm^{-1}$ while $\delta(CH_3)$ absorptions can be distinguished at 1488, 1418 and 1384 $cm^{-1}$. A weak band at 721 $cm^{-1}$ corresponds to $CH_2$ rocking vibration (Günzler and Gremlich, 2002). The MIR spectrum of octadecylammonium ($ODA^+$) Mt (Fig. 13.4A,c) shows, in addition to the $CH_3$ and $CH_2$ bands, also the vibrational modes related to $NH_3^+$ groups. The bands at 3255 and 3187 $cm^{-1}$ are assigned to the $\nu_{as}(NH_3^+)$ and $\nu_s(NH_3^+)$, respectively. The identification of the $NH_3^+$ deformation modes is less straightforward because they are overlapped with C–H or H–O–H deformation vibrations. The second derivative analysis reveals that a band at 1619 $cm^{-1}$ involves the $\delta(H_2O)$ at 1633 $cm^{-1}$ and $\delta_{as}(NH_3^+)$ at 1614 $cm^{-1}$. The $\delta_s(NH_3^+)$ mode overlaps the $\delta_{as}CH_3$ band ($\sim$1486 $cm^{-1}$).

In addition to the MIR spectra, the NIR region also provides useful information about organo-clays (OC). The NIR spectrum of $Na^+$-JP shows a broad complex band at 7073 $cm^{-1}$ due to the $2\nu(OH)$ of structural OH groups and $2\nu(HOH)$ of weakly H-bonded water molecules, a shoulder near 6830 $cm^{-1}$ due to the $2\nu(HOH)$ of strongly H-bonded $H_2O$, the intense band at 5249 $cm^{-1}$ related to $(\nu+\delta)H_2O$ and the band at 4523 $cm^{-1}$ to the $(\nu+\delta)OH$ of the structural OH groups (Fig. 13.4B,a). The presence of the $ODTMA^+$ cations gives rise to the $2\nu(CH)$ and $(\nu+\delta)CH$ bands appearing in the 6100–5500 $cm^{-1}$ and 4500–4000 $cm^{-1}$ regions, respectively (Fig. 13.4B,b). The complex band with a maximum at 5784 $cm^{-1}$ involves the contributions of several CH overtones, the accurate position of which was obtained using the second derivative analysis (see insert in Fig. 13.4B). Two weak bands at 6041 and 5998 $cm^{-1}$ are due to $2\nu_{as}(CH_3\text{–}N)$ and weak inflections at 5910 and 5865 $cm^{-1}$ to $2\nu_s(CH_3\text{–}N)$ and $2\nu_{as}(CH_3\text{–}C)$, respectively. The most pronounced peak appearing at 5784 $cm^{-1}$ belongs to $2\nu_{as}(CH_2)$, and a broad shoulder at 5677 $cm^{-1}$ to both $2\nu_s(CH_3\text{–}C)$ and $2\nu_s(CH_2)$. The spectra in the 4400–4000 $cm^{-1}$ region show several bands related to the combinations of C–H stretching and deformation modes of the $CH_3$ and $CH_2$ group. The band at 4460 $cm^{-1}$ corresponds to both $(\nu+\delta)_{as}CH_3$ and $(\nu+\delta)_{as}CH_3\text{-}N$ vibrations. The bands at 4330 $cm^{-1}$ and 4253 $cm^{-1}$ are due to $(\nu+\delta)_{as}CH_2$ and $(\nu+\delta)_sCH_2$, respectively (Workman and Weyer, 2008; Madejová et al., 2011, 2016).

The presence of $NH_3^+$ groups modifies the shape of $ODA^+$-JP spectrum (Fig. 13.4B,c) compared to $ODTMA^+$-JP (Fig. 13.4B,b). The $2\nu_{as}NH_3^+$ and $2\nu_sNH_3^+$ overtones occur at 6366 and 6220 $cm^{-1}$, respectively. The combination region of $ODA^+$-JP reveals a complex band with two overlapping maxima at 4850 and 4769 $cm^{-1}$ related to $(\nu+\delta)_{as}NH_3^+$ and $(\nu+\delta)_sNH_3^+$ modes, respectively. The high intensity of the $NH_3^+$ combination band confirms a special potential of this region in the identification of N–H vibrations (Madejová et al., 2011).

The $ODTMA^+$ and $ODA^+$ cations also influence the shape and the intensities of the bands of parent Mt (Fig. 13.4B). The $2\nu(OH)$ band shows a shoulder near 7200 $cm^{-1}$ reflecting a different environment of the structural OH groups after organic cations intercalation (Madejová et al., 2011). Though the content of water molecules in organo-Mts (O-Mt) significantly decreased, as confirmed the intensity of the $(\nu+\delta)H_2O$ of both, $ODTMA^+$-, and $ODA^+$-JP, some water molecules remained trapped in the interlayer spaces. The basal oxygen atoms of the tetrahedral sheets thus can interact either with $H_2O$ or with the tails of the alkyl chains, which consequently affects also the vibrations of OH groups. A more detailed analysis of the shape of the $2\nu(OH)$ band of O-Mts is given in (Madejová et al., 2011).

A different characteristic of the alkylammonium cation headgroup significantly affects the positions of the $2\nu(OH)$ and $(\nu+\delta)H_2O$ bands (Fig. 13.4B). While the $2\nu(OH)$ of $ODTMA^+$-JP is at almost the same position as $Na^+$-JP (7075 and 7073 $cm^{-1}$, respectively), the OH band is shifted to 7085 $cm^{-1}$ for $ODA^+$-JP. The reason for such shift is still unknown. It is possible that the proton of the $NH_3$ group can partly enter the ditrigonal cavities of the tetrahedral sheet and repulse the O–H group from its original position, thus influencing the vibrational energy of the O–H group. A different position is also observed for the $(\nu+\delta)H_2O$ band. An upward shift to 5262 $cm^{-1}$ for $ODTMA^+$-JP indicates the presence of mainly weakly H-bonded water molecules to the hydrophobic surface of the $ODTMA^+$-JP. The $ODA^+$ cation, however, contains not only a hydrophobic part, such as octadecylammonium chain but also a hydrophilic part, like $NH_3^+$ headgroup capable to create strong H-bond with water molecules. As a result, the position of the $(\nu+\delta)H_2O$ band of $ODA^+$-JP and $Na^+$-JP is nearly the same (Fig. 13.4B,a and c).

The arrangement of the alkylammonium cations (cationic surfactants) in smectites interlayer spaces affects the properties of OC and thus also their utilization. Vaia et al. (1994) was the first to show that the positions of the bands related to the $\nu_{as}(CH_2)$, $\nu_s(CH_2)$ and $\delta(CH_2)$ modes were sensitive to the organic cations configuration. With the increasing amount of surfactants present in the smectite interlayer spaces, the alkyl chain length and decreasing temperature the $\nu(CH_2)$ bands were shifted from a higher position characteristic for disordered (*gauche*) conformers to lower wavenumbers characteristic for ordered (all-*trans*) conformers. Many authors used MIR spectroscopy to study the arrangement of surfactants in clay minerals (e.g., Osman et al., 2004; He et al., 2004a; Zhu et al., 2007a,b). A detailed review of their work is presented in Chapter 10 in this volume.

The overtone ($2\nu$) and combination ($\nu+\delta$) modes of $CH_2$ groups appearing in the near-IR spectra of OC are directly related to the fundamental $CH_2$ vibrations; therefore, their positions should also reflect the modification of the alkyl chain arrangements. The potential of the NIR spectroscopy to follow the conformational changes of alkylammonium surfactants was briefly introduced by Madejová et al. (2011). A more detailed methodological NIR study was recently published (Madejová et al., 2016). The NIR spectra of JP Mt saturated with alkyl trimethylammonium cations of varying alkyl chain lengths (C6-JP up to C18-JP) were analyzed and the changes in the positions of the $2\nu_{as}(CH_2)$, $(\nu+\delta)_{as}CH_2$ and $(\nu+\delta)_{s}CH_2$ bands were compared with $\nu_{as}(CH_2)$, $\nu_{s}(CH_2)$ occurring in the corresponding MIR spectra. The bands found in the 5811–5802 $cm^{-1}$ range reflected predominance of the *gauche* conformers, characteristic for short alkyl chains, while wavenumbers below 5790 $cm^{-1}$ implied an increase of ordered all-*trans* conformers for surfactants with longer chains. The magnitude of the $2\nu_{as}(CH_2)$ shift from C6-JP to C18-JP (28 $cm^{-1}$) was significantly higher than that observed in the MIR region for $\nu_{as}(CH_2)$ ($\Delta\nu = 8\ cm^{-1}$) or $\nu_{s}(CH_2)$ ($\Delta\nu = 10\ cm^{-1}$). The combination $(\nu+\delta)_{as}CH_2$ and $(\nu+\delta)_{s}CH_2$ modes were also shifted towards lower wavenumbers by increasing the alkyl chain length. The overall decrease (by 12 and 18 $cm^{-1}$, respectively) was higher compared to MIR region, but less pronounced than in the C–H overtone region. The results show that NIR spectroscopy is a useful method that provides information on the configuration of alkylammonium surfactants intercalated into Mt interlayer spaces.

The majority of works dealing with OC were addressed to study the interactions between swelling clay minerals and quaternary ammonium salts. Polymer-modified clays, a novel class of hybrid organic–inorganic materials, have also attracted industrial and academic interest due to their wide range of applications, such as in the fields of catalysis, nanofillers, controlled drug release systems, or as a solid matrix for the encapsulation of photoactive molecules, such as organic dyes. For proper characterization of polymer-intercalated clays the MIR spectroscopy, in addition to XRD and TG, is most often used (Zampori et al., 2010; Ma et al., 2010; Finocchio et al., 2014). The utilization of NIR spectroscopy in studies of clay-polymer nanocomposites (CPN), however, is less frequent (Finocchio et al., 2011; Czímerová et al., 2013; Pentrák et al., 2014).

Czímerová et al. (2013) used polydiallyl dimethylammonium (PDDA) chloride to prepare an organo-Mt complex. The content of PDDA was 50, 100, 200, 300, 500 and 1000 mg per 1 g of Mt. PDDA-Mt were further modified with a laser dye Rhodamine 6G (R6G) to obtain photo-functional hybrid materials and the effect of the amount of PDDA on the luminescence properties of complex hybrid system was followed. The PDDA loading into Mt was verified by XRD and NIR spectroscopy using the complex $2\nu(CH)$ band. With a growing amount of PDDA used for CPN preparation, the intensity of the $2\nu(CH)$ band increases (Fig. 13.5A). The values of the areas of the complex $2\nu(CH)$ band (Fig. 13.5B), reflecting the amount of PDDA in the sample, show a pronounced increase up to 500 mg PDDA used per 1 g of Mt. The increase of the PDDA content at 1000 mg loading is less steep but still clearly visible. Because the XRD pattern reveals only negligible expansion of the Mt basal spacing for PDDA loading above 300 mg per 1 g of Mt (Fig. 13.5C) the increasing area of the $2\nu(CH)$ band (Fig. 13.5B) indicates that not all

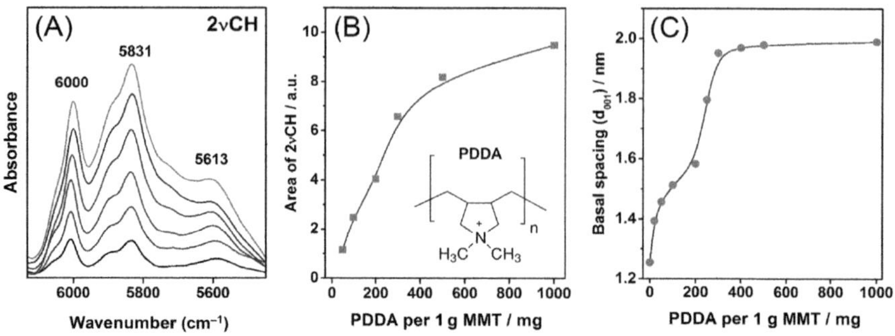


**FIG. 13.5** (A) The first CH overtone band of montmorillonite intercalated with increasing amount of PDDA (from the bottom to the top), (B) the area of the 2$\nu$(CH) band, (C) basal spacing of PDDA-montmorillonite.

polycations are intercalated in the Mt interlayer spaces. Instead, the PDDA molecules are more likely attached onto the outer surface of Mt particles.

Finocchio et al. (2011) used MIR and NIR spectroscopy to investigate the ability of the $Ca^{2+}$-Mt STx-1 to interact with three different nonionic polymers: two poly(ethylene glycol)-based surfactants (PEG 1500 and PEG 4000), having similar molecular structure but different weight, and poly(ethylene oxide) surfactant (BRIJ78), characterized by the presence of an aliphatic chain. The polymer intercalation within interlayer spaces was confirmed by the increasing of the $d_{001}$ in XRD. In addition to CH stretching and deformation modes, the CH overtone and combination bands were analyzed. The presence of polymers in samples confirmed the appearance of 2$\nu$(CH) bands: a broad band with two maxima near 5787 and 5624 $cm^{-1}$ for PEG samples and several components band (5786, 5670, 5589, 5498 $cm^{-1}$) for the sample treated with BR1J78. Upon polymer intercalation the reduction of the 2$\nu$(OH) (near 7075 $cm^{-1}$) band intensity and a shift of the $(\nu+\delta)H_2O$ from 5245 $cm^{-1}$ to 5230–5235 $cm^{-1}$ was observed. These authors explain the $(\nu+\delta)H_2O$ shift as a result of H-bonding between $H_2O$ and hydrophilic polymer fragments. Reduction of the intensity, on the other side was mainly attributed to the partial substitution of cations' hydration water molecules with hydrophilic polymer.

NIR spectroscopy can also help to study the intercalation of potassium acetate (KAc) into non-swelling 1:1 kaolinite (Zhang et al., 2011) and halloysite (Cheng et al., 2011a,b). Both, MIR and NIR spectra provided the significant differences and the obvious structural change between kaolinite and the kaolinite-KAc and halloysite and halloysite-KAc intercalation complexes. The NIR spectrum of the kaolinite-KAc intercalation complex showed a decrease in the intensities of the structural OH overtone (7300–7000 $cm^{-1}$) and the combination (4600–4300 $cm^{-1}$) bands. Moreover, the presence of KAc in both complexes confirmed the bands in the 6000–5500 $cm^{-1}$ region assigned to the overtones of the $CH_3$ stretching band observed at 2994 $cm^{-1}$. Both studies confirm that the application of NIR spectroscopy for the study of intercalated kaolinite/halloysite has a great potential for the understanding of the interactions between the surface hydroxyls of 1:1 clay minerals and the inserting ions.

## 13.5.2 Interaction of Organoclays With Water and Other Organic Species

The exchange of the inorganic by organic, most often alkylammonium, cations significantly modifies the surface properties of smectites, including their capacity for water uptake. The ability of OC to interact with water molecules is governed by the nature of the organic cation, such as a type of surfactant, alkyl chain lengths or the size of cation (Giese and van Oss, 2002). The MIR spectroscopy is often used to examine the hydration of OC (e.g., Fripiat et al., 1969; Stevens and Anderson, 1996; Ma et al., 2010). The utilization of the NIR spectroscopy, however, was presented only recently (Pálková et al., 2015). The authors compared the extent of the hydration of SAz-1 Mt modified with tetramethylammonium cations ($TMA^+$-S) and hexadecyl trimethylammonium cations ($HDTMA^+$-S). The size of the cation significantly influences the intensity and position of the $(\nu+\delta)H_2O$ band (Fig. 13.6A). A weak band at 5258 $cm^{-1}$ observed for dried $HDTMA^+$-S indicates very low content of water molecules bonded to Mt surface by weak H-bonds. Slightly higher intensity and a lower wavenumber (5246 $cm^{-1}$), found for dried $TMA^+$-S, reflects more water involved in stronger H-bonds. After exposure of the samples to water vapor under 52 and 100% relative humidities (52RH and 100RH, respectively) the $(\nu+\delta)H_2O$ band becomes broader and its intensity is significantly increased due to enlarged water content. The position found for $TMA^+$-S (5241 $cm^{-1}$) hydrated at 100RH reflects the adsorption of water molecules between pillars of $TMA^+$ cations preventing the formation of an extensive network of strongly H-bonded water molecules. In contrast, the band at 5234 $cm^{-1}$

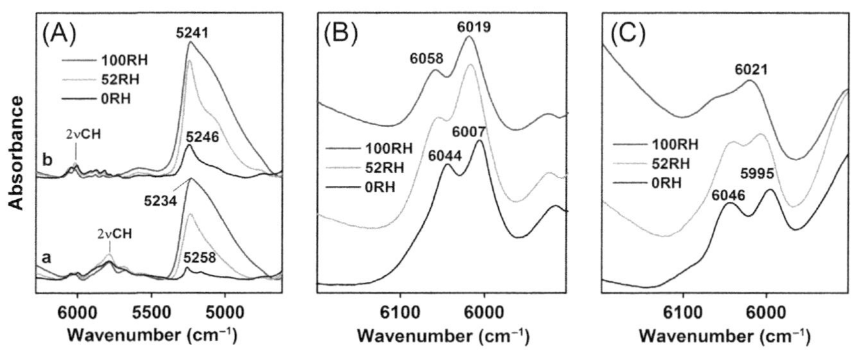


**FIG. 13.6** (A) NIR spectra of (a) HDTMA-SAz and (b) TMA-SAz hydrated at 0RH, 50RH and 100RH, (B) detail of the CH overtone of TMA-SAz, (C) detail of the CH overtone of HDTMA-SAz.

of HDTMA$^+$-S indicates a presence of more strongly H-bonded water molecules. The interlayer space of Mt is almost fully occupied by HDTMA$^+$ and only limited content of water molecules is supposed to be trapped in the interlayer spaces. The majority of $H_2O$ is adsorbed as pore water on the outer surfaces of O-Mts, also revealing the effect of hydration on the vibration modes of $CH_3$ groups. The $2\nu_{as}(CH_3–N)$ bands observed at 6044 and 6007 cm$^{-1}$ for dried TMA$^+$-S (Fig. 13.6B) and at 6046 and 5995 cm$^{-1}$ for dried HDTMA$^+$-S (Fig. 13.6C) are shifted to higher wavenumbers upon hydration at 100% RH. The water molecules penetrate closer to the headgroup of the organic cation, therefore changing their near environment, which affects the positions of the $CH_3$ bands.

OC are widely used as adsorbents for organic contaminants from soils and industrial effluents. Although there have been many studies focused on the adsorption of organic pollutants by OC using MIR spectra, only few papers report the use of NIR spectroscopy (Zhou et al., 2008; Liu et al., 2009; Tomić et al., 2012; Gamba et al., 2015). Zhou et al. (2008), for example, utilized the NIR spectroscopy to measure adsorption of *para*-nitrophenol on untreated and HDTMA$^+$-exchanged Mt. *P*-nitrophenol was characterized by intense NIR bands at 8890 and 6052 cm$^{-1}$, which shifted to 8840 and 6017 cm$^{-1}$, respectively, upon adsorption on HDTMA$^+$-Mt. This indicates a strong interaction between the *p*-nitrophenol and the OC. The bands were not observed for *p*-nitrophenol adsorbed on untreated Mt. The authors concluded that *p*-nitrophenol was adsorbed to significantly greater amounts on the OC compared with the untreated Mt.

Liu et al. (2009) used NIR spectroscopy, in addition to MIR and XRD, to measure the adsorption of a phenol, a common component of contaminated soil, groundwater or leachate, on untreated Na$^+$-Mt and on HDTMA$^+$-Mt. This study demonstrated that phenol was adsorbed and penetrated the interlayer space of the HDTMA$^+$-modified Mt, resulting in the expansion of the interlayer space. The NIR spectra showed a prominent band at 8260 cm$^{-1}$, assigned to the combination result of the C–H stretching vibrations of HDTMA$^+$ and adsorbed phenol. A detailed analysis of the first C–H overtones and combination modes of HDTMA$^+$-Mt and OC with adsorbed phenol was provided. The higher the value of the surfactant (HDTMA$^+$) concentration, the more phenol was adsorbed by the O-Mt. Changes in both the wavenumber and intensity of MIR and NIR spectra revealed that phenol was not only bonded to the water in the Na hydration sphere but also to the HDTMA$^+$ in the interlayer space. As the loading of the surfactant increased the amount of available water decreased. The phenol then interacted with the clay mineral surface and with the surfactant molecules. Both studies (Zhou et al., 2008; Liu et al., 2009) demonstrate that NIR is an effective technique to determine the adsorption of the organic pollutants on clay minerals. It is of high importance to understand the adsorption mechanism of clay materials to organic pollutants due to the application of clay-based materials in pollution prevention and environmental remediation.

### 13.5.3 Acid Treatment of Organoclays

The properties of clay minerals can be considerably improved when using a combination of different chemical treatments. For example, acid activation of clay minerals is one of the most effective methods to produce active materials for adsorption and catalysis due to their high SSA and acidity. On the other hand, the progressive replacement of inorganic by organic cations modifies the surface properties of a clay mineral from highly hydrophilic to increasingly organophilic (hydrophobic). These materials (OC) exhibit high adsorption capacity towards hydrophobic molecules. The combination of acid-activation with alkylammonium intercalation provides a potential enhancement of the clay properties. In the first instance, the clay mineral is H-saturated by a cation exchange in acid solution (acidifying process), resulting in a proton rich environment when presented to organic molecules. In the second instance, the catalytic abilities can be further improved by providing organic cations in the interlayer space, which permit access to and interact effectively with organic molecules. The possibilities for catalysis in selected systems (e.g., $TMA^+$-clays) are therefore greater than initial clay mineral because both the internal and the external surfaces are available for reaction (Breen et al., 1997; Moronta et al., 2002).

The size and the amount of alkylammonium cations, the intensity of acid treatment, and the type of clay mineral play a key role for the extent of OC dissolution. Breen et al. (1997) compared the extent of dissolution of three dioctahedral smectites saturated with tetramethylammonium, dodecyl trimethylammonium and octadecyl trimethylammonium cations to different CEC level. The MIR spectra clearly showed that the long-chain alkylammonium cations restricted the access of protons to the layers and the extent of the acid attack was decreased. In addition to MIR, NIR spectra are also used for acid-treated OC studies. Madejová et al. (2009) performed detailed analysis of the NIR spectra of $Li^+$, $TMA^+$ and $HDTMA^+$ saturated SAz-1 Mt dissolved in 6 M HCl at 80°C. Both, MIR and NIR spectra proved that dissolution of the $TMA^+$-SAz-1 was only partially reduced compared to Li-SAz-1, while the presence of bigger $HDTMA^+$ cations significantly protected the Mt layers from the acid attack. Only NIR region, however, could clearly monitor the creation of the SiOH groups, which are an important indicator of the Mt surface acidification. The OH overtone region in the spectra of $Li^+$-SAz-1 and $TMA^+$-SAz-1 revealed that the SiOH band near 7315 $cm^{-1}$ increased in intensity with enhanced acid treatment. In contrast, no SiOH groups were identified in the NIR spectra of HCl-treated $HDTMA^+$-SAz-1. This observation indicates that $HDTMA^+$ cations fully cover the inner and outer surfaces of Mt and hinder the access of protons to the Si–$O^-$ bonds.

Tomić et al. (2012) used NIR and MIR spectroscopies for the characterization of hydrophilic (Na-saturated) and hydrophobic (HDTMA-saturated) bentonite from Petrovac (Serbia) treated with sulphuric acid of different concentrations. Spectra of acid-treated Na-samples showed a slight decrease in the intensities of the OH overtone (near 7100–7000 $cm^{-1}$) and combination bands (4600–4300 $cm^{-1}$) with increasing acid concentration reflecting a fewer number of octahedral atoms. The lack of a SiOH band near 7315 $cm^{-1}$, which usually confirms the presence of protonated amorphous silica, demonstrated that the conditions selected for acid treatment were more moderate than used by Madejová et al. (2009). Even a lower extent of decomposition was observed for $HDTMA^+$-Mt. Only subtle wavenumber changes and a low decrease of intensity of $2\nu(OH)$ and $(\nu+\delta)H_2O$ bands in the spectrum of the $HDTMA^+$-Mt treated in 6 M acid (i.e. the highest acid concentration used) indicated partial change in the structure of hydrophobic Mt. Modest changes in the region of CH overtone and combination bands were observed as well. Minimal changes of O-Mt layers upon $H_2SO_4$ treatment resulted from a complete covering of inner and outer surface of Mt by $HDTMA^+$ cations, which blocked the access of protons to the layers. The study confirmed that the NIR region provides results that can be comparable to the MIR spectroscopy outcome and also showed a greater sensitivity of the structural changes in Mt, moisture content and impact of $H_2SO_4$ to inorganic–organic Mt.

The NIR spectra of acid treated O-Mt prepared from five symmetric short alkyl-chain cations with a progressively increasing number of carbons in one single alkyl chain: tetramethylammonium ($Me_4N^+$), tetraethylammonium ($Et_4N^+$), tetrapropylammonium ($Pr_4N^+$), tetrabutylammonium ($But_4N^+$), tetrapentylammonium ($Pe_4N^+$) and SAz-1 Mt were reported in Madejová et al. (2012). This study was an extension of the work of Pálková et al. (2011) in which mainly surface and textural properties of acid-treated O-Mt, prepared using the same surfactants (except of $Pe_4N^+$), were investigated. The NIR spectra of selected samples from this series, namely $Me_4N^+$-SAz, $Pr_4N^+$-SAz and $Pe_4N^+$-SAz are given in Fig. 13.7.

The NIR spectrum of acid-untreated $Me_4N^+$-SAz shows $2\nu(OH)$ at 7055 $cm^{-1}$, $(\nu+\delta)H_2O$ at 5246 $cm^{-1}$ and $(\nu+\delta)OH$ at 4514 $cm^{-1}$ (Fig. 13.7A,a). The presence of methyl groups confirms a complex C–H overtone band with a maximum at 6013 $cm^{-1}$ assigned to $2\nu_{as}(CH_3–N)$ and several sharp bands below 4450 $cm^{-1}$ related to the combination modes of $CH_3$ groups. A detailed assignment of the individual CH overtone and combination components of $Me_4N^+$ as well as other short-chain alkylammonium cations is given in Madejová et al. (2012). A shift and reduced intensity of the $2\nu(OH)$ band upon $Me_4N^+$-SAz decomposition in HCl reflects a release of the central atoms from the octahedral sheets and lower amount of water molecules as confirms also the intensity of the $(\nu+\delta)H_2O$. An upward shift of both $2\nu(OH)$ and $(\nu+\delta)H_2O$ vibrational modes provides evidence that the $H_2O$ molecules adsorbed on the surface of destroyed layers and/or amorphous silica are less hydrogen bonded than those in untreated clay minerals. A new $2\nu(SiOH)$ band near 7315 $cm^{-1}$ appears after 2 hours of treatment and its intensity increases with the extent of Mt decomposition and creation of

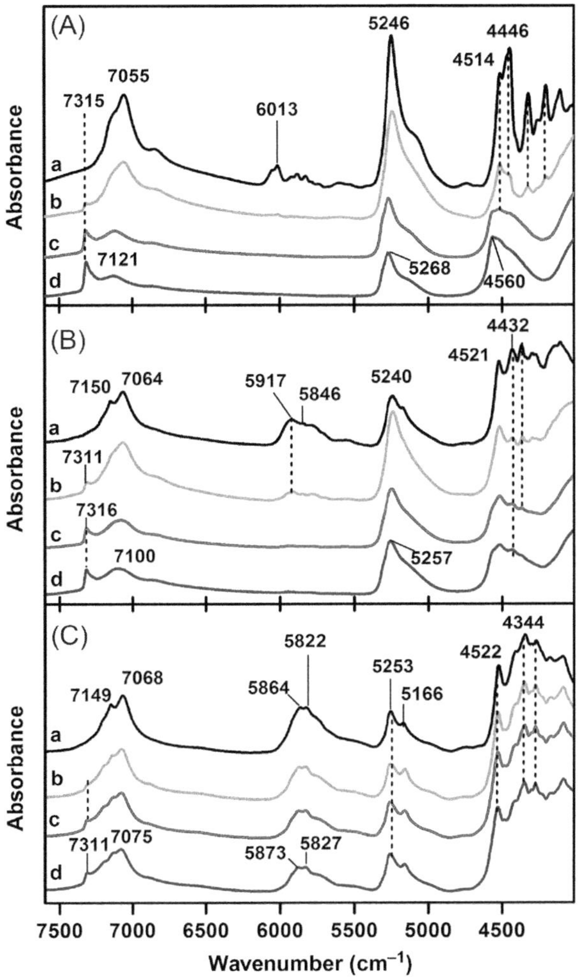


**FIG. 13.7** NIR spectra of (A) $Me_4N$-SAz, (B) $Pr_4N$-SAz and (C) $Pe_4N$-SAz treated with 6 M HCl at 80°C for (a) 0 h, (b) 2 h, (c) 6 h, and (d) 8 h.

protonated amorphous silica (Fig. 13.7A). The C–H overtone and combination regions of acid-treated $Me_4$-SAz confirm gradual de-intercalation (desorption) of $Me_4N^+$ cations. While the $2\nu(CH_3)$ band cannot be recognized in the spectrum of $Me_4N^+$-SAz-2 h, the $(\nu+\delta)CH_3$ bands occurring in the 4450–4000 cm$^{-1}$ region confirm the presence of methyl groups in this sample. Only the bands related to the protonated amorphous silica can the identified in the spectra of $Me_4N^+$-SAz-6 h and $Me_4N^+$-SAz-8 h (Fig. 13.7A).

The NIR spectrum of acid-untreated $Pr_4N^+$-SAz shows similar basic features as that of $Me_4N^+$-SAz, however, several differences are clearly recognized (Fig. 13.7B,a). Due to the interactions of the structural OH groups with $Pr_4N^+$, the $2\nu(OH)$ band is split into two components at 7150 and 7064 $cm^{-1}$ reflecting different nearest environment of the OH groups (Madejová et al., 2011). Higher hydrophobicity of the $Pr_4N^+$-SAz surface is reflected in the pronounced decrease of the $(\nu+\delta)H_2O$ band intensity. The presence of organic cations confirms a broad complex band centered at 5917 $cm^{-1}$. The maximum of the band corresponds to $2\nu_{as}(CH_3)$ vibration while the weaker band at 5846 $cm^{-1}$ is due to $2\nu_{as}(CH_2)$ vibration. The $(\nu+\delta)_{as}CH_2$ appears at 4432 and $(\nu+\delta)_{s}CH_2$ at 4368 $cm^{-1}$. Dissolution of $Pr_4N^+$-SAz in HCl modifies the shape of the NIR spectrum. After 2 hours of treatment the $2\nu OH$ component near 7150 $cm^{-1}$ is transformed into a weak inflection (Fig. 13.7B,b). The complete disappearance of this overtone upon 6 hours of treatment confirms its relation to the OH groups interacting with alkylammonium cations. As expected, the presence of the $2\nu(SiOH)$ band at 7311 $cm^{-1}$ proves partial decomposition of the Mt layers in the $Pr_4N^+$-SAz-2 h sample. Its intensity increases with time of treatment. The CH overtone and combination regions of $Pr_4N^+$-SAz-2 h confirm a substantial decrease of the surfactant content after two hours of treatment. While the $2\nu(CH)$ bands are not clearly visible in the spectrum of the 6 hour treated sample, the $(\nu+\delta)CH$ modes can be identified even in $Pr_4N^+$-SAz-8 h proving very high sensitivity of this region for the identification of organic cations.

The release of organic cations affects the surface properties of $Pr_4N^+$-SAz. Presence of protons partially replacing $Pr_4N^+$ imparts more hydrophilic character on the $Pr_4N^+$-SAz-2 h surfaces. As a result, the amount of adsorbed water significantly increases with the intensity of the $(\nu+\delta)H_2O$ band at 5240 $cm^{-1}$. The 6 hour and 8 hour dissolution generates more silica in the solid reaction product with more hydrophobic surface; therefore the combination water band decreases in intensity (Fig. 13.7B). Although the NIR spectra of 8 hour treated $Me_4N^+$-SAz and $Pr_4N^+$-SAz show similar patterns the comparison of the intensities of the $2\nu OH$ (related to Mt) and $2\nu SiOH$ (related to reaction product) bands demonstrates different degree of the Mt decomposition. The lower intensity of the 7316 $cm^{-1}$ band and higher intensity of the 7100 $cm^{-1}$ in $Pr_4N^+$-SAz-2 h demonstrates the higher stability of $Pr_4N^+$SAz than $Me_4N^+$-SAz.

The shape of the NIR spectrum of $Pe_4N^+$-SAz (Fig. 13.7C,a) resembles $Pr_4N^+$-SAz (Fig. 13.7B,a) though the positions and intensities of selected bands differ substantially. For example, due to the higher number of $CH_2$ groups the complex $2\nu(CH)$ band is shifted to a lower wavenumber and its intensity increased. Larger $Pe_4N^+$ cations impart strong hydrophobic character to $Pe_4N^+$-SAz surface. The content of adsorbed water molecules is reduced significantly as confirmed by the low intensity of the $(\nu+\delta)H_2O$ band. Splitting of the band into two components indicates the presence of two groups of water molecules. The band at 5253 $cm^{-1}$ corresponds to residual $H_2O$ molecules weakly H-bonded to the siloxane surface while the 5166 $cm^{-1}$ is a characteristic position for “liquid” water (Workman and

Weyer, 2008), such as water molecules H-bonded to each other, but not directly associated with cations or surfaces. The NIR spectra of acid-treated $Pe_4N^+$-SAz show a considerably lower degree of their decomposition than $Pr_4N^+$-SAz. While no 2$\nu$CH bands can be recognized in the spectrum of $Pr_4N^+$-SAz-8 h, the $Pe_4N^+$-SAz-8 h shows a well-resolved complex band near 5870 $cm^{-1}$. Moreover, the manner in which the Mt layers decomposed was also different. The NIR spectra of $Pr_4N^+$-SAz indicate a gradual release of octahedral atoms and the creation of Si–OH groups with the time of treatment. In contrast, the NIR spectra of $Pe_4N^+$-SAz show structural modification for a 2 hour treated sample, but longer dissolution in HCl evokes only small changes indicating retardation of the acid-dissolution process.

The results reported in Madejová et al. (2012) show that the size of the organic cation plays a significant role in the modification of the Mt structure upon decomposition in HCl. Organic cations with short tetraalkylammonium chains are located on the Mt surface as discrete units isolated by more or less uncovered mineral surface that can act as potential access for protons. The size of $Me_4N^+$ is small enough to leave a sufficient part of the siloxane surface free for protons attack. Due to the increased size of $Et_4N^+$ and $Pr_4N^+$ less siloxane area is available for the penetration of protons to the layers and the $Et_4N^+$-SAz and $Pr_4N^+$-SAz are more stable in HCl at short treatment times. Upon longer dissolution the organic cations are removed and the extent of Mt destruction almost reaches that of $Me_4N^+$-SAz. Mt saturated with the largest organo-cations within the studied series ($Bu_4N^+$, $Pe_4N^+$) shows the highest stability upon the acid-treatment. Bulky alkylammonium surfactants covering its inner and outer surfaces prevent the ingress of the protons towards the layers and thus protect the Mt from the acid attack even after 8 hours of treatment. These results also prove that short-chain alkylammonium cations prevent the access of protons to the layers protecting the mineral from degradation in the acid.

All of the reports given above were focused on the acid-treatment of Mt modified with alkylammonium cations. Commonly utilized quaternary alkylammonium cations, however, show limited thermal stability, which may create problems when they are used as fillers for polymers. Phosphonium cations have also been used as clay mineral modifiers for CPN as they show higher thermal stability compared to their ammonium counterparts. Pálková et al. (2017) applied the IR spectroscopy to compare the thermal stability of acid-untreated and acid-treated SAz-1 Mt saturated with tetrabutylammonium ($Bu_4N^+$) and tetrabutylphosphonium ($Bu_4P^+$) cations. A thermogravimetric analyzer connected with MIR spectrometer for the analysis of evolved gasses and an NIR spectrometer equipped with a special temperature cell allowing the in-situ measurements of NIR spectra were used to follow the effect of central atom (N and P) of surfactant headgroup on the thermal stability of OC. Upon organic phase liberation the $\nu$(CH) absorption bands of gaseous products (aliphatic hydrocarbons) appeared in the MIR spectra. The intensity of the absorption bands reached maxima at 277°C and 446°C for $Bu_4N^+$-S and $Bu_4P^+$-S, respectively. The NIR spectra of acid-untreated samples showed that a decrease of the intensity of the first 2ν(CH) overtones bands due to organic cation release was initiated at lower temperature for $Bu_4N^+$-S (above 380°C) than for $Bu_4P^+$-S (above 500°C). The acid treatment of $Bu_4N^+$-S and $Bu_4P^+$-S in 6 M HCl for 12 hours caused the removal of a part of organic cation content and partial decomposition of the structure of Mt. Upon heating, the content of the surfactants further dropped down. A clear decrease of the 2ν(CH) band intensity was observed at 320°C and 420°C for $Bu_4N^+$-S and $Bu_4P^+$-S, respectively. The first overtone of the 2ν(SiOH) units, due to the presence of hydrated amorphous silica, the product of Mt decomposition, was detected for both acid treated samples. The intensity of the band increased significantly with increasing temperature, remaining in the spectra even after heating for 540°C. The IR spectra of both, acid-untreated and acid-treated Mt confirmed higher stability of the samples containing phosphonium as the central atom of the surfactant headgroup.

### 13.5.4 Adsorption of Pyridine on Acid-Treated Samples

The interaction of Mt with inorganic acid significantly changes the surface acidity of a mineral. A powerful technique capable to distinguish different types of clay mineral acid sites is MIR spectroscopy. For example, pyridine as a base of moderate strength can be physically and/or chemically adsorbed on the Mt surface. Chemisorbed pyridine can either form H bonds between the pyridine nitrogen atom and OH groups of clay minerals, can be protonated to pyridinium cation at strong Brønsted acid (BA) sites (e.g., from polarized water molecules), or can be coordinately bound through pair of electrons on nitrogen atom to Lewis acid centre (e.g., $Al^{3+}$ cations). These various species are easily identified and distinguished by examining the 1700–1400 $cm^{-1}$ region, where the absorption bands related to ring stretching vibrations (skeletal modes) of pyridine are located.

Pálková et al. (2013) used MIR spectroscopy and pyridine to characterize the acid sites on $Na^+$-saturated Cheto Mt surface before ($Na^+$-SAz) and after dissolution in 6 M HCl at 80°C for 2, 4, and 8 h (SAz-2 h, SAz-4 h, and SAz-8 h). After pyridine adsorption the spectrum of $Na^+$-SAz showed the intense bands at 1441 and 1594 $cm^{-1}$ corresponding to the H-bonded pyridine, a band near 1487 $cm^{-1}$ indicating the presence of weak BA sites and the vibration at 1573 $cm^{-1}$ attributed to the physisorbed pyridine (Yariv, 2002). The modification of the $Na^+$-SAz structure in HCl affected the bonds between pyridine and the sample. The spectrum of SAz-2 h showed a decrease of the intensities of the bands related to H-bonded and physi-sorbed pyridine and a new well-developed band at 1541 $cm^{-1}$ assigned to $N^+$-H deformation vibration of pyridinium cation. The pyridine molecules accepted protons from $H_3O^+$, present in the sample after HCl treatment, thus confirming the existence of strong BA sites on the mineral surface. The spectrum of SAz-4 h revealed a pronounced decrease of the intensity of the 1541 $cm^{-1}$ band reflecting the reduction of the amount of the strong BA sites. After 8 hours of HCl treatment, no band was found in the SAz-8 h spectrum due to pyridinium cations. Only bands corresponding to the H-bonded and/or physisorbed pyridine trapped on the amorphous silica surface were seen in the 1700–1400 $cm^{-1}$ region.

The same set of samples was used by Madejová et al. (2015) to examine the potential of NIR spectroscopy to distinguish different pyridine species adsorbed on SAz-1 Mt surface (Fig. 13.8). The NIR spectrum of acid-untreated Na-SAz shows characteristic bands at 7057 and 4515 $cm^{-1}$ related to the structural OH groups and a combination water band at 5239 $cm^{-1}$ (Fig. 13.8A,a). The spectra of acid treated samples show a gradual decrease in the intensities of the structural OH bands due to the release of the central atoms from the octahedral sheets and the appearance of the SiOH overtone at 7315 $cm^{-1}$ confirming the formation of protonated silica (Fig. 13.8A,b–d).

The NIR spectra of the samples after pyridine vapors adsorption are presented in Fig. 13.8B. Due to partial replacement of water molecules by pyridine the environment of the structural OH groups of $Na^+$-SAz was changed and the $2\nu(OH)$ band is split into two components at 7156 $cm^{-1}$ and 7059 $cm^{-1}$ (13.8.B,a). A decreased intensity of the $(\nu+\delta)H_2O$ band also confirms the partial replacement of water by pyridine. The first C–H overtone of the aromatic ring and combination bands in the 6200–5800 $cm^{-1}$ and 4700–4000 $cm^{-1}$ regions, respectively, correspond to physi-sorbed and/or

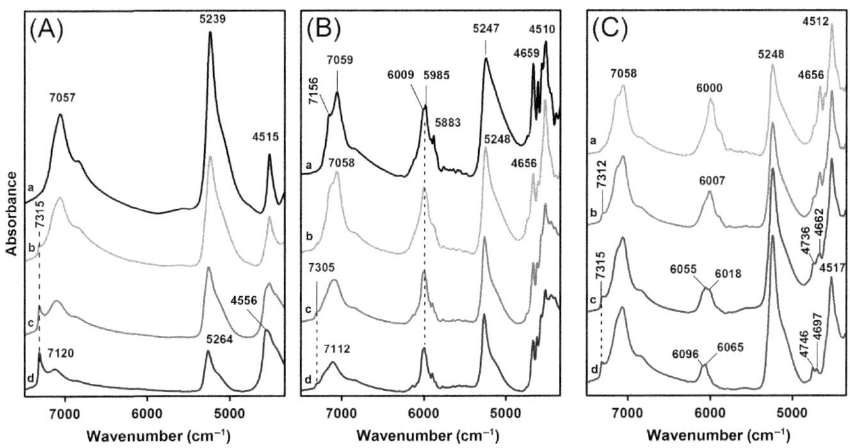


**FIG. 13.8** NIR spectra of (A) Na-SAz treated with 6 M HCl at 80°C for (a) 0 h, (b) 2 h, (c) 4 h and (d) 8 h, (B) the same samples as in (A) but after pyridine adsorption, (C) SAz treated with HCl for 2 h with adsorbed pyridine (a) unheated and heated at (b) 110°C, (c) 170°C and (d) 230°C.

H-bonded pyridine (Fig. 13.8B,a). A detailed assignment of the individual CH bands/components of pyridine is given in Madejová et al. (2015). After pyridine adsorption on the acid-treated samples, the pronounced modification of the $2\nu$(SiOH) band is observed. No band at 7315 $cm^{-1}$ can be identified for SAz-2 h (Fig. 13.8B,b) and only a very weak shoulder near 7305 $cm^{-1}$ is recognized for SAz-4 h and SAz-8 h (Fig. 13.8B,c and d). The majority of the silanol groups, as weak Brønsted acid sites, form H bonds with pyridine-nitrogen. The $2\nu$(SiOH) band of H-bonded SiOH groups was shifted to lower wavenumbers and contributed to the broad complex band near 7100 $cm^{-1}$. The shape and position of the $2\nu$(CH) band of pyridine is modified only slightly comparing to the $Na^+$-SAz, i.e. acid-untreated sample. In contrast to the MIR region no feature/band due to pyridinium cation can be identified in the NIR spectra presented in Fig. 13.8B.

To examine the strength of the pyridine bonding the samples were heated at 110, 170 and 230°C prior to IR analysis. Fig. 13.8C shows the NIR spectra of heated SAz-2 h sample. The re-appearance of the 7315 $cm^{-1}$ band assigned to the free Si–OH groups and the increased intensity of the $(\nu+\delta)H_2O$ band confirmed that pyridine H-bonded to silanol groups and water molecules was released upon heating. The intensity of the $2\nu$(CH) overtone decreased due to desorption of H-bonded pyridine and the band was considerably shifted to the higher wavenumbers (Fig. 13.8C,c and d). Pálková et al. (2013) showed that the strong BA sites in SAz-2 h provide protons for the generation of pyridinium cations absorbing at 1539 $cm^{-1}$. Moreover, only pyridine molecules adsorbed on these types of acid sites persisted at the heating of the sample at 230°C. Thus the bands at 6096 and 6065 $cm^{-1}$ in the NIR spectrum of SAz-2 h heated at 230°C were assigned to C–H overtones of the pyridinium cations. Similar C–H positions were observed for SAz Mt directly saturated with pyridinium cations. For SAz-2 h unheated and heated at 110°C the pyridinium bands could not be observed because they were covered with stronger bands of physi-sorbed and/or H-bonded pyridine molecules. The position of the C–H overtone with peaks at 6055 and 6018 $cm^{-1}$ found for SAz-2 h heated at 170°C indicated the overlapping contribution of both the H-bonded pyridine molecules and the pyridinium cations.

The NIR spectra of heated SAz-4 h (not shown) confirmed the gradual release of physically adsorbed and/or weakly H-bonded pyridine leaving only those molecules that interacted with strong BA sites. Although the intensity of the $2\nu$(CH) band in the samples heated at 170°C and 230°C was significantly lower than in the corresponding SAz-2 h, the peaks at 6097 and 6066 $cm^{-1}$ were compatible with the presence of strongly bonded pyridinium cations. The reaction product obtained after 8 hours of decomposition of $Na^+$-SAz in HCl contained about 97% of protonated amorphous silica (Madejová et al., 2015). The acidity of the Si–OH groups was not strong enough to provide protons for the creation of a pyridinium cation. After SAz-8 h heating at 230°C the absence of the $2\nu$(CH) bands confirmed almost complete desorption of pyridine from the sample. The obtained results showed that in addition to MIR, NIR spectroscopy can also distinguish different pyridine species on the Mt surface.

## 13.6 CLAY-BASED HETEROSTRUCTURES

Porous clay heterostructures (PCH) belong to the silica based mesoporous materials, the extensive research of which began in the 1992 when a new family of ordered mesoporous molecular sieves was synthesized via templating approach (Beck et al., 1992). In 1995, Galarneau et al. (1995) performed the successful synthesis of mesoporous solids from fluorohectorite. The synthesis of PCH begins with an expansion of clay mineral interlayer space (usually smectite) with quaternary alkylammonium cation with a long alkyl chain (e.g., hexadecyl trimethylammonium) followed by the introduction of neutral amine as the co-surfactant. The addition of Si-alkoxide as a silica source leads to the self-assembly of surfactants into micelles and the formation of a condensed silica network around the surfactant micelles template. Removal of the template leaves porous nanocomposite materials formed from the open silica framework between and/or around the clay layers. The resulting materials possess not only large SSA and combined micro/mesoporosity, but also acidity due to the presence of clay mineral component (Galarneau et al., 1995; Polverejan et al., 2000; Benjelloun et al., 2002). The surface of PCH may be further modified by incorporating of heteroatoms within and/or on the external position of the Si-network (e.g., Ahenach et al., 2000; Cool et al., 2000; Chmielarz et al., 2009; Zimowska et al., 2013, 2016). These obtained materials, compared to the initial PCH, showed superior acidity, improved efficiency in catalytic reactions and variations in the selectivity.

The IR spectroscopy has been used to follow the PCH structure evolution (Pálková et al., 2010; Belver et al., 2012; Zimowska et al., 2013, 2016), to assess the PCH surface acidities utilizing their interaction with molecules such as acetonitrile and pyridine (Ahenach et al., 2000; Chmielarz et al., 2009; Zimowska et al., 2013), but also to characterize the PCH adsorption properties (Pires et al., 2008). Pálková et al. (2010) used the IR spectroscopy in both MIR and NIR regions to describe in detail the individual steps of PCH synthesis from synthetic hectorite (Laponite RD®). The appearance of the C–H stretching and deformation bands in the MIR spectrum of OC confirmed the modification of Laponite (Lap) with $HDTMA^+$ surfactant.

Further, adding dodecylamine and tetraethylorthosilicate (TEOS) to OC resulted in the creation of a three dimensional silica network around the surfactant micelles, which was confirmed by the broadening of the complex Si–O band of the PCH-precursor compared to the initial Lap or OC. In the final step of synthesis, the organic templating phase was removed by calcination and the spectrum showed just absorption bands of initial Lap and those assigned to siliceous matrix. The layered structure of clay mineral resisted the heat treatment. Only the thermal damage was connected with the possible partial dehydroxylation of the octahedral sheets reflected by an intensity decrease of the 3688 and 653 $cm^{-1}$ bands corresponding to the stretching and bending modes of the structural OH groups, respectively.

The NIR spectra of studied samples provide the complementary information to the MIR spectra. The individual steps of synthesis of PCH prepared from trioctahedral Lap (Pálková et al., 2010) and from dioctahedral SAz-1 Mt (SAz) are compared in Fig. 13.9. The preparation of PCH from SAz (PCH-SAz) was carried out using the same experimental procedure as reported in Pálková et al. (2010). The NIR spectra of initial smectites show the characteristic bands for trioctahedral (Fig. 13.9A,a) and dioctahedral (Fig. 13.9B,a) smectite. The $2\nu(OH)$ overtone region of Lap shows the complex band with maximum at 7214 $cm^{-1}$ and a shoulder at 7280 $cm^{-1}$ corresponding to the $2\nu(Mg_3OH)$ and $2\nu(LiMg_2OH)$ overtones, respectively. The band at 7080 $cm^{-1}$ is assigned to the $2\nu(H_2O)$ band of water molecules weakly H-bonded to the basal oxygen atoms of the tetrahedral sheets, while a shoulder near 6840 $cm^{-1}$ is a first overtone of stretching vibrations of $H_2O$ engaged in strong H bonds. The spectrum of SAz shows only one complex band at 7064 $cm^{-1}$ consisting of the contributions of the first O–H overtone of the structural OH groups of Mt and water molecules involved either in weak or strong H-bonds. The complex band near 5200 $cm^{-1}$, observed in the spectra of both smectites, belongs to the combination of the stretching and bending vibrations of water molecules. The $(\nu+\delta)$OH region of Lap shows the band at 4361 and 4325 $cm^{-1}$ corresponding to $(\nu+\delta)Mg_2LiOH$ and $(\nu+\delta)Mg_3OH$,

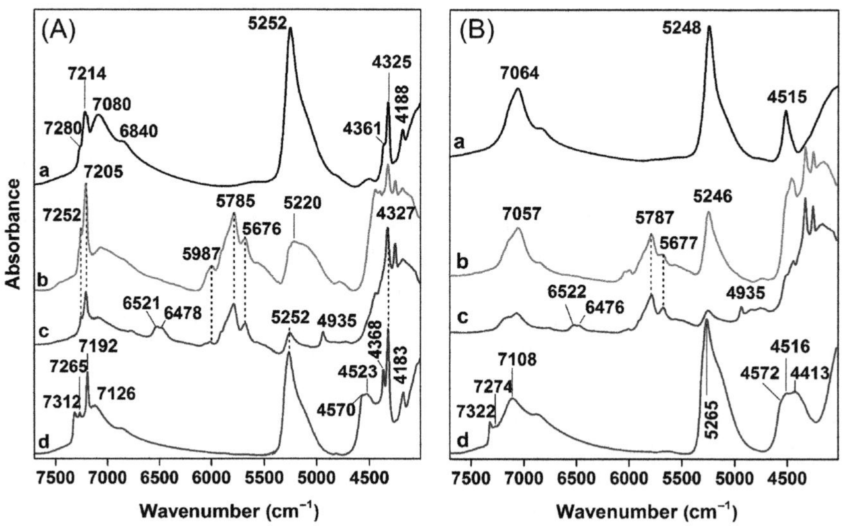


**FIG. 13.9** NIR spectra of (A): (a) Na-LAP, (b) HDTMA-LAP, (c) PCH-LAP-precursor, (d) PCH-LAP and (B): (a) Na-SAz, (b) HDTMA-SAz, (c) PCH-SAz-precursor, (d) PCH-SAz.

respectively. Only a single band at 4515 $cm^{-1}$, mainly due to $(\nu+\delta)Al_2OH$, is present in the combination region of SAz Mt.

The presence of $HDTMA^+$ in both O-Mt confirms the complex $2\nu(CH)$ band occurring in the 6100–5400 $cm^{-1}$ region and the $(\nu+\delta)CH$ bands observed below 4500 $cm^{-1}$. The hydrophobic character of the OC surfaces illustrates the significantly reduced intensity of the combination water band (Fig. 13.9A,b and B,b). New bands at 6521 and 6478 $cm^{-1}$ for PCH-Lap and at 6522 and 6476 $cm^{-1}$ for PCH-SAz attributed to asymmetric and symmetric $2\nu(NH)$ modes give evidence of the presence of co-surfactant dodecylamine in both PCH-precursors. Combination $(\nu+\delta)NH$ mode occurs at 4935 $cm^{-1}$ (Fig. 13.9A,c and B,c). As was shown previously (Madejová et al., 2011) the NIR region is more sensitive for detecting the presence of amines compared to MIR, because corresponding N–H stretching and bending vibrations absorbing in 3400–3300 $cm^{-1}$ region and near 1620 $cm^{-1}$ are often overlapped with absorption bands of molecular water adsorbed on the sample and/or KBr used as matrix for IR measurements (Pálková et al., 2010; Madejová et al., 2011).

The calcination of the PCH-precursors considerably modifies the shape of their NIR spectra (Fig. 13.9A,d and B,d). The bands related to $HDTMA^+$ and dodecylamine disappeared and the spectra consist of the bands of clay mineral and OH groups formed as a part of developed Si-network. The OH overtone region of PCH-LAP shows three sharp bands in addition to the $2\nu(H_2O)$ overtone at 7126 $cm^{-1}$. The band at 7192 $cm^{-1}$ is due to $2\nu(OH)$ vibrations of $Mg_3OH$ groups of Lap, the band at 7312 $cm^{-1}$ is due to $2\nu(SiOH)$ mode of free (not H-bonded) Si–OH groups, and the 7265 $cm^{-1}$ band corresponds to silanol groups bonded to water vapor or to coke species created upon calcinations. A more detailed assignment of the $2\nu(SiOH)$ bands is given in (Pálková et al., 2010).

The OH overtone region of PCH-SAz shows a complex band at 7108 $cm^{-1}$ including the $2\nu(OH)$ and $2\nu(H_2O)$ components (Fig. 13.9B,d). Several factors can induce a shift of the $2\nu(OH)$ band position compared to spectra of SAz or $HDTMA^+$-SAz. Due to high temperature treatment during the template removal, the local environment of the structural OH groups is changed. Moreover, the amount of water molecules involved in weak H-bonds to create a silica network has increased, which moves the complex $2\nu(OH)$ band to higher wavenumbers. Similar upward shift can be observed also for $(\nu+\delta)H_2O$ band. A sharp band detected near 7322 $cm^{-1}$ is assigned to the $2\nu(SiOH)$ of free (not-bonded) Si–OH groups. The second derivative analysis revealed also a component at 7274 $cm^{-1}$ corresponding to H-bonded Si–OH groups (Fig. 13.9B,d).

The 4600–4000 $cm^{-1}$ region of PCH-Lap and PCH-SAz reveals the bands of both the clay mineral and silica network. The 4368, 4325 and 4183 $cm^{-1}$ bands related to the combination modes of the structural OH group confirm the persistence of the Lap structure during calcination. A broad complex band with components at 4570, 4523 and 4480 cm$^{-1}$, obtained by second derivative analysis, reflects wide diversity of the Si–OH groups present in PCH-Lap (Fig. 13.9A,d). Based on the results reported by Christy (2010) and Pentrák et al. (2012a) the 4570 cm$^{-1}$ is assigned to $(\nu+\delta)$SiOH mode while the other components are assigned to the H bonded silanol groups. Even the broader OH combination band is seen in the spectrum of PCH-SAz (Fig. 13.9A,d). The second derivative analysis revealed the positions of the components at 4572, 4516 and 4413 cm$^{-1}$ resembling those reported for silanol groups on silica gel surface by Christy (2010). The width and intensity of the band, however, indicates that the $(\nu+\delta)$OH vibrations of the structural OH groups of Mt (near 4515 cm$^{-1}$) also contribute to this band.

The pyridine adsorption/desorption experiments and the IR spectroscopy are frequently used to assess PCH acidity, a property that is important for their utilization for catalysis. As discussed in Section 13.5.4, pyridine can either form H bonds between pyridine nitrogen atom and OH groups of clay mineral or can be protonated to pyridinium cation at strong Brønsted acid (BA) sites. To detect the acid sites on the surface of PCH-Lap and PCH-SAz the MIR spectra in the 1700–1450 cm$^{-1}$ and NIR in the 6200–5800 cm$^{-1}$ were examined after adsorption of pyridine on the samples and subsequent heating at 200°C to remove weakly bonded pyridine (Fig. 13.10). The bands at 1594, 1490 and 1446 cm$^{-1}$ confirm the presence of H-bonded pyridine in both PCH samples. The spectrum of PCH-SAz also contains the 1548 cm$^{-1}$ band due to the $N^+$-H deformation vibration of pyridinium cation diagnostic for strong Brønsted acid sites (Fig. 13.10B). The presence of pyridine in the heated samples also confirms the first $2\nu$CH overtone of pyridine found at 6015 cm$^{-1}$ for PCH-Lap and at 6019 cm$^{-1}$ for PCH-SAz (Fig. 13.10A). The $2\nu$CH of PCH-SAz is not only shifted to higher wavenumbers, but is also

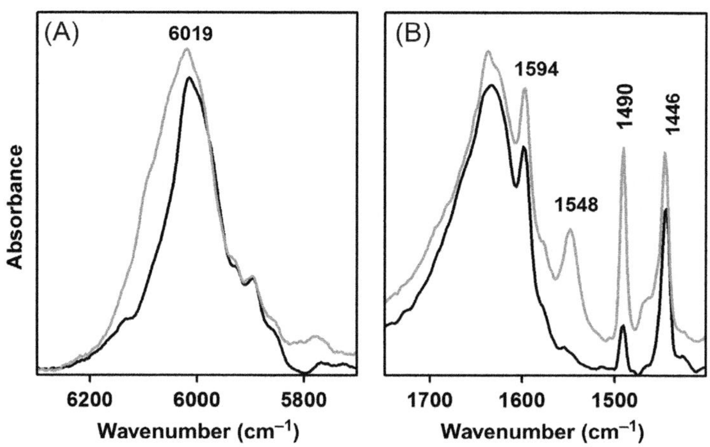


**FIG. 13.10** (A) NIR spectra and (B) MIR spectra of PCH-LAP (bottom spectrum) and PCH-SAz (upper spectrum) after pyridine adsorption and consequent heating at 200°C.

broader. Madejová et al. (2015) related the upward shift of this band to the presence of pyridinium cation. It appears that the presence of aluminum in octahedral sheets of SAz Mt is responsible for the higher acidity of PCH prepared from this Mt.

## 13.7 CONCLUDING REMARKS

NIR spectroscopy represents an analytical method that is enjoying an increase in popularity in the clay sciences due to its low running costs and also because it does not require sample preparation. NIR has proven to be a powerful technique to investigate the structure and selected properties of clay minerals modified by mechanical, thermal or chemical treatment. Due to the sensitivity of the structural OH groups of clay minerals to the changes in their environment, the first O–H overtone and combination regions allow, for example, the identification of fixed $Li^+$ cations in the previously vacant octahedra, creation of silanol groups upon clay minerals acid treatment and during PCH synthesis, or dehydroxylation of the octahedral sheets upon acid treatment and dry grinding. The intensity and position of the combination band of water molecules provides information not only on the amount of water adsorbed on the clay mineral surface but also on the strength of H-bonding in the system. NIR spectroscopy is extremely useful in OC studies. The presence of organic species can often be more easily identified in the C–H and N–H overtone and combination regions than in their MIR spectra. The changes in the position of the first overtone of the $CH_2$ groups provide information on the gauche/trans arrangement of alkylammonium surfactants intercalated into Mt interlayer spaces. The different fields of application of NIR spectroscopy reported in the present chapter highlight its enormous versatility.

The potential of this technique is certainly not exhausted. On the contrary, new advances in its utilization are expected in the near future. The application of two main mathematical techniques for analyzing complex bandshapes, namely band decomposition and derivative analysis, allows the extraction of more information from the NIR spectra of clay minerals. Significant progress has occured also in portable IR spectrometers operating in the NIR region to allow complex characterization of soils and clay deposits. To correctly evaluate the data obtained from the field, a well-interpreted database of the spectra of original but also modified clay minerals is needed. It follows clearly that NIR spectroscopy, together with MIR, will remain one of the most important experimental methods for clay minerals investigation in the upcoming period.

Chapter 14

# Remote Detection of Clay Minerals

J.L. Bishop, J.R. Michalski and J. Carter

## 14.1 PRESENCE OF CLAY MINERALS IN OUR SOLAR SYSTEM

Clay minerals are not only found on Earth, but also on our neighbouring planet Mars and in meteorites, asteroids and comets. These minerals can be identified through several techniques on Earth, most commonly by X-ray diffraction (XRD) or visible/near-infrared (VNIR) reflectance spectroscopy. However, identification and characterisation of clay minerals on Mars is largely carried out using orbital VNIR spectra from ~0.4 to 5 μm. Thermal infrared (TIR) spectra collected in the mid-IR (MIR) region from ~200 to 2000 $cm^{-1}$ (~5–50 μm) also provide orbital information about mineralogy including clay minerals. The Mars Science Laboratory (MSL) rover uses an XRD instrument called CheMin to identify clay minerals on the surface at the landing site inside Gale Crater (Blake et al., 2013) and several martian surface missions have measured surface chemistry and this data was then used to infer mineralogy (Baird et al., 1977; Toulmin III et al., 1977; McSween et al., 2009). Detections of clay minerals on asteroids such as Ceres (De Sanctis et al., 2015) and Jupiter-family comets (Lisse et al., 2006) are typically achieved through NIR and MIR spectroscopy because these small bodies are too far away for other techniques. Many more detections of clay minerals in meteorites have been made than in asteroids, likely due to the ability to study meteorites in the lab and challenges resulting from space weathering on asteroid surfaces. Clay minerals are frequently identified in meteorites using microprobe analyses of thin sections.

The history of clay minerals on the planet Mars has experienced a rather disjointed path towards our current understanding. Elemental analyses of surface material by Viking first led scientists to believe clay minerals were likely present on Mars. Modelling of the major element data was found to be consistent with 60–80 wt.% smectite (Baird et al., 1977; Toulmin III et al., 1977). This was based on the major element chemistry of the surface material and

Developments in Clay Science, Vol. 8. https://doi.org/10.1016/B978-0-08-100355-8.00014-X

also the expectation that alteration of volcanic ash and tephra on Mars would form clay minerals as observed on Earth. Telescopic VNIR and TIR spectra of Mars followed in the 1980s and 1990s without clear evidence of any clay minerals (McCord et al., 1982; Singer, 1985; Bibring et al., 1990; Bell et al., 1994). As the signal to noise ratio of these data improved and the spot size became smaller and no clay minerals were found, scepticism about clay minerals on Mars arose. Further remote sensing of Mars with the Thermal Emission Spectrometer (TES) instrument on Mars Global Surveyor beginning in the late 1990s (Christensen et al., 2001) mapped the martian surface at 3–5 km spot size and still did not find evidence of clay minerals. It was not until the Observatoire pour la Minéralogie, l'Eau, les Glaces et l'Activité (OMEGA) VNIR imaging spectrometer on Mars Express started mapping Mars at 1–3 km spot sizes in 2004 that clay minerals could be definitively identified on that planet (Poulet et al., 2005). As the Compact Reconnaissance Imaging Spectrometer for Mars (CRISM) instrument on the Mars Reconnaissance Orbiter began collecting targeted high resolution (18 m/pixel) hyperspectral imagery in 2006, windows of small and large clay mineral-bearing outcrops were discovered nearly everywhere that the ancient rocks were exposed on the surface (Mustard et al., 2008; Murchie et al., 2009).

## 14.2 REMOTE DETECTION OF CLAY MINERALS

Remote sensing of clay minerals is primarily performed using VNIR and TIR instruments. Similar instruments are used on Earth, Mars and other planetary bodies. Remote detection of clay minerals in the MIR region with TIR spectra uses Si–O stretching ($\nu$(SiO)) and bending ($\delta$(SiO)) vibrations, and M–OH bending $\delta$(MOH) vibrations (where M is a metal cation such as Al, $Fe^{3+}$, $Fe^{2+}$ or Mg), while NIR clay mineral detections focus mostly on M–OH stretching plus bending combinations (($\nu+\delta$)MOH) and stretching overtones (2$\nu$(OH)). Similar modes also occur for $H_2O$ in phyllosilicates. These include a stretching plus bending combination band (($\nu+\delta$)$H_2O$) and a stretching overtone (2$\nu$ ($H_2O$)). The extended visible region covers iron absorptions and is particularly helpful in discriminating between $Fe^{3+}$ and $Fe^{2+}$ in clay minerals. The MIR region has the advantage that the spectral features are fundamental vibrations and the bands are strong; however, a disadvantage is that other silicates exhibit bands in the same wavelength range and identifying clay minerals in the presence of quartz, feldspar and pyroxene can be challenging. An advantage of the NIR region is that there are few other spectral features at these wavelengths so detection is easier. In addition, there are bands near 1.4, 1.9 and 2.1–2.5 μm that can be evaluated jointly for added confidence in the clay mineral detection. Band assignments were confirmed using comparison of the NIR overtones and combinations with the MIR fundamental stretching and bending vibrations from MIR transmittance spectra (Bishop et al., 2002c; Petit et al., 2004b). While MIR transmission spectroscopy uses an active radiation source, MIR remote sensing is passive. The thermal emission spectra used in remote sensing studies includes components from absorption as well as scattering, so they are more complex than transmittance spectra, which include only the absorption component. Reflectance spectra are often easier to measure in the laboratory, but they are not always directly comparable to thermal emission data. Further complicating matters is the reality that all VNIR and MIR spectra of planetary surfaces contain particle size scattering effects that cannot be completely reproduced in any laboratory setting. However, the spectral contrast, shapes and absorption positions measured on planets provide strong clues regarding the effects of particle size, lending confidence to clay mineral detections using purely IR techniques.

### 14.2.1 VNIR Bands Used for Detection of Clay Minerals

Detection of clay minerals in the VNIR region varies with mineral structure and cation type (Fig. 14.1). Several studies have investigated the VNIR reflectance

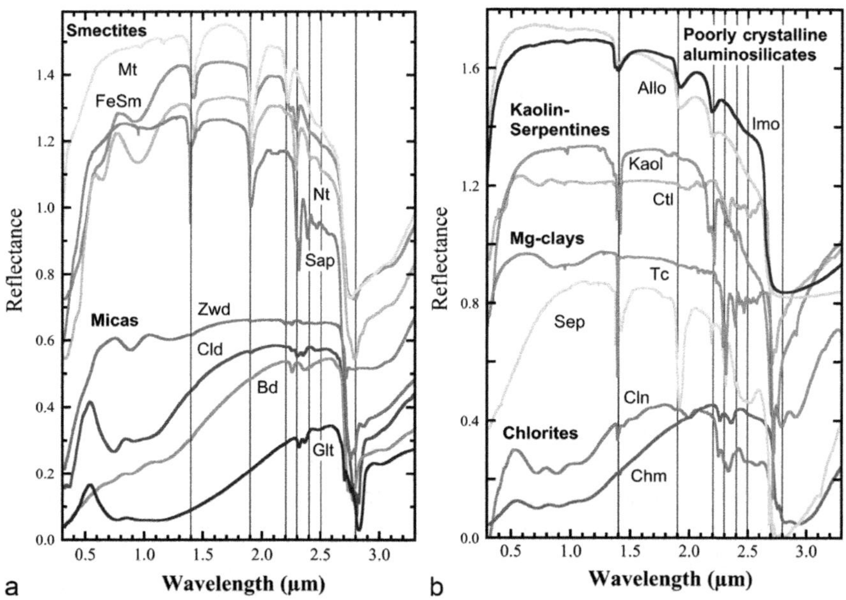


**FIG. 14.1** Example VNIR reflectance spectra from 0.3 to 3.3 μm of clay minerals found in remote sensing studies. The spectra are offset for clarity and grouped by type. (A) Smectites: montmorillonite (Mt), Fe-smectite (FeSm), nontronite (Nt), saponite (Sap); Micas: zinnwaldite (Zwd), celadonite (Cld), biotite (Bt), glauconite (Glt). (B) Poorly crystalline aluminosilicates: allophane (Allo), imogolite (Imo); Kaolinite-serpentines: kaolinite (Kaol), chrysotile (Ctl); Other Mg-rich clays: talc (Tc), sepiolite (Sep); Chlorites: clinochlore (Cln), chamosite (Chm). Grey lines mark features near 1.4, 1.9, 2.2, 2.3, 2.4, 2.5 and 2.8 μm. These spectra were measured as particulate samples at Brown University's RELAB for previous studies (Bishop et al., 2008a, 2013a).

spectra of pure clay minerals for remote sensing studies. These include studies of smectites (Cariati et al., 1981; Post, 1984; Clark et al., 1990; Post and Noble, 1993; Bishop et al., 1994, 1999, 2002a,c, 2008a, 2011; Post et al., 1997; Gates, 2005; Decarreau et al., 2008), kaolin-serpentine group minerals (King and Clark, 1989; Clark et al., 1990; Petit et al., 1999b; Bishop et al., 2002c, 2008a,b), chlorites (King and Clark, 1989; Bishop et al., 2008a), micas (Bishop et al., 2008a), talc (Petit et al., 2004a), and nanophase/poorly crystalline clay minerals such as allophane and imogolite (Bishop et al., 2013a). In remote sensing studies the materials measured are always mixtures of multiple components. Thus, studies of the spectral properties of natural mixtures of clay minerals (Cuadros et al., 2015; Michalski et al., 2015), mixtures prepared from clay minerals and other minerals (e.g. Madejová et al., 2002b; McKeown et al., 2011), and clay mineral-bearing rocks (e.g. Bishop et al., 2002d; Hamilton et al., 2008b; Ehlmann et al., 2012) are important as well.

### 14.2.1.1 VNIR Characterisation of Smectites

Smectites have received the most study of clay minerals in this spectral region and have clearly identifiable bands for their identification. All smectite spectra exhibit an $H_2O$ stretching overtone ($2\nu(H_2O)$) at 1.41 μm and an asymmetric $H_2O$ combination ($(\nu+\delta)_{as}H_2O$) band centred at 1.91 μm (Fig. 14.1, Bishop et al., 1994). The MOH features depend on the presence of Al, $Fe^{3+}$ or Mg cations in the octahedral sites, as well as Al and $Fe^{3+}$ substitutions in the tetrahedral sites, as observed in spectra of the *Clay Mineral Society* standards SAz-1, SWy-1, SWa-1, SHCa-1, SapCa-1 and others. Montmorillonites with high Al abundance in the octahedral sites (e.g. SAz-1) exhibit an OH combination band near 2.20 μm, but this can be extended toward 2.21 μm for montmorillonite with some $Fe^{3+}$ or Mg in the octahedral sites (e.g. SWy-1). The AlOH stretching overtone ($2\nu(Al_2OH)$) lies at 1.41 μm, at the same wavelength as the $2\nu(H_2O)$, which was a source of confusion for decades in making band assignments. Beidellite spectra exhibit $2\nu(Al_2OH)$ vibrations at 1.40 and 2.18 μm (Bishop et al., 2011), which are shorter wavelengths than the bands observed in montmorillonite spectra due to increased tetrahedral Al substitution for Si. Nontronite spectra include a $2\nu(Fe^{3+}{}_2OH)$ band at 1.43 μm and a $(\nu+\delta)Fe^{3+}{}_2OH$ band at 2.29 μm. Spectra of nontronite containing some $Al^{3+}$ in octahedral sites (e.g. SWa-1) have an additional shoulder or weak $(\nu+\delta)AlFe^{3+}OH$ band near 2.23 μm. Hectorite (e.g. SHCa-1) and saponite (e.g. SapCa-1) spectra include an MgOH stretching overtone ($2\nu(Mg_3OH)$) band at 1.39 μm and an MgOH combination ($(\nu+\delta)Mg_3OH$) band at 2.31 μm. For spectra of Fe- and Mg-rich smectites, the $2\nu(OH)$ bands are distinct from the $2\nu(H_2O)$ bands, which aides in their identification in remote sensing data. Smectites also include an additional OH band at longer wavelengths (~2.4–2.5 μm) that depends on the octahedral cations. This occurs at 2.38 μm for saponite, 2.39 μm for hectorite, 2.41 for nontronite, 2.44 μm for beidellite, and 2.45 μm for montmorillonite.

#### 14.2.1.2 VNIR Characterisation of Kaolinites-Serpentines

VNIR spectra of kaolinite and serpentine are readily distinguished from spectra of smectites because the kaolin-serpentine group clays do not have $H_2O$ bands, except for halloysite. Kaolinite spectra exhibit a sharp pair of bands due to the $2\nu(Al_2OH)$ stretching overtone at 1.396 and 1.416 μm and a $(\nu+\delta)Al_2OH$ doublet at 2.17 and 2.21 μm. The halloysite spectrum is similar to that of kaolinite, but also includes a water band near 1.9 μm due to $H_2O$ in the structure (e.g. Clark et al., 1990). Serpentine spectra include a $2\nu(Mg_3OH)$ overtone at 1.39 μm and a $(\nu+\delta)Mg_3OH$ combination band near 2.33 μm. Many serpentine spectra also include bands near 2.52–2.57 μm.

#### 14.2.1.3 VNIR Characterisation of Chlorites

VNIR spectra of chlorites include the expected OH bands near 1.39 and 2.33–2.35 μm and the band positions vary depending on the relative abundance of $Fe^{3+}$ and Mg; however, they are distinguished from serpentines by the presence of an additional $(\nu+\delta)AlFe^{2+}MgOH$ band near 2.25 μm. The Mg-rich chlorites such as clinochlore have spectral bands near 1.39, 2.25 and 2.33 μm, while $Fe^{2+}$-rich chlorites such as chamosite have spectral bands near 1.42, 2.26 and 2.36 μm. Also, the $2\nu(Fe^{2+}{}_3OH)$ band for chamosite is weaker in intensity, likely because of the strongly absorbing $Fe^{2+}$ absorptions from ~0.7 to 1.2 μm. Sometimes $H_2O$ bands are observed near 2.0 μm in chlorite spectra, rather than near 1.9 μm as in most hydrated clay minerals. Spectra of some chlorites also contain a band near 2.48–2.57 μm.

#### 14.2.1.4 VNIR Characterisation of Micas, Illites and Other Clay Minerals

Micas tend to exhibit much weaker NIR bands from 1.4 to 2.5 μm than smectites, kaolin-serpentines, and chlorites, and the bands near 1.4 and 1.9 μm are generally too weak to be useful for remote detection. The OH combination vibrations typically occur as multiple bands in micas and vary with the octahedral cation composition. Al-rich micas like zinnwaldite have spectral bands near 2.20 and 2.25 μm, while celadonite has bands near 2.25, 2.30 and 2.35 μm due to multiple cations. Biotite with Al and $Fe^{3+}$ cations has bands near 2.25 and 2.35 μm, while glauconite with $Fe^{3+}$ and $Fe^{2+}$ cations includes spectral bands near 2.30 and 2.36 μm. Additional clay minerals such as illite, palygorskite and sepiolite all exhibit spectral features due to both OH and $H_2O$ in their structures. Illite spectra have bands near 1.41, 1.91, 2.22, 2.35 and 2.44 μm. Palygorskite spectra contain bands near 1.41–1.42, 1.91, a doublet at 2.18 and 2.22, and a weak band near 2.25 μm. The VNIR spectra of sepiolite include features near 1.38, 1.42, 1.91, 2.18 and 2.31 μm due to both Al and Mg in its structure. The VNIR spectra of

Mg-talc include a band at 1.39 μm, a doublet at 2.29 and 2.31 μm, and additional bands near 2.39 and 2.46 μm.

#### *14.2.1.5 VNIR Characterisation of Poorly Crystalline Clay Minerals*

The importance of poorly crystalline clay minerals including allophane and imogolite as soil components and proto-clays was recognised long ago (Wada et al., 1972; Farmer et al., 1979, 1983; Parfitt and Furkert, 1980). More recently, the importance of these for remote sensing on Mars has been recognised (Rampe et al., 2012; Bishop and Rampe, 2016). The nanophase aluminosilicates allophane and imogolite have VNIR spectral features similar to those of Al-rich smectites (AlSm) and opal. The primary difference is broadened spectral features for the nanophase clay minerals due to a distribution of Al–OH and Si–OH sites in the structure. Another difference is that the $(\nu+\delta)$ $H_2O$ band is centred at 1.92 μm instead of 1.91 μm. The corresponding MOH bands are also shifted. The $2\nu(Al_2OH)$ bands occur at 1.38 and 1.40 μm for allophane and at 1.37 and 1.39 μm for imogolite, while the $(\nu+\delta)Al_2OH$ band occurs near 2.19 μm for both of these materials.

### 14.2.2 MIR Bands Used for Detection of Clay Minerals in TIR Spectra

The properties of transmittance spectra of clay minerals are described in detail in previous chapters. Transmittance spectra of fine powders depend on the imaginary component of the complex index of refraction $k$, while reflectance and emission spectra depend on both the real $n$ and imaginary $k$ components of the index of refraction, and are thus susceptible to the effects of surface scattering. Because MIR remote sensing data are measured as TIR, that is the best technique for comparison of lab data to remote sensing data. Lab emissivity spectra have been acquired of numerous samples including clay minerals for comparison with the Thermal Emission Spectrometer (TES) data of Mars (Christensen et al., 2000a). Detailed lab studies of clay minerals (Michalski et al., 2005, 2006a,b) and related materials such as zeolites (Ruff, 2004), glass (Minitti et al., 2002; Byrnes et al., 2007) and amorphous phases (Kraft et al., 2003; Rampe et al., 2012) were performed, although the majority of TIR lab studies focus on other minerals. Hemispherical reflectance spectra can often be used to calculate emission spectra using Kirchhoff's Law: $E=1-R$; however, this does not always hold true for small particle sizes or if the surface is not a Lambertian scatterer (e.g. Salisbury et al., 1994). Reflectance spectra collected using a biconical system can provide only an approximation of emission spectra. For finely particulate samples it is often easier to obtain high-quality reflectance spectra in the lab than emissivity spectra. Clay minerals are typically pressed into powders in order to improve the band strength in emission spectra (Michalski et al., 2005).

Pressing finely grained materials into powders removes pore space and reduces internal scatter, which strengthens the Reststrahlen bands (reflectance maximum or emission minimum). Most importantly, clay minerals occurring naturally within rocks (e.g. mesostatic replacement in volcanic rocks or compacted clays in clastic rocks) behave spectroscopically like coarse materials or pressed pellets whereas clays in dust behave more like fine particulates. TIR spectral studies of clay-bearing rocks were also performed (e.g. Wyatt et al., 2001; Michalski et al., 2004; Hamilton et al., 2008b; Ehlmann et al., 2012; Rampe et al., 2013) in order to support identification of clays on planetary surfaces.

#### *14.2.2.1 Si–O Stretching Vibrations*

The strongest bands for identification of clay minerals in thermal remote sensing typically result from vibrations of the tetrahedral sheet $Si(Al,Fe)O_4$ (Michalski et al., 2006b). However, related $SiO_4$ vibrations are observed in spectra of other silicates as well. Salisbury (1993) describes the Christiansen Feature (CF) as a reflectance minimum or emission maximum located at the short-wavelength edge of the *ν*(SiO) Reststrahlen feature. The CF occurs where the imaginary component of the index of refraction approaches 0 and the real component approaches 1. Salisbury (1993) noted shifts in the wavelength of this feature with mineral composition. Typically the CF occurs at longer wavelengths for clay minerals compared to mafic silicates (Salisbury, 1993). Michalski et al. (2005) observed a shift in the *ν*(SiO) in spectra of clay minerals, aluminosilicate glasses and altered phases, such that the band near 1000 $cm^{-1}$ transitions towards lower wavenumbers (longer wavelengths) with increasing abundance of tetrahedral Al and $Fe^{3+}$ and decreasing abundance of Si. This is consistent with the trends observed by Salisbury et al. (1991b) for a large collection of silicate minerals and with observed transmittance spectra of clay minerals (Farmer, 1974b). For example, Michalski et al. (2005) noted emissivity minima of 8.8 μm (1135 $cm^{-1}$) for SWy-1 montmorillonite with an Si/O ratio of 0.399, and of 9.5 μm (1056 $cm^{-1}$) for NAu-1 nontronite with an Si/O ratio of 0.349.

The $Si(Al,Fe)O_4$ vibrations observed in emission spectra of several clay minerals are shown in Fig. 14.2. The *ν*(SiO) band centre exhibits a trend of increasing wavelength across the smectites as the octahedral cation changes from Al to $Fe^{3+}$ to Mg. The Si–O stretching band position for Al-rich smectite is similar to that observed for kaolinite and aluminosilicate gel. The Si–O stretching band occurs at longer wavelengths for the poorly crystalline allophane and imogolite and also for clay minerals rich in $Fe^{3+}$ and Mg (Fig. 14.2).

#### *14.2.2.2 Si–O Bending Vibrations*

The bending vibrations for tetrahedral $SiO_4$ groups as well as lattice deformation modes such as Si–O-$M_{oct}$ absorptions, occur near 18–25 μm (550–400 $cm^{-1}$) in emission spectra when little or no tetrahedral substitution occurs

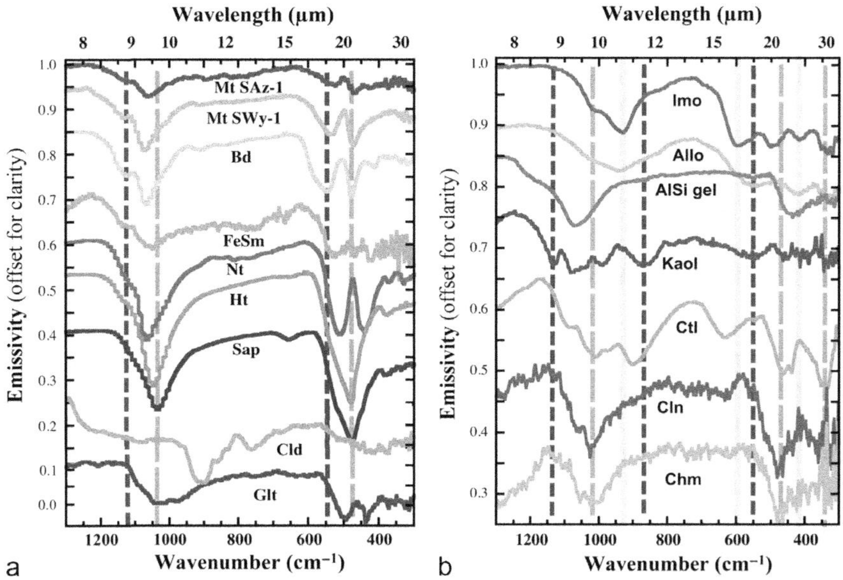


**FIG. 14.2** Example TIR spectra from 1300 to 300 $cm^{-1}$ (~8–30 μm) of clay minerals found in remote sensing studies. The spectra are offset for clarity and grouped by type. (A) Smectites: montmorillonite (Mt), beidellite (Bd), Fe-smectite (FeSm), nontronite (Nt), hectorite (Ht), saponite (Sap); Micas: celadonite (Cld), glauconite (Glt). (B) Poorly crystalline aluminosilicates: allophane (Allo), imogolite (Imo) and aluminosilicate gel (Al/Si gel); Kaolinite-serpentines: kaolinite (Kaol), chrysotile (Ctl); Chlorites: clinochlore (Cln), chamosite (Chm). In panel A medium grey dashed lines mark the Si–O stretching band of saponite near 1035 $cm^{-1}$ and the bending band near 480 $cm^{-1}$, while dark grey dashed lines mark bands near 1125 and 545 $cm^{-1}$ found for Bd. In panel B light grey lines mark features near 930, 595 and 415 $cm^{-1}$ found in Imo, medium grey dashed lines mark features near 1015, 470 and 340 $cm^{-1}$ found in serpentines and chlorites, and dark grey dashed lines mark features near 1130, 865 and 550 $cm^{-1}$ found in Kaol. These spectra were measured as pressed powders or particulate samples at the Mars Space Flight Facility at Arizona State University for previous studies (Michalski et al., 2005; Bishop et al., 2008a, 2013a; Rampe et al., 2012).

(Michalski et al., 2005; Salisbury et al., 1991b). Spectra of serpentines, chlorites and biotite contain a strong $SiO_4$ bending vibration centred near 20.6–21.3 μm (485–470 $cm^{-1}$). In TOT (tetrahedral-octahedral-tetrahedral configuration) clay minerals, bands associated with Si–O–$M_{oct}$ absorptions ($\delta(SiOMg_{oct})$) partially overlap with the Si–O bending ($\delta(SiO)$) absorption. In trioctahedral clay minerals, the overlap between $\delta(SiO)$ and $\delta(SiOMg_{oct})$ is significant, resulting in a single, strong band located near 20.8 μm (480 $cm^{-1}$), the components of which can only be resolved through modeling. However, for dioctahedral clay minerals, the $\delta(SiOAl_{oct})$ and $\delta(SiOFe^{3+}{}_{oct})$ bands shift to shorter wavelengths (18.6–19.6 μm or 540–510 $cm^{-1}$), separating from the $\delta(SiO)$ bands. These absorptions are key because they can be used to distinguish different clay mineral compositions using long-wavelength thermal emission remote sensing data (Michalski et al., 2010a). Coordinated bending also occurs along the Si–O–$M_{oct}$ bonds from the tetrahedral Si–O into octahedral O–M bonds and is responsible for the weaker bands that often occur near 15–18 μm (670–550 $cm^{-1}$) for Al and $Fe^{3+}$ cations and near 22–24 μm (450–400 $cm^{-1}$) for Mg cations. Si–O–$Mg_{oct}$ vibrations are particularly well-resolved in the reflectance spectra of the Mg-rich minerals chrysotile and clinochlore (Fig. 14.2).

## 14.3 CHARACTERISATION OF CLAY MINERALS ON EARTH

It is out of necessity that remote sensing (VNIR and MIR) is used to study clay minerals beyond Earth. However, remote sensing is also a useful technique for clay mineral identification and mapping on our home planet as well. In fact, the ability to identify clay minerals and interpret their compositions remotely on Earth lends confidence in the capability to do so on other planets. Just as is the case in planetary exploration, remote sensing provides a tool for mapping distant, rugged terrains that are difficult or impossible to access in the field. Remote sensing data, when calibrated by careful sample analysis and field measurements, can allow for extrapolation of detailed information from a local area to mapping of vast study sites. In addition, remote sensing provides opportunities to identify and map alteration mineralogy, which might be obvious in IR data, but difficult to identify in the field.

### 14.3.1 Challenges of Remote Detection of Clay Minerals on Earth

IR spectroscopy is a powerful tool for clay mineral analyses in the laboratory; in part because significant amounts of radiation can be supplied using an active source in either reflectance or transmission. Having large numbers of photons means that the measured spectrum can be split into many channels, achieving the high spectral resolution required to resolve the precise position of specific absorptions without sacrificing much in terms of signal-to-noise-ratio (SNR). This is generally not the case in a remote sensing setting where a sensor can measure only reflected sunlight or emitted heat radiation. Therefore, remote sensing instruments are designed to maximise the science return through judicious use of available photons.

Remote sensing instruments are therefore designed as either 'multispectral' (a few to a few tens of channels) or 'hyperspectral' (hundreds of channels). Hyperspectral instruments typically offer spectral channels comparable to the spectral sampling of laboratory instruments, but with much lower SNR than what is achieved in the lab setting. Multispectral instruments have fewer channels, but they are strategically located to collect as much relevant information as possible. The photon budget also governs tradeoffs in spatial sampling (pixel size); all else being equal, smaller pixels result in lower SNR.

Spatial sampling is an important factor in remote sensing. In the laboratory, the spot size of an IR measurement can be from $<1\ mm^2$ to $1\ cm^2$ or more. While significantly larger than the clay mineral grains of interest, this ‘pixel size’ is small enough to provide confidence in exactly what part of the sample is being measured. In the case of remote sensing a single pixel can contain several surfaces as different as bedrock, regolith, soil, vegetation, water and man-made materials. An advantage of smaller pixel size is to minimise this effect; the best-case scenario in IR remote sensing of Earth is typically $\sim 10\ m^2$.

IR spectral analysis of clay mineral-bearing samples is also affected by spectral sampling as a function of depth, in addition to spatial resolution. Just as in the laboratory, reflected NIR or emitted TIR radiation samples only the uppermost 1–100 μm in most cases. Because of this effect, thin coatings can obscure the dominant mineralogy of rocks in remote sensing studies. Clay minerals compose a significant fraction of some siliceous coatings present on Earth (e.g. desert varnish) and could be important in some remote sensing settings beyond Earth. Clay mineral-bearing rocks are also in some cases masked by coatings and not detected via remote sensing.

The second major difference between laboratory and remote sensing IR studies is that environmental conditions are controlled in the laboratory setting to minimise or eliminate absorptions and scattering from gases and aerosols. The Earth’s atmosphere contains abundant water vapour, most of which occurs in the troposphere and is unavoidable in any remote IR measurement. The most significant absorptions are due to $H_2O$ and occur at 0.718, 0.810, 0.935, 1.13, 1.38, 1.88 and 2.68 μm (e.g. Gao et al., 1993; Farrand et al., 1994; Kruse, 2004). Other major absorptions affecting IR remote sensing are $CO_2$ absorptions at ~1.4, 1.6 and 2.0 μm, the $O_3$ absorption at 9.6 μm (e.g. Gao et al., 1993), and aerosol scattering throughout. Complex physical models exist for separation of atmospheric components from remote sensing data in order to retrieve spectra of the surface. Due to the complexity of Earth’s atmosphere, remote spectral observations of other planetary bodies can be more straightforward to interpret. A hyperspectral image of Meteor Crater, Arizona collected by the Airborne VIR Imaging Spectrometer (AVIRIS) (e.g. Porter and Enmark, 1987; Green et al., 1990) illustrates the power of spectral imaging over a planetary analogue site (Fig. 14.3). The atmospherically corrected data show the limitations imposed by the atmospheric windows.

Many other important differences between laboratory and remote sensing analyses exist, but they are all secondary to the issues of SNR, spectral sampling, spatial sampling and atmospheric effects. Despite all of the challenges and realities outlined above, remote sensing provides a tremendous opportunity for studying clay mineralogy on Earth. Remotely collected IR data reveal the presence of clay minerals in sites where they were previously unrecognised in the field, they show patterns of the occurrence of clay minerals that

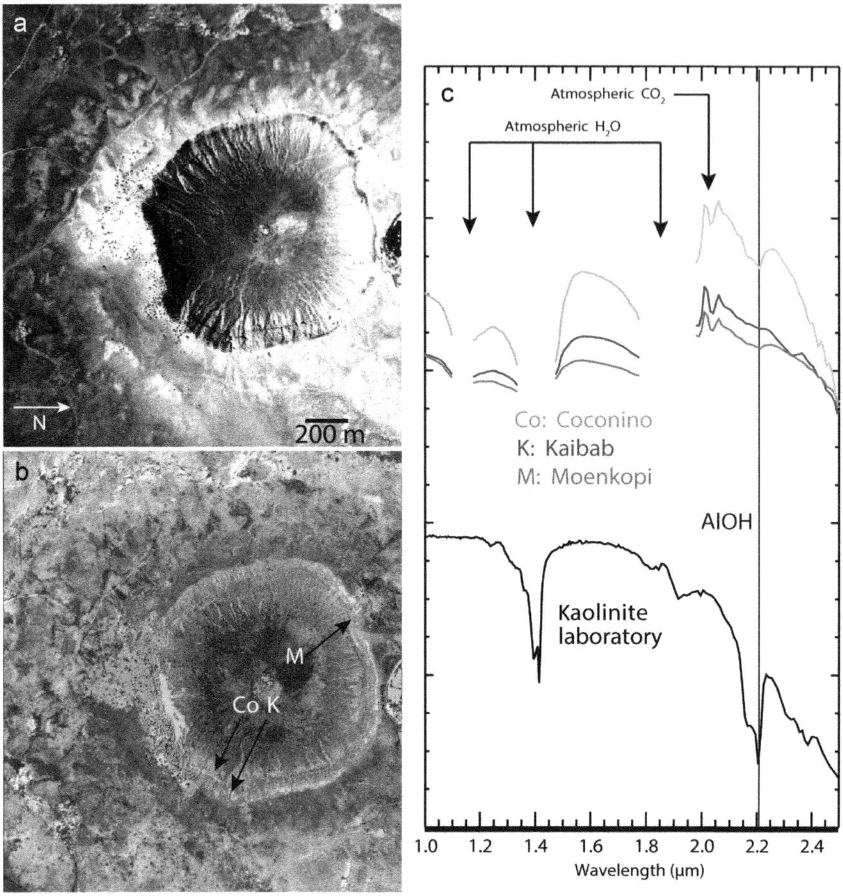


**FIG. 14.3** An example of clay mineral detection in a terrestrial impact crater (Meteor Crater) using hyperspectral AVIRIS data. (A) Approximate true colour image of the 1.2 km diameter crater. (B) A principal component image where the fourth, third and second principal components are displayed as RGB, respectively. The colours demonstrate real compositional differences, though they do not have literal meaning. Pink colours in the uppermost crater rim correspond to the Moenkopi (M) Formation. Yellow-red colours underlying the Moenkopi Formation in the mid-crater wall correspond to the Kaibab (K) formation. Green colours refer to the Coconino (Co) Sandstone. Blue colours in the crater correspond to alluvium and lacustrine sediments. (C) Atmospherically corrected spectra of several sites around the crater compared with a lab spectrum of kaolinite. Black arrows mark the positions of atmospheric $H_2O$ bands that were removed. A grey line marks the position of the $(\nu+\delta)Al_2OH$ kaolinite band at 2.21 μm. (For interpretation of the references to colour in this figure legend, the reader is referred to the web version of this article.)

are surprising, and that challenge geological assumptions. The data provide a powerful tool for characterising clay mineralogy and constraining crystal chemistry over large areas. In addition, remote sensing of clay minerals on Earth lends tremendous confidence to the remote detection of clay minerals on Mars and other planets—environments where atmospheric effects are lacking or minimal, and where surface vegetative cover, water bodies, and the presence of man-made materials are not concerns.

### 14.3.2 Instruments and Datasets Available for IR Remote Sensing of Clay Minerals on Earth

Terrestrial remote sensing can be carried out from orbital or airborne platforms (Table 14.1). While an orbital platform provides significant advantages in terms of global coverage, there are substantial tradeoffs with respect to spatial resolution and SNR. Most orbital IR instruments to date have been multispectral, with the exception of Hyperion, a hyperspectral imager aboard the EO–1 satellite (e.g. Kruse et al., 2003; Pearlman et al., 2003). Hyperion was able to image nearly anywhere on the globe at 30 m/pixel spatial resolution, but at the expense of very low SNR. These data can be used for qualitative maps showing compositional diversity, but they are not useful for clay

**TABLE 14.1** Instruments and Datasets Available and Planned for Remote Analysis of Clay Minerals on Earth

| Sensor Name | # Bands | Wavelength Range (μm) | Spectral Resolution | Spatial Resolution (m/pixel) |
|---|---|---|---|---|
| *Orbital* | | | | |
| Landsat 8 | 9 | 0.43–2.3 | Multi | 15–30 |
| ASTER | 14 | 0.52–11.65 | Multi | 15–90 |
| Hyperion | 220 | 0.4–2.5 | Hyper | 30 |
| HyspIRI | 212 | 0.38–2.5 | Hyper | 60 |
| HyspIRI | 8 | 3–12 | Multi | 60 |
| *Airborne* | | | | |
| AVIRIS | 224 | 0.4–2.5 | Hyper | 5–50 |
| NS001 | 8 | 0.46–12.3 | Multi | 5–50 |
| TIMS | 6 | 8.2–12.2 | Multi | 5–50 |
| MASTER | 50 | 0.4–13 | Multi | 5–50 |
| HyMap | 128 | 0.45–2.5 | Hyper | 3–10 |
| HyTES | 256 | 7.5–12 | Hyper | 2–10 |

*Note*: only the most recent Landsat imager is included in this table; the spatial resolution of airborne sensors depends on the height at which they are flown; HyspIRI is in the planning stage at the time of this writing.

mineral identification. ASTER and LANDSAT satellites have provided global multispectral NIR/short-wave IR (SWIR) coverage of the Earth's surface at 30 m/pixel resolution. Due to the multispectral nature of these data, spectral bands associated with specific minerals cannot be identified. However, they can provide useful information through broad spectral patterns and these data can be used for mapping alteration, and potentially outcrops of clay minerals through coordination with hyperspectral sensors or ground testing. Such techniques have been developed (Viviano and Moersch, 2012) for detection of clay minerals on Mars using the Mars Odyssey thermal emission imaging system (THEMIS) multispectral sensor (Christensen et al., 2003).

Sensors flown on airborne platforms can generally achieve higher SNR and smaller pixel size by flying at low altitude. The major tradeoff is that the data need to be deliberately collected, which involves planning flights. In other words, the data are only available when targeted for a specific site. Over the last 25 years, AVIRIS has been the primary remote sensing instrument for IR clay mineral mapping (e.g. Kruse et al., 1993; Boardman et al., 1995; Clark et al., 1995; Clark et al., 2002; Crowley and Zimbleman, 1997; Chabrillat et al., 2002; Guinness et al., 2007). A more recently developed instrument, HyMap provides similar data (e.g. Cocks et al., 1998) and has been used as well for detection of clays (e.g. Brown et al., 2005; Nunez et al., 2015).

### 14.3.3 Remote Characterisation of Clay Minerals on Earth

A number of authors have used remote sensing to study clay minerals on Earth (e.g. Crowley and Zimbleman, 1997; Chabrillat et al., 2002; Clark et al., 2003; Guinness et al., 2007; Seelos et al., 2009; Swayze et al., 2014). A driving factor in remote sensing of clay minerals has been to identify alteration patterns linked to economic mineralisation in rugged, complex terrains (e.g. van der Meer et al., 2012). The global reach of multispectral datasets such as ASTER has proven fruitful for the identification and basic characterisation of hydrothermal alteration associated with ore deposits (e.g. Bakardjiev and Popov, 2015), but hyperspectral mapping by airborne sensors has revealed a tremendous amount of diversity in hydrothermally altered, mineralised terrains (e.g. Swayze et al., 2014; van Ruitenbeek et al., 2012). Clay mineral-bearing soils were mapped along the Front Range Urban Corridor in Colorado using AVIRIS and HyMap imagery (Chabrillat et al., 2002). This study illustrated the utility of hyperspectral imaging for identification of distinct clay minerals such as smectite, inner stratifications of smectite with illite and kaolinite-bearing soils. A set of band shape algorithms termed Tetracorder have also been developed for identification of clays and clay minerals and other minerals using band centre and asymmetry (Clark et al., 2003).

While hydrothermal deposits associated with economic mineral outcrops might be a driver for remote exploration, studies of alteration relevant to environmental issues are also carried out via remote sensing. For example, Crowley and Zimbleman (1997) used AVIRIS data to map various clay minerals on Mount Rainier, in areas that are extremely difficult to access through field work alone. These highly altered terrains are most susceptible to slope failure, leading to landslide hazards.

Remote sensing studies of clay minerals have also been carried out in order to better understand clay minerals detections beyond Earth. For example, Mustard and Pieters (1987) used data from an airborne hyperspectral sensor to identify and map alteration units within serpentinised terrain on the Colorado Plateau. Studies of weathering patterns and occurrences of clay minerals in igneous rocks on Earth has helped to understand how the breakdown of primary minerals impacts interpretations of bedrock lithology on the Earth and Mars (Michalski et al., 2004). Spectral analyses of alteration associated with Hawaiian cinder cones and calderas shed light on the detection of silica and clay minerals in analogous environments on Mars (Guinness et al., 2007; Seelos et al., 2009).

#### *14.3.3.1 Clays and Clay Minerals Detected in Hyaloclastites in Askja, Iceland*

The central Icelandic desert contains youthful, well-preserved examples of landforms that occur when volcanoes interact with glacial ice. These include tuyas (table mountains) and tindars or moberg ridges that form beneath thick ice (Allen, 1979; Chapman and Smellie, 2007). The products of these subice eruptions contain intensely fragmented volcanic glass that has largely been altered to smectite, oxides and oxyhydroxides, and minor carbonates and sulphates (Honnorez, 1981; Furnes, 1978; Jakobsson, 1978). Examples of palagonitised hyalotuffs were investigated in situ at Hlöðufell tuya and Thórólfsfell ridge (Bishop et al., 2002d) and examples of clay mineral-bearing altered basalts were characterised from flows near Hvalfjordur and Berufjordur (Ehlmann et al., 2012).

The remoteness of the Askja region make it difficult to reach with airborne sensors, but ASTER is well suited for clay mineral mapping in the vast (>5000 km$^2$), largely unvegetated, rugged landscape (Graettinger et al., 2013). An ASTER SWIR principal component image of the Askja region (Fig. 14.4) shows strongly altered components of the tuya and moberg in bright blue and slightly less altered parts (most of the structure) in purple. Extracted reflectance spectra illustrate strong absorptions in ASTER bands 7 and 8, which correspond approximately to 2.24–2.37 μm, where trioctahedral clays and $Fe^{3+}$-rich dioctahedral clay minerals absorb. TIR spectra (in the MIR region) of the same surfaces (Fig. 14.4d) show a strong absorption at 9.3 μm but little absorption from 10 to 11 μm. This observation suggests that much of the mafic glass with absorptions from 10 to 11 μm has been converted to smectite, which has strong Si–O absorptions at shorter wavelengths (9–9.5 μm). Taken together,

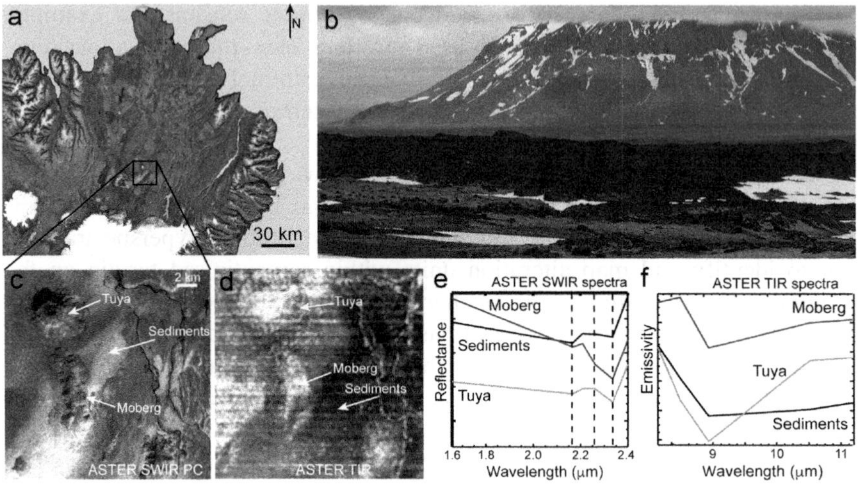


**FIG. 14.4** Remote sensing of clay minerals in the Askja region of central Iceland (A). The landscape includes lavas, reworked volcanics, moberg ridges and tuyas (B). ASTER SWIR (C) and TIR (D) principal component images show colour variations related to volcanism and hydrothermal alteration. Extracted reflectance (E) and thermal emission (F) spectra show features related to octahedral cation–OH and Si–O vibrational absorptions in clays.

the SWIR and TIR results provide a strong case for the formation of nontronite and/or $Fe^{3+}$-rich trioctahedral smectite in the Askja hyaloclastites.

Analyses of materials sampled from these areas confirm the above hypothesis and reveal more information about the altered hyaloclastites (Michalski and Bleacher, 2014). NIR reflectance spectra collected in the lab contain $2\nu$ (OH) absorptions near 1.4 μm, $2\nu(H_2O)$ absorptions at 1.41 μm, $(\nu+\delta)$ $(H_2O)$ absorptions at 1.91 μm, and structural $(\nu+\delta)$MOH bands at 2.305 consistent with the presence of Fe/Mg-smectite.

### *14.3.3.2 Clays and Clay Minerals Detected at Swansea, Arizona*

Swansea, Arizona contains extremely, structurally complex Micocene volcanic and sedimentary rocks juxtaposed against Precambrian basement along the Buckskin-Rawhide detachment fault. The Miocene was a period of extreme and active extension in southwestern North America, where movement along the fault is estimated at 65 km or more (Reynolds and Spencer, 1985). As the crust extended, a number of shallow basins containing alkaline lakes formed. These surface and subsurface basin brines pervasively metasomatised volcanic ash and lava and debris flows composed of fragments of both (Chapin and Lindley, 1986). The K-metasomatism is remarkable because it has resulted in 100% chemical and mineralogical replacement of the protoliths, which preserved the primary textures. Despite the fact that some of the altered rocks have K/Na ratios >500, this alteration is very difficult to identify in the field.

Remote sensing data provide valuable insight into the alteration patterns in the Swansea area. While it was conceivable that the alteration might have been limited to fluid conduits along the Buckskin-Rawhide detachment fault or associated normal faults, ASTER clay mineral maps (Fig. 14.5) show that this is not the case. In fact, analysis of ASTER imagery reveals that the

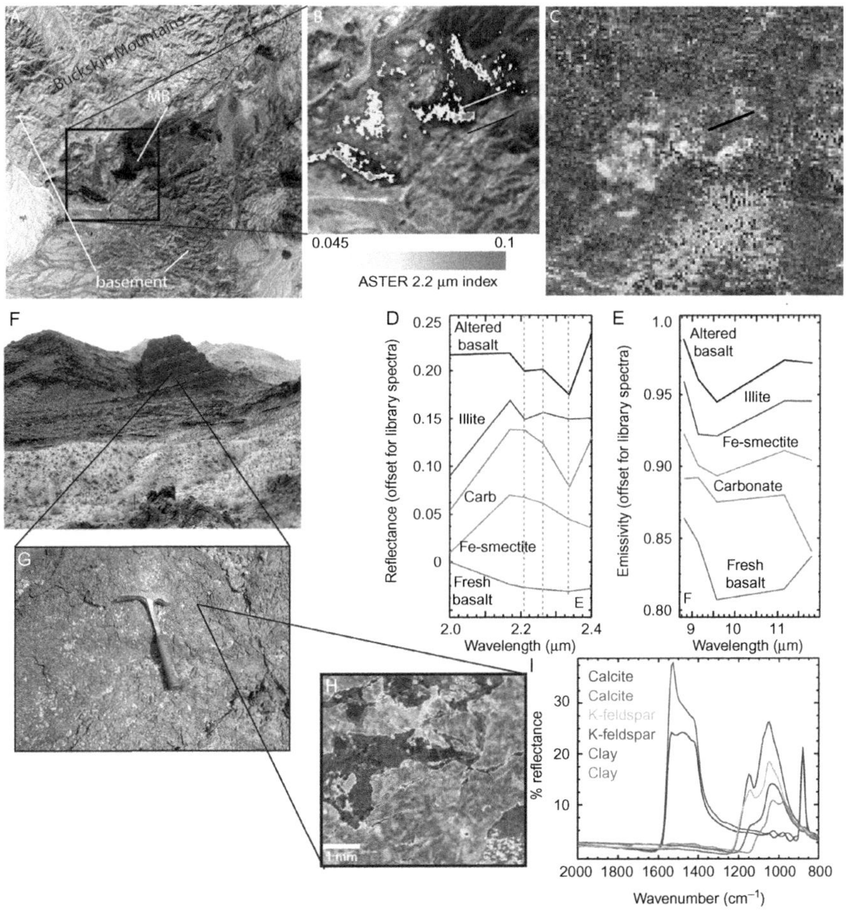


**FIG. 14.5** ASTER visible data show Miocene volcanic and sedimentary rocks near Swansea, Arizona, within the NE-SW-trending Buckskin Mountains (A). An ASTER 2.2 μm clay index map shows the occurrence of Al-rich clay minerals in the region (B). TIR emissivity data (C) show colour variations related to alteration mineralogy. Extracted reflectance (D) and emission (E) data show absorptions related to octahedral cation–OH and Si–O absorptions in clay minerals, described in the main text. The basalt unit (dark rock in 'F') contains patchy carbonate replacements, but shows no evidence of clay mineral-rich material in the field (G). However, remote sensing and laboratory data show that the rock has been entirely replaced by secondary minerals associated with K-metasomatism. Micro-FTIR reflectance data show original plagioclase lathes that have been replaced by K-feldspar and mafic groundmass that has been totally replaced by illite (H and I).

volcanic units (mostly basaltic lava protolith) are altered everywhere they are observed. All of the rock units were exposed to basin brines as the basins were buried during rapid extension. However, the alteration only affected the mafic lithologies; it was driven entirely by susceptibility of the protolith to alteration by the K-rich brines.

ASTER reflectance data suggest the presence of Al-rich clays based on the 2.2 μm absorption band (Fig. 14.5). TIR ASTER data show narrow Si–O absorptions shifted to shorter wavelengths, consistent with the occurrence of clay minerals rather than mafic minerals in basalt exposures. Laboratory analyses of the bulk rock, as well as the <2 mm fraction of the crushed and sieved rock particles, show that the altered basalt contains illite. Micro-FTIR mapping of the alteration at the scale of hundreds of microns reveals how the plagioclase has been replaced by K-feldspar and the pyroxene and amphibole have been completely replaced by illite (Michalski et al., 2007).

#### *14.3.3.3 Clays and Clay Minerals Detected at the Painted Desert, Arizona*

The Painted Desert in Arizona is another region rich in clay minerals and low in vegetation with coverage by the HyMap imager. HyMap has a spectral range of 0.45–2.48 μm with a sampling of 13–17 nm per channel, and a spatial resolution of 4 m per pixel. Spectra are shown in Fig. 14.6 from sites 1-1 and 1-6 (McKeown et al., 2009b) that include mixtures of montmorillonite, calcite, dolomite, analcime, quartz and plagioclase. The montmorillonite bands are present in the lab, field and aerial data of these sites, although the relative intensity of the montmorillonite and carbonate bands varies. XRD indicates a higher abundance of montmorillonite at site 1-6, which corresponds well with the stronger montmorillonite signatures in those data.

## 14.4 CHARACTERISATION OF CLAYS AND CLAY MINERALS ON MARS

The primary source of data for martian clay minerals are the OMEGA (Bibring et al., 2005) and CRISM (Murchie et al., 2009) orbital imaging spectrometers that collect reflected sunlight of the surface in the VNIR spectral region. The near entirety of Mars has been mapped at VNIR wavelengths at the km to ~200 m resolution, while selected spot observations have been carried out at tens of metres resolution. TES orbital imaging provides context for investigations of clay minerals with near global coverage at 1–5 km resolution (Christensen et al., 2001). Clay minerals frequently occur in small outcrops on Mars that are less than 1 km across. Such occurrences are not possible to detect with TES. However, modelling developed by the TES team enables estimations of major and minor surface components including clay minerals and related altered aluminosilicate-bearing materials (Ramsey and Christensen,

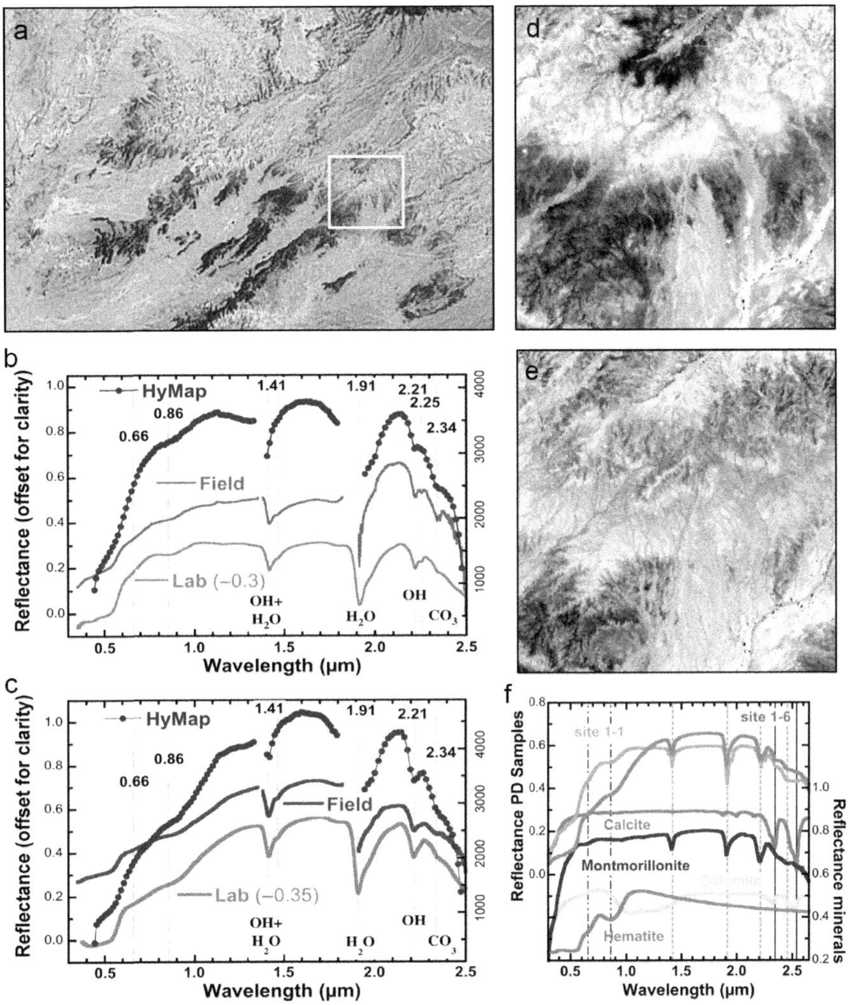


**FIG. 14.6** Example remote sensing analyses from the Painted Desert, AZ (modified from McKeown et al., 2009b). (A) Aerial photograph of the region showing clay mineral-bearing units in white/grey or red where the topsoil has been eroded away, (B) lab, field and HyMap spectra of site 1-1 from the study, (C) lab, field and HyMap spectra of site 1-6 from the study, (D) visible HyMap image highlighting variations in iron content in the lower-left corner in red/purple tones (colours mapped as R: 0.65 μm, G: 0.55 μm, B: 0.45 μm), (E) NIR HyMap image highlighting variations in mineralogy in the upper half of the image in blue and purple tones (colours mapped as R: 2.48 μm, G: 1.50 μm, B: 1.01 μm) and (F) lab spectra of minerals compared to lab spectra of Painted Desert samples. (For interpretation of the references to colour in this figure legend, the reader is referred to the web version of this article.)

1998; Rogers and Christensen, 2007; McDowell and Hamilton, 2009). Band parameters have also been developed for detection of Fe/Mg-clay minerals using THEMIS imagery (Viviano and Moersch, 2012), which has better spatial resolution but only a few channels (Christensen et al., 2003). Detecting clay minerals on Mars with TES has been challenging due to spot size, surface morphology and the nature of the MIR clay mineral bands in relation to other silicate features. Despite these challenges, investigations of selected regions have identified components due to altered basalt, glass, zeolite, high-Si phases and poorly crystalline aluminosilicates (Michalski et al., 2013b; Michalski and Fergason, 2009; Rogers and Christensen, 2007; Rampe et al., 2012; Ruff, 2004). Recently, coordinated studies of CRISM and TES analyses have found that including poorly crystalline aluminosilicates in the model enables detection of clay minerals in TES data as well for Mawrth Vallis, one of the most clay mineral-rich sites of Mars (Bishop and Rampe, 2016).

An early assessment of clay mineral detections on Mars using OMEGA data that indicated they are largely found in the ancient Noachian terrains (Bibring et al., 2006) continues to hold for most of the planet. More recently formed clay-rich materials have been observed in isolated regions such as Noctis Labyrinthus (e.g. Weitz et al., 2011), Coprates Chasma (e.g. Weitz et al., 2014), and some impact craters. Early investigations of outcrops bearing clays and clay minerals at Mawrth Vallis revealed that Al-rich clays and clay minerals always occurred in strata above Fe/Mg-rich smectite (Bishop et al., 2008c). A more recent study found that this trend is common across most of the planet where clay mineral exposures are observed (Carter et al., 2015). Detections of clay minerals from orbit (Wray et al., 2009; Milliken et al., 2010) have even guided detections of clay minerals by surface rovers at Endeavour crater (Arvidson et al., 2014; Fox et al., 2016) and Gale crater (Ming et al., 2014; Vaniman et al., 2014), where more detailed analyses of these ancient aqueous outcrops were performed. Characterisation of clay minerals by instruments on the Curiosity rover at Gale crater support formation of smectites in Hesperian rocks in low water to rock ratio environments (e.g. McLennan et al., 2014; Grotzinger et al., 2014; Grotzinger et al., 2015; Bristow et al., 2015).

### 14.4.1 Global Mapping of Clays and Clay Minerals and Aqueous Alteration on Mars

Mars exhibits several thousands to tens of thousands of clay mineral exposures at the coarse kilometric scale across its surface (Figs. 14.7 and 14.8). The distribution of these clay mineral-bearing outcrops is not homogeneous, with a majority of detections originating from the southern, Noachian-aged highlands ($>$3.8 Gyrs old). Terrains younger than the Late Hesperian ($\sim$3.5–3.8 Gyrs) are nearly devoid of these signatures, indicating that the bulk of the clay mineral-forming environments occurred during and before that time. Little is known about approximately one third of the martian surface

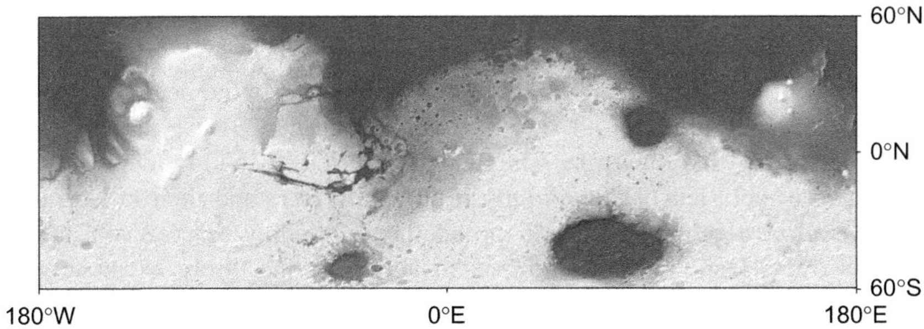


**FIG. 14.7** Global map of aqueous minerals on Mars, based on data from the OMEGA and CRISM VNIR spectrometers. The background is Mars topography from MOLA and colours indicate mineral composition: red for Fe/Mg-clay minerals (80% smectites, 20% chlorites), green for hydrated sulphates and cyan for Al clay minerals. Yellow indicates mixtures of Fe/Mg-smectites and sulphates. Most aqueous mineral exposures, especially non-Fe/Mg-clay minerals detections are not visible at this scale. (For interpretation of the references to colour in this figure legend, the reader is referred to the web version of this article.)

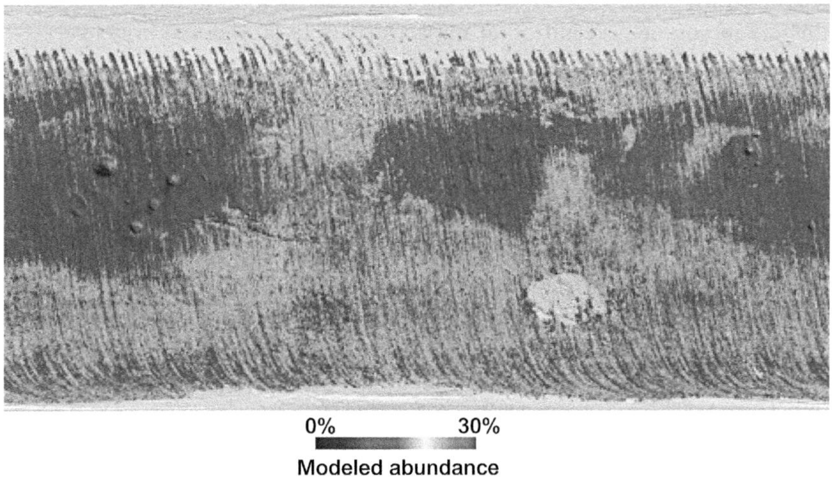


**FIG. 14.8** TES global abundances of clay minerals on Mars. Modelled abundances correspond to clay minerals and other poorly crystalline aluminosilicates (Michalski et al., 2006b). *(modified from Bandfield, 2002.)*

as it is masked to VNIR instruments, either by coatings of anhydrous dust or by the cryosphere (ice, recurring ice clouds and seasonal frosts). Clay mineral exposures are typically small in size (a few 100 m to a few km) and often take the form of discontinuous patches, owing to the degraded state of the ancient surface. Their geological context is often too disrupted to be uniquely inferred from orbit. Recent reviews of orbital remote sensing at Mars describe the current understanding of the martian surface composition including clay minerals (Murchie et al., 2018; Hamilton et al., 2018).

#### *14.4.1.1 VNIR Observations of Clays and Clay Minerals and Aqueous Alteration on Mars*

About half of the clay mineral exposures are observed in the peaks, walls and ejecta of impact craters of all sizes (e.g. Carter et al., 2013; Ehlmann et al., 2013, indicating that clay minerals are often buried at shallow depths, up to a few km, usually less). The ubiquity of impact craters and their broad range of excavation depths are used to probe the mineralogy of the subsurface. Impacts over Hesperian and Amazonian aged terrains rarely excavate clay minerals, with the exception of larger craters (typically >20–200 km), which are thought to have penetrated the deeply buried and altered Noachian crust. This applies in particular to the northern lowland hemisphere of Mars, where the few clay mineral signatures are restricted to remnant Noachian units and the largest craters. In the south, consistent trends have been found wherein clay minerals are replaced at depth by higher temperature and lower water activity species that include chlorites, zeolites and prehnite (e.g. Ehlmann et al., 2009; Marzo et al., 2010). Low-grade metamorphic facies have been recognised for some sites. A few craters are also thought to have formed clay minerals in situ, as a result of an impact-driven hydrothermal system.

Non impact-related clay mineral exposures are numerous and include several deposits of vast sizes (up to hundreds of km across) spread across the southern highlands (e.g. Poulet et al., 2005; Bishop et al., 2008c, 2013c; Mustard et al., 2008; Murchie et al., 2009; McKeown et al., 2009a; Milliken et al., 2010; Loizeau et al., 2012a; Carter et al., 2015; Weitz et al., 2012; Weitz et al., 2015; Weitz and Bishop, 2016). Those studied in detail have uncovered over 10 types of clay minerals (smectites including nontronite, saponite, beidellite and montmorillonite, vermiculites, corrensites, chlorites, K/Al-rich micas, $Fe^{2+}$-smectite/mica mixtures, kaolinite, mixed layered kaolinite-smectite, allophane, talc and serpentines). These are summarised by Ehlmann and Edwards (2014). Correlations between the clay mineral and related mineralogy and morphological context across the planet have been interpreted as originating from a number of surface to near-surface formation and depositional environments (e.g. Ehlmann et al., 2011). These include detrital and authigenic lacustrine clay minerals, climate-mediated surface weathering (involving rain or snow fall), evaporitic playa environments and hydrothermal systems (deep crustal, epithermal and possibly spring deposits). Geochemical conditions proposed for these environments range from highly acidic (pH$\sim$2) to mildly alkaline, near freezing to >100°C, water-dominated to rock-dominated, mostly oxidising, and with varying ion activities (e.g. Ehlmann et al., 2013; Carter et al., 2015; Murchie et al., 2018).

Beyond this preserved diversity, global mineralogical trends exist. The bulk of Mars alteration (>80% of the deposits) appears to have formed Fe/Mg-rich clay minerals with a strong smectitic contribution (Ehlmann et al., 2011; Carter et al., 2013; Michalski et al., 2015). For some of these, there is evidence for mixed-layering with chlorites and micas, while chlorites are found in a spectrally pure form less than 20% of the time. Hydrated sulphates and silica are the second and third most abundant aqueously altered species (accounting for ~40% of the aqueous deposits on the surface), while aluminium clay minerals (both smectites and kaolinites) are found in about a third of the mineral deposits. It should be noted that these relative abundances may be significantly biased by a number of factors: (1) the selective detection sensitivity to the diagnostic spectral absorptions of different clay minerals, (2) the selective survivability of some species to later chemical transformations (diagenesis) or to disruptive processes such as impact gardening or aeolian erosion, (3) the lack of access to a third of the surface of Mars, (4) the poor penetration depth of VNIR spectroscopy (typically hundreds of microns or less), which largely restrict the study to surficial layers and erosional windows.

The dominance of Fe/Mg-rich clay minerals at Mars is consistent with water-limited alteration of basaltic and ultramafic protoliths, but their associated geological context and alteration process remains elusive in some cases. It is likely that a large fraction of the alteration occurred in the martian subsurface, but it is also clear that surface alteration has occurred throughout ancient Mars. The biggest mystery over climate remains: was Mars episodically warm and wet or did most of the alteration occur in cold, snow or ice-dominated environments?

#### *14.4.1.2 TIR Observations of Clays and Clay Minerals and Aqueous Alteration on Mars*

TIR emission spectral provide a global view of martian surface mineralogy, albeit at coarse spatial resolution (hundreds of km$^2$) (Christensen et al., 2001; Hamilton et al., 2018a). Martian dark regions are spectrally dominated by basaltic materials in the thermal IR (Christensen et al., 2000b). Linear spectral unmixing of data from the TES spectrometer provides estimates of mineral abundances in the dark regions. This approach suggests that dark volcanic surfaces are composed of approximately 25–40% plagioclase feldspar, 25–40% pyroxene and 5–10% olivine (Rogers and Christensen, 2007). The data also show, however, that martian dark regions contain 15–35% clay minerals or high-silica phases (Fig. 14.8) (Michalski et al., 2006b), with 10–15% occurring at low latitudes and 20–30% occurring at higher northern latitudes (Wyatt et al., 2004).

TES spectra provide an interesting perspective on the occurrence and abundance of clay minerals on Mars. The spectral unmixing approach that has often been applied to TES data depends heavily on the spectral shape in the 900–1300 cm$^{-1}$ region, where Si–O absorptions occur in all silicate minerals and amorphous phases. In this spectral range, clays and clay minerals exhibit nonunique absorptions related to Si–O vibrations. However, these somewhat broad unremarkable features are shared by other poorly crystalline materials with similar Si/O molar ratios (Michalski et al., 2005; Rampe et al., 2012).

Early results from TES showing 10–15% of clay minerals or similar materials in the dark, low-latitude regions were difficult to interpret in the pre-OMEGA era, but now these results seem reasonable as an average composition of Noachian martian crust. Such detections might also include allophane, imogolite and other poorly ordered aluminosilicates, possibly associated with weathered volcanic ash in the ancient crust (Rampe et al., 2012; Bishop and Rampe, 2016).

Increased abundances of modelled clay-like, high-silica materials at high latitudes are more complicated to interpret because, while NIR data suggest the presence of mafic glass in some parts of the high latitudes (Horgan and Bell, 2012), neither OMEGA nor CRISM has detected such widespread deposits consistent with high-silica materials. It is possible that the high-latitude deposits detected by TES correspond to thin, but highly absorbing coatings and rinds (Kraft et al., 2003; Michalski et al., 2006a) similar to those that form as immature weathering products in low water/rock ratio settings on Earth.

## 14.4.2 Regional Mapping of Clays and Clay Minerals and Aqueous Alteration on Mars

At the regional scale, Mars exhibits variations in the clay mineralogy that can sometimes be linked to a particular morphological context or at least a broad geological unit. Detailed investigations often yield additional morphological and mineralogical diversity and allow refining interpretations of aqueous environments. The main limitation is often the lack of deep erosional windows to assess the stratigraphy, obscuration by dust or other mantles (e.g. volcanic capping) or lack of submetre resolution context imagery. Another potential complicating factor is the lack of a direct analogue on Earth for some of the alteration processes and geologic contexts observed on Mars, which provides a great challenge to interpretation of martian geology. A few examples of mineral provinces at Mars are described in detail below that are representative of the clay context globally and/or have been identified as targets of particular interest, including for future exploration.

### *14.4.2.1 The Clay-Laden Eastern Margin of Chryse Planitia*

The Chryse Planitia basin provides a diversity of terrains with ages ranging from the middle Noachian to the Amazonian (Tanaka et al., 2005). Since its early emplacement, the basin has been in-filled by mass wasting, catastrophic flood deposits, and aeolian and volcanic resurfacing processes. The margin of Chryse Planitia retains, however, the signatures of several aqueous environments present during the Noachian to early Hesperian time period. The near entirety of the preserved margin exhibits clay-rich units (Fig. 14.7), the most prominent of which are located toward the east in the Mawrth Vallis plateau and Oxia Planum deposits (Fig. 14.9). The former has been recognised as the best example of past pedogenesis on Mars based on its stratigraphy and composition (Michalski et al., 2010b; Bishop et al., 2013b). The lower altered strata consist of hundreds of meters-thick layered $Fe^{3+}$-rich smectite (mapped in red) that may have formed by the syn-depositional alteration of volcanic sediments involving meteoritic water (e.g. Michalski and Noe Dobrea, 2007; Wray et al., 2008; Bishop et al., 2008c; McKeown et al., 2009a; Loizeau et al., 2010; Loizeau et al., 2012b; Bishop and Rampe, 2016). Upper, highly leached strata consist of mixtures dominated by Al-smectite and kaolinite (mapped in cyan), sulphates and poorly crystallised phases such as allophane and hydrated silica (Bishop et al., 2016). Changes in climate conditions from near neutral to mildly acidic are proposed to have caused this chemical gradient in the Mawrth Vallis region. Although the Oxia Planum deposits appear to pertain to the same circum-Chryse clay mineral unit as Mawrth Vallis, their composition and stratigraphy differ. The aqueous deposits at Oxia Planum consist of thinner strata (tens of metres) of layered clay minerals with a remarkably homogeneous composition, at least as viewed from orbit, and the clay minerals identified at Oxia Planum are either ferroan saponite or vermiculite-bearing. An alluvial fan or delta postdates this Late Noachian

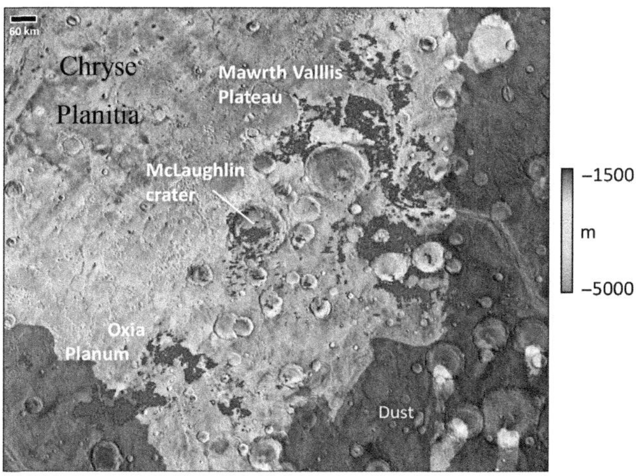


**FIG. 14.9** Regional scale, clay mineral-rich units on the eastern margin of Chryse Planitia. Background is THEMIS IR imagery colour coded (green-brown tones) with altimetry from MOLA. Darkened areas correspond to a dust-mantle partly shrouding the surface to VNIR spectroscopy (mapped with OMEGA). Fe/Mg-clay minerals are overlain in red and vary from $Fe^{3+}$-nontronite at the Mawrth Vallis Plateau, $Fe^{3+/2+}$-saponite/vermiculite at Oxia Planum, to serpentine and carbonate bearing at McLaughlin crater. Aluminous clay minerals are mapped in blue, and are comprised of kaolinites, smectites and allophane. Spectrally pure deposits of sulphates, carbonates and minor phases are not visible at this scale. Map is centred approximately at −22°E, 22°N. (For interpretation of the references to colour in this figure legend, the reader is referred to the web version of this article.)

unit, and contains hydrated silica and kaolinite. These are interpreted as authigenic phases formed during a shallow lake phase overlying the clay minerals (Carter et al., 2016; Quantin et al., 2016). Located in between Mawrth Vallis and Oxia Planum is McLaughlin crater, where mixtures of clay minerals with carbonate and serpentines have been found (Michalski et al., 2013a). These may originate from a combination of inherited, deeply buried material excavated by a nearby 30 km-large crater, while the carbonates may have formed authigenically within an alkaline, ground-water fed crater lake (Michalski et al., 2013a). The high scientific potential of these three deposits at the margin of the Chryse Plateau has led to multiple landing site proposals for current and upcoming rover missions.

Layered outcrops in the Mawrth Vallis region of Mars contain the greatest diversity of aqueous alteration materials that provide an opportunity to infer past aqueous environments. Numerous orbital investigations have documented aluminous and siliceous clay-bearing units overlying an Fe/Mg-smectite-rich unit below (e.g. Bishop et al., 2008c; Wray et al., 2008; McKeown et al., 2009a; Michalski et al., 2010b; Noe Dobrea et al., 2010; Loizeau et al., 2012b). Many different secondary minerals have been identified in the upper aluminous and siliceous clay-rich units (e.g. Bishop et al., 2013b), but the presence of poorly crystalline phases was reported only recently (Bishop and Rampe, 2016). Further, allophane and imogolite comprise a significant portion of the uppermost stratum of the aluminous and siliceous clay-rich units, on top of the montmorillonite- and opal-rich unit. These results signify a change in climate on Mars from a warm and wet environment to one where water was sporadic and likely depleted rapidly (Bishop and Rampe, 2016). A more complete stratigraphy including five units was developed recently for the Mawrth Vallis region using newly created CRISM parameters, HRSC DTMs across hundreds of kms and HiRISE DTMs across hundreds of metres (Bishop et al., 2016). This stratigraphy is illustrated in Fig. 14.10. The distinct mineralogic units were characterised using spectral properties from CRISM and surface morphology from HiRISE. Bishop et al. (2016) describe an ancient wet and warm geologic record that formed the thick nontronite unit, a period of wet/dry cycling to create acid alteration, followed by leaching or pedogenesis to result in aluminous clay minerals, and finally a drier, colder climate that left the altered ash in the form of nanophase clay minerals, rather than crystalline clay minerals.

#### *14.4.2.2 The Clay-Bearing Region West and South of Isidis Planitia*

After the Chryse Planitia basin the regions surrounding the Isidis basin have the next greatest abundance of clay and clay mineral-rich material (Figs. 14.7 and 14.11). The Nili Fossae region northwest of Isidis hosts pristine olivine and pyroxene outcrops (Mustard et al., 2009) and abundant and varied clay

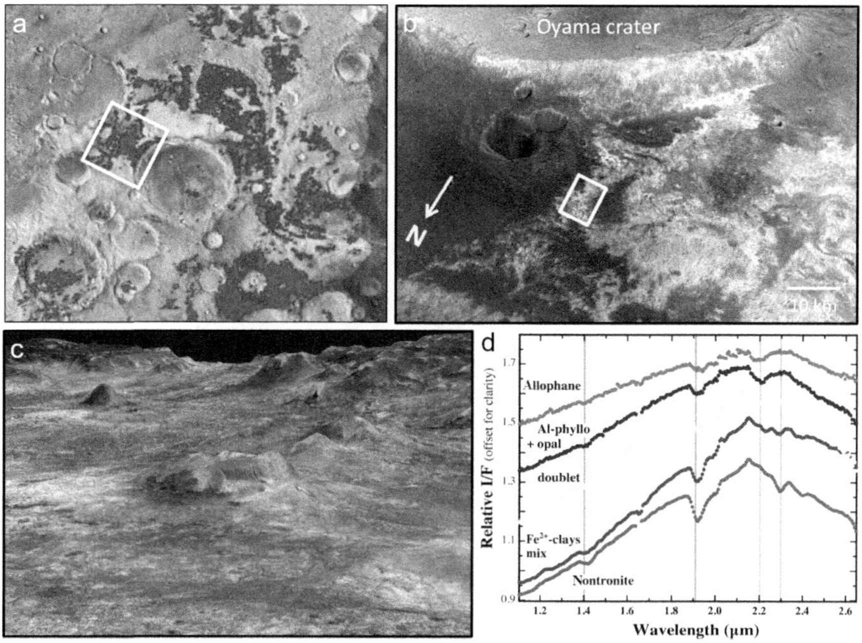


**FIG. 14.10** Views of clay-bearing outcrops at Mawrth Vallis. (A) Mawrth Vallis region from Fig. 14.9 with a white box showing the location of the region shown in B. (B) Close-up view of northwestern region in CTX imagery over an HRSC DTM with colours from CRISM overlain. Nontronite (Nt) is mapped in red, ferroan clay minerals (FeSm mix) in purple, a sulphate-bearing and/or acidic altered unit in yellow, Al-clay minerals (AlSm) and opal in blue, and poorly crystalline aluminosilicates (allophane, Allo) in green. The white box shows the location of C. (C) HiRISE DTM with colours from CRISM, view towards N. (D) Example spectra from the five stratigraphic units identified recently in the region. *(modified from Bishop et al., 2016.) (For interpretation of the references to colour in this figure legend, the reader is referred to the web version of this article.)*

mineral-bearing units in several provinces with distinct mineral assemblages (Mustard et al., 2008; Ehlmann et al., 2009). Fe/Mg-smectite is found throughout the Nili Fossae region in the ancient Noachian rocks. This Fe/Mg-smectite unit occurs in eroded terrains in Eastern Nili Fossae where it is stratigraphically covered by either kaolinite or magnesite (Ehlmann et al., 2009, Fairén et al., 2010). In other outcrops the Fe/Mg-smectite is mixed with mica, illite or chlorite. The presence of zeolites and prehnite together with Fe/Mg-smectite in and around impact craters in the west and south are indicators of low-grade metamorphic or aqueous hydrothermal alteration (Ehlmann et al., 2009; Fairén et al., 2010; Marzo et al., 2010). Sulphates are observed together with Fe/Mg-smectite, magnesite and other clay minerals along the southwestern part of Nili Fossae between Jezero crater and the Syrtis volcanics region (Ehlmann and Mustard, 2012). Jarosite ridges were observed on layered material containing polyhydrated sulphates and Fe/Mg-smectite. The diversity in clay minerals observed

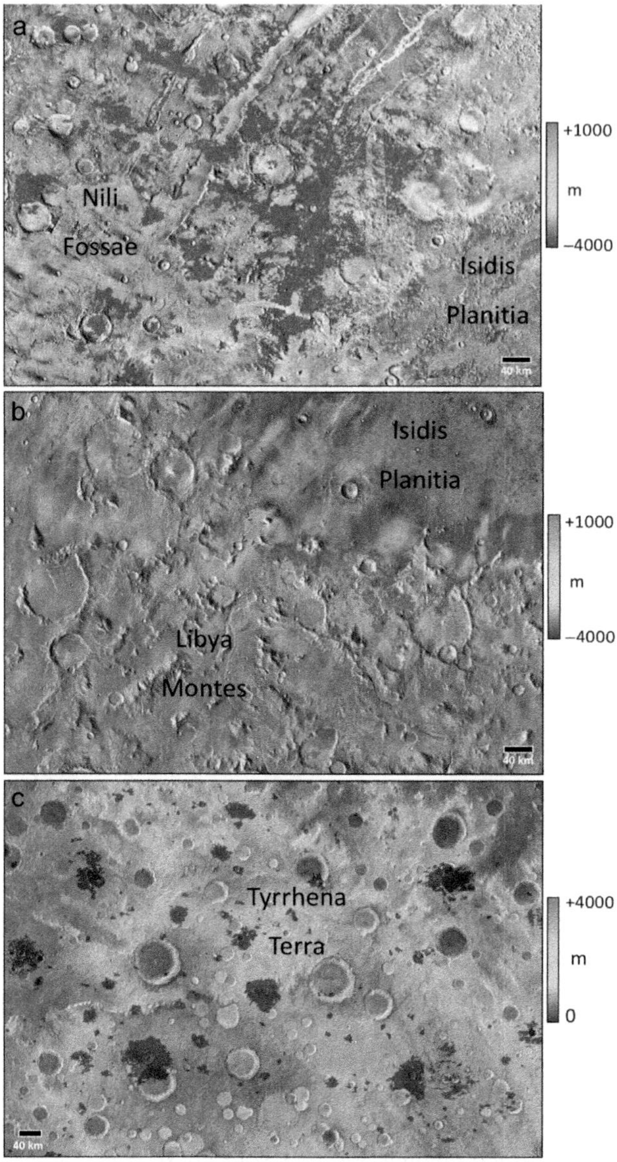


**FIG. 14.11** Regional scale clay mineral-rich units near Isidis Planitia. The background is THEMIS IR as in Fig. 14.9. Red indicates ferroan-magnesian clay minerals (80% smectites, 20% chlorites), green hydrated sulphates, cyan for Al-rich clay minerals, and yellow mixtures of ferroan-magnesian smectites and sulphates. (A) Nili Fossae region northwest of Isidis Planitia centred near 75°E, 20°N, (B) Libya Montes region south of Isidis Planitia centred near 84°E, 2°N and (C) Tyrrhena Terra region centred near 88°E, 10°S between Isidis Planitia, and Hellas Basin.

across the Nili Fossae region, as well as the geologic context of these clay mineral-bearing outcrops indicates that several episodes of aqueous alteration took place (Ehlmann et al., 2009).

The Libya Montes region, located at the southern rim of the Isidis impact basin, hosts a variety of geological processes that have shaped the martian surface. Evidence for fluvial, lacustrine, aeolian, volcanic and hydrothermal processes have been documented, resulting in a variety of landforms and ample evidence for chemical alteration of the local rocks (Crumpler and Tanaka, 2003; Tornabene et al., 2008; Jaumann et al., 2010; Erkeling et al., 2012). The Libya Montes region contains widespread pyroxene- and olivine-bearing units (Tornabene et al., 2008) and a variety of smectites and Fe/Mg-carbonate similar to dolomite (Bishop et al., 2013c). The ancient Noachian basaltic crustal materials experienced extensive aqueous alteration at the time of the Isidis impact, during which the montes were also formed, followed by emplacement of a rough olivine-rich lava or melt and finally the smooth pyroxene-bearing caprock unit. The Fe/Mg-smectite in this region vary from nontronite with an OH band at 2.29 μm to Fe/Mg-smectite with a band at 2.30 μm and saponite with a band at 2.31 μm (Fig. 14.12). Beidellite is observed here due to an $(\nu+\delta)Al_2OH$ band at 2.19 μm rather than montmorillonite, which would have a band at 2.21 μm. Beidellite is a higher temperature smectite and this beidellite likely formed via low temperature hydrothermal alteration from the Isidis impact. Abundant fluvial features dating from the Noachian to Amazonian time periods traverse the Libya Montes region; however, clay minerals are primarily observed in the ancient Noachian rocks.

The Tyrrhena Terra region lies south of Libya Montes and north of Hellas basin and exhibits some of the highest and most ancient crust of Mars, ranging from the Early to Late Noachian. It consists of highland plateaus, at times dissected, but rarely resurfaced by the accumulation of later sediments or volcanic deposits. It displays a large number of impact craters of all sizes (up to ~100 km in diameter), which provide natural probes to the upper crust from the near surface to several kms deep. Older craters have experienced erosion so that their depth-to-diameter ratio has decreased, or because they have been locally infilled by younger material. Fresh craters however have a well identifiable rim, central peak (or rings) and ejecta blanket, which are necessary to explore the excavated vertical stratification and disentangle contributions from postimpact, clay mineral-forming hydrothermal environments. As shown in Fig. 14.11C, most fresh craters, regardless of their size, display significant clay mineral abundances in their ejecta blanket, while simultaneously exhibiting clay minerals in intracrater deposits (Loizeau et al., 2012a; Carter et al., 2013). Most of these signatures are interpreted as previously altered excavated material, while postimpact formation is considered minor in this region (Loizeau et al., 2012a). A trend is found such that craters typically smaller than about 10 km excavate smectitic clay minerals, while larger craters have stronger

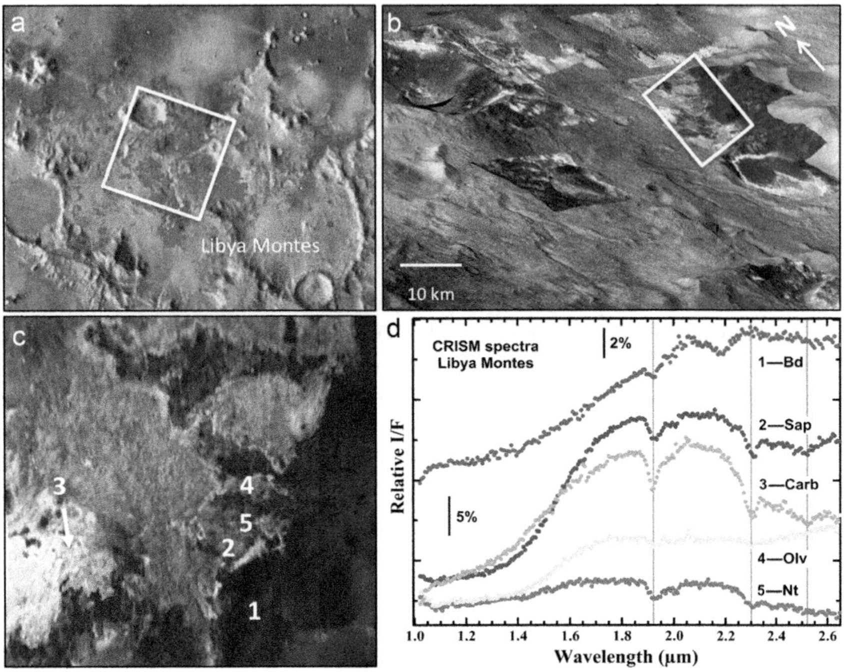


**FIG. 14.12** Views of clay mineral-bearing outcrops at Libya Montes. (A) A portion of the Libya Montes region from Fig. 14.11 with a white box showing the location of B. (B) Close-up view of the central region in CTX imagery over an HRSC DTM with colours from CRISM overlain. Ferroan-magnesian smectite is mapped in red/pink, olivine in green and pyroxene in blue. The white box shows the location of C. (C) CRISM parameter map showing outcrops of multiple units and the locations of example spectra shown in D: 1-beidellite (Bd), 2-saponite (Sap), 3-Fe/Mg -carbonate mixed with smectite (Carb), 4-olivine (Olv) and 5-nontronite (Nt). (D) Example spectra illustrating the features observed for the geologic units containing smectites, carbonate and olivine. Grey lines at 1.92, 2.30 and 2.53 μm mark the spectral features. *(modified from Bishop et al., 2013c.) (For interpretation of the references to colour in this figure legend, the reader is referred to the web version of this article.)*

contributions from chlorites. Zeolites and prehnite are observed locally as excavated material within the largest craters (except in a few small craters where they are thought to have been excavated from an underlying ejecta blanket). The vertical stratigraphy of smectites to chlorites ± zeolites to chlorites ± prehnite is interpreted as the signature of burial diagenesis and low-grade metamorphism, while direct precipitation in hot and deep aquifers can also contribute to the observed mineralogy (Ehlmann et al., 2011; Carter et al., 2013). The metamorphic mineral epidote is found at a few sites, but its original context/stratigraphy has not been inferred yet. Carbonates and talc have also been detected in some regions of Tyrrhena Terra associated with clay minerals, especially near the Hellas rim (Viviano-Beck et al., 2016).

## 14.5 CHARACTERISATION OF CLAYS IN METEORITES

Meteorites serve as delivery systems for extraterrestrial clay minerals to Earth. They are classified by petrologic type and origination, with some coming from the Moon and Mars and many others coming from the asteroid belt or further sources. The majority of meteorites are stony type (chondrites and achondrites), while fewer are iron meteorites or stony-iron meteorites (Wasson, 1985; Lauretta and McSween Jr., 2006; Papike, 1989). The lunar and martian meteorites fall into the achondrite category. Attempts have been made to connect meteorites to asteroid types through their spectral properties (Chapman and Salisbury, 1973; Gaffey, 1976; Sandford, 1984; Salisbury and Hunt, 1974; Salisbury et al., 1991a), but these have not been entirely successful, likely due to weathering of meteorites and because the meteorites are not well represented on the surfaces of their parent bodies. Clay minerals are rare in meteorites, but are present in some martian meteorites and some chondrites (Zolensky and McSween, 1988).

Early detections of clay minerals and altered materials in carbonaceous chondrites were reported for Allende (Dominik et al., 1978), Plainview (Nozette and Wilkening, 1982), and Mokoia (Cohen et al., 1983), although it was greatly debated in the community whether these clay minerals were formed by terrestrial alteration or had an extraterrestrial origin. Carbonaceous chondrites of the Ivuna and Orgueil types contain clay minerals intimately mixed with a carbon-rich matrix attributed to preterrestrial alteration by aqueous solutions (Morlok et al., 2006). The C2 Tagish Lake meteorite includes mixed clay minerals, carbon, carbonate and sulphides, which are also thought to be preterrestrial due to the brief environmental exposure of this meteorite (Brown et al., 2000). The spectral properties of both $Fe^{3+/2+}$-rich (Calvin and King, 1997) and $Mg^{2+}$-rich (King and Clark, 1989) clay minerals were investigated in comparison with the spectral properties of meteorites containing Ivuna and Mighei type chondrites. Mixtures including Fe-rich serpentines and chamosite were found to be most consistent with these meteorites (Calvin and King, 1997).

The majority of martian meteorites follow three categories: shergottites, nakhlites and chassignites, named after meteorites from these locations (McSween and Treiman, 1998). Whether or not clay minerals detected in these meteorites were formed by terrestrial weathering or preterrestrial processes was hotly debated for years, and only became widely accepted after abundant clay minerals were observed on Mars by the orbiters. Smectites together with carbonate and sulphate were detected in interior regions of the Nakhla meteorite and attributed to preterrestrial alteration because of their location in the rock and the morphology of the grains (Gooding et al., 1991). Further investigation of additional shergottite, nakhlite and chassignite meteorites revealed preterrestrial clay minerals and associated aqueous alteration minerals (Gooding, 1992). Continuing investigation of clay minerals in these meteorites discovered Fe-rich smectite (Treiman et al., 1993). Modelling of alteration pathways for formation of these clays, carbonates and sulphates suggest evaporation from low temperature brines in contact with parent igneous rocks (Bridges et al., 2001). One important martian meteorite that does not fall in one of these categories is the famous orthopyroxenite ALH 84001 (Mittlefehldt, 1994), which received great attention following reports of tiny globules containing preterrestrial organics, carbonates, magnetite and amorphous materials (McKay et al., 1996). Clays in this meteorite are found exclusively in the carbonates and not in the glass, thus strengthening support for their preterrestrial origin (Brearley, 2000). A recent review of clay mineral detections in martian meteorites suggests that broadly similar clay minerals are observed in the meteorites and on the surface of Mars (Velbel, 2012). Unfortunately, the clay minerals present in meteorites tend to occur in extremely tiny aliquots on the scale of nm or smaller, thus challenging researchers to identify and characterise them. The nakhlite meteorite group contains the largest quantity of clay minerals and for this reason they have been characterised in more detail. The most abundant clay minerals present include $Fe^{3+/2+}$-enriched trioctahedral saponite and serpentine (Hicks et al., 2014). Modelling of the geochemical conditions required for formation of the observed carbonate and Fe/Mg-smectite in nakhlites indicates there was a $CO_2$-rich hydrothermal fluid at 150–200°C with pH ~ 6–8 and a water/rock ratio ≤300 (Bridges and Schwenzer, 2012).

## 14.6 CHARACTERISATION OF CLAY MINERALS AT ASTEROID 1-CERES

The asteroid 1-Ceres is the largest body in the asteroid belt and is composed of silicate minerals (McCord and Gaffey, 1974; McCord and Sotin, 2005). Spectral features near 3 μm in telescopic data first led to the idea of clay minerals on the Dwarf planet Ceres (Lebofsky et al., 1981). Ammoniated clay minerals were detected on Ceres through telescopic measurements using the spectral bands near 3.07 μm (King et al., 1992). Their data were most consistent with $NH_4$-saponite. Further comparison of ammoniated saponite with emission spectra of Ceres found consistencies near 6.2 and 6.9 μm, but a difference near 9.5 μm (Cohen et al., 1998), suggesting additional components are present as well. A lab study of $NH_4$-treated smectite showed shifts in the 3.07 μm band depending on how the $NH_4$ was bound to the smectite (Bishop et al., 2002b). For tightly bound $NH_4$ this band was observed at 3.06 μm, while for smectite washed in $NH_4$ solution this band occurred at 3.07–3.1 μm. Additional analyses of improved telescopic data of Ceres reported the presence of carbonate and Fe-smectite (Rivkin et al., 2006). Further analyses of this new data led to the interpretation that Ceres contains brucite, carbonate and serpentine (Milliken and Rivkin, 2009).

More recent spectral data were acquired for Ceres by the Dawn spacecraft from 0.4 to 5 μm (De Sanctis et al., 2015). These spectra are consistent with widespread $NH_4$-bearing minerals across the surface of Ceres. De Sanctis et al. (2015) support formation of these $NH_4$-clay minerals during differentiation, perhaps through incorporation of material from the outer Solar System. The type of ammoniated clay or clay mineral present on Ceres has not yet been determined, but it is thought to be mixed with magnetite, antigorite and carbonate (De Sanctis et al., 2015). Brucite was ruled out as contributing to the 3.05–3.1 μm feature in Ceres' spectrum through laboratory mixture experiments (De Angelis et al., 2016). A recent review of telescopic spectra of asteroids and $NH_4$-bearing minerals determined that 1-Ceres, 10-Hygiea and 324-Bamberga all exhibit spectral features near 3.05–3.07 μm attributed to ammoniated clays (Berg et al., 2016). Several other spectral bands are present for $NH_4$-treated clay minerals, but these are more difficult to resolve in spectra of asteroids. Thus, the current understanding of Ceres supports the presence of ammoniated clay minerals, but the specific type of clay minerals cannot yet be uniquely assigned.

## 14.7 CHARACTERISATION OF CLAY MINERALS IN COMETS

The Spitzer Space Telescope has collected TIR spectra (5–40 μm) of several Jupiter family comets including 9B-Tempel 1 (Werner et al., 2004). Spitzer spectra were collected of dust ejected from the coma of 9B-Tempel 1 before and after the collision of the Deep Impact mission with its nucleus. Features due to clay minerals, carbonate, water ice and organics were detected in the Spitzer spectra of the dust after impact (Lisse et al., 2006). Fine grains of crystalline and amorphous olivine and pyroxene were observed in the dust around 9B-Tempel 1 before and after impact and were modelled together with the clay minerals, carbonate, water ice and organics (Kelley and Wooden, 2009). Lisse et al. (2006) found nontronite to correspond best to the clay mineral signatures and estimated its presence as 5–10% of the silicate material in the dust. Lisse et al. (2007) compared the Spitzer data for comet 9B-Tempel 1 with International Space Observatory (ISO) spectra of comet Hale-Bopp following its 1995–1996 apparition resulting in strong outflow of dust (Williams et al., 1997) and an ISO spectrum of HD100546, a young star often compared to comets (Waelkens et al., 1997). Coordinated spectral analyses of these bodies indicate similar components including nontronite (Lisse et al., 2007). They calculated weighted surface areas for nontronite, carbonate, water ice, organics, sulphides, olivine and pyroxene for these objects. Their results included weighted (unit-less) surface areas of 0.14, 0.08, 0.05 and 0.04 for nontronite, carbonate, water ice and polyaromatic hydrocarbons, respectively, for 9B-Tempel 1, compared to 0.18, 0.33, 0.34 and 0.14 for Hale-Bopp, and 0.14, 0.08, 0.46 and 0.51 for HD 100546 (Lisse et al., 2007). The variations in clay minerals and volatile components suggest different formation histories for these planetary bodies.

## 14.8 SUMMARY OF REMOTE OBSERVATIONS OF PLANETARY CLAY MINERALS

Spectral remote sensing in the VNIR and MIR regions has enabled detection and characterisation of multiple clay minerals on Earth and elsewhere in the Solar System. Remote sensing on Earth has the greatest challenge due to atmospheric absorptions that interfere with detection of surface minerals. Still, a greater variety of clay minerals has been observed on Earth than other planets due to the maturity of this planet. Clay minerals have arguably been mapped in more detail on Mars because they are not masked by vegetation and the atmosphere is thinner and primarily composed of $CO_2$ and thus easier to process. Fe/Mg-smectite is the most abundant clay mineral identified on the surface of Mars, is present (with $Fe^{2+}$/Mg-enriched clay minerals) in some meteorites, and is also the most common clay mineral in comets where detected. The Fe/Mg-smectite on Mars is almost always found in ancient rocks ~4 Gy old. Could this clay mineral have been a component of the material formed early in the life of the Solar System? Perhaps this question could be addressed as more spectrometers are sent on missions to other planetary bodies.

# Bibliography


Abkowicz-Bienko, A.J., Latajka, Z., Bienko, D.C., Michalska, D., 1999. Theoretical infrared spectrum and revised assignment for para-nitrophenol. Density functional theory studies. Chem. Phys. 250, 123–129.

Adams, J.M., 1979. The crystal structure of a dickite: N-methylformamide intercalate. Acta Crystallogr. B 35, 1084–1087.

Adams, J.M., Jefferson, D.A., 1976. The crystal structure of a dickite: formamide intercalate $Al_2Si_2O_5(OH)_4 \cdot HCONH_2$. Acta Crystallogr. B 32, 1180–1183.

Afseth, N.K., Kohler, A., 2012. Extended multiplicative scatter correction in vibrational spectroscopy, a tutorial. Chemom. Intell. Lab. Syst. 117, 92–99.

Ahenach, J., Cool, P., Vansant, E.F., 2000. Enhanced Brønsted acidity created upon Al-grafting of porous clay heterostructures via aluminium acetylacetonate adsorption. Phys. Chem. Chem. Phys. 2, 5750–5755.

Ahlrichs, J.L., Serna, C.J., Serratosa, J.M., 1975. Structural hydroxyls in sepiolites. Clays Clay Miner. 23, 119–124.

Akcay, M., 2006. Characterization and adsorption properties of tetrabutylammonium montmorillonite (TBAM) clay: thermodynamic and kinetic calculations. J. Colloid Interface Sci. 296, 16–21.

Akcay, G., Akcay, M., Yurdakoc, K., 2006. The characterization of prepared organomontmorillonite (DEDMAM) and sorption of phenoxyalkanoic acid herbicides from aqueous solution. J. Colloid Interface Sci. 296, 428–433.

Akyuz, S., Akyuz, T., Davies, J.E.D., Esmer, K., Erbolukbas Ozel, A., 1995. Fourier transform Raman and Fourier transform IR spectroscopic investigation of pyrazine adsorbed by sepiolite and bentonite from Anatolia. J. Raman Spectrosc. 26, 883–888.

Alabarse, F.G., Conceição, R.V., Balzaretti, N.M., Schenato, F., Xavier, A.M., 2011. In situ FTIR analyses of bentonite under high pressure. Appl. Clay Sci. 51, 202–208.

Alba, M.D., Castro, M.A., Naranjo, M., Pavon, E., 2006. Hydrothermal reactivity of Na-n-micas ($n = 2$, 3, 4). Chem. Mater. 18, 2867–2872.

Alba, M.D., Castro, M.A., Orta, M.M., Pavon, E., Pazos, M.C., Rios, J.S.V., 2011. Formation of organo-highly charged mica. Langmuir 27, 9711–9718.

Alcalà, M., Blanco, M., Moyano, D., Broad, N.W., O'Brian, N., Friedrich, D., Pfeifer, F., Siesler, H.W., 2013. Qualitative and quantitative pharmaceutical analysis with a novel hand-held miniature near infrared spectrometer. J. Near Infrared Spectrosc. 21, 445–457.

Allen, C.C., 1979. Volcano-ice interactions on Mars. J. Geophys. Res. 84, 8048–8060.

Alver, O., Alver, B.E., 2013. Fourier transform infrared, solid state nuclear magnetic resonance, and X-ray diffraction studies of natural and acid-treated bentonites from Turkey. Spectrosc. Lett. 46, 128–132.

Alves, A.P.M., Araujo, A.S., Bezerra, F.A., Sousa, K.S., Lima, S.J.G., Fonseca, M.G., 2014. Kinetics of dehydration and textural characterization of selectively leached vermiculites. J. Therm. Anal. Calorim. 117, 19–26.

Ambers, C.P., Murray, H.H., 1995. The role of carbonate bedrock in the formation of Indianaite halloysitic clays. Indiana Geol. Surv. Bull. 65, 1–29.

Anbalagan, G., Sivakumar, G., Prabakaran, A.R., Gunasekaran, S., 2010. Spectroscopic characterization of natural chrysotile. Vib. Spectrosc. 53, 122–127.

Andrades, M.S., Rodriguez-Cruz, M.S., Sanchez-Martin, M.J., Sanchez-Camazano, M., 2004. Effect of the modification of natural clay minerals with hexadecylpyridinium cation on the adsorption-desorption of fungicides. Int. J. Environ. Anal. Chem. 84, 133–141.

Andrés, J.M., Bona, M.T., 2005. Analysis of coal by diffuse reflectance near-infrared spectroscopy. Anal. Chim. Acta 535, 123–132.

Andrieux, P., Petit, S., 2010. Hydrothermal synthesis of dioctahedral smectites: the Al–Fe chemical series. Part I: influence of experimental conditions. Appl. Clay Sci. 48, 5–17.

Angel, B.R., Hall, P.L., 1972. Electron spin resonance studies of kaolins. In: Serratosa, J.M. (Ed.), Proc. Int. Clay Conf. Div. Ciencias. C.S.I.C, Madrid, Spain, pp. 47–59.

Anirudhan, T.S., Bringle, C.D., Rijith, S., 2010. Removal of uranium(VI) from aqueous solutions and nuclear industry effluents using humic acid-immobilized zirconium-pillared clay. J. Environ. Radioact. 101, 267–276.

Annabi-Bergaya, F., Cruz, M.I., Gatineau, L., Fripiat, J.J., 1980. Adsorption of alcohols by smectites. 3. Nature of the bonds. Clay Miner. 15, 225–237.

Arab, M., Bougeard, D., Smirnov, K.S., 2002. Experimental and computer simulation study of the vibrational spectra of vermiculite. Phys. Chem. Chem. Phys. 4, 1957–1963.

Arvidson, R.E., Squyres, S.W., Bell, J.F., Catalano, J.G., Clark, B.C., Crumpler, L.S., De Souza, P.A., Fairén, A.G., Farrand, W.H., Fox, V.K., Gellert, R., Ghosh, A., Golombek, M.P., Grotzinger, J.P., Guinness, E.A., Herkenhoff, K.E., Jolliff, B.L., Knoll, A.H., Li, R., Mclennan, S.M., Ming, D.W., Mittlefehldt, D.W., Moore, J.M., Morris, R.V., Murchie, S.L., Parker, T.J., Paulsen, G., Rice, J.W., Ruff, S.W., Smith, M.D., Wolff, M.J., 2014. Ancient aqueous environments at endeavour crater, Mars. Science 343, https://doi.org/10.1126/science.1248097.

Aufort, J., Ségalen, L., Gervais, C., Brouder, C., Balan, E., 2016. Modeling the attenuated total reflectance infrared (ATR-FTIR) spectrum of apatite. Phys. Chem. Miner. 43, 615–626.

Aung, L.L., Tertre, E., Petit, S., 2015. Effect of the morphology of synthetic kaolinites on their sorption properties. J. Colloid Interface Sci. 443, 177–186.

Auzende, A.L., Daniel, I., Reynard, B., Lemaire, C., Guyot, F., 2004. High-pressure behaviour of serpentine minerals: a Raman spectroscopic study. Phys. Chem. Miner. 31, 269–277.

Bahranowski, K., Wlodarczyk, W., Wisla-Walsh, E., Gawel, A., Matusik, J., Klimek, A., Gil, B., Michalik-Zym, A., Dula, R., Socha, R.P., Serwicka, E.M., 2015. [Ti, Zr]-pillared montmorillonite—a new quality with respect to Ti- and Zr-pillared clays. Microporous Mesoporous Mater. 202, 155–164.

Bailey, S.W., 1980a. Structures of layer silicates. In: Brindley, G.W., Brown, G. (Eds.), Crystal Structures of Clay Minerals and Their X-Ray Identification. Mineralogical Society, London, pp. 1–123. Monograph 5.

Bailey, S.W., 1980b. Summary of recommendations of AIPEA Nomenclature Committee. Clays Clay Miner. 28, 73–78.

Bailey, S.W., 1989. Halloysite—a critical assessment. Sci. Geol. Mem. 86, 89–98.

Baird, A.K., Castro, A.J., Clark, B.C., Toulmin P. III, Rose Jr., H., Keil, K., Gooding, J.L., 1977. The Viking X ray fluorescence experiment: sampling strategies and laboratory simulations. J. Geophys. Res. 82, 4595–4624.

Bakardjiev, D., Popov, K., 2015. ASTER spectral band ratios for detection of hydrothermal alterations and ore deposits in the Panagyurishte Ore Region, Central rednogorie, Bulgaria. Rev. Bulg. Geol. Soc. 76, 79–88.

Balan, E., Saitta, A.M., Mauri, F., 2001. First-principles modeling of the infrared spectrum of kaolinite. Am. Mineral. 86, 1321–1330.

Balan, E., Bish, D.L., Calas, G., 2014. Kaolin-group minerals: from hydrogen-bonded layers to environmental recorders. Elements 10, 183–188.

Balan, E., Lazzeri, M., Morin, G., Mauri, F., 2006. First-principles study of the OH stretching modes of gibbsite. Am. Mineral. 91, 115–119.

Balan, E., Blanchard, M., Hochepied, J.-F., Lazzeri, M., 2008. Surface modes in the infrared spectrum of hydrous minerals: the OH stretching modes of bayerite. Phys. Chem. Miner. 35, 279–285.

Balan, E., Delattre, S., Guillaumet, M., Salje, E.K.H., 2010. Low-temperature infrared spectroscopic study of OH-stretching modes in kaolinite and dickite. Am. Mineral. 95, 1257–1266.

Balan, E., Saitta, A.M., Mauri, F., Lemaire, C., Guyot, F., 2002b. First-principles calculation of the infrared spectrum of lizardite. Am. Mineral. 87, 1286–1290.

Balan, E., Lazzeri, M., Saitta, A.M., Allard, T., Fuchs, Y., Mauri, F., 2005. First-principles study of OH-stretching modes in kaolinite, dickite, and nacrite. Am. Mineral. 90, 50–60.

Balan, E., Lazzeri, M., Delattre, S., Meheut, M., Refson, K., Winkler, B., 2007. Anharmonicity of inner-OH stretching modes in hydrous phyllosilicates: assessment from first-principles frozen-phonon calculations. Phys. Chem. Miner. 34, 621–625.

Balan, E., Mauri, F., Lemaire, C., Brouder, C., Guyot, F., Saitta, A.M., Devouard, B., 2002a. Multiple ionic-plasmon resonances in naturally occurring multiwall nanotubes: infrared spectra of chrysotile asbestos. Phys. Rev. Lett. 89. 177401-1–177401-4.

Balan, E., Fritsch, E., Allard, T., Morin, G., Guillaument, M., Delattre, S., Blanchard, M., Calas, G., 2011a. Spectroscopic investigation and theoretical modeling of kaolinite-group minerals and other low-temperature phases. Compt. Rendus Geosci. 343, 177–187.

Balan, E., Refson, K., Blanchard, M., Delattre, S., Lazzeri, M., Ingrin, J., Mauri, F., Wright, K., Winkler, B., 2009. Theoretical infrared absorption coefficient of OH groups in minerals. Am. Mineral. 93, 950–953.

Balan, E., Delattre, S., Roche, D., Segalen, L., Morin, G., Guillaumet, M., Blanchard, M., Lazzeri, M., Brouder, C., Salje, E.K.H., 2011b. Line-broadening effects in the powder infrared spectrum of apatite. Phys. Chem. Miner. 38, 111–122.

Baldermann, A., Dohrmann, R., Kaufhold, S., Nickel, C., Letofsky-Papst, I., Dietzel, M., 2014. The Fe-Mg-saponite solid solution series—a hydrothermal synthesis study. Clay Miner. 49, 391–415.

Bancroft, G.M., Burns, R.G., 1969. Mössbauer and absorption spectral study of alkali amphiboles. In: Papike, J.J. (Ed.), Pyroxenes and Amphiboles: Crystal Chemistry and Phase Petrology. Mineralogical Society of America, Special Paper 2, pp. 137–148.

Bandfield, J.L., 2002. Global mineral distributions on Mars. J. Geophys. Res. 107, https://doi.org/10.1029/2001JE001510.

Bank, S., Bank, J.F., Ellis, P.D., 1989. Solid-state $^{113}Cd$ nuclear magnetic resonsnce study of exchanged montmorillonites. J. Phys. Chem. 93, 4847–4855.

Baraldi, P., 1990. Infrared emission spectroscopy (IRES): a tool in research and application. Chimicaoggi, 53–57.

Bard, D., Yarwood, J., Tylee, B., 1997. Asbestos fiber identification by Raman microspectroscopy. J. Raman Spectrosc. 28, 803–809.

Bardhan, R., Zarick, H.F., Schwartzberg, A., Pint, C.L., 2013. Size-dependent phononicproperties of PdO nanocrystals probed by nanoscale optical thermometry. J. Phys. Chem. C 117, 21558–21568.

Barman, S., Venkataraman, N.V., Vasudevan, S., Seshadri, R., 2003. Phase transitions in the anchored organic bilayers of long-chain alkylammonium lead iodides ($(C_nH_{2n+1}NH_3)_2PbI_4$; $n=12$, 16, 18. J. Phys. Chem. B 107, 1875–1883.

Barnes, R.J., Dhanioa, M.S., Lister, S.J., 1989. Standard normal variate transformation and de-trending of near-infrared diffuse reflectance spectra. Appl. Spectrosc. 43, 772–777.

Baron, F., Petit, S., 2016. Interpretation of the infrared spectra of the lizardite-nepouite series in the near- and mid-infrared range. Am. Mineral. 101, 423–430.

Baron, F., Petit, S., Tertre, E., Decarreau, A., 2016a. Influence of aqueous Si and Fe speciation on tetrahedral Fe(III) substitution in nontronites: a clay synthesis approach. Clays Clay Miner. 64, 230–244.

Baron, F., Pushparaj, S.S.C., Fontaine, C., Sivaiah, M.V., Decarreau, A., Petit, S., 2016b. Microwave-assisted hydrothermal synthesis of Ni-Mg layered silicate clays. Curr. Microwave Chem. 3, 85–89.

Baroni, S., De Gironcoli, S., Dal Corso, A., Giannozzi, P., 2001. Phonons and related crystal properties from density-functional perturbation theory. Rev. Mod.. Phys. 73, 515–561.

Barrios, J., Plançon, A., Cruz, M.I., Tchoubar, C., 1977. Qualitative and quantitative study of stacking faults in a hydrazine treated kaolinite-relationship with infrared spectra. Clays Clay Miner. 25, 422–429.

Barron, P.F., Frost, R.L., Skjemstad, J.O., 1983. $^{29}Si$ spin-lattice relaxation in aluminosilicates. J. Chem. Soc., Chem. Commun. 581–583.

Bautier, M.D., Potdevin, J.L., Lopez, M., Petit, S., 1996. Occurance of nacrite in the Lodeve Permian basin (France). Eur. J. Mineral. 8, 847–852.

Basak, U.K., Ghosh, S.K., 1996. Preparation and characterization of synthetic allophane-organic complexes. Clay Res. 15, 28–37.

Beaufort, D., Cassagnabère, A., Petit, S., Lanson, B., Berger, G., Lacharpagne, J.C., Johansen, H., 1998. Kaolinite-to-dickite reaction in sandstone reservoirs. Clay Miner. 33, 297–316.

Beck, J.S., Vartuli, J.C., Roth, W.J., Leonowicz, M.E., Kresge, C.T., Schmitt, K.D., Chu, C.T.W., Olson, D.H., Sheppard, E.W., Mccullen, S.B., Higgins, J.B., Schlenker, J.L., 1992. A new family of mesoporous molecular-sieves prepared with liquid-crystal templates. J. Am. Chem. Soc. 114, 10834–10843.

Bell J.F. III, Pollack, J.B., Geballe, T.R., Cruikshank, D.P., Freedman, R., 1994. Spectroscopy of Mars from 2.04 to 2.44 μm during the 1993 opposition: absolute calibration and atmospheric vs. mineralogic origin of narrow absorption features. Icarus 111, 106–123.

Bellamy, L.J., 1975. The Infra-Red Spectra of Complex Molecules, vol. 1. Chapman and Hall, London. 433 p.

Belver, C., Bañarez Muñoz, M.A., Vicente, M.A., 2002. Chemical activation of a kaolinite under acid and alkaline conditions. Chem. Mater. 14, 2033–2043.

Belver, C., Aranda, P., Martín-Luengo, M.A., Ruiz-Hitzky, E., 2012. New silica/alumina–clay heterostructures: properties as acid catalysts. Microporous Mesoporous Mater. 147, 157–166.

Ben Haj Amara, A., Ben Rhaiem, H., Plancon, A., 2000. Structural evolution of nacrite as a function of the nature of intercalated organic molecules. J. Appl. Crystallogr. 33, 1351–1359.

Benco, L., Tunega, D., Hafner, J., Lischka, H., 2001a. Orientation of OH groups in kaolinite and dickite: ab initio molecular dynamic study. Am. Mineral. 86, 1057–1065.

Benco, L., Tunega, D., Hafner, J., Lischka, H., 2001b. Upper limit of the O-H···O hydrogen bond. Ab initio study of the kaolinite structure. J. Phys. Chem. B 105, 10812–10817.

Benjelloun, M., Cool, P., Van Der Voort, P., Vansant, E.F., 2002. Template extraction from porous clay heterostructures: influence on the porosity and the hydrothermal stability of the materials. Phys. Chem. Chem. Phys. 4, 2818–2823.

Benndorf, J., 2015. Making use of online production data: sequential updating of mineral resource models. Math. Geosci. 47, 547–563.

Bentabol, M., Ruiz Cruz, M.D., 2013. Chemistry, morphology and structural characteristics of synthetic Al–Ni and Al–Co-lizardites. Appl. Clay Sci. 77–78, 68–78.

Bentabol, M., Ruiz Cruz, M.D., Huertas, F.J., 2009. Isomorphous substitution vs. defect density in hydrothermally synthesized (200°C) $Fe^{3+}$, $Ga^{3+}$ and $Cr^{3+}$-substituted kaolinites. Appl. Clay Sci. 45, 36–43.

Bentabol, M., Cruz, M.D.R., Sobrados, I., 2010. Chemistry, morphology and structural characteristics of synthetic Al-lizardite. Clay Miner. 45, 131–143.

Beran, A., Libowitzky, E., 1999. IR Spectroscopy and Hydrogen Bonding in Minerals. Springer, Dordrecht.

Berg, B.L., Cloutis, E.A., Beck, P., Vernazza, P., Bishop, J.L., Driss, T., Reddy, V., Applin, D., Mann, P., 2016. Reflectance spectroscopy (0.35–25 μm) of ammonium-bearing minerals and comparison to Ceres family asteroids. Icarus 265, 218–237.

Bergaoui, L., Lambert, J.-F., Franck, R., Suquet, H., Robert, J.-L., 1995. Al-pillared saponites. Part 3. Effect of parent clay layer charge on the intercalation-pillaring mechanism and structural properties. J. Chem. Soc., Faraday Trans. 91, 2229–2239.

Bergaya, F., Lagaly, G., 2001. Surface modification of clay minerals. Appl. Clay Sci. 19, 1–3.

Bergaya, F., Lagaly, G., 2011. Intercalation processes of layered minerals. In: Brigatti, M.F., Mottana, A. (Eds.), Layered Mineral Structures and Their Application in Advanced Technology. EMU Notes in Mineralogy, vol. 11. EMU, London, pp. 259–284.

Bergaya, F., Lagaly, G. (Eds.), 2013a. Handbook of Clay Science. Part A: Fundamentals, second ed. Elsevier, Amsterdam, NL.

Bergaya, F., Lagaly, G., 2013b. General Introduction: Clays, Clay Minerals, And Clay Science. In: Bergaya, F., Lagaly, G. (Eds.), Handbook of Clay Science. Developments in Clay Science, vol. 5A. Elsevier, Amsterdam, pp. 1–19.

Bergaya, F., Mandalia, T., Amigouet, P., 2005. A brief survey on CLAYPEN and nanocomposites based on unmodified PE and organo-pillared clays. Colloid Polym. Sci. 283, 773–782.

Bergaya, F., Jaber, M., Lambert, J.-F., 2011b. Clays and clay minerals. In: Galimberti, M. (Ed.), Rubber-clay Nanocomposites: Science, Technology, and Applications. John Wiley & Sons, Canada, pp. 3–44.

Bergaya, F., Jaber, M., Lambert, J.-F., 2011a. Organophilic clay minerals. In: Galimberti, M. (Ed.), Rubber-clay Nanocomposites: Science, Technology, and Applications. John Wiley & Sons, Canada, pp. 45–86.

Bergaya, F., Jaber, M., Lambert, J.-F., 2012. Clays and clay minerals as layered nanofillers for (bio) polymers. In: Averous, L., Pollet, E. (Eds.), Environmental Silicate Nano-Biocomposites. Springer-Verlag, London, pp. 41–75.

Bergaya, F., Hassoun, N., Barrault, J., Gatineau, L., 1993. Pillaring of synthetic hectorite by mixed $[Al_{13-x}Fe_x]$ pillars. Clay Miner. 28, 109–122.

Bergaya, F., Detellier, C., Lambert, J.-F., Lagaly, G., 2013. Introduction on clay polymer nanocomposites (CPN). In: Bergaya, F., Lagaly, G. (Eds.), Handbook of Clay Science. Developments in Clay Science, vol. 5A. Elsevier, Amsterdam, pp. 655–677.

Bertaut, E.F., 1950. Raies de Debye-Scherrer et Repartition des Dimensions des Domaines de Bragg Dans les Poudres Polycristallines. Acta Crystallogr. 3, 14–18.

Bertella, F., Pergher, S.B.C., 2015. Pillaring of bentonite clay with Al and Co. Microporous Mesoporous Mater. 201, 116–123.

Bertoluzza, A., Bonora, S., Fini, G., Battaglia, M.A., Monti, P., 1981. Hydrogen bonding in dimethylsulphoxide-proton donor interactions: a Raman and infrared study of the DMSO-HCl system. J. Raman Spectrosc. 11, 430–436.

Besson, G., Drits, V.A., 1997a. Refined relationship between chemical composition of dioctahedral fine-grained mica minerals and their infrared spectra within the OH stretching region. Part 1. Identification of the OH stretching bands. Clays Clay Miner. 45, 158–169.

Besson, G., Drits, V.A., 1997b. Refined relationship between chemical composition of dioctahedral fine-grained mica minerals and their infrared spectra with the OH stretching region. Part 2. Main factors affecting OH vibrations and quantitative analysis. Clays Clay Miner. 45, 170–183.

Beyer, K.A., Zhao, H., Borkiewicz, O.J., Newton, M.A., Chupas, P.J., Chapman, K.W., 2013. Simultaneous diffuse reflection infrared spectroscopy and X-ray pair distribution function measurements. J. Appl. Crystallogr. 47, 95–101.

Bibring, J.-P., Langevin, Y., Mustard, J.F., Poulet, F., Arvidson, R., Gendrin, A., Gondet, B., Mangold, N., Pinet, P., Forget, F., 2006. Global mineralogical and aqueous Mars history derived from OMEGA/Mars Express data. Science 312, 400–404.

Bibring, J.-P., Langevin, Y., Gendrin, A., Gondet, B., Poulet, F., Berthé, M., Soufflot, A., Arvidson, R., Mangold, N., Mustard, J., Drossart, P., 2005. Mars surface diversity as revealed by the OMEGA/Mars Express observations. Science 307, 1576–1581.

Bibring, J.-P., Combes, M., Langevin, Y., Cara, C., Drossart, P., Encrenaz, T., Erard, S., Forni, O., Gondet, B., Ksanfomality, L., Lellouch, E., Masson, P., Moroz, V., Rocard, F., Rosenqvist, J., Sotin, C., Soufflot, A., 1990. ISM observations of Mars and Phobos: First Results. In: Proc. Lunar Planet. Sci. Conf. XX, pp. 461–471.

Bickley, R.I., Edwards, H.G.M., Rose, S.J., Gustar, R., 1990. A raman spectroscopic study of nickel(II) acetate, $Ni(CH_3COO)_2$ and its aqueous and methanolic solutions. J. Mol. Struct. 238, 15–26.

Billault, V., Beaufort, D., Patrier, P., Petit, S., 2002. Crystal chemistry of Fe-sudoites from uranium deposits in the Athabasca basin (Saskatchewan, Canada). Clays Clay Miner. 50, 70–81.

Bineesh, K.V., Kim, D.-K., Cho, H.-J., Park, D.-W., 2010a. Synthesis of metal-oxide pillared montmorillonite clay for the selective catalytic oxidation of $H_2S$. J. Ind. Eng. Chem. (Amsterdam, Neth.) 16, 593–597.

Bineesh, K.V., Kim, D.-K., Kim, M.-I., Park, D.-W., 2011. Selective catalytic oxidation of $H_2S$ over $V_2O_5$ supported on $TiO_2$-pillared clay catalysts in the presence of water and ammonia. Appl. Clay Sci. 53, 204–211.

Bineesh, K.V., Kim, D.-K., Kim, D.-W., Cho, H.-J., Park, D.-W., 2010b. Selective catalytic oxidation of $H_2S$ to elemental sulfur over $V_2O_5$/Zr-pillared montmorillonite clay. Energy Environ. Sci. 3, 302–310.

Binitha, N.N., Sugunan, S., 2006. Preparation, characterization and catalytic activity of titania pillared montmorillonite clays. Microporous Mesoporous Mater. 93, 82–89.

Bish, D.L., 1993. Rietveld refinement of the kaolinite structure at 1.5. K. Clays Clay Miner. 41, 738–744.

Bish, D.L., Johnston, C.T., 1993. Rietveld refinement and fourier-transform infrared spectroscopic study of the dickite structure at low-temperature. Clays Clay Miner. 41, 297–304.

Bishop, J.L., Murad, E., 2004. Characterization of minerals and biogeochemical markers on Mars: A Raman and IR spectroscopic study of montmorillonite. J. Raman Spectrosc. 35, 480–486.

Bishop, J.L., Rampe, E.B., 2016. Evidence for a changing Martian climate from the mineralogy at Mawrth Vallis. Earth Planet. Sci. Lett. 448, 42–48.

Bishop, J.L., Pieters, C.M., Edwards, J.O., 1994. Infrared spectroscopic analyses on the nature of water in montmorillonite. Clays Clay Miner. 42, 702–716.

Bishop, J.L., Murad, E., Dyar, M.D., 2002c. The influence of octahedral and tetrahedral cation substitution on the structure of smectites and serpentines as observed through infrared spectroscopy. Clay Miner. 37, 617–628.

Bishop, J.L., Schiffman, P., Southard, R.J., 2002d. Geochemical and mineralogical analyses of palagonitic tuffs and altered rinds of pillow lavas on Iceland and applications to Mars. In: Smellie, J.L., Chapman, M.G. (Eds.), Volcano-Ice Interactions on Earth and Mars. Geological Society, London. Special Publication No. 202.

Bishop, J.L., Banin, A., Mancinelli, R.L., Klovstad, M.L., 2002b. Detection of soluble and fixed $NH_4^+$ in clay minerals by DTA and IR reflectance spectroscopy: a potential tool for planetary surface exploration. Planet. Space Sci. 50, 11–19.

Bishop, J.L., Madejová, J., Komadel, P., Fröschl, H., 2002a. The influence of structural Fe, Al, and Mg on the infrared OH bands in spectra of dioctahedral smectites. Clay Miner. 37, 607–616.

Bishop, J.L., Dyar, M.D., Sklute, E.C., Drief, A., 2008b. Physical alteration of antigorite: a Mössbauer spectroscopy, reflectance spectroscopy and TEM study with applications to Mars. Clay Miner. 43, 55–67.

Bishop, J.L., Lane, M.D., Dyar, M.D., Brown, A.J., 2008a. Reflectance and emission spectroscopy study of four groups of phyllosilicates: smectites, kaolinite-serpentines, chlorites and micas. Clay Miner. 43, 35–54.

Bishop, J.L., Gates, W.P., Makarewicz, H.D., Mckeown, N.K., Hiroi, T., 2011. Reflectance spectroscopy of beidellites and their importance for Mars. Clays Clay Miner. 59, 376–397.

Bishop, J.L., Murad, E., Madejová, J., Komadel, K., Wagner, U., Scheinost, A.C., 1999. Visible, Mössbauer and infrared spectroscopy of dioctahedral smectites: structural analysis of the Fe-bearing smectites Sampor, SWy-1 and SWa-1. ICC In: Kodoma, H., Mermut, A.R., Torrance, J.K. (Eds.), Clays for Our Future. Proceedings of the 11th International Clay Conference, Ottawa, Canada, 1997, pp. 413–419.

Bishop, J.L., Rampe, E.B., Bish, D.L., Abidin, Z., Baker, L.L., Matsue, N., Henmi, T., 2013a. Spectral and hydration properties of allophane and imogolite. Clays Clay Miner. 61, 57–74.

Bishop, J.L., Loizeau, D., Mckeown, N.K., Saper, L., Dyar, M.D., Des Marais, D., Parente, M., Murchie, S.L., 2013b. What the ancient phyllosilicates at Mawrth Vallis can tell us about possible habitability on early Mars. Planet. Space Sci. 86, 130–149.

Bishop, J.L., Noe Dobrea, E.Z., McKeown, N.K., Parente, M., Ehlmann, B.L., Michalski, J.R., Milliken, R.E., Poulet, F., Swayze, G.A., Mustard, J.F., Murchie, S.L., Bibring, J.-P., 2008c. Phyllosilicate diversity and past aqueous activity revealed at Mawrth Vallis, Mars. Science 321, 830–833. https://doi.org/10.1126/science.1159699.

Bishop, J.L., Gross, C., Rampe, E.B., Wray, J.J., Parente, M., Horgan, B., Loizeau, D., Viviano-Beck, C.E., Clark, R.N., Seelos, F.P., Ehlmann, B.L., Murchie, S.L., 2016. Mineralogy of layered outcrops at Mawrth Vallis and implications for early aqueous geochemistry on Mars. In: Lunar Planet. Sci. Conf. XLVII, 2016 The Woodlands, TX. Abstract #1332.

Bishop, J.L., Tirsch, D., Tornabene, L.L., Jaumann, R., McEwen, A.S., Mcguire, P.C., Ody, A., Poulet, F., Clark, R.N., Parente, M., Voigt, J., Aydin, Z., Bamberg, M., Petau, A., Mckeown, N.K., Mustard, J.F., Hash, C., Murchie, S.L., Swayze, G., Neukum, G., Seelos, F., 2013c. Mineralogy and morphology of geologic units at Libya Montes, Mars: ancient aqueous outcrops, mafic flows, fluvial features and impacts. J. Geophys. Res. 118, 487–513.

Bisio, C., Gatti, G., Boccaleri, E., Marchese, L., Superti, G.B., Pastore, H.O., Thommes, M., 2008. Understanding physico-chemical properties of saponite synthetic clays. Microporous Mesoporous Mater. 107, 90–101.

Blaha, J.J., Rosasco, G.J., 1978. Raman microprobe spectra of individual microcrystals and fibers of talc, tremolite, and related silicate minerals. Anal. Chem. 50, 892–896.

Blake, D.F., Morris, R.V., Kocurek, G., Morrison, S.M., Downs, R.T., Bish, D., Ming, D.W., Edgett, K.S., Rubin, D., Goetz, W., Madsen, M.B., Sullivan, R., Gellert, R., Campbell, I., Treiman, A.H., Mclennan, S.M., Yen, A.S., Grotzinger, J., Vaniman, D.T., Chipera, S.J., Achilles, C.N., Rampe, E.B., Sumner, D., Meslin, P.-Y., Maurice, S., Forni, O., Gasnault, O., Fisk, M., Schmidt, M., Mahaffy, P., Leshin, L.A., Glavin, D., Steele, A., Freissinet, C., Navarro-González, R., Yingst, R.A., Kah, L.C., Bridges, N., Lewis, K.W., Bristow, T.F., Farmer, J.D., Crisp, J.A., Stolper, E.M., Des Marais, D.J., Sarrazin, P., Team, M.S., 2013. Curiosity at Gale Crater, Mars: characterization and analysis of the Rocknest sand shadow. Science 341, https://doi.org/10.1126/science.1239505.

Blanc, P., Vieillard, P., Gailhanou, H., Gaboreau, S., 2013. Chapter 6—Thermodynamics of clay minerals. In: Bergaya, F., Lagaly, G. (Eds.), Developments in Clay Science. Elsevier.

Blanchard, M., Lazzeri, M., Mauri, F., Balan, E., 2008. First-principles calculation of the infrared spectrum of hematite. Am. Mineral. 93, 1019–1027.

Blanchard, M., Balan, E., Giura, P., Béneut, K., Yi, H., Morin, G., Pinilla, C., Lazzeri, M., Floris, A., 2013. Infrared spectroscopic properties of goethite: anharmonic broadening, long-range electrostatic effects and Al substitution. Phys. Chem. Miner. 41, 289–302.

Blanco, C., González, F., Pesquera, C., Benito, I., Mendioroz, S., Pajares, J.A., 1989. Differences between one aluminic palygorskite and another magnesic by infrared spectroscopy. Spectrosc. Lett. 22, 659–673.

Bloise, A., Belluso, E., Fornero, E., Rinaudo, C., Barrese, E., Capella, S., 2010. Influence of synthesis conditions on growth of Ni-doped chrysotile. Microporous Mesoporous Mater. 132, 239–245.

Bloise, A., Critelli, T., Catalano, M., Apollaro, C., Miriello, D., Croce, A., Barrese, E., Liberi, F., Piluso, E., Rinaudo, C., Belluso, E., 2014. Asbestos and other fibrous minerals contained in the serpentinites of the Gimigliano-Mount Reventino Unit (Calabria, S-Italy). Environ. Earth Sci. 71, 3773–3786.

Boardman, J.W., Kruse, F.A., Green, R.O., 1995. Mapping target signatures via partial unmixing of AVIRIS data. In: Green, R.O. (Ed.), Summaries of Fifth Annual JPL Airborne Earth Science Workshop.

Bodoardo, S., Chiappetta, R., Onida, B., Figueras, F., Garrone, E., 1998. Ammonia interaction and reaction with Al-pillared montmorillonite: an IR study. Microporous Mesoporous Mater. 20, 187–196.

Bona, M.T., Andrés, J.M., 2007. Coal analysis by diffuse reflectance near-infrared spectroscopy: hierarchical cluster and linear discriminant analysis. Tatlanta 72, 1423–1431.

Bonnin, D., Muller, S., Calas, G., 1982. Le fer dans les kaolins. Etude par spectrométries RPE, Mössbauer, EXAFS (Iron in kaolins. EPR, Mössbauer, EXAFS spectrometric studies). Bull. Minéral. 105, 467–475.

Booij, E., Kloprogge, J.T., Van Veen, J.A.R., 1996a. Preparation, structural characteristics and catalytic properties of large-pore rare earth element (Ce/La)/Al-pillared smectites. Clays Clay Miner. 44, 774–782.

Booij, E., Kloprogge, J.T., Van Veen, J.A.R., 1996b. Large pore REE/Al pillared bentonites: preparation, structural aspects and catalytic properties. Appl. Clay Sci. 11, 155–162.

Borade, R.B., Clearfield, A., 1992. Characterization of acid sites in Beta and ZSM-20 zeolites. J. Phys. Chem. 96, 6729–6737.

Borges, R., Macedo Dutra, L., Barison, A., Wypych, F., 2016. MAS NMR and EPR study of structural changes in talc and montmorillonite induced by grinding. Clay Miner. 51, 69–80.

Borges, R., Brunatto, S.F., Leitao, A.A., De Carvalho, G.S.G., Wypych, F., 2015. Solid-state mechano-chemical activation of clay minerals and soluble phosphate mixtures to obtain slow release fertilizers. Clay Miner. 50, 153–162.

Born, M., Huang, K., 1954. Dynamical Theory of Crystal Lattices. Oxford University Press, Oxford. 420 pp.

Botella, V., Timon, V., Escamilla-Roa, E., Hernandez-Laguna, A., Sainz-Diaz, C.I., 2004. Hydrogen bonding and vibrational properties of hydroxy groups in the crystal lattice of dioctahedral clay minerals by means of first principles calculations. Phys. Chem. Miner. 31, 475–486.

Böttger, H., 1983. Principles of the Theory of Lattice Dynamics. Akademie-Verlag, Berlin.

Boudjema, S., Vispe, E., Choukchou-Braham, A., Mayoral, J.A., Bachir, R., Fraile, J.M., 2015. Preparation and characterization of activated montmorillonite clay supported 11-molybdo-vanado-phosphoric acid for cyclohexene oxidation. RSC Adv. 5, 6853–6863.

Bougeard, D., Smirnov, K.S., Geidel, E., 2000. Vibrational spectra and structure of kaolinite: a computer simulation model. J. Phys. Chem. B 104, 9210–9217.

Bowers, G.M., Singer, J.W., Bish, D.L., Kirkpatrick, R.J., 2011. Alkali metal and $H_2O$ dynamics at the smectite/water interface. J. Phys. Chem. C 115, 23395–23407.

Boyd, S.A., Johnston, C.T., Laird, D.A., Teppen, B.J., Li, H., 2011. Comprehensive Study of Organic Contaminant Adsorption by Clays: Methodologies, Mechanisms, and Environmental Implications. Biophysico-Chemical Processes of Anthropogenic Organic Compounds in Environmental Systems. John Wiley & Sons, Inc.

Boyer, M., Paineau, E., Bacia-Verloop, M., Thill, A., 2014. Aqueous dispersion state of amphiphilic hybrid aluminosilicate nanotubes. Appl. Clay Sci. 96, 45–49.

Bradley, S.M., Kydd, R.A., 1993. Comparison of the species formed upon base hydrolyses of gallium(III) and iron(III) aqueous solutions: the possibility of existence of an $[FeO_4Fe_{12}(OH)_{24}(H_2O)_{12}]^{7+}$ polyoxocation. J. Chem. Soc., Dalton Trans. 2407–2413.

Bradley, S.M., Kydd, R.A., Yamdagni, R., 1990a. Comparison of the hydrolysis of gallium(III) and aluminium(III) solutions by nuclear magnetic resonance spectroscopy. J. Chem. Soc., Dalton Trans. 9, 2653–2656.

Bradley, S.M., Kydd, R.A., Yamdagni, R., 1990b. Detection of a new polymeric species formed through the hydrolysis of gallium (III) salt solutions. J. Chem. Soc., Dalton Trans. 9, 413–417.

Brauns, E.B., Meier, R.J., 2009. Issues in curve fitting vibrational spectra: bandshape justification and the influence of background correction. Vib. Spectrosc. 49, 303–304.

Brearley, A.J., 2000. Hydrous phases in ALH 84001: further evidence for preterrestrial alteration and a shock-induced thermal overprint. In: Lunar Planet Sci. Conf., 2000 LPI, Houston. Abstract #1203.

Breen, C., Illés, J., Yarwood, J., Skuse, D.R., 2007. Variable temperature diffuse reflectance infrared Fourier transform spectroscopic investigation of the effect of ball milling on the water sorbed to kaolin. Vib. Spectrosc. 43, 366–379.

Breen, C., Watson, R., Madejová, J., Komadel, P., Klapyta, Z., 1997. Acid-activated organoclays: preparation, characterization and catalytic activity of acid-treated tetraalkylammonium-exchanged smectites. Langmuir 13, 6473–6479.

Bridges, J., 2014. Comparing clays from Mars and Earth: implications for martian habitability. Am. Mineral. 99, 2163–2164.

Bridges, J.C., Schwenzer, S.P., 2012. The nakhlite hydrothermal brine on Mars. Earth Planet. Sci. Lett. 359–360, 117–123.

Bridges, J.C., Catling, D.C., Saxton, J.M., Swindle, T.D., Lyon, I.C., Grady, M.M., 2001. Alteration assemblages in Martian Meteorites: implications for near-surface processes. Space Sci. Rev. 96, 365–392.

Brigatti, M.F., Galan, E., Theng, B.K.G., 2013. Structure and mineralogy of clay minerals. Chapter 2, In: Bergaya, F., Lagaly, G. (Eds.), Handbook of Clay Science, Part A: Fundamentals, second ed. Elsevier, The Netherlands, pp. 21–82.

Brindley, G.W., Sempels, R.E., 1977. Preparation and properties of some hydroxy-aluminium beidellites. Clay Miner. 12.

Brindley, G.W., Brown, G., 1980. Crystal Structures of Clay Minerals and their X-ray Identification. Mineralogical Society, London, p. 495.

Brindley, G.W., Kao, C.C., Harrison, J.L., Lipsicas, M., Raythatha, R., 1986. Relation between structural disorder and other characteristics of kaolinites and dickites. Clays Clay Miner. 34, 239–249.

Bristow, T.F., Bish, D.L., Vaniman, D.T., Morris, R.V., Blake, D.F., Grotzinger, J.P., Rampe, E.B., Crisp, J.A., Achilles, C.N., Ming, D.W., Ehlmann, B.L., King, P.L., Bridges, J.C., Eigenbrode, J.L., Sumner, D.Y., Chipera, S.J., Moorokian, J.M., Treiman, A.H., Morrison, S.M., Downs, R.T., Farmer, J.D., Marais, D.D., Sarrazin, P., Floyd, M.M., Mischna, M.A., McAdam, A.C., 2015. The origin and implications of clay minerals from Yellowknife Bay, Gale crater, Mars. Am. Mineral. 100, 824–836.

Bro, R., Smilde, A.K., 2014. Principal component analysis. Anal. Methods 6, 2812–2831.

Brown, G.E., Henrich, V.E., Casey, W.H., Clark, D.L., Eggleston, C., Felmy, A., Goodman, D.W., Gratzel, M., Maciel, G., McCarthy, M.I., Nealson, K.H., Sverjensky, D.A., Toney, M.F., Zachara, J.M., 1999. Metal oxide surfaces and their interactions with aqueous solutions and microbial organisms. Chem. Rev. 99 (1), 77–174.

Brown, G.E., Catalano, J.G., Templeton, A.S., Trainor, T.P., Farges, F., Bostick, B.C., Kendelewicz, T., Doyle, C.S., Spormann, A.M., Revill, K., Morin, G., Juillot, F., Calas, G., 2005. Environmental interfaces, heavy metals, microbes, and plants: applications of XAFS spectroscopy and related synchrotron radiation methods to environmental science. Phys. Scr. T115, 80–87.

Brown, P.G., Hildebrand, A.R., Zolensky, M.E., Grady, M., Clayton, R.N., Mayeda, T.K., Tagliaferri, E., Spalding, R., Macrae, N.D., Hoffman, E.L., Mittlefehldt, D.W., Wacker, J.F., Bird, J.A., Campbell, M.D., Carpenter, R., Gingerich, H., Glatiotis, M., Greiner, E., Mazur, M.J., McCausland, P.J., Plotkin, H., Mazur, T.R., 2000. The fall, recovery, orbit, and composition of the Tagish Lake Meteorite: a new type of carbonaceous chondrite. Science 290, 320–325.

Brtáňová, A., Madejová, J., Bizovská, V., Komadel, P., 2014. Utilization of infrared spectroscopy for studying solvation properties of Cu-montmorillonites. Spectrochim. Acta A 123, 385–391.

Bruce, L.A., Sanders, J.V., Turney, T.W., 1986. Hydrothermal synthesis and characterization of cobalt clays. Clays Clay Miner. 34, 25–36.

Bruesch, P., 1982. Phonons: Theory and Experiments I: Lattice dynamics and models of interatomic forces. In Solid-State Sciences 34. Springer Series, Berlin. 262 pp.

Bruesch, P., 1986. Phonons: Theory and Experiments II: Experiments and Interpretation of Experimental Results. In Solid-State Sciences 65. Springer Series, Berlin. 275 pp.

Bruggeman, D.A.G., 1935. Berechnung verschiedener physikalischer Konstanten von heterogenen Substanzen. I. Dielektrizitätskonstanten und Leitfähigkeiten der Mischkörper aus isotropen Substanzen. Ann. Phys. 416, 665–679.

Brydon, J.E., Kodama, H., 1966. The nature of aluminum hydroxide-montmorillonite complexes. Am. Mineral. 51, 875–889.

Bujdák, J., Iyi, N., Fujita, T., 2002. The aggregation of methylene blue in montmorillonite dispersions. Clay Miner. 37, 121–133.

Bukas, V.J., Tsampodimou, M., Gionis, V., Chryssikos, G.D., 2013. Synchronous ATR infrared and NIR-spectroscopy investigation of sepiolite upon drying. Vib. Spectrosc. 68, 51–60.

Bukowska, J., 1979. On the assignment of the $NH_2$ stretching modes in the vibrational spectrum of formamide. Spectrochim. Acta A: Mol. Biomol. Spectrosc. 35, 985–988.

Burgess, J., 1978. Metal Ions in Solution. Ellis Horwood Limited, Chichester.

Buswell, A.M., Krebs, K., Rodebush, W.H., 1937. Infrared studies. III. Absorption bands of hydrogels between 2.5 and 3.5 micrometers. J. Am. Chem. Soc. 59, 2603–2605.

Butler, I.S., Frost, R.L., 2006. An overview of the high-pressure vibrational spectra of clays and related minerals. Appl. Spectrosc. Rev. 41, 449–471.

Butruille, J.R., Pinnavaia, T.J., 1996. Alumina pillared clays. In: Albert, G., Bein, T. (Eds.), Comprehensive Supramolecular Chemistry. Solid-State Supramolecular Chemistry: Two and Three Dimensional Inorganic Networks. Pergamon Press, New York, pp. 219–250.

Butruille, J.R., Michot, L.J., Barres, O., Pinnavaia, T.J., 1993. Fluorine-mediated acidity of alumina-pillared fluorohectorite. J. Catal. 139, 664–678.

Byrnes, J.M., Ramsey, M.S., King, P.L., Lee, R.J., 2007. Thermal infrared reflectance and emission spectroscopy of quartzofeldspathic glasses. Geophys. Res. Lett. 34 (L01802), 1–5.

Bzdek, B.R., Mcguire, M.M., 2009. Polarized ATR-FTIR investigation of Fe reduction in the Uley nontronites. Clays Clay Miner. 57, 227–233.

Caglar, B., Cubuk, O., Demir, E., Coldur, F., Catir, M., Topcu, C., Tabak, A., 2015. Characterization of AlFe-pillared Unye bentonite: a study of the surface acidity and catalytic property. J. Mol. Struct. 1089, 59–65.

Calvet, R., Prost, R., 1971. Cation migration into empty octahedral sites and surface properties of clays. Clays Clay Miner. 19, 175–186.

Calvin, W.M., King, T.V.V., 1997. Spectral characteristics of Fe-bearing phyllosilicates: comparison to Orgueil (C11), Murchison and Murray (CM2). Meteorit. Planet. Sci. 32, 693–701.

Cancela, G.D., Huertas, F.J., Taboada, E.R., Sanchezrasero, F., Laguna, A.H., 1997. Adsorption of water vapor by homoionic montmorillonites. Heats of adsorption and desorption. J. Colloid Interface Sci. 185, 343–354.

Čapková, P., Driessen, R.A.J., Numan, M., Schenk, H., Weiss, Z., Klika, Z., 1998. Molecular simulations of montmorillonite intercalated with aluminum complex cations. Part I: Intercalation with $[Al_{13}0_4(OH)_{24+x}(H_2O)_{12-x}]^{(7-x)+}$. Clays Clay Miner. 46, 232–239.

Cariati, F., Erre, L., Gessa, C., Micera, G., Piu, P., 1981. Water molecules and hydroxyl groups in montmorillonites as studied by near infrared spectroscopy. Clays Clay Miner. 29, 157–159.

Cariati, F., Erre, L., Micera, G., Piu, P., Gessa, C., 1983a. Effects of layer charge on the near infrared spectra of water molecules in smectites and vermiculites. Clays Clay Miner. 31, 447–449.

Cariati, F., Erre, L., Micera, G., Piu, P., Gessa, C., 1983b. Polarization of water molecules in phyllosilicates in relation to exchange cations as studied by near infrared spectroscopy. Clays Clay Miner. 31, 155–157.

Carrado, K.A., Komadel, P., 2009. Acid activation of bentonites and polymer-clay nanocomposites. Elements 5, 111–116.

Carrizosa, M.J., Rice, P.J., Koskinen, W.C., Carrizosa, I., Hermosin, M.D., 2004. Sorption of isoxaflutole and DKN on organoclays. Clays Clay Miner. 52, 341–349.

Carter, J., Poulet, F., Bibring, J.P., Mangold, N., Murchie, S., 2013. Hydrous minerals on Mars as seen by the CRISM and OMEGA imaging spectrometers: updated global view. J. Geophys. Res.: Planets 118, 831–858.

Carter, J., Loizeau, D., Mangold, N., Poulet, F., Bibring, J.-P., 2015. Widespread surface weathering on early Mars: a case for a warmer and wetter climate. Icarus 248, 373–382.

Casal, H.L., Mantsch, H.H., Cameron, D.G., Snyder, R.G., 1982. Interchain vibrational coupling in phase-II (hexagonal) n-alkanes. J. Chem. Phys. 77, 2825–2830.

Cases, J.M., Grillet, Y., François, M., Michot, L.J., Villieras, F., Yvon, J., 1991. Evolution of the porous structure and surface area of palygorskite under vacuum thermal treatment. Clays Clay Miner. 39, 191–201.

Cashion, J.D., Gates, W.P., Thomson, A., 2008. Mössbauer and IR analysis of iron sites in four ferruginous smectites. Clay Miner. 43, 83–93.

Cashion, J.D., Gates, W.P., Riley, G.M., 2009. Origin of the two quadrupole doublets in NAu-1 nontronite. ICMAE. In: J. Phys., conference Series No. 217, 012065, https://doi.org/10.1088/1742-6596/217/1/012065. IOP Publishing.

Cashion, J.D., Gates, W.P., Greaves, T.L., Dorjkhaidav, O., 2011. Identification of $Fe^{3+}$ site coordinations in NAU-2 nontronite. In: Proceedings of the 35th A&NZ Condensed Matter Materials Meeting, Wagga Wagga, NSW, Australia. http://www.aip.org.au/wagga2011/Wagga2011_6.pdf.

Catalano, M., Belluso, E., Miriello, D., Barrese, E., Bloise, A., 2014. Synthesis of Zn-doped talc in hydrothermal atmosphere. Cryst. Res. Technol. 49, 283–289.

Čavajda, V., Uhlík, P., Derkowski, A., Čaplovičová, M., Madejová, J., Mikula, M., Ifka, T., 2015. Influence of grinding and sonification on the crystal structure of talc. Clays Clay Miner. 63, 311–327.

Cecilia, J.A., 2013. Montmorillonite based porous clay heterostructures: influence of Zr in the structure and acidic properties. Microporous Mesoporous Mater. 176, 95–102.

Chabrillat, S., Goetz, A.F.H., Krosley, L., Olsen, H.W., 2002. Use of hyperspectral images in the identification and mapping of expansive clay soils and the role of spatial resolution. Remote Sens. Environ. 82, 431–445.

Chae, H.J., Nam, I.-S., Ham, S.-W., Hong, S.B., 2004. Characteristics of vanadia on the surface of $V_2O_5$/Ti-PILC catalyst for the reduction of $NO_x$ by $NH_3$. Appl. Catal. B 53, 117–126.

Chahi, A., Petit, S., Decarreau, A., 2002. Infrared evidence of dioctahedral-trioctahedral site occupancy in palygorskite. Clays Clay Miner. 50, 306–313.

Chang, C.-W., Laird, D., Mausbach, M.J., Jurburgh Jr., C.R., 2001. Near-infrared reflectance spectroscopy—principal components regression analysis of soil properties. Soil Sci. Soc. Am. J. 65, 480–490.

Chang, F.-R.C., Skipper, N.T., Sposito, G., 1995. Computer simulation of interlayer molecular structure in sodium montmorillonite hydrates. Langmuir 11, 2734–3741.

Chapin, C.E., Lindley, J.I., 1986. Potassium metasomatism of igneous and sedimentary rocks in detachment terranes and other sedimentary basins: economic implications. In: Beatty, B., Wilkinson, P.A.K. (Eds.), Frontiers in Geology and Ore Deposits of Arizona and the Southwest. Arizona Geological Society Digest, Tucson, AZ.

Chapman, C.R., Salisbury, J.W., 1973. Comparisons of meteorite and asteroid spectral reflectivities. Icarus 19, 507–522.

Chapman, M.G., Smellie, J.L., 2007. Mars interior layered deposits and terrestrial sub-ice volcanoes compared: observations and interpretations of similar geomorphic characteristics. In: Chapman, M.G. (Ed.), The Geology of Mars: Evidence from Earth-based Analogs. Cambridge University Press, UK.

Chase, D.B., 1986. Fourier transform Raman spectroscopy. J. Am. Chem. Soc. 108, 7485–7488.

Chatterjee, A., Iwasaki, T., Ebina, T., Hayashi, H., 1996. Relationship between infrared spectra and isomorphous substitution in smectites: a computer simulation study. J. Mol. Graph. Model. 302–305.

Che, C., Glotch, T.D., Bish, D.L., Michalski, J.R., Xu, W., 2011. Spectroscopic study of the dehydration and/or dehydroxylation of phyllosilicate and zeolite minerals. J. Geophys. Res. 116 (E05007), 1–23.

Chemmi, A., Brendle, J., Marichal, C., Lebeau, B., 2013. A novel fluoride route for the synthesis of aluminosilicate nanotubes. Nanomaterials 3, 117–125.

Chen, B.L., Zhu, L.Z., Zhu, J.X., 2005. Configurations of the bentonite-sorbed myristylpyridinium cation and their influences on the uptake of organic compounds. Environ. Sci. Technol. 39, 6093–6100.

Chen, D., 2008. Modification of microporous mineral materials and adsorption mechanism of organic contaminants onto thematerials (Ph.D. Thesis). Graduate University of the Chinese Academy of Sciences, China.

Chen, K., Vyazovkin, S., 2006. Mechanistic differences in degradation of polystyrene and polystyrene-clay nanocomposite: thermal and thermo-oxidative degradation. Macromol. Chem. Phys. 207, 587–595.

Chen, Q., Wu, P., Li, Y., Zhu, N., Dang, Z., 2009. Heterogeneous photo-Fenton photodegradation of reactive brilliant orange X-GN over iron-pillared montmorillonite under visible irradiation. J. Hazard Mater. 168, 901–908.

Cheng, H.F., Yang, J., Liu, Q.F., Zhang, J.S., Frost, R.L., 2010. A spectroscopic comparison of selected Chinese kaolinite, coal bearing kaolinite and halloysite-A mid-infrared and near-infrared study. Spectrochim. Acta A 77, 856–861.

Cheng, H.F., Liu, Q.F., Yang, J., Zhang, J., Frost, R.L., Du, X.M., 2011a. Infrared spectroscopic study of halloysite-potassium acetate intercalation complex. J. Mol. Struct. 990, 21–25.

Cheng, H.F., Yang, J., Frost, R.L., Liu, Q., Zhang, Z., 2011b. Thermal analysis and infrared emission spectroscopic study of kaolinite-potassium acetate intercalate complex. J. Therm. Anal. Calorim. 103, 507–513.

Cheng, H.F., Liu, Q., Liu, J., Sun, B., Kang, Y., Frost, R.L., 2014. TG–MS–FTIR (evolved gas analysis) of kaolinite–urea intercalation complex. J. Therm. Anal. Calorim. 116, 195–203.

Chevallier, S., Franck, R., Suquet, H., Lambert, J.F., Barthomeuf, D., 1994. Al-pillared saponites. Part 1. IR studies. J. Chem. Soc., Faraday Trans. 90, 667–674.

Chigwada, G., Jiang, D.D., Wilkie, C.A., 2005. Polystyrene nanocomposites based on carbazole-containing surfactants. Thermochim. Acta 436, 113–121.

Chmielarz, L., Kuśtrowski, P., Dziembaj, R., Cool, P., Vansant, E.F., 2007. Selective catalytic reduction of NO with ammonia over porous clay heterostructures modified with copper and iron species. Catal. Today 119, 181–186.

Chmielarz, L., Piwowarska, Z., Kuśtrowski, P., Gil, B., Adamski, A., Dudek, B., Michalik, M., 2009. Porous clay heterostructures (PCHs) intercalated with silica-titania pillars and modified with transition metals as catalysts for the DeNOx process. Appl. Catal. B Environ. 91, 449–459.

Choy, J.H., Kwak, S.Y., Han, Y.S., Kim, B.W., 1997. New organo-montmorillonite complexes with hydrophobic and hydrophilic functions. Mater. Lett. 33, 143–147.

Christensen, P.R., Bandfield, J.L., Smith, M.D., Hamilton, V.E., Clark, R.N., 2000b. Identification of basaltic component on the Martian surface from thermal emission spectrometer data. J. Geophys. Res. 105, 9609–9621.

Christensen, P.R., Bandfield, J.L., Hamilton, V.E., Lane, M.D., Piatek, J.L., Ruff, S.W., Stefanov, W.L., 2000a. A thermal emission spectral library of rock-forming minerals. J. Geophys. Res. 105, 9735–9739.

Christensen, P.R., Bandfield, J.L., Bell J.F. III, Gorelick, N., Hamilton, V.E., Ivanov, A., Jakosky, B.M., Kieffer, H.H., Lane, M.D., Malin, M.C., McConnochie, T., McEwen, A.S., McSween Jr., H.Y., Mehall, G.L., Moersch, J.E., Nealson, K.H., Rice Jr., J.W., Richardson, M.I., Ruff, S.W., Smith, M.D., Titus, T.N., Wyatt, M.B., 2003. Morphology and composition of the surface of Mars: Mars Odyssey THEMIS results. Science 300, 2056–2061.

Christensen, P.R., Bandfield, J.L., Hamilton, V.E., Ruff, S.W., Kieffer, H.H., Titus, T.N., Malin, M.C., Morris, R.V., Lane, M.D., Clark, R.L., Jakosky, B.M., Mellon, M.T., Pearl, J.C., Conrath, B.J., Smith, M.D., Clancy, R.T., Kuzmin, R.O., Roush, T., Mehall, G.L., Gorelick, N., Bender, K., Murray, K., Dason, S., Greene, E., Silverman, S., Greenfield, M., 2001. Mars global surveyor thermal emission spectrometer experiment: investigation description and surface science results. J. Geophys. Res. 106, 23,823–23,871.

Christidis, G.E., Makri, P., Perdikatsis, V., 2004. Influence of grinding colour properties of on the structure and talc, bentonite and calcite white fillers. Clay Miner. 39, 163–175.

Christy, A.A., 2010. New insights into the surface functionalities and adsorption evolution of water molecules on silica gel surface: a study by second derivative near infrared spectroscopy. Vib. Spectrosc. 54, 42–49.

Chryssikos, G.D., Gionis, V., Kacandes, G.H., Stathopoulou, L.T., Suárez, M., García-Romero, E., Sánchez Del Río, M., 2009. Octahedral cation distribution in palygorskite. Am. Mineral. 94, 200–203.

Chukhrov, F.V., Zvyagin, B.B., Drits, V.A., Gorshkov, A.I., Ermilova, L.P., Goilo, E.A., Rudnitskaya, E.S., 1979. The ferric analogue of pyrophyllite and related phases. In: Proc. Int. Clay Conf., Oxford, 1978, pp. 55–64.

Churchman, G.J., Carr, R.M., 1972. Stability fields of hydration states of halloysite. Am. Mineral. 57, 914–923.

Churchman, G.J., Carr, R.M., 1973. Dehydration of the washed potassium acetate complex of halloysite. Clays Clay Miner. 21, 423–424.

Churchman, G.J., Aldridge, L.P., Carr, R.M., 1972. The relationship between the hydrated and dehydrated states of an halloysite. Clays Clay Miner. 20, 241–246.

Churchman, G.J., Whitton, J.S., Claridge, G.G.C., Theng, B.K.G., 1984. Intercalation method using formamide for differentiating halloysite from kaolinite. Clays Clay Miner. 32, 241–248.

Číčel, B., Komadel, P., Bednáriková, E., Madejová, J., 1992. Mineralogical composition and distribution of Si, Al, Fe, Mg, and Ca in the fine fractions of some Czech and Slovak bentonites. Geol. Carph. Clays 1, 3–7.

Clark, R.N., King, T.V.V., Klejwa, M., Swayze, G.A., 1990. High spectral resolution reflectance spectroscopy of minerals. J. Geophys. Res. 95, 12653–12680.

Clark, R.N., Swayze, G.A., Heidebrecht, K.B., Green, R.O., Goetz, A.F.H., 1995. Calibration to surface reflectance of terrestrial imaging spectrometry data: comparison of methods. In: Green, R.O. (Ed.), Proceedings of the Fifth Airborne Visible/Infrared Imaging Spectrometer (AVIRIS) Workshop, JPL, Pasadena, pp. 41–42.

Clark, R.N., Swayze, G.A., Livo, K.E., Kokaly, R.F., Sutley, S.J., Dalton, J.B., Mcdougal, R.R., Gent, C.A., 2003. Imaging spectroscopy: Earth and planetary remote sensing with the USGS Tetracorder and expert systems. J. Geophys. Res. 108, 5131. https://doi.org/10.1029/2002JE001847.

Clark, R.N., Swayze, G.A., Livo, K.E., Kokaly, R.F., King, T.V.V., Dalton, J.B., Vance, J.S., Rockwell, B.W., Hoefen, T., McDougal, R.R., 2002. Surface reflectance calibration of terrestrial imaging spectroscopy data: a tutorial using AVIRIS. In: Proceedings of the Eleventh Annual JPL Airborne Earth Science Workshop, JPL, Pasadena.

Clark, S.J., Segall, M.D., Pickard, C.J., Hasnip, P.J., Probert, M.J., Refson, K., Payne, M.C., 2005. First principles methods using CASTEP. Z. Kristallogr. 220, 567–570.

Coblentz, W.W., 1910. Note on water of crystallization. Phys. Rev. 30, 322–327.

Coblentz, W.W., 1911. The role of water in minerals. J. Franklin Inst. 172, 309–335.

Coblentz, W.W., 1908. IR Spectra, Supplementary Investigations of Infra-red Spectra. The Carnegie Institute of Washington Publications, Washington, DC, 183 pp.

Coblentz, W.W., 1912. Selective radiation from various substances, IV. Bull. Bur. Stand. 9, 81–117.

Cocks, T., Jenssen, R., Stewart, A., Wilson, I., Shields, T., 1998. The HyMap airborne hyperspectral sensor: the system, calibration and performance. In: Schaepman, M., Schläpfer, D., Itten, K.I. (Eds.), Proceedings of 1st EARSeL Workshop on Imaging Spectroscopy. EARSeL Zurich, Paris, pp. 37–43.

Cohen, M., Witteborn, F.C., Roush, T., Bregman, J., Wooden, D., 1998. Spectral Irradiance Calibration in the Infrared. VIII. 5-14 Micron Spectroscopy of the Asteroids Ceres, Vesta, and Pallas. Astron. J. 115, 1671.

Cohen, R.E., Kornacki, A.S., Wood, J.A., 1983. Mineralogy and petrology of chondrules and inclusions in the Mokoia CV3 chondrite. Geochim. Cosmochim. Acta 47, 1739–1757.

Cole, K.C., 2009. The use of attenuated total reflection infrared spectroscopy to study the intercalation of molten polymer into layer silicates in real time. Appl. Spectrosc. 63, 1343–1350.

Cool, P., Ahenach, J., Collart, O., Vansant, E.F., 2000. Al-modified porous clay heterostructures with combined micro- and mesoporosity. Stud. Surf. Sci. Catal. 129, 409–416.

Corona, J.C., Jenkins, D.M., Dyar, M.D., 2015. The experimental incorporation of Fe into talc: a study using X-ray diffraction, Fourier transform infrared spectroscopy, and Mossbauer spectroscopy. Contrib. Mineral. Petrol. 170, 1–15.

Costenaro, D., Gatti, G., Carniato, F., Paul, G., Bisio, C., Marchese, L., 2012. The effect of synthesis gel dilution on the physico-chemical properties of acid saponite clays. Microporous Mesoporous Mater. 162, 159–167.

Cowe, I.A., McNicol, J.W., 1985. The use of principal components in the analysis of near-infrared spectra. Appl. Spectrosc. 39, 257–266.

Craciun, C., 1984. Influence of the $Fe^{3+}$ for $Al^{3+}$ octahedral substitutions on the IR spectra of montmorillonite minerals. Spectrosc. Lett. 17, 579–590.

Cradwick, P.D.G., Farmer, V.C., Russell, J.D., Masson, C.R., Wada, K., Yoshinaga, N., 1972. Imogolite, a hydrated aluminium silicate of tubular structure. Nat. Phys. Sci. 240, 187–189.

Cristovan, F.H., Nascimento, C.M., Bell, M.J.V., Laureto, E., Duarte, J.L., Dias, I.F.L., Cruz, W.O., Marletta, A., 2006. Synthesis and optical characterization of poly(styrene sulfonate) films doped with Nd(III). Chem. Phys. 326, 514–520.

Crowley, J.K., Zimbleman, D.R., 1997. Mapping hydrothermally altered rocks on Mount Rainier, Washington, with Airborne Visible/Infrared Imaging Spectrometer (AVIRIS) data. Geology 25, 559–562.

Crumpler, L.S., Tanaka, K.L., 2003. Geology and MER target site characteristics along the southern rim of Isidis Planitia, Mars. J. Geophys. Res. 108. https://doi.org/10.1029/2002JE002040.

Cuadros, J., Dekov, V.M., Fiore, S., 2008. Crystal-chemistry of the mixed-layer sequence talc-talc-smectite-smectite from submarine hydrothermal vents. Am. Mineral. 93, 1338–1348.

Cuadros, J., Vega, R., Toscano, A., 2015. Mid-infrared features of kaolinite-dickite. Clays Clay Miner. 63, 73–84.

Cuadros, J., Michalski, J.R., Dekov, V., Bishop, J.L., 2016. Octahedral chemistry of 2:1 clay minerals and hydroxyl band position in the near-infrared: application to Mars. Am. Mineral. 101, 554–563.

Cutler, D.J., 1990. The development of Fourier Transform Raman Spectroscopy. Spectrochim. Acta A: Mol. Biomol. Spectrosc. 46, 123–129.

Cygan, R.T., Liang, J.J., Kalinichev, A.G., 2004. Molecular models of hydroxide, oxyhydroxide, and clay phases and the development of a general force field. J. Phys. Chem. B 108, 1255–1266.

Cygan, R.T., Daemen, L.L., Ilgen, A.G., Krumhansl, J.L., Nenoff, T.M., 2015. Inelastic neutron scattering and molecular simulation of the dynamics of inter layer water in smectite clay minerals. J. Phys. Chem. C 119, 28005–28019.

Czímerová, A., Jankovič, Ľ., Madejová, J., Čeklovský, A., 2013. Unique photoactive nanocomposites based on rhodamine 6G/polymer/montmorillonite hybrid systems. J. Polym. Sci, B, Polym Phys. 51, 1672–1679.

Damen, T.C., Porto, S.P.S., Tell, B., 1966. Raman effect in zinc oxide. Phys. Rev. 142, 570–574.

Dawley, M.M., Scott, A.M., Hill, F.C., Leszczynski, J., Orlando, T.M., 2012. Adsorption of formamide on kaolinite surfaces: a combined infrared experimental and theoretical study. J. Phys. Chem. C 116, 23981–23991.

de Angelis, S., Manzari, P., de Sanctis, M.C., Ammannito, E., di Iorio, T., 2016. VIS-IR study of brucite–clay–carbonate mixtures: implications for Ceres surface composition. Icarus 280, 315–327.

De La Calle, C., Suquet, H., 1988. Vermiculite. In: Baily, S.W. (Ed.), Hydrous Phyllosilicates (Exclusive of Micas), Reviews in Mineralogy, vol. 19. Mineralogical Society of America, Washington, DC, pp. 455–492.

De León, M.A., De Los Santos, C., Latronica, L., Cesio, A.M., Volzone, C., Castiglioni, J., Sergio, M., 2014. High catalytic activity at low temperature in oxidative dehydrogenation of propane with Cr-Al pillared clay. Chem. Eng. J. (Amsterdam, Neth.) 241, 336–343.

De Ligny, D., Guillaud, E., Gailhanou, H., Blanc, P., 2013. Raman spectroscopy of adsorbed water in clays: first attempt at band assignment. Proc. Earth Planet. Sci. 7, 203–206.

De Paiva, L.B., Morales, A.R., Valenzuela Díaz, F.R., 2008. Organoclays: properties, preparation and applications. Appl. Clay Sci. 42, 8–24.

De Sanctis, M.C., Ammannito, E., Raponi, A., Marchi, S., McCord, T.B., McSween, H.Y., Capaccioni, F., Capria, M.T., Carrozzo, F.G., Ciarniello, M., Longobardo, A., Tosi, F., Fonte, S., Formisano, M., Frigeri, A., Giardino, M., Magni, G., Palomba, E., Turrini, D., Zambon, F., Combe, J.P., Feldman, W., Jaumann, R., Mcfadden, L.A., Pieters, C.M., Prettyman, T., Toplis, M., Raymond, C.A., Russell, C.T., 2015. Ammoniated phyllosilicates with a likely outer Solar System origin on (1) Ceres. Nature 528, 241–244.

Deblase, F.J., Compton, S., 1991. Infrared emission spectroscopy: a theoretical and experimental review. Appl. Spectrosc. 45, 611–618.

Decarreau, A., Bonnin, D., 1986. Synthesis and crystallogenesis at low temperature of iron(III)-smectites by evolution of coprecipitated gels: experiments in partially reducing conditions. Clay Miner. 21, 861–877.

Decarreau, A., Grauby, O., Petit, S., 1992. The actual distribution of octahedral cations in 2:1 clay minerals: results from clay synthesis. Appl. Clay Sci. 7, 147–167.

Decarreau, A., Bonnin, D., Badaut-Trauth, D., Couty, R., Kaiser, P., 1987a. Synthesis and crystallogenesis of ferric smectite by evolution of silicon-iron coprecipitates in oxidizing conditions. Clay Miner. 22, 207–223.

Decarreau, A., Petit, S., Martin, F., Farges, F., Vieillard, P., Joussein, E., 2008. Hydrothermal synthesis, between 75 and 150°C, of high-charge, ferric nontronites. Clays Clay Miner. 56, 322–337.

Decarreau, A., Colin, F., Herbillon, A., Manceau, A., Nahon, D., Paquet, H., Trauth-Badaut, D., Trescases, J.J., 1987b. Domain segregation in Ni-Fe-Mg-smectites. Clays Clay Miner. 35, 1–10.

Deeds, C.T., van Olphen, H., Bradley, W.F., 1996. Intercalation and interlayer hydration of minerals in rhe kaolinite group. In: Heller, L., Weiss, A. (Eds.), Proc. Int. Clay Conf., Jerusalem.

Dékány, I., Szanto, F., Nagy, L.G., Foti, G., 1975. Adsorption of liquid-mixtures on bentonite and organophilic bentonite. J. Colloid Interface Sci. 50, 265–271.

Delattre, S., Balan, E., Lazzeri, M., Blanchard, M., Guillaumet, M., Beyssac, O., Haussühl, E., Winkler, B., Salje, E.K.H., Calas, G., 2012. Experimental and theoretical study of the vibrational properties of diaspore (α-AlOOH). Phys. Chem. Miner. 39, 93–102.

Delineau, T., Allard, T., Muller, J.-P., Barres, O., Yvon, J., Cases, J.-M., 1994. FTIR reflectance vs EPR studies of structural iron in kaolinites. Clays Clay Miner. 42, 308–320.

Deng, Y., White, G.N., Dixon, J.B., 2002. Effect of structural stress on the intercalation rate of Kaolinite. J. Colloid Interface Sci. 250, 379–393.

Dentel, S.K., Jamrah, A.I., Sparks, D.L., 1998. Sorption and cosorption of 1,2,4-trichlorobenzene and tannic acid by organo-clays. Water Res. 32, 3689–3697.

Diaz, M., Farmer, V.C., Prost, R., 2000. Characterization and assignment of far infrared absorption bands of $K^+$ in muscovite. Clays Clay Miner. 48, 433–438.

Diaz, M., Robert, J.-L., Schroeder, P.A., Prost, R., 2010. Far-infrared study of the influence of the octahedral sheet composition on the $K^+$-layer interactions in synthetic phlogopites. Clays Clay Miner. 58, 263–271.

DiFoggio, R., 2000. Guidelines for applying chemometrics to spectra: feasibility and error propagation. Appl. Spectrosc. 54, 94A–113A.

Dill, H.G., Dohrmann, R., Kaufhold, S., 2011. Disseminated and faultbound autohydrothermal ferroan saponite in Late Paleozoic andesites of the Saar-Nahe Basin, SW Germany: implications for the economic geology of intermediate (sub)volcanic rocks. Appl. Clay Sci. 51, 226–240.

Ding, C., Guo, B.C., He, H., Jia, D.M., Hong, H.Q., 2005. Preparation and structure of highly confined intercalated polystyrene/montmorillonite nanocomposite via a two-step method. Eur. Polym. J. 41, 1781–1786.

Ding, H., Wang, Y.B., Liang, Y., Qin, F.X., 2014. Preparation and characterization of cetyl trimethylammonium intercalated sericite. Adv. Mater. Sci. Eng. 1–8.

Ding, Z., Kloprogge, J.T., Frost, R.L., Lu, G.Q., Zhu, H.Y., 2001. Smectites and porous pillared clay catalysts. Part 2: A review of the catalytic and molecular sieve applications. J. Porous Mater. 8, 273–293.

Dominik, B., Jessberger, E.K., Staudacher, T., Nagel, K., El Goresy, A., 1978. A new type of white inclusion in allende—petrography, mineral chemistry, $^{40}Ar - ^{39}Ar$ ages and genetic implications. In: Proceedings 9th Lunar Planet Science Conf., Houston, TZ. Pergamon Press, Inc., pp.1249–1266.

Doner, H.E., Mortland, M.M., 1969. Benzene complexes with copper(II) montmorillonite. Science 166, 1406–1407.

Donini, J.C., Michaelian, K.H., 1986. Near-infrared photoacoustic FTIR spectroscopy of clay minerals and coal. Infrared Phys. 26, 135–140.

Dontsova, K.M., Norton, L.D., Johnston, C.T., 2005. Calcium and magnesium effects on ammonia adsorption by soil clays. Soil Sci. Soc. Am. J. 69, 1225–1232.

Dontsova, K.M., Norton, L.D., Johnston, C.T., Bigham, J.M., 2004. Influence of exchangeable cation on water adsorption by soil clays. Soil Sci. Soc. Am. J. 68, 1218–1227.

Dovesi, R., Orlando, R., Erba, A., Zicovich-Wilson, C.M., Civalleri, B., Casassa, S., Maschio, L., Ferrabone, M., De La Pierre, M., D'Arco, P., Noel, Y., Causa, M., Rerat, M., Kirtman, B., 2014. CRYSTAL14: a program for the ab initio investigation of crystalline solids. Int. J. Quantum Chem. 114, 1287–1317.

Drits, V.A., Besson, G., Muller, F., 1995. An improved model for structural transformations of heat-treated aluminous dioctahedral 2:1 layer silicates. Clays Clay Miner. 43, 718–731.

Drits, V.A., McCarty, D.K., Zviagina, B.B., 2006. Crystal-chemical factors responsible for the distribution of octahedral cations over trans- and cis-sites in dioctahedral 2.1 layer silicates. Clays Clay Miner. 54, 131–152.

Du, C., Zhou, J., Wang, H., Chen, X., Zhu, A., Zhang, J., 2008. Determination of soil properties using Fourier transform mid-infrared photoacoustic spectroscopy. Vib. Spectrosc. 49, 32–37.

Dudik, J.M., Johnson, C.R., Asher, S.A., 1985. UV resonance Raman studies of acetone, acetamide, and N-methylacetamide: models for the peptide bond. J. Phys. Chem. 89, 3805–3814.

Dumas, A., Mizrahi, M., Martin, F., Requejo, F.G., 2015. Local and extended-order evolution of synthetic talc during hydrothermal synthesis: extended X-ray absorption fine structure, X-ray diffraction, and Fourier transform infrared spectroscopy studies. Cryst. Growth Des. 15, 5451–5463.

Dumas, A., Martin, F., Le Roux, C., Micoud, P., Petit, S., Ferrage, E., Brendle, J., Grauby, O., Greenhill-Hooper, M., 2013. Phyllosilicates synthesis: a way of accessing edges contributions in NMR and FTIR spectroscopies. Example of synthetic talc. Phys. Chem. Miner. 40, 361–373.

Duong, L.V., Wood, B.J., Kloprogge, J.T., 2005. XPS study of basic aluminium sulphate and basic aluminium nitrate. Mater. Lett. 59, 1932–1936.

Durig, J.R., Zheng, C., 2002. On the vibrational spectra and structural parameters of hydrazine and some methyl substituted hydrazines. Vib. Spectrosc. 30, 59–67.

Eberly Jr., P.E., 1967. High-temperature infrared spectroscopy of olefins adsorbed on faujasites. J. Phys. Chem. B 71, 1717–1722.

Edge, J.S., Skipper, N.T., Fernandez-Alonso, F., Lovell, A., Srinivas, G., Bennington, S.M., Sakai, V.G., Youngs, T.G.A., 2014. Structure and dynamics of molecular hydrogen in the interlayer pores of a swelling 2:1 clay by neutron scattering. J. Phys. Chem. C 118, 25740–25747.

Ehlmann, B., Berger, G., Mangold, N., Michalski, J., Catling, D., Ruff, S., Chassefière, E., Niles, P., Chevrier, V., Poulet, F., 2013. Geochemical consequences of widespread clay mineral formation in Mars' ancient crust. Space Sci. Rev. 174, 329–364.

Ehlmann, B.L., Mustard, J.F., 2012. An in-situ record of major environmental transitions on early Mars at Northeast Syrtis Major. Geophys. Res. Lett. 39. L11202. https://doi.org/10.1029/2012GL051594.

Ehlmann, B.L., Edwards, C.S., 2014. Mineralogy of the Martian Surface. Annu. Rev. Earth Planet. Sci. 42, 291–315.

Ehlmann, B.L., Bish, D.L., Ruff, S.W., Mustard, J.F., 2012. Mineralogy and chemistry of altered Icelandic basalts: application to clay mineral detection and understanding aqueous environments on Mars. J. Geophys. Res.: Planets 117, E00J16. doi:10.1029/2012JE004156.

Ehlmann, B.L., Mustard, J.F., Murchie, S.L., Bibring, J.P., Meunier, A., Fraeman, A.A., Langevin, Y., 2011. Subsurface water and clay mineral formation during the early history of Mars. Nature 479, 53–60.

Ehlmann, B.L., Mustard, J.F., Swayze, G.A., Clark, R.N., Bishop, J.L., Poulet, F., Marais, D.J.D., Roach, L.H., Milliken, R.E., Wray, J.J., Barnouin-Jha, O., Murchie, S.L., 2009. Identification of hydrated silicate minerals on Mars using MRO-CRISM: geologic context near Nili Fossae and implications for aqueous alteration. J. Geophys. Res. 114. https://doi.org/10.1029/2009JE003339.

El-Eswed, B.I., Zughul, M.B., Derwish, G.A.W., 1997. Infrared spectroscopic study of the role of water in crown ethers and their molecular complexes with 3- and 4-nitrophenol. J. Inclusion Phenom. Mol. Recognit. Chem. 28, 245–258.

Emmerich, K., 2013. Full characterization of smectites. In: Bergaya, F., Lagaly, G. (Eds.), Handbook of Clay Science. Developments in Clay Science, vol. 5B, second ed. (Chapter 13).

Emmerich, K., Wolters, F., Kahr, G., Lagaly, G., 2009. Clay profiling: the classification of montmorillonites. Clays Clay Min. 57, 104–114.

Erkeling, G., Reiss, D., Hiesinger, H., Poulet, F., Carter, J., Ivanov, M.A., Hauber, E., Jaumann, R., 2012. Valleys, paleolakes and possible shorelines at the Libya Montes/Isidis boundary: implications for the hydrologic evolution of Mars. Icarus 219, 393–413.

Esbensen, K.H., Geladi, P., 2010. Principles of proper validation. J. Chemometr. 24, 168–187.

Escamilla-Roa, E., Hernandez-Laguna, A., Sainz-Diaz, C.I., 2014. Theoretical study of the hydrogen bonding and infrared spectroscopy in the cis-vacant polymorph of dioctahedral 2:1 phyllosilicates. J. Molec. Model 20, Article Number: 2404.

Evans, B.W., Guggenheim, S., 1988. Talc, pyrophyllite and related minerals. In: Bailey, S.W. (Ed.), Hydrous Phyllosilicates (Exclusive of Micas). Mineralogical Society of America, Washington, DC, pp. 225–294.

Fabbri, G., Baraldi, P., 1972. Infrared emission spectra of solids. Appl. Spectrosc. 26, 593–598.

Faghihian, H., Mohammadi, M.H., 2012. A novel catalyst for alkylation of benzene. C. R. Chim. 15, 962–968.

Faghihian, H., Mohammadi, M.H., 2013. Surface properties of pillared acid-activated bentonite as catalyst for selective production of linear alkylbenzene. Appl. Surf. Sci. 264, 492–499.

Fairén, A.G., Chevrier, V., Abramov, O., Marzo, G.A., Gavin, P., Davila, A.F., Tornabene, L.L., Bishop, J.L., Roush, T.L., Gross, C., Kneissl, T., Uceda, E.R., Dohm, J.M., Schulze-Makuch, D., Rodríguez, J.A.P., Amils, R., McKay, C.P., 2010. Phyllosilicates in impact craters on mars: pre- or post-impact genesis. In: Proceedings of the National Academy of Sciences, 107, pp. 12,095–12,100. https://doi.org/10.1073/pnas.1002889107.

Falini, G., Foresti, E., Lesci, G., Roveri, N., 2002. Structural and morphological characterization of synthetic chrysotile single crystals. Chem. Commun. (Cambridge, UK), 1512–1513.

Farmer, V.C., 1958. The infrared spectra of talc, saponite and hectorite. Mineral. Mag. 31, 829–845.

Farmer, V.C., 1964. Infrared absorption of hydroxyl groups in kaolinite. Science 145, 1189–1190.

Farmer, V.C., 1968. Infrared spectroscopy in clay mineral studies. Clay Miner. 7, 373–387.

Farmer, V.C., 1971. The characterization of adsorption bonds in clays by infrared spectroscopy. Soil Sci. 112, 62–68.

Farmer, V.C., 1974a. The Infrared Spectra of Minerals. Mineralogical Society Monograph 4, London, 539 pp.

Farmer, V.C., 1974b. The layer silicates. In: Farmer, V.C. (Ed.), The Infrared Spectra of Minerals. Mineralogical Society, London, pp. 331–363.

Farmer, V.C., 1979. Infrared spectroscopy. In: van Olphen, H., Fripiat, J.J. (Eds.), Data Handbook for Clay Materials and Other Non-metallic Minerals. Pergamon Press, Oxford, pp. 285–337.

Farmer, V.C., 1997. Conversion of ferruginous allophanes to ferruginous beidellites at 95°C under alkaline conditions with alternating oxidation and reduction. Clays Clay Miner. 45, 591–597.

Farmer, V.C., 1998. Differing effects of particle size and shape in the infrared and Raman spectra of kaolinite. Clay Miner. 33, 601–604.

Farmer, V.C., 2000. Transverse and longitudinal crystal modes associated with OH stretching vibrations in single crystals of kaolinite and dickite. Spectrochim. Acta A 56, 927–930.

Farmer, V.C., Russell, J.D., 1964. The infra-red spectra of layer silicates. Spectrochim. Acta 20, 1140–1173.

Farmer, V.C., Russell, J.D., 1966. Effects of particle size and structure on the vibrational frequencies of layer silicates. Spectrochim. Acta 22, 389–398.

Farmer, V.C., Russell, J.D., 1967. Infrared absorption spectrometry in clay studies. Clays Clay Miner. 15, 121–142.

Farmer, V.C., Russell, J.D., 1971. Interlayer complexes in layer silicates. The structure of water in lamellar ionic solutions. Trans. Faraday Soc. 67, 2737–2749.

Farmer, V.C., Velde, B., 1973. Effects of structural order and disorder on the infrared spectra of brittle mica. Mineral. Mag. 39, 282–288.

Farmer, V.C., Lazarev A.N., 1974. Symmetry and crystal vibrations. In: V.C. Farmer (Ed.), The Infrared Spectra of Minerals. Mineralogical Society, London, pp. 51–67.

Farmer, V.C., Russell, J.D., Ahlrichs, J.D., Velde, B., 1967. Vibration of the hydroxyl group in layer silicates. Bull. Grpe Fr. Argiles 19, 5–10.

Farmer, V.C., Fraser, A.R., Tait, J.M., 1977b. Synthesis of imogolite: a tubular aluminium silicate polymer. J. Chem. Soc., Chem. Comm., 462–463.

Farmer, V.C., Fraser, A.R., Tait, J.M., 1979. Characterization of the chemical structures of natural and synthetic aluminosilicate gels and sols by infrared spectroscopy. Geochim. Cosmochim. Acta 43, 1417–1420.

Farmer, V.C., Krishnamurti, G.S.R., Huang, P.M., 1991a. Synthetic allophane and layer-silicate formation in $SiO_2$-$Al_2O_3$-FeO-$Fe_2O_3$-MgO-$H_2O$ systems at 23°C and 89°C in a calcareous environment. Clays Clay Miner. 39, 561–570.

Farmer, V.C., Fraser, A.R., Russell, J.D., Yoshinaga, N., 1977a. Recognition of imogolite structures in allophanic clays by infrared spectroscopy. Clay Miner. 12, 55–57.

Farmer, V.C., Adams, M.J., Fraser, A.R., Palmieri, F., 1983. Synthetic imogolite: properties, synthesis, and possible applications. Clay Miner. 18, 459–472.

Farmer, V.C., Mchardy, W.J., Elsass, F., Robert, M., 1994. HK-ordering in aluminous nontronite and saponite synthesized near 90-degrees C—effect of synthesis conditions on nontronite composition and ordering. Clays Clay Miner. 42, 180–186.

Farmer, V.C., Smith, B.F.L., Wilson, M.J., Loveland, P.J., Payton, R.W., 1988. Readily-extractable hydroxyaluminium interlayers in clay-sized and silt-sized vermiculite. Clay Miner. 23, 271–277.

Farmer, V.C., Mchardy, W.J., Palmieri, F., Violante, A., Violante, P., 1991b. Synthetic allophanes formed in calcareous environments: nature, conditions of formation, and transformations. Soil Sci. Soc. Am. J. 55, 1162–1166.

Farmer, V.C., Russell, J.D., Mchardy, W.J., Newman, A.C.D., Ahlrichs, J.L., Rimsaite, J.Y.H., 1971. Evidence of loss of protons and octahedral iron from oxidized biotites and vermiculites. Mineral. Mag. 38, 121–137.

Farrand, W.H., Singer, R.B., Merényi, E., 1994. Retrieval of apparent surface reflectance from AVIRIS data: a comparison of empirical line, radiative transfer, and spectral mixture methods. Remote Sens. Environ. 47, 311–321.

Fatimah, I., Huda, T., 2013. Preparation of cetyltrimethylammonium intercalated Indonesian montmorillonite for adsorption of toluene. Appl. Clay Sci. 74, 115–120.

Favre, H., Lagaly, G., 1991. Organo-bentonites with quaternary alkylammonium ions. Clay Miner. 26, 19–32.

Fehervari, A., Gates, W.P., Turney, T.W., Patti, A.F., Bouazza, A., 2016. Cyclic organic carbonate modification of sodium bentonite for enhanced containment of hyper saline leachates. Appl. Clay Sci. 134, 2–12.

Feng, X.Y., Hu, G.J., Meng, X.F., Ding, Y.F., Zhang, S.M., Yang, M.S., 2009. Influence of ethanol addition on the modification of montmorillonite by hexadecyl trimethylammonium bromide. Appl. Clay Sci. 45, 239–243.

Ferrage, E., Lanson, B., Sakharov, B.A., Drits, V.A., 2005. Investigation of smectite hydration properties by modeling experimental X-ray diffraction patterns: Part I. Montmorillonite hydration properties. Am. Mineral. 90, 1358–1374.

Ferrage, E., Lanson, B., Michot, L.J., Robert, J.-L., 2010. Hydration properties and interlayer organization of water and ions in synthetic Na-smectite with tetrahedral layer charge. Part 1. Results from X-ray diffraction profile modeling. J. Phys. Chem. C. 114, 4515–4526.

Ferrage, E., Martin, F., Micoud, P., Petit, S., De Parseval, P., Beziat, D., Ferret, J., 2003a. Cation site distribution in clinochlorites: a NIR approach. Clay Miner. 38, 329–338.

Ferrage, E., Martin, F., Petit, S., Pejo-Soucaille, S., Micoud, P., Fourty, G., Ferret, J., Salvi, S., De Parseval, P., Fortune, J.P., 2003b. Evaluation of talc morphology using FTIR and H/D substitution. Clay Miner. 38, 141–150.

Ferrage, E., Martin, F., Boudet, A., Petit, S., Fourty, G., Jouffret, F., Micoud, P., De Parseval, P., Salvi, S., Bourgerette, C., Ferret, J., Saint-Gerard, Y., Buratto, S., 2002. Talc as nucleating agent of polypropylene: morphology induced by lamellar particles addition and interface mineral-matrix modelization. J. Mater. Sci. 37, 1561–1573.

Ferrante, F., Armata, N., Lazzara, G., 2015. Modeling of the halloysite spiral nanotube. J. Phys. Chem. C 119, 16700–16707.

Ferraro, J.R., Nakamoto, K., Brown, Ch.W., 2003. Introductory Raman Spectroscopy, second ed. Academic Press, Elsevier Science, San Diego.

Fialips, C.-I., Petit, S., Decarreau, A., Beaufort, D., 2000. Influence of synthesis pH on kaolinite "crystallinity" and surface properties. Clays Clay Miner. 48, 173–184.

Fialips, C.-I., Huo, D., Yan, L., Wu, J., Stucki, J.W., 2002. Infrared study of reduced and reduced-reoxidized ferruginous smectite. Clays Clay Miner. 50, 455–469.

Figueras, F., 1988. Pillared clays as catalysts. Catal. Rev. Sci. Eng. 30, 457–499.

Filgueiras, P.R., Sad, C.M.S., Loureiro, A.R., Santos, M.F.P., Castro, E.V.R., Dias, J.C.M., Poppi, R.J., 2014. Determination of API gravity, kinematic viscosity and water content in petroleum by ATR-FTIR spectroscopy and multivariate calibration. Fuel 116, 123–130.

Finocchio, E., Cristiani, C., Dotelli, G., Stampino, P.G., Zampori, L., 2014. Thermal evolution of PEG-based and BRIJ-based hybridorgano-inorganic materials FT-IR studies. Vib. Spectrosc. 71, 47–56.

Finocchio, E., Baccini, I., Cristiani, C., Dotelli, G., Gallo Stampino, P., Zampori, L., 2011. Hybrid organo-inorganic clay with nonionic inter layers. Mid- and near-IR spectroscopic studies. J. Phys. Chem. A 115, 7484–7493.

Flach, C.R., Gericke, A., Mendelsohn, R., 1997. Quantitative determination of molecular chain tilt angles in monolayer films at the air/water interface: infrared reflection/absorption spectroscopy of behenic acid methyl ester. J. Phys. Chem. B. 101, 58–65.

Flego, C., Galasso, L., Millini, R., Kiricsi, I., 1998. The influence of the composition on the thermal and acid characteristics of multi-component oxide pillared montmorillonite. Appl. Catal., A 168, 323–331.

Foresti, E., Fornero, E., Lesci, I.G., Rinaudo, C., Zuccheri, T., Roveri, N., 2009. Asbestos health hazard: A spectroscopic study of synthetic geoinspired Fe-doped chrysotile. J. Hazard. Mater. 167, 1070–1079.

Foresti, E., Hochella, M.F., Kornishi, H., Lesci, I.G., Madden, A.S., Roveri, N., Xu, H., 2005. Morphological and chemical-physical characterization of Fe-doped synthetic chrysotile nanotubes. Adv. Funct. Mater. 15, 1009–1016.

Fourdrin, C., Allard, T., Balan, E., Boukari, C., Calas, G., 2008. Induced modifications of kaolinite under ionizing radiation: an infrared spectroscopic study. Phys. Chem. Miner. 36, 291–299.

Fox, V.K., Arvidson, R.E., Guinness, E.A., Mclennan, S.M., Catalano, J.G., Murchie, S.L., Powell, K.E., 2016. Smectite deposits in Marathon Valley, Endeavour Crater, Mars, identified using CRISM hyperspectral reflectance data. Geophys. Res. Lett. 43, 4885–4892.

Franco, F., Ruiz Cruz, M.D., 2003. Thermal behaviour of dickite-dimethyl sulfoxide intercalation complex. J. Therm. Anal. Calorim. 73, 151–165.

Fraser, A.R., Russell, J.D., 1994. Infrared methods. In: Wilson, M.J. (Ed.), Clay Mineralogy: Spectroscopic and Chemical Determinative Methods. Chapman & Hall, London, pp. 11–67.

Fraser, H.J., Mccoustra, M.R.S., Williams, D.A., 2002. The molecular universe. Astron. Geophys. 43, 10–18.

Friedlander, L., Glotch, T.D., Bish, D.L., Dyar, M.D., Sharp, T.G., Sklute, E.C., Michalski, J.R., 2015. Structural and spectroscopic changes to natural nontronite induced by experimental impacts between 10 and 40 GPa. J. Geophys. Res.: Planets 120, 888–912.

Friedrich, F., Weidler, P.G., 2010. Contact pressure effects on vibrational bands of kaolinite during infrared spectroscopic measurements in a diamond attenuated total reflection cell. Appl. Spectrosc. 64, 500–506.

Fripiat, J.J., 1982. Application of Far-infrared spectroscopy to the study of clay minerals and zeolites. In: Fripiat, J.J. (Ed.), Advanced Techniques for Clay Mineral Analysis. Elsevier, Amsterdam, pp. 191–210.

Fripiat, J.J., 1988. High resolution solid state NMR study of pillared clays. In: Burch, R. (Ed.), Catal. Today 2–3, 281–295.

Fripiat, J.J., Tousaint, F., 1963. Dehydroxylation of kaolinite. II. Conductometric measurements and infrared spectroscopy. J. Phys. Chem. 67, 30–36.

Fripiat, J.J., Chaussidon, J., Touillaux, R., 1960. Study of dehydration of montmorillonite and vermiculite by infrared spectroscopy. J. Phys. Chem. 64, 1234–1241.

Fripiat, J.J., Pennequin, M., Poncelet, G., Cloos, P., 1969. Influence of the van der Waals force on the infrared spectra of short aliphatic alkylammonium cations held on montmonllonite. Clay Miner. 8, 119–134.

Frost, R.L., 1995. Fourier transform Raman spectroscopy of kaolinite, dickite and halloysite. Clays Clay Miner. 43, 191–195.

Frost, R.L., 1997. The structure of the kaolinite minerals—a FT-Raman study. Clay Miner. 32, 65–77.

Frost, R.L., 1998. Hydroxyl deformation in kaolins. Clays Clay Miner. 46, 280–289.

Frost, R.L., Kloprogge, J.T., 1999. Raman spectroscopy of the low-frequency region of kaolinite at 298 and 77 K. Appl. Spectrosc. 53, 1610–1616.

Frost, R.L., Kloprogge, J.T., 2000a. Raman spectroscopy of kaolinite hydroxyls between 25°C and 500°C. J. Raman Spectrosc. 31, 415–420.

Frost, R.L., Kloprogge, J.T., 2000b. Raman spectroscopy of nacrite single crystals at 298 and 77 K. Spectrochim. Acta A 56, 931–939.

Frost, R.L., Kloprogge, J.T., 2000c. Vibrational spectroscopy of ferruginous smectite and nontronite. Spectrochim. Acta A 56, 2177–2189.

Frost, R.L., Kloprogge, J.T., 2000d. Raman spectroscopy of nontronites. Appl. Spectrosc. 54, 402–405.

Frost, R.L., Kloprogge, J.T., 2000e. Raman spectroscopy of the acetates of sodium, potassium and magnesium at liquid nitrogen temperature. J. Mol. Struct. 526, 131–141.

Frost, R.L., Kloprogge, J.T., 2001. Towards a single crystal Raman spectrum of kaolinite at 77 K. Spectrochim. Acta A 57, 163–175.

Frost, R.L., Kloprogge, J.T., 2012. Raman and infrared spectroscopic study of the modification of kaolinite surfaces by intercalation with organic molecules. Encycl. Surf. Colloid Sci. (2nd Ed.), 5361–5372.

Frost, R.L., Kristof, J., 1997. Intercalation of halloysite: a Raman spectroscopic study. Clays Clay Miner. 45, 551–563.

Frost, R.L., Rintoul, L., 1996a. Lattice vibrations of montmorillonite: an FT Raman and X-ray diffraction study. Appl. Clay Sci. 11, 171–183.

Frost, R.L., Rintoul, L., 1996b. FT Raman spectroscopy of montmorillonites. Int. Ceram. Monogr. 2, 325–331.

Frost, R.L., Shurvell, H.F., 1997. Raman microprobe spectroscopy of halloysite. Clays Clay Miner. 45, 68–72.

Frost, R.L., Van Der Gaast, S.J., 1997. Kaolinite hydroxyls—a Raman microscopy study. Clay Miner. 32, 471–484.

Frost, R.L., Vassallo, A.M., 1997. Fourier-transform infrared emission spectroscopy of kaolinite dehydroxylation. Mikrochim. Acta Suppl. 14, 789–791.

Frost, R.L., Fredericks, P.M., Bartlett, J.R., 1993. Fourier transform Raman spectroscopy of kandite clays. Spectrochim. Acta Part A 49A, 667–674.

Frost, R.L., Fredericks, P.M., Shurvell, H.F., 1996. Raman microscopy of some kaolinite clay minerals. Can. J. Appl. Spectrosc. 41, 10–14.

Frost, R.L., Tran, T.H.T., Kristóf, J., 1997a. FT-Raman spectroscopy of the lattice region of kaolinite and its intercalates. Vib. Spectrosc. 13, 175–186.

Frost, R.L., Tran, T.H., Kristóf, J., 1997b. The structure of an intercalated ordered kaolinite—a Raman microscopy study. Clay Miner. 32, 587–596.

Frost, R.L., Kristof, J., Tran, T.H., 1998f. Kinetics of deintercalation of potassium acetate from kaolinite—a Raman spectroscopic study. Clay Miner. Bull. 33, 605–617.

Frost, R.L., Kloprogge, J.T., Tran, T.H.T., Kristóf, J., 1998c. The effect of pressure on the intercalation of an ordered kaolinite. Am. Mineral. 83, 1182–1187.

Frost, R.L., Kristóf, J., Paroz, G.N., Kloprogge, J.T., 1998a. Modification of the kaolinite hydroxyl surfaces through intercalation with potassium acetate under pressure. J. Colloid Interface Sci. 208, 478–486.

Frost, R.L., Kristóf, J., Paroz, G.N., Kloprogge, J.T., 1998d. Molecular structure of dimethyl sulfoxide intercalated kaolinites. J. Phys. Chem. B 102, 8519–8532.

Frost, R.L., Kristóf, J., Paroz, G.N., Kloprogge, J.T., 1998b. Role of water in the intercalation of kaolinite with hydrazine. J. Colloid Interface Sci. 208.

Frost, R.L., Tran, T.H., Rintoul, L., Kristof, J., 1998g. Raman microscopy of dickite, kaolinite and their intercalates. Analyst 123, 611–616.

Frost, R.L., Kloprogge, J.T., Kristóf, J., Horváth, E., 1999f. Deintercalation of hydrazine-intercalated low-defect kaolinite. Clays Clay Miner. 47, 732–741.

Frost, R.L., Kristóf, J., Horváth, E., Kloprogge, J.T., 1999c. Deintercalation of dimethylsulfoxide intercalated kaolinites—a DTA/TGA and Raman spectroscopic study. Thermochim. Acta 327, 155–166.

Frost, R.L., Kristóf, J., Horváth, E., Kloprogge, J.T., 1999a. Modification of kaolinite surfaces with cesium acetate at 25, 120 and 220°C. Langmuir 15, 8787–8794.

Frost, R.L., Kristóf, J., Horváth, E., Kloprogge, J.T., 1999b. Molecular structure of dimethy sulfoxide in DMSO-intercalated kaolinites at 298 and 77 K. J. Phys. Chem. A 103, 9654–9660.

Frost, R.L., Kristóf, J., Paroz, G.N., Kloprogge, J.T., 1999e. Intercalation of kaolinite with acetamide. Phys. Chem. Miner. 26, 257–263.

Frost, R.L., Lack, D.A., Paroz, G.N., Tran, T.H.T., 1999d. New techniques for studying the intercalation of kaolinites from Georgia with formamide. Clays Clay Miner. 47, 297–303.

Frost, R.L., Kristóf, J., Horváth, E., Kloprogge, J.T., 2000a. Effect of water on the formamide-intercalation of kaolinite. Spectrochim. Acta A 56, 1711–1729.

Frost, R.L., Kristóf, J., Horváth, E., Kloprogge, J.T., 2000d. Kaolinite hydroxyls in dimethysulphoxide-intercalated kaolinites at 77 K—a Raman spectroscopic study. Clay Miner. 35, 443–454.

Frost, R.L., Kristóf, J., Horváth, E., Kloprogge, J.T., 2000c. Vibrational spectroscopy of formamide-intercalated kaolinites. Spectrochim. Acta A 56, 1191–1204.

Frost, R.L., Kristóf, J., Makó, É., Kloprogge, J.T., 2000b. Modification of the hydroxyl surface in potassium-acetate-intercalated kaolinite between 25 and 300°C. Langmuir 16, 7421–7428.

Frost, R.L., Kristóf, J., Rintoul, L., Kloprogge, J.T., 2000e. Raman spectroscopy of urea and urea-intercalated kaolinites at 77 K. Spectrochim. Acta A 56, 1681–1691.

Frost, R.L., Fredericks, P.M., Kloprogge, J.T., Hope, G.A., 2001a. Raman spectroscopy of kaolinites using different excitation wavelengths. J. Raman Spectrosc. 32, 657–663.

Frost, R.L., Forsling, W., Holmgren, A., Kloprogge, J.T., Kristóf, J., 1998e. Raman spectroscopy at temperatures between 298 and 423 K and at 77 K of kaolinites intercalated with formamide. J. Raman Spectrosc. 29, 1065–1069.

Frost, R.L., Locos, O.B., Ruan, H., Kloprogge, J.T., 2001b. Near-infrared and mid-infrared spectroscopic study of sepiolites and palygorskites. Vib. Spectrosc. 27, 1–13.

Frost, R.L., Kristóf, J., Horváth, E., Kloprogge, J.T., 2001c. Raman microscopy of formamide-intercalated kaolinites treated by controlled-rate thermal analysis technology. J. Raman Spectrosc. 32, 873–880.

Frost, R.L., Kristóf, J., Horváth, E., Kloprogge, J.T., 2001d. Separation of adsorbed formamide and intercalated formamide using controlled rate thermal analysis methodology. Langmuir 17, 3216–3222.

Frost, R.L., Kristóf, J., Horváth, E., Kloprogge, J.T., 2001e. Raman spectroscopy of potassium acetate-intercalated kaolinites over the temperature range 25–300°C. J. Raman Spectrosc. 32, 271–277.

Frost, R.L., Kristóf, J., Horváth, E., Kloprogge, J.T., 2001f. The modification of hydroxyl surfaces of formamide-intercalated kaolinites synthesized by controlled rate thermal analysis. J. Colloid Interface Sci. 239, 126–133.

Frost, R.L., Kristóf, J., Kloprogge, J.T., Horváth, E., 2001g. Modification of the hydroxyl surface in cesium acetate intercalated kaolinite. Langmuir 17, 4067–4073.

Frost, R.L., Kristóf, J., Horváth, E., Martens, W.N., Kloprogge, J.T., 2002a. Complexity of intercalation of hydrazine into kaolinite–controlled rate thermal analysis and DRIFT spectroscopic study. J. Colloid Interface Sci. 251, 350–359.

Frost, R.L., Kloprogge, J.T., Ding, Z., 2002b. Near-infrared spectroscopic study of nontronites and ferruginous smectite. Spectrochim. Acta Part A 58, 1657–1668.

Frost, R.L., Makó, É., Kristóf, J., Kloprogge, J.T., 2002c. Modification of kaolinite surfaces through mechanochemical treatment—a mid-IR and near-IR spectroscopic study. Spectrochim. Acta A 58, 2849–2859.

Frost, R.L., Zhou, Q., He, H.P., Xi, Y.F., 2008. An infrared study of adsorption of para-nitrophenol on mono-, di- and tri-alkyl surfactant intercalated organoclays. Spectrochim. Acta A: Mol. Biomol. Spectrosc. 69, 239–244.

Fukamachi, C.R.B., Wypych, F., Mangrich, A.S., 2007. $Fe^{3+}$ ion probe to study the stability of urea-intercalated kaolinite by electron paramagnetic resonance. J. Colloid Interface Sci. 313, 537–541.

Furnes, H., 1978. Element mobility during palagonitization of a subglacial hyaloclastite in Iceland. Chem. Geol. 22, 249–264.

Gaffey, M.J., 1976. Spectral reflectance characteristics of the meteorite classes. J. Geophys. Res. 81, 905–920.

Galán, E., Carretero, M.I., 1999. A new approach to compositional limits for sepiolite and palygorskite. Clays Clay Miner. 47, 399–409.

Galarneau, A., Barodawalla, A., Pinnavaia, T.J., 1995. Porous clay heterostructures formed by gallery-templated synthesis. Nature 374, 529–531.

Gamba, M., Flores, F.M., Madejová, J., Sanchez, R.M.T., 2015. Comparison of imazalil removal onto montmorillonite and nanomontmorillonite and adsorption surface sites involved: an approach for agricultural wastewater treatment. Ind. Eng. Chem. Res. 54, 1529–1538.

Gammoudi, S., Frini-Srasra, N., Srasra, E., 2012. Influence of exchangeable cation of smectite on HDTMA adsorption: equilibrium, kinetic and thermodynamic studies. Appl. Clay Sci. 69, 99–107.

Gan, H., Low, P.F., 1993. Spectroscopic study of ionic adjustments in the electric double layer of montmorillonite. J. Colloid Interface Sci. 161, 1–5.

Ganeshsrinivas, E., Sathyanarayana, D.N., Machida, K., Miwa, Y., 1996. Simulation of the infrared spectra of acetamide by an extended molecular mechanics method. J. Mol. Struct. (Theochem) 361, 217–227.

Gao, B.-C., Heidebrecht, K.B., Goetz, A.F.H., 1993. Derivation of scaled surface reflectances from AVIRIS data. Remote Sens. Environ. 44, 165–178.

Gao, H., Zhao, B.-X., Luo, J.-C., Wu, D., Ye, W., Wang, Q., Zhang, X.-L., 2014. Fe-Ni-Al pillared montmorillonite as a heterogeneous catalyst for the Catalytic Wet Peroxide Oxidation degradation of Orange Acid II: Preparation condition and properties study. Microporous Mesoporous Mater. 196, 208–215.

Garnett, J.C.M., 1904. Colours in metal glasses and metallic films. Philos. Trans. R. Soc. A 203, 385–420.

Gashti, M.P., Eslami, S., 2012. Structural, optical and electromagnetic properties of aluminum-clay nanocomposites. Superlattices Microstruct. 51, 135–148.

Gates, W.P., 2004. Crystalline swelling of organo-modified clays in ethanol-water solutions. Appl. Clay Sci. 27, 1–12.

Gates, W.P., 2005. Infrared spectroscopy and the chemistry of dioctahedral smectites. In: Klopproge, J.T. (Ed.), The Application of Vibrational Spectroscopy to Clay Minerals and Layered Double Hydroxides, CMS Workshop Lectures, vol. 13. The Clay Mineral Society, Aurora, CO, pp. 125–168.

Gates, W.P., 2008. Cation mass–valence sum (CM-VS) approach to assigning OH-bending bands in dioctahedral smectites. Clays Clay Miner. 56, 10–22.

Gates, W.P., Cashion J.D., 2014. Nearest neighbour distributions of site occupancies in nontronites. In: Proceedings of the 23rd Biennial Conference of the Australian Clay Minerals Society, Perth, Australia. Session 3, p. 81; http://www.smectech.com.au/ACMS/ACMS_Conferences/ACMS23/Program/Paper_By_Author/Papers_Author.htm.

Gates, W.P., Slade, P.G., Manceau, A., Lanson, B., 2002. Site occupancies by iron in nontronites. Clays Clay Miner. 50, 223–239.

Gates, W.P., Shaheen, U., Turney, T.W., Patti, A.F., 2016. Cyclic carbonate-sodium smectite intercalates. Appl. Clay Sci. 124–125, 94–101.

Gates, W.P., Bordallo, H.N., Aldridge, L.P., Seydel, T., Jacobsen, H., Marry, V., Churchman, G.J., 2012. Neutron time-of-flight quantification of water desorption isotherms of montmorillonite. J. Phys. Chem. C 116, 5558–5570.

Geladi, P., Kowalski, B.R., 1986. Partial least squares regression: a tutorial. Anal. Chim. Acta 185, 1–17.

Gervais, F., Piriou, B., 1974. Anharmonicity in several-polar-mode crystals: adjusting phonon self-energy of LO and TO modes in Al2O3 and TiO2 to fit infrared reflectivity J. Phys. C: Solid State Phys. 7, 2374–2386.

Ghabru, S.K., Mermut, A., Arnoud, R.J.S., 1989. Layer charge and cation-exchange characteristics of vermiculite (weathered biotite) isolated from a Gray Luvisol in northeastern Saskatchewan. Clays Clay Miner. 37, 1–11.

Ghosez, P.H., Michenaud, J.-P., Gonze, X., 1998. Dynamical atomic charges: the case of ABO3 compounds. Phys. Rev. B 58, 6224–6240.

Giannozzi, P., Baroni, S., Bonini, N., Calandra, M., Car, R., Cavazzoni, C., Ceresoli, D., Chiarotti, G.L., Cococcioni, M., Dabo, I., Dal Corso, A., De Gironcoli, S., Fabris, S., Fratesi, G., Gebauer, R., Gerstmann, U., Gougoussis, C., Kokalj, A., Lazzeri, M., Martin-Samos, L., Marzari, N., Mauri, F., Mazzarello, R., Paolini, S., Pasquarello, A., Paulatto, L., Sbraccia, C., Scandolo, S., Sclauzero, G., Seitsonen, A.P., Smogunov, A., Umari, P., Wentzcovitch, R.M., 2009. Quantum ESPRESSO: a modular and open-source software project for quantum simulations of materials. J. Phys.: Condens. Matter 21, 395502.

Giese, R.F., 1988. Kaolin minerals: structures and stabilities. In: Baily, S.W. (Ed.), Reviews in Mineralogy. Hydrous Phyllosilicates, vol. 19. Mineralogical Society of America, BookCrafters Inc, Chelsea, Michigan.

Giese, R.F., 1990. Kaolin minerals: structures and stabilities. In: Bailey, S.W. (Ed.), Hydrous Phyllosilicates (Exclusive of Micas). Reviews in Mineralogy, vol. 19. Mineralogical Society of America, Washington, DC.

Giese, R.F., van Oss, C.J., 1993. The surface thermodynamic properties of silicates and their interactions with biological materials. In: Guthrie, C.D., Mossman, B.T. (Eds.), Health Effects of Mineral Dusts. Mineralogical Society of America, Washington, DC, pp. 327–346.

Giese, R.F., van Oss, C.J., 2002. Organophilicity and hydrophobicity of organo-clays. In: Yariv, S., Cross, H. (Eds.), Organo-Clay Complexes and Interactions. Marcel Dekker, Inc., New York, pp. 175–189 (Chapter 4).

Gil, A., Gandia, L.M., Vicente, M.A., 2000. Recent advances in the synthesis and catalytic applications of pillared clays. Catal. Rev. Sci. Eng. 42, 145–212.

Gionis, V., Kacandes, G.H., Kastritis, I.D., Chryssikos, G.D., 2006. On the structure of palygorskite by mid- and near-infrared spectroscopy. Am. Mineral. 91, 1125–1133.

Gionis, V., Kacandes, G.H., Kastritis, I.D., Chryssikos, G.D., 2007. Combined near-infrared and X-ray diffraction investigation of the octahedral sheet composition of palygorskite. Clays Clay Miner. 55, 543–553.

Goetz, A.F.H., Curtiss, B., Shiley, D.A., 2009. Rapid gangue mineral concentration measurement over conveyors by NIR reflectance spectroscopy. Miner. Eng. 22, 490–499.

Goh, T.B., Huang, P.M., 1986. Influence of citric and tannic acids on hydroxy-Al interlayering in montmorillonite. Clays Clay Miner. 34, 37–44.

Gomez, C., Viscarra Rossel, R.A., McBratney, A.B., 2008. Soil organic carbon prediction by hyperspectral remote sensing and field vis-NIR spectroscopy: an Australian case study. Geoderma 146, 43–411.

Gonzalez, F., Pesquera, C., Benito, I., Mendioroz, S., Poncelet, G., 1992. High conversion and selectivity for cracking of n-heptane on cerium-aluminium montmorillonite catalysts. J. Chem. Soc., Chem. Comm., 491-493.

González-Martín, I., González-Pérez, C., Hernández-Méndez, J., Alvarez-García, N., 2002. Mineral analysis (Fe, Zn, Ca, Na, K) of fresh Iberian pork loin by near infrared reflectance spectrometer: determination of Fe, Na and K with a remote fibre-optic reflectance probe. Anal. Chim. Acta 468, 293–301.

Gonze, X., Lee, C., 1997. Dynamical matrices, Born effective charges, dielectric permittivity tensors, and interatomic force constants from density-functional perturbation theory. Phys. Rev. B 55, 10355–10368.

Gonze, X., Rignanese, G.-M., Verstraete, M., Beuken, J.-M., Pouillon, Y., Caracas, R., Jollet, F., Torrent, M., Zerah, G., Mikami, M., Ghosez, P.H., Veithen, M., Raty, J.-Y., Olevano, V., Bruneval, F., Reining, L., Godby, R., Onida, G., Hamann, D.R., Allan, D.C., 2005. A brief introduction to the ABINIT software package. Z. Kristallogr. 220, 558–562.

Gooding, J.L., 1992. Soil mineralogy and chemistry on Mars: possible clues from salts and clays in SNC meteorites. Icarus 99, 28–41.

Gooding, J.L., Wentworth, S.J., Zolensky, M.E., 1991. Aqueous alteration of the Nakhla meteorite. Meteoritics 26, 135–143.

Goodman, B.A., Russell, J.D., Fraser, A.R., 1976. A Mössbauer and I.R. spectroscopy study of the structure of nontronite. Clays Clay Miner. 24, 53–59.

Goodman, A.L., Campus, L.M., Schroeder, K.T., 2005. Direct evidence of carbon dioxide sorption in Argonne premium coals using attenuated total reflectance—Fourier transform spectroscopy. Energy Fuels 19, 471–476.

Gordon, R., 1956. The emissivity of transparent materials. J. Am. Ceram. Soc. 39, 278–287.

Graettinger, A.H., Ellis, M.K., Skilling, I.P., Reath, K., Ramsey, M.S., Lee, R.J., Hughes, C.G., Mcgarvie, D.W., 2013. Remote sensing and geologic mapping of glaciovolcanic deposits in the region surrounding Askja (Dyngjufjöll) volcano, Iceland. Int. J. Remote Sens. 34, 7178–7198.

Grauby, O., Petit, S., Decarreau, A., Baronnet, A., 1993. The beidellite-saponite series: an experimental approach. Eur. J. Mineral. 5, 623–635.

Grauby, O., Petit, S., Decarreau, A., Baronnet, A., 1994. The nontronite-saponite series: an experimental approach. Eur. J. Mineral. 6, 99–112.

Greathouse, J.A., Cygan, R.T., 2013. Chapter 3—Molecular simulation of clay minerals. In: Bergaya, F., Lagaly, G. (Eds.), Developments in Clay Science. Elsevier.

Greathouse, J.A., Cygan, R.T., Fredrich, J.T., Jerauld, G.R., 2016. Molecular dynamics simulation of diffusion and electrical conductivity in montmorillonite interlayers. J. Phys. Chem. C 120, 1640–1649.

Green, R.O., Conel, J.E., Carrere, V., Bruegge, C.J., Margolis, J.S., Rast, M., Hoover, G., 1990. Determination of the in-flight spectral and radiometric characteristics of the Airborne Visible/Infrared Imaging Spectrometer (AVIRIS). In: Green, R.O. (Ed.), Proc. 3rd AVIRIS Workshop, JPL Publication 91-28. Jet Propulsion Laboratory, Pasadena, CA.

Griffiths, P.R., 1972. Infrared emission spectroscopy. I. Basic considerations. Appl. Spectrosc. 26, 73–76.

Griffiths, P.R., 2002. Beer's Law. In: Chalmers, J.M., Griffiths, P.R. (Eds.), Handbook of Vibrational Spectroscopy, vol. 3. John Wiley & Sons, Chichester, pp. 2225–2234.

Griffiths, P.R., De Haseth, J.A., 2007. Fourier Transform Infrared Spectrometry, second ed., John Wiley and Sons. ISBN: 978 0 471 19404 0.

Grootveld, M., 2015. Introduction to the applications of chemometric techniques in 'Omics' research: common pitfalls, misconceptions and 'rights and wrongs'. In: Grootveld, M., Issues in Toxicology No.21. Metabolic Profiling: Disease and Xenobiotics. The Royal Society of Chemistry, pp. 1–33.

Groppo, C., Rinaudo, C., Cairo, S., Gastaldi, D., Compagnoni, R., 2006. Micro-Raman spectroscopy for a quick and reliable identification of serpentine minerals from ultramafics. Eur. J. Mineral. 18, 319–329.

Grotzinger, J.P., Gupta, S., Malin, M.C., Rubin, D.M., Schieber, J., Siebach, K., Sumner, D.Y., Stack, K.M., Vasavada, A.R., Arvidson, R.E., Calef, F., Edgar, L., Fischer, W.F., Grant, J.A., Griffes, J., Kah, L.C., Lamb, M.P., Lewis, K.W., Mangold, N., Minitti, M.E., Palucis, M., Rice, M., Williams, R.M.E., Yingst, R.A., Blake, D., Blaney, D., Conrad, P., Crisp, J., Dietrich, W.E., Dromart, G., Edgett, K.S., Ewing, R.C., Gellert, R., Hurowitz, J.A., Kocurek, G., Mahaffy, P., McBride, M.J., Mclennan, S.M., Mischna, M., Ming, D., Milliken, R., Newsom, H., Oehler, D., Parker, T.J., Vaniman, D., Wiens, R.C., Wilson, S.A., 2015. Deposition, exhumation, and paleoclimate of an ancient lake deposit, Gale crater, Mars. Science 350.

Grotzinger, J.P., Sumner, D.Y., Kah, L.C., Stack, K., Gupta, S., Edgar, L., Rubin, D., Lewis, K., Schieber, J., Mangold, N., Milliken, R., Conrad, P.G., Desmarais, D., Farmer, J., Siebach, K., Calef, F., Hurowitz, J., Mclennan, S.M., Ming, D., Vaniman, D., Crisp, J., Vasavada, A., Edgett, K.S., Malin, M., Blake, D., Gellert, R., Mahaffy, P., Wiens, R.C., Maurice, S., Grant, J.A., Wilson, S., Anderson, R.C., Beegle, L., Arvidson, R., Hallet, B., Sletten, R.S., Rice, M., Bell, J., Griffes, J., Ehlmann, B., Anderson, R.B., Bristow, T.F., Dietrich, W.E., Dromart, G., Eigenbrode, J., Fraeman, A., Hardgrove, C., Herkenhoff, K., Jandura, L., Kocurek, G., Lee, S., Leshin, L.A., Leveille, R., Limonadi, D., Maki, J., McCloskey, S., Meyer, M., Minitti, M., Newsom, H., Oehler, D., Okon, A., Palucis, M., Parker, T., Rowland, S., Schmidt, M.,

Squyres, S., Steele, A., Stolper, E., Summons, R., Treiman, A., Williams, R., Yingst, A., Team, M.S., Kemppinen, O., Bridges, N., Johnson, J.R., Cremers, D., Godber, A., Wadhwa, M., Wellington, D., McEwan, I., Newman, C., Richardson, M., Charpentier, A., Peret, L., King, P., Blank, J., Weigle, G., Li, S., Robertson, K., Sun, V., Baker, M., Edwards, C., Farley, K., Miller, H., Newcombe, M., Pilorget, C., Brunet, C., Hipkin, V., Léveillé, R., et al., 2014. A habitable Fluvio-Lacustrine environment at Yellowknife Bay, Gale Crater, Mars. Science 343.

Gualtieri, A.F., Ferrari, S., Leoni, M., Grathoff, G., Hugo, R., Shatnawi, M., Paglia, G., Billinge, S., 2008. Structural characterization of the clay mineral illite-1M. J. Appl. Crystallogr. 41, 402–415.

Guggenheim, S., Eggleton, R.A., 1988. Crystal chemistry, classification, and identification of modulated layer silicates. In: Bailey, S.W. (Ed.), Hydrous Phyllosilicates (Exclusive of the Micas). Reviews in Mineralogy, vol. 20. Mineralogical Society of America, Washington, DC, pp. 675–725.

Guinness, E.A., Arvidson, R.E., Bradley, J.P., Jolliff, B.L., Seelos, K.D., Seelos, F.P., Ming, D.W., Morris, R.V., Graff, T.G., 2007. Hyperspectral reflectance mapping of cinder cones at the summit of Mauna Kea and implications for equivalent observations on Mars. J. Geophys. Res. 112. doi:1029/2006JE002822.

Gunnarsson, I., Arnorsson, S., Jakobsson, S., 2005. Precipitation of poorly crystalline antigorite under hydrothermal conditions. Geochim. Cosmochim. Acta 69, 2813–2828.

Günzler, H., Gremlich, H.-U., 2002. IR Spectroscopy. An Introduction. Wiley Verlag Gmbh, Weinheim, Germany. p. 361.

Gupta, G.C., Malik, W.U., 1969a. Fixation of hydroxy-aluminum by montmorillonite. Am. Mineral. 54, 1625–1634.

Gupta, G.C., Malik, W.U., 1969b. Chloritization of montmorillonite by its coprecipitation with magnesium hydroxide. Clays Clay Miner. 17, 331–338.

Gupta, G.C., Malik, W.U., 1969c. Transformation of montmorillonite to nickel-chlorite. Clays Clay Miner. 17, 233–239.

Ha, J.-H., Kim, Y.S., Hochstrasser, R.M., 2006. Vibrational dynamics of N–H, C–D, and C·O modes in formamide. J. Chem. Phys. 124. 064508.

Hadni, A., 1974. The interaction of infrared radiation with crystals. In: Farmer, V.C. (Ed.), The Infrared Spectra of Minerals. Mineralogical Society, London, pp. 27–49.

Haley, L.V., Wylie, I.W., Koningstein, J.A., 1982. An investigation of the lattice and interlayer water vibrational spectral regions of muscovite and vermiculite using Raman microscopy. A Raman microscopic study of layer silicates. J. Raman Spectrosc. 13, 203–205.

Hamilton, V.E., Morris, R.V., Gruener, J.E., Mertzman, S.A., 2008b. Visible, near-infrared, and middle infrared spectroscopy of altered basaltic tephras: spectral signatures of phyllosilicates, sulfates, and other aqueous alteration products with application to the mineralogy of the Columbia Hills of Gusev Crater, Mars. J. Geophys. Res.: Planets, 113. https://doi.org/10.1029/2007JE003049.

Hamilton, V.E., Christensen, P.R., Bandfield, J.L., Rogers, A.D., Edwards, C.S., 2018a. Chapter 25. Thermal infrared spectral analyses of Mars from orbit using TES and THEMIS. In: Bishop, J.L., Moersch, J.E., Bell J.F. III (Eds.), Remote Compositional Analysis: Techniques for Understanding Spectroscopy, Mineralogy, and Geochemistry of Planetary Surfaces. Cambridge University Press, Cambridge, UK.

Han, B., Dong, Q., Chen, J., Feng, W., Liu, S., Wang, H.-I., 2015. Photocatalytic degradation of methyl orange over $Y^{3+}$ doped $TiO_2$ pillared montmorillonite. J. Adv. Oxid. Technol. 18, 98–104.

Hanesch, M., 2009. Raman spectroscopy of iron oxides and (oxy)hydroxides at low laser power and possible applications in environmental magnetic studies. Geophys. J. Int. 177, 941–948.

Hansen, E.L., Larsen, N.W., Nicolaisen, F.-M., 1980. An infrared investigation of formamide, acetamide and thioacetamide in the vapour phase. Inversion of the amino group. Chem. Phys. Lett. 69, 327–331.

Hara, M., Inukai, J., Miyatake, K., Uchida, H., Watanabe, M., 2011. Temperature dependence of the water distribution inside a Nafion membrane in an operating polymer electrolyte fuel cell. A micro-Raman study. Electrochim. Acta 58, 449–455.

Hargreaves, M.D., Page, K., Munshi, T., Tomsett, R., Lynch, G., Edwards, H.G.M., 2008. Analysis of seized drugs using portable Raman spectroscopy in an airport environment—a proof of principle study. J. Raman Spectrosc. 39, 873–880.

Harrick, N.J., 1967. Internal Reflection Spectroscopy. Interscience Publishers, New York, London, Sydney. ISBN 0470352507.

Hasegawa, T., 2006. Principal component regression and partial least squares modeling. In: Chalmers, J.M., Griffiths, P.R. (Eds.), Handbook of Vibrational Spectroscopy, vol. 3. John Wiley & Sons, Chichester, pp. 2293–2312.

Haskouri, J.E., Cabrera, S., Gutierrez, M., Porter, A.B., Porter, D.B., Marcos, M.D., Amoros, P., 2001. Very high titanium content mesoporous silicas. Chem. Commun. 309–310.

Hassan, M.S., Salem, S.M., 2001. Distribution and influence of iron phases on the physico-chemical properties of phyllosilicates. Chin. J. Geochem. 20, 120–129.

Hassan, M.S., Baioumy, H.M., 2006. Structural and chemical alteration of glauconite under progressive acid treatment. Clays Clay Miner. 54, 491–499.

Hayashi, S., 1995. NMR study of dynamics of dimethyl sulfoxide molecules in kaolinite/dimethyl sulfoxide intercalation compound. J. Phys. Chem. 99, 7120–7129.

Hayashi, S., Akiba, E., 1994. Interatomic distances in layered silicates and their intercalation compounds as studied by cross polarization NMR. Chem. Phys. Lett. 226, 495–500.

Hazen, R.M., 2014. Data-driven abductive discovery in mineralogy. Am. Mineral. 99, 2165–2170.

Hazen, R.M., Ferry, J.M., 2010. Mineral evolution: mineralogy in the fourth dimension. Elements 6, 9–12.

He, H.P., Frost, R.L., Zhu, J.X., 2004a. Infrared study of HDTMA+ intercalated montmorillonite. Spectrochim. Acta Part A 60, 2853–2859.

He, H.P., Frost, R.L., Xi, Y.F., Zhu, J.X., 2004b. Raman spectroscopic study of organo-montmorillonites. J. Raman Spectrosc. 35, 316–323.

He, H.P., Ma, Y.H., Zhu, J.X., Yuan, P., Qing, Y.H., 2010. Organoclays prepared from montmorillonites with different cation exchange capacity and surfactant configuration. Appl. Clay Sci. 48, 67–72.

He, H.P., Frost, R.L., Deng, F., Zhu, J.X., Wen, X.Y., Yuan, P., 2004c. Conformation of surfactant molecules in the interlayer of montmorillonite studied by $^{13}C$ MAS NMR. Clays Clay Miner. 52, 350–356.

He, H.P., Tao, Q., Zhu, J.X., Yuan, P., Shen, W., Yang, S.Q., 2013. Silylation of clay mineral surfaces. Appl. Clay Sci. 71, 15–20.

He, H.P., Ma, L.Y., Zhu, J.X., Frost, R.L., Theng, B.K.G., Bergaya, F., 2014. Synthesis of organoclays: a critical review and some unresolved issues. Appl. Clay Sci. 100, 22–28.

He, H.P., Ding, Z., Zhu, J.X., Yuan, P., Xi, Y.F., Yang, D., Frost, R.L., 2005. Thermal characterization of surfactant-modified montmorillonites. Clays Clay Miner. 53, 287–293.

He, H.P., Frost, R.L., Bostrom, T., Yuan, P., Duong, L., Yang, D., Yunfel, X.F., Kloprogge, J.T., 2006a. Changes in the morphology of organoclays with $HDTMA^+$ surfactant loading. Appl. Clay Sci. 31, 262–271.

He, H.P., Zhou, Q., Martens, W.N., Kloprogge, T.J., Yuan, P., Yunfei, X.F., Zhu, J.X., Frost, R.L., 2006b. Microstructure of $HDTMA^+$-modified montmorillonite and its influence on sorption characteristics. Clays Clay Miner. 54, 689–696.

Heinz, H., Vaia, R.A., Krishnamoorti, R., Farmer, B.L., 2007. Self-assembly of alkylammonium chains on montmorillonite: effect of chain length, head group structure, and cation exchange capacity. Chem. Mater. 19, 59–68.

Heller-Kallai, L., 2013. Thermally modified clay minerals. In: Bergaya, F., Lagaly, G. (Eds.), Handbook of Clay Science, Developments in Clay Science Series, vol. 5A, second ed. Elsevier, Oxford, pp. 411–433.

Heller-Kallai, L., Lapides, I., 2015. Dehydroxylation of muscovite: study of quenched samples.

Helsen, J., 1982. Clay minerals as solid acids and their catalytic properties. J. Chem. Educ. 59, 1063–1065.

Hendricks, S.B., 1929. Diffraction of X-radiation from some crystalline aggregates. Z. Kristallogr. 71, 273–275.

Hembree, D.M., Smyrl, H.R., 1989. Anomalous dispersion effects in diffuse reflectance infrared Fourier transform spectroscopy: a study of optical geometries. Appl. Spectrosc. 43, 267–274.

Henmi, T., Tange, K., Minagawa, T., Yoshinaga, N., 1981. Effect of $SiO_2$-$Al_2O_3$ ratio on the thermal-reactions of allophane. 2. Infrared and X-ray-powder diffraction data. Clays Clay Miner. 29, 124–128.

Herbillon, A.J., Mestdagh, M.M., Vielvoye, L., Derouane, E.G., 1976. Iron in kaolinite with special reference to kaolinite from tropical soils. Clay Miner. 11, 201–220.

Herrmann, W., Blake, M., Doyle, M., Huston, D., Kamprad, J., Merry, N., Pontual, S., 2001. Short wavelength infrared (SWIR) spectral analysis of hydrothermal alteration zones associated with base metal sulfide deposits at Rosebery and Western Tharsis, Tasmania, and Highway-Reward, Queensland. Econ. Geol. 96, 939–955.

Hicks, L.J., Bridges, J.C., Gurman, S.J., 2014. Ferric saponite and serpentine in the nakhlite martian meteorites. Geochim. Cosmochim. Acta 136, 194–210.

Higashi, S., Miki, K., Komarneni, S., 2002. Hydrothermal synthesis of Zn-smectites. Clays Clay Miner. 50, 299–305.

Hillier, S., Brydson, R., Delbos, E., Fraser, T., Gray, N., Pendlowski, H., Phillips, I., Robertson, J., Wilson, I., 2016. Correlations among the mineralogical and physical properties of halloysite nanotubes (HNTs). Clay Miner. 51, 325–350.

Hirschfeld, T., Chase, B., 1986. FT-Raman spectroscopy—development and justification. Appl. Spectrosc. 40, 133–137.

Hochella, M.F., 2008. Nanogeoscience: from origins to cutting-edge applications. Elements 4, 373–379.

Hochella, M.F., Lower, S.K., Maurice, P.A., Penn, R.L., Sahai, N., Sparks, D.L., Twining, B.S., 2008. Nanominerals, mineral nanoparticles, and Earth systems. Science 319, 1631–1635.

Hofmann, U., Klemen, R., 1950. Verlust der Austauschfähigkeit von Lithiumionen an Bentonit durch Erhitzung. Z. Anorg. Allg. Chem. 262, 95–99.

Hofmeister, A.M., Bowey, J.E., 2006. Quantitative infrared spectra of hydrosilicates and related minerals. Mon. Not. R. Astron. Soc. 367, 577–591.

Hofmeister, A.M., Keppel, E., Speck, A.K., 2003. Absorption and reflection spectra of MgO and other diatomic compounds. Mon. Not. R. Astron. Soc. 345, 16–38.

Hogg, C.S., Malden, P.J., Meads, R.E., 1975. Identification of iron-containing impurities in natural kaolinites using the Mössbauer effect. Mineral. Mag. 40, 89–96.

Holtz, M., Solin, S.A., Pinnavaia, T.J., 1993. Effect of pressure on the Raman vibrational modes of layered aluminosilicate compounds. Phys. Rev. B Condens. Matter 48, 13312–13317.

Holtz, M., Solin, S.A., Pinnavaia, T.J., 1994. Pressure-Raman studies of layered alumino-silicate compounds. AIP Conf. Proc. 309, 1539–1542.

Hongo, T., Yoshino, S., Yamazaki, A., Yamasaki, A., Satokawa, S., 2012. Mechanochemical treatment of vermiculite in vibration milling and its effect on lead (II) adsorption ability. Appl. Clay Sci. 70, 74–78.

Honnorez, J., 1981. The aging of the oceanic crust at low temperature. In: Emiliani, C. (Ed.), The Oceanic Lithosphere. John Wiley, New York.

Horgan, B., Bell, J.F., 2012. Widespread weathered glass on the surface of Mars. Geology. https://doi.org/10.1130/G32755.1.

Horváth, E., Frost, R.L., Makó, É., Kristóf, J., Cseh, T., 2003. Thermal treatment of mechanochemically activated kaolinite. Thermochim. Acta 404, 227–234.

Hotelling, H., 1931. The generalization of Student's ratio. Ann. Math. Stat. 2 (3), 360–378.

Hou, M.-F., Ma, C.-X., Zhang, W.-D., Tang, X.-Y., Fan, Y.-N., Wan, H.-F., 2011. Removal of rhodamine B using iron-pillared bentonite. J. Hazard. Mater. 186, 1118–1123.

House, J.E., 2013. Chapter 7—Ionic bonding and structures of solids. In: House, J.E. (Ed.), Inorganic Chemistry, second ed. Academic Press.

Hrachová, J., Komadel, P., Fajnor, V.Š., 2007a. The effect of mechanical treatment on the structure of montmorillonite. Mater. Lett. 61, 3361–3365.

Hrachová, J., Madejová, J., Billik, P., Komadel, P., Fajnor, V.Š., 2007b. Dry grinding of Ca and octadecyltrimethylammonium montmorillonite. J. Colloid Interface Sci. 316, 589–595.

Hrobáriková, J., Madejová, J., Komadel, P., 2001. Effect of heating temperature on Li-fixation, layer charge and properties of fine fractions of bentonites. J. Mater. Chem. 11, 1452–1457.

Hu, L., Ji, S., Jiang, Z., Song, H., Wu, P., Liu, Q., 2007. Direct synthesis and structural characteristics of ordered SBA-15 mesoporous silica containing tungsten oxides and tungsten carbides. J. Phys. Chem. C 111, 15173–15184.

Huang, J., Romero-Torres, S., Moshgbar, M., 2010. Practical considerations in data pretreatment for NIR and Raman spectroscopy. American Pharmaceutical Review, http://www.americanpharmaceuticalreview.com/Featured-Articles/116330-Practical-Considerations-in-Data-Pre-treatment-for-NIR-and-Raman-Spectroscopy/.

Huertas, F.J., Huertas, F., Linares, J., 1993. Hydrothermal synthesis of kaolinite: method and characterization of synthetic materials. Appl. Clay Sci. 7, 345–356.

Huertas, F.J., Cuadros, J., Huertas, F., Linares, J., 2000. Experimental study of the hydrothermal formation of smectite in the beidellite-saponite series. Am. J. Sci. 300, 504–527.

Huheey, J.E., Keiter, E.A., Keiter, R.L., 1997. Inorganic Chemistry: Principles of Structure and Reactivity. Prentice Hall, New York.

Hunt, G.R., 1980. Modification of integrating sphere accessory to allow spectroscopic measurements of horizontal surfaces from above. Appl. Opt. 19, 1746–1747.

Hunt, G.R., Ross, H.P., 1967. A bidirectional reflectance accessory for spectroscopic measurements. Appl. Opt. 6, 1687–1690.

Hunt, G.R., Salisbury, J.W., 1970. Visible and infrared spectra of minerals and rocks: I Silicate minerals. Mod. Geol. 1, 283–300.

Hunt, G.R., Salisbury, J.W., Lenhoff, C.J., 1973. Visible and near infrared spectra of minerals and rocks: VI. Additional silicates. Mod. Geol. 4, 85–106.

Hvistendhal, J., Rytter, E., Øye, H.A., 1983. IR emission spectroscopy of molten salts and other liquids using thick samples as reference. Appl. Spectrosc. 37, 182–187.

Iglesias, J.E., Ocana, M., Serna, C.J., 1990. Aggregation and matrix effects on the infrared spectrum of microcrystalline powders. Appl. Spectrosc. 44, 418–426.

Iishi, M., Shimanouchi, T., Nakahira, M., 1967. Far infrared absorption spectra of layer silicates. Inorg. Chim. Acta 1, 387–392.

Iriarte, I., Petit, S., Huertas, F.J., Fiore, S., Grauby, O., Decarreau, A., Linares, J., 2005. Synthesis of kaolinite with a high level of Fe3+ for Al substitution. Clays Clay Miner. 53, 1–10.

Ishida, K., 1990. Identification of infrared hydroxyl librational bands of talc-willemseite solid solutions and aluminum(IV)-free amphiboles through deuteration. Mineral. J. 15, 93–104.

Ishii, R., Nakatsuji, M., Ooi, K., 2005. Preparation of highly porous silica nanocomposites from clay mineral: a new approach using pillaring method combined with selective leaching. Microporous Mesoporous Mater. 79, 111–119.

Itoh, K., Shimanouchi, T., 1972. Vibrational spectra of crystalline formamide. J. Mol. Spectrosc. 42, 86–99.

Jaber, M., Komarneni, S., Zhou, C.-H., 2013. Synthesis of clay minerals. In: Bergaya, F., Lagaly, G. (Eds.), Handbook of Clay Science, second ed. Elsevier, Oxford, pp. 223–241.

Jackson, R.S., 2002. Continuous scanning interferometers for mid-infrared spectrometry. In: Chalmers, J.M., Griffiths, P.R. (Eds.), Handbook of Vibrational Spectroscopy, vol. 1. John Wiley & Sons, Chichester, pp. 264–282.

Jakobsson, S.P., 1978. Environmental factors controlling the palagonitization of the Surtsey tephra, Iceland. Bull. Geol. Soc. Den. 27, 91–105.

Janek, M., Lagaly, G., 2003. Interaction of a cationic surfactant with bentonite: a colloid chemistry study. Colloid Polym. Sci. 281, 293–301.

Janek, M., Bugár, I., Lorenc, D., Szöcs, V., Velič, D., Chorvát, D., 2009. Terahertz time-domain spectroscopy of selected layered silicates. Clays Clay Miner. 57, 416–424.

Janek, M., Matejdes, M., Szöcs, V., Bugár, I., Gaál, A., Velič, D., Darmo, J., 2010. Dielectric properties of micaceous clays determined by terahertz time-domain spectroscopy. Philos. Mag. 90, 14–28.

Jaumann, R., Nass, A., Tirsch, D., Reiss, D., Neukum, G., 2010. The Western Libya Montes Valley System on Mars: evidence for episodic and multi-genetic erosion events during the Martian history. Earth Planet. Sci. Lett. 294, 272–290.

Jaynes, W.F., Bigham, J.M., 1987. Charge reduction, octahedral charge, and lithium retention in heated. Li-saturated smectites. Clays Clay Miner. 35, 440–448.

Jaynes, W.F., Boyd, S.A., 1991. Hydrophobicity of siloxane surfaces in smectites as revealed by aromatic hydrocarbon adsorption from water. Clays Clay Miner. 39, 428–436.

Jaynes, W.F., Traina, S.J., Bigham, J.M., Johnston, C.T., 1992. Preparation and characterization of reduced-charge hectorites. Clays Clay Miner. 40, 397–404.

Jie, D., Xie, L., Fu, X., Rao, X., Ying, Y., 2013. Variable selection for partial least squares analysis of soluble solids content in watermelon using near-infrared diffuse transmission technique. J. Food Eng. 118, 387–392.

Johansson, U., Frost, R.L., Forsling, W., Kloprogge, J.T., 1998. Raman spectroscopy of the kaolinite hydroxyls at 77 K. Appl. Spectrosc. 52, 1277–1282.

John, W.D., Sen Gupta, P.K., 1967. Vermiculite-alkylammonium complexes. Am. Mineral. 52, 1706–1724.

Johnston, C.T., 1989. Raman and FT-IR spectra of the kaolinite-hydrazine intercalate. ACS Symp. Ser. 415, 432–454.

Johnston, C.T., 2010. Probing the nanoscale architecture of clay minerals. Clay Miner. 45, 245–279.

Johnston, C.T., Stone, D.A., 1990. Influence of hydrazine on the vibrational modes of kaolinite. Clays Clay Miner. 38, 121–128.

Johnston, C.T., Aochi, Y.O., 1996. Fourier transform infrared and Raman spectroscopy. In: Sparks, D.L. (Ed.), Methods of Soil Analysis, Part 3: Chemical Methods. SSSA Book Series No. 5, third ed. Soil Science Society of America, Madison, pp. 269–321.

Johnston, C.T., Premachandra, G.S., 2001. Polarized ATR-FTIR study of smectite in aqueous suspension. Langmuir 17, 3712–3718.

Johnston, C.T., Wang, S.L., 2002. Applications of vibrational spectroscopy in soil and environmental sciences. In: Chalmers J.M., Griffiths P.R. (Eds.), Handbook of Vibrational Spectroscopy. Applications in Industry, Materials and the Physical Sciences. J. Wiley & Sons, UK, pp. 3192–3204.

Johnston, C.T., Sposito, G., Birge, R.R., 1985. Raman spectroscopic study of kaolinite in aqueous suspension. Clays Clay Miner. 33, 483–489.

Johnston, C.T., Agnew, S.F., Bish, D.L., 1990. Polarized single-crystal Fourier-transform infrared microscopy of Ouray dickite and Keokuk kaolinite. Clays Clay Miner. 38, 573–583.

Johnston, C.T., Sposito, G., Erickson, C., 1992a. Vibrational probe studies of water interactions with montmorillonite. Clays Clay Miner. 40, 722–730.

Johnston, C.T., Sposito, G., Earl, W.L., 1993. Surface spectroscopy of environmental particles by Fourier transform infrared and nuclear magnetic resonance spectroscopy. In: Buffle, J., Van Leeuwen, H.P. (Eds.) Environmental Particles. Volume 2 in the Environmental Analytical and Physical Chemistry Series. Lewis Publ, Boca Raton.

Johnston, C.T., Sposito, G., Bocian, D.F., Birge, R.R., 1984. Vibrational spectroscopic study of the interlamellar kaolinite-dimethyl sulfoxide complex. J. Phys. Chem. 88, 5959–5964.

Johnston, C.T., Helsen, J., Schoonheydt, R.A., Bish, D.L., 1998. Single-crystal Raman spectroscopic study of dickite. Am. Mineral. 83, 75–84.

Johnston, C.T., Bish, D.L., Eckert, J., Brown, L.A., 2000. Infrared and inelastic neutron scattering study of the 1.03- and 0.95-nm kaolinite-hydrazine intercalation complexes. J. Phys. Chem. B 104, 8080–8088.

Johnston, C.T., Tipton, T., Stone, D.A., Erickson, C., Trabue, S.L., 1991. Chemisorption of p-dimethoxybenzene on Cu-montmorillonite. Langmuir 7, 289–296.

Johnston, C.T., Tipton, T., Trabue, S.L., Erickson, C., Stone, D.A., 1992b. Vapor phase sorption of p-xylene on Co and Cu exchanged SAz-1 montmorillonite. Environ. Sci. Technol. 26, 382–390.

Johnston, C.T., Elzea Kogel, J., Bish, L., Kogure, T., Murray, H., 2008. Low-temperature FTIR study of kaolin-group minerals. Clays Clay Miner. 56, 470–485.

Johnston, C.T., Wang, S.L., Bish, D.L., Dera, P., Agnew, S.F., Kenney, J.W., 2002a. Novel pressure-induced phase transformations in hydrous layered materials. Geophys. Res. Lett. 29, art-1770.

Johnston, C.T., Wang, S.L., Bish, D.L., Dera, P., Agnew, S., Kenney J.W. III 2002b. Novel pressure-induced phase transformations in hydrous layered materials. Geophys. Res. Lett. 29, 17-1–17-4.

Jolicoeur, C., 1981. Infrared and thermogravimetric studies of the thermal degradation of chrysotile asbestos fibres: evidence for matrix effects. Can. J. Chem. 59, 1521–1526.

Jones, B.F., Galan, E., 1988. Sepiolite and palygorskite. Rev. Miner. 19, 631–674.

Jonson, B., Rebenstorf, B., Larsson, R., Primet, M., 1986. Infrared emission spectroscopic studies of metal oxide catalysts. Appl. Spectrosc. 40, 798–803.

Jourdan, S., Bernet, M., Schwartz, S., Guillot, S., Tricart, P., Chauvel, C., Dumont, T., Montagnac, G., Bureau, S., 2012. Tracing the Oligocene-Miocene evolution of the western

Alps drainage divide with pebble petrology, geochemistry, and Raman spectroscopy of foreland basin deposits. J. Geol. 120, 603–624.

Joussein, E., Petit, S., Decarreau, A., 2001. Une nouvelle méthode de dosage des minéraux argileux en mélange par spectroscopie IR. C. R. Académie des Sciences, Paris. Sciences de la terre et des planètes/Earth Planet. Sci. 332, 83–89.

Joussein, E., Petit, S., Churchman, G.J., Theng, B., Righi, D., Delvaux, B., 2005. Halloysite clay minerals—a review. Clay Miner. 40, 383–426.

Jovic-Jovicic, N., Milutinovic-Nikolic, A., Bankovic, P., Mojovic, Z., Zunic, M., Grzetic, I., Jovanovic, D., 2010. Organo-inorganic bentonite for simultaneous adsorption of Acid Orange 10 and lead ions. Appl. Clay Sci. 47, 452–456.

Kaklhana, M., Kotaka, M., Okamoto, M., 1983. Vibrational analysis of the acetate ion molecules and estimation of equilibrium constants for their hydrogen isotope exchange reactions. J. Phys. Chem. A 87, 2526–2535.

Kamitsos, E.I., 2015. Infrared spectroscopy of glasses. In: Affatigato, M. (Ed.), Modern Glass Characterization. John Wiley and Sons, Chapter 2, pp. 32–73.

Kanesaka, I., Mitsuhashi, E., 2008. A damped oscillator model for infrared spectra of formamide in the liquid state. Bull. Chem. Soc. Jpn. 81, 248–253.

Kapff, S.F., 1946. Infrared emission spectra of liquids. Science 105, 274–275.

Kapff, S.F., 1948. Infrared emission spectra of hot liquids. J. Chem. Phys. 16, 446–453.

Kar, P., Nayak, A., Bhoi, Y.P., Mishra, B.G., 2016. Preparation and catalytic application of sulfonated PVA-Zr-pillared clay nanocomposite materials towards one pot synthesis of hexahydropyrimidines. Microporous Mesoporous Mater. 223, 176–186.

Karakassides, M.A., Petridis, D., Gournis, D., 1997. Infrared reflectance of thermally treated Li- and Cs-montmorillonites. Clays Clay Miner. 45, 649–658.

Karakassides, M.A., Madejová, J., Arvaiová, B., Bourlinos, A., Petrides, D., Komadel, P., 1999. Location of Li(I), Cu(II) and Cd(II) in heated montmorillonite: evidence from specular reflectance infrared and electron spin resonance spectroscopies. J. Mater. Chem. 9, 1553–1558.

Karstang, T.V., Christy, A.A., Dahl, B., Kvalheim, O.M., 1991. Diffuse reflectance Fourier-transformed infrared spectroscopy in petroleum exploration: a multivariate approach to maturity determination. J. Geol. Explorat. 41, 213–226.

Katdare, S.P., Ramaswamy, V., Ramaswamy, A.V., 1999. Ultrasonication: a competitive method of intercalation for the preparation of alumina pillared montmorillonite catalyst. Catal. Today 49, 313–320.

Kaufhold, S., Hein, M., Dohrmann, R., Ufer, K., 2012. Quantification of the mineralogical composition of clays using FTIR spectroscopy. Vib. Spectrosc. 59, 29–39.

Kawano, K., Madhu, S.B., Mizutamari, M., Nakata, R., 2002. Paramagnetic properties of the spinel-type $Al_{1-x}Mn_x$-sulphate salts for its pillared intercalation into a clay mineral. J. Phys. Chem. Solids 64, 477–489.

Kawata, S., 2002. Instrumentation for NIR spectroscopy. In: Siesler, H.W., Ozaki, Y., Kawata, S., Heise, H.M. (Eds.) Near-Infrared Spectroscopy. Principles, Instruments, Applications. Wiley-VCH Verlag, Weinheim, Germany, Ch. 3, 43–73.

Keeling, J., Raven, M., Gates, W.P., 2000. Geology and preliminary characterisation of two nontronites from Uley graphite mine, South Australia. Clays Clay Miner. 48, 537–548.

Keiluweit, M., Kleber, M., 2009. Molecular-level interactions in soil and sediments: the role of aromatic π-systems. Environ. Sci. Technol. 43, 3421–3429.

Kelley, M.S., Wooden, D.H., 2009. The composition of dust in Jupiter-family comets inferred from infrared spectroscopy. Planet. Space Sci. 57, 1133–1145.

Kendrick, J., Burnett, A.D., 2016. PDielec: The calculation of infrared and terahertz absorption for powdered crystals. J. Comp. Chem. 16, 1491–1505.

Kerr, A., Rafuse, H., Sparkes, G., Hinchey, J., Sandeman, H., 2011. Visible/infrared spectroscopy (VIRS) as a research tool in economic geology: background and pilot studies from Newfoundland and Labrador. Current Research, Newfoundland and Labrador Department of Natural Resources, Geological Survey Report 11–1, 145–166.

Keuleers, R., Desseyn, H.O., Rousseau, B., Van Alsenoy, C., 1999. Vibrational analysis of urea. J. Phys. Chem. A 103, 4621–4630.

Khedr, M.Z., Arai, S., 2013. Origin of Neoproterozoic ophiolitic peridotites in south Eastern Desert, Egypt, constrained from primary mantle mineral chemistry. Miner. Petrol. 107, 807–828.

Kim, T.H., Lim, S.T., Lee, C.H., Choi, H.J., Jhon, M.S., 2003. Preparation and rheological characterization of intercalated polystyrene/organophilic montmorillonite nanocomposite. J. Appl. Polym. Sci. 87, 2106–2112.

King, T.V.V., Clark, R.N., 1989. Spectral characteristics of chlorites and Mg-serpentines using high-resolution reflectance spectroscopy. J. Geophys. Res. 94, 13,997–14,008.

King, T.V.V., Clark, R.N., Calvin, W.M., Sherman, D.M., Brown, R.H., 1992. Evidence for ammonium-bearing minerals on Ceres. Science 255, 1551–1553.

Kiricsi, I., Molnar, A., Palinko, I., Lazar, K., 1995. Synthesis of iron-containing montmorillonite by various methods. Characterization of the intercalants and the behavior of the intercalated substances in acid-catalyzed reactions. Stud. Surf. Sci. Catal. 94, 63–70.

Kiricsi, I., Molnar, A., Palinko, I., Fudala, A., Nagy, J.B., 1997. Nanoscale redox catalysts: Cr- and Cr, Al-pillared layer clays: characterization and catalytic activity. Solid State Ionics 101–103, 793–797.

Kirkpatrick, R.J., Kalinichev, A.G., Wang, J., Hou, X., Amonette, J.E., 2005. Molecular modeling of the vibrational spectra of interlayer and surface species of layered double hydroxides. In: Kloprogge, J.T. (Ed.), The Application of Vibrational Spectroscopy to Clay Minerals and Layered Double Hydroxides. CMS Workshop Lectures, vol. 13, The Clay Mineral Society, Aurora, CO, pp. 239–285.

Kitagawa, Y., Watanabe, Y., 1970. Preparation of dioctahedral vermiculite from muscovite. Clay Sci. 4, 31–36.

Kitamura, R., Pilon, L., Jonasz, M., 2007. Optical constants of silica glass from extreme ultraviolet to far infrared at near room temperature. Appli. Opt. 46, 8118–8133.

Kjeldhal, K., Bro, R., 2010. Some common misunderstandings in chemometrics. J. Chemometr. 24, 558–564.

Klapyta, Z., Fujita, T., Iyi, N., 2001. Adsorption of dodecyl- and octadecyltrimethylammonium ions on a smectite and synthetic micas. Appl. Clay Sci. 19, 5–10.

Klatte, S.J., Beck, T.L., 1993. Molecular-dynamics of tethered alkanes-temperature-dependent behavior in a high-density chromatographic system. J. Phys. Chem. 97, 5727–5734.

Kleppe, A.K., Jephcoat, A.P., Welch, M.D., 2003. The effect of pressure upon hydrogen bonding in chlorite: a Raman spectroscopic study of clinochlore to 26.5 GPa. Am. Mineral. 88, 567–573.

Klinkenberg, M., Dohrmann, R., Kaufhold, S., Stanjek, H., 2006. A new method for identifying Wyoming bentonite by ATR-FTIR. Appl. Clay Sci. 195–206.

Kloprogge, J.T., 1998. Synthesis of smectites and porous pillared clay catalysts: a review. J. Porous Mater. 5, 5–41.

Kloprogge, J.T., 2005. The Application of Vibrational Spectroscopy to Clay Minerals and Layered Double Hydroxides. CMS Workshop Lectures, vol. 13. The Clay Mineral Society, Aurora, CO, 285 pp.

Kloprogge, J.T., 2006. Spectroscopic studies of synthetic and natural beidellites: a review. Appl. Clay Sci. 31, 165–179.

Kloprogge, J.T., Frost, R.L., 1998. The dehydroxylation of basic aluminum sulfate: an Infrared Emission Spectroscopic Study. Thermochim. Acta 320, 245–252.

Kloprogge, J.T., Frost, R.L., 1999a. Raman microscopy study of basic aluminum sulfate. Part II. Raman microscopy at 77 K. J. Mater. Sci. 34, 4367–4374.

Kloprogge, J.T., Frost, R.L., 1999b. Raman microscopy study of basic aluminum sulfate. J. Mater. Sci. 34, 4199–4202.

Kloprogge, J.T., Frost, R.L., 1999c. Infrared emission spectroscopy of Al-pillared beidellite. Appl. Clay Sci. 15, 431–445.

Kloprogge, J.T., Frost, R.L., 1999d. Raman microprobe spectroscopy of hydrated halloysite from a Neogene cryptokarst from Southern Belgium. J. Raman Spectrosc. 30, 1079–1085.

Kloprogge, J.T., Frost, R.L., 1999e. An infrared emission spectroscopic study of synthetic and natural pyrophyllite. Neues Jahrb. Mineral., Monatsh. 62–74.

Kloprogge, J.T., Frost, R.L., 2000. The effect of synthesis temperature on the FT-Raman and FT-IR spectra of saponites. Vib. Spectrosc. 23, 119–127.

Kloprogge, J.T., Frost, R.L., 2001. Infrared emission spectroscopic study of the dehydroxylation of some natural and synthetic saponites. Neues Jahrb. Mineral., Monatsh. 446–463.

Kloprogge, J.T., Frost, R.L., 2005. Infrared emission spectroscopy of clay minerals. In: Kloprogge, J.T. (Ed.), The Application of Vibrational Spectroscopy to Clay Minerals and Layered Double Hydroxides. CMS Workshop Lectures, vol. 13. The Clay Mineral Society, Aurora, CO, pp. 99–124.

Kloprogge, J.T., Jansen, J.B.H., Geus, J.W., 1990a. Characterization of synthetic Na-beidellite. Clays Clay Miner. 38, 409–414.

Kloprogge, J.T., Frost, R.L., Hickey, L., 1999a. Infrared absorption and emission study of synthetic mica-montmorillonite in comparison to rectorite, beidellite and paragonite. J. Mater. Sci. Lett. 18, 1921–1923.

Kloprogge, J.T., Frost, R.L., Kristóf, J., 1999b. Complex expansion of kaolinite with hydrazine; some preliminar observations. Neues Jahrb. Mineral, Monatsh. 49–61.

Kloprogge, J.T., Frost, R.L., Rintoul, L., 1999c. Single crystal Raman microscopic study of the asbestos mineral chrysotile. Phys. Chem. Chem. Phys. 1, 2559–2564.

Kloprogge, J.T., Fry, R., Frost, R.L., 1999d. An infrared emission spectroscopic study of the thermal transformation mechanisms in $Al_{13}$-pillared clay catalysts with and without tetrahedral substitutions. J. Catal. 184, 157–171.

Kloprogge, J.T., Hickey, L., Frost, R.L., 1999e. Infrared study of some synthetic saponites: effect of $NH_4$/Al and $H_2O$/(Si+Al) ratios during the crystallization. J. Mater. Sci. Lett. 18, 1401–1403.

Kloprogge, J.T., Komarneni, S., Amonette, J.E., 1999f. Synthesis of smectite clay minerals: a critical review. Clays Clay Miner. 47, 529–554.

Kloprogge, J.T., Hickey, L., Frost, R.L., 2000a. The effect of increasing layer charge on the infrared absorption spectra of synthetic beidellites. J. Mater. Sci. Lett. 19, 1131–1134.

Kloprogge, J.T., Ruan, H., Frost, R.L., 2000b. Near-infrared spectroscopic study of synthetic and natural pyrophyllite. Neues Jahrb. Mineral., Monatsh. 337–347.

Kloprogge, J.T., Ruan, H., Frost, R.L., 2001a. Near-infrared spectroscopic study of basic aluminum sulfate and nitrate. J. Mater. Sci. 36, 603–607.

Kloprogge, J.T., van der Eerden, A.M.J., Jansen, J.B.H., Geus, J.W., 1990b. Hydrothermal synthesis of sodium-beidellite. Geol. Mijnbouw 69, 351–357.

Kloprogge, J.T., Geus, J.W., Jansen, J.B.H., Seykens, D., 1992a. Thermal stability of basic aluminum sulfate. Thermochim. Acta 209, 265–276.

Kloprogge, J.T., Jansen, J.B.H., Schuiling, R.D., Geus, J.W., 1992b. The interlayer collapse during dehydration of synthetic Na0.7-beidellite: a 23Na solid state Magic-Angle spinning NMR study. Clays Clay Miner. 40, 561–566.

Kloprogge, J.T., Breukelaar, J., Jansen, J.B.H., Geus, J.W., 1993. Development of ammonium-saponites from gels with variable ammonium concentration and water content at low temperatures. Clays Clay Miner. 41, 103–110.

Kloprogge, J.T., Dirken, P.J., Jansen, J.B.H., Geus, J.W., 1995. One- and two-dimensional $^{27}Al$ MAS NMR study of basic aluminum sulfate. J. Non-Cryst. Solids 181, 151–156.

Kloprogge, J.T., Komarneni, S., Yanagisawa, K., Frost, R.L., 1999g. Infrared emmission spectroscopic study of the dehydroxylatino via surface silanol groups of synthetic and natural beidellite. J. Colloid Interface Sci. 212, 562–569.

Kloprogge, J.T., Van Veen, J.A.R., Booij, E., Frost, R.L., 2001b. Comments on "Enhanced thermal stability of Al-pillared smectites modified with Ce and La" by J.L. Valverde, P. Canizares, M.R. Sun Kou, and C.B. Molina. Clays Clay Miner. 49, 183–185.

Kloprogge, J.T., Evans, R., Hickey, L., Frost, R.L., 2002. Characterisation and Al-pillaring of smectites from Miles, Queensland (Australia). Appl. Clay Sci. 20, 157–163.

Kloprogge, J.T., Komarneni, S., Yanagisawa, K., Frost, R.L., Fry, R., 1998. Infrared study of some synthetic and natural beidellites. J. Mater. Sci. Lett. 17, 1853–1855.

Kloprogge, J.T., Duong, L.V., Martens, W.N., van der Eerden, A., Frost, R.L., 2007. Mid- and near-infrared spectroscopy of hydrothermally synthesized 2:1 phyllosilicates in the system $Na_2O$-$Al_2O_3$-$SiO_2$-$H_2O$. In: Buckley, R.W. (Ed.), Progress in Solid State Chemistry Research. Nova Science Publishers, Hauppauge, NY, pp. 285–299.

Kodama, T., Komarneni, S., 1999. Na-4-mica: $Cd^{2+}$, $Ni^{2+}$, $Co^{2+}$, $Mn^{2+}$ and $Zn^{2+}$ ion exchange. J. Mater. Chem. 9, 533–539.

Kodama, T., Higuchi, T., Shimizu, T., Shimizu, K., Komarneni, S., Hoffbauer, W., Schneider, H., 2001. Synthesis of Na-2-mica from metakaolin and its cation exchange properties. J. Mater. Chem. 11, 2072–2077.

Kogure, T., Elzea-Kogel, J., Johnston, C.T., Bish, D.L., 2010. Stacking disorder in a sedimentary kaolinite. Clays Clay Miner. 58, 62–71.

Kohyama, N., Fukushima, K., Fukami, A., 1978. Observation of hydrated form of tubular halloysite by an electron-microscope equipped with an environmental cell. Clays Clay Miner. 26, 25–40.

Komadel, P., 2003. Chemically modified smectites. Clay Miner. 38, 127–138.

Komadel, P., 2016. Acid activated clays: materials in continuous demand. Appl. Clay Sci. 131, 84–99.

Komadel, P., Madejová, J., 2013. In: Bergaya, F., Lagaly, G. (Eds.), Handbook of Clay Science Part A: Fundamentals, second ed., Chap. 10.1. Elsevier Ltd., Amsterdam, pp. 386–409.

Komadel, P., Madejová, J., Stucki, J.W., 1995. Reduction and reoxidation of nontronite: question of reversibility. Clays Clay Miner. 39, 467–477.

Komadel, P., Madejová, J., Bujdák, J., 2005. Preparation and properties of reduced-charge smectites—a review. Clays Clay Miner. 53, 313–334.

Komadel, P., Bujdák, J., Madejová, J., Šucha, V., Elsass, F., 1996a. Effect of non-swelling layers on the dissolution of reduced-charge montmorillonite in hydrochloric acid. Clay Miner. 31, 333–345.

Komadel, P., Madejová, J., Laird, D.A., Xia, Y., Stucki, J.W., 2000. Reduction of Fe(III) in griffithite. Clay Miner. 35, 625–634.

Komadel, P., Madejová, J., Janek, M., Gates, W.P., Kirkpatrick, R.J., Stucki, J.W., 1996b. Dissolution of hectorite in inorganic acids. Clay Clay Min. 44, 228–236.

König, I.R., Malley, J.D., Weimar, C., Diener, H.-C., Ziegler, A., 2007. Practical experiences on the necessity of external validation. Stat. Med. 26, 5499–5511.

Konrad, F., Stalder, R., Tessadri, R., 2015. Quantitative phase analysis of lateritic bauxite with NIR-spectroscopy. Miner. Eng. 77, 117–120.

Kooli, F., Jones, W., 1997. Characterization and catalytic properties of a saponite clay modified by acid activation. Clay Miner. 32, 633–643.

Košařová, V., Hradil, D., Němec, I., Bezdička, P., Kanický, V., 2013. Microanalysis of clay-based pigments in painted artworks by the means of Raman spectroscopy. J. Raman Spectrosc. 44, 1570–1577.

Köster, H.M., Ehrlicher, U., Gilg, H.A., Jordan, R., Murad, E., Onnich, K., 1999. Mineralogical and chemical characteristics of five nontronites and Fe-rich smectites. Clay Miner. 34, 579–599.

Koswojo, R., Utomo, R.P., Ju, Y.H., Ayucitra, A., Soetaredjo, F.E., Sunarso, J., Ismadji, S., 2010. Acid Green 25 removal from wastewater by organo-bentonite from Pacitan. Appl. Clay Sci. 48, 81–86.

Kozai, T., Yamashita, S., Hirochi, K., Miyagawa, H., Tsurumachi, N., Koshiba, S., Nakanishi, S., Itoh, H., 2012. Molecular vibrational dynamics in polyvinyl alcohol studied by femtosecond coherent anti-stokes Raman spectroscopy. Chem. Phys. Lett. 553, 26–29.

Kraft, M.D., Michalski, J.R., Sharp, T.G., 2003. Effects of pure silica coatings on thermal emission spectra of basaltic rocks: considerations for Martians surface mineralogy. Geophys. Res. Lett. 30, 1–4.

Kramer, R., 1998. Chemometric Techniques for Quantitative Analysis. Marcel Dekker, Inc., New York.

Kresse, G., Furthmüller, J., 1996. Efficient iterative scheme for ab initio total-energy calculations using a plane-wave basis set. Phys. Rev. B 54, 11169–11186.

Kristóf, J., Frost, R.L., Felinger, A., Mink, J., 1997. FTIR spectroscopic study of intercalated kaolinite. J. Mol. Struct. 410–411, 119–122.

Kirstóf, J., Frost, R.L., Kloprogge, J.T., Horvath, E., Gabor, M., 1999. Thermal behaviour of kaolinite intercalated with formamide. dimethyl sulphox ide and hydrazine. J. Therm. Anal. 56, 885–891.

Krukowski, E.G., Goodman, A., Rother, G., Ilton, E.S., Guthrie, G., Bodnar, R.J., 2015. FT-IR study of $CO_2$ interaction with $Na^+$ exchanged montmorillonite. Appl. Clay Sci. 114, 61–68.

Kruse, F.A., 1996. Identification and mapping of minerals in drill core using hyperspectral image analysis of infrared reflectance data. Int. J. Remote Sens. 17, 1623–1632.

Kruse, F.A., 2004. Comparison of ATREM, ACORN, and FLAASH atmospheric corrections using low-altitude AVIRIS Data of Boulder, Colorado. In: Proceedings of 13th JPL Airborne Geoscience Workshop, Pasadena, CA.

Kruse, F.A., Lefkoff, A.B., Dietz, J.B., 1993. Expert system-based mineral mapping in northern death valley, California/Nevada, using the Airborne Visible/Infrared Imaging Spectrometer (AVIRIS). Remote Sens. Environ. 44, 309–336.

Kruse, F.A., Boardman, J.W., Huntington, J.F., 2003. Comparison of airborne hyperspectral data and EO-1 Hyperion for mineral mapping. IEEE Trans. Geosci. Remote Sens. 41, 1388–1400.

Kubota, M.M., Sacco, B.L., Bento, D.C., De Santana, H., 2015. Synthesis and spectroscopic analysis of polydiphenylamine via oxidation with bentonite clay in the solid state. Spectrochim. Acta A: Mol. Biomol. Spectrosc. 151, 80–88.

Kuligiewicz, A., Derkowski, A., Szczerba, M., Gionis, V., Chryssikos, G.D., 2015b. Revisiting the infrared spectrum of the water-smectite interface. Clays Clay Miner. 63, 15–29.

Kuligiewicz, A., Derkowski, A., Emmerich, K., Christidis, G.E., Tsiantos, C., Gionis, V., Chryssikos, G.D., 2015a. Measuring the layer charge of dioctahedral smectite by O-D vibrational spectroscopy. Clays Clay Miner. 63, 443–456.

Kung, K.H.S., Hayes, K.F., 1993. Fourier-transform infrared spectroscopic study of the adsorption of cetyltrimethylammonium bromide and cetylpyridinium chloride on silica. Langmuir 9, 263–267.

Kydd, R.A., Dunham, A.P.C., 1980. The infrared spectra and structure of acetamide and acrylamide. J. Mol. Struct. 69, 79–88.

Lagaly, G., 1981. Characterization of clays by organic compounds. Clay Miner. 16, 1–21.

Lagaly, G., 1982. Layer charge heterogeneity in vermiculites. Clays Clay Miner. 30, 215–222.

Lagaly, G., 1984. Clay organic reactions. Philos. Trans. R. Soc. Lond. Ser. A 311, 315–332.

Lagaly, G., Dékány, I., 2005. Adsorption on hydrophobized surfaces: clusters and self-organization. Adv. Colloid Interface Sci. 114, 189–204.

Lagaly, G., Ogawa, M., Dékány, I., 2013. Clay mineral–organic interactions. In: Bergaya, F., Lagaly, G. (Eds.), Handbook of Clay Science Part A: Fundamentals, second ed., Chap. 7.3. Elsevier Ltd., Amsterdam, pp. 437–505.

Lahav, N., 1990. Preparation of stable suspensions of de laminated kaolinite by combined dimethylsulfoxide-ammonium fluoride treatment. Clays Clay Miner. 38, 219–222.

Laird, D.A., Fleming, A.M., 1999. Mechanisms for adsorption of organic bases on hydrated smectite surfaces. Environ. Toxicol. Chem. 18 (8), 1668–1672.

Laird, D.A., Barriuso, E., Dowdy, R.H., Koskinen, W.C., 1992. Adsorption of atrazine on smectites. Soil Sci. Soc. Am. J. 56, 62–67.

Laird, D.A., Yen, P.Y., Koskinen, W.C., Steinheimer, T.R., Dowdy, R.H., 1994. Sorption of atrazine on soil clay components. Environ. Sci. Technol. 26, 1054–1061.

Lambert, J.F., Poncelet, G., 1997. Acidity in pillared clays origin and catalytic manifestations. Top. Catal. 4, 43–56.

Lambert, J.F., Bergaya, F., 2013. Smectites-polymer nanocomposites. In: Bergaya, F., Lagaly, G. (Eds.), Handbook of Clay Science. Developments in Clay Science, vol. 5A. Elsevier, Amsterdam, pp. 699–706.

Lambert, J.F., Chevalier, S., Franck, R., Suquet, H., Barthomeuf, D., 1994. Al pillared saponites Part 2-NMR studies. J. Chem. Soc. Faraday Trans. 90, 675–682.

Landauer, R., 1978. Electrical conductivity in inhomogeneous media. AIP Conf. Proc. 40, 2–45.

Lantenois, S., Beny, J.M., Muller, F., Champallier, R., 2007. Integration of Fe in natural and synthetic Al-pyrophyllites: an infrared spectroscopic study. Clay Miner. 42, 129–141.

Lantenois, S., Champallier, R., Beny, J.M., Muller, F., 2008a. Hydrothermal synthesis and characterization of dioctahedral smectites: a montmorillonites series. Appl. Clay Sci. 38, 165–178.

Lantenois, S., Muller, F., Beny, J.M., Mahiaoui, J., Champallier, R., 2008b. Hydrothermal synthesis of beidellites: characterization and study of the cis- and trans-vacant character. Clays Clay Miner. 56, 39–48.

Laperche, V., Prost, R., 1991. Assignment of the far-infrared absorption bands of K in micas. Clays Clay Miner. 39, 281–289.

Lapides, I., Michaelian, K.H., Yariv, S., 1997. X-ray study of the thermal intercalation of alkali halides into kaolinite. J. Therm. Anal. Calorim. 49, 1423–1432.

Larentzos, J.P., Greathouse, J.A., Cygan, R.T., 2007. An ab initio and classical molecular dynamics investigation of the structural and vibrational properties of talc and pyrophyllite. J. Phys. Chem. C 111, 12752–12759.

Larson, S.J., Pardoe, G.W.F., Gebbie, H.A., Larson, E.E., 1972. Use of far infrared interferometric spectroscopy for mineral identification. Am. Mineral. 57, 998–1002.

Lauretta, D.S., McSween Jr., H.Y. (Eds.), 2006. Meteorites and the Early Solar System II. The University of Arizona Press, Tucson, AZ.

Lazarev, A.N., 1974. The dynamics of crystal lattices. In: Farmer, V.C. (Ed.), The Infrared Spectra of Minerals. Mineralogical Society, London, pp. 69–85.

Lazzeri, M., Mauri, F., 2003. First principles calculation of vibrational Raman spectra in large systems: signature of small rings in crystalline $SiO_2$. Phys. Rev. Lett. 90. 036401.

Lazzeri, M., Calandra, M., Mauri, F., 2003. Anharmonic phonon frequency shift in $MgB_2$. Phys. Rev. B 68, 220509 (R).

Lebofsky, L.A., Feierberg, M.A., Tokunaga, A.T., Larson, H.P., Johnson, J.R., 1981. The 1.7- to 4.2-micron spectrum of asteroid 1 Ceres—evidence for structural water in clay minerals. Icarus 48, 453–459.

Ledoux, R.L., White, J.L., 1964a. Infrared study of selective deuteration of kaolinite and halloysite at room temperature. Science 145, 47–49.

Ledoux, R.L., White, J.L., 1964b. Infrared study of the OH group in expanded kaolinite. Science 143, 244–246.

Ledoux, R.L., White, J.L., 1966. Infrared studies of hydrogen bonding interaction between kaolinite surfaces and intercalated potassium acetate, hydrazine, formamide and urea. J. Colloid Interface Sci. 21, 127–152.

Lee, J.J., Choi, J.Y., Park, J.W., 2002. Simultaneous sorption of lead and chlorobenzene by organobentonite. Chemosphere 49, 1309–1315.

Lee, S.S., Kim, S.J., 2004. Surface modification of clay and its effect on the intercalation behavior of the polymer/clay nanocomposites. J. Polym. Sci. B Polym. Phys. 42, 2367–2372.

Lee, S.Y., Kim, S.J., 2002. Expansion characteristics of organoclay as a precursor to nanocomposites. Colloid Surf. A:Physicochem. Eng. Asp. 211, 19–26.

Lee, S.Y., Kim, S.J., Chung, S.Y., Jeong, C.H., 2004. Sorption of hydrophobic organic compounds onto organoclays. Chemosphere. 55, 781–785.

Lerot, L., Low, P.F., 1976. Effect of swelling on infrared-absorption spectrum of montmorillonite. Clays Clay Miner. 24, 191–199.

Levard, C., Basile-Doelsch, I., 2016. Geology and mineralogy of imogolite-type marerials. Chapter 3 Nanosized Tubular Clay Minerals, vol. 7. In: Thill, A., Yuan, P., Bergaya, F. (Eds.), Developments in Clay Science, pp. 49–65.

Levard, C., Doelsch, E., Basile-Doelsch, I., Abidin, Z., Miche, H., Masion, A., Rose, J., Borschneck, D., Bottero, J.Y., 2012. Structure and distribution of allophanes, imogolite and proto-imogolite in volcanic soils. Geoderma 183–184, 100–108.

Levitan, D.M., Hammarstrom, J.M., Gunter, M.E., Seal R.R. II, Chou, I.M., Piatak, N.M., 2009. Mineralogy of mine waste at the Vermont Asbestos Group mine, Belvidere Mountain. Vermont. Am. Mineral. 94, 1063–1066.

Lewis, I.R., Chaffin, N.C., Gunter, M.E., Griffiths, P.R., 1996. Vibrational spectroscopic studies of asbestos and comparison of suitability for remote analysis. Spectrochim. Acta Part A 52, 315–328.

Leyden, D.E., Murthy, R.S.S., 1988. Diffuse reflectance Fourier transform IR spectroscopy. Trends Anal. Chem. 7, 154–169.

Li, B., Liu, Z., Han, C., Ma, W., Zhao, S., 2012. In situ synthesis, characterization, and catalytic performance of tungstophosphoric acid encapsulated into the framework of mesoporous silica pillared clay. J. Colloid Interface Sci. 377, 334–341.

Li, J.F., Li, Y.M., Lu, J.H., 2009. Adsorption of herbicides 2, 4-D and acetochlor on inorganic-organic bentonites. Appl. Clay Sci. 46, 314–318.

Li, L., Liu, X., Ge, Y., Xu, R., Rocha, J., Klinowski, J., 1993. Structural studies of pillared saponites. J. Phys. Chem. B 97, 10389–10393.

Li, S.-Z., Wu, P.-X., 2010. Characterization of sodium dodecyl sulfate modified iron pillared montmorillonite and its application for the removal of aqueous Cu(II) and Co(II). J. Hazard Mater. 173, 62–70.

Li, W., Yue, H.H., Valle-Cervantes, S., Qin, S.J., 2000. Recursive PCA for adaptive process monitoring. J. Process Control 10, 471–486.

Li, Y.Q., Ishida, H., 2003a. Concentration-dependent conformation of alkyl tail in the nanoconfined space: hexadecylamine in the silicate galleries. Langmuir 19, 2479–2484.

Li, Y.Q., Ishida, H., 2003b. Solution intercalation of polystyrene and the comparison with poly(ethyl methacrylate). Polymer 44, 6571–6577.

Li, Y.Q., Ishida, H., 2005. A study of morphology and intercalation kinetics of polystyrene-organoclay nanocomposites. Macromolecules 38, 6513–6519.

Li, Z.H., Jiang, W.T., Hong, H.L., 2008. An FTIR investigation of hexadecyltrimethylammonium intercalation into rectorite. Spectrochim. Acta A. 71, 1525–1534.

Liang, Y., Ding, H., Wang, Y.B., Liang, N., Wang, G.L., 2013. Intercalation of cetyl trimethylammonium ion into sericite in the solvent of dimethyl sulfoxide. Appl. Clay Sci. 74, 109–114.

Lidner, G.-G., Nakazawa, H., Hayashi, S., 1998. Hollow nanospheres, allophanes 'all-organic' synthesis and characterization. Microporous Mesoporous Mater. 21, 381–386.

Libowitzky, E., 1999. Correlation of O-H stretching frequencies and O-H···O hydrogen bond lengths in minerals. Monatsh. Chem. 130, 1047–1059.

Li-Chan, E., Chalmers, J., Griffiths, P., 2011. Applications of Vibrational Spectroscopy in Food Science. John Wiley and Sons. ISBN: 0470742992. 872p.

Likos, W.J., Bowders, J.J., Gates, W.P., 2010. Mineralogy and engineering properties of Bentonite. In: Bouazza, A., Bowders, J.J. (Eds.) Geosynthetic Clay Liners for Waste Containment Facilities. CRC Press, 31–53.

Lin, J.J., Cheng, I.J., Wang, R., Lee, R.J., 2001. Tailoring basal spacings of montmorillonite by poly(oxyalkylene)diamine intercalation. Macromolecules 34, 8832–8834.

Lindberg, J.D., Snyder, D.G., 1972. Diffuse reflectance spectra of several clay minerals. Am. Mineral. 57, 485–493.

Lisse, C.M., Kraemer, K.E., Nuth J.A. III, Li, A., D., Joswiak, 2007. Comparison of the composition of the Tempel 1 ejecta to the dust in Comet C/Hale–Bopp 1995 O1 and YSO HD 100546. Icarus 187, 69–86.

Lisse, C.M., Vancleve, J., Adams, A.C., A'Hearn, M.F., Fernández, Y.R., Farnham, T.L., Armus, L., Grillmair, C.J., Ingalls, J., Belton, M.J.S., Groussin, O., Mcfadden, L.A., Meech, K.J., Schultz, P.H., Clark, B.C., Feaga, L.M., Sunshine, J.M., 2006. Spitzer spectral observations of the deep impact ejecta. Science 313, 635–640.

Liu, G.-H., Ko, A.-N., Chang, Y.-C., 1995. Synthesis and properties of pillared montmorillonite formed by intercalation of transition metal macrocyclic complexes. Microporous Mater. 5, 61–67.

Liu, H.M., Yuan, P., Qin, Z.H., Liu, D., Tan, D.Y., Zhu, J.X., He, H.P., 2013. Thermal degradation of organic matter in the interlayer clay-organic complex: a TG-FTIR study on a montmorillonite/12-aminolauric acid system. Appl. Clay Sci. 80–81, 398–406.

Liu, J., Li, X., Zuo, S., Yu, Y., 2007. Preparation and photocatalytic activity of silver and $TiO_2$ nanoparticles/montmorillonite composites. Appl. Clay Sci. 37, 275–280.

Liu, N., Wang, M.X., Liu, M.M., Liu, F., Weng, L.P., Koopal, L.K., Tan, W.F., 2012. Sorption of tetracycline on organo-montmorillonites. J. Hazard. Mater. 225, 28–35.

Liu, Q., Zhang, S., Cheng, H., Wang, D., Li, X., Hou, X., Frost, R.L., 2014. Thermal behavior of kaolinite-urea intercalation complex and molecular dynamics simulation for urea molecule orientation. J. Therm. Anal. Calorim. 117, 189–196.

Liu, R., Frost, R.L., Martens, W.N., 2009. Near infrared and mid infrared investigations of adsorbed phenol on HDTMAB organoclays. Mater. Chem. Phys. 113, 707–713.

Liu, X.D., Lu, X.C., 2006. A thermodynamic understanding of clay-swelling inhibition by potassium ions. Angew. Chem. Int. Ed. 45, 6300–6303.

Loh, E., 1973. Optical vibrations in sheet silicates. J. Phys. C 6, 1091–1104.

Loizeau, D., Werner, S.C., Mangold, N., Bibring, J.P., Vago, J.L., 2012b. Chronology of deposition and alteration in the Mawrth Vallis region, Mars. Plan. Space Sci. 72, 31–43.

Loizeau, D., Mangold, N., Poulet, F., Ansan, V., Hauber, E., Bibring, J.P., Gondet, B., Langevin, Y., Masson, P., Neukum, G., 2010. Stratigraphy in the Mawrth Vallis region through OMEGA, HRSC color imagery and DTM. Icarus 205, 396–418.

Loizeau, D., Carter, J., Bouley, S., Mangold, N., Poulet, F., Bibring, J.P., Costard, F., Langevin, Y., Gondet, B., Murchie, S.L., 2012a. Characterization of hydrated silicate-bearing outcrops in Tyrrhena Terra, Mars: implications to the alteration history of Mars. Icarus 219, 476–497.

Long, D.A., 1977. Raman Spectroscopy. McGraw-Hill international Book Company, 292 pp.

Long, D.A. 2002. The Raman effect. A unified treatment of the theory of Raman scattering by molecules. John Wiley and Sons. ISBN: 0 471 49028 8.

Low, M.J.D., 1965. Infra-red emission spectra of minerals. Nature 206, 1089–1090.

Ma, L.Y., Chen, Q.Z., Zhu, J.X., Xi, Y.F., He, H.P., Zhu, R.L., Tao, Q., Ayoko, G.A., 2016. Adsorption of phenol and Cu(II) onto cationic and zwitterionic surfactant modified montmorillonite in single and binary systems. Chem. Eng. J. 283, 880–888.

Ma, Y.H., 2010. Factors Controlling and Affecting the Interlayer Structure and Properties of Organo (Pillared) Montmorillonites. (Master's Thesis) Graduate University of the Chinese Acadamy of Sciences, China.

Ma, Y.H., Zhu, J.X., He, H.P., Yuan, P., Shen, W., Liu, D., 2010. Infrared investigation of organo-montmorillonites prepared from different surfactants. Spectrochim. Acta A: Mol. Biomol. Spectrosc. 76, 122–129.

MacEwan, D.M.C., 1961. Montmorillonite minerals. In: Brown, G. (Ed.), The X-ray Identification and Crystal Structures of Clay. Mineralogical Society, London, pp. 43–207.

Mackenzie, R.C., 1957. The Differential Thermal Analysis of Clays. Mineralogical Society, London. 456 pp.

Maddams, W.F., Mead, W.L., 1982a. The measurement of derivative I.R. spectra—I. Background studies. Spectrochim. Acta 38A, 437–444.

Maddams, W.F., Southon, M.J., 1982b. III. The effect of band width and band shape on resolution enhancement by derivative spectrocopy. Spectrochim. Acta 38A, 459–466.

Madejová, J., 2003. FTIR techniques in clay mineral studies. Vib. Spectrosc. 31, 1–10.

Madejová, J., 2005. Studies of reduced-charge smectites by near infrared spectroscopy. In: Kloprogge, J.T. (Ed.), The Application of Vibrational Spectroscopy to Clay Minerals and Layered Double Hydroxides, CMS Workshop Lectures, vol. 13. The Clay Mineral Society, Aurora, CO, pp. 169–202.

Madejová, J., Komadel, P., 2001. Baseline studies of the clay minerals society source clays: infrared methods. Clays Clay Miner. 49, 410–432.

Madejová, J., Komadel, P., 2005. Information available from infrared spectra of the fine fraction of bentonites. In: Klopproge, J.T. (Ed.), The Application of Vibrational Spectroscopy to Clay Minerals and Layered Double Hydroxides, CMS Workshop Lectures, vol. 13. The Clay Mineral Society, Aurora, CO, pp. 65–98.

Madejová, J., Komadel, P., Čičel, B., 1994. Infrared study of octahedral site populations in smectites. Clay Miner. 29, 319–326.

Madejová, J., Pálková, H., Komadel, P., 2006. Behaviour of $Li^+$ and $Cu^{2+}$ in heated montmorillonite: evidence from far-, mid-, and near-IR regions. Vib. Spectrosc. 40, 80–88.

Madejová, J., Pálková, H., Komadel, P., 2010. IR spectroscopy of clay minerals and clay nanocomposites. In: Yarwood, J. (Ed.), Spectroscopic Properties of Inorganic and Organometallic Compounds, vol. 41. RSC Publishing, UK, pp. 22–71.

Madejová, J., Balan, E., Petit, S., 2011. Application of vibrational spectroscopy to the characterization of phyllosilicates and other industrial minerals. In Christidis, G.E. (Ed.)

Advances in the Characterization of Industrial Minerals. EMU Notes in Mineralogy, **9**, European Mineralogical Union and the Mineralogical Society of Great Britain and Ireland, 171-226.

Madejová, J., Pálková, H., Jankovič, Ľ., 2012. Degradation of surfactant-modified montmorillonites in HCl. Mater. Chem. Phys. 134, 768–776.

Madejová, J., Pálková, H., Jankovič, Ľ., 2015. Near-infrared study of the interaction of pyridine with acid-treated montmorillonite. Vib. Spectrosc. 76, 22–30.

Madejová, J., Bujdák, L., Gates, W., Komadel, P., 1996. Preparation and infrared spectroscopic characterization ofreauceu-charge montmorillonite with various Li contents. Clay Miner. 31, 233–241.

Madejová, J., Kraus, I., Tunega, D., Šamajová, E., 1997. Fourier transform infrared spectroscopic characterisation of kaolinite group minerals from the main Slovak deposits. Geol. Carpathi. Clays 6, 3–10.

Madejová, J., Bujdák, J., Janek, M., Komadel, P., 1998. Comparative FT-IR study of the structural modifications during acid treatment of dioctahedral smectites and hectorite. Spectrochim. Acta A 54, 1397–1406.

Madejová, J., Bujdák, J., Petit, S., Komadel, P., 2000a. Effects of chemical composition and temperature of heating on the infrared spectra of Li-saturated dioctahedral smectites. (I) Mid-infrared region. Clay Miner. 35, 739–751.

Madejová, J., Bujdák, J., Petit, S., Komadel, P., 2000b. Effects of chemical composition and temperature of heating on the infrared spectra of Li-saturated dioctahedral smectites. Part (II) Near-infrared region. Clay Miner. 35, 753–761.

Madejová, J., Kečkéš, J., Pálková, H., Komadel, P., 2002b. Identification of components in smectite-kaolinite mixtures. Clay Miner. 37, 377–388.

Madejová, J., Pentrák, M., Pálková, H., Komadel, P., 2009. Near-infrared spectroscopy: a powerful tool in studies of acid-treated clay minerals. Vib. Spectrosc. 49, 211–218.

Madejová, J., Jankovič, Ľ., Pentrák, M., Komadel, P., 2011. Benefits of near-infrared spectroscopy for characterization of selected organo-montmorillonites. Vib. Spectrosc. 57, 8–14.

Madejová, J., Janek, M., Komadel, P., Herbert, H.J., Moog, H.C., 2002a. FTIR analyses of water in MX-80 bentonite compacted from high salinary salt solution systems. Appl. Clay Sci. 20, 255–271.

Madejová, J., Sekeráková, Ľ., Bizovská, V., Slaný, M., Jankovič, Ľ., 2016. Near-infrared spectroscopy as an effective tool for monitoring the conformation of alkylammonium surfactants in montmorillonite interlayers. Vib. Spectrosc. 84, 44–52.

Madejová, J., Andrejkovičová, S., Bujdák, J., Čeklovský, A., Hrachová, J., Valúchová, J., Komadel, P., 2007. Characterization of products obtained by acid leaching of Fe-bentonite. Clay Miner. 42, 527–540.

Maes, A., Schoonheydt, R.A., Cremers, A., Uytterhoeven, J.B., 1980. Spectroscopy of $Cu(en)_2^{2+}$ on clay surfaces. Surface and charge-density effects. J. Phys. Chem. 84, 2795–2799.

Maes, N., Vansant, E.F., 1995. Study of $Fe_2O_3$-pillared clays synthesized using the trinuclear Fe(III)-acetato complex as pillaring precursor. Microporous Mater. 4, 43–51.

Maes, N., Molinard, A., Vansant, E.F., Roy, G.D., 1994. Fe2O3-pillared clays: the interaction of the tri-nuclear Fe(III)-acetato complex with montmorillonite clay. Process Technol. Proc. 11, 403–412.

Mahadevaiah, N., Vijayakumar, B., Hemalatha, K., Prakash, B.S.J., 2011. Uptake of permanganate from aqueous environment by surfactant modified montmorillonite batch and fixed bed studies. Bull. Mater. Sci. 34, 1675–1681.

Maiti, M., Bhowmick, A.K., 2009. Synthesis and properties of new fluoroelastomer nanocomposites from tailored anionic layered magnesium silicates (hectorite). J. Appl. Polym. Sci. 111, 1094–1104.

Makó, É., Kristóf, J., Horváth, E., Vágvölgyi, V., 2009. Kaolinite–urea complexes obtained by mechanochemical and aqueous suspension techniques—a comparative study. J. Colloid Interface Sci. 330, 367–373.

Makó, É., Kristóf, J., Horváth, E., Vágvölgyi, V., 2013. Mechanochemical intercalation of low reactivity kaolinite, Appl. Clay Sci. 83–84, 24–31.

Maksimović, Z., White, J.L., 1973. Infrared study of chromium-bearing halloysites. In: Proc. Int. Clay Conf., Madrid, Spain, pp. 61–73.

Maksimović, Z., White, J.L., Logar, M., 1981. Chromium-bearing dickite and chromium-bearing kaolinite from Teslić, Yugoslavia. Clays Clay Miner. 29, 213–218.

Maleki, M.R., Mouazen, A.M., Ramon, H., De Baerdemaeker, J., 2007. Multiplicative Scatter Correction during on-line measurement with near-infrared spectroscopy. Biosyst. Eng. 96, 427–433.

Manceau, A., Calas, G., 1985. Heterogeneous distribution of nickel in hydrous silicates from New caledonia ore deposits. Am. Mineral. 70, 549–558.

Manceau, A., Calas, G., 1986. Nickel–bearing clay minerals: II Intracrystalline distribution of nickel: a X-ray absorption study. Clay Miner. 21, 341–360.

Manceau, A., Chateigner, D., Gates, W.P., 1998. Polarized EXAFS, distance-valence least-squares modeling (DLVS), and quantitative texture analysis approaches to the structural refinement of Garfield nontronite. Phys. Chem. Miner. 25, 347–365.

Mandalia, T., Bergaya, F., 2006. Organo clay mineral-melted polyolefin nanocomposites—effect of surfactant/CEC ratio. J. Phys. Chem. Solids 67, 836–845.

Mandalia, T., Crespin, M., Messad, D., Bergaya, F., 1998. Large interlayer repeat distance observed for montmorillonites treated by mixed Al-Fe pillaring solutions. Chem. Commun. 19, 2111–2112.

Mano, E.S., Caner, L., Petit, S., Chaves, A.P., Mexias, A.S., 2014. Mineralogical characterization of Ni-bearing smectites from Niquelandia, Brazil. Clays Clay Miner. 62, 325–336.

Manohar, D.M., Noeline, B.F., Anirudhan, T.S., 2006. Adsorption performance of Al-pillared bentonite clay for the removal of cobalt(II) from aqueous phase. Appl. Clay Sci. 31, 194–206.

Mao, H., Gao, X., Yang, J., Li, B., 2011. A novel one-step synthesis of mesostructured silica-pillared clay with highly ordered gallery organic-inorganic hybrid frame. Appl. Surf. Sci. 257, 4655–4662.

Mao, H., Li, B., Li, X., Liu, Z., Ma, W., 2009. Mesoporous nickel (or cobalt)-doped silica-pillared clay: Synthesis and characterization studies. Mater. Res. Bull. 44, 1569–1575.

Mao, H., Zhu, K., Li, B., Yao, C., Kong, Y., 2014a. Synthesis of titania modified silica-pillared clay (SPC) with highly ordered interlayered mesoporous structure for removing toxic metal ion Cr(VI) from aqueous state. Appl. Surf. Sci. 292, 1009–1019.

Mao, H., Liu, X., Yang, J., Li, B., Chen, Q., Zhong, J., 2014b. Fabrication of magnetic silica-pillared clay (SPC) nanocomposites with ordered interlayer mesoporous structure for controlled drug release. Microporous Mesoporous Mater. 184, 169–176.

Mao, H., Zhu, K., Lu, X., Zhang, G., Yao, C., Kong, Y., Liu, J., 2015. Restructuring of silica-pillared clay (SPC) through posthydrothermal treatment and application as phosphotungstic acid supports for cyclohexene oxidation. J. Colloid Interface Sci. 446, 141–149.

Maqueda, C., Santas Romero, A., Morillo, E., Pérez-Rodrígez, J.L., Lerf, A., Wagner, F.E., 2008. The behaviour of Fe in ground and acid-treated vermiculite from Santa Olalla. Spain. Clays Clay Miner. 56, 380–388.

Mardia, K.V., Kent, J.T., Bibby, J.M., 1979. Multivariate Analysis. Academic Press.

Mark, H., Workman Jr., J., 2003. Derivatives in spectroscopy. Part I—The behavior of the derivative. Spectroscopy 18 (4), 32–37.

Marosi, T., Dékány, I., Lagaly, G., 1994. Displacement processes on hydrophilic hydrophobic surfaces in 1-propanol-water mixtures. Colloid Polym. Sci. 272, 1136–1142.

Marsal, A., Bautista, E., Ribosa, I., Pons, R., García, M.T., 2009. Adsorption of polyphenols in wastewater by organo-bentonites. Appl. Clay Sci. 44, 151–155.

Marsh, S.E., McKeon, J.B., 1983. Integrated analysis of high-resolution field and airborne spectroradiometer data for alteration mapping. Econ. Geol. 78, 618–632.

Martens, H., 1979. Factor analysis of chemical mixtures. Anal. Chim. Acta 112, 423–448.

Martens, H., Nielson, J.P., Engelsen, S.B., 2003. Light scattering and light absorbance separated by extended multiplicative signal correction. Application of near-infrared transmission analysis of powder mixtures. Anal. Chem. 75, 394–404.

Martens, W.N., Frost, R.L., Kristóf, J., Kloprogge, J.T., 2002a. Raman spectroscopy of dimethyl sulphoxide and deuterated dimethyl sulphoxide at 298 and 77 K. J. Raman Spectrosc. 33, 84–91.

Martens, W.N., Ding, Z., Frost, R.L., Kristóf, J., Kloprogge, J.T., 2002b. Raman spectroscopy of hydrazine-intercalated kaolinite at 77, 298, 323, 343 and 358 K. J. Raman Spectrosc. 33, 31–36.

Martin, F., Petit, S., Grauby, O., Lavie, M.P., 1999a. Gradual H D substitution in synthetic germanium bearing talcs: a method for infrared band assignment. Clay Mine. 34, 365–374.

Martin, F., Petit, S., Decarreau, A., Grauby, O., Hazemann, J.L., Noack, Y., 1992. Experimental study of silicon-germanium tetrahedral solid solution in nickel-cobalt-magnesium talcs. Thin Solid Films 222, 189–195.

Martin, F., Petit, S., Decarreau, A., Ildefonse, P., Grauby, O., Beziat, D., Parseval, P., Noack, Y., 1998. Ga/Al substitution in synthetic kaolinites and smectites. Clay Miner. 33, 231–241.

Martin, F., Micoud, P., Delmotte, L., Marichal, C., Le Dred, R., De Parseval, P., Mari, A., Fortune, J.P., Salvi, S., Beziat, D., Grauby, O., Ferret, J., 1999b. The structural formula of talc from the Trimouns deposit, Pyrenees, France. Can. Mineral. 37, 997–1006.

Martínez-Alonso, S., Rustad, J.R., Goetz, A.F.H., 2002a. Ab initio quantum mechanical modelling of infrared vibrational frequencies of the OH group in dioctahedral phyllosilicates. Part 1: Methods, results and comparison to experimental data. Am. Mineral. 87, 1215–1223.

Martínez-Alonso, S., Rustad, J.R., Goetz, A.F.H., 2002b. Ab initio quantum mechanical modeling of infrared vibrational frequencies of the OH group in dioctahedral phyllosilicates. Part II: Main physical factors governing the OH vibrations. Am. Mineral. 87, 1224–1234.

Marx, D., Hutter, J., 2009. Ab Initio Molecular Dynamics: Basic Theory and advanced Methods. Cambridge University Press, 578 pp.

Marzo, G.A., Davila, A.F., Tornabene, L.L., Dohm, J.M., Fairén, A.G., Gross, C., Kneissl, T., Bishop, J.L., Roush, T.L., McKay, C.P., 2010. Evidence for Hesperian impact-induced hydrothermalism on Mars. Icarus 208, 667–683.

McBride, J.R., Hass, K.C., Weber, W.H., 1991. Resonance-Raman and lattice-dynamics studies of single-crystal PdO. Phys. Rev. B 44, 5016–5028.

McBride, M.B., Farmer, V.C., Russell, J.D., Tait, J.M., Goodman, B.A., 1984. Iron substitution in aluminosilicate sols synthesized at low pH. Clay Miner. 19, 1–8.

McCabe, R.W., Adams, J.M., 2013. Chapter 4.3—Clay minerals as catalysts. In: Faïza, B., Gerhard, L. (Eds.), Developments in Clay Science. Elsevier.

McCauley, J.R., 1988. Stable intercalated clays and preparation method, International Patent Application PCT/US88/00567.

McClelland, J.F., Jones, R.W., Luo, S., Seaverson, L.M., 1993. A practical guide to FTIR photoacoustic spectroscopy. In: Coleman, P.B. (Eds) Practical Sampling Techniques for Infrared Analysis. CRC Press. Ch. 5, 107–144.

McCord, T., Gaffey, M.J., 1974. Asteroids: surface composition from reflectance spectroscopy. Science 186, 352–355.

McCord, T., Sotin, C., 2005. Ceres: evolution and current state. J. Geophys. Res. 110E05009.

McCord, T.B., Clark, R.N., Singer, R.B., 1982. Mars: near-infrared spectral reflectance of surface regions and compositional implications. J. Geophys. Res. 87, 3021–3032.

McCutcheon, A., Hu, J., Kannangara, G.S.K., Wilson, M.A., Reddy, N., 2005. $^{29}Si$ labelled nanoaluminosilicate imogolite. J. Non-Cryst. Solids 351, 1967–1972.

McDowell, M.L., Hamilton, V.E., 2009. Seeking phyllosilicates in thermal infrared data: a laboratory and Martian data case study. J. Geophys. Res. https://doi.org/10.1029/2008JE003317. 114.

McKay, D.S., Gibson, Jr., E.K., Thomas-Keprta, K.L., Vali, H., Romanek, C.S., Clemett, S.J., Chillier, X.D.F., Maechling, C.R., Zare, R.N., 1996. Search for past life on Mars: possible relic biogenic activity in Martian meteorite ALH 84001. Science 273, 924-930.

Mckeown, D.A., Post, J.E., Etz, E.S., 2002. Vibrational analysis of palygorskite and sepiolite. Clays Clay Miner. 50, 667–680.

McKeown, N.K., Noe Dobrea, E.Z., Bishop, J.L., Silver, E.A., 2009b. Coordinated lab, field, and aerial study of the Painted Desert, AZ, as a potential analog site for phyllosilicates at Mawrth Vallis, Mars. Lunar and Planetary Science Conference XL, 2009, Houston, TX. LPS XL, abs. #2509.

McKeown, N.K., Bishop, J.L., Cuadros, J., Hillier, S., Amador, E., Makarewicz, H.D., Parente, M., Silver, E., 2011. Interpretation of reflectance spectra of clay mineral-silica mixtures: implications for Martian clay mineralogy at Mawrth Vallis. Clays Clay Miner. 59, 400–415.

McKeown, N.K., Bishop, J.L., Noe Dobrea, E.Z., Ehlmann, B.L., Parente, M., Mustard, J.F., Murchie, S.L., Swayze, G.A., Bibring, J.-P., Silver, E., 2009a. Characterization of phyllosilicates observed in the central Mawrth Vallis region, Mars, their potential formational processes, and implications for past climate. J. Geophys. Res. 114. doi:10.1029/2008JE003301.

McLennan, S.M., Anderson, R. B., Bell, J. F., Bridges, J. C., Calef, F., Campbell, J. L., Clark, B. C., Clegg, S., Conrad, P., Cousin, A., Des Marais, D. J., Dromart, G., Dyar, M. D., Edgar, L. A., Ehlmann, B. L., Fabre, C., Forni, O., Gasnault, O., Gellert, R., Gordon, S., Grant, J. A., Grotzinger, J. P., Gupta, S., Herkenhoff, K. E., Hurowitz, J. A., King, P. L., Le Mouélic, S., Leshin, L. A., Léveillé, R., Lewis, K. W., Mangold, N., Maurice, S., Ming, D. W., Morris, R. V., Nachon, M., Newsom, H. E., Ollila, A. M., Perrett, G. M., Rice, M. S., Schmidt, M. E., Schwenzer, S. P., Stack, K., Stolper, E. M., Sumner, D. Y., Treiman, A. H., Vanbommel, S., Vaniman, D. T., Vasavada, A., Wiens, R. C., Yingst, R. A., Kemppinen, O., Bridges, N., Johnson, J. R., Minitti, M., Cremers, D., Farmer, J., Godber, A., Wadhwa, M., Wellington, D., McEwan, I., Newman, C., Richardson, M., Charpentier, A., Peret, L., Blank, J., Weigle, G., Li, S., Milliken, R., Robertson, K., Sun, V., Baker, M., Edwards, C., Farley, K., Griffes, J., Miller, H., Newcombe, M., Pilorget, C., Siebach, K., Brunet, C., Hipkin, V., Marchand, G., Sánchez, P. S., Favot, L., Cody, G., Steele, A., Flückiger, L., Lees, D., Nefian, A., Martin, M., Gailhanou, M., Westall, F., Israël, G., Agard, C., Baroukh, J., Donny, C., Gaboriaud, A., Guillemot, P., Lafaille, V., Lorigny, E., Paillet, A., et al., 2014. Elemental geochemistry of sedimentary rocks at Yellowknife Bay, Gale Crater, Mars. Science 343.

McMahon, H.O., 1950. Thermal radiation from partially transparent reflecting bodies. J. Opt. Soc. Am. 410, 376–380.

McMillan, P. 1985. Vibrational spectroscopy in the mineral sciences. In: Kieffer, S.W., Navrotsky, A. (Eds.), Reviews in Mineralogy, vol. 14. Microscopic to macroscopic. Atomic Environments to Mineral Thermodynamics. Mineralogical Society of America, Washington, DC.

McMillan, P.F., Hofmeister, A.M., 1988. Infrared and Raman Spectroscopy. In: Hawthorne, F.C. (Ed.), Spectroscopic Methods in Mineralogy and Geology. Reviews in Mineralogy, Mineralogical Society of America, vol.18. 99–159.

McSween Jr., H.Y.J., Treiman, A.H., 1998. Chapter 6: Martian meteorites. In: Papike, J.J. (Ed.), Reviews in Mineralogy, vol. 36. Planetary Materials, Mineralogical Society of America.

McSween Jr., H.Y.J., Taylor, G.J., Wyat, M.B., 2009. Elemental composition of the Martian crust. Science 324, 736–739.

Meads, R.E., Malden, P.J., 1975. Electron spin resonance in natural kaolinites containing iron(3+) and other transition metal ions. Clay Miner. 10, 313–345.

Mellini, M., 1982. The crystal structure of lizardite-1T: hydrogen bonds and polytypism. Am. Mineral. 67, 587–598.

Mellini, M., Viti, C., 1994. Crystal structure of lizardite-1T from Elba. Italy. Am. Mineral. 79, 1194–1198.

Mendelovici, E., Yariv, S., Villalba, R., 1979. Iron-bearing kaolinite in Venezuelan laterites: I. Infrared spectroscopy and chemical dissolution evidence. Clay Miner. 14, 323–331.

Mendelsohn, R., Brauner, J.W., Gericke, A., 1995. External infrared reflection absorption spectrometry of monolayer films at the air-water interface. Annu. Rev. Phys. Chem. 46, 305–334.

Menéndez, J., Cardona, M., 1984. Temperature dependence of the first-order Raman scattering by phonons in Si, Ge, and α-Sn: anharmonic effects. Phys. Rev. B 29, 2051–2059.

Mennucci, B., Martínez, J.M., 2005. How to model solvation of peptides? Insights from a quantum-mechanical and molecular dynamics study of N-methylacetamide. 1. Geometries, infrared, and ultraviolet spectra in water. Phys. Chem. B 109, 9818–9829.

Mermut, A.R., 1994. Problems associated with layer charge characterization of phyllosilicates. In: Mermut, A.R. (Ed.), Layer Charge Characteristics of Clays, CMC Workshop Lectures. The Clay Minerals Society, 6. Boulder, CO, pp. 106–122.

Mermut, A.R., Cano, A.F., 2001a. Baseline studies of the Clay Minerals Society Source Clays: chemical analyses of major elements. Clays Clay Miner. 49, 381–386.

Mermut, A.R., Lagaly, G., 2001b. Baseline studies of the clay minerals society source clays: layer-charge determination and characteristics of those minerals containing 2.1 layers. Clays Clay Miner. 49, 393–397.

Mestdagh, M.M., Vielvoye, L., Herbillon, A.J., 1980. Iron in kaolinite: II. The relationship between kaolinite crystallinity and iron content. Clay Miner. 15, 1–13.

Mestdagh, M.M., Herbillon, A.J., Rodique, L., Rouxhet, P.G., 1982. Evaluation du rôle du fer structural sur la cristallinité des kaolinites. Bull. Minéral. 105, 457–466.

Meunier, A., 2005. Clays. Springer, Berlin Heidelberg New York.

Michaelian, K.H., 1986. The Raman spectrum of kaolinite #9 at 21°C. Can. J. Chem. 64, 285–289.

Michalik-Zym, A., Dula, R., Duraczynska, D., Krysciak-Czerwenka, J., Machej, T., Socha, R.P., Wlodarczyk, W., Gawel, A., Matusik, J., Bahranowski, K., Wisla-Walsh, E., Litynska-Dobrzynska, L., Serwicka, E.M., 2015. Active, selective and robust Pd and/or Cr catalysts supported on Ti-, Zr- or [Ti, Zr]-pillared montmorillonites for destruction of chlorinated volatile organic compounds. Appl. Catal. B 174–175, 293–307.

Michálková Scott, A., Dawley, M.M., Orlando, T.M., Hill, F.C., Leszczynski, J., 2012. Theoretical study of the roles of $Na^+$ and water on the adsorption of formamide on kaolinite surfaces. J. Phys. Chem. C 116, 23992–24005.

Michalski, J.R., Noe Dobrea, E.Z., 2007. Evidence for a sedimentary origin of clay minerals in the Mawrth Vallis region, Mars. Geology 35, 951–954.

Michalski, J.R., Fergason, R.L., 2009. Composition and thermal inertia of the Mawrth Vallis region of Mars from TES and THEMIS data. Icarus 199, 25–48.

Michalski, J.R., Bleacher, J.E., 2014. Volcano-ice interactions in the Askja Region, Central Iceland: High-resolution remote sensing analyses. GSA Annual Meeting, 2014 Vancouver, BC. Abstract #234-2.

Michalski, J. R., Reynolds, S. J., Sharp, T. G. & Christensen, P. R. 2004. Thermal infrared analysis of weathered granitic rock compositions in the Sacaton Mountains, Arizona: implications for petrologic classifications from thermal infrared remote-sensing data. J. Geophys. Res.: Planets 109 (E03005), 1–18.

Michalski, J.R., Kraft, M.D., Sharp, T.G., Christensen, P.R., 2006a. Effects of chemical weathering on infrared spectra of Columbia River Basalt and spectral interpretations of martian alteration. Earth Planet. Sci. Lett. 248, 822–829.

Michalski, J., Poulet, F.S., Bibring, J.-P., Mangold, N., 2010a. Analysis of phyllosilicate deposits in the Nili Fossae region of Mars: comparison of TES and OMEGA data. Icarus 206, 269–289.

Michalski, J.R., Niles, P.B., Cuadros, J., Baldridge, A.M., 2013b. Multiple working hypotheses for the formation of compositional stratigraphy on Mars: Insights from the Mawrth Vallis region. Icarus 226, 816–840.

Michalski, J.R., Kraft, M.D., Sharp, T.G., Williams, L.B., Christensen, P.R., 2005. Mineralogical constraints on the high-silica martian surface component observed by TES. Icarus 174, 161–177.

Michalski, J.R., Kraft, M.D., Sharp, T.G., Williams, L.B., Christensen, P.R., 2006b. Emission spectroscopy of clay minerals and evidence for poorly crystalline aluminosilicates on Mars from Thermal Emission Spectrometer data. J. Geophys. Res. 111 (E03004), 1–14. https://doi.org/10.1029/2005JE002438.

Michalski, J.R., Reynolds, S.J., Niles, P.B., Sharp, T.G., Christensen, P.R., 2007. Alteration mineralogy in detachment zones: insights from Swansea, Arizona. Geosphere 3, 184–198.

Michalski, J.R., Cuadros, J., Niles, P.B., Parnell, J., Deanne Rogers, A., Wright, S.P., 2013a. Groundwater activity on Mars and implications for a deep biosphere. Nature Geosci 6, 133–138.

Michalski, J.R., Cuadros, J., Bishop, J.L., Darby Dyar, M., Dekov, V., Fiore, S., 2015. Constraints on the crystal-chemistry of Fe/Mg-rich smectitic clays on Mars and links to global alteration trends. Earth Planet. Sci. Lett. 427, 215–225.

Michalski, J.R., Poulet, F., Loizeau, D., Mangold, N., Noe Dobrea, E.Z., Bishop, J.L., Wray, J.J., McKeown, M.K., Parente, M., Hauber, E., Altieri, F., Carrozzo, F.G., Niles, P.B., 2010b. The Mawrth Vallis region of mars: a potential landing site for the Mars Science Laboratory (MSL) mission. Astrobiology 10, 687–703.

Michot, L.J., Villieras, F., 2002. Assessment of surface energetic heterogeneity of synthetic Na-saponites. The role of layer charge. Clay Miner. 37, 39–57.

Michot, L.J., Ferrage, E., Delville, Jimenez-Ruiz, M., 2016. Influence of layer charge, hydration state and cation nature on the collective dynamics of interlayer water in synthetic swelling clay minerals. Appl. Clay Sci. 119, 375–384.

Michot, L.J., Villieras, F., Francois, M., Yvon, J., Ledred, R., Cases, J.M., 1994. The structural microscopic hydrophilitcy of talc. Langmuir 10, 3765–3773.

Milliken, R.E., Rivkin, A.S., 2009. Brucite and carbonate assemblages from altered olivine-rich materials on Ceres. Nat. Geosci. 2, 258–261.

Milliken, R.E., Grotzinger, J.P., Thomson, B.J., 2010. The paleoclimate of Mars as captured by the stratigraphic record in Gale Crater. Geophys. Res. Lett. 37 (L04201), 1–6. https://doi.org/10.1029/2009GL041870.

Milosevic, M., 2012. Internal Reflection and ATR Spectroscopy, first ed., John Wiley and Sons. ISBN: 978 0 470 27832 1.

Ming, D.W., Archer, P.D., Glavin, D.P., Eigenbrode, J.L., Franz, H.B., Sutter, B., Brunner, A.E., Stern, J.C., Freissinet, C., McAdam, A.C., Mahaffy, P.R., Cabane, M., Coll, P., Campbell, J.L., Atreya, S.K., Niles, P.B., Bell, J.F., Bish, D.L., Brinckerhoff, W.B., Buch, A., Conrad, P.G., Des Marais, D.J., Ehlmann, B.L., Fairén, A.G., Farley, K., Flesch, G.J., Francois, P., Gellert, R., Grant, J.A., Grotzinger, J.P., Gupta, S., Herkenhoff, K.E., Hurowitz, J.A., Leshin, L.A., Lewis, K.W., McLennan, S.M., Miller, K.E., Moersch, J., Morris, R.V., Navarro-González, R., Pavlov, A.A., Perrett, G.M., Pradler, I., Squyres, S.W., Summons, R.E., Steele, A., Stolper, E.M., Grotzinger, J.P., Gupta, S., Herkenhoff, K.E., Hurowitz, J.A., Leshin, L.A., Lewis, K.W., McLennan, S.M., Miller, K.E., Moersch, J., Morris, R.V., Navarro-González, R., Pavlov, A.A., Perrett, G.M., Pradler, I., Squyres, S.W., Summons, R.E., Steele, A., Stolper, E.M., Sumner, D.Y., Szopa, C., Teinturier, S., Trainer, M.G., Treiman, A.H., Vaniman, D.T., Vasavada, A.R., Webster, C.R., Wray, J.J., Yingst, R.A., Team, M.S., 2014. Volatile and organic compositions of sedimentary rocks in Yellowknife Bay, Gale Crater, Mars. Science 343 (6169), 1245267. https://doi.org/10.1126/science.1245267.

Minitti, M.E., Mustard, J.F., Rutherford, M.J., 2002. The effects of glass content and oxidation on the spectra of SNC-like basalts: applications to mars remote sensing. J. Geophys. Res. 107 (6), 1–16.

Mink, J., Keresztury, G., 1993. FT-IR emission spectroscopy and its applications. Appl. Spectrosc. 47, 1446–1451.

Mittal, V., 2012. Surface modification of layered silicates. II. Factors affecting thermal stability. Philos. Mag. 92, 4518–4535.

Mittlefehldt, D.W., 1994. ALH84001, a cumulate orthopyroxeneite member of the Martian meteorite clan. Meteoritics 29, 214–221.

Miwa, Y., Drews, A.R., Schlick, S., 2006. Detection of the direct effect of clay on polymer dynamics: The case of spin-labeled poly(methyl acrylate)/clay nanocomposites studied by ESR, XRD, and DSC. Macromolecules 39, 3304–3311.

Miyawaki, R., Tomura, S., Inukai, K., Shibasaki, Y., Okazaki, M., Samejima, S., Satokawa, S., 1992. Formation process of kaolinite from the amorphous mixture of silica and alumina. Clay Sci. 8, 273–284.

Miyawaki, R., Tomura, S., Inukai, K., Okazaki, M., Toriyama, K., Shibasaki, Y., Kamori, M., 1993. Formation process of kaolinite from amorphous calcium silicate and aluminum chloride. Clay Sci. 9, 21–32.

Mohammadnejad, S., Provis, J.L., van Deventer, J.S.J., 2014. Effects of grinding on the pregrobbing behaviour of pyrophyllite. Hydrometallurgy 146, 154–163.

Mojović, Z., Jović-Jovičić, N., Milutinović-Nikolić, A., Banković, P., Rabi-Stanković, A.A., Jovanović, D., 2011. Phenol determination on HDTMA-bentonite-based electrodes. J. Hazard. Mater. 194, 178–184.

Molu, Z.B., Yurdakoc, K., 2009. Preparation and characterization of aluminum pillared K10 and KSF for adsorption of trimethoprim. Microporous Mesoporous Mater. 127, 50–60.

Montarges-Pelletier, E., Bogenez, S., Pelletier, M., Razafitianamaharavo, A., Ghanbaja, J., Lartiges, B., Michot, L.J., 2005. Synthetic allophane-like particles: textural properties. Colloids Surf. A: Physicochem. Eng. Asp. 255, 1–10.

Morlok, A., Bischoff, A., Stephan, T., Floss, C., Zinner, E., Jessberger, E.K., 2006. Brecciation and chemical heterogeneities of CI chondrites. Geochim. Cosmochim. Acta 70, 5371–5394.

Moronta, A., Ferrer, V., Quero, J., Arteaga, G., Choren, E., 2002. Influence of preparation method on the catalytic properties of acid-activated tetramethylammonium-exchanged clays. Appl. Catal. A:Gen. 230, 127–135.

Mortensen, A., Nielsen, O.F., Yarwood, J., Shelley, V., 1994. Vibrational spectra of mixtures of isotopomers of formamide. Anomalies in the carbonyl stretching region. J. Phys. Chem. 98, 5221–5226.

Mortensen, A., Nielsen, O.F., Yarwood, J., Shelley, V., 1995a. Interactions in liquid mixtures of formamide isotopomers studied by Raman spectroscopy. J. Raman Spectrosc. 26, 669–674.

Mortensen, A., Nielsen, O.F., Yarwood, J., Shelley, V., 1995b. Anomalies in the Isotropic Raman spectra of liquid mixtures of isotopomers of formamide. Intermolecular interactions in the carbonyl stretching region. J. Phys. Chem. 99, 4435–4440.

Mortland, M.M., Raman, K.V., 1968. Surface acidity of smectites in relation to hydration, exchangeable-cation and structure. Clays Clay Miner. 16, 393–398.

Mortland, M.M., Pinnavaia, T.J., 1971. Formation of copper(II) arene complexes on the interlamellar surfaces of montmorillonite. Nature 229, 75–77.

Mosser, C., Petit, S., Mestdagh, M., 1993. ESR and IR evidence for chromium in kaolinites. Clay Miner. 28, 353–364.

Mosser, C., Michot, L.J., Villieras, F., Romeo, M., 1997. Migration of cations in copper(II)-exchanged montmorillonite and laponite upon heating. Spectrochim. Acta A 55, 2467–2476.

Mottl, M.J., Glazer, B.T., Kaiser, R.I., Meech, K.J., 2007. Water and astrobiology. Chem Erde-Geochem. 67, 253–282.

Mouazen, A.M., Kuang, B., De Baierdemaeker, J., Ramon, H., 2010. Comparison among principal component, partial least squeares and back propogation network analyses for accuracy of measurement of selected soil properties with visible and near infrared spectroscopy. Geoderma 158, 23–31.

Mulla, D.J., Low, P.F., Roth, C.B., 1985. Measurement of the specific surface area of clays by internal reflectance spectroscopy. Clays Clay Miner. 33, 391–396.

Muller, F., Besson, G., Manceau, A., Drits, V.A., 1991. Distribution of isomorphous cations with octahedral sheets in montmorillonite from Camp-Bertaux. Phys. Chem. Miner. 24, 159–166.

Muller, F., Drits, V.A., Plançon, A., Robert, J.-L., 2000. Structural transformation of 2:1 dioctahedral layer silicates during dehydroxylation-rehydroxylation reactions. Clays Clay Miner. 48, 572–585.

Müller, C.M., Pejcic, B., Esteban, L., Delle Piane, C., Raven, M., Mizaikoff, B., 2014. Infrared attenuated total reflectance spectroscopy: an innovative strategy for analyzing mineral components in energy relevant systems. Sci. Rep. 4 (6764), 1–11.

Murad, E., 1997. Identification of minor amounts of anatase in kaolins by Raman spectroscopy. Am. Mineral. 82, 203–206.

Murchie, S.L., Bibring, J.P., Arvidson, R.E., Bishop, J.L., Carter, J., Ehlmann, B.L., Langevin, Y., Mustard, J.F., Poulet, F., Riu, L., Seelos, K.D., Viviano-Beck, C.E., 2018. Chapter 24. VSWIR spectral analyses of Mars from orbit using CRISM and OMEGA. In: Bishop, J.L., Moersch, J.E., Bell J.F. III (Eds.), Remote Compositional Analysis: Techniques for Understanding Spectroscopy, Mineralogy, and Geochemistry of Planetary Surfaces. Cambridge University Press, Cambridge, UK.

Murchie, S.L., Mustard, J.F., Ehlmann, B.L., Milliken, R.E., Bishop, J.L., Mckeown, N.K., Noe Dobrea, E. Z., Seelos, F. P., Buczkowski, D. L., Wiseman, S. M., Arvidson, R. E., Wray, J. J., Swayze, G. A., Clark, R. N., Des Marais, D. J., Mcewen, A. S. & Bibring, J. P., 2009. A synthesis of Martian aqueous mineralogy after 1 Mars year of observations from the Mars Reconnaissance Orbiter. J. Geophys. Res. 114. https://doi.org/10.1029/2009JE003342.

Murray, H.H., 2000. Traditional and new applications for kaolin, smectite, and palygorskite: a general overview. Appl. Clay Sci. 17, 207–221.

Mustard, J.F., Pieters, C.M., 1987. Abundance and distribution of ultramafic microbreccia in moses rock dike: quantitative application of mapping spectroscopy. J. Geophys. Res.: Solid Earth 92, 10376–10390.

Mustard, J.F., Ehlmann, B.L., Murchie, S.L., Poulet, F., Mangold, N., Head, J.W., Bibring, J.-P., Roach, L.H., 2009. Composition, morphology, and stratigraphy of Noachian crust around the Isidis basin. J. Geophys. Res. 114. https://doi.org/10.1029/2009JE003349.

Mustard, J.F., Murchie, S.L., Pelkey, S.M., Ehlmann, B.L., Milliken, R.E., Grant, J.A., Bibring, J.-P., Poulet, F., Bishop, J.L., Noe Dobrea, E.Z., Roach, L.A., Seelos, F., Arvidson, R.E., Wiseman, S., Green, R., Hash, C., Humm, D., Malaret, E., McGovern, J.A., Seelos, K., Clancy, R.T., Clark, R.N., Des Marais, D., Izenberg, N., Knudson, A.T., Langevin, Y., Martin, T., McGuire, P., Morris, R.V., Robinson, M., Roush, T., Smith, M., Swayze, G.A., Taylor, H., Titus, T.N., Wolff, M., 2008. Hydrated silicate minerals on mars observed by the mars reconnaissance orbiter CRISM instrument. Nature 454, 305–309. https://doi.org/10.1038/nature07097.

Naes, T., Martens, H., 1988. Principal component regression in NIR analysis: viewpoints, background details and selection of components. J. Chemometr. 2, 155–167.

Naes, T., Isaksson, T., Fearn, T., Davies, T., 2002. A user-friendly guide to multivariate calibration and classification. NIR Publications, Chichester.

Nakakuki, T., Fujimura, K., Aisawa, S., Hirahara, H., Narita, E., 2004. Synthesis and physicochemical properties of Zn-hectorite. Clay Sci. 12, 285–291.

Nasdala, L., Smith, D., Kaindl, R., Ziemann, M., 2004. Raman spectroscopy: analytical perspectives in mineralogical research. In: Beran, A., Libowitzky, E. (Eds.), Spectroscopic Methods in Mineralogy. EMU Notes in Mineralogy, vol. 6. 281–343.

Neder, R.B., Burghammer, M., Grasl, T., Schulz, H., Bram, A., Fiedler, S., 1999. Refinement of the kaolinite structure from single-crystal synchrotron data. Clays Clay Miner. 47, 487–494.

Neumann, A., Petit, S., Hofstetter, T.B., 2011. Evaluation of redox-active iron sites in smectites using middle and near infrared spectroscopy. Geochim. Cosmochim. Acta 75, 2336–2355.

Newman, A.C.D., 1987. The interaction of water with clay mineral surfaces. In: Newman, A.C.D. (Ed.), Chemistry of clays and clay minerals. Monograph No. 6. London: Minerlogical Society.

Newton, M.A., 2009. Applying dynamic and synchronous DRIFT/EXAFS to the structural reactive behaviour of dilute ($\leq$1 wt%) supported $Rh/Al_2O_3$ catalysts using quick and energy dispersive EXAFS. Top. Catal. 52, 1410–1424.

Newton, M.A., van Beek, W., 2010. Combining synchrotron-based X-ray techniques with vibrational spectroscopies for the *in situ* study of heterogeneous catalysts: a view from a bridge. Chem. Soc. Rev. 39, 4845–4863.

Nguyen, T.T., 1986. Infrared spectroscopic study of the formamide-Na-montmorillonite complex. Conversion of s-triazine to formamide. Clays Clay Miner. 35, 521–528.

Ni, Y.C., Skinner, J.L., 2015. IR and SFG vibrational spectroscopy of the water bend in the bulk liquid and at the liquid-vapor interface, respectively. J. Chem. Phys. 143, 12.

Nicolini, K.P., Fukamachi, C.R.B., Wypych, F., Mangrich, A.S., 2009. Dehydrated halloysite intercalated mechanochemically with urea: thermal behavior and structural aspects. J. Colloid Interface Sci. 338, 474–479.

Nikolopoulou, A., Papoulis, D., Komarneni, S., Tsolis-Katagas, P., Panagiotaras, D., Kacandes, G.H., Zhang, P., Yin, S., Sato, T., 2009. Solvothermal preparation of $TiO_2$/saponite nanocomposites and photocatalytic activity. Appl. Clay Sci. 46, 363–368.

Nkoumbou, C., Villieras, F., Njopwouo, D., Ngoune, C.Y., Barres, O., Pelletier, M., Razafitianamaharavo, A., Yvon, J., 2008. Physicochemical properties of talc ore from three deposits of Lamal Pougue area (Yaounde Pan-African Belt, Cameroon), in relation to industrial uses. Appl. Clay Sci. 41 (113–132), 2008.

Noe Dobrea, E.Z., Bishop, J.L., Mckeown, N.K., Fu, R., Rossi, C.M., Michalski, J. R., Heinlein, C., Hanus, V., Poulet, F., Arvidson, R., Mustard, J.F., Ehlmann, B.L., Murchie, S., Mcewen, A.S., Swayze, G., Bibring, J.-P., Malaret, J.F.E., Hash, C., 2010. Mineralogy and stratigraphy of phyllosilicate-bearing and dark mantling units in the greater Mawrth Vallis/west Arabia Terra area: constraints on geological origin. J. Geophys. Res. 115. doi:10.1029/2009JE003351.

Nozette, S., Wilkening, L.L., 1982. Evidence for aqueous alteration in a carbonaceous xenolith from the Plainview (H5) chondrite. Geochim. Cosmochim. Acta 46, 557–563.

Nunez, J.I., Bridges, N.T., Seelos, F.P., Hook, S.J., Baldridge, A.M., Thompson, B.J., 2015. Comparison of simulated CRISM observations using airborn hyperspectral images with ground truth: implications for mars. Lunar Planet. Sci. Conf. XLVII, 2015 The Woodlands, TX. Abstract #1133.

Ocelli, M.L., Finseth, D.H., 1986. Preparation and characterization of pillared hectorite catalysts. J. Catal. 99, 316–326.

Ogawa, M., Matsutomo, T., Okada, T., 2009. Preparation of iron-containing hectorite-like swelling silicate. Bull. Chem. Soc. Jpn. 82, 408–412.

Ogloza, A., Malhorta, V., 1989. Dehydroxylation induced structural transformations in montmorillonite: an isothermal FTIR study. Phys. Chem. Miner. 16, 378–385.

Okoye, I.P., Obi, C., 2011a. Synthesis and characterization of titanium pillared bentonite clay mineral. Res. J. Appl. Sci. 6, 443–446.

Okoye, I.P., Obi, C., 2011b. Synthesis and characterization of Al-pillared bentonite clay minerals. Res. J. Appl. Sci. 6, 447–450.

Okoye, I.P., Obi, C., Maduakolam, A.I., 2012. The preparation and application of environmentally benign titanium pillared clay catalyst for esterification of ethanol and acetic acid. Am. Chem. Sci. J. 2, 45–59.

Olejnik, S., Posner, A.M., Quirk, J.P., 1971. Infrared spectra of interlamellar kaolinite-amide complexes. II. Acetamide, N-methylacetamide, and dimethylacetamide. J. Colloid Interface Sci 37, 536–547.

Onikata, M., Kondo, M., Hayashi, N., Yamanaka, S., 1999. Complex formation of cation-exchanged montmorillonites with propylene carbonate: osmotic swelling in aqueous electrolyte solutions. Clays Clay Miner. 47, 672–677.

Ookawa, M., Inoue, Y., Watanabe, M., Suzuki, M., Yamaguchi, T., 2006a. Synthesis and characterization of Fe containing imogolite. Clay Sci. 12, 280–284.

Ookawa, M., Hirao, Y., Watanabe, M., Maekawa, T., Inukai, K., Miyamoto, S., Yamaguchi, T., 2006b. Synthesis of aluminum germanate tubular material using germanium oxide as the source of germanium. Clay Sci. 13, 69–73.

Osman, M.A., 2006. Organo-vermiculites: synthesis, structure and properties. Platelike nanoparticles with high aspect ratio. J. Mater. Chem. 16, 3007–3013.

Osman, M.A., Ploetze, M., Skrabal, P., 2004. Structure and properties of alkylammonium monolayers self-assembled on montmorillonite platelets. J. Phys. Chem. B 108, 2580–2588.

Paiva, L., Morales, A., Valenzueladiaz, F., 2008. Organoclays: properties, preparation and applications. Appl. Clay Sci. 42, 8–24.

Pajcini, V., Dhamelincourt, P., 1994. Raman study of OH-stretching vibrations in kaolinite at low temperature. Appl. Spectrosc. 48, 638–641.

Palinko, I., Lazar, K., Kiricsi, I., 1997. Cationic mixed pillared layer clays: infrared and Mossbauer characteristics of the pillaring agents and pillared structures in Fe, Al and Cr, Al pillared montmorillonites. J. Mol. Struct. 410–411, 547–550.

Palinko, I., Lazar, K., Hannus, I., Kirisci, I., 1996. Step toward nanoscale Fe moieties: intercalation of simple and Keggin-type iron-containing ions in-between the layers of Na-montmorillonite. J. Phys. Chem. Solids 57, 1067–1072.

Pálková, H., Madejová, J., Righi, D., 2003. Acid dissolution of reduced-charge Li- and Ni-montmorillonites. Clays Clay Miner. 51, 133–142.

Pálková, H., Madejová, J., Zimowska, M., Serwicka, E.M., 2010. Laponite-derived porous clay heterostructures: II. FTIR study of the structure evolution. Microporous Mesoporous Mater. 127, 237–244.

Pálková, H., Jankovič, Ľ., Zimowska, M., Madejová, J., 2011. Alterations of the surface and morphology of tetraalkyl-ammonium modified montmorillonites upon acid treatment. J. Colloid Interface Sci. 363, 213–222.

Pálková, H., Hronský, V., Jankovič, Ľ., Madejová, J., 2013. The effect of acid treatment on the structure and surface acidity of tetraalkylammonium-montmorillonites. J. Colloid Interface Sci. 395, 166–175.

Pálková, H., Hronský, V., Bizovská, V., Madejová, J., 2015. Spectroscopic study of water adsorption on $Li^{+}$, $TMA^{+}$ and $HDTMA^{+}$ exchanged montmorillonite. Spectrochim. Acta A 149, 751–761.

Pálková, H., Zimowska, M., Jankovič, Ľ., Sulikowski, B., Serwicka, E.M., Madejová, J., 2017. Thermal stability of tetrabutyl-phosphonium and -ammonium exchanged montmorillonite: influence of acid treatment. Appl. Clay Sci. 138, 63–73.

Pampuch, R., Wilkos, K., 1963. Zuordnung der I.R.-Banden in Gebiet der Hydroxylschwingungen für die Gruppe der Kaolinminerale und kaolinitische Tone, In: Tamas, F. (Ed.), Proc. 7th Conf. Silic. Ind., Akadémiai Kiadó, Budapest.

Papike, J.J. (Ed.), 1989. Planetary Materials: Mineralogical Society of America.

Parfitt, R.L., Furkert, R.J., 1980. Identification and structure of two types of allophane from volcanic ash soils and tephra. Clays Clay Miner. 28, 328–334.

Parfitt, R.L., Russell, J.D., Farmer, V.C., 1976. Confirmation of the surface structures of geothite (alpha-FeOOH) and phosphated goethite by infrared spectroscopy. J. Chem. Soc. Faraday Trans. I 72, 1082–1087.

Park, M., Lee, D.H., Choi, C.L., Kim, S.S., Kim, K.S., Choi, J., 2002. Pure Na-4-mica: synthesis and characterization. Chem. Mater. 14, 2582–2589.

Parker, R.W., Frost, R.L., 1996. The application of DRIFT spectroscopy to the multicomponent analysis of organic chemicals adsorbed on montmorillonite. Clays Clay Miner. 44, 32–40.

Parker, T.W., 1969. Classification of kaolinites by infrared spectroscopy. Clay Miner. 8, 135–141.

Paroz, G.N., Frost, R.L., Cash, G.A., 1998. Dehydroxylation of intercalated kaolinite. An infrared emission spectroscopic study. In: de Haseth, J.A. (Ed.), 11th Inr. Conf. Fourier Transform Spectroscopy. American Institute of Physics (AIP), Woodburg. New York, pp. 678–680.

Parry, S.A., Pawley, A.R., Jones, R.L., Clark, S.M., 2007. An infrared spectroscopic study of the OH stretching frequencies of talc and 10-A phase to 10 Gpa. Am. Mineral. 92, 525–531.

Pascale, F., Tosoni, S., Zicovich-Wilson, C., Ugliengo, P., Orlando, R., Dovesi, R., 2004. Vibrational spectrum of brucite, $Mg(OH)_2$, a periodic ab initio quantum mechanical calculation including OH anharmonicity. Chem. Phys. Lett. 396, 308–315.

Pasquarello, A., Car, R., 1997. Dynamical charge tensors and the infrared spectrum of amorphous $SiO_2$. Phys. Rev. Lett. 79, 1766–1769.

Pauling, L., 1929. The principles determining the structure of complex ionic crystals. J. Am. Chem. Soc. 51, 1010–1026.

Pauling, L., 1930. The structure of chlorites. Proc. Natl. Acad. Sci. U. S. A. 16, 578–582.

Pawly, A.R., Jones, R.L., 2011. Hydroxyl stretching in phyllosilicates at high pressures and temperatures: An infrared spectroscopic study. Phys. Chem. Miner. 38, 753–765.

Pazos, M.C., Cota, A., Osuna, F.J., Pavon, E., Alba, M.D., 2015. Self-assembling of tetradecylammonium chain on swelling high charge micas (Na-mica-3 and Na-mica-2): effect of alkylammonium concentration and mica layer charge. Langmuir 31, 4394–4401.

Pazos, M.C., Castro, M.A., Orta, M.M., Pavon, E., Rios, J.S.V., Alba, M.D., 2012. Synthetic high-charge organomica: effect of the layer charge and alkyl chain length on the structure of the adsorbed surfactants. Langmuir 28, 7325–7332.

Pearlman, J.S., Barry, P.S., Segal, C.C., Shepanski, J., Beiso, D., Carman, S.L., 2003. Hyperion, a space-based imaging spectrometer. IEEE Trans. Geosci. Remote Sens. 41, 1160–1173.

Pecharroman, C., Iglesias, J.E., 1996. Modeling particle size and clumping effects in the IR absorbance spectra of dilute powders. Appl. Spectrosc. 50, 1553–1562.

Pecharroman, C., Iglesias, J.E., 2000. Effect of particle shape on the IR reflectance spectra of pressed powders of anisotropic materials. Appl. Spectrosc. 54, 634–638.

Pecharroman, C., Ocana, M., Tartaj, P., Serna, C.J., 1994. Infrared optical properties of zircon. Mater. Res. Bull. 29, 417–442.

Pelletier, M., Michot, L.J., Barres, O., Humbert, B., Petit, S., Robert, J.-L., 1999. Influence of KBr conditioning on the infrared hydroxyl-stretching region of saponites. Clay Miner. 34, 439–445.

Pelletier, M., Michot, L.J., Humbert, B., Barres, O., d'Espinose De La Caillerie, J.-B., Robert, J.-L., 2003. Influence of layer charge on the hydroxyl stretching of trioctahedral clay minerals: a vibrational study of synthetic Na- and K-saponites. Am. Mineral. 88, 1801–1808.

Pentrák, M., Madejová, J., Komadel, P., 2009. Acid and alkali treatment of kaolins. Clay Miner. 44, 511–523.

Pentrák, M., Madejová, J., Komadel, P., 2010. Effect of chemical composition and swelling on acid dissolution of 2:1 clay minerals. Philos. Mag. 90, 2387–2397.

Pentrák, M., Bizovská, V., Madejová, J., 2012a. Near-IR study of water adsorption on acid-treated montmorillonite. Vib. Spectrosc. 63, 360–366.

Pentrák, M., Madejová, J., Andrejkovičová, S., Uhlík, P., Komadel, P., 2012b. Stability of kaolin sands from Vyšný Petrovec deposit (south Slovakia) in an acid environment. Geol. Carpath. 63, 503–512.

Pentrák, M., Pentráková, L., Radian, A., Mishael, Y.G., Stucki, J.W., 2014. Nitrate reduction by redox-modified smectites exchanged with chitosan. Clays Clay Miner. 62, 403–414.

Petit, S., 2005. Crystal-chemistry of talcs: A NIR and MIR spectroscopic approach. In: Kloprogge, J.T. (Ed.), The Application of Vibrational Spectroscopy to Clay Minerals and Layered Double Hydroxides, CMS Workshop Lectures. The Clay Mineral Society, 13. Aurora, CO, pp. 41–64.

Petit, S., 2006. Fourier Transform Infrared Spectroscopy. Chapter 12.6. In: Bergaya, F., Theng, B.K.G., Lagaly, G. (Eds.), Handbook of Clay Science. Developments in Clay Science, vol. I. Elsevier, The Netherlands, pp. 909–918.

Petit, S., Decarreau, A., 1990. Hydrothermal (200°C) synthesis and crystal chemistry of iron-rich kaolinites. Clay Miner. 25, 181–196.

Petit, S., Madejová, J., 2013. Fourier transform infrared spectroscopy. Chapter 2.7. In: Bergaya, F., Lagaly, G. (Eds.), Handbook of Clay Science, second ed., Part B: Techniques and Applications. Elsevier, The Netherlands, pp. 213–231.

Petit, S., Righi, D., Madejová, J., 2006. Infrared spectroscopy of $NH_4^+$ - bearing and saturated clay minerals: a review of the study of layer charge. Appl. Clay Sci. 34, 22–30.

Petit, S., Righi, D., Decarreau, A., 2008. Transformation of synthetic Zn-stevensite to Zn-talc induced by the hofmann-klemen effect. Clays Clay Miner. 56, 645–654.

Petit, S., Decarreau, A., Eymery, J.P., Thomassin, J.H., 1988. Synthesis of ferric kaolinites at 200°C. Comparison with kaolinite of supergene origin: $Fe^{3+}$ content, morphology, crystallinity. C.R. Acad Sci. Ser. II, t307.

Petit, S., Madejová, J., Decarreau, A., Martin, F., 1999b. Characterization of octahedral substitutions in kaolinites using near infrared spectroscopy. Clays Clay Miner. 47, 103–108.

Petit, S., Righi, D., Madejová, J., Decarreau, A., 1999a. Interpretation of the infrared $NH_4^+$ spectrum of the $NH_4^+$-clays: application to the evaluation of the layer charge. Clay Miner. 34, 543–549.

Petit, S., Decarreau, A., Martin, F., Buchet, R., 2004b. Refined relationship between the position of the fundamental OH stretching and the first overtones for clays. Phys. Chem. Miner. 31, 585–592.

Petit, S., Baron, F., Graby, O., Decarreau, A., 2016. Revisiting the cation mas-valence sum approach to assigning infrared OH-bands in dioctrahedral smectites in the light of new data from synthetic Ga-$Fe^{3+}$ smectites. Vib. Spectrosc. 87, 137–142.

Petit, S., Decarreau, A., Mosser, C., Ehret, G., Grauby, O., 1995. Hydrothermal synthesis (250°C) of copper-substituted kaolinites. Clays Clay Miner. 43, 482–494.

Petit, S., Martin, F., Wiewiora, A., De Parseval, Ph., Decarreau, A., 2004a. Crystal-chemistry of talcs: a NIR and MIR spectroscopic approach. Am. Mineral. 89, 319–326.

Petit, S., Decarreau, A., Gates, W.P., Andrieux, P., Grauby, O., 2015. Hydrothermal synthesis of dioctahedral smectites: The Al-$Fe^{3+}$ chemical series. Part II. Crystal-chemistry. Appl. Clay Sci. 104, 96–105.

Petit, S., Caillaud, J., Righi, D., Madejová, J., Elsass, F., Köster, H.M., 2002. Characterisation and crystal chemistry of a Fe-rich montmorillonite. Clay Miner. 37, 283–297.

Petra, L., Billik, P., Komadel, P., 2015. Preparation and characterization of hybrid materials consisting of high-energy ground montmorillonite and α-amino acids. Appl. Clay Sci. 115, 174–178.

Petriglieri, J.R., Salvioli-Mariani, E., Mantovani, L., Tribaudino, M., Lottici, P.P., Laporte-Magoni, C., Bersani, D., 2015. Micro-Raman mapping of the polymorphs of serpentine. J. Raman Spectrosc. 46, 953–958.

Pinnavaia, T.J., Mortland, M.M., 1971. Interlamellar metal complexes on layer silicates. I. Copper(II) arene complexes on montmorillonite. J. Phys. Chem. 75, 3957–3962.

Pinnavaia, T.J., Landau, S.D., Tzou, M.S., Johnson, I.D., 1985. Layer cross-linking in pillared clays. J. Am. Chem. Soc. 107, 7222–7224.

Pires, J., Pinto, M., Estella, J., Echeverría, J.C., 2008. Characterization of the hydrophobicity of mesoporous silicas and clays with silica pillars by water adsorption and DRIFT. J. Colloid Interface Sci. 317, 206–213.

Pironon, J., Pelletier, M., De Donato, P., Mosser-Ruck, R., 2003. Characterization of smectite and illite by FTIR spectroscopy of $NH_4^+$ interlayer cations. Clay Miner. 38, 201–211.

Plachá, D., Martynková, G.S., Rummeli, M.H., 2008. Preparation of organovermiculites using HDTMA: structure and sorptive properties using naphthalene. J. Colloid Interface Sci. 327, 341–347.

Planck, M., 1901. Ueber das gesetz der Energieverteilung im Normalspektrum. Ann. Phys. 4th series (vierte folge), 553-563.

Plée, D., Borg, F., Gatineau, L., Fripiat, J.J., 1985. High resolution solid state $^{27}Al$ and $^{29}Si$ nuclear magnetic resonance study of pillared clays. J. Am. Chem. Soc. 107, 2362–2369.

Poinsignon, C., Cases, J.M., Fripiat, J.J., 1978. Electrical-polarization of water molecules adsorbed by smectites. An infrared study. J. Phys. Chem. 82, 1855–1860.

Polverejan, M., Pauly, T.R., Pinnavaia, T.J., 2000. Acidic porous clay heterostructures (PCH): intragallery assembly of mesoporous silica in synthetic saponite clays. Chem. Mater. 12, 2698–2704.

Porter, W.M., Enmark, H.T., 1987. A System Overview Of The Airborne Visible/Infrared Imaging Spectrometer (Aviris). Proc. SPIE 0834 Imaging Spectroscopy II, San Diego, CA, pp. 22–31.

Post, J.L., 1984. Saponite from near Ballarat, California. Clays Clay Miner. 32, 147–153.

Post, J.L., Noble, P.N., 1993. The near-infrared combination band frequencies of dioctahedral smectites, micas, and illites. Clays Clay Miner. 41, 639–644.

Post, J.L., Borer, L., 2000. High-resolution infrared spectra, physical properties, and micromorphology of serpentines. Appl. Clay Sci. 16, 73–85.

Post, J.L., Cupp, B.L., Madsen, F.T., 1997. Beidellite and associated clays from the DeLamar mine and Florida mountain area, Idaho. Clays Clay Miner. 45, 240–250.

Post, J.E., Bish, D.L., Heaney, P.J., 2007. Synchrotron powder X-ray diffraction study of the structure and dehydration behavior of sepiolite. Am. Mineral. 92, 91–97.

Poulet, F., Bibring, J.-P., Mustard, J.F., Gendrin, A., Mangold, N., Langevin, Y., Arvidson, R.E., Gondet, B., Gomez, C., Omega, 2005. Phyllosilicates on Mars and implications for early Martian climate. Nature 438, 623–627.

Prencipe, M., Noel, Y., Bruno, M., Dovesi, R., 2009. The vibrational spectrum of lizardite-1T $[Mg_3Si_2O_5(OH)_4]$ at the $\Gamma$ point: a contribution from an ab initio periodic B3LYP calculation. Am. Mineral. 94, 986–994.

Presinger, A., 1963. Sepiolite and related compounds: its stability and applications. Clays Clay Miner. 10, 365–371.

Prieto, A.C., Dubessy, J., Cathelineau, M., 1991. Structure-composition relationships in trioctahedral chlorites: a vibrational spectroscopy study. Clays Clay Miner. 39, 531–539.

Prosanov, I.Y., Matvienko, A.A., 2010. Study of PVA thermal destruction by means of IR and Raman spectroscopy. Phys. Solid State 52, 2203–2206.

Prost, R., 1975a. Étude de l' hydratation des argiles: Interactions eau-minéral et mécanisme de la rétention de l'eau. I. Étude d' une zéolite (natrolite) et de deux minéraux fibreux (attapulgite et sépiolite). Ann. Agron. 26, 401–461.

Prost, R., 1975b. Étude de l' hydratation des argiles: interactions eau-minéral et mécanisme de la rétention de l'eau. II. Étude d' une smectite (hectorite). Ann. Agron. 26, 462–535.

Prost, R., Dameme, A., Huard, E., Driard, J., 1987. Low temperature (300-5K) IR study of structural OH groups of kaolinite, dickite, and nacrite. In: Schultz, L.G., van Olphen, H., Mumpton, F.A. (Eds.) Proceedings of the International Clay Conference. Denver 1985. Clay Minerals Society, Boulder, CO.

Prost, R., Koutit, T., Benchara, A., Huard, E., 1998. State and location of water adsorbed on clay minerals: consequences of the hydration and swelling-shrinkage phenomena. Clays Clay Miner. 46, 117–131.

Prost, R., Damene, A., Huard, E., Driard, J., Leydecker, J.P., 1989. Infrared study of structural OH in kaolinite, dickite, nacrite and poorly crystalline kaolinite at 5 to 600 K. Clays Clay Miner. 37, 464–468.

Putiš, M., Koppa, M., Snarska, B., Koller, F., Uher, P., 2012. The blueschist-associated perovskite-andradite-bearing serpentinized harzburgite from Dobsina (the Meliata Unit), Slovakia. J. Geosci. 57, 221–240.

Qu, X.L., Liu, P., Zhu, D.Q., 2008. Enhanced sorption of polycyclic aromatic hydrocarbons to tetra-alkyl ammonium modified smectites via cation-π interactions. Environ. Sci. Technol. 42, 1109–1116.

Quantin, C., Carter, J., Thollot, P., Broyer, J., Lozach, L., Davis, J., Grindrod, P. M., Pajola, M., Baratti, E., Rossato, S., Allemand, P., Bultel, B., Leyrat, C., Fernando, J., Ody, A., 2018. Oxia Planum is a wide noachian clay bearing plain hosting younger fluvio-deltaic deposits which has been chosen by ESA as the landing site for Exomars 2018. Lunar Planet. Sci. Conf. XLVII, 2016 The Woodlands, TX. Abstract #2863.

Raghuvanshi, G.S., Khandelwal, D.P., Bist, H.D., 1985. Fermi resonance in $M(CH_3COO)_2 \cdot 4H_2O$ (M = Mg, Co and Ni). Spectrochim. Acta A 41, 391–398.

Raghuvanshi, G.S., Pal, M., Patel, M.B., Bist, H.D., 1983. Vibrational spectra and phase transitions in $Ni(CH_3COO)_2.4X_2O$ (X = H, D). J. Mol. Struct. 101, 7–21.

Raman, C.V., Krishnan, K.S., 1928. A new type of secondary radiation. Nature 121, 501–502.

Rampe, E.B., Kraft, M.D., Sharp, T.G., 2013. Deriving chemical trends from thermal infrared spectra of weathered basalt: Implications for remotely determining chemical trends on Mars. Icarus 225, 749–762.

Rampe, E.B., Kraft, M.D., Sharp, T.G., Golden, D.C., Ming, D.W., Christensen, P.R., 2012. Allophane detection on Mars with Thermal Emission Spectrometer data and implications for regional-scale chemical weathering processes. Geology 40. https://doi.org/10.1130/G33215.1.

Ramsey, M.S., Christensen, P.R., 1998. Mineral abundance determination: quantitative deconvolution of thermal emission spectra. J. Geophys. Res. 103, 577–596.

Rana, K., Boyd, S.A., Teppen, B.J., Li, H., Liu, C., Johnston, C.T., 2009. Probing the microscopic hydrophobicity of smectite surfaces. A vibrational spectroscopic study of dibenzo-*p*-dioxin sorption to smectite. Phys. Chem. Chem. Phys. 11, 2976–2985.

Randolph, T.W., 2006. Scale-based normalization of spectral data. Cancer Biomarkers 2, 135–144.

Ras, R.H.A., Schoonheydt, R.A., Johnston, C.T., 2007. Relation between s-polarized and p-polarized internal reflection spectra: application for the spectral resolution of perpendicular vibrational modes. J. Phys. Chem. A 111, 8787–8791.

Ras, R.H.A., Németh, J., Johnston, C.T., Dékány, I., Schoonheydt, R.A. 2004. Infrared reflection absorption spectroscopy study of smectite clay monolayers. Thin Solid Films, 291–294.

Ras, R.H.A., Umemura, Y., Johnston, C.T., Yamagishi, A., Schoonheydt, R.A., 2006. Ultrathin hybrid films of clay minerals. Phys. Chem. Chem. Phys. 9, 918–932.

Ras, R.H.A., Johnston, C.T., Franses, E.I., Ramackers, R., Maes, G., Foubert, P., De Schryver, F.C., Shoonheydt, R.A., 2003. Polarized infrared study of hydrid Langmuir-Blodgett monolayers containing clay minerals nanoparticles. Langmuir 19, 4295–4302.

Raupach, M., Barron, P.F., Thompson, J.G., 1987. Nuclear magnetic resonance, infrared, and x-ray powder diffraction study of dimethylsulfoxide and dimethylselenoxide intercalates with kaolinite. Clays Clay Miner. 35, 208–219.

Ray, S.S., Okamoto, M., 2003. Polymer/layered silicate nanocomposites: a review from preparation to processing. Prog. Polym. Sci. 28, 1539–1641.

Reddy, B.M., Ganesh, I., Chowdhury, B., 1999. Design of stable and reactive vanadium oxide catalysts supported on binary oxides. Catal. Today 49, 115–121.

Reddy, T.R., Reddy, S.L., Endo, T., 2016. Structural characterization studies on the natural mineral pyrophyllite. Radiat. Eff. Defects Solids 171, 307–315.

Reeves, J.B., 2010. Near- versus mid-infrared diffuse reflectance spectroscopy for soil analysis emphasizing carbon and laboratory versus on-site analysis: where are we and what needs to be done? Geoderma 158, 3–14.

Reeves, J.B., Smith, D.B., 2009. Potential of mid- and near-infrared diffuse reflectance spectroscopy for determining major- and trace-element concentrations in soils from a geochemical survey of North America. Appl. Geochem. 24, 1472–1481.

Reeves, J.B., McCarthy, G.W., Hively, W.D., 2010. Mid-versus near-infrared spectroscopy for on-site analysis of soil. P 133-142 IN R.A. Viscarra Rossel. In: McBratney, A.B., Minasny, B. (Eds.), Proximal Soil Sensing, Progress in Soil Science, Springer.

Reich, G., 2005. Near-infrared spectroscopy and imaging: basic principles and pharmaceutical applications. Adv. Drug Deliv. Rev. 57, 1109–1143.

Reinert, L., Ohashi, F., Kehal, M., Bantignies, J.-L., Goze-Bac, C., Duclaux, L., 2011. Characterization and boron adsorption of hydrothermally synthesised allophanes. Appl. Clay Sci. 54, 274–280.

Rendon, J.L., Serna, C.J., 1981. IR spectra of powder hematite: effects of particle size and shape. Clay Miner. 16, 375–381.

Reynard, B., Wunder, B., 2006. High-pressure behavior of synthetic antigorite in the $MgO$-$SiO_2$-$H_2O$ system from raman spectroscopy. Am. Mineral. 91, 459–462.

Reynolds, S.J., Spencer, J.E., 1985. Evidence for large-scale transport on the Bullard detachment fault, west-central Arizona. Geology 13, 353–356.

Rinaudo, C., Gastaldi, D., Belluso, E., 2003. Characterization of chrysotile, antigorite, and lizardite by FT-Raman spectroscopy. Can. Mineral. 41, 883–890.

Rinaudo, C., Roz, M., Boero, V., Franchini-Angela, M., 2004. FT-Raman spectroscopy on several di- and tri-octahedral T-O-T phyllosilicates. Neues Jahrb. Mineral., Monatsh. 537-554.

Ringwald, S.C., Pemberton, J.E., 2000. Adsorption interaction of aromatics and heteroaromatics with hydrated and dehydrated silica surfaces by Raman and FTIR spectrocopies. Environ. Sci. Technol. 34, 259–265.

Rinnan, A., van den Berg, F., Balling Engelsen, S., 2009. Review of the most common preprosessing techniques for near-infrared spectra. Trends Anal. Chem. 28, 1201–1222.

Rinnert, E., Carteret, C., Humbert, B., Fragneto-Cusani, G., Ramsay, J.D.F., Delville, A., Robert, J.-L., Bihannic, I., Pelletier, M., Michot, L.J., 2005. Hydration of a synthetic clay with tetrahedral charges: a multidisciplinary experimental and numerical study. J. Phys. Chem. B 109, 23745–23759.

Ritz, M., Zdralková, J., Valášková, M., 2014. Vibrational spectroscopy of acid treated vermiculites. Vib. Spectrosc. 70, 63–69.

Ritz, M., Vaculíková, L., Kupková, J., Plevová, E., Bartoňová, L., 2016. Different level of fluorescence in Raman spectra of montmorillonites. Vib. Spectrosc. 84, 7–15.

Rivkin, A.S., Volquardsen, E.L., Clark, B.E., 2006. The surface composition of Ceres: discovery of carbonates and iron-rich clays. Icarus 185, 563–567.

Robert, J.-L., Kodama, H., 1988. Generalization of the correlations between hydroxyl-stretching wavenumbers and composition of micas in the system $K_2O$-$MgO$-$Al_2O_3$-$SiO_2$-$H_2O$: a single model for trioctahedral and dioctahedral micas. Am. J. Sci. 288A, 196–212.

Robin, V., Petit, S., Beaufort, D., Prêt, D., 2013. Mapping kaolinite and dickite in sandstone thin sections using infrared microscopy. Clays Clay Miner. 61, 141–151.

Rocchiccioli-Deltcheff, C., Foumier, M., Frank, R., 1983. Vibrational investigations of polyoxometalates. 2. Evidence for anion-anion interactions in molybdenum(VI) and tungsten(VI) compounds related to the Keggin structure. Inorg. Chem. 22, 207–216.

Rogers, A.D., Christensen, P.R., 2007. Surface mineralogy of Martian low-albedo regions from MGS-TES data: implications for upper crustal evolution and surface alteration. J. Geophys. Res. 112E01003. https://doi.org/10.1029/2006JE002727.

Rosasco, G.J., Blaha, J.J., 1980. Raman microprobe spectra and vibrational mode assignments of talc. Appl. Spectrosc. 34, 140–144.

Ross-Medgaarden, E.I., Wachs, I.E., 2007. Structural determination of bulk and surface tungsten oxides with UV–vis diffuse reflectance spectroscopy and Raman spectroscopy. J. Phys. Chem. C 111, 15089–15099.

Ruan, H.D., Frost, R.L., Kloprogge, J.T., 2001. Comparison of Raman spectra in characterizing gibbsite, bayerite, diaspore and boehmite. J. Raman Spectrosc. 32, 745–750.

Ruff, S.W., 2004. Spectral evidence for zeolite in the dust on Mars. Icarus 168, 131–143.

Ruiz-Garcia, C., Pérez-Carvajal, J., Berenguer-Murcia, A., Darder, M., Aranda, P., Cazorla-Amorós, D., Ruiz-Hitzky, E., 2013. Clay-supported graphene materials: application to hydrogen storage. Chem. Phys. Phys. Chem. 15, 18635–18641.

Russell, J.D., 1974. Instrumentation and techniques. In: Farmer, V.C. (Ed.), The infrared spectra of minerals. Mineralogical Society Monograph 4, Mineralogical Society, London, pp. 11–25.

Russell, J.D., 1979. Infrared spectroscopy of ferrihydrite: evidence for the presence of structural hydroxyl groups. Clay Miner. 14, 109–114.

Russell, J.D., 2008. A Tribute to Victor Colin Farmer. Opening lecture of the ALUSIV conference, organized by the Clay Minerals Group of the Mineralogical Society and the British Society of Soil Science, Aberdeen 2008, unpublished. Courtesy of Steve Hillier.

Russell, J.D., Farmer, V.C., 1964. Infra-red spectroscopic study of the dehydration of montmorillonite and saponite. Clay Miner. Bull. 5, 443–464.

Russell, J.D., Fraser, A., 1994. Infrared methods. In: Wilson, M.J. (Ed.), Clay Mineralogy: Spectroscopic and Chemical Determinative Methods. Chapman and Hall, London, pp. 11–67.

Russell, J.D., Farmer, V.C., Velde, B., 1970. Replacement of OH by OD in layer silicates and identification of the vibration of these groups in infrared spectra. Mineral. Mag. 37, 869–879.

Russell, J.D., Parfitt, R.L., Fraser, A.R., Farmer, V.C., 1974. Surface structures of gibbsite goethite and phosphated goethite. Nature 213, 220–221.

Rutkai, G., Makó, É., Kristóf, T., 2009. Simulation and experimental study of intercalation of urea in kaolinite. J. Colloid Interface Sci. 334, 65–69.

Ryskin, YA. I., 1974. The vibrations of protons in minerals: hydroxyl, water and ammonium. In: V.C. Farmer (Ed.), The Infrared Spectra of Minerals. Mineralogical Society, London, pp. 137–181.

Ryu, K.-W., Jang, Y.-N., Chae, S.-C., 2010. Hydrothermal synthesis of kaolinite and its formation mechanism. Clays Clay Miner. 58, 44–51.

Ryu, K.-W., Jang, Y.-N., Chae, S.-C., Bae, I.-K., Choi, S.-H., 2006. Hydrothermal synthesis of smectite from dickite. Clays Clay Miner. 54, 80–86.

Rywak, A.A., Burtlich, J.M., 1996. The crystal chemistry and thermal stability of sol-gel prepared fluorine-substituted talc. Phys. Chem. Miner. 23, 418–431.

Saboya, R.M.A., Cecilia, J.A., Garcia-Sancho, C., Luna, F.M.T., Rodriguez-Castellon, E., Cavalcante Jr., C.L., 2016. $WO_3$-based catalysts supported on porous clay heterostructures (PCH) with Si-Zr pillars for synthetic esters production. Appl. Clay Sci. 124–125, 69–78.

Sainz-Diaz, C.I., Timon, V., Botella, V., Andez-Laguna, A.H., 2000. Isomorphous substitution effect on the vibration frequencies of hydroxyl groups in molecular cluster models of the clay octahedral sheet. Am. Mineral. 85, 1038–1045.

Saito, M., D'Amico, F., Camisasca, G., Bencivenga, F., Cucini, R., Gessini, A., Principi, E., Ogura, T., Masciovecchio, C., 2015. Resonance raman spectroscopy with chemical state selectivity on histidine and acetamide using synchrotron radiation. Bull. Chem. Soc. Jpn 88, 591–596.

Salerno, P., Asenjo, M.B., Mendioroz, S., 2001. Influence of preparation method on thermal stability and acidity of Al-PILCs. Thermochim. Acta 379, 101–109.

Salisbury, F.B., Wald, A., D'Aria, D.M., 1994. Thermal-infrared remote sensing and Kirchhoff's law 1. Laboratory measurements. J. Geophys. Res. 99, 11897–11911.

Salisbury, J.W., 1993. Mid-infrared spectroscopy: laboratory data. In: Pieters, C.M., Englert, P.A.J. (Eds.), Remote Geochemical Analysis: Elemental and Mineralogical Composition. Cambridge University Press, Cambridge.

Salisbury, J.W., Hunt, G.R., 1974. Meteorite spectra and weathering. J. Geophys. Res. 79, 4439–4441.

Salisbury, J.W., D'Aria, D.M., Jarosewich, E., 1991a. Midinfrared (2.5-13.5 μm) reflectance spectra of powdered stony meteorites. Icarus 92, 280–297.

Salisbury, J.W., Walter, L.S., Vergo, N., D'aria, D.M., 1991b. Infrared (2.1–25 μm) Spectra of Minerals, Baltimore, Johns Hopkins University Press.

Salje, E.K.H., Bismayer, U., 1997. Hard mode spectroscopy: the concept and applications. Phase Trans. 63, 1–75.

Salje, E.K.H., Carpenter, M.A., Malcherek, T., Boffa-Ballaran, T., 2000. Autocorrelation analysis of infrared spectra from minerals. Eur. J. Mineral. 12, 503–519.

Salles, F., Douillard, J.M., Bildstein, O., El Ghazi, S., Prelot, B., Zajac, J., Van Damme, H., 2015. Diffusion of interlayer cations in swelling clays as a function of water content: case of montmorillonites saturated with alkali cations. J. Phys. Chem. C 119, 10370–10378.

Samuels, A.C., Zhu, C., Williams, B.R., Ben-David, A., Miles, R.W., Hulet, M., 2006. Improving the linearity of infrared diffuse reflection spectroscopy data for quantitative analysis: an application in quantifying organophosphorus contaminated soil. Anal. Chem. 78, 408–415.

san Cristóbal, A.G., Castelló, R., Luengo, M.A.M., Vizcayno, C., 2009. Acid activation of mechanically and thermally modified kaolins. Mater. Res. Bull. 44, 2103–2111.

Sánchez Del Río, M., Boccaleri, E., Milanesio, M., Croce, G., van Beek, W., Tsiantos, C., Chryssikos, G.D., Gionis, V., Kacandes, G.H., Suárez, M., García-Romero, E., 2009. A combined synchrotron powder diffraction and vibrational study of the thermal treatment of palygorskite-indigo to produce Maya Blue. J. Mater. Sci. 44, 5524–5536.

Sanchez-Camazano, M., Sanchez-Martin, M.J., 1994. Organo-clays as adsorbents for azinphosmethyl and dichlorvos in aqueous-medium. Water Air Soil Pollut. 74, 19–28.

Sandford, S.A., 1984. Infrared transmission spectra from 2.5 to 25 μm of various meteorite classes. Icarus 60, 115–126.

Santagata, M.C., Premachandra, G.S. & Johnston, C.T. Nanoconfined water in clay minerals. Book of Abstracts for XIV International Clay Conference "Mico et Nano Scientiae Mare Magnum", 2009 Castellaneta Marina (TA), Italy. AIPEA, vol. 1, p. 428.

Sarkar, B., Xi, Y.F., Megharaj, M., Krishnamurti, G.S.R., Bowman, M., Rose, H., Naidu, R., 2012. Bioreactive organoclay: a new technology for environmental remediation. Crit. Rev. Env. Sci. Technol. 42, 435–488.

Savitzky, A., Golay, M.J.E., 1964. Smoothing and differentiation of data by simplified least squares procedures. Anal. Chem. 36, 1629–1639.

Sawant, S.Y., Pawar, R.R., Somani, R.S., Bajaj, H.C., 2014. Facile hard template approach for synthetic hectorite hollow microspheres. Mater. Lett. 128, 121–124.

Sawhney, B.L., 1968. Aluminum interlayers in layer silicates: effect of OH/Al ratio of Al solution, time of reaction, and type of structure. Clays Clay Miner. 16, 157–163.

Sayin, M., von Reichenbach, H.G., 1979. X-ray and infrared investigations of a synthetic dioctahedral vermiculite saturated with alkali and alkaline earth cations. Clay Miner. 14, 211–228.

Schaber, P.M., Colson, J., Higgins, S., Thielen, D., Anspach, B., Brauer, J., 2004. Thermal decomposition (pyrolysis) of urea in an open reaction vessel. Thermochim. Acta 424, 131–142.

Scheuing, D.R., Weers, J.G., 1990. A Fourier-transform infrared spectroscopic study of dodecyltrimethylammonium chloride sodium dodecyl-sulfate surfactant mixtures. Langmuir 6, 665–671.

Schirmann, J.-P., Bourdauducq, P., 2000. Hydrazine, Ullmann's Encyclopedia of Industrial Chemistry. Wiley-VCH Verlag GmbH & Co, KGaA.

Schlens, 2014. A tutorial on principal component analysis. On-line Resource, Cornell University, arXiv:1404.1100[cs.LG].

Schmidt, R.L., 1980. Applications of photoacoustic spectroscopy to the study of soils and clay minerals. In: Stucki, J.W., Banwart, W.L. (Eds.), Advanced chemical methods for soil and clay mineral research. NATO Advanced Study Institutes Series, vol. 63, D. Reidel Publishing Company, Dordrecht, Holland, pp. 451–465.

Schnetzer, F., Thissen, P., Giraudo, N., Emmerich, K., 2016. Unraveling the coupled processes of (de)hydration and structural changes in $Na^{+}$-saturated montmorillonite. J. Phys. Chem. C 120, 15282–15287.

Scholtzová, E., Tunega, D., Nagy, L.T., 2003. Theoretical study of cation substitution in trioctahedral sheet of phyllosilicates. An effect on inner OH group. J. Mol. Struct. (Theochem) 620, 1–8.

Scholtzová, E., Madejová, J., Tunega, D., 2014. Structural properties of montmorillonite intercalated with tetraalkylammonium cations—computational and experimental study. Vib. Spectrosc. 74, 120–126.

Scholtzová, E., Tunega, D., Madejová, J., Pálková, H., Komadel, P., 2013. Theoretical and experimental study of montmorillonite intercalated with tegramethylammonium cation. Vib. Spectrosc. 66, 123–131.

Schoonheydt, R.A., 2016. Reflections on the material science of clay minerals. Appl. Clay Sci. 131, 107–112.

Schoonheydt, R.A., Johnston, C.T., 2011. The Surface Properties of Clay Minerals. In: Brigatti, M.F., Mottana, A. (Eds.), Layered Mineral Structures and their Applications in Advanced Technologies, EMU Notes in Mineralogy, vol 11. The European Mineralogical Union and The Mineralogical Society of Great Britain & Ireland, London (Chapter 10).

Schoonheydt, R.A., Johnston, C.T., 2013. Surface and interface chemistry of clay minerals. In: Bergaya, F., Theng, B.K.G., Lagaly, G. (Eds.), Handbook of Clay Science. Part A. Fundamentals, second ed. Elsevier, Amsterdam.

Schoonheydt, R.A., Pinnavaia, T., Lagaly, G., Gangas, N., 1999. Pillared clays and pillared layered solids. Pure Appl. Chem. 71, 2367–2371.

Schrader, M.E., Yariv, S., 1990. Wettability of clay minerals. J. Colloid Interface Sci. 136, 85–94.

Schroeder, P.A., 1992. Far-infrared study of the interlayer torsional-vibrational mode of mixed-layer illite/smectites. Clays Clay Miner. 40, 81–91.

Schroeder, P.A., 2002. Infrared spectroscopy in clay science. In: Rule, A., Guggenheim, S. (Eds.), CMS Workshop Lectures. Teaching Clay Science, vol. 11. The Clay Minerals Society, Aurora, CO, pp. 181–206.

Schroeder, P.A., Melear, N.D., Pruett, R.J., 2003. Quantitative analysis of anatase in Georgia kaolins using Raman spectroscopy. Appl. Clay Sci. 23, 299–308.

Schuttlefield, J.D., Cox, D., Grassian, V.H., 2007b. An investigation of water uptake on clay minerals using ATR-FTIR spectroscopy coupled with quartz microbalance measurements. J. Geophys. Res. 112 (D21303), 1–14.

Schuttlefield, J.D., Al-Hosney, H., Zachariah, A., Grassian, V.H., 2007a. Attenuated total reflection spectroscopy to investigate water uptake and phase transitions in atmospherically relevant particles. Appl. Spectrosc. 61, 283–292.

Schutz, A., Stone, W.E.E., Poncelet, G., Fripiat, J.J., 1987. Preparation and characterization of bidimensional zeolitic structures obtained from synthetic beidellite and hydroxy-aluminum solutions. Clays Clay Miner. 35, 251–261.

Schwartz, S., Guillot, S., Tricart, P., Bernet, M., Jourdan, S., Dumont, T., Montagnac, G., 2012. Source tracing of detrital serpentinite in the Oligocene molasse deposits from the western Alps (Barrême basin): implications for relief formation in the internal zone. Geol. Mag. 149, 841–856.

Schwartz, S., Guillot, S., Reynard, B., Lafay, R., Debret, B., Nicollet, C., Lanari, P., Auzende, A.L., 2013. Pressure-temperature estimates of the lizardite/antigorite transition in high pressure serpentinites. Lithos 178, 197–210.

Seelos, K.D., Arvidson, R.E., Jolliff, B.L., Chemtob, S.M., Morris, R.V., Ming, D.W., Swayze, G.A., 2009. Silica in a Mars Analog Environment: Ka'u Desert, Kilauea Volcano. Hawaii. J. Geophys. Res. 114. https://doi.org/10.1029/2009JE003347.

Serna, C.J., Van Scoyoc, G.E., Ahlrichs, J.L., 1977a. Hydroxyl groups and water in palygorskite. Am. Mineral. 62, 784–792.

Serna, C.J., Velde, B.D., White, J.L., 1977b. Infrared evidence of order-disorder in amesites. Am. Mineral. 62, 296–303.

Serna, C.J., Ocaña, M., Iglesias, J.E., 1987. Optical properties of a-$Fe_2O_3$ microcrystals in the infrared. J. Phys. C: Solid State Phys. 20, 473–484.

Serratosa, J.M., 1960. Dehydration studies by infrared spectroscopy. Am. Mineral. 45, 1101–1104.

Servagent-Noinville, S., Revault, M., Quiquampoix, H., Baron, M.H., 2000. Conformational changes of bovine serum albumin induced by adsorption on different clay surfaces: FTIR analysis. J. Colloid Interface Sci. 221, 273–283.

Shannon, R.D., 1976. Revised effective ionic radii and systematic studies of interatomic distances in halides and chalcogenides. Acta Crystallogr. A 32 (5), 751–767.

Shao, J., 1993. Linear model selection by cross-validation. J. Am. Stat. Assoc. 88, 486–494.

Sharma, S.K., Misra, A.K., Clegg, S.M., Barefield, J.E., Wiens, R.C., Acosta, T., 2010. Time-resolved remote Raman study of minerals under supercritical CO2 and high temperatures relevant to Venus exploration. Philos. Trans. R. Soc., A 368, 3167–3191.

Shen, B., Ma, H., Yao, Y., 2012. Mn-CeOx/Ti-PILCs for selective catalytic reduction of NO with NH3 at low temperature. J. Environ. Sci. (China) 24, 499–506.

Sheng, G.Y., Xu, S.H., Boyd, S.A., 1999. A dual function organoclay sorbent for lead and chlorobenzene. Soil Sci. Soc. Am. J. 63, 73–78.

Sherman, D.M., Waite, T.D., 1985. Electronic spectra of $Fe^{3+}$ oxides and oxide hydroxides in the near IR to near UV. Am. Mineral. 70, 1262–1269.

Shewring, N.I.E., Jones, T.G.J., Maitland, G., Yarwodd, J., 1995. Fourier transform Infrared spectroscopic techniques to investigate surface hydration processes on bentonite. J. Colloid Interface Sci. 176, 308–317.

Shirai, M., Aoki, K., Torii, K., Arai, M., 1999. Acidity and 1-butene isomerization of synthesized smectite-type catalysts containing different divalent cations. Appl. Catal. A 187, 141–146.

Shirozu, H., 1980. Cation distribution, sheet thickness, and O-OH space in trioctahedral chlorites—an X-ray and infrared study. Mineral. J. 10, 14–34.

Shirozu, H., Momoi, H., 1972. Synthetic magnesium chlorite in relation to natural chlorite. Miner. J. 6, 464–476.

Shirozu, H., Ishida, K., 1982. Infrared study of some 7 Å and 14 Å layer silicates by deuteration. Mineral. J. 11, 161–171.

Shoval, S., Yariv, S., Michaelian, K.H., Boudeulle, M., Panczer, G., 2002. Hydroxyl-stretching bands in polarized micro-Raman spectra of oriented single-crystal Keokuk kaolinite. Clays Clay Miner. 50, 56–62.

Shylesh, S., Singh, A.P., 2004. Synthesis, characterization and catalytic activity of vanadium-incorporated, -grafted, and -immobilized mesoporous MCM-41 in the oxidation of aromatics. J. Catal. 228, 333–346.

Sihvola, A.H., Kong, J.A., 1988. Effective permittivity of dielectric mixtures. IEEE Trans. Geosci. Remote Sens. 26, 420–429.

Singer, R.B., 1985. Spectroscopic observation of Mars. Adv. Space Res. 5, 59–68.

Singh, N., Megharaj, M., Gates, W.P., Churchman, G.J., Naidu, R., Kookana, R.S., 2004. Sorption-desorption of fenamiphos in surfactant-modified clays. Bull. Environ. Contam. Toxicol. 72, 276–282.

Singh, N., Megharaj, M., Gates, W.P., Churchman, G.J., Anderson, J., Kookana, R.S., Naidu, R., Chen, Z., Slade, P.G., Sethunathan, N., 2003. Bioavailability of an organophosphate pesticide, fenamiphos, sorbed on an organo-clay. J. Agric. Food Chem. 51, 2653–2658.

Singh, S.S., Kodama, H., 1994. Effect of the presence of aluminum ions in iron solutions on the formation of iron oxyhydroxides (FeOOH) at room temperature under acidic environment. Clays Clay Miner. 42, 606-.

Singha, M., Singh, L., 2016. Vibrational spectroscopic study of muscovite and biotite layered phyllosilicates. Ind. J Pure Appl. Phys. 54, 116–122.

Siqueira, L.J.A., Constantino, V.R.L., Camilo, F.F., Torresi, R.M., Temperini, M.L.A., Ribeiro, M.C.C., Izumi, C.M.S., 2014. Probing the local environment of hybrid materials designed from ionic liquids and synthetic clay by Raman spectroscopy. Spectrochim. Acta A: Mol. Biomol. Spectrosc. 122, 469–475.

Sivaraman, B., Raja Sekhar, B.N., Nair, B.G., Hatode, V., Mason, N.J., 2013. Infrared spectrum of formamide in the solid phase. Spectrochim. Acta A: Mol. Biomol. Spectrosc. 105, 238–244.

Skoubris, E.N., Chryssikos, G.D., Christidis, G.E., Gionis, V., 2013. Structural characterization of reduced-charge montmorillonites—evidence based on FTIR spectroscopy, thermal behavior, and layer-charge systematics. Clays Clay Miner. 61, 83–97.

Slade, P.G., Gates, W.P., 2004a. The ordering of HDTMA in the interlayers of vermiculite and the influence of solvents. Clays Clay Miner. 52, 204–210.

Slade, P.G., Gates, W.P., 2004b. The swelling of HDTMA smectites as influenced by their preparation and layer charges. Appl. Clay Sci. 25, 93–101.

Slonimskayá, M.V., Besson, G., Dainyak, L.G., Tchoubar, C., Drits, V.A., 1986. Interpretation of the IR spectra of celadonites and glauconites in the region of OH-stretching frequencies. Clay Miner. 21, 115–149.

Slosiariková, H., Bujdák, J., Hlavatý, V., 1992. IR-spectra of octadecylammonium-montmorillonite in the range of the Si-O vibrations. J. Incl. Phenom. Mol. Recognit. Chem. 13, 267–272.

Smekal, A., 1923. Zur Quantentheorie der Dispersion. Naturwissenschaften 43, 873–875.

Smrčok, Ľ., Gyepesová, D., Chmielová, M., 1990. New X-ray Rietveld refinement of kaolinite from Keokuk, Iowa. Cryst. Res. Technol. 25, 105–110.

Smrčok, L., Tunega, D., Ramirez-Cuesta, A.J., Scholtzová, E., 2010. The combined inelastic neutron scattering and solid state DFT study of hydrogen atoms dynamics in a highly ordered kaolinite. Phys. Chem. Miner. 37, 571–579.

Snyder, R.G., 1979. Vibrational correlation splitting and chain packing for the crystalline n-alkanes. J. Chem. Phys. 71, 3229–3235.

Solc, R., Gerzabek, M.H., Lischka, H., Tunega, D., 2011. Wettability of kaolinite (001) surfaces—molecular dynamic study. Geoderma 169, 47–54.

Soler, J.M., Artacho, E., Gale, J.D., García, A., Junquera, J., Ordejón, P., Sánchez-Portal, D., 2002. The Siesta method for ab initio order-N materials simulation. J. Phys.: Condens. Matter 14, 2745–2779.

Solihin, Z.Q., Tongamp, W., Saito, F., 2011. Mechanochemical synthesis of kaolin–$KH_2PO_4$ and kaolin–$NH_4H_2PO_4$ complexes for application as slow release fertilizer. Powder Technol. 212, 354–358.

Šontevska, V., Jovanovski, G., Makreski, P., 2007. Minerals from Macedonia. Part XIX. Vibrational spectroscopy as identificational tool for some sheet silicate minerals. J. Mol. Struct. 834–836, 318–327.

Spanier, J.E., Herman, I.P., 2000. Use of hybrid phenomenological and statistical effective-medium theories of dielectric functions to model the infrared reflectance of porous SiC films. Phys. Rev. B 61, 10437–10450.

Sposito, G., Prost, R., 1982. Structure of water adsorbed on smectites. Chem. Rev. 82, 553–573.

Sposito, G., Prost, R., Gaultier, J.P., 1983. Infrared spectroscopic study of adsorbed water on reduced-charge Na/Li montmorillonites. Clays Clay Miner. 31, 9–16.

Srasra, E., Bergaya, F., Fripiat, J.J., 1994. Infrared spectroscopy study of tetrahedral and octahedral substitutions in an interstratified illite-smectite clay. Clays Clay Miner. 42, 237–241.

Środoń, J., 1984. X-ray powder diffraction identification of illitic materials. Clay Clay Miner. 32, 337–349.

Stathopoulou, E.T., Psycharis, V., Chryssikos, G.D., Gionis, V., Theodorou, G., 2008. Bone diagenesis: New data from infrared spectroscopy and X-ray diffraction. Palaeogeogr. Palaeoclimatol. Palaeoecol. 266, 168–174.

Stathopoulou, E.T., Suárez, M., García-Romero, E., Sánchez Del Río, M., Kacandes, G.H., Gionis, V., Chryssikos, G.D., 2011. Trioctahedral entities in palygorskite: near-infrared evidence for sepiolite-palygorskite polysomatism. Eur. J. Mineral. 23, 567–576.

Steinier, J., Termonia, Y., Deltour, J., 1972. Smoothing and differentiation of data by simplified least square procedure. Anal. Chem. 44, 1906–1909.

Sterte, J., 1991a. Preparation and properties of large pore LaAl-pillared montmorillonite. Clays Clay Miner. 39, 167–173.

Sterte, J., 1991b. Preparation and properties of large pore RE/Al-pillared montmorillonite. A comparison of RE-cations. In: Poncelet, V.G., Jacobs, P.A., Grange, P., Delmon, B. (Eds.), Preparation of Catalysts. Elsevier, Amsterdam, pp. 301–310.

Steudel, A., Heinzmann, R., Indris, S., Emmerich, K., 2015. CEC and Li-7 MAS NMR study of interlayer $Li^+$ in the montmorillonite-beidellite series at room temperature and after heating. Clay Clay Miner. 63, 337–350.

Stevens, J.J., Anderson, S.J., 1996. An FTIR study of water sorption on TMA- and TMPA-montmorillonites. Clays Clay Miner. 44, 142–150.

Steyerberg, E.W., Bleeker, S.E., Moll, H.A., Grobbee, D.E., Moons, K.G.M., 2003. Internal and external validation of predictive models: a simulation study of bias and precision in small samples. J. Clin. Epidemiol. 56, 441–447.

Stockmeyer, M.R., 1991. Adsorption of organic compounds on organophilic bentonites. Appl. Clay Sci. 6, 39–57.

Stone, W.E.E., Torres-Sanchez, R.M., 1988. Nuclear magnetic resonance spectroscopy applied to minerals. 6. Structural iron in kaolinites as viewed by proton magnetic resonance. J. Chem. Soc., Faraday Trans. 1 84, 117-132.

Stubičan, V., Roy, R., 1961. A new approach to the assignment of infrared absorption bands in layer silicates. Z. Kristallogr. 115, 200–214.

Stucki, J.W., Roth, C.B., 1976. Interpretation of infrared spectra of oxidized and reduced nontronite. Clays Clay Miner. 24, 293–296.

Stucki, J.W., Bish, D.L., 1990. Thermal analysis in clay science. CMS Workshop Lectures, 3. Boulder, The Clay Minerals Society, 192 pp.

Su, X.L., Ma, L.Y., Wei, J.M., Zhu, R.L., 2016. Structure and thermal stability of organo-vermiculite. Appl. Clay Sci. 132–133, 261–266.

Šucha, V., Kraus, I., Madejová, J., 1994. Ammonium illite from anchimetamorphic shales associated with anthracite in the Zemplinicum of the Western Carpathians. Clay Miner. 29, 369–377.

Šucha, V., Elsass, F., Eberl, D.D., Kuchta, L., Madejová, J., Gates, W.P., Komadel, V., 1998. Hydrothermal synthesis of ammonium illite. Am. Mineral. 83, 58–67.

Sudduth, K.A., Hummel, J.W., 1993. Soil organic matter, CEC and moisture sensing with a portable NIR spectrophotometer. Trans. ASAE 36, 1571–1582.

Sun, S., Jiang, Y., Yu, L., Li, F., Yang, Z., Hou, T., Hu, D., Xia, M., 2006. Enhanced photocatalytic activity of microwave treated $TiO_2$ pillared montmorillonite. Mater. Chem. Phys. 98, 377–381.

Sun, Y., Seccombe, P.K., Yang, K., 2001. Application of short-wave infrared spectroscopy to define alteration zones associated with the Elura zinc-lead-silver deposit, NSW, Australia. J. Geochem. Explorat. 73, 11–26.

Suquet, H., Prost, R., Pezerat, H., 1977. Étude par spectroscopie infrarouge de l' eau adsorbée par la saponite-calcium. Clay Miner. 12, 113–126.

Swartzen-Allen, S.L., Matijevi, E., 1974. Surface and colloid chemistry of clays. Chem. Rev. 74, 385–400.

Swayze, G.A., Clark, R.N., Goetz, A.F.H., Livo, K.E., Breit, G.N., Kruse, F.A., Sutley, S.J., Snee, L.W., Lowers, H.A., Post, J.L., Stoffregen, R.E., Ashley, R.P., 2014. Mapping advanced argillic alteration at Cuprite, Nevada, using imaging spectroscopy. Econ. Geol. 109, 1179–1221.

Swoboda, A.R., Kunze, G.W., 1964. Infrared study of pyridine adsorbed on montmorillonite surfaces. Clays Clay Miner. 13, 277–288.

Szczerba, M., Kuligiewicz, A., Derkowski, A., Gionis, G., Chryssikos, G.D., Kalinichev, A.G., 2016. Structure and dynamics of water–smectite interfaces: hydrogen bonding and the origin of the sharp O Dw/O Hw infrared band from molecular simulations. Clays Clay Miner. 64, 452–471.

Tahani, A., Karroua, M., Van Damme, H., Levitz, P., Bergaya, F., 1999. Adsorption of a cationic surfactant on Na-montmorillonite: inspection of adsorption layer by X-ray and fluorescence spectroscopies. J. Colloid Interface Sci. 216, 242–249.

Tahir, M., Amin, N.S., 2013. Photocatalytic reduction of carbon dioxide with water vapors over montmorillonite modified $TiO_2$ nanocomposites. Appl. Catal. B 142–143, 512–522.

Tanaka, K. L., Skinner Jr., J. A., Hare, T. M., 2005. Geologic map of the northern plains of Mars, Map 2888. Flagstaff, AZ: U.S. Geol. Surv.

Tang, M.J., Cziczo, D.J., Grassian, V.H., 2016. Interactions of water with mineral dust aerosol: water adsorption, hygroscopicity, cloud condensation, and ice nucleation. Chem. Rev. 116, 4205–4259.

Tarasevi, Y.I., Gribina, I.A., 1972. Infrared spectroscopic study of state of water in halloysite. Colloid J. Russ. Acad. Sci. 34, 346–350.

Tauler, R., Lacorte, S., Barceló, D., 1996. Application of multivariate curve self-modeling curve resolution for the quantitation of trace levels of organophosphorous pesticides in natural waters from interlaboratory studies. J. Chromatogr. A. 730, 177–183.

Teixeira-Neto, Â.A., Izumi, C.M.S., Temperini, M.L.A., Da C. Ferreira, A.M., Constantino, V.R. L., 2012. Hybrid materials based on smectite clays and nutraceutical anthocyanins from the Açaí fruit. Eur. J. Inorg. Chem. 32, 5411–5420.

Tennakoon, D.T.B., Jones, W., Thomas, J.M., 1986. Structural aspect of metal-oxide-pillared sheet silicates. J. Chem. Soc. Faraday Trans. 82, 3081–3095.

Tennakoon, D.T.B., Jones, W., Thomas, J.M., Ballantine, J.H., Purnell, J.H., 1987. Characterization of clay and pillared clay catalysts. Solid State Ionics 24, 205–212.

Theng, B.K.G., 1974. The Chemistry of Clay–Organic Reactions. Adam Hilger, London, 343 pp.

Theng, B.K.G., 2012. Formation and Properties of Clay-Polymer Complexes. Developments in Clay Science, vol. 4. Elsevier, Amsterdam, 511 pp.

Theng, B.K.G., Greenland, D.J., Quirk, J.P., 1967. Adsorption of alkylammonium cations by montmorillonite. Clay Miner. 7, 1–17.

Theng, B.K.G., Churchman, G.J., Gates, W.P., Yuan, G., 2008. Organically modified clays for pollutant uptake and environmental protection. In: Huang, Q., Huang, P.M., Violante, A. (Eds.), Soil Mineral-Microbe-Organic Interactions: Theories and Applications. Springer-Verlag, Berlin, pp. 145–174.

Thill, A., 2016. Characterisation of imogolite by microscopic and spectroscopic methods. Chapter 10. In: Yuan, P., Thill, A., Bergaya, F. (Eds.), Nanosized Tubular Clay Minerals—Halloysite and Imogolite. Developments in Clay Science, vol. 7, pp. 223–253.

Thomas, C.L., Hickey, J., Stecker, G., 1950. Chemistry of clay cracking catalysts. Ind. Eng. Chem. 42, 866–871.

Thomas, M., Walter, M.R., 2002. Application of hyperspectral infrared analysis of hydrothermal alteration on Earth and Mars. Astrobiology 2, 335–351.

Thompson, J.G., Cuff, C., 1985. Crystal structure kaolinite:dimethyl sulfoxide intercalate. Clays Clay Miner. 33, 490–500.

Thompson, J.G., Gabbitas, N., Unwins, P.J.R., 1993. The intercalation of kaolinite by alkali halides in the solid state: a systematic study of the intercalates and their derivatives. Clays Clay Miner. 41, 73–86.

Tichit, D., Fajula, F., Figueras, F., Ducourant, B., Mascherpa, G., Gueguen, C., Bousquet, J., 1988. Sintering of montmorillonites pillared by hydroxy-aluminum species. Clays Clay Miner. 36, 369–375.

Timofeeva, M.N., Panchenko, V.N., Gil, A., Chesalov, Y.A., Sorokina, T.P., Likholobov, V.A., 2011. Synthesis of propylene glycol methyl ether from methanol and propylene oxide over alumina-pillared clays. Appl. Catal. B 102, 433–440.

Tipton, T., Johnston, C.T., Trabue, S.L., Erickson, C., Stone, D.A., 1993. Gravimetric/FTIR apparatus for the study of vapor sorption on clay films. Rev. Sci. Inst. 64, 1091–1092.

Titulaer, M.K., Van Miltenburg, J.C., Jansen, J.B.H., Geus, J.W., 1993. Characterization of tubular chrysotile by thermoporometry, nitrogen sorption, DRIFTS, and TEM. Clays Clay Miner. 41, 496–513.

Tomić, Z.P., Asanin, D., Antic-Mladenovic, S., Poharc-Logar, V., Makreski, P., 2012a. NIR and MIR spectroscopic characteristics of hydrophilic and hydrophobic bentonite treated with sulphuric acid. Vib. Spectrosc. 58, 95–103.

Tomić, Z.P., Ašanin, D., Đurović, R., Đorđević, A., Makreski, P., 2012b. Near-infrared spectroscopy study for determination of adsorbed acetochlor in the organic and inorganic bentonites. Spectrochim. Acta Part A 98, 47–52.

Tomita, K., 1970. Syntheses of montmorillonites and vermiculitelike minerals from sericite and pyrophyllite. Ganseki Kobutsu Kosho Gakkaishi 63, 109–121.

Tomita, K., Yamane, H., Kawano, M., 1993. Synthesis of smectite from volcanic glass at low temperature. Clays Clay Miner. 41, 655–661.

Tomul, F., Balci, S., 2009. Characterization of Al. Cr-pillared clays and CO oxidation. Appl. Clay Sci. 43, 13–20.

Tomul, F., Turgut Basoglu, F., Canbay, H., 2016. Determination of adsorptive and catalytic properties of copper, silver and iron contain titanium-pillared bentonite for the removal bisphenol A from aqueous solution. Appl. Surf. Sci. 360, 579–593.

Torii, H., Tasumi, M., 1998. Liquid structure, infrared and isotropic/anisotropic Raman noncoincidence of the amide I band, and low-wavenumber vibrational spectra of liquid formamide: Molecular dynamics and ab initio molecular orbital studies. J. Phys. Chem. B 102, 315–321.

Torii, K., Iwasaki, T., 1988. Synthesis of novel nickel-hectorite inorganic complexes. Chem. Lett. 2045–2048.

Tornabene, L.L., Moersch, J.E., McSween, H.Y., Hamilton, V.E., Piatek, J.L., Christensen, P.R., 2008. Surface and crater-exposed lithologic units of the Isidis Basin as mapped by coanalysis of THEMIS andd TES derived data products. J. Geophys. Res., 113. https://doi.org/10.1029/2007JE002988.

Torrie, B.H., Brown, B.A., 1994. Raman and far-infrared spectra of formamide at temperatures down to 20 K. J. Raman Spectrosc. 25, 183–187.

Tosoni, S., Doll, K., Ugliengo, P., 2006. Hydrogen bond in layered materials: structural and vibrational properties of kaolinite by a periodic B3LYP approach. Chem. Mater. 18, 2135–2143.

Tosoni, S., Pascale, F., Ugliengo, P., Orlando, R., Saunders, V.R., Dovesi, R., 2005. Quantum mechanical calculation of the OH vibrational frequency in crystalline solids. Mol. Phys. 103, 2549–2558.

Toulmin P. III, Baird, A.K., Clark, B.C., Keil, K., Rose, J.H.J., Christian, R.P., Evans, P.H., Kelliher, W.C., 1977. Geochemical and mineralogial interpretation of the Viking inorganic chemical results. J. Geophys. Res. 82, 4625–4634.

Treiman, A.H., Barrett, R.A., Gooding, J.L., 1993. Preterrestrial aqueous alteration of the Lafayette (SNC) meteorite. Meteoritics 28, 86–97.

Trillo, J.M., Alba, M.D., Alvero, R., Castro, M.A., Munoz, A., Poyato, J., 1993. Montmorillonite intercalated with Al(III), La(III) and alumina pillars: structural aspects and reactivity. Solid State Ionics 63–65, 457–463.

Trittschack, R., Grobety, B., 2013. The dehydroxylation of chrysotile: a combined in situ micro-Raman and micro-FTIR study. Am. Mineral. 98, 1133–1145.

Trittschack, R., Grobety, B., Koch-Muller, M., 2012. In situ high-temperature Raman and FTIR spectroscopy of the phase transformation of lizardite. Am. Mineral. 97, 1965–1976.

Trombetta, M., Busca, G., Lenarda, M., Storaro, L., Ganzerla, R., Piovesan, L., Jimenez Lopez, A., Alcantara-Rodriguez, M., Rodriguez-Castellon, E., 2000. Solid acid catalysts from clays. Evaluation of surface acidity of mono- and bi-pillared smectites by FT-IR spectroscopy measurements, $NH_3$-TPD and catalytic tests. Appl. Catal. A 193, 55–69.

Trujillano, R., Rico, E., Vicente, M.A., Herrero, M., Rives, V., 2010. Microwave radiation and mechanical grinding as new ways for preparation of saponite-like materials. Appl. Clay Sci. 48, 32–38.

Tsampodimou, M., Bukas, V.J., Stathopoulou, E.T., Gionis, V., Chryssikos, G.D., 2015. Near-infrared investigation of folding sepiolite. Am. Mineral. 100, 195–202.

Tsiantos, C., Tsampodimou, M., Kacandes, G.H., Sánchez Del Río, M., Gionis, V., Chryssikos, G.D., 2012. Vibrational investigation of indigo–palygorskite association(s) in synthetic Maya blue. J. Mater. Sci. 47, 3415–3428.

Tunega, D., Benco, L., Haberhauer, G., Gerzabek, M.H., Lischka, H., 2002. Ab initio molecular dynamics study of adsorption sites on the (0 0 1) surfaces of 1:1 dioctahedral clay minerals. J. Phys. Chem. B 106, 11515–11525.

Tuo, B., Wang, J., Yao, Y., Liu, Z., 2014. Performance study of Ti-pillared montmorillonite nanocomposites. Adv. Mater. Res. 1004-1005, 85-88, 85 pp.

Uhlík, P., Šucha, V., Eberl, D.D., Puškelová, L., Čaplovičová, M., 2000. Evolution of pyrophyllite particle sizes during dry grinding. Clay Miner. 35, 423–432.

Vahedi-Faridi, A., Guggenheim, S., 1997. Crystal structure of tetramethylammonium-exchanged vermiculite. Clays Clay Miner. 45, 859–866.

Vaia, R.A., Teukolsky, R.K., Giannelis, E.P., 1994. Interlayer structure and molecular environment of alkylammonium layred silicates. Chem. Mater. 6, 1017–1022.

Valverde, J.L., Canizares, P., Sun Kou, M.R., Molina, C.B., 2000. Enhanced thermal stability of Al-pillared smectites modified with Ce and La. Clays Clay Miner. 48, 424–432.

Valverde, J.L., Canizares, P., Kou, M.R.S., Molina, C.B., 2001. Reply to the comment by J. T. Kloprogge, J. A. Rob van Veen, E. Booij, and Ray L. Frost on "Enhanced thermal stability of Al-pillared smectites modified with Ce and La" Clays Clay Miner. 49, 186–187.

Valverde, J.L., De Lucas, A., Dorado, F., Romero, A., Garcia, P.B., 2005. Study by in situ FTIR of the SCR of NO by propene on $Cu^{2+}$ ion-exchanged Ti-PILC. J. Mol. Catal. A Chem. 230, 23–28.

van der Marel, H.W., Beutelspacher, H., 1974. Atlas of Infrared Spectroscopy of Clay Minerals and their Admixtures. Elsevier, Amsterdam.

van der Meer, F.D., van der Werff, H.M.A., van Ruitenbeek, F.J.A., Hecker, C.A., Bakker, W.H., Noomen, M.F., van der Meijde, M., Carranza, E.J.M., Smeth, J.B.D., Woldai, T., 2012. Multi- and hyperspectral geologic remote sensing: a review. Int. J. Appl. Earth Observ. Geoinform. 14, 112–128.

van Oss, C.J., Giese, R.F., Li, Z., Murphy, K., Norris, J., Charudhury, M.K., Good, R.J., 1992. Determination of contact angles and pore sizes of porous-media by column and thinlayer wicking. J. Adhes. Sci. Technol. 6, 413–428.

van Ruitenbeek, F.J.A., Cudahy, T.J., van der Meer, F.D., Hale, M., 2012. Characterization of the hydrothermal systems associated with Archean VMS-mineralization at Panorama, Western Australia, using hyperspectral, geochemical and geothermometric data. Ore Geol. Rev. 45, 33–46.

van der Gaast, S.J., Wada, K., Wada, S.-I., Kakuto, Y., 1985. Small-angle X-ray powder diffraction, morphology, and structure of allophane and imogolite. Clays Clay Miner. 33, 237–243.

Vandenabeele, P., Edwards, H.G.M., Jehlička, J., 2014. The role of mobile instrumentation in novel applications of Raman spectroscopy: archaeometry, geosciences, and forensics. Chem. Soc. Rev. 43, 2628–2649.

Vaniman, D.T., Bish, D.L., Ming, D.W., Bristow, T.F., Morris, R.V., Blake, D.F., Chipera, S.J., Morrison, S.M., Treiman, A.H., Rampe, E.B., Rice, M., Achilles, C.N., Grotzinger, J.P., Mclennan, S.M., Williams, J., Bell, J.F., Newsom, H.E., Downs, R.T., Maurice, S., Sarrazin, P., Yen, A.S., Morookian, J.M., Farmer, J.D., Stack, K., Milliken, R.E., Ehlmann, B.L., Sumner, D.Y., Berger, G., Crisp, J.A., Hurowitz, J.A., Anderson, R., Des Marais, D.J., Stolper, E.M., Edgett, K.S., Gupta, S., Spanovich, N., Team, M.S., 2014. Mineralogy of a Mudstone at Yellowknife Bay, Gale Crater, Mars. Science 343. https://doi.org/10.1126/science.1243480.

Vantelon, D., Pelletier, M., Michot, L.J., Barres, O., Thomas, F., 2001. Fe Mg and Al distribution in the octahedral sheet of montmorillonites. An infrared study in the OH-bending region. Clay Miner. 36, 369–379.

Vantelon, D., Montarges-Pelletier, E., Michot, L.J., Briois, V., Pelletier, M., Thomas, F., 2003. Iron distribution in the octahedral sheet of dioctahedral smectites. An Fe K-edge Xray absorption spectroscopy study. Phys. Chem. Miner. 30, 44–53.

Vassallo, A.M., Cole-Clarke, P.A., Palmisano, A.J., 1992. Infrared emission spectroscopy of coal minerals and their thermal. Appl. Spectrosc. 46, 73–78.

Vaughan, D.J., Pattrick, R.A.D., 1995. Mineral Surfaces, vol. 5. Springer, 370 pp.

Vaughan, D.E.W., 1987. Multimetallic pillared interlayered clay products and processes of making them. U.S. Patent 4, 666,877.

Vaughan, D.E.W., Lussier, R.J., Magee Jr., S.J., 1979. Pillared interlayered clay materials useful as catalysts and sorbents, US Patent 4 176 090.

Vedder, W., 1964. Correlations between infrared spectrum and chemical composition in mica. Am. Mineral. 49, 736–768.

Vedder, W., McDonald, R.S., 1963. Vibrations of the OH ions in muscovite. J. Chem. Phys. 38, 1583–1590.

Veithen, M., Gonze, X., Ghosez, P.H., 2005. Nonlinear optical susceptibilities, Raman efficiencies, and electro-optic tensors from first-principles density functional perturbation theory. Phys. Rev. B 71, 125107.

Velbel, M.A., 2012. Aqueous alteration in Martian meteorites: comparing mineral relations in igneous-rock weathering of Martian meteorites and in the sedimentary cycle of Mars. In: Grotzinger, J.P., Milliken, R.E. (Eds.), Sedimentary Geology of Mars. SEPM Special Pub. #102.

Velde, B., 1980. Ordering in synthetic aluminous serpentines; infrared spectra and cell dimensions. Phys. Chem. Miner. 6, 209–220.

Velde, B., 1983. Infrared OH-stretch bands in potassic micas, talcs and saponites; influence of electronic configuration and site of charge compensation. Am. Mineral. 68, 1169–1173.

Venkataraman, N.V., Vasudevan, S., 2002. Characterization of alkyl chain conformation in an intercalated cationic lipid bilayer by IR spectroscopy. J. Phys. Chem. B. 106, 7766–7773.

Venyaminov, S.Y., Prenderast, F.G., 1997. Water (H2O and D2O) molar absorptivity in the 1000–4000 cm–1 range and quantitative infrared spectroscopy of aqueous solutions. Anal. Biochem. 248, 234–245.

Vertuccio, L., Gorrasi, G., Sorrentino, A., Vittoria, V., 2009. Nano clay reinforced PCL/starch blends obtained by high energy ball milling. Carbohydr. Polym. 75, 172–179.

Vianna, M.M.G.R., Dweck, J., Kozievitch, V.F.J., Valenzuela-Diaz, F.R., Buchler, P.M., 2005. Characterization and study of sorptive properties of differently prepared organoclays from a Brazilian natural bentonite. J. Therm. Anal. Calorim. 82, 595–602.

Vicente, M.A., Gil, A., Bergaya, F., 2013. Pillared clays and clay minerals. In: Bergaya, F., Lagaly, G. (Eds.), Handbook of Clay Science, Developments in Clay Science Series, second ed, 5A. Elsevier, Oxford, pp. 523–557.

Vicente, I., Salagre, P., Cesteros, Y., Medina, F., Sueiras, J.E., 2010. Microwave-assisted synthesis of saponite. Appl. Clay Sci. 48, 26–31.

Vidal, N., Volzone, C., 2012. Influence of organobentonite structure on toluene adsorption from water solution. Mater. Res. 15, 944–953.

Villar, S.E.J., Edwards, H.G.M., 2005. Near-infrared Raman spectra of terrestrial minerals: relevance for the remote analysis of Martian spectral signatures. Vib. Spectrosc. 39, 88–94.

Viscarra Rossel, R.A., Walvoort, D.J.J., McBratney, A.B., Janik, L.J., Skjemstad, J.O., 2006. Visible, near infrared, mid infrared or combined diffuse reflectance spectroscopy for simultaneous assessment of various soil properties. Geoderma 131, 59–75.

Vítek, P., Ali, E.M.A., Edwards, H.G.M., Jehlička, J., Cox, R., Page, K., 2012. Evaluation of portable Raman spectrometer with 1064 nm excitation for geological and forensic applications. Spectrochim. Acta A. 86, 320–327.

Viviano, C.E., Moersch, J.E., 2012. A technique for mapping Fe/Mg-rich phyllosilicates on Mars using THEMIS multispectral thermal infrared images. J. Geophys. Res.: Planets 117. https://doi.org/10.1029/2011JE003985.

Viviano-Beck, C., Seelos, K.D., Murchie, S.L., Brown, A.J., 2016. Alteration associated with large impact basins on Mars. Lunar Planet. Sci. Conf. XLVII, 2016 The Woodlands, TX. Abstract #1738.

Vogels, R.M.J., Kloprogge, J.T., Geus, J.W., 2005. Synthesis and characterization of saponite clays. Am. Mineral. 90, 931–944.

Vogels, R.M.J., Breukelaar, J., Kloprogge, J.T., Jansen, J.B.H., Geus, J.W., 1997. Hydrothermal crystalization of ammonium-saponite at 200°C and autogenous water pressure. Clays Clay Miner. 45, 1–7.

Volzone, C., Rinaldi, J.O., Ortiga, J., 2006. Retention of gases by hexadecyltrimethylammonium-montmorillonite clays. J. Environ. Manage. 79, 247–252.

Voncken, J.H.L., Wevers, J.M.A.R., van der Eerden, A.M.J., Bos, A.M.J., Jansen, J.B.H., 1987. Hydrothermal synthesis of tobelite, $NH_4Al_2Si_3AlO_{10}(OH)_2$, from various starting materials and implications for its occurrence in nature. Geol. Mijnbouw 66, 259–269.

Wada, K., 1961. Lattice expansion of kaolinite minerals by treatment with potassium acetate. Am. Mineral. 46, 78–91.

Wada, K., 1965. Intercalation of water in kaolin minerals. Am. Mineral. 50, 924–941.

Wada, K., 1967. A study of hydroxyl groups in kaolin minerals utilizing selective-deuteration and infrared spectroscopy. Clay Miner. 7, 51–61.

Wada, N., 1987. Experimental studies on physical properties of vermiculite, muscovite and kaolinite clays. AIP Conf. Proc. 154, 195–205.

Wada, K., Kubo, H., 1975. Precipitation of amorphous aluminosilicates from solutions containing monomeric silica and aluminium ions. J. Soil Sci. 26, 100–111.

Wada, K., Henmi, T., Yoshinaga, N., Patterson, S.H., 1972. Imogolite and allophane formed in saprolite of basalt on Maui, Hawaii. Clays Clay Miner. 20, 375–380.

Wada, N., Kamitakahara, W.A., 1991. Inelastic neutron- and Raman-scattering studies of muscovite and vermiculite layered silicates. Phys. Rev. B Condens. Matter 43, 2391–2397.

Wada, N., Hines, D.R., Ahrenkiel, S.P., 1990. X-ray-diffraction studies of hydration transitions in sodium vermiculite. Phys. Rev. B Condens. Matter 41, 12895–12901.

Wada, S.I., Eto, A., Wada, K., 1979. Synthetic allophane and imogolite. J. Soil Sci. 30, 347–355.

Wada, S., Wada, K., 1982. Effects of substitution of germanium for silicon in imogolite. Clays Clay Miner. 30, 123–128.

Waelkens, C., Malfait, K., Waters, L.B.F.M., 1997. Comet Hale-Bopp, Circumstellar Dust, and the Interstellar Medium. Earth Moon Planets 79, 265–274.

Walker, G.F., 1961. Vermiculite minerals. In: Brown, G. (Ed.), The X-ray Identification and Crystal Structures of Clay. Mineralogical Society, London, pp. 297–324.

Walker, G.F., 1967. Interactions of n-alkylammonuim ions with mica-type layer lattices. Clay Miner. 7, 129–143.

Walrafen, G.E., Hokmabadi, M.S., Yang, W.H., 1988. Raman investigation of the temperature-dependence of the bending $\nu_2$ and combination $\nu_2+\nu_1$ bands from liquid water. J. Phys. Chem. 92, 2433–2438.

Wang, A., Jolliff, B.L., Haskin, L.A., 1999. Raman spectroscopic characterization of a highly weathered basalt: igneous mineralogy, alteration products, and a microorganism. J. Geophys. Res.: Planets 104, 27067-27077.

Wang, A., Freeman, J.J., Jolliff, B.L., 2015. Understanding the Raman spectral features of phyllosilicates. J. Raman Spectrosc. 46, 829–845.

Wang, L., Zhang, Z., 1997a. Principles and methods of quantitative analysis on b-axis disorder in 2:1 dioctahedral phyllosilicate. Chin. Sci. Bull. 42, 1908–1911.

Wang, L., Zhang, Z., 1997b. Orientating structure of hydroxyls in 2:1 phyllosilicates. Chin. Sci. Bull. 42, 321–324.

Wang, L., Xie, X., Wu, D., 2007. The spectroscopy characterization of kaolinite-organic intercalation materials. Key Eng. Mater. 353–358, 1362–1365.

Wang, L., Zhang, M., Redfern, S.A.T., Zhang, Z., 2002. Dehydroxylation and transformations of the 2:1 phyllosilicate pyrophyllite at elevated temperatures: an infrared spectroscopic study. Clays Clay Miner. 50, 272–283.

Wang, L., Chen, Z.Y., Wang, X., Yan, S.Y., Wang, J.H., Fan, Y.W., 2011. Preparations of organo-vermiculite with large interlayer space by hot solution and ball milling methods: a comparative study. Appl. Clay Sci. 51, 151–157.

Wang, L.Q., Liu, J., Exarhos, G.J., Flanigan, K.Y., Bordia, R., 2000. Conformation heterogeneity and mobility of surfactant molecules in intercalated clay minerals studied by Solid-State NMR. J. Phys. Chem. B 104, 2810–2816.

Wang, R., Li, Y., 2011. Multiexcitation Raman spectroscopy in identification of Chinese Jade. Spectrosc. Lett. 44, 432–439.

Wang, S.L., Johnston, C.T., Bish, D.L., White, J.L., Hem, S.L., 2003. Water-vapor adsorption and surface area measurement of poorly crystalline boehmite. J. Colloid Interface Sci. 260, 26–35.

Wang, Y.Q., Zhang, Z.B., Li, Q., Liu, Y.H., 2012a. Adsorption of thorium from aqueous solution by $HDTMA^+$-pillared bentonite. J. Radioanal. Nucl. Chem. 293, 519–528.

Wang, Y.Y., Gan, F.X., Zhao, H.X., 2012b. Nondestructive analysis of Lantian jade from Shaanxi Province. China. Appl. Clay Sci. 70, 79–83.

Wang, Z., Pinnavaia, T.J., 2003. Intercalation of poly(propyleneoxide) amines (Jeffamines) in synthetic layered silicas derived from ilerite, magadiite, and kenyaite. J. Mater. Chem. 13, 2127–2131.

Waruru, B.K., Shepherd, K.D., Ndegwa, G.M., Kamoni, P.T., Sila, A.W., 2014. Rapid estimation of soil engineering properties using diffuse reflectance near infrared spectroscopy. Biosyst. Eng. 121, 177–185.

Wasson, J.T., 1985. Meteorites: Their Record of Early Solar System History. W.H. Freeman, New York.

Weckhuysen, B.M., Wachs, I.E., Schoonheydt, R.A., 1996. Surface chemistry and spectroscopy of chromium in inorganic oxides. Chem. Rev. 96, 3327–3349.

Weers, J.G., Scheuing, D.R., 1990. Micellar sphere to rod transitions. In: Scheuing, D.R. (Ed), Fourier Transform Infrared Spectroscopy in Colloid and Interface Science. ACS Symposium series, American Chemical Society, Washington, DC, pp. 87–122.

Weiss, A., Orth, H., 1973. Zur Kenntnis der Intercalationsverbindungen von Kaolinit, Nakrit, Dickit and Halloysit mit Pyridin-N-oxid and Picolin-N-oxid. Z. Naturforsch. 28b, 252–254.

Weiss, A., Thielepape, W., Orth, H., 1966. Intercalation into kaolinite minerals. In: Heller, L., Weiss, A. (Eds.), Proc. Int. Clay Conf. Israel Programme for Scientific Translations, Jerusalem, Israel.

Weiss, A., Ruthard, R., Orth, H., 1973. Neue Einlagerungsverbindungen von Kaolinit, Nakrit, Dickit, Halloysit und Titandisulfid mit Imidazol und Methylimidazol. Z. Naturforsch. 28b, 446–449.

Weiss, C.A., Kirkpatrick, R.J., Altaner, S.P., 1990. The structural environments of cations adsorbed onto clays: Cesium-133 variable temperature MAS NMR spectroscopy of hectorite. Geochim. Cosmochim. Acta 54, 1655–1669.

Weitz, C.M., Bishop, J.L., 2016. Stratigraphy and formation of clays, sulfates, and hydrated silica within a depression in Coprates Catena, Mars. J. Geophys. Res. 121, 805–835.

Weitz, C.M., Noe Dobrea, E., Wray, J.J., 2015. Mixtures of clays and sulfates within deposits in western Melas Chasma, Mars. Icarus 251, 291–314.

Weitz, C. M., Noe Dobrea, E. Z., Lane, M. D. & Knudson, A. T. 2012. Geologic relationships between gray hematite, sulfates, and clays in Capri Chasma. J. Geophys. Res.: Planets 117, E00J09.

Weitz, C.M., Bishop, J.L., Baker, L.L., Berman, D.C., 2014. Fresh exposures of hydrous Fe-bearing amorphous silicates on Mars. Geophys. Res. Let. 41, 8744–8751.

Weitz, C.M., Bishop, J.L., Thollot, P., Mangold, N., Roach, L.H., 2011. Diverse mineralogies in two troughs of Noctis Labyrinthus, Mars. Geology 39, 899–902. https://doi.org/10.1130/G32045.1.

Welch, M., Montgomery, W., Balan, E., Lerch, P., 2012. Insights into the high-pressure behavior of kaolinite from infrared spectroscopy and quantum-mechanical calculations. Phys. Chem. Miner. 39, 143–151.

Welch, M.D., Barras, J., Klinowski, J., 1995. A multinuclear NMR study of clinochlore. Am. Mineral. 80, 441–447.

Werner, M.W., Roellig, T.L., Low, F.J., Rieke, G.H., Rieke, M., Hoffmann, W.F., Young, E., Houck, J.R., Brandl, B., Fazio, G.G., Hora, J.L., Gehrz, R.D., Helou, G., Soifer, B.T., Stauffer, J., Keene, J., Eisenhardt, P., Gallagher, D., Gautier, T.N., Irace, W., Lawrence, C.R., Simmons, L., Cleve, J.E.V., Jura, M., Wright, E.L., Cruikshank, D.P., 2004. The Spitzer Space Telescope Mission. Astrophys. J. Suppl. Ser. 154, 1.

Wersin, P., Jenni, A., Mader, U.K., 2015. Interaction of corroding iron with bentonite in the ABM1 experiment at Äspö, Sweden: a microscopic approach. Clays Clay Miner. 63, 51–68.

Westad, F., Marini, F., 2015. validation of chemometric models—a tutorial. Anal. Chim. Acta 893, 14–24.

Wiewiora, A., Wieckowski, T., Sokolowska, A., 1979. The Raman spectra of kaolinite sub-group minerals and of pyrophyllite. Arch. Mineral. 35, 5–12.

Wight, J.P., Ashworth, A.J., Allen, F.L., 2016. Organic substrate, clay type, texture, and water influence on NIR carbon measurements. Geoderma 261, 36–43.

Wilke, I., Ramanathan, V., Lachance, J., Tamalonis, A., Aldersley, M., Joshi, P.C., Ferris, J., 2014. Characterization of the terahertz frequency optical constants of montmorillonite. Appl. Clay. Sci. 87, 61–65.

Wilkins, R.W.T., Ito, J., 1967. Infrared spectra of some synthetic talcs. Am. Mineral. 52, 1649–1661.

Williams, D.M., Mason, C.G., Gehrz, R.D., Jones, T.J., Charles, E.W., Harker, D.E., Hanner, M.S., Wooden, D.H., Witteborn, F.C., Butner, H.M., 1997. Measurement of Submicron Grains in the Coma of Comet Hale-Bopp C/1995 O1 during 1997 February 15-20 UT. Astrophys. J. Lett. 489, L91–L94.

Wold, S., Sjöström, M., Eriksson, L., 2001. PLS-regression: a basic tool of chemometrics. Chemom. Intell. Lab. Syst. 58, 109–130.

Wolters, F., Lagaly, G., Kahr, G., Nüesch, R., Emmerich, K., 2009. A comprehensive characterization of dioctahedral smectites. Clays Clay Miner. 57, 115–133.

Workman, J., Weyer, L., 2008. Practical Guide to Interpretive Near-Infrared Spectroscopy. Taylor & Francis Group, Boca Raton, FL.

Wray, J. J., Ehlmann, B. L., Squyres, S. W., Mustard, J. F. & Kirk, R. L. 2008. Compositional stratigraphy of clay-bearing layered deposits at Mawrth Vallis, Mars. Geophys. Res. Lett. 35. doi:10.1029/2008GL034385.

Wray, J. J., Noe Dobrea, E. Z., Arvidson, R. E., Wiseman, S. M., Squyres, S. W., McEwen, A. S., Mustard, J. F. & Murchie, S. L. 2009. Phyllosilicates and sulfates at Endeavour Crater, Meridiani Planum, Mars. Geophys. Res. Lett. 36. doi:10.1029/2009GL040734.

Wu, N., Li, B., Liu, Z., Han, C., 2014. Synthesis of Keggin-type lacunary 11-tungstophosphates encapsulated into mesoporous silica pillared in clay interlayer galleries and their catalytic performance in oxidative desulfurization. Catal. Commun. 46, 156–160.

Wu, N., Wu, L.M., Liao, L.B., Lv, G.C., 2015. Organic intercalation of structure modified vermiculite. J. Colloid Interface Sci. 457, 264–271.

Wunder, B., Baronnet, A., Schreyer, W., 1997. Ab-initio synthesis and TEM confirmation of antigorite in the system $MgO-SiO_2-H_2O$. Am. Mineral. 82, 760–764.

Wunder, B., Wirth, R., Gottschalk, M., 2001. Antigorite: pressure and temperature dependence of polysomatism and water content. Eur. J. Mineral. 13, 485–495.

Wyatt, M.B., McSween, H.Y., Tanaka, K.L., Head, J.W., 2004. Global geologic context for rock types and surface alteration on Mars. Geology 32, 645–648.

Wyatt, M.B., Hamilton, V.E., McSween, H.Y., Christensen, J.R., Taylor, L.A., 2001. Analysis of terrestrial and martian volcanic compositions using thermal emission spectroscopy, 1. Determination of mineralogy, chemistry, and classification strategies. J. Geophys. Res. 106, 14,711–14,732.

Xi, Y.F., Frost, R.L., He, H.P., 2007. Modification of the surfaces of Wyoming montmorillonite by the cationic surfactants alkyl trimethyl, dialkyl dimethyl, and trialkyl methyl ammonium bromides. J. Colloid Interface Sci. 305, 150–158.

Xi, Y.F., Ding, Z., He, H.P., Frost, R.L., 2005. Infrared spectroscopy of organoclays synthesized with the surfactant octadecyltrimethylammonium bromide. Spectrochim. Acta A: Mol. Biomol. Spectrosc. 61, 515–525.

Xie, W., Xie, R.C., Pan, W.P., Hunter, D., Koene, B., Tan, L.S., Vaia, R., 2002. Thermal stability of quaternary phosphonium modified montmorillonites. Chem. Mater. 14, 4837–4845.

Xin, X.D., Wang, J., Yu, H.Q., Du, B., Wei, Q., Yan, L.G., 2011. Removal of o-nitrobenzoic acid by adsorption on to a new organoclay: montmorillonite modified with HDTMA microemulsion. Environ. Technol. 32, 447–454.

Xu, H., Chen, X., 2011. A thermal study of Al-pillared montmorillonites. Adv. Mater. Res. 187, 112-116.

Xu, S.H., Boyd, S.A., 1995. Cationic surfactant adsorption by swelling and nonswelling layer silicates. Langmuir 11, 2508–2514.

Xu, W., Johnston, C.T., Parker, P., Agnew, S.F., 2000. Infrared study of water sorption on Na-, Li-, Ca- and Mg-exchanged (SWy-1 and SAz-1) montmorillonite. Clays Clay Miner. 48, 120–131.

Xu, X., Pan, Y., Cui, X., Suo, Z., 2004. Catalytic combustion of methane over Ti-pillared clay supported copper catalysts. J. Nat. Gas Chem. 13, 204–208.

Yan, H.Q., Chen, X.Q., Wu, T.T., Feng, Y.H., Wang, C.X., Li, J.C., Lin, Q., 2014. Mechanochemical modification of kaolin surfaces for immobilization and delivery of pesticides in alginate-chitosan composite beads. Polym Bull. 71, 2923–2944.

Yan, L.B., Stucki, J.W., 1999. Effects of structural Fe oxidation state on the coupling of interlayer water and structural Si-O stretching vibrations in montmorillonite. Langmuir 15, 4648–4657.

Yan, L.B., Roth, C.B., Low, P.F., 1996a. Changes in the Si-O vibrations of smectite layers accompanying the sorption of interlayer. Langmuir 12, 4421–4429.

Yan, L.B., Roth, C.B., Low, P.F., 1996b. Effects of monovalent, exchangeable cations and electrolytes on the infrared vibrations of smectite layers and interlayer water. J. Colloid Interface Sci. 184, 663–670.

Yan, L.G., Wang, J., Yu, H.Q., Wei, Q., Du, B., Shan, X.Q., 2007. Adsorption of benzoic acid by CTAB exchanged montmorillonite. Appl. Clay Sci. 37, 226–230.

Yan, P.Y., Tani, M., Usami, M., Kono, S., Kersting, R., Zhang, X.-C., 2001. J. Appl. Phys. 89, 2357–2359.

Yang, K., Huntington, J.F., Browne, P.R.L., Ma, C., 2000. An infrared spectral reflectance study of hydrothermal alteration minerals from the Te Mihi sector of the Wairakei geothermal system, New Zealand. Geothermics 29, 377–392.

Yang, K., Browne, P.R.L., Huntington, J.F., Walshe, J.L., 2001. Characterizing the hydrothermal alteration of the Broadlands-Ohaaki geothermal system, New Zealand, using short-wave infrared spectroscopy. J. Volcanol. Geotherm. Res. 106, 53–65.

Yang, L., Kruse, B., 2004. Revised Kubelka-Muck theory. I. Theory and applications. J. Opt. Soc. Am. 21, 1933–1944.

Yang, L., Miklavicic, S.J., 2005. Revised Kubelka-Munk theory. III. A general theory of light propagation in scattering and absorptive media. J. Opt. Soc. Am. A 22, 1868–1873.

Yang, P.W., Mantsch, H.H., Baudis, F., 1986. A critical evaluation of three types of diffuse reflectance infrared accessories. Appl. Spectrosc. 40, 974–978.

Yang, X., Zhu, H., Liu, J., Gao, X., Martens, W.N., Frost, R.L., Shen, Y., Yuan, Z., 2008. A mesoporous structure for efficient photocatalysts: anatase nanocrystals attached to leached clay layers. Microporous Mesoporous Mater. 112, 32–44.

Yao, H.Y., Zhu, J., Morgan, A.B., Wilkie, C.A., 2002. Crown ether-modified clays and their polystyrene nanocomposites. Polym. Eng. Sci. 42, 1808–1814.

Yariv, S., 1992a. The effect of tetrahedral substitution of Si by Al on the surface-acidity of the oxygen plane of clay-minerals. Int. Rev. Phys. Chem. 11, 345–375.

Yariv, S., 1992b. Wettability of clay minerals. In: Schrader, M.E., Loeb, G. (Eds.), Modern Approaches to Wettability: Theory and Applications. Plenum Press, New York, pp. 279–326.

Yariv, S., 2002. IR spectroscopy and thermo-IR spectroscopy in the study of the fine structure of organo-clay complexes. In: S. Yariv, H. Cross (Eds.), Organo-clay complexes and interactions. Marcel Dekker, Inc., New York, 2002, pp. 345–462.

Yariv, S., Shoval, S., 1975. Nature of interaction between water-molecules and kaolin-like layers in hydrated halloysite. Clays Clay Miner. 23, 473–474.

Yariv, S., Lapides, I., 2008. Thermo-infrared-spectroscopy analysis of dimethylsulfoxide-kaolinite intercalation complexes. J. Therm. Anal. Calorim. 94, 433–440.

Yasunari, T.J., Stohl, A., Hayano, R.S., Burkhart, J.F., Eckhardt, S., Yasunari, T., 2011. Cesium-137 deposition and contamination of Japanese soils due to the Fukushima nuclear accident. Proc. Natl. Acad. Sci. U. S. A. 108, 19530–19534.

Yatsimirksii, K.B., Vasil'ev, V.P., 1960. Instability Constants of Complex Compounds. Pergamon, Elmsford, NY.

Yei, D.R., Kuo, S.W., Fu, H.K., Chang, F.C., 2005. Enhanced thermal properties of PS nanocomposites formed from montmorillonite treated with a surfactant/cyclodextrin inclusion complex. Polymer. 46, 741–750.

Yilmaz, N., Yapar, S., 2004. Adsorption properties of tetradecyl- and hexadecyl trimethylammonium bentonites. Appl. Clay Sci. 27, 223–228.

Yu, J., Yang, Q.-X., 2010. Magnetization improvement of Fe-pillared clay with application of polyetheramine. Appl. Clay Sci. 48, 185–190.

Yu, X.F., Zhao, L.Y., Gao, X.X., Zhang, X.P., Wu, N.Z., 2006. The intercalation of cetyltrimethylammonium cations into muscovite by a two-step process: II. The intercalation of cetyltrimethylammonium cations into Li-muscovite. J. Solid State Chem. 179, 1525–1535.

Zachara, J.M., Serne, J., Freshley, M., Mann, F., Anderson, F., Wood, M., Jones, T., Myers, D., 2007. Geochemical processes controlling migration of tank wastes in Hanford's vadose zone. Vadose Zone J. 6, 985–1003.

Zaghouane-Boudiaf, H., Boutahala, M., 2011. Preparation and characterization of organo-montmorillonites. Application in adsorption of the 2, 4, 5-trichlorophenol from aqueous solution. Adv. Powder Technol. 22, 735–740.

Zamama, M., Burneau, A., Mokhlisse, R., 1995. IR study of deuterated dickite-DMSO intercalate: ($Al_2Si_2O_5(OH)_4$-$CH_3SOCH_3$). Effect of particle size on hydroxyl stretching frequencies. Spectrochim. Acta Part A 51A, 101–108.

Zampori, L., Gallo Stampino, P., Cristiani, C., Dotelli, G., Cazzola, P., 2010. Synthesis of organoclays using non-ionic surfactants: effect of time, temperature and concentration. Appl. Clay Sci. 48, 97–102.

Zecchina, A. & Arean, C.O. 1996. Diatomic molecular probes for mid-IR studies of zeolites. Chem. Soc. Rev. 25, 187–197.

Zha, W., Choi, S., Lee, K.M., Han, C.D., 2005. Dispersion characteristics of organoclay in nanocomposites based on end-functionalized homopolymer and block copolymer. Macromolecules 38, 8418–8429.

Zhai, S., Ito, E., Yoneda, A., 2008. Effects of pre-heated pyrophyllite gaskets on high-pressure generation in the Kawai-type multi-anvil experiments. High Pressure Res. 28, 265–271.

Zhang, B., Ren, J., Liu, X., Guo, Y., Guo, Y., Lu, G., Wang, Y., 2010. Novel sulfonated carbonaceous materials from p-toluenesulfonic acid/glucose as a high-performance solid-acid catalyst. Catal. Commun. 11, 629–632.

Zhang, J., Yan, J., Sheng, J., 2015. Dry grinding effect on pyrophyllite–quartz natural mixture and its influence on the structural alternation of pyrophyllite. Micron 71, 1–6.

Zhang, J., Cheng, H., Liu, Q., He, J., Frost, R.L., 2011. Mid-infrared and near-infrared spectroscopic study of kaolinite-potassium acetate intercalation complex. J. Mol. Struct. 994, 55–60.

Zhang, J.G., Wilkie, C.A., 2004. A carbocation substituted clay and its styrene nanocomposite. Polym. Degrad. Stab. 83, 301–307.

Zhang, M., Hui, Q., Lou, X.-J., Redfern, S.A.T., Salje, E.K.H., Tarantino, S., 2006. Dehydroxylation, proton migration, and structural changes in heated talc: an infrared spectroscopic study. Am. Mineral. 91, 816–825.

Zhang, M., Salje, E.K.H., Carpenter, M.A., Wang, J.Y., Groat, L.A., Lager, G.A., Wang, L., Beran, A., Bismayer, U., 2007. Temperature dependence of IR absorption of hydrous/hydroxyl species in minerals and synthetic materials. Am. Mineral. 92, 1502–1517.

Zhang, W.F., He, Y.L., Zhang, M.S., Yin, Z., Chen, Q., 2000. Raman scattering study of anatase $TiO_2$ nanocrystals. J. Phys. D: Appl. Phys. 33, 912–916.

Zhang, Y., Liu, Q., Wu, Z., Zheng, Q., Cheng, H., 2012. Thermal behavior analysis of kaolinite-dimethylsulfoxide intercalation complex. J. Therm. Anal. Calorim. 110, 1167–1172.

Zhao, B., Dang, L., Zhang, X., Yang, N., Sun, Y., 2012. Preparation of $TiO_2$-pillared montmorillonite as photocatalyst and photocatalytic degradation of methyl orange. Appl. Mech. Mater. 190–191, 534–538.

Zhao, D., Wang, G., Yang, Y., Guo, X., Wang, Q., Ren, J., 1993. Preparation and characterization of hydroxy-FeAl pillared clays. Clays Clay Miner. 41, 317–327.

Zhao, H.T., Vance, G.F., 1998. Sorption of trichloroethylene by organo-clays in the presence of humic substances. Water Res. 32, 3710–3716.

Zhao, H.X., Li, Q.H., Liu, S., 2015. Investigation of some Chinese jade artifacts (5000 BC to 771 BC) by confocal laser micro-Raman spectroscopy and other techniques. J. Raman Spectrosc, Ahead of Print.

Zhao, X.S., Lu, G.Q., Whittaker, A.K., Millar, G.J., Zhu, H.Y., 1997. Comprehensive study of surface chemistry of MCM-41 Using $^{29}Si$ CP/MAS NMR, FTIR, pyridine-TPD, and TGA. J. Phys. Chem. B 101, 6525–6531.

Zhen, W., Lu, C., Li, C., Liang, M., 2012. Structure and properties of thermoplastic saponite/poly(vinyl alcohol) nanocomposites. Appl. Clay Sci. 57, 64–70.

Zheng, L., Hao, Y., Tao, L., Zhang, Y., Xue, Z., 1992. MAS NMR and IR studies of pillared clays. Zeolites 12, 374–379.

Zheng, S.M., Wang, K.M., 2014. Preparation and characterization of lizardite. Appl. Mech. Mater. 556–562, 109–112.

Zhou, J., Zheng, W., Xu, J., Chen, L., Zhang, Z., Li, Y., Ma, N., Du, P., 2013. Formation of 0.84 nm hydrated kaolinite as an environmentally friendly precursor of a kaolinite intercalation compound. Clays Clay Miner. 61, 416–423.

Zhou, Q., Frost, R.L., He, H.P., Xi, Y.F., 2007a. Changes in the surfaces of adsorbed p-nitrophenol on methyltrioctadecylammonium bromide organoclay—an XRD, TG, and infrared spectroscopic study. J. Colloid Interface Sci. 314, 405–414.

Zhou, Q., He, H.P., Frost, R.L., Xi, Y.F., 2007b. Adsorption of p-nitrophenol on mono-, di-, and trialkyl surfactant-intercalated organoclays: a comparative study. J. Phys. Chem. C. 111, 7487–7493.

Zhou, Q., Xi, Y., He, H.P., Frost, R.L., 2008. Application of near infrared spectroscopy for the determination of adsorbed p-nitrophenol on HDTMA organoclay-implications for the removal of organic pollutants from water. Spectrochim. Acta A: Mol. Biomol. Spectrosc. 69, 835–841.

Zhou, Q., Frost, R.L., He, H.P., Xi, Y.F., Liu, H.W., 2007c. Adsorbed para-nitrophenol on HDTMAB organoclay—a TEM and infrared spectroscopic study. J. Colloid Interface Sci. 307, 357–363.

Zhou, Q., Deng, S.B., Yu, Q., Zhang, Q.Y., Yu, G., Huang, J., He, H.P., 2010. Sorption of perfluorooctane sulfonate on organo-montmorillonites. Chemosphere 78, 688–694.

Zhu, L.H., Zhu, R.L., Xu, L.H., Ruan, X.X., 2007a. Influence of clay charge densities and surfactant loading amount on the microstructure of CTMA–montmorillonite hybrids. Colloids Surf. A Physicochem. Eng. Asp. 304, 41–48.

Zhu, L.Z., Chen, B.L., Shen, X.Y., 2000. Sorption of phenol, p-nitrophenol, and aniline to dual-cation organobentonites from water. Environ. Sci. Technol. 34, 468–475.

Zhu, L.Z., Tian, S.L., Zhu, J.X., Shi, Y., 2007b. Silylated pillared clay (SPILC): a novel bentonite-based inorgano–organo composite sorbent synthesized by integration of pillaring and silylation. J. Colloid Interface Sci. 315, 191–199.

Zhu, J.X., Zhu, L.Z., Zhu, R.L., Chen, B.L., 2008. Microstructure of organo-bentonites in water and the effect of steric hindrance on the uptake of organic compounds. Clays Clay Miner. 56, 144–154.

Zhu, J.X., He, H.P., Guo, J.G., Yang, D., Xie, X.D., 2003. Arrangement models of alkylammonium cations in the interlayer of $HDTMA^+$ pillared montmorillonites. Chin. Sci. Bull. 48, 368–372.

Zhu, J.X., He, H.P., Zhu, L.Z., Wen, X.Y., Deng, F., 2005. Characterization of organic phases in the interlayer of montmorillonite using FTIR and $^{13}C$ NMR. J. Colloid Interface Sci. 286, 239–244.

Zhu, J.X., Qing, Y.H., Wang, T., Zhu, R.L., Wei, J.M., Tao, Q., Yuan, P., He, H.P., 2011. Preparation and characterization of zwitterionic surfactant-modified montmorillonites. J. Colloid Interface Sci. 360, 386–392.

Zhu, R.L., Wang, T., Ge, F., Chen, W.X., You, Z.M., 2009. Intercalation of both CTMAB and $Al_{13}$ into montmorillonite. J. Colloid Interface Sci. 335, 77–83.

Zhu, R.L., Zhou, Q., Zhu, J.X., Xi, Y.F., He, H.P., 2015. Organo-clays as sorbents of hydrophobic organic contaminants: sorptive characteristics and approaches to enhancing sorption capacity. Clays Clay Miner. 63, 199–221.

Zhuang, G.Z., Zhang, Z.P., Guo, J.S., Liao, L.B., Zhao, J.L., 2015. A new ball milling method to produce organo-montmorillonite from anionic and nonionic surfactants. Appl. Clay Sci. 104, 18–26.

Zidelkheir, B., Abdelgoad, M., 2008. Effect of surfactant agent upon the structure of montmorillonite. J. Therm. Anal. Calorim. 94, 181–187.

Zimowska, M., Pálková, H., Madejová, J., Dula, R., Pamin, K., Olejniczak, Z., Gil, B., Serwicka, E.M., 2013. Laponite-derived porous clay heterostructures: III The effect of alumination. Microporous Mesoporous Mater. 175, 67–75.

Zimowska, M., Gurgul, J., Pálková, H., Olejniczak, Z., Łątka, K., Lityńska-Dobrzyńska, L., Matachowski, L., 2016. Structural rearrangements in Fe-porous clay heterostructures composites derived from Laponite®—Influence of preparation methods and Fe source. Microporous Mesoporous Mater. 231, 66–81.

Zolensky, M.E., McSween Jr., H.Y., 1988. Aqueous alteration. In: Kerridge, J.F., Matthews, M.S. (Eds.), Meteorites and the Early Solar System. University of Arizona Press, Tucson, AZ.

Zuo, S., Zhou, R., Qi, C., 2011. Synthesis and characterization of aluminum and Al/REE pillared clays and supported palladium catalysts for benzene oxidation. J. Rare Earths 29, 52–57.

Zviagina, B.B., McCarty, D.K., Środón, J., Drits, V.A., 2004. Interpretation of infrared spectra of dioctahedral smectites in the region of OH-stretching vibrations. Clays Clay Miner. 52, 399–410.

Zviagina, B.B., Drits, D.A., Środón, J., McCarty, D.K., Dorzhieva, O.V., 2015. The illite-aluminoceladonite series: distinguishing features and identification criteria from X-ray diffraction and infrared spectroscopy data. Clays Clay Miner. 63, 378–394.

# Index

Note: Page numbers followed by "*f*" indicate figures, and "*t*" indicate tables.

## N

# Q

# T

## U



## V

## X



## Z

9780081003558